$ 2-

THE DISCOVERERS

THE
DISCOVERERS

DANIEL J. BOORSTIN

And take upon 's the mystery of things,
As if we were God's spies.

—SHAKESPEARE, *King Lear*, V, 3

VINTAGE BOOKS
A DIVISION OF RANDOM HOUSE
NEW YORK

Credit for copyright page:
Cover art: after an early 16th century woodcut,
courtesy of the Bettmann Archive

Cover design: Robert Aulicino

FIRST VINTAGE BOOKS EDITION, February 1985

Copyright © 1983 by Daniel J. Boorstin
All rights reserved under International and Pan-American Copyright Conventions. Published in the United States by Random House, Inc., New York, and simultaneously in Canada by Random House of Canada Limited, Toronto. Originally published by Random House, Inc., in 1983.

Library of Congress Cataloging in Publication Data

Boorstin, Daniel J. (Daniel Joseph), 1914–
The discoverers.

Bibliography: p.
Includes index.
1. Civilization—History. 2. Discoveries (in geography) 3. Science—History. I. Title.
CB69.B66 1985   909   84-40056
ISBN 0-394-72625-1 (pbk.)

Manufactured in the United States of America
3579C8642

FOR RUTH

Nay, the same Solomon the king, although he excelled in the glory of treasure and magnificent buildings, of shipping and navigation, of service and attendance, of fame and renown, and the like, yet he maketh no claim to any of those glories, but only to the glory of inquisition of truth; for so he saith expressly, "The glory of God is to conceal a thing, but the glory of the king is to find it out"; as if, according to the innocent play of children, the Divine Majesty took delight to hide his works, to the end to have them found out; and as if kings could not obtain a greater honour than to be God's play-fellows in that game.

—FRANCIS BACON, *The Advancement of Learning* (1605)

CONTENTS

BOOK THREE: NATURE

BOOK FOUR: SOCIETY

A Personal Note to the Reader

MY hero is Man the Discoverer. The world we now view from the literate West—the vistas of time, the land and the seas, the heavenly bodies and our own bodies, the plants and animals, history and human societies past and present—had to be opened for us by countless Columbuses. In the deep recesses of the past, they remain anonymous. As we come closer to the present they emerge into the light of history, a cast of characters as varied as human nature. Discoveries become episodes of biography, unpredictable as the new worlds the discoverers opened to us.

The obstacles to discovery—the illusions of knowledge—are also part of our story. Only against the forgotten backdrop of the received common sense and myths of their time can we begin to sense the courage, the rashness, the heroic and imaginative thrusts of the great discoverers. They had to battle against the current "facts" and dogmas of the learned. I have tried to recapture those illusions—about the earth, the continents and the seas before Columbus and Balboa, Magellan and Captain Cook, about the heavens before Copernicus and Galileo and Kepler, about the human body before Paracelsus and Vesalius and Harvey, about plants and animals before Ray and Linnaeus, Darwin and Pasteur, about the past before Petrarch and Winckelmann, Thomsen and Schliemann, about wealth before Adam Smith and Keynes, about the physical world and the atom before Newton and Dalton and Faraday, Clerk Maxwell and Einstein.

I have asked some unfamiliar questions. Why didn't the Chinese "discover" Europe or America? Why didn't the Arabs circumnavigate Africa and the world? Why did it take so long for people to learn that the earth goes around the sun? Why did people begin to believe that there are "species" of plants and animals? Why were the facts of prehistory and the discovery of the progress of civilization so slow in coming?

I have included the story of only a few crucial inventions—the clock, the compass, the telescope and the microscope, the printing press and movable type—which have been essential instruments of discovery. I have not told the story of the shaping of governments, the waging of wars, the rise and fall of empires. I have not chronicled culture, the story of Man the Creator, of architecture, painting, sculpture, music, and literature, much as these have multiplied the delights of human experience. My focus remains on mankind's need to *know*—to know what is out there.

The plan of the book as a whole is chronological. In detail it is a shingle scheme. Each of the fifteen Parts overlaps chronologically with its predecessor as the story advances from antiquity to the present. I begin with Time, the most elusive and mysterious of the primitive dimensions of experience. Then I turn to Western man's widening vistas of the Earth and the Seas. On to Nature—the physical objects in the heavens and on earth, plants and animals, the human body and its processes. Finally to Society, finding that the human past was not what had been imagined, to the self-discovery of Man the Discoverer, and to the Dark Continents in the atom.

This is a story without end. All the world is still an America. The most promising words ever written on the maps of human knowledge are *terra incognita*—unknown territory.

BOOK ONE

TIME

Time is the greatest innovator.

—FRANCIS BACON, "OF INNOVATIONS" (1625)

The first grand discovery was time, the landscape of experience. Only by marking off months, weeks, and years, days and hours, minutes and seconds, would mankind be liberated from the cyclical monotony of nature. The flow of shadows, sand, and water, and time itself, translated into the clock's staccato, became a useful measure of man's movements across the planet. The discoveries of time and of space would become one continuous dimension. Communities of time would bring the first communities of knowledge, ways to share discovery, a common frontier on the unknown.

PART ONE

THE HEAVENLY EMPIRE

God did not create the planets and stars with the intention that they should dominate man, but that they, like other creatures, should obey and serve him.

—PARACELSUS, *Concerning the Nature of Things*
(c. 1541)

1

The Temptations of the Moon

FROM far-northwest Greenland to the southernmost tip of Patagonia, people hail the new moon—a time for singing and praying, eating and drinking. Eskimos spread a feast, their sorcerers perform, they extinguish lamps and exchange women. African Bushmen chant a prayer: "Young Moon! . . . Hail, hail, Young Moon!" In the light of the moon everyone wants to dance. And the moon has other virtues. The ancient German communities, Tacitus reported nearly two thousand years ago, held their meetings at new or full moon, "the seasons most auspicious for beginning business."

Everywhere we find relics of mythic, mystic, romantic meanings—in "moonstruck" and "lunatic" (Latin *luna* means moon), in "moonshine," and in the moonlight setting of lovers' meetings. Even deeper is the primeval connection of the moon with measuring. The word "moon" in English and its cognate in other languages are rooted in the base *me* meaning measure (as in Greek *metron,* and in the English *meter* and *measure*), reminding us of the moon's primitive service as the first universal measurer of time.

Despite or because of its easy use as a measure of time, the moon proved to be a trap for naïve mankind. For while the phases of the moon were convenient worldwide cycles which anybody could see, they were an attractive dead end. What hunters and farmers most needed was a calendar of the seasons—a way to predict the coming of rain or snow, of heat and cold. How long until planting time? When to expect the first frost? The heavy rains?

For these needs the moon gave little help. True, the cycles of the moon had an uncanny correspondence with the menstrual cycle of women, because a sidereal month, or the time required for the moon to return to the same position in the sky, was a little less than 28 days, and a pregnant woman could expect her child after ten of these moon-months. But a solar year—the proper measure of days between returning seasons—measures 365¼ days. The cycles of the moon are caused by the moon's movement around the earth at the same time that the earth is moving around the sun. The moon's orbit is elliptical, and departs by an angle of about five degrees from the earth's orbit about the sun. This explains why eclipses of the sun do not occur every month.

The discomfiting fact that the cycles of the moon and the cycles of the sun are incommensurate would stimulate thinking. Had it been possible to calculate the year, the round of seasons, simply by multiplying the cycles of the moon, mankind would have been saved a lot of trouble. But we might also have lacked the incentive to study the heavens and to become mathematicians.

The seasons of the year, as we now know, are governed by the movements of the earth around the sun. Each round of the seasons marks the return of the earth to the same place in its circuit, a movement from one equinox (or solstice) to the next. Man needed a calendar to find his bearings in the seasons. How to begin?

The ancient Babylonians started with the lunar calendar and stayed with it. Their obstinacy in sticking with moon cycles for their calendar-making had important consequences. In search of a way to measure the cycle of the seasons in multiples of moon cycles, they eventually discovered, probably around 432 B.C., the so-called Metonic cycle (after an astronomer Meton) of nineteen years. They found that if they used a nineteen-year cycle, assigned to seven of the years thirteen months, and assigned to the other twelve years only twelve months, they could continue to use the conveniently visible phases of the moon as the basis of their calendar. Their "intercalation," or insertion of extra months, avoided the inconvenience of a "wandering" year in which the seasons wandered gradually through the lunar months, so that there was no easy way of knowing which month would bring the new season. This Metonic calendar with its nineteen-year clusters was too complicated for everyday use.

The Greek historian Herodotus, writing in the fifth century B.C., illustrated these complications in a famous passage when he reported how the wise Solon answered the rich and irascible Croesus, who asked him who was the happiest of mortals. To impress on Croesus the vast unpredictability of fortune, he calculated according to the Greek calendar then in use the number of days in the seventy years which he regarded as the limit of the life of man. "In these seventy years," he observed, "are contained, without reckoning intercalary months, 25,200 days. Add an intercalary month to every other year, that the seasons may come round at the right time, and there will be, besides the seventy years, thirty-five such months, making an addition of 1,050 days. The whole number of the days contained in the seventy years will thus be 26,250, whereof not one but will produce events unlike the rest. Hence man is wholly accident. For yourself, Croesus, I see that you are wonderfully rich, and the lord of many nations; but with respect to your question, I have no answer to give, until I hear that you have closed your life happily."

The Egyptians somehow escaped the temptations of the moon. So far as we know, they were the first to discover the length of the solar year and to define it in a useful, practical fashion. As with many other crucial human achievements, we know the *what,* but remain puzzled still about the *why,* the *how,* and even the *when.* The first puzzle is why it was the Egyptians. They had no astronomical instruments not already well known to the ancient world. They showed no special genius for mathematics. Their astronomy remained crude compared with that of the Greeks and others in the Mediterranean and was dominated by religious ritual. But it seems that by about 2500 B.C. they had figured out how to predict when the rising or setting sun would gild the tip of any particular obelisk, which helped them add a glow to their ceremonies and anniversaries.

The Babylonian scheme, which kept the lunar cycles and tried to adjust them to the seasonal or solar year by "intercalation," was inconvenient. Local whims prevailed. In Greece, fragmented by mountains and bays and fertile in landscape loyalties, each city-state made its own calendar, arbitrarily "intercalating" the extra month to mark a local festival or to suit political needs. The result was to defeat the very purpose of a calendar—a time scheme to hold people together, to ease the making of common plans, such as agreements on the planting of crops and the delivery of goods.

The Egyptians, even without the Greek yen for mathematics, solved the practical problem. They invented a calendar that served everyday needs throughout their land. As early as 3200 B.C., the whole Nile Valley was united with the Nile Delta into a single kingdom which lasted for three thousand years, until the Age of Cleopatra. Political unity was reinforced by nature. Like the heavenly bodies themselves, the Nile displayed a regular but more melodramatic natural rhythm. The longest river in Africa, the Nile stretches four thousand miles from its remote headstream, gathering the rainfall and snowmelt of the Ethiopian highlands and all the northeastern continent into a single grand channel to the Mediterranean. The pharaoh's realm was aptly called the Empire of the Nile. The ancients, taking Herodotus' cue, called Egypt "the gift of the Nile." The search for the sources of the Nile, like the search for the Holy Grail, had mystic overtones, which stirred death-defying explorers into the nineteenth century.

The Nile made possible the crops, the commerce, and the architecture of Egypt. Highway of commerce, the Nile was also a freightway for the materials of colossal temples and pyramids. A granite obelisk of three thousand tons could be quarried at Aswan and then floated two hundred miles down the river to Thebes. The Nile fed the cities that clustered along its banks. No wonder that the Egyptians called the Nile "the sea" and in the Bible it is "the river."

The rhythm of the Nile was the rhythm of Egyptian life. The annual rising of its waters set the calendar of sowing and reaping with its three seasons: inundation, growth, and harvest. The flooding of the Nile from the end of June till late October brought down rich silt, in which crops were planted and grew from late October to late February, to be harvested from late February till the end of June. The rising of the Nile, as regular and as essential to life as the rising of the sun, marked the Nile year. The primitive Egyptian calendar, naturally enough, was a "nilometer"—a simple vertical scale on which the flood level was yearly marked. Even a few years' reckoning of the Nile year showed that it did not keep in step with the phases of the moon. But very early the Egyptians found that twelve months of thirty days each could provide a useful calendar of the seasons if another five days were added at the end, to make a year of 365 days. This was the "civil" year, or the "Nile year," that the Egyptians began to use as early as 4241 B.C.

Avoiding the seductively convenient cycle of the moon, the Egyptians had found another sign to mark their year: Sirius, the Dog Star, the brightest star in the heavens. Once a year Sirius rose in the morning in direct line with the rising sun. This "heliacal rising" of Sirius, which occurred every year in the midst of the Nile's flood season, became the beginning of the Egyptian year. It was marked by a festival, the five "epagomenal days" (days outside the months), celebrating in turn the birthday of Osiris, of his son Horus, of his Satanic enemy, Set, of his sister and wife, Isis, and of Nephthys, the wife of Set.

Since the solar year, of course, is not precisely 365 days, the Egyptian year of 365 days would, over the centuries, become a "wandering year" with each named month gradually occurring in a different season. The discrepancy was so small that it took many years, far longer than any one person's lifetime, for the error to disturb daily life. Each month moved through all the seasons in fourteen hundred and sixty years. Still, this Egyptian calendar served so much better than any other known at the time that it was adopted by Julius Caesar to make his Julian calendar. It survived the Middle Ages and was still used by Copernicus in his planetary tables in the sixteenth century.

While the Egyptians for their everyday calendar succeeded in declaring their independence of the moon, the moon retained a primeval fascination. Many peoples, including the Egyptians themselves, kept the lunar cycle to guide religious festivals and mystic anniversaries. Even today people dominated by their religion let themselves be governed by the cycles of the moon. The daily inconvenience of living by a lunar calendar becomes a daily witness to religious faith.

The Jews, for example, preserve their lunar calendar, and each Jewish month still begins with the appearance of a new moon. To keep their lunar calendar in step with the seasonal year the Jews have added an extra month for each leap year, and the Jewish calendar has become a focus of esoteric rabbinical learning. The Jewish year was made to comprise twelve months, each of 29 or 30 days, totaling some 354 days. In order to fill out the solar year, Jewish leap years—following the Metonic cycle of Babylonia—add an extra month in the third, sixth, eighth, eleventh, fourteenth, seventeenth, and nineteenth year of every nineteen-year period. Other adjustments are required occasionally to make festivals occur in their proper seasons—for example, to ensure that Passover, the spring festival, will come after the vernal equinox. In the Bible most of the months retain their Babylonian, rather than the Hebrew, names.

Christianity, following Judaism for most religious anniversaries, has kept its tie to the lunar calendar. "Movable feasts" in the Church were moved around in the solar calendar because of the effort to keep festivals in step with the cycles of the moon. They still remind us of the primeval charm of the most conspicuous light in the night sky. The most important of these Christian, moon-fixed festivals is, of course, Easter, which celebrates the resurrection of Jesus. "Easter-Day," prescribes the English Book of Common Prayer, "is always the first Sunday after the Full Moon which happens upon, or next after the Twenty-first day of March; and if the Full Moon happens upon a Sunday, Easter-Day is the Sunday after." At least a dozen other Church festivals are fixed by reference to Easter and its lunar date, with the result that Easter controls about seventeen weeks in the ecclesiastical calendar. The fixing of the date of Easter—in other words, the calendar—became a great issue and a symbol. Since the New Testament recounted that Jesus was crucified on the Passover, the anniversary of the Easter resurrection would obviously be tied to the Jewish calendar. The inevitable result was that the date of Easter would depend on the complicated lunar calculations by which the highest Jewish council, the Sanhedrin, defined the Passover.

Many of the early Christians, following their own literal interpretation of the Bible, fixed the death of Jesus on a Friday, and the Easter resurrection on the following Sunday. But if the anniversary of the festival was to follow the Jewish lunar calendar, there was no assurance that Easter would occur on a Sunday. The bitter quarrel over the calendar led to one of the earliest schisms between the Eastern Orthodox Church and the Church of Rome. The Eastern Christians, holding to the lunar calendar, continued to observe Easter on the fourteenth day of the lunar month, regardless of the day of the week. At the very first ecumenical (worldwide) council of the Christian Church, held at Nicaea in Asia Minor in 325, one of the world-unifying questions to be decided was the date of Easter. A uniform date was fixed in such a way as both to stay with the traditional lunar calendar and to assure that Easter would always be observed on Sunday.

But this did not quite settle the matter. For community planning someone still had to predict the phases of the moon and locate them on a solar calendar. The Council of Nicaea had left this task to the bishop of Alexandria. In that ancient center of astronomy he was to forecast the phases of the moon for all future years. Disagreement over how to predict those specified cycles led to a division in the Church, with the result that different parts of the world continued to observe Easter on different Sundays.

The reform of the calendar by Pope Gregory XIII was needed because the year that Julius Caesar had borrowed from the Egyptians, and which had ruled Western civilization since then, was not a precise enough measure of the solar cycle. The actual solar year—the time required for the earth to complete an orbit around the sun—is 365 days, 5 hours, 48 minutes, and 46 seconds. This was some 11 minutes and 14 seconds less than the 365¼ days in the Egyptian year. As a result, dates on the calendar gradually lost their intended relation to solar events and to the seasons. The crucial date, the vernal equinox, from which Easter was calculated, had been fixed by the First Council of Nicaea at March 21. But the accumulating inaccuracy of the Julian calendar meant that by 1582 the vernal equinox was actually occurring on March 11.

Pope Gregory XIII, though notorious now for his public Thanksgiving for the brutal massacre of Protestants in Paris on Saint Bartholomew's Day (1572), was in some matters an energetic reformer. He determined to set the calendar straight. Climaxing a movement for calendar reform which had been developing for at least a century, in 1582 Pope Gregory ordained that October 4 was to be followed by October 15. This meant, too, that in the next year the vernal equinox would occur, as the solar calendar of seasons required, on March 21. In this way the seasonal year was restored to what it had been in 325. The leap years of the old Julian calendar were readjusted. To prevent the accumulation of another 11-minute-a-year discrepancy, the Gregorian calendar omitted the leap day from years ending in hundreds, unless they were divisible by 400. This produced the modern calendar by which the West still lives.

Simply because the reform had come from Rome, Protestant England and the Protestant American colonies obstinately refused to go along. Not until 1752 were they persuaded to make the change. The Old Style calendar year that governed them till then had begun on March 25, but the New Style year began on January 1. When the necessary eleven days were added, George Washington's birthday, which fell on February 11, 1731, Old Style, became February 22, 1732, New Style.

Back in 1582, when Pope Gregory took ten days out of the calendar, there had been grumbling and confusion. Servants demanded their usual full monthly pay for the abridged month; employers refused. People objected to having their lives shortened by papal decree. But when Britain and the

American colonies finally got around to making the change, Benjamin Franklin, aged forty-six when he lost the ten days of his life, with his usual cheery ingenuity gave readers of his *Poor Richard's Almanack* something to be thankful for:

> Be not astonished, nor look with scorn, dear reader, at such a deduction of days, nor regret as for the loss of so much time, but take this for your consolation, that your expenses will appear lighter and your mind be more at ease. And what an indulgence is here, for those who love their pillow to lie down in peace on the second of this month and not perhaps awake till the morning of the fourteenth.

The world never entirely accepted the Gregorian reform. The Eastern Orthodox Church, wary of subjecting itself to any Romish rule, has kept the Julian calendar for its own calculation of Easter. And so the Christian world, supposedly held together by a Prophet of Peace, has not been able to agree even on the date to celebrate the resurrection of their Savior.

Still, for everyday secular affairs, the whole Christian world has shared a solar calendar which serves the convenience of the farmer and the merchant. But Islam, insisting on literal obedience to the words of the prophet Muhammad and to the dictates of the holy Koran, continues to live by the cycles of the moon.

The crescent, the sign of the new moon, appears on the flag of Muslim nations. Despite scholarly dispute about the origin of the crescent symbol, there can be no doubt of its appropriateness for the peoples who have obediently submitted the schedule of their lives to the divinely commanded measure of the moon. And it is doubly significant as a conspicuous exception to the Muslim ban on representing natural objects. At least as early as the thirteenth century, the crescent became the military and religious symbol of the Ottoman Turks. There is reason to believe that its adoption and its survival as a sign of Islam came from the dominance of the new moon, which not only is a signal of the beginning and end of the month-long Muslim season of fasting, but is the regular punctuation for the whole calendar.

The new moons, declares the Koran, "are fixed times for the people and for the pilgrimage." The Muslim world, with orthodox scrupulosity, has tried to live by the moon. Just as Caesar had decisively committed his world to the convenience of the solar year, with the months serving as indices of the seasons, so Muhammad committed his everyday world to the cycles of the moon. These lunar cycles would guide the faithful to the divinely ordained dates for the prime religious duties—the pilgrimage to Mecca and the Ramadan month of fasting. The Muslim year consists of twelve lunar months, of alternately 29 and 30 days. The fractional correction to keep the months in step with the moon was secured by varying the length of the twelfth month in the year. A cycle of thirty Muslim years was defined, in nineteen of which the final month had 29 days, with 30 days in the others.

Since the Muslim calendar contains only 354 or 355 days, the months have no regular relation to the seasons. Ramadan, the ninth month—the month of fasting, the observance which marks the true Muslim—and Dhu'l-Hijja, the twelfth month, during the first two weeks of which the faithful are to make their pilgrimage to Mecca, may occur in summer or in winter. In each year the festival of Ramadan and the Pilgrimage occur ten or eleven days earlier than in the previous year. The everyday inconveniences of this kind of calendar are simply another reminder of the good Muslim's surrender to the will of Allah. The calendar itself, for others a mere schedule of worldly affairs, the Muslim makes an affirmation of faith.

The Muslim's literal submission to the moon cycle has had some interesting consequences. To live by the God-given visible phases of the moon (and not by some human calculation of when the new moon is expected) has meant, of course, that celebration of a festival must await the actual sight of the moon. Most Muslims hold to this view, following a traditionally accepted utterance of the prophet Muhammad, "Do not fast until you see the new moon, and do not break the fast until you see it; but when it is hidden from you [by cloud or mist] give it its full measure." If clouds or mist prevent the new moon from being seen in certain villages, those villages will observe the beginning and the end of Ramadan at different times from their neighbors.

One of the most hotly debated issues in Islam is whether it is permissible to define the beginning and end of festivals not by observation but "by resorting to calculation." The members of the Ismaili sect, who separated themselves in this way, failed to persuade most of their fellow Muslims, who still stand by the need to *observe,* that is, actually to see the new moon. Strict adherence to the lunar calendar has become a touchstone of loyalty to traditional Islam. "Resort to calculation"—the appeal to the sophisticated mathematics of a solar year rather than to the simple, visible dictates of the lunar cycle—has marked the modern revolts against tradition. In 1926, when Kemal Atatürk (Mustapha Kemal) proclaimed the end of the sultanate in Turkey and "modernized" the nation by adopting a new code of laws, by making civil marriage compulsory, and by abolishing the fez for men and the veil for women, he also abandoned the lunar calendar of Islam and adopted the solar calendar of the West.

While for many in the West the calendar may seem nothing more than a system of chronological bookkeeping, it has proved to be one of the most rigid of human institutions. That rigidity comes partly from the potent mystic aura of sun and moon, partly from the fixed boundaries of the seasons. Revolutionaries have frequently tried to remake the calendar, but their success has been short-lived. The National Convention of the French Revolution set up a committee on calendar reform—made up of mathematicians, an educator, a poet, and the great astronomer Laplace—which produced a new calendar of charming rational symmetry. In 1792 their decimal calendar replaced the 7-day week by a 10-day week called a *décade,* each day of which was given a Latin numerical name, three of which comprised a month. The day was divided into ten hours, each consisting of 100 minutes, each minute of 100 seconds. In addition to the 360 days of these twelve months, the extra 5 or 6 days were given edifying names: Les Vertus, Le Génie, Le Travail, L'Opinion, Les Récompenses, with a leap day called Sans-culottide dedicated to holidays and sports. This calendar, designed to loosen the grip of the Church on daily life and thought, lasted uneasily for only thirteen years. When Napoleon became ruler of France, he restored the Gregorian calendar with its traditional saint's days and holidays, for which he received the Pope's blessing.

In China the Revolution of 1911 brought a reform, which introduced the calendar of the West, alongside the traditional Chinese calendar.

In 1929 the Soviet Union, aiming to dissolve the Christian year, replaced the Gregorian with a Revolutionary calendar. The week was to have 5 days, 4 for work, the fifth free, and each month would consist of six weeks. The extra days needed to make up each year's complement of 365 or 366 would be holidays. The Gregorian names of the months were kept, but the days of the week were simply numbered. By 1940 the Soviet Union had returned to the familiar Gregorian calendar.

2

The Week: Gateway to Science

So long as man marked his life only by the cycles of nature—the changing seasons, the waxing or waning moon—he remained a prisoner of nature. If he was to go his own way and fill his world with human novelties, he would have to make his own measures of time. And these man-made cycles would be wonderfully varied.

The week—or something very like it—was probably the earliest of these artificial time clusters. Our English word "week" seems to come from an Old High German word meaning to change, or turn about (like the English "vicar" and the German *Wechsel*). But the week is no Western invention, nor has it everywhere been a cluster of seven days. Around the world, people have found at least fifteen different ways, in bunches of 5 to 10 days each, of clustering their days together. What is planet-wide is not any particular bouquet of days but the need and the desire to make some kind of bouquet. Mankind has revealed a potent, pressing desire to play with time, to make more of it than nature has made.

Our own Western 7-day week, one of the most arbitrary of our institutions, came into being from popular need and spontaneous agreement, not from a law or the order of any government. How did it happen? Why? When?

Why a *seven*-day week?

The ancient Greeks, it seems, had no week. Romans lived by an 8-day week. Farmers who worked in the fields for 7 days came to town for the eighth day—the market day (or *nundinae*). This was a day of rest and festivity, a school holiday, the occasion for public announcements and for entertaining friends. When and why the Romans fixed on 8 days and why they eventually changed to a 7-day week is not clear. The number seven almost everywhere has had a special charm. The Japanese found seven gods of happiness, Rome was set on seven hills, the ancients counted seven wonders of the world, and medieval Christians enumerated seven deadly sins. The Roman change from eight to seven seems not to have been accomplished by any official act. By the early third century A.D. Romans were living with a 7-day week.

There must have been some popular new ideas afloat. One of these was the idea of the Sabbath, which appears to have come to Rome through the Jews. "Remember the sabbath day, to keep it holy," says the Fourth Commandment. "Six days shalt thou labor, and do all thy work but the seventh day is the sabbath of the Lord thy God; in it thou shalt not do any work, thou, nor thy son, nor thy daughter, thy manservant, nor thy maidservant, nor thy cattle, nor the stranger that is within thy gates. For in six days the Lord made heaven and earth, the sea, and all that in them is, and rested the seventh day; wherefore the Lord blessed the sabbath day, and hallowed it" (Exodus 20:8–11). Every week God's creatures reenacted His Creation. The Jews made their week a memorial too of their liberation from slavery. "And remember that thou wast a servant in the land of Egypt, and that the Lord thy God brought thee out thence through a mighty hand and by a stretched out arm: therefore the Lord thy God commanded thee to keep the sabbath day" (Deuteronomy 5:15).

When the Jews observed the Sabbath, they dramatized the Again-and-Again quality of their world.

There were other, less theological forces at work, such as the human need to refresh body and mind. The idea of a seventh day of rest, the very name Sabbath (from the Babylonian *Sabattu*), appears to have survived from the years when the Jews were in Babylonian captivity. The Babylonians observed certain enumerated days—the seventh, fourteenth, nineteenth, twenty-first, and twenty-eighth days of the month—when specific activities were forbidden to their king.

We find another clue in the name Saturday, which the Jews and the Romans and others after them came to use for their Sabbath day. Among the Romans, Saturn's Day, or Saturday, was a day of evil omen when all tasks were ill-starred, a day when battles should not be fought, nor journeys begun. No prudent person would want to risk the mishaps that Saturn might bring. According to Tacitus, the Sabbath was observed in honor of Saturn because "of the seven stars which rule human affairs Saturn has the highest sphere and the chief power."

By the third century the seven-day week had become common in private life throughout the Roman Empire. Each day was dedicated to one of the seven planets. Those seven, according to the current astronomy, included the sun and the moon, but not the earth. The order in which planets governed the days of the week was: sun, moon, Mars, Mercury, Jupiter, Venus, and Saturn. This order was *not* that of their then supposed distance from the earth, which was the "normal" order in which Dante, for example, later described the zones in the heavens, and in which the names of the planets were recited in the schools down to the time of Copernicus.

Our familiar order of the names for the days of the week came from this order of the planets that the Romans thought "governed" the first hour of each day in turn. The astrologers of the day did make use of the "order" of the planets according to their supposed distance from the earth, to calculate the "influence" of each planet on worldly affairs. They believed that each planet would govern an hour, then in the next hour would give way to the influence of the next planet nearer the earth, and so on through the cycle of all seven planets. After each cycle of seven hours, the planetary influences would begin all over again in the same order. The "governing" planet for each day, then, was the planet that happened to preside over the first hour of that day, and each day of the week thus took its name from the planet that governed its first hour. The result of this way of calculating was to name the days of the week in their now familiar order.

The days of our week remain a living witness to the early powers of astrology. We easily forget that our days of the week really are named after the "planets" as they were known in Rome two thousand years ago. The days of the week in European languages are still designated by the planets' names. The survival is even more obvious in languages other than English. Here, with the dominant planet, are some examples:

English	French	Italian	Spanish
Sunday (Sun)	dimanche	domenica	domingo
Monday (Moon)	lundi	lunedì	lunes
Tuesday (Mars)	mardi	martedì	martes
Wednesday (Mercury)	mercredi	mercoledì	miércoles
Thursday (Jupiter)	jeudi	giovedì	jueves
Friday (Venus)	vendredi	venerdì	viernes
Saturday (Saturn)	samedi	sabato	sábado

When peoples have tried to extinguish ancient idolatry, they have replaced the planetary names with simple numbers. So the Quakers call their days First Day, Second Day, and so on up to Seventh Day. They hold their religious meetings not on Sunday but on First Day. In modern Israel, too, the days of the week are given ordinal numbers.

One of the more unexpected examples of the power of the planetary idea is the Christian change of the Sabbath from Saturday, or Saturn's day, to Sunday, or Sun's day. When Christianity first took root in the Roman Empire, pious Church Fathers worried over the survival of the pagan gods in the names of the planets that governed the Christian week. The Eastern Church had some success in exterminating this pagan influence: the names of the days in both modern Greek and Russian ceased to be planetary. But Western Christianity proved more willing to turn Roman beliefs and prejudices to their own purpose. The Church Father Justin Martyr (c. 100–c. 165) shrewdly explained to the Emperor Antoninus Pius and his sons (c. 150) why the Christians chose their particular day for Gospel reading and to celebrate the Eucharist. "It is on what is called the Sun's day that all who abide in the town or the country come together. . . . and we meet on the Sun's day because it is the first day on which God formed darkness and mere matter into the world and Jesus Christ our saviour rose from the dead. For on the day before Saturn's day they crucified him, and on the day after Saturn's day which is the Sun's day he appeared to his apostles and disciples and taught them."

Saturn's day, the traditionally unlucky day when the Jews found it wise to abstain from work, somehow remained the pivot around which the weekly auspices would revolve. But there were still other influences. The Mithraists, followers of the Persian mystery religion who worshipped the sun-god Mithras, which was one of the strongest competitors of Christianity in the Roman Empire, adopted a seven-day week. They naturally felt special reverence for what everybody then called the Sun's day.

The Christians fixed their Lord's day, then, so that every week's passing would relive the drama of Jesus Christ. By taking Communion, every Christian would somehow become one of the Disciples at the Last Supper. The script for this mystic drama was, of course, the liturgy of the mass. The Eucharist, like the other sacraments, became a repeat performance of a crucial symbolic event in the history of the Church. What a happy coincidence that the Sun's day was already known as a day of joy and renewal! "The Lord's day is reverenced by us," a Church Father, Maximus of Turin, explained in the fifth century, "because on it the Savior of the world like the rising sun, dispelling the darkness of hell, shone with the light of resurrection, and therefore is the day called by men of the world the Sun's day, because Christ the sun of righteousness illumines it." The Sun's day, like the first David, prefigured the dazzling light of the sun in the true Savior. The Church Fathers made this coincidence further evidence that the world had long been preparing itself for the Savior's coming.

The making of our week was another forward step in man's mastery of the world, in his reach toward science. The week was man's own cluster, not dictated by the visible forces of nature, for the planetary influences were invisible. By seeking astral regularities, by imagining that regularly recurring forces at a distance, forces that could be judged only by their effects, might govern the world, mankind was preparing a new arsenal of thought, an escape from the prison of Again-and-Again. The planets, unworldly forces, would lead mankind out into the world of history.

The planetary week was a path into astrology. And astrology was a step toward new kinds of prophecy. The earlier forms of prophecy can give us a hint of why astrology was a step forward into the world of science. Ancient rituals brought with them a complicated "science" for using parts of a sacrificed animal to foretell the future of the person who offered the sacrifice. Osteomancy, for example, prophesied by examining a bone of the sacrificed animal. In the mid-nineteenth century, Sir Richard Burton reported from Sindh, in the valley of the Indus, the elaborate technique still widely used for divining from the shoulder blade of a sacrificed sheep. The osteomancers divided the bone into twelve areas, or "houses," each answering a different question about the future. If in the first "house" the bone was clear and smooth, the omen was propitious and the consulter would prove to be a good man. If, in the second "house," which pertained to herds, the bone was clear and clean, the herds would thrive, but if there were layers of red and white streaks, robbers must be expected.

And so it went. Hepatoscopy, which predicted by examining the sacrificed animal's liver, was one of the earliest popular techniques of prophecy among the Assyro-Babylonians. It seems to have been used in China in the Bronze Age. Then the Romans and many others continued the practice. The liver impressed the diviners by its large size, its interesting shape, and its heavy burden of blood. An elaborate bronze model of a sheep's liver, which survives from Piacenza, Italy, is covered with inscriptions indicating what was to be foretold by the condition of each part. Every conceivable human activity or experience—from the knotting of strings to the interpretation of dreams—has become an oracle, witnessing man's desperate eagerness for clues to his future.

By contrast with these other kinds of prophecy, astrology was progressive. Astrology differed in asserting the continuous, regular force of a power at a distance. The influences of heavenly bodies on the events on earth it described as periodic, repetitious, *invisible* forces like those that would rule the scientific mind.

It is not surprising that earliest man was awed by the heavens and enticed by the stars. These first night-lights which inspired the priests of ancient Babylonia also sparked the popular fancy. The changelessness of the rhythm of life on earth made the shifting fireworks of the sky into melodrama. The coming and going of the stars, their rising and falling, their moving about the heavens, became the conflicts and the adventures of the gods.

If the rising and setting of the sun made so much difference on earth, why not also the movement of the other heavenly bodies? The Babylonians made the whole sky a stage for their mythological imagination. Like the rest of nature, the heavens were a scene of living drama. Like the entrails of sacrificial victims, the heavens were divided into zones and then peopled with fantastic figures. The evening star, later called the planet Venus—the brightest heavenly object next to the sun and moon—became a luminous lion roaming the sky from east to west. The great god El, jealous of so bright and high-rising a luminary, put the lion to death again at every dawn. The Old Testament presents this fantasy in the vision of Lucifer cut down for his pride: "How art thou fallen from heaven, O Lucifer, son of the morning! . . . For thou hast said in thine heart, I will ascend into heaven, I will exalt my throne above the stars of God. . . . I will ascend above the heights of the clouds; I will be like the most High" (Isaiah, 14:12–14) This diurnal assassination was accomplished by the messenger of El, Michael (meaning "Who is like El"?). In the sky the gods fought battles, made love, formed alliances, and hatched conspiracies. How inconceivable that these cosmic events should not affect life on earth! Every farmer knew that the clouds in the sky, the warmth of the sun, and the heavenly gift of rain decided the fate of crops and so really governed his own life. Of course, the subtler, more obscure heavenly events required proper interpretation by priests.

This lure of the skies produced a fertile lore of the skies. The powers of sun and rain, the *correspondence* between happenings in the heavens and happenings on earth, stirred the search for other correspondences. The Babylonians were among the first who elaborated a mythological frame for these universal correspondences. Their vivid imaginings would be perpetuated by Greeks, Jews, Romans, and others over the following centuries.

The theory of correspondences became astrology, which sought new links between space and time, between the movements of physical bodies and the unfolding of all human experience. The growth of science would depend on man's willingness to believe the improbable, to cross the dictates of common sense. With astrology man made his first great scientific leap into a scheme for describing how unseen forces from the greatest distance, from the very depth of the heavens, shaped everyday trivia. The heavens, then, were the laboratory of mankind's first science, just as the interior of the human body, the intimate inward realm of his consciousness, and the Dark Continents in the atom, would be the scenes for his latest sciences. Man sought to use his growing knowledge of the patterns of repeating experience in his neverending struggle to break the iron ring of repetition.

Social or wholesale prophecy flourished in Babylonia. It forecast the large events—battles, droughts, plagues, harvests—which affected the whole community. For centuries such astrology remained a lore rather than a doctrine. The Greeks made a science of it. Personal astrology—"judicial" astrology, or genethlialogy—which cast *a person's* fortune from the position of the heavenly bodies at the moment of his birth, developed more slowly. The person was called the "native," and the prophecy came to be called a "nativity," or horoscope.

The Greeks, too, were torn between wanting to know the good news and fearing to know the bad. Their medical astrologers divided the whole sky according to the signs of the zodiac, then assigned a particular stellar force to each part of the body. Then Greek anti-astrologers attacked the whole dogma of astral forces with arguments that would last into modern times. The names assigned to the stars, the anti-astrologers argued, were quite accidental. Why should this planet be called Mars and that Saturn or Venus? And why did the astrologers limit their horoscopes to human beings? Would not the same astral fortunes govern all animals? And how could astrologers explain the different fortunes of twins? The Epicureans, whose philosophy was built on the belief in each man's freedom to shape his destiny, attacked astrology as a way of making men think they were mere slaves of the stars.

In ancient Rome, astrology attained an influence seldom equaled in later centuries. Astrologers—called *Chaldaei* after the Chaldean or Babylonian origins of the science, or *Mathematici* from their astronomical calculations

—were a recognized profession whose repute varied with the turbulent times. Under the Roman Republic they became so powerful and so unpopular that in 139 B.C. they were expelled not only from Rome but from all Italy. Afterwards, under the empire, when their dangerous prophecies had brought several astrologers to trial for treason, they were repeatedly banished. But the same emperor who would banish some astrologers for their ominous forebodings, would employ others to guide his imperial household. Some areas were declared off limits. In the late empire, even when astrologers were tolerated or encouraged, they were forbidden to make prophecies about the life of the emperor.

Christian emperors failed in their efforts to discourage astrology. "There are many," the historian Ammianus Marcellinus recounted in the late fourth century after Constantine officially converted the empire to Christianity, "who do not presume either to bathe, or to dine, or to appear in public, till they have diligently consulted, according to the rules of astrology, the situation of Mercury and the aspect of the moon. It is singular enough that this vain credulity may often be discovered among the profane sceptics, who impiously doubt or deny the existence of a celestial power." By that time the powers of the seven planets were attested by the quiet transformation of the eight-day week into the week of seven days, with each day now subordinate to one of the seven planets. When Romans attended the imperial circus, the astral powers were seen everywhere. Over each of the twelve stalls from which the chariots started their race, there appeared a sign of one of the twelve constellations of the zodiac. Each of the seven tracks of the racecourse was intended to represent a heavenly circuit of one of the seven planets.

3

God and the Astrologers

ASTROLOGY married the human needs which later centuries would divorce into science and religion. Was astrology in ancient Rome, as historians commonly say, simply a superstitious fatalism, a triumph of the irrational? There is no denying that awe of the stars—those "visible gods"—inspired an awe of astrologers. "He to whom the gods themselves reveal the future," observed Arellius Fuscus, an eminent rhetorician of the Augustan Age, "who imposes his will even on kings and peoples, cannot be fashioned by the same womb which bore us ignorant men. His is a superhuman rank. Confident of the gods, he is himself divine. . . . let us lift up our minds by means of the science which reveals to us the future, and before the appointed hour of death let us taste the pleasures of the Blest."

But astral religion was not to be separated from astral science. The leading scientists took for granted the influence of the stars on human events. They disagreed only over *how* the stars exerted their powers. The great scientific encyclopedia of the age, Pliny's *Natural History,* propagated the rudiments of astrology by showing the influence of the stars everywhere. Seneca's only complaint was that the astrologers were not comprehensive enough. "What? Think you so many thousand stars shine on in vain? What else, indeed, is it which causes those skilled in nativities to err than that they assign us to a few stars, although all those that are above us have a share in the control of our fate? . . . But even those stars that are motionless, or because of their speed keep equal pace with the rest of the universe and seem not to move, are not without rule and dominion over us."

The most influential of all ancient Roman scientists proved to be the most durable authority on astrology. Ptolemy (Claudius Ptolemaeus of Alexandria) provided the solid treatise which would give substance and respectability to this science for the next thousand years. But his reputation has suffered from the overadvertised fate of two of his crucially erroneous theories. Both were current in his time, both were developed and perpetuated in his writings. The geocentric, or Ptolemaic, theory of the universe would become a modern byword for astronomical error. Similarly, the land-dominant view, the view that dry land comprised most of the earth's surface, became a byword for geographic error. These two popular misconceptions were destined to obscure Ptolemy's colossal achievements. But never since Ptolemy has anyone provided so comprehensive a survey of the whole scientific knowledge of an age.

Yet the life of this encyclopedic genius remains a mystery. Probably descended from Greek immigrants, Ptolemy (90–168) lived in Egypt during the reigns of Emperors Hadrian and Marcus Aurelius. His Alexandria had continued to be a great center of learning even after its famous library was burned by Caesar in 48 B.C.

Ptolemy dominated the popular and literary view of the universe throughout the Middle Ages. The world as depicted in Dante's *Divine Comedy* came right out of Ptolemy's *Almagest.* In many ways Ptolemy spoke as a prophet. For he enlarged the uses of mathematics in the service of science. While he drew on the best observations made before him, he emphasized the need for repeated, increasingly precise observation. In fact,

Ptolemy was a harbinger of the scientific spirit, an unsung pioneer of the experimental method. In trigonometry, for example, his table of chords has been found accurate to five decimal places. In spherical geometry he offered an elegant solution to the problems of sundials, which had special significance in that age before mechanical clocks. There was no branch of physical science that he did not survey and organize into newly usable forms. Geography, astronomy, optics, harmonics—each he expounded in a system. The best known of these was his treatise on astronomy, the *Almagest.* His *Geography,* which aimed to map the whole known world, pioneered in listing places systematically by latitude and longitude. There, too, he offered his own improved method for projecting spherical surfaces on flat maps. In view of the scanty factual reports available to him, Ptolemy's maps of his "known world," the Roman Empire, were a remarkable accomplishment. He showed the crucial scientific talents—shaping theories to fit available facts and testing old theories by new facts.

The Arabs appreciated the grandeur of Ptolemy's work and brought him to the West. His astronomy was destined to bear an Arabic name (*Almagest,* from *al majisti,* "the greatest compilation"), and his *Geography* was translated into Arabic early in the ninth century. The *Tetrabiblos,* his four books on astrology, which Ptolemy regarded as the companion to his *Almagest,* also came West through the Arabic.

"Mortal as I am," wrote Ptolemy, "I know that I am born for a day, but when I follow the serried multitude of the stars in their circular course, my feet no longer touch the earth; I ascend to Zeus himself to feast me on ambrosia, the food of the gods." He helped others find a refuge from earthly cycles into heavenly mysteries. Ptolemy's *Tetrabiblos* became the leading textbook of astrology, one of our best guides to science in the Middle Ages. While his *Almagest* predicted the changing positions of heavenly bodies, his astrology predicted their influences on earthly events. Did not the cycles of the sun and the moon obviously affect happenings on earth? Then why should not lesser celestial bodies also affect events down here? If illiterate sailors could forecast the weather from the sky, could not learned astrologers use the data of the heavens to forecast human events? Ptolemy saw the astral influence as purely physical, as only one of many forces. He conceded that astrology was of course no more foolproof than any other science. But that was no reason why careful observation of the correspondence of earthly events with celestial events should not produce some useful, though not mathematically certain, predictions.

In this practical spirit, Ptolemy laid the foundation of the most durable of the occult sciences. Of the four books of his *Tetrabiblos,* the first two, on "astrological geography" and weather prediction, cover the influences of celestial bodies on earthly physical events, and the last two cover their influence on human events. Ptolemy expounds the science of horoscopes, the prediction of human destinies from the position of the stars at the moment of a person's birth. Although Ptolemy's work did become the leading textbook of astrology for a thousand years, because he ignored "catarchic"—the technique of answering questions about the future by the position of heavenly bodies at the time the question was asked—his work did not fully satisfy the needs of astrological practitioners.

Ptolemy's adventure into the occult world of astrology would even outlive his magisterial works in the more familiar realms of modern science. Copernicus' epoch-making *De Revolutionibus* (1543), which changed the center of the solar system, still confirmed in its form and much of its content the dominant influence of Ptolemy's *Almagest.* Not until a half century later, when Tycho Brahe's *Astronomiae Instauratae Mechanica* (1598) displaced Ptolemy's star catalogue with a new catalogue based entirely on independent observations, did Ptolemy's data, as well as his theory, finally become obsolete. Ptolemy's geographic speculations about *terra incognita,* and about other parts of the world remote from Europe, had already begun to be out of date when his *Geography,* translated from Greek into Latin, reached Western Europe in 1406. Still, Ptolemy's *Geography* long remained popular as the best guide to the "known world" in the West. Maps published in the fifteenth and sixteenth centuries, including Mercator's own great map of Europe that appeared in 1554, were commonly based on Ptolemy, whose technique of map projection continued to stimulate cartography throughout the sixteenth century. Meanwhile, Ptolemy's *Tetrabiblos* remained the bible of astrology. It was republished twice—once in England, once in Germany—during World War II, when it still was thought useful.

Astrology expressed a crucial transformation of human feelings in the land-bound age. There was a world of difference between the old wine-inspired Dionysiac intoxication that Euripides describes in his *Bacchae* and the new star-inspired ecstasy, the new astral mysticism. Now, as the historian of religion Franz Cumont observes, "it is with pure light that reason quenches her thirst for truth; and 'the abstemious intoxication,' which exalts her to the stars, kindles in her no ardour save a passionate yearning for divine knowledge. The source of mysticism is transferred from earth to heaven."

The popular claims of pagan astrologers disturbed the early prophets of Christianity. Church Fathers who declared their own power to forecast everyman's fate in the next world begrudged the powers of prophecy to those who pretended to know any man's destiny on earth. If the astrologers' horoscopes meant what they said, where was the room for free will, for freedom to choose good over evil, to forsake Mammon or Caesar for Jesus Christ?

The very struggle to become a Christian—to abandon pagan superstition for Christian free will—seemed to be a struggle against astrology. Saint Augustine (354–430) recalls in his *Confessions:* "Those impostors then, whom they style Mathematicians [Astrologers], I consulted without scruple; because they seemed to use no sacrifice, nor to pray to any spirit for their divinations. ' And he was tempted by the astrologers' counsel: "The cause of thy sin is inevitably determined in heaven; this did Venus, or Saturn, or Mars: That man, forsooth, flesh and blood, and proud corruption, might be blameless; while the Creator and Ordainer of heaven and the stars is to bear the blame."

Saint Augustine struggled to reject "the lying divinations and impious dotages of the astrologers." Two acquaintances reminded him "that there was no such art whereby to foresee things to come, but that men's conjectures were a sort of lottery, and that out of many things which they said should come to pass, some actually did, unawares to them who spake it, who stumbled upon it, through their oft speaking."

At this providential moment of his burgeoning doubts, God "provided then a friend for me, no negligent consulter of the astrologers; nor yet well skilled in those arts, but . . . a curious consulter with them, and yet knowing something which he said he had heard of his father, which how far it went to overthrow the estimation of that art, he knew not."

A story told by this friend, Firminus, shook the young Augustine from his pagan faith. Firminus' father, an earnest experimenter in astrology, always noted the positions of the stars and even "took care with the most exact diligence to know the birth of his very puppies." Firminus' father learned that one of his women-servants was to be delivered of a child at about the same time that Firminus' mother was expecting. "Both were delivered at the same instant; so that both were constrained to allow the same constellations, even to the minutest points, the one for his son, the other for his new-born slave. For so soon as the women began to be in labour, they each gave notice to the other what was fallen out in their houses, and had messengers ready to send to one another, so soon as they had notice of the actual birth, of which they had easily provided, each in his own province, to give instant intelligence. Thus then the messengers of the respective parties met, he averred, at such an equal distance from either house, that neither of them could make out any difference in the position of the stars, or any other minutest points; and yet Firminus, born in a high estate in his parents' house, ran his course through the gilded paths of life, was increased in riches, raised to honours; whereas that slave continued to serve his masters, without any relaxation of his yoke, as Firminus, who knew him, told me." The diverse destiny of twins struck Augustine as the most obvious and appealing argument against astrology.

Not only in the autobiographical *Confessions* but even in his great theoretical work, *The City of God,* he labors lengthily against the astrologers. The Roman Empire, and all other kingdoms, he warns, have their destiny shaped not by the stars but by the will of God. His clinching Biblical argument is the example of Jacob and Esau—"two twins born so near together that the second held the first by the heel; yet in their lives, manners, and actions, was such a disparity, that that very difference made them enemies one to another." And he elaborates on the cases of other twins.

Astrology remained the *bête noire* of the Christian Church Fathers. Faith in a star-written destiny had dissuaded Romans, such as Emperor Tiberius, from paying homage even to their pagan gods. Tertullian (c. 160–c. 230) warned against astrology because "men, presuming that we are disposed of by the immutable arbitrament of the stars, think on that account that God is not to be sought after."

Shrewd Christian theologians of the Middle Ages managed to find holy uses for the widespread belief in astral powers. Both Albertus Magnus and Saint Thomas Aquinas admitted the strong governing influence of the stars, but they insisted that man's freedom was his very power to resist that influence. Even if astrologers often did make true predictions, Aquinas explains, these usually concerned events in which large numbers of men were involved. In such cases the passions of the many would prevail against the rational wisdom of the few; but the individual Christian's free will had not been exercised.

The great medieval theologians earnestly enlisted the prevailing belief in astrology to reinforce the truths of Christianity. They liked to recall the astrological prediction of the Virgin Birth of Jesus Christ. If Jesus Christ was not Himself subject to the rule of the stars, the stars did give signs of His coming. What else was the Star of Bethlehem? Was it not likely that those wise enough to follow that star, the *Magi,* were actually learned astrologers?

PART TWO

FROM
SUN TIME
TO
CLOCK TIME

The gods confound the man who first found out
How to distinguish hours! Confound him, too,
Who in this place set up a sun-dial,
To cut and hack my days so wretchedly
Into small portions.

—PLAUTUS (c. 200 B.C.)

4

Measuring the Dark Hours

So long as mankind lived by raising crops and herding animals there was not much need for measuring small units of time. The seasons were all-important—to know when to expect the rain, the snow, the sun, the cold. Why bother with hours and minutes? Daylight time was the only important time, the only time when men could work. To measure useful time, then, was to measure the hours of the sun.

No change in daily experience is more emptying than the loss of the sense of contrast between day and night, light and dark. Our century of artificial light tempts us to forget the meaning of night. Life in a modern city is always a time of mixed light and darkness. But for most of the human centuries night was a synonym for the darkness that brought all the menace of the unknown. "Never greet a stranger in the night," warned the Talmud (c. 200 B.C.), "for he may be a demon." "I must work the works of him that sent me, while it is day," announced Jesus (John 9:4–5), "the night cometh, when no man can work. As long as I am in the world, I am the light of the world." Few subjects have been more enticing to the literary imagination. "In the dead vast and middle of the night" was when Shakespeare and other dramatists placed their crimes.

> O comfort-killing night, image of Hell;
> Dim register and notary of shame;
> Black stage for tragedies and murders fell;
> Vast sin-concealing chaos, nurse of blame.

The first step in making night more like day was taken long before people became accustomed to artificial lighting. It came when man, playing with time, began measuring it off into shorter slices.

While the ancients measured the year and the month, and set the pattern for our week, the shorter units of time remained vague and played little part in the common human experience until the last few centuries. Our precise uniform hour is a modern invention, while the minute and the second are still more recent. Naturally enough, when the working day was the sunlit day, the first efforts to divide time measured the passing of the sun across the heavens. For this purpose sundials, or shadow clocks, were the first measuring devices. The original meaning of our English word "dial," which since has taken on so many other meanings (from the Latin *dies,* or day; medieval Latin *dialis*), was sundial. Primitive societies noticed that the shadow of an upright post (or *gnomon,* from the Greek "to know") became shorter as the sun rose in the heavens, and lengthened again as the sun set. The ancient Egyptians used such a device, and we can still see one that survives from the time of Thutmose III (c. 1500 B.C.). A horizontal bar about a foot long had a small T-shaped structure at one end which would cast a shadow on the calibration on the horizontal bar. In the morning the bar was set with the T facing east; at noon the device was reversed with the T facing west. When the prophet Isaiah promised to cure King Hezekiah by making time go backward, he announced that he would do so by making the sun's shadow recede.

For centuries the sun's shadow remained the universal measure of time. And this was a handy measure, since a simple sundial could be made anywhere by anybody without special knowledge or equipment. But the cheery boast "I count only the sunny hours," inscribed on modern sundials, announces the obvious limitation of the sundial for measuring time. A sundial measures the sun's shadow: no sun, no shadow. A shadow clock was useful only in those parts of the world where there was lots of sunlight, and then it served only when the sun was actually shining.

Even when the sun shone bright, the movement of the sun's shadow was so slow that it would be small help in marking minutes and useless for signaling seconds. The dial that marked the day's passing in any one place was ill suited to measure a universal standard unit, such as our hour of sixty minutes. For everywhere except on the equator the length of daylight hours varies from one day to the next and around the seasons. To use the sun's shadow in any place to define the hour according to Greenwich Mean Time requires a combined knowledge of astronomy, geography, mathematics, and mechanics. Not until about the sixteenth century could sundials be marked with these true hours. When this "science of dialing" was developed, it became fashionable to carry a pocket sundial. But by then the clock and the watch existed and were more convenient and useful in every way.

The early sundials had other limitations. The horizontal instrument of Thutmose III could not record the earliest hour in the morning or the latest hour of the afternoon because the shadow of the horizontal T-bar would stretch to infinite length and so not register on the scale. The great ancient advance in sundial design, while no help in describing universal time, did make it easier to divide the daylight hours into equal parts. This was a sundial shaped into a hemicycle, the interior of a half-sphere, with the pointer extended from one side to the center and the opening facing upward. The shadow path during any day, then, would be a perfect replica of the sun's path in the hemisphere of the sky above. This arc traced by the sun and marked inside the hemicycle was divided into twelve equal parts. After paths were drawn for different dates, the twelve "hour" divisions for each date were joined with curves, indicating the varying twelfth of the daylight hours.

The Greeks, adept at geometry, succeeded in making many advances in sundial design. One delightful example survives in the Tower of the Winds in Athens. On this eight-sided tower, each of the eight principal directions is personified in its winds, and each face bears a sundial, so that an Athenian could read the time on at least three faces at once. The sundial became so common a Roman fixture that the architect Vitruvius, in the first century B.C., could list thirteen kinds of sundials. But the many beautiful monumental sundials that the Romans looted abroad to decorate their villas were nearly useless as timepieces in the Roman latitude. If we believe Plautus (d. 184 B.C.), the Romans relied on sundials to fix their mealtimes:

> The gods confound the man who first found out
> How to distinguish hours! Confound him, too,
> Who in this place set up a sun-dial,
> To cut and hack my days so wretchedly
> Into small portions. When I was a boy,
> My belly was my sun-dial; one more sure,
> Truer, and more exact than any of them.
> This dial told me when 'twas proper time
> To go to dinner, when I had aught to eat.
> But now-a-days, why even when I have,
> I can't fall-to, unless the sun give leave.
> The greater part of its inhabitants,
> Shrunk up with hunger, creep along the streets!

Even after the sundial was designed to divide the daylight time into equal segments, it was not much help in comparing times from one season to another. Summer days were long, so too were summer hours. Roman soldiers, in the reign of Emperor Valentinian I (364–375), were drilled to march "at the rate of 20 miles in five *summer* hours." An "hour"—one-twelfth of the daylight—on a particular day in one place would be quite different from that on another day or in another place. The sundial was an elastic yardstick.

How was mankind liberated from the sun? How did we conquer the night, making it part of the intelligible world? Only by escaping the sun's tyranny would we ever learn to measure out our time in universally uniform spoonfuls. Only then could the recipes for action, for making and doing be understood anywhere anytime. Time was, in Plato's phrase, "a moving image of eternity." No wonder that measuring its course tantalized mankind all over the planet.

Anything that would flow, that would be consumed or would consume has been tried somewhere or another as a measure of time. All were efforts to escape the tyranny of the sun, to grasp time more firmly, more predictably and bring it into the service of man. The simple universal measure needed every day, the measure of life itself, would have to be something better than the whimsical, fleeting, slow-moving, often obscured shadow. Man must find something better than the Greeks' "timepiece they call Hunt-the-Shadow."

Water, that wonderful, flowing medium, the luck of the planet—which would serve humankind in so many ways, and which gives our planet a special character—made possible man's first small successes in measuring the dark hours. Water, which could be captured in any small bowl, was more manageable than the sun's shadow. When mankind began to use water to serve him for a timepiece, he took another small step forward in making the planet into his household. Man could make the captive water flow fast or slow, day and night. He could measure out its flow in regular, constant units, which would be the same at the equator or on the tundra, winter or summer. But perfecting this device was long and difficult. By the time the water clock was elaborated into a more or less precision instrument, it had already begun to be supplanted by something far more convenient, more precise, and more interesting.

Yet, for most of history, water provided the measure of time when the sun was not shining. And until the perfection of the pendulum clock about 1700, the most accurate timepiece was probably the water clock. During all those centuries, the water clock ruled man's daily—or, rather, his nightly —experience.

Man discovered very early that he could measure the passage of time by the amount of water that dripped from a pot. Within five hundred years after their first sundials, the ancient Egyptians were using water clocks. In their sunny country, the sundial well served their needs by day, but they needed the water clock to measure off hours of night. Their god of the night, Thoth, who was also god of learning, of writing, and of measurement, presided over both outflow and inflow models of water clocks. The outflow type was an alabaster vessel with a scale marked inside and a single hole near the bottom from which the water dripped. By noting the drop in the water level inside from one mark to the next mark below on the scale, the passage of time was measured. The later inflow type, which marked the passing time by the rise of water in the vessel, was more complicated, as it required a constant source of regulated supply. Even such simple devices were not without their problems. In cold climates the changing viscosity of the water would be troublesome. But, in any climate, to keep the clock at constant speed, the hole through which the water poured must not become clogged or worn larger. The outflow clocks posed another minor problem because the speed of flow depended on the water pressure, and that always varied with the amount of water left in the bowl. The Egyptians therefore slanted the walls of the vessel so that the height declined by the same amount for equal periods of time.

The problem of designing a useful water clock was simple enough so long as its only purpose was like that of a modern egg timer, which marks the passage of short uniform units of time. But to use the water clock as an instrument for dividing the hours of daylight or of darkness into equal segments posed a difficult problem of calibration. The winter night in Egypt was, of course, longer than the night in summer. The water clock at Thebes required, according to Egyptian measures, that the summer night be measured by only twelve fingers of water, while the winter night required fourteen. These variant "hours"—equal subdivisions of the total hours of daylight or darkness—were not really chronometric hours. They came to be called "temporary hours" or "temporal" hours, for they had meaning that was only temporary and did not equal an hour the next day. It would have been much simpler to make a water clock measure off a fixed, unchanging unit. But centuries passed before abstracted time was captured by a machine that measured something other than a fragment of daylight or darkness.

The Greeks, who had perfected the sundial to measure their daylight, also used the water clock as an everyday timer. Their picturesque name, *klepsydra*, meaning water-thief, would designate the device for centuries to come. They used their water clock to limit the times for pleading in Athenian courts. The surviving court clocks flow for about six minutes. Demosthenes in his legal speeches, referring to the time running out in the water clock, often asked that the flow of the water be stopped while he read from laws or depositions so that his speaking time would not be used up. The elegant Tower of the Winds had attached to it a circular cistern as a reservoir for a water clock. Ctesibius of Alexandria (second century B.C.), the ingenious Greek physicist and inventor, who also devised a hydraulic organ and an air gun, contrived a water clock with a floating indicator to mark the time on a vertical scale set above.

Adept though they were at engineering and mechanics, the Romans relied on the water clock as their only mechanical device, apart from the sundial, for measuring time. Latinizing *klepsydra* to *clepsydra*—or the *horologium ex aqua*—they elaborated and popularized it into an everyday convenience. They made miniature sundials measuring only one and a half inches across for carrying about in the pocket. At the same time the Roman feeling for the grandiose was displayed on the Campus Martius in the great obelisk of Montecitorio, serving as gnomon for a giant sundial whose shadow was measured on lines of bronze in the surrounding marble pavement.

The Romans showed a similar versatility with their water clocks. Like other practical, commercial people, they were alert to the value of time. But only gradually, and then only crudely, did they divide their day into smaller parts. They never invented a mechanical clock that would conveniently subdivide the hours. Even at the end of the fourth century B.C. they formally divided their day into only two parts: before midday (*ante meridiem*, A.M.) and after midday (*post meridiem*, P.M.). An assistant to the consul was assigned to notice when the sun crossed the meridian, and to announce it in the Forum, since lawyers had to appear in the courts before noon. Eventually they made finer subdivisions. First by dividing each half of the day into two parts: early morning (*mane*) and forenoon (*ante meridiem*); afternoon (*de meridie*) and evening (*suprema*). Then they took to marking the "temporary" hours according to a sundial that had been brought from Catana in Sicily. Made for a different latitude, it was hardly precise. Finally in 164 B.C. the censor Q. Marcius Philippus earned popularity by setting up a sundial properly oriented for Rome. Beside the sundial a waterclock was installed to tell time on foggy days and at night.

The Romans used their sundials to calibrate and set the water clocks, which had become the common timepiece in imperial Rome. The water clocks still offered only "temporary" hours, with daylight and darkness measurements for all the days of a month lumped together, though these really varied from day to day. Since no one in Rome could know the exact hour, promptness was an uncertain, and uncelebrated, virtue. It was as impossible to find agreement among the clocks of Rome, the wit Seneca (c. 4 B.C.–A.D. 65) observed, as to find agreement among Roman philosophers.

The "hours" of their daily lives—their temporary "hour" was one-twelfth the time of daylight or of darkness on that day—were more elastic than we can now imagine. At the winter solstice, even if the sun shone all day there would be, by our modern measures, only 8 hours, 54 minutes of sunlight, leaving a long night of 15 hours, 6 minutes. At the summer solstice the time, by our modern hours, was exactly reversed. But from the Romans' point of view both day and night always had precisely 12 hours year round. In Rome at the winter solstice the *first* hour of their day (*hora prima*) began at what we would call 7:33 A.M. and lasted only till 8:17 A.M., while the *twelfth* hour (*hora duodecima*) began at 3:42 P.M. and expired at 4:27 P.M., when the longer night hours began. What a problem for the clockmakers! We must be amazed not that they did not provide a more precise timepiece, but that under these circumstances they were able to provide an instrument that served daily needs at all.

By elaborate systems of calibration they made their water clocks indicate the shifting lengths of hours from month to month. It was far too complicated to mark the shifting increments from day to day. This meant, too, that there was no accepted way of subdividing each day's passing hours.

When daily needs required shorter standard units, a simple water clock served them with all the precision of an egg timer. In the Roman courtroom, for example, where the lawyers of opposing parties were supposed to have equal time, the simple water clock worked well. For this purpose they followed the Athenian example, using a bowl with a hole near the bottom. This timer emptied in about twenty minutes. A lawyer might ask the judge to grant him an additional "six clepsydrae," or about two of our modern hours, to make his case. A particularly long-winded advocate was once actually granted sixteen water clocks—five hours! While the Romans doubtless shared our view that "time is money," they often equated time with water. In Rome the phrase *aquam dare,* "to grant water," meant to allot time to a lawyer, while *aquam perdere,* "to lose water," meant to waste time. If a speaker in the Senate spoke out of turn or talked too long, his colleagues would shout that his water should be taken away. Under other circumstances they might petition that more water be allowed.

Lawyers were no less wordy then than they are today. One especially tiresome advocate inspired the Roman wit Martial (c. 40–c. 102) to suggest:

> Seven water-clocks' allowance you asked for in loud tones, Caecilianus, and the judge unwillingly granted them. But you speak much and long, and with back-tilted head, swill tepid water out of glass flasks. That you may once for all sate your oratory and your thirst, we beg you, Caecilianus, now to drink out of the water-clock!

For every bowlful of water the lawyer drank he would reduce the judge's boredom by twenty minutes.

The simple water clock challenged the Romans' ingenuity. To prevent the escape hole from wearing away or being clogged, the hole was fashioned from a gem, much as later mechanical clockmakers used "jewels." Some of the Roman water clocks described by the architect Vitruvius were fitted with elaborate floats which announced the Roman "hour" by tossing pebbles—or eggs—into the air, or by blowing a whistle. The water clock, like the piano in middle-class European households in the nineteenth century, became a symbol of status. "Has he not got a clock in his dining-room?" observed the admirers of the parvenu Trimalchio in the Age of Nero. "And a uniformed trumpeter to keep telling him how much of his life is lost and gone?"

In later centuries people everywhere, after their fashion, found ways to use water to mark off the portions of life. The Saxons in the ninth century characteristically used a bowl of strong and rustic elegance. The bowl, with a small hole in the bottom, was floated on water and sank as it filled, always marking the same period. The Chinese, who had their own simple water clocks from remote antiquity, returned home from Western travels with astonishing tales of complicated striking water clocks. They especially admired a giant water clock that ornamented the east gate of the Great Mosque at Damascus. At each "hour" of the day or night two weights of brightly shining brass fell from the mouths of two brazen falcons into brazen cups, perforated to allow the balls to return into position. Above the falcons was a row of open doors, one for each "hour" of the day, and above each door was an unlighted lamp. At each hour of the day, when the balls fell, a bell was struck and the doorway of the completed hour was closed. Then, at nightfall, the doors all automatically opened. As the balls fell announcing each "hour" of the night, the lamp of that hour was lit, giving off a red glow, so that finally by dawn all the lamps were illuminated. With the coming of daylight the lamps were extinguished and the doors of daytime hours resumed their cycle. It required the full time of eleven men to keep this machine in working order.

Not the flowing waters of time but the falling sands of time have given modern poets their favorite metaphor for the passing hours. In England, sandglasses were frequently placed in coffins as a symbol that life's time had run out. "The sands of time are sinking," went the hymn. "The dawn of heaven breaks."

But the hourglass, measuring time by dripping sand, comes late in our story. Sand was, of course, less fluid than water, and hence less adapted to the subtle calibration required by the variant "hours" of day and night in early times. You could not float an indicator on it. But sand would flow in climates where water would freeze. A practical and precise sandglass required the mastery of the glassmaker's art.

We hear of sand hourglasses in Europe in the eighth century, when legend credits a monk at Chartres with their invention. As glassmaking progressed it became possible to seal the hourglass to keep out the moisture that slowed the fall of the sand. Elaborate processes dried the sand before it was inserted in the glass. A medieval treatise prescribed in place of sand a fine-ground black-marble dust, boiled nine times in wine. At each boiling, the scum was skimmed off, and finally the dust was dried in the sun.

Sandglasses were ill adapted for daylong timekeeping. Either they had to be made too large for convenience—like the sandglass Charlemagne ordered which was so large that it had to be turned only once in twelve hours —or, if they were small, they had to be turned frequently at the precise moment when the last grain had dropped. Some had a little dial attached with a pointer that could be advanced with each turn of the glass. But the sandglass did serve better than a water clock for measuring the shortest intervals when no other device was yet known. Columbus on his ships noted the passing time by a half-hour sandglass that was turned as it emptied to keep track of the seven "canonical" hours. By the sixteenth century the sandglass was already being used to measure short intervals in the kitchen. Or to help a preacher (and his congregation!) regulate the length of his sermon. An English law of 1483 was said to require clocks to be placed *over* pulpits, since congregations could not otherwise see the "sermon-glass." The House of Commons kept a two-minute glass to time the ringing of bells to announce divisions for voting. Stonemasons and other craftsmen used a glass to count their hours of work. Teachers brought their hourglass along to measure the duration of their lecture or the length of the students' prescribed study period. An Oxford don in Elizabethan times once threatened his idle pupils "that if they did not doe their exercise better he would bring an Hower-glasse two howers long."

The unique use of the sandglass, after the sixteenth century, was measuring a ship's speed. Knots were tied at seven-fathom intervals on a line tied to a log chip that would float astern. A sailor dropped the log chip off the end of the speeding ship and counted off the number of knots paid out while a small sandglass measured a half-minute. If five knots passed in the interval, the ship was making five nautical miles an hour. Throughout the nineteenth century, sailing vessels still "heaved the log" every hour to keep track of the speed.

In the long run the sandglass was not much use in measuring the hours of the night, for it was inconvenient to keep the glass turning. From time to time, as a way of solving this problem, people tried to combine a timepiece with a lighting device. For centuries, ingenuity was lavished on the effort to use the fire that illuminated the night also to measure the passing hours of darkness. The inventions, however original, were not practical. They were costly, sometimes dangerous, and never succeeded in aligning night hours with day hours. So long as "hours" remained elastic, a fire clock, like a sandglass, would measure off a short, fixed unit but could not be widely used for daylong timekeeping.

A famous candle clock was that reputedly designed to help the pious Alfred the Great (849–899), king of the West Saxons, keep the vow he made when he was a fugitive from his native country. He swore that if his kingdom was restored, he would devote a full third of each day to the service of God. According to legend, when he was back in England he ordered a candle clock. From seventy-two pennyweight of wax, six twelve-inch candles were made, all uniform in thickness, each marked in one-inch divisions. The candles were lit in rotation, and the six candles were said to last a full twenty-four hours. They were protected by transparent horn panels set into wooden frames to prevent the light from being extinguished by a draft. If King Alfred devoted the time of two full candle lengths to his religious duties, he could be assured that he was fulfilling his vow.

Other sovereigns who could afford to use candles or lamp oil for the purpose of timekeeping—King Alfonso X of Castile (c. 1276), King Charles V of France (the Wise, 1337–1380), King Philip I of Spain (1478–1506)—experimented with clock lamps. The search for a practical portable clock lamp led a Milanese physician, Girolamo Cardano (1501–1576), to invent a fountain-feeding device that used the principle of the vacuum to draw in a constant flow of oil. Cardano's lamp provided a convenient and popular lighting device until the late eighteenth century.

Even after mechanical clocks came into general use, restless inventors continued to try all sorts of expedients—some using the flame of an oil lamp to propel the mechanism of a clock, others using the consumption of oil indicated on a calibrated transparent container, still others using the changing shadow of a diminishing candle cast on a scale that marked the changing night hours—to conquer both night and time with the same device.

In China, Japan, and Korea, the use of fire to measure time took quite another turn. The custom of burning incense gave them a clue to a range of ingenious and beautiful devices. These produced a pleasing aroma while a continuous trail of powdered incense was burned in an elaborate seal. The time was indicated by the place within the seal that was reached by the fire. One of the most intricate of these—the "hundred-gradations incense seal" —was invented in China in 1073, when a drought had dried up the wells and so made it impossible to use the customary water clocks. The Chinese aromatic clock, in turn, inspired later generations to find newly vivid and elaborate ways to use a fire clock to measure time into temporary hours, which varied with the seasons. The charming intricacy of the Chinese designs was a delightful by-product of the effort to make a virtue of the variant hour.

There seems no end to the desperate ingenuity spent on ways to count the passing hours of the night before inexpensive artificial illumination became universal. After the invention of the mechanical clock the striking hour was the obvious way of conquering darkness. A clever French inventor in the late seventeenth century, M. de Villayer, tried using the sense of taste. He designed a clock so arranged that when he reached for the hour hand at night, it guided him to a small container with a spice inserted in place of numbers, a different spice for each hour of the night. Even when he could not see the clock, he could always taste the time.

5

The Rise of the Equal Hour

WHILE man allowed his time to be parsed by the changing cycles of daylight he remained a slave of the sun. To become the master of his time, to assimilate night into the day, to slice his life into neat, usable portions, he had to find a way to mark off precise small portions—not only equal hours, but even minutes and seconds and parts of seconds. He would have to make a machine. It is surprising that machines to measure time were so long in coming. Not until the fourteenth century did Europeans devise mechanical timepieces. Until then, as we have seen, the measuring of time was left to the shadow clock, the water clock, the sandglass, and the miscellaneous candle clocks and scent clocks. While there was remarkable progress five thousand years ago in measuring the year, and useful week clusters of days were long in use, the subdivided day was another matter. Only in modern times did we begin to live by the hour, much less by the minute.

The first steps toward the mechanical measurement of time, the beginnings of the modern clock in Europe, came not from farmers or shepherds, nor from merchants or craftsmen, but from religious persons anxious to perform promptly and regularly their duties to God. Monks needed to know the times for their appointed prayers. In Europe the first mechanical clocks were designed not to *show* the time but to *sound* it. The first true clocks were alarms. The first Western clockworks, which set us on the way to clockmaking, were weight-driven machines which struck a bell after a measured interval. Two kinds of clocks were made for this purpose. Probably the earlier were small monastic alarms, or chamber clocks—called *horologia excitatoria,* or awakening clocks—for the cell of the *custos horologii,* or guardian of the clock. These rang a small bell to alert a monk to summon the others to prayer. He would then go up to strike the large bell, usually set high in a tower, so that all could hear. About the same time much larger turret clocks began to be made and placed in the towers, where they would ring the large bell automatically.

These monastic clocks announced the canonical hours, the times of day prescribed by the Church's canons, or rules, for devotion. The number of these hours varied, of course, with the changing canons of the Church, with the varied customs from place to place, and with the rules of particular orders. In the sixth century, after Saint Benedict, the canonical hours were standardized at seven. Distinct prayers were specified to be said at the first light or dawn (Matins or Lauds), with the sunrise *(Hora Prima),* at mid-morning *(Hora Tertia),* at noon *(Hora Sexta* or *Meridies),* at midafternoon *(Hora Nona),* at sunset (Vespers, or *Hora Vesperalis*), and at nightfall (Compline, or *Completorium*). The number of strokes of the bell varied from four at sunrise to one at noon and back to four again at nightfall. The precise hour, by our modern calculation, for each of these prayers depended in any given place on the latitude and the season. Despite the complexity of the problem, monastic clocks were adjusted to vary the time between bells according to the season.

Efforts to adapt earlier timekeeping devices to the making of sounds had never been quite successful. A clever Parisian fitted a lens into his sundial to act as a burning glass which precisely at noon focused on the touchhole of a small cannon, and so automatically saluted the sun at its apex. This elegant cannon clock, installed by the Duke of Orléans in the garden of the Palais Royal in 1786, is said to have fired the shot that started the French Revolution. Centuries before, complicated water clocks had been designed to mark passing time by tossing pebbles or blowing whistles. Some such devices were probably tried in monasteries.

But a new kind of timepiece, a mechanical timepiece that was a true clock, would be much better adapted to the new mechanical needs. The very word "clock" bears the mark of its monastic origins. The Middle English *clok* came from the Middle Dutch word for bell and is a cognate of the German *Glocke,* which means bell. Strictly speaking, in the beginning a timepiece was not considered to be a clock unless it rang a bell. It was only later that it came to mean any device that measured passing time.

These first mechanical clocks came into an age when sunlight circum-scribed the times of life and movement, when artificial lights had not yet begun to confuse night with day. Medieval striking clocks remained silent during the dark hours. After the four strokes which announced Compline, the prayers at nightfall, the next bell was not sounded until the time for Matins, the prayers at dawn the next morning. But in the long run the unintended consequence of the making of mechanical clocks, and a hidden imperative of the machine itself, was to incorporate both hours of darkness and hours of sunlight into a single equal-houred twenty-four-hour day. The monastic clock, specially designed for *sounding* the time, pointed the way to a new way of thinking about time.

The sundial, water clock, and hourglass were all designed primarily to *show* the passing time, by the visible gradual flow of a shadow across a dial, of water from a bowl, of sand through a glass. But the mechanical clock, in its monastic origins, was made for a decisive mechanical act, a stroke of a hammer on a bell. The needs of mechanical timekeeping, the logic of the machine itself, imposed a new feeling. Instead of being synonymous with repeated cycles of the sun, which varied as the cycles of the seasons commanded, or with the shorter cycles of other flowing media, time now was to be measured by the staccato of a machine. Making a machine to *sound* the canonical hours required, and achieved, mechanical novelties which would be the foundation of clockmaking for centuries to come.

The force that moved the arm that struck the bell was provided by falling weights. What made the machine truly novel was the device that prevented the free fall of the weights and interrupted their drop into regular intervals. The sundial had shown the uninterrupted movement of the sun's shadow, and the sandglass operated by the free-falling of water or sand. What gave this new machine a longer duration and measured off the units was a simple enough device, which has remained almost uncelebrated in history. It was called an escapement, since it was a way of regulating the "escape" of the motive power into the clock, and it held revolutionary import for human experience.

With the simplicity of the greatest inventions, the "escapement" was nothing more than an arrangement that would regularly interrupt the force of a falling weight. The interruptor was so designed that it would alternately check and then release the force of the weight on the moving machinery of the clock. This was the basic invention that made all modern clocks possible. Now a weight falling only a short distance could keep a clock going for hours as the regular downward pull of the falling weights was translated into the interrupted, staccato movement of the clock's machinery.

The earliest simple form was the "verge" escapement. An unknown mechanical genius first imagined a way of connecting the falling weight by intersecting cogged wheels to a vertical axle which carried a horizontal bar, or verge, with weights attached. These weights regulated the movement. When they were moved outward, the clock beat slower; when moved inward, the clock went faster. The back-and-forth movement of the bar (moved by the large falling weights) would alternately engage and disengage the cogs on the clock's machinery. These interrupted movements eventually measured off the minutes and, later, the seconds. When, in due course, clocks became common, people would think of time no longer as a flowing stream but as the accumulation of discrete measured moments. The sovereign time that governed daily lives would no longer be the sunlight's smooth-flowing elastic cycles. Mechanized time would no longer flow. The tick-tock of the clock's escapement would become the voice of time.

Such a machine plainly had nothing to do with the sun or the movements of the planets. Its own laws provided an endless series of uniform units. The "accuracy" of a clock—which meant the uniformity of its measured units —would depend on the precision and regularity of the escapement.

The canonical hours, which had measured out the daylight into the appropriate elastic units between divine services, were registered on clocks until about the fourteenth century. It was around 1330 that the hour became our modern hour, one of twenty-four equal parts of a day. This new "day" included the night. It was measured by the time between one noon and the next, or, more precisely, what modern astronomers call "mean solar time." For the first time in history, an "hour" took on a precise, year-round, everywhere meaning.

There are few greater revolutions in human experience than this movement from the seasonal or "temporary" hour to the equal hour. Here was man's declaration of independence from the sun, new proof of his mastery over himself and his surroundings. Only later would it be revealed that he had accomplished this mastery by putting himself under the dominion of a machine with imperious demands all its own.

The first clocks did not have dials or hands at all. They did not need them, since their use was simply to *sound* the hour. An illiterate populace that might have trouble reading a dial would not mistake the sound of bells. With the coming of the "equal" hour, replacing the "temporary" or "canonical" hour, the sounding of hours was ideally adapted for measurement by a simple machine. Sun time was translated into clock time.

By the fourteenth century in Europe large turret clocks in the belfries of churches and town halls were sounding the equal hours, heralding a new time-consciousness. Church towers, built to salute God and to mark man's heavenward aspirations, now became clock towers. The *torre* became the campanile. As early as 1335, the campanile of the Chapel of the Blessed Virgin in Milan was admired by the chronicler Galvano della Fiamma for its wonderful clock with many bells. "A very large hammer . . . strikes one bell twenty-four times according to the number of the twenty-four hours of the day and night; so that at the first hour of the night it gives one sound, at the second, two strokes, at the third, three, and at the fourth, four; and thus it distinguishes hour from hour, which is in the highest degree necessary for all conditions of men." Such equal-hour clocks became common in the towns of Europe. Now serving the whole community, they were a new kind of public utility, offering a service each citizen could not afford to provide himself.

People unwittingly recognized the new era when, noting the time of day or night, they said it was nine "o'clock"—a time "of the clock." When Shakespeare's characters mentioned the time, "of the clock," they recalled the hour they had heard last struck. Imogen, Cymbeline's daughter, explains that a faithful lover is accustomed "to weepe 'twixt clock and clock" for her beloved. While the populace now began to know the "hour," several centuries passed before they could talk of "minutes." During the whole fourteenth century, dials were seldom found on clocks, for the clocks' function was still to *sound* the hours. They are not found on the Italian campaniles, though there may have been one on St. Paul's Cathedral (1344) in London. Early dials were not like ours. Some showed hours only from I to VI with hands that moved around the dial four times in twenty-four hours. Others, like the famous work of Giovanni de' Dondi (1318–1389), enumerated the full twenty-four hours.

It was not too difficult to improve clocks that already struck the hour so that they could strike the quarter-hour. A dial, marked 1 to 4, was sometimes added to indicate the quarters. Later these were replaced by the figures 15, 30, 45, and 60 to indicate minutes. There was still no minute hand.

By 1500 the clock at Wells Cathedral in England was striking the quarter-hours, but had no way to mark the minutes. To measure minutes you still had to use a sandglass. A separate concentric minute hand, in addition to the hour hand, did not come into use until the pendulum was successfully applied to clocks. The pendulum also made it possible to indicate seconds. By 1670 it was not unusual for clocks to have a second hand whose movements were controlled by a 39-inch pendulum with a period of just one second.

More than any earlier invention the mechanical clock began to incorporate the dark hours of night into the day. In order to show the right time at daybreak this time machine had to be kept going continuously all night.

When does a "day" begin? Answers to this question have been almost as numerous as those to the question of how many days should be in a week. "The evening and the morning were the first day," we read in the first chapter of Genesis. The very first "day" then was really a night. Perhaps this was another way of describing the mystery of Creation, leaving God to perform his miraculous handiwork in the dark. The Babylonians and the early Hindus calculated their day from sunrise. The Athenians, like the Jews, began their "day" at sunset, and carried on the practice through the nineteenth century. Orthodox Muslims, literally following Holy Script, continue to begin their day at sunset, when they still set their clocks at twelve.

As we have seen, for most of history, mankind did not think of a day as a unit of twenty-four hours. Only with the invention and diffusion of the mechanical clock did this notion become common. The early Saxons divided their day into "tides"—"morningtide," "noontide," and "eveningtide"—and some of the earliest English sundials are so marked.

Other widespread ways of dividing the day were much simpler than the system of "temporary" hours, which subdivided daylight and darkness. The seven canonical hours marked the passing time for Columbus and his crew.

Even after the arrival of the mechanical clock the sun left its mark on the measuring of the hours. The "double-twelve" system, by which Americans count the hours, is such a relic. When the *daylight* hours were measured off and subdivided, in contrast to the hours of night, the hours of each of the two parts were numbered separately. And so it remained, even after a machine required that time be measured continuously. The first 24-hour clocks—while substituting equal mechanical hours for the elastic canonical or "temporary" hours—still remained curiously tied to the sun. They normally used sunset as the end of the twenty-fourth hour.

To ask how we came to our day, hour, minute, and second takes us deep into the archaeology of everyday life. Our English word "day" (no relation to the Latin *dies*) comes from an old Saxon word "to burn," which also meant the hot season, or the warm time. Our "hour" comes from Latin and Greek words meaning season, or time of day. It meant one-twelfth part of the sunlight or the darkness—the "temporary," or seasonal, hour—varying with season and latitude, long before it acquired its modern meaning of one twenty-fourth part of the equinoctial day.

Why the twenty-four? Historians do not help us much. The Egyptians did divide their day into twenty-four "hours"—"temporary," of course. Apparently they chose this number because they used the sexagesimal system of numbers, based on multiples of sixty, which had been developed by the Babylonians. This pushes the mystery back into earlier centuries, for we have no clear explanation of why the Babylonians built their arithmetic as they did. But their use of the number sixty seems to have had nothing to do with astronomy or the movement of heavenly bodies. We have seen how the Egyptians fixed 360 days as the regular days of their year—12 months of 30 days each, supplemented by 5 additional days at the end of each year. They also marked off 360 degrees in a circle, perhaps by analogy to the yearly circuit of the sun. Sixty, being one-sixth of the 360 and so a natural subdivision in their sexagesimal system, became a convenient subdivision of the circle, and also of each "degree" or each hour. Perhaps the Chaldean Babylonians, noticing five planets—Mercury, Venus, Mars, Jupiter, and

Saturn—multiplied 12 (the number of the months, and a multiple of 6) by the planetary 5, and so arrived at the significant 60.

An everyday relic of the primitive identification of the circuit of the sun with the full circle is our sign for a "degree." The tiny circle we now use to designate a degree is probably a hieroglyph for the sun. If the degree sign ° was a picture of the sun, then 360°—a full circle—would also properly mean a cycle of 360 days, or a full year. The degree as a way of dividing the circle was first applied by ancient Babylonian and Egyptian astronomers to the circle of the zodiac to designate the stage or distance traveled by the sun each day, just as a *sign* described for them the astronomical space passed through in a month.

Our "minute," from the medieval Latin *pars minuta prima* (first minute or small part), originally described the one-sixtieth of a unit in the Babylonian system of sexagesimal fractions. And "second," from *partes minutae secundae,* was a further subdivision on the base of sixty. Since the Babylonian arithmetic was based on that unit, it was their version of a decimal and was easier to handle in their scientific calculations than other "vulgar fractions" *(minutae)* would have been. Ptolemy used this sixty-unit system for subdividing the circle, and he also used it to divide the day. Not until much later, perhaps in the thirteenth century with the arrival of the mechanical clock, did the minute become a division of the hour. The language, again, is a clue to the needs and capacities of time-keeping machinery. The "second" was at first an abbreviation for "second minute," and originally described the unit resulting from the second operation of sexagesimal subdivision. Long used for subdivisions of a circle, seconds were not applied to timekeeping until clockmaking was refined in the late sixteenth century.

The clock did not entirely liberate itself from the sun, from the dictates of light and darkness. In Western Europe the hours of the clock continued to be numbered from noon, when the sun was at the meridian, or from midnight midway between two noons. In most of Europe and in America a new day still begins at midnight by the clock.

The archaeology of our everyday life leads us all over the world. The 365 days of our year acknowledge our debt to ancient Egyptian priests, while the names of months—January, February, March—and of the days of our seven-day week—Saturday, Sunday, Monday—remain our tie to the early Hebrews and to Greek and Roman astrologers. When we mark each hour of our 24-hour day, and designate the minutes after the hour, we are living, as a historian of ancient science reminds us, by "the results of a Hellenistic modification of an Egyptian practice combined with Babylonian numerical procedures."

The broadcasting medium of the medieval town was bells. Since the human voice could not reach all who needed to hear a civic announcement, bells told the hours, summoned help to extinguish a fire, warned of an approaching enemy, called men to arms, brought them to work, sent them to bed, knelled public mourning at the death of a king, sounded public rejoicing at the birth of a prince or a coronation, celebrated the election of a pope or a victory in war. "They may ring their bells now," Sir Robert Walpole observed in 1739 on hearing bells rung in London to announce the declaration of war against Spain, "before long they will be wringing their hands." Americans treasure a relic of that age of bells in the Liberty Bell, which announced Independence in Philadelphia.

There was supposed to be power and therapy in the sound of the bells that were rung to ward off an epidemic or to prevent a storm. Citizens of Lyons, France, in 1481 petitioned their town council that they "sorely felt the need for a great clock whose strokes could be heard by all citizens in all parts of the town. If such a clock were to be made, more merchants would come to the fairs, the citizens would be very consoled, cheerful and happy and would live a more orderly life, and the town would gain in decoration."

Community pride was a pride of bells. Churches, monasteries, and whole towns were judged by the reach and resonance of the peals from their towers. An inscription on an old bell boasted, "I mourn death, I disperse the lightening, I announce the Sabbath, I rouse the lazy, I scatter the winds, I appease the bloodthirsty" *(Funera plango, fulmina frango, Sabbath pango, Excito lentos, dissipo ventos, paco cruentos).* Paul Revere, the messenger of the American Revolution, made a reputation, and a fortune, as a caster of bells for proud New England towns. The art of bell-casting and experiments with bell-ringing devices advanced the art of the clockmaker and encouraged the elaboration of clocks.

Widespread illiteracy helps explain why dials were slow to appear on the exterior of public clocks. Not everybody could read even the simple numbers on a clockface. The very same factors that delayed the production of calibrated dials also encouraged experiment, ingenuity, and playfulness with clockwork performances. The great public clocks of the Middle Ages did not much advance the precision of clockworks, which, before the pendulum, lost or gained as much as an hour a day. It was technically difficult to improve the escapement hidden inside the machinery, and that regulated the accuracy of the movement. But it was easy to add wheels to wheels to improve the automated public display.

Nowadays the calendrical or astronomical indicators on antique clocks seem superfluous ornaments on a machine that should only show us hours, minutes, and seconds. For at least two centuries after the great mechanical clocks began to be built in Europe, it was quite otherwise. The magnificent clock made about 1350 for the Cathedral of Strasbourg served the public as both a calendar and an aid to astrology. Also an instructive and entertaining toy, it performed a variety show as it tolled the hours. In addition to a moving calendar and an astrolabe with pointers marking the movement of sun, moon, and planets, in its upper compartment the Three Magi bowed in procession before a statue of the Virgin Mary while a tune played on the carillon. At the end of the procession of the Magi, an enormous cock made of wrought iron with a copper comb and set on a gilded base opened its beak and stuck out its tongue, crowing as it flapped its wings. When rebuilt in 1574, the Strasbourg clock included a calendar showing movable feasts, a Copernican planetarium with revolutions of the planets, phases of the moon, eclipses, apparent and sidereal times, precession of the equinoxes, and equations for translating sun and moon indicators into local time. A special dial showed the saint's days. Each of the four quarters of each hour was struck by a figure showing one of the Four Ages of Man: Infancy, Adolescence, Manhood, and Old Age. Every day at noon the twelve Apostles passed before Christ to receive His blessing. The days of the week were indicated by chariots among clouds, each carrying the appropriate pagan god. The burghers of Strasbourg boasted that they had produced one of the Seven Wonders of Germany. In the late nineteenth century, German immigrant clockmakers in the Pennsylvania Dutch countryside produced Americanized versions of these "apostolic clocks" which added to the traditional procession of Magi and Apostles a patriotic parade of presidents of the United States.

The most popular dramas of the Middle Ages did not occur on a theater stage, nor even at the fairs or in the courtyards of churches, but were broadcast from clock towers. When the great turret clocks were in full display, they performed every hour on the hour, and every day, including Sundays and holidays. The Wells Cathedral clock, first built in 1392 and improved in the following centuries, offered a widely appealing show. Dials indicated the hour, the age and phases of the moon. Opposite the moon was a figure of Phoebus, for the sun, weighted to remain upright. Another dial showed a minute hand concentric with an hour hand that carried an image of the sun which made a full circle each twenty-four hours. In a niche above, two pairs of armored knights circled around in combat in opposite directions. As the bell struck the hour one of them was unhorsed and then, when out of view, regained his saddle. A conventional uniformed figure, "Jack Blandifet," struck each hour with a hammer but sounded the quarter-hours on two smaller bells with his heels.

Clockmakers lost no opportunity for drama. In place of a clapper hidden in the bell, they preferred vivid automata to strike the hours and the quarters. The striking figure became personified as "Jack," derived from Jacquemart, a shortened form of Jacques combined with *marteau* (hammer). This word later became generalized into "jack," meaning a tool that saved labor. A pair of such Jacks, two robust men of bronze, dating from 1499, still perform for us on the Piazza San Marco in Venice. Here was something for everyone. As the chronicler at Parma observed in 1431, to the whole populace *("al popolo")* the town clock told only the simple hours, while to the few who could understand *("agli intelligenti")* it showed the phases of the moon and all sorts of astronomical subtleties.

The clock dial, a convenience for the literate, and the first mechanical device for registering time visibly rather than aurally, is said to have been invented by Jacopo de' Dondi of Chioggia, Italy, in 1344. For this he was honored with the title of the Horologist *(Del Orologio),* which became his family name. "Gracious Reader," his epitaph boasted, "advised from afar from the top of a high tower how you tell the time and the hours, though their number changes, recognize my invention. . . ." His son, Giovanni de' Dondi, completed in 1364 one of the most complicated clockworks ever built, combining a planetarium and a timepiece. Although the clock itself has disappeared, Dondi left detailed descriptions and complete drawings from which this famous "astrarium" has been reconstructed, and can now be viewed in the Smithsonian Institution in Washington, D.C. An elegant heptagonal machine of brass activated by falling weights, it stands about five feet high. In many respects it was centuries ahead of its time, for it took account of such subtleties as the slightly elliptical orbit of the moon. On its numerous dials, it recorded the mean hour and minute, the times of the setting and rising of the sun, conversion of mean time to sidereal time, the "temporary" hours, the day of the month and month of the year, the fixed feasts of the Church, the length of daylight for each day, the dominical letter of the year, the solar and lunar cycles, the annual movement of the sun and moon in the ecliptic, and the annual movements of the five planets. In addition, Dondi provided the means to predict eclipses, indicated the movable feasts of the Church, and devised a perpetual calendar for Easter. People from everywhere came to Padua to see the clock and meet the genius who had spent sixteen years making it.

In that age the boundary was much less sharp than it would become later between the data of the heavens and the needs of everyday life. Night was more threatening, and darker, and the modern mechanical antidotes to darkness, heat, and cold had not yet been invented. For people on the seacoast or on a river the tide times were crucial. Over everybody and everything, the influence of the planets—the astral powers—governed. The

Strasbourg clock of 1352 used the data of the heavens to provide the community also with medical advice. A conventional human figure was surrounded by the signs of the zodiac. Lines were drawn from each sign to the parts of the body over which it ruled and which should be treated only when that sign was dominant. The clock then offered information about the changing dominance of the signs, helping citizens and doctors to choose the best times for medical treatment. The astrological indications on the public clock in Mantua, Italy, impressed a visitor in 1473 with its display of "the proper time for phlebotomy, for surgery, for making dresses, for tilling the soil, for undertaking journeys, and for other things very useful in this world."

Making Time Portable

In 1583 Galileo Galilei (1564–1642), a youth of nineteen attending prayers in the baptistery of the Cathedral of Pisa, was, according to tradition, distracted by the swinging of the altar lamp. No matter how wide the swing of the lamp, it seemed that the time it took the lamp to move from one end to the other was the same. Of course Galileo had no watch, but he checked the intervals of the swing by his own pulse. This curious everyday puzzle, he said, enticed him away from the study of medicine, to which his father had committed him, to the study of mathematics and physics. In the baptistery he had discovered what physicists would call the isochronism, or equal time of the pendulum—that the time of a pendulum's swing varies not with the width of the swing but with the length of the pendulum.

This simple discovery symbolized the new age. Astronomy and physics at the University of Pisa, where Galileo was enrolled, had consisted of lectures on the texts of Aristotle. But Galileo's own way of learning, from observing and measuring what he saw, expressed the science of the future. His discovery, although never fully exploited by Galileo himself, opened a new era in timekeeping. Within three decades after Galileo's death the average error of the best timepieces was reduced from fifteen minutes to only ten seconds per day.

A clock that kept perfect step with countless other clocks elsewhere made time a measure transcending space. Citizens of Pisa could know what time it was in Florence or in Rome at that very moment. Once such clocks were synchronized they would stay synchronized. No longer a mere local convenience for measuring the craftsman's hours or fixing the time for worship or the town council's meeting, henceforth the clock was a universal yardstick. Just as the equal hour standardized the units of day and night, summer and winter, in any particular town, so now the precision clock standardized the units of time all over the planet.

Certain peculiarities of our planet made this magic possible. Because the earth turns on its axis, every place on earth experiences a 24-hour day with each full 360-degree turn. The meridians of longitude mark off these degrees. As the earth turns, it brings noon successively to different places. When it is noon in Istanbul, it is still only 10 A.M. westward at London. In one hour the earth turns 15 degrees. Therefore we can say that London is 30 degrees longitude, or two hours, west of Istanbul, which makes those degrees of longitude measures of both space and time. If you have an accurate clock set to the time at London and carry it to Istanbul, by comparing the time on the clock you have carried with the local time in Istanbul, you will also know precisely how far you have traveled eastward, or how far east Istanbul is from London.

If you are a long-distance traveler and want to know precisely where you are, you are apt to find it much harder on sea than on land. On land you can get your bearings by landmarks—mountains and rivers, buildings, roads and towns. But seamarks are few, and mostly discernible only to the skilled observer who has been there before. The emptiness and homogeneity of the sea, the vast sameness of the oceans on the surface naturally drove sailors to seek their bearings in the heavens, in the sun and moon and stars and constellations. They sought skymarks to serve for seamarks. It is no wonder that astronomy became the handmaiden of the sailor, that the Age of Columbus ushered in the Age of Copernicus. With the aid of the newly invented telescope fixed on the heavens, with Galileo's new vision of the moon, Jupiter, and Venus, men discovered the seas, charted the oceans, and defined new continents.

When people set out to explore the oceans, they found it more than ever necessary to know the heavens. They had to locate themselves both on latitude north or south of the equator and on longitude east or west of some agreed point. But it was always much more difficult to determine longitude (east-west relations) than latitude (north-south relations)—which helps us understand why the New World lay so long undiscovered, why Columbus had the courage to go on his voyage of discovery, and why "East" and "West" were so long separated. To locate his east-west position on the planet the navigator had to measure the difference between the time, for example, when the sun was at noon in different places.

Defining latitude is much simpler, for the altitude of the sun above the horizon is the crucial fact. At the equator at the equinoxes the sun at noon is directly above or at altitude 90 degrees, while at the North Pole the sun is totally invisible in winter and always visible in summer. In between, the altitude of the sun above the horizon at noon can be noted, then compared with astronomical tables in national almanacs that tell us how far we are north or south of the equator. For this purpose the only device needed is a simple sighting instrument to measure in degrees the altitude of the sun above the horizon. The Greeks' way of determining latitude—by merely noting the elevation of the circumpolar stars—actually required no instrument at all. The astronomical tables in medieval nautical manuals were already so accurate that a person who had properly determined the declination of the sun could fix his latitude to within half a degree or less.

For the necessary latitude observations, medieval sailors used the simple cross-staff, or Jacob's Staff. Two rods hinged at one end could measure the angle of declination when the observer leveled the bottom rod on the horizon and the upper on the sun or star. The principle of the cross-staff, known to the ancient Greeks as *dioptra* and to the Arabs as *kamal,* was applied in western Europe at least as early as 1342. An Englishman, John Davis, devised in 1595 the handier backstaff, or English quadrant, which allowed the observer to stand with the sun behind him and avoid being dazzled.

Seafarers, reaching out into the vast oceans, began to discover how little they knew of the planet. They had to solve the problem of longitude. From the States General of the United Provinces of the Netherlands Galileo heard about the urgent need of that seagoing people to solve this problem. As early as 1610 he suggested to the States General that the longitude could be determined at sea by observation of the four satellites of Jupiter that he had discovered earlier that year. But this required observations over an extended period through a long telescope resting on the shifting deck of a ship at sea, which obviously made it impractical. He then devised a helmet with a telescope attached which the observer could wear while seated in a chair set in gimbals, like those used to keep the ship's compass horizontal. Although this method would eventually prove practical for surveying on land, it never worked at sea. Finally he recommended the making of an accurate seagoing timepiece. After he had found the pendulum to be a simple natural timekeeper, he reflected that if it could time the human pulse, perhaps it could also provide an accurate marine clock. Not until the last ten years of his life in enforced retirement did Galileo himself explore this possibility, and then his blindness prevented him from putting together the clock he had designed.

The Dutch, who now had outposts far eastward on Asian shores, more than ever felt the need for better defining longitude, the need for a seafaring clock. The brilliant young Christiaan Huygens (1629–1695), set out to solve the problem. From the age of twenty-seven when he devised his first pendulum clock, he tried again and again. But he never quite succeeded, because a pendulum would not keep accurate time on a surging, rocking ship.

Before there was an accurate seafaring clock, the sailor seeking his bearings had to be a trained mathematician. The accepted way to find longitude at sea was by precise observations of the moon, which required refined instruments and subtle calculations. An error as small as 5′ in observing the moon meant an error of 2½ degrees of longitude, which on the ocean could be as much as 150 miles—enough to wreck a ship on treacherous shoals. Fatal miscalculation might come from a crude instrument, from an error in the nautical tables, or from the rocking motion of the ship.

This made the problem of longitude an educational as well as a technological problem. The great seafaring nations optimistically organized mathematics courses for common sailors. When Charles II set up a mathematics course for forty pupils at Christ's Hospital, the famous "Bluecoat" charity school in London, teachers found it hard to satisfy both the sailors and the mathematicians. The governors of the school, noting that Drake, Hawkins, and other great sailors had done well enough without mathematics, asked whether future sailors really needed it. On the side of mathematics, Sir Isaac Newton argued that the old rule of thumb was no longer good enough. "The Mathematicall children, being the flower of the Hospitall, are capable of much better learning, and when well instructed and bound out to skilful Masters may in time furnish the Nation with a more skilful sort of Sailors, builders of Ships, Architects, Engineers, and Mathematicall Artists of all sorts, both by Sea and Land, than France can at present boast of." Samuel Pepys, then Secretary to the Admiralty, had already set up a naval lieutenant's examination which included navigation and, following Newton's advice, naval schoolmasters were actually put on board ships to instruct the crew in mathematics.

Still, the calculations to find longitude from the moon were discouragingly complicated. Some method, preferably a machine, had to be found to give semiliterate crews their bearings. In 1604 King Philip III of Spain offered a prize of 10,000 ducats for a solution, and later Louis XIV of France offered 100,000 florins. And the Dutch States General announced the prize to which Galileo had responded.

In England the urgent impulse to solve the problem of longitude came not from the needs of sailors on distant oceans but from an avoidable catastrophe just off Land's End on the southwest coast. In 1707 an English fleet foundered on the rocks of the Scilly Islands, a cluster of a hundred and forty islets not forty miles offshore. Down with the crew went their admiral, Sir Clowdisley Shovell, the very model of a heroic sea captain. In the glorious heyday of the British navy, the loss of so many seamen so close to home, and not from enemy action, was humiliating. The public conscience was stricken. Two eminent mathematicians declared publicly that the wreck could have been prevented if only the seamen had not been ignorant of their longitude. What was needed, then, was some way to find longitude that was "easy to be understood and practis'd by ordinary seamen, without the necessity of any puzzling calculations in astronomy."

Stirred by this event, Parliament in 1714 passed an act "For providing a Publick Reward for such Person or Persons as shall discover the Longitude at Sea." A Board of Longitude, including seamen and scholars, would grant sums up to £2,000 to support promising experiments and then could award a prize of £20,000 for solving the problem. This was, of course, an invitation to eccentrics. As late as 1736 in the madhouse that Hogarth shows us in his *Rake's Progress*, we see an inmate earnestly trying to solve the puzzle of longitude. One proposal was to locate sunken ships in known positions all over the world and then send up signals from them. Another was to publish a worldwide table of the tides, then use a portable barometer so that a seaman could locate his position by the expected rise and fall at that particular place. Still another suggested using lighthouses to flash the needed time signals on to clouds. Many claimed to have techniques that they dared not reveal publicly for fear someone else would get the reward. To qualify for the prize the method had to prove its success on a voyage to the West Indies and back with a round-trip error of less than 30' or 2 minutes of time.

It was obvious that the prize could not be won by a weight-driven pendulum-escapement clock. To keep precisely measured pace on a pitching, dipping, rolling ship there must be some other way. The clock must be freed of weights and pendulums.

It had occurred to someone that if a thin piece of metal was coiled into a spring, it could propel the machine as it unwound. The Italian architect Brunelleschi seems to have made a spring-driven clock about 1410. Within a century a German locksmith was making small clocks driven by springs. But the spring had its own problems. While a descending weight exerted the same force whether at the beginning or at the end of its descent, an uncoiling spring exerted a decreasing force as it unwound. An ingenious solution to this problem was the "fusee," a conical spool so designed that as the spring unwound and so uncoiled the connecting cord, the shape of the spool itself exerted an increasing force on the machine. This device was borrowed from the military engineers who had invented a conical axle to ease the job of spanning the cord on a heavy crossbow. And before clock-makers began to call it a fusee, military men, drawing on their own wide experience, had already christened it "the virgin" because, they said, it offered the least resistance when the bow was slack and the most when it was taut.

At first the shapes of these newly portable clocks were as whimsical as the dramatic performances of the early public clocks. They took the forms of skulls, eggs, prayer books, crucifixes, dogs, lions, or pigeons. And some were designed to provide astronomical calendars, showing the movements of the sun, the moon, and the stars.

But these early spring-driven clocks were no more accurate than the weight-driven models they replaced. At first the dial was put horizontally on top, and there was only an hour hand. Until the early seventeenth century the works were not enclosed or protected from dust and moisture. After the works were set upright, with the dial facing out, they began to be enclosed in cases made of brass. Still, as long as they relied on the primitive verge escapement, their inaccuracy was proverbial. While Cardi-nal Richelieu was showing his collection of clocks, a careless visitor dropped two of them at once on the floor. The imperturbable Cardinal observed, "That's the first time they've both gone together!"

A more accurate portable clock would require a more accurate regulator. Neither the old verge escapement nor Galileo's pendulum escapement would serve the new purpose. An ingenious anchorsmith turned clock-maker, William Clement, devised an "anchor" escapement adapted from the flukes of an inverted anchor. He used the pendulum to move these flukes back and forth, releasing one tooth at a time the teeth of an escape-wheel. He made a clock of this design for King's College, Cambridge, in 1671. The verge escapement had required a swing of 40 degrees, but the anchor limited the swing of the pendulum to a small arc of 3 or 4 degrees. In this small arc the pendulum's free swing exactly coincided with the truly isochronous cycloidal arc. Still, Clement's anchor escapement could not solve the sea-farer's problem.

A seafaring clock would have to be independent of gravity not only for its motive force but also for its regulator. If the force of a spring could be used to drive the clock, could not the resilience of a spring also be used in place of a pendulum to regulate the works? This was Robert Hooke's simple idea.

Before Robert Hooke (1635–1703) was ten, he saw a clock taken apart and then built one for himself out of wood. At Christ Church, Oxford, where he was older than other students, he joined a scientific discussion group which included the pioneer economist William Petty, the architect Christopher Wren, and the physicist Robert Boyle. Hooke made the machines to test the theories that the others were developing. When the Royal Society was founded in 1662, they wisely chose Hooke, still only twenty-seven, for the novel post of Curator of Experiments, assigned to try the experiments suggested by members. "The truth is"—Hooke in his *Micrographia* (1665) sounded the keynote of the new age—"the science of Nature has been already too long made only a work of the brain and the fancy: It is now high time that it should return to the plainness and soundness of observations on material and obvious things."

In 1658, when he was only twenty-three, Hooke had already conjectured that the regulator for a marine clock might be made by the "use of Springs instead of Gravity for the making of a Body vibrate in any Posture." A spring attached to a balance wheel could make the wheel oscillate back and forth around its own center of gravity, thus providing the periodic movement needed to stop and start the clockworks, and so mark the units of time. This crucial insight would make possible the marine clock.

If his device had been patented, it might have made Hooke's fortune. Fellow scientists, including Robert Boyle and William Brouncker, the first president of the Royal Society, both wealthy men, wanted to back the project. But Hooke retreated when they could not satisfy every last one of his demands. In 1674, when his Dutch competitor, Huygens, actually made a watch with a balance spring, Hooke was outraged and accused Huygens of stealing his invention. To assert his priority, Hooke made a watch the next year to present to the king, inscribed with the assertion that Hooke had invented the crucial device back in 1658. Hooke became the undisputed author of Hooke's Law: *Ut tensio sic vis*—a spring when stretched resists with a force proportionate to its extension. But his primacy for nearly every one of his many specific inventions, including the balance spring, was widely disputed. Whoever may have been their "first inventor," the combination of a mainspring for motive power and a balance spring as regulator eventually provided a marine clock.

Government prizes and the evolving law of patents were beginning to give public and profitable rewards to the *first* inventor. One of the most effective uses of public funds to advance science and technology was the prize that, as we have seen, was announced by the British Parliament in 1714, for a practical way of finding longitude at sea. The lucky winner was John Harrison (1693–1776), the skillful and persistent son of a Yorkshire carpenter. He had promptly responded to the British Parliament's offer, and after repeated efforts financed by an interest-free loan from George Graham, the famous London instrument maker, Harrison finally succeeded. In 1761 his Model No. 4 seemed to meet the test. On a nine-week trip to Jamaica his clock lost only five seconds, or about 1.25 minutes of longitude, which was well within the margin of 30 minutes of longitude which the Board of Longitude had allowed. A second trial confirmed the clock's accuracy, and brought him £10,000, which was half the prize.

Until cheaper marine clocks were produced, captains continued to use the lunar method. But in the long run it would be easier to provide cheap clocks than to turn out mathematically educated sailors. It was not only sailors who would have convenient new access to time, for Harrison's marine clock was really a large watch. The spring mechanism brought portability on land as well as on sea. The new sense of time which came with the portable clock—ever smaller, gravity-free, in a pocket or on the wrist —would fill all the interstices of life.

PART THREE

THE MISSIONARY CLOCK

One need not be astonished that the Chinese sages did not make these steps. The astonishing thing is that these discoveries were made at all.

—ALBERT EINSTEIN (1953)

Open Sesame to China

NOW it was technically possible for people everywhere to find their bearings on the planet, and return to the new places they would discover. But what was possible technically was not always possible socially. Tradition, custom, institutions, language, a thousand little habits—ways of thinking and behaving—became barriers. The drama of the clock in the West was not reenacted in the East.

In 1577, at the Jesuit College in Rome, when the twenty-five-year-old priest Matteo Ricci (1552–1610) met a Father returning from the Jesuit outpost in India, he decided to join the missions on that side of the world "to garner into the granaries of the Catholic Church a rich harvest from this sowing of the gospel seed." The young Ricci had already shown the independence of spirit which would make him one of the greatest of missionaries. His father had sent him to Rome at the age of seventeen to study law. Fearing that Matteo might be tempted into the priesthood, his father expressly ordered him to avoid religious subjects. Despite his father's efforts he entered the Society of Jesus before he was twenty, then wrote his father for approval. When the elder Ricci, hastening to Rome to withdraw his son from the Jesuit novitiate, fell sick en route, he was persuaded that this was a sign from God that Matteo should be allowed to follow his call. Matteo Ricci then left Rome for Genoa, where he sailed for Portugal, to take passage in one of the annual trading ships for India. Arriving in Goa, a Portuguese enclave on the southwest coast, in September 1578, he spent four years studying and teaching theology. Then his Jesuit superiors assigned him to the mission at Macao, where he set about learning Chinese. Across the bay from the great commercial city of Canton, Macao seemed an ideal jumping-off place for missionaries.

Ricci and his fellow priest, Michele Ruggieri, stayed for seven years in Chao-ch'ing, a town west of Canton. They built a mission house, and despite popular suspicion and occasional hails of rocks from the hostile populace, they were accepted as men of learning. On the wall of the mission's reception room Ricci mounted his map of the world. As Ricci himself reported:

Of all the great nations, the Chinese have had the least commerce, indeed, one might say that they have had practically no contact whatever, with outside nations, and consequently they are grossly ignorant of what the world in general is like. True, they had charts somewhat similar to this one, that were supposed to represent the whole world, but their universe was limited to their own fifteen provinces, and in the sea painted around it they had placed a few islands to which they gave the names of different kingdoms they had heard of. All of these islands put together would not be as large as the smallest of the Chinese provinces. With such a limited knowledge, it is evident why they boasted of their kingdom as being the whole world, and why they call it Thienhia, meaning, everything under the heavens. When they learned that China was only a part of the great east, they considered such an idea, so unlike their own, to be something utterly impossible, and they wanted to be able to read about it, in order to form a better judgment. . . .

We must mention here another discovery which helped to win the good will of the Chinese. To them the heavens are round but the earth is flat and square, and they firmly believe that their empire is right in the middle of it. They do not like the idea of our geographies pushing their China into one corner of the Orient. They could not comprehend the demonstrations proving that the earth is a globe, made up of land and water, and that a globe of its nature has neither beginning nor end. The geographer was therefore obliged to change his design and, by omitting the first meridian of the Fortunate Islands, he left a margin on either side of the map, making the Kingdom of China to appear right in the center. This was more in keeping with their ideas and it gave them a great deal of pleasure and satisfaction. Really, at that time and in the particular circumstances, one could not have hit upon a discovery more appropriate for disposing this people for the reception of the faith. . . .

Because of their ignorance of the size of the earth and the exaggerated opinion they have of themselves, the Chinese are of the opinion that only China among the nations is deserving of admiration. Relative to the grandeur of empire, of public administration and of reputation for learning, they look upon all other people not only as barbarous but as unreasoning animals. To them there is no other place on earth that can boast of a king, of a dynasty, or of culture. The more their pride is inflated by this ignorance, the more humiliated they become when the truth is revealed.

Ricci's learning and tact did not allay the fears of the townspeople. One night they stoned the mission, denounced Ricci for conspiring to bring in the Portuguese to sack the town, accused him of knowing alchemy yet refusing to tell the people its secrets, and then destroyed his house. Father Ricci moved on northward, toward Peking and the headquarters of the Emperor.

By tradition, the Chinese emperors kept themselves hidden from their subjects. In these last decadent years of the Ming dynasty, Emperor Wan-li, a pathological recluse, had imprisoned himself inside the "Great Within," the Forbidden City, with his wives and scores of concubines, attended by uncounted eunuchs. Even his highest officials seldom saw him, and had to send messages by the palace eunuchs.

As Ricci and his fellow Jesuits approached Peking, they were arrested and their possessions seized. The magistrate especially warned Ricci "to reduce to dust, and if possible to annihilate every sample they had in their possession of the man who was nailed to the cross." Chinese officials were horrified by the bleeding figure being crucified, which they feared as an instrument of black magic. For six months the imprisoned Jesuit Fathers, seeing no other hope, "turned their thoughts toward God and prepared themselves resolutely and joyfully to meet any difficulty, even death itself, in the cause they had undertaken."

For twenty years Father Ricci had been trying to reach the Emperor, who alone could open the way for the Gospel, and he began to fear that his mission would end in a prison cell in Peking. Then out of the blue came a call from the Emperor to approach the palace, and be sure to bring his presents from Europe. The surprising explanation, according to Ricci, was that "one day, the King, of his own prompting, suddenly remembered a certain petition that had been sent in to him and said, 'Where is that clock, I say, where is that clock that rings of itself; the one the foreigners were bringing here to me, as they said in their petition?' "

So Ricci was released from prison, his presents were delivered to the palace, and a gun was fired announcing that the Emperor had just received tribute. The gifts had been sent first to the Board of Rites for their advice, which was as follows:

> The western ocean countries have had no relations with us, and do not accept our laws. The images and paintings of the Lord of Heaven and of a virgin which Li Ma-tou [Ricci] offers as tribute are not of great value. He offers a purse in which he says there are the bones of immortals, as if the immortals, when they ascend to heaven did not take their bones with them. On a similar occasion Han Yü [an anti-Buddhist scholar, who had been counseled on the offering of a reputed finger of Buddha] said that one should not allow such novelties to be introduced into the palace for fear of bringing misfortune. We advise, therefore, that his presents should not be received, and he should not be permitted to remain in the capital. He should be sent back to his country.

In spite of this, the Emperor received the gifts and summoned Ricci to the Great Within.

Ricci's presents included two elegant timepieces of the latest Italian design—a large clock driven by weights and a small clock driven by a spring. Both had been delivered to the palace a few days before Ricci's arrival, and when Ricci was summoned, the smaller clock was still running. The large clock had stopped because its weights had reached bottom. The "self-ringing bells," which had so delighted the Emperor, ceased to ring.

The Emperor, a child with a broken toy, through his chief eunuch gave Ricci three days to get the clock running.

Luckily, when Ricci, back in Rome, had trained for his exotic mission, he took care to be well instructed in the clockmaker's craft. Now he was prepared to give a short course in clock repair.

By dint of hard work, the four mathematicians who were assigned to the clocks finally acquired sufficient knowledge to regulate them and for fear that something might go wrong, they wrote out every detail of the instruction and of the mechanism of the clocks. For a eunuch to make a mistake in the presence of the King is equivalent to placing his life in danger. They say the sovereign is so rigid with them in this respect that even for a slight fault the poor unfortunates are sometimes beaten to death. Their first care was to ask for the names, in Chinese, of all the wheels, springs and accessories, all of which Ricci gave them in Chinese characters, because if any parts were missing, the names of the parts would be readily forgotten. . . .

The three days assigned for instructions had not passed before the King called for the clocks. They were brought to him at his order and he was so pleased with them that he immediately promoted the eunuchs and raised their wages. This they were delighted to report to the Fathers and particularly so because, from that day on, two of them were permitted to enter the presence of the King, to wind the small clock, which he always kept before him, because he liked to look at it and to listen to it ringing the time. These two became very important figures in the royal palace.

The Emperor ordered a wooden tower built for the big clock in one of the inner courtyards where only His Majesty and a few privileged dignitaries were admitted.

The Emperor wanted to see the strangers who had brought these machines with self-ringing bells. Yet he dared not break his custom of never appearing in the company of anyone save his immediate family, his wives, concubines, and eunuchs. Least of all could he favor foreigners over his own magistrates. Instead of calling the Fathers into his presence, he sent two of his best artists to paint their full-length portraits.

During the next nine years Father Ricci became a quite different kind of emissary than he had intended. The Emperor's clock "struck all the Chinese dumb with astonishment," Ricci explained, simply because it was "a work the like of which had never been seen, nor heard, nor even imagined, in Chinese history." But Ricci was wrong. Although the Fathers did not know it, the mechanical clock already had a long and remarkable history in China. Five hundred years before the Jesuits arrived there a few privileged Chinese courtiers were dazzled by a spectacular astronomical clock. By the time the Jesuit Fathers arrived in China that Heavenly

Clockwork survived only as a legend known to a handful of antiquarian scholars.

The building of Su Sung's Heavenly Clockwork was itself a saga. In 1077 Su Sung, a learned civil servant, was sent by the Sung emperor of North China to offer birthday congratulations to a "barbarian" emperor in the farther north. That year the barbarian emperor's birthday happened to fall on the winter solstice. When Su Sung arrived at his destination, he found to his dismay that he was a day early. The barbarian calendar, it seemed, was more accurate than the Chinese. Since Su Sung dared not admit the inferiority of his own emperor's calendar, he persuaded his hosts to allow him to accomplish his diplomatic mission on the day he had originally intended.

Back home in China, the issuing of a calendar, like the royal minting of coins in Europe, declared a new dynasty's authority. To counterfeit the calendar or use an unauthorized calendar was lèse majesté. An inaccurate calendar could also spell disaster for the farmer. Astronomy and mathematics had to be confined to authorized persons, for others might use their astrology with help from these sciences to note when the stars were propitious to overthrow the regime. The Emperor was responsible for pleasing heaven by his ordering of earthly events.

When the Emperor asked his returned emissary whether the Chinese or the barbarian calendar was right, "Su Sung told him the truth," reads the Chinese chronicle, "with the result that the officials of the Astronomical Bureau were all punished and fined." Su Sung then received the Emperor's order to design an astronomical clock more useful and more beautiful than any before seen.

Su Sung's purpose was not to make a timepiece for public convenience but to create a calendar machine, a private "heavenly clockwork," for the Son of Heaven:

> According to your servant's opinion there have been many systems and designs for astronomical instruments during past dynasties all differing from one another in minor respects. But the principle of the use of water-power for the driving mechanism has always been the same. The heavens move without ceasing but so also does water flow [and fall]. Thus if the water is made to pour with perfect evenness, then the comparison of the rotary movements [of the heavens and the machine] will show no discrepancy or contradiction; for the unresting follows the unceasing.

His "New Design for a Mechanized Armillary Sphere and Celestial Globe" was so detailed that working drawings have been prepared from it and functioning models have been made.

The thirty-foot-high astronomical clock tower was a five-story pagodalike structure. On the topmost platform reached by a separate outside staircase was a huge bronze power-driven armillary sphere within which there rotated automatically a celestial globe. Outside each of the five stories a procession of manikins carrying bells and gongs was programmed to ring at appropriate hours. Inside the main tower, reaching three-stories upward from the ground, was a huge clockwork, driven by water flowing at ground level and alternately filling and emptying the scoops on a vertical rotating wheel.

Every quarter-hour the whole structure reverberated to bells and gongs, the splashing of water, the creaking of giant wheels, the marching of manikins. The escapement that stopped and started the machine as it marked off the units of time was, of course, the crucial element. Su Sung's ingenious water escapement made use of the fluid qualities of water—as Hooke and Huygens later would use the elastic properties of metal—to provide the staccato motion required in a mechanical timepiece.

Those few who were privileged to view Su Sung's Heavenly Clockwork witnessed a mechanical fantasy, which Su Sung himself described:

> There are ninety-six jacks. They are arranged to correspond in timing with the sounding of "quarters" on the bell-and-drum floor of this belfry. . . .
> At sunset a jack wearing red appears to report, and then after two and a half "quarters" there comes another in green to report darkness. The night-watches each contain five subdivisions. A jack wearing red appears at the beginning of the night-watch, marking the first subdivision, while for the remaining four subdivisions the jacks are all in green. In this way there are twenty-five jacks for the five night-watches. When the time of waiting for dawn comes, with its ten "quarters," a jack in green comes out to report this. Then dawn with its two and a half "quarters" is marked by another jack wearing green, and sunrise is reported by a jack wearing red. All these jacks appear in the central doorway.

In 1090 this machine was ready to entertain and instruct the Emperor and a few high officials in the palace grounds.

When a new emperor came to power in 1094, his minions, following custom, declared the previous emperor's calendar to be faulty. No longer under the imperial protection, Su Sung's Heavenly Clockwork became a quarry of bronze for vandals, and it dissolved from the memories of the learned. When Ricci arrived in Peking, Chinese scholars of the court were dazzled by the wonderful "European" invention, which they considered something new under the sun.

Ricci and Jesuit missionaries who came after him used their knowledge of astronomy and the calendrical sciences to secure influence within the Chinese government. On his arrival, Ricci noted that the Chinese lunar calen-

dar was in error, as it had been for centuries. The Imperial Astronomers
had repeatedly erred in predicting the eclipse of the sun, which, of course,
cast suspicion on the Emperor's power to obey the will of heaven.

The great opportunity for the Jesuits came when an eclipse was expected on the morning of June 21, 1629. The Imperial Astronomers predicted that the eclipse would occur at 10:30 and would last for two hours.
The Jesuits forecast that the eclipse would not come until 11:30 and would
last only two minutes. On the crucial day, as 10:30 came and went the sun
shone in full brilliance. The Imperial Astronomers were wrong, but were
the Jesuits right? Then, just at 11:30, the eclipse began and lasted for a
brief two minutes, as the Jesuits had predicted. Their place in the Emperor's confidence was now secure, and the door of China which Ricci
had set ajar was opened to the science of the West. The Imperial Board of
Rites begged the Emperor to command revision of the calendar, and on
September 1 the Emperor ordered the Jesuits to begin the work. Incidentally, with their Chinese collaborators they translated Western books on
mathematics, optics, hydraulics, and music, and they began building telescopes for China. At the very moment when, in Rome, Galileo was being
tried by the Pope for his heresies, Jesuits in Peking were preaching the
Galilean gospel.

The skill and tact of a succession of Jesuit missionary-astronomers made
them intimates of the Sons of Heaven. They attained a power never equaled
by foreigners there until the European advisers of Asian potentates came
in the nineteenth century. It was their understanding of the calendar that
opened the Jesuits' way to the imperial court. Yet it was not the calendar
but the clock that opened a new commerce between West and East. In the
West the clock soon became an everyday convenience, while in China it
long remained a toy.

During the eighteenth century, clocks, watches, and clockwork toys
became a treasured currency in European dealings with the Chinese imperial court. The young Emperor K'ang-hsi, the patron of Father Ferdinand
Verbiest, was delighted to receive from the Jesuit Father Gabriel de Magal-
haen a soldier-automaton motivated by clockwork, which flourished a
sword in one hand while holding a shield in the other, and another clock,
which played a melody after each hour was struck. The brilliant French
missionary Father Jean Matthieu de Ventavon produced for the Emperor
a legendary clockwork-automaton that could write in Manchu, Mongolian,
and Tibetan. By the 1760's the Jesuit Father in charge of the Emperor's
collection reported that the imperial palace was "stuffed with clocks . . .
watches, carillons, repeaters, organs, spheres and astronomical clocks of all
kinds and description—there are more than four thousand pieces from the
best masters of Paris and London."

Chinese emperors established their own factories and workshops for these charming toys. By the mid-eighteenth century the imperial clockworks employed a hundred workers, but their product was not up to European standards. Because the Chinese could not make high-quality clock springs they remained back in the age of weight-driven clocks. The first manual of clockmaking finally appeared in Chinese in 1809, when there were enough secondhand clocks in China to occupy hundreds of repairmen.

As soon as Europeans learned of the seductive appeal of clockwork in China, they set about to meet the demand. Clockwork toys of every conceivable shape and performing every imaginable trick poured into China. "I have been appointed by the Emperor as clockmaker," the Jesuit Father Jean Matthieu de Ventavon complained in 1769, "but I should rather say that I am here as a machinist because the Emperor expects me to produce not really clocks but curious machines and automata."

The ambassador of the Dutch East India Company in Peking at the end of the eighteenth century sent back to Holland advice that "one should bring to Peking especially those playthings that European boys use to amuse themselves. Such objects will be received here with much greater interest than scientific instruments or *objets d'art.*"

The situation stirred the ingenuity—and the knavery—of European merchants and sounded the keynote for future Western relations with China. John Barrow (1764–1848), founder of the Royal Geographical Society, a self-made man who became one of the great explorers of his age, in his *Travels in China* (1804) helps us understand why Western merchants would be held in low esteem in China.

The gaudy watches of indifferent workmanship, fabricated purposely for the China market and once in universal demand, are now scarcely asked for. One gentleman in the Honourable East India Company's employ took it into his head that cuckoo clocks might prove a saleable article in China, and accordingly laid in a large assortment, which more than answered his most sanguine expectations. But as these wooden machines were constructed for sale only, and not for use, the cuckoo clocks became all mute long before the second arrival of this gentleman with another cargo. His clocks were now not only unsaleable, but the former purchasers threatened to return theirs upon his hands, which would certainly have been done, had not a thought entered his head, that not only pacified his former customers but procured him also other purchasers for his second cargo; he convinced them by undeniable authorities, that the cuckoo was a very odd kind of bird which sung only at certain seasons of the year, and assured them that whenever the proper time arrived, all the cuckoos they had purchased would once again "tune their melodious throats." After this it would only be fair to allow the Chinese sometimes to trick the European purchaser with a wooden ham instead of a real one.

In China the man who could afford to gratify his taste for these "curious baubles" was not satisfied to have only one. If he possessed a clock at all, he was likely to be a collector. And he was unlikely to use it as a time marker. When there were so few public clocks and so few people carrying watches, a timepiece was not much use in daily intercourse. One Jesuit watchmaker stationed in Peking reported the Chinese gentlemen's custom of carrying several watches, which they were anxious should always harmonize. A grandson of the Emperor, in Barrow's time, had collected at least a dozen watches. In the mid-nineteenth century a British doctor in China noted that clocks were seldom seen, except in public offices, where it was not uncommon to see as many as a half-dozen clocks in a row—with few if any in working order.

8

Mother of Machines

PRECISELY because the clock did not start as a practical tool shaped for a single purpose, it was destined to be the mother of machines. The clock broke down the walls between kinds of knowledge, ingenuity, and skill, and clockmakers were the first consciously to apply the theories of mechanics and physics to the making of machines. Progress came from the collaboration of scientists—Galileo, Huygens, Hooke, and others—with craftsmen and mechanics.

Since clocks were the first modern measuring machines, clockmakers became the pioneer scientific-instrument makers. The enduring legacy of the pioneer clockmakers, though nothing could have been further from their minds, was the basic technology of machine tools. The two prime examples are the gear (or toothed wheel) and the screw. The introduction of the pendulum, by Galileo and then by Huygens, made it possible for clocks to be ten times more accurate than they had been, but this could be accomplished only by precisely divided and precisely cut toothed wheels. Clockmakers developed new, simpler, more precise techniques both for dividing the circumference of a circular metal plate into equal units and for cutting the gear teeth with an efficient profile. Clocks also required precision screws, which in turn required the improvement of the metal lathe.

Gears were, of course, the essential connective tissue in a mechanical clock. The teeth in the wheels within the clock were not apt to be accurately spaced or cleanly cut if they were hand-hewn. The first gear-cutting machine of which we have any record is the work of an Italian craftsman, Juanelo Torriano of Cremona (1501–1575), who went to Spain in 1540 to make an elegant large planetary clock for Emperor Charles V. Torriano spent twenty years planning a timepiece with eighteen hundred gear wheels and then three and a half years building it. "So every day (not counting holidays)," his friend reported, "he had to make . . . more than three wheels that were different in size, number and shape of teeth, and in the way in which they are placed and engaged. But in spite of the fact that this speed is miraculous, even more astounding is a most ingenious lathe that he invented . . . to carve out with a file iron wheels to the required dimension and degree of uniformity of the teeth . . . no wheel was made twice because it always came out right the first time." During Torriano's lifetime his "lathe" was already being used by other clockmakers. It appears to have become the model for the "wheel-cutting engines" used by English and French clockmakers in the seventeenth century, when timepieces were reaching a wider market. Without such a device it would have been impossible for clocks to be made in large numbers for the commercial market. With such a gear-cutting machine, it was possible to make countless other machines and scientific instruments.

The screw, like the gear, was essential for a new world of machines. Its prototypes, like those of the gear, go back to the time of Archimedes or before. An ancient Greek scientist, Hero, may have devised a screw-cutting tool. But making a simple screw long remained a difficult operation. Until the mid-nineteenth century, when finally screws were made with points, it was always necessary in advance to prepare a hole for the full length of the screw.

During the Middle Ages, metal screws were rare. For centuries the screw was used in the winepress and for irrigation before it was applied to printing and to the pressing of coins. The wooden-threaded barrels of winepresses were laboriously made: first the diagonal channels were marked, then cut by hand. The earliest recorded mechanical device for threading a screw was the work of clockmakers. By about 1480 a German clockmaker had devised a remarkable little lathe, operated by a crank handle and equipped with what was later known as a compound slide rest.

The first all-metal lathes were made by clockmakers and for clockmakers. Later lathes, basic for the machine-tool industry, were little more than elaborations of the clockmakers' early design. The pioneer clockmakers of the seventeenth and eighteenth centuries proved to be the pioneer lathe-makers.

Improvements in the clock required improvements in the lathe. For example, the spring-driven clock used a special kind of conical screw device —a fusee—as the gear to compensate for the varying force exerted by the driving spring as it unwound. But the spiral groove in this fusee was difficult to cut by hand with the required accuracy. By 1741 French clockmakers had invented a "fusee machine" that cut the spiral groove on the tapered barrel, and so made it possible to produce clocks and watches in larger numbers for the commercial market. This fusee cutting machine still required skill to direct the tool that cut the channel as the barrel was turned. In 1763 the famous Swiss clockmaker Ferdinand Berthoud made a fully automatic fusee machine, which, incidentally, was one of the earliest examples of "building the skill into the machine." The next step was taken by Jesse Ramsden (1735–1800), an English instrument-maker with a celebrated passion for precision. He drew on the techniques of the early clockmakers to make a lathe that produced the master screw required for his "dividing engine," which made it possible to turn out countless other new scientific instruments—sextants, theodolites, micrometers, balances, barometers, microscopes, and telescopes.

The great Salisbury clock (1380), the oldest clock still operating in England and said to be the oldest surviving mechanical clock in the world, was made without the use of a single screw thread. Its iron frame was held together by rivets or wedges. Much of its construction was the work of a blacksmith who must have made the holes for the arbor pivots, rivets, and wedge bolts by punching through the hot metal. The diffusion of clocks came only with the making of smaller, portable clocks. If clocks were to be supplied not only to monasteries and town halls and noble palaces but to individual citizens, they had to be made in sizes suitable for a modest house or a craftsman's shop. This required scaling down the design, which brought a whole new technology of precision-machine craftsmanship.

A small clock obviously could not be hammered together or forged by a blacksmith. To assemble it without knocking it apart, screws were needed. Smallness required screws, and then screws made possible a host of other portable machines. Of course smaller clocks appealed to a wider market, and the demands of this wider market in turn were incentives to making cheaper clocks that people could afford to buy.

By the seventeenth century clockmakers had advanced spectacularly ahead of the other technology of the age, and had begun to apply the principle of division of labor. In 1763 Ferdinand Berthoud could list sixteen different sorts of workmen involved in producing clocks, and twenty-one making watches. Among these were makers of the movement. finishers, borers, makers of springs, engravers of brass needles, pendulum makers, engravers of dials, polishers of brass parts, enamelers of dials, silverers of brass dials, engravers of cases, bronze gilders, painters to imitate gilding in colors, founders of wheels, turners and polishers of bells.

The clock stimulated new talents, new kinds of understanding and imagination. In the age of the French Revolution, Condorcet, the mathematician-philosopher and encyclopedist, while praising an inventor of improvements in silk-weaving machinery, observed:

> Generally speaking, people have a very erroneous idea of the type of talent proper to the ideal mechanician. He is not a geometrician who, delving into the theory of movement and the categories of phenomena, formulates new mechanical principles or discovers unsuspected laws of nature. . . .
>
> In most other branches of science are to be found constant principles; a multitude of methods offer to the genius inexhaustible possibilities. If a scholar poses himself a new problem, he can attack it fortified by the pooled knowledge of all his predecessors. No elementary textbook contains the principles of this [new] science; no one can learn its history. The workshops, the machines themselves, show what has been achieved, but results depend on individual effort. To understand a machine it has to be divined. This is the reason why talent for mechanics is so rare, and can so easily go astray, and this is why it is hardly ever manifested without that boldness and the errors which, in the infancy of science, characterize genius.

The clock enticed men across boundaries of religion, language, and politics. Even before the vast colonial migrations and the settlement of the New World, the movements of skilled craftsmen exerted an influence far out of proportion to their numbers. Before power-driven transportation, before the rise of mass production, it was often the craftsmen themselves, rather than their product, that traveled. When clocks were gigantic machines built into the tops of public towers, these had to be erected where they would be used. At first there was no demand for more than one clock to a community, which meant, of course, that the professional clockmaker had to be a traveler. A clockmaker came from Basel to build the Strasbourg cathedral clock in the early 1350's, then went on to construct the first public clock in Lucerne. A German clockmaker made the first clock in the royal palace in Paris. A Paris clockmaker journeyed to Avignon in 1374 to make a clock for the Pope. While it was unusual to build a large clock in one place for use in another, there were a few famous exceptions. Genoa's first mechanical public clock was actually made in Milan in 1353 and the elegant clock we can still see on the Piazza San Marco in Venice was brought there from Reggio. After clocks were scaled down into small fragile machines, there were new reasons for making them near their customers.

Centuries passed in Europe before there were enough clockmakers in any one place to form a guild to protect their monopoly. The earliest clockmak-

ers were drawn from blacksmiths, locksmiths, or gun founders. A clockmakers' guild was begun in 1544 in Paris and in 1601 in Geneva. Not until
1630 was a Clockmakers' Company chartered in England—a result of complaints in 1627 that "the Clockmakers Freemen of this Cittie . . . are
exceedingly oppressed by intrusion of Clockmakers straingers." By the
sixteenth and seventeenth centuries the local market in European metropolises supported settled groups of clockmakers who organized to enforce
their monopoly against foreigners.

The market for their product attracted some. Political and religious
upheaval and epidemics pushed out others. In the fifteenth century Italy
had been a Mecca for craftsmen who came to serve Lorenzo the Magnificent
(de' Medici; 1449–1492) in Florence, and others in Milan, Genoa, Rome, or
Naples. In the sixteenth century skilled craftsmen, victims of religious
persecution in Germany, went out to enrich metropolises elsewhere. In
France it seems that a considerable number of clockmakers had become
Protestants, and so were targets of Catholic regimes that forced them to
move on.

Before the end of the fifteenth century there were no great centers of
clockmaking in Europe. For a while Augsburg and Nuremberg flourished
by building on their traditions of metalworking. Peter Henlein, clockmaker
of Augsburg, was given legendary credit for inventing the pocket watch. But
the chaos of the Thirty Years War (1618–1648) dispersed Augsburg and
Nuremberg clockmakers all over Europe. By the eighteenth century the
most precise and most elegant timepieces were being made in Geneva and
in London. The future of the pioneer machine lay in two islands—in Switzerland, isolated by mountain barriers, and in seagirt England. Here were
safe gathering places where mobile craftsmen from all Europe could meet,
combine, and exchange their talents. London and Geneva would reap the
fruits of persecution by others.

The Protestant Reformation, which divided Western Christendom, had
brought a new era of upheaval, persecution—and mobility. In 1517 Luther
nailed up his provocative 95 Theses on the door of All Saints Church in
Wittenberg, and so opened the Reformation in Germany. Within two years
Zwingli was preaching reformation in Zurich. Within another decade Calvin announced the Reformation in France. Banished from Paris, Calvin
took refuge in Basel, where he published his *Institutes of the Christian
Religion* (1536), the first textbook of Protestant Christianity. In the next
decades the thousands who followed Calvin to Geneva made that city a
center for European refugees.

Just as, four centuries later, the Nazi and Fascist persecutions would
make the United States a world center of atomic physics, so independent
Geneva profited in the science and technology of its day, and speedily

became a world center of clockmaking. In both eras a few refugees with specialized skills could make all the difference. In 1515, when the clock in Geneva's Cathedral of St. Pierre needed repairing, there still was no qualified clockmaker in the city. But soon after 1550, as the persecution of Protestants heated up in France and elsewhere, the clockmakers arrived. To Geneva they came not only from France but also from the Low Countries, from Germany, and from Italy. By 1600 Geneva counted some twenty-five master clockmakers, in addition to an unrecorded number of apprentices and workmen. Before the seventeenth century was over, more than one hundred masters and some three hundred workers there were turning out five thousand clocks a year.

Protestant England also became a place of refuge, and the rise of English clockmaking was a measure of the persecutions across the channel. In the earliest era of mechanical clocks, England was not a land of pioneers. But it offered a vacuum of skills into which the enterprising foreigners gladly flowed. When Henry VIII needed repairs on the clock in Nonsuch Palace, he still had to import Frenchmen. Queen Elizabeth, too, used French clockmakers. There is no record of an English-made watch before 1580, and those made in the next two decades were slavish copies of French or German models. In due time the growing number of native English clockmakers complained of "the manie straingers invadinge this Realme," and argued that they needed a royally chartered guild monopoly, which was finally granted in 1631.

One of these troublesome strangers was Lewis Cuper, whose family moved from Germany to become eminent among the clockmakers of Blois, in France, before emigrating to London about 1620. In the early seventeenth century, England continued to live on borrowed talents. The Fromanteel family of London, still famous among clock collectors, came from the Netherlands. They were the first to make pendulum clocks in England— an art that John Fromanteel had gone to the Netherlands to learn from Huygens and Coster. But before the end of the century an Englishman, Robert Hooke, was making epochal improvements in clock design, and by the early eighteenth century London took its place beside Geneva.

In England watchmaking was showing the advantages of specialization and the division of labor. The Clerkenwell district of London housed various groups of workmen who called themselves escapement makers, engine turners, fusee cutters, secret springers, or finishers. The Clockmakers' Company reported to the Board of Trade in 1786 that they were exporting about eighty thousand clocks and watches each year to Holland, Flanders, Germany, Sweden, Denmark, Norway, Russia, Spain, Portugal, Italy, Turkey, the East and West Indies, China, and elsewhere.

In the tolerant interval after the Edict of Nantes (1598) granted the

French Huguenots freedom of worship, and before it was revoked in 1685, French clockmaking seems to have been growing. But the French guilds still excluded new talent, inhibited movement into new lines of work, and enforced countless narrow monopolies. While in England, makers of new scientific instruments—depending on whether their main interests were mechanical or optical—could join either the Clockmakers' Company or the Spectacle-makers' Company, some managed to pursue their craft without joining either, and many joined the Grocers' Company. In France, guild membership was strictly enforced. When the making of mathematical instruments was first listed among French monopolies, in 1565, it was assigned to the cutlery engravers, with King Charles IX's express prohibition that, except for members of that guild, "no one may make shears or scissors . . . nor surgical instruments of metal, nor cases for falconry or any other cases furnished with astrological and geometrical instruments." As it turned out, fine scientific instruments had to be made of brass, and the smelting of copper necessary for making brass was reserved only for the founders' guild. The result was lengthy and acrimonious disputes over who held this particular monopoly. At the end of the seventeenth century, when barometers and thermometers were being made commercially in France, they came under the monopoly of the enamelers simply because the graduated scale was inscribed on an enameled metal plate. During the eighteenth century, an age of innovation in Europe, while French guilds were imposing exorbitant dues on their own members, and restricting the numbers of apprentices and of workshops, the guilds themselves were also being exorbitantly taxed by a domineering government. French instrument-makers, at the same time, were snubbed by men of science, who thought them no better than manual workers or tradesmen, and kept them out of the learned societies.

From time to time, the French government made strenuous but futile efforts to invigorate the craft. The famous English clockmaker and horologist Henry Sully (1680–1728) was imported to be clockmaker to the Duke of Orléans. But all of Sully's efforts, even his importation of sixty skilled English craftsmen, could not overbalance the retarding forces in French society, and his workshops at Versailles and St. Germain were soon abandoned.

Life in England—looser guild restrictions, and the increasing demands of a growing prosperous middle class for clocks and watches—was much more favorable to the clockmakers' crafts. Competition increased and the market widened. It is not surprising that when seamen of the expanding worldwide empires required seafaring clocks and better scientific instruments of all sorts, English clockmakers were pioneers.

Philosophers were always looking for new handles on the universe—new similes, new metaphors, new analogies. Despite their scorn for those who cast the Creator of the Universe in man's image, the theologians never ceased to scrutinize man's own handiwork as their clues to God. Now man was a proud clockmaker, a maker of self-moving machines. Once set in motion, the mechanical clock seemed to tick with a life of its own. Might not the universe itself be a vast clock made and set in motion by the Creator Himself? This interesting possibility, not conceivable until the mechanical clock was on the scene, would be a main way station toward modern physics.

The older view of the movement of physical bodies, as expounded by Aristotle, was that nothing moved unless it was constantly being pushed by some outside force. But by the time the first mechanical clocks were striking in the town belfries of Europe, an interest in predictable regularities was growing—toward a new theory of motion. Now, it was argued, things kept moving because of forces originally imprinted on them *(vis impresa)* that simply continued to operate. De' Dondi's elegant model of a clockwork universe, recently completed, was already astounding the scholarly world. In the late fourteenth century an influential French popularizer of science, Bishop Nicole d'Oresme (1330?–1382), created the unforgettable metaphor: a clockwork universe, God the perfect clockmaker! And "if anyone should make a mechanical clock," Oresme asked, "would he not make all the wheels move as harmoniously as possible?"

This metaphor guided and inspired scientists like the great astronomer Johannes Kepler (1571–1630). "My aim," he observed in 1605, "is to show that the celestial machine is to be likened not to a divine organism but rather to a clockwork." And Descartes, too, the philosopher-mathematician, made the clock his prototypical machine. His doctrine of dualism—that mind and body operated independently—was explained in a famous clock metaphor. Suppose there are two clocks, Descartes's Dutch disciple Geulincx suggested, both of which keep perfect time. When one points to the hour, the other always strikes. If you did not know about their machinery and how they were made, you might mistakenly assume that the movements of the one caused the other to strike. This is the way both the mind and the body function. God the Clockmaker created each quite independently of the other, then wound up both of them and set them going so they are in perfect harmony. When I decide to lift my arm, I may think that my mind is acting on my body. But really both move independently, each a part of God's perfectly harmonized clockwork.

This fertile mother of machines was the missing link between man's own efforts to master his physical universe and his awed reverence before his Creator. In the seventeenth century the pioneer Puritan physicist and founder of the Royal Society, Robert Boyle (1627–1691), saw the universe as "a great piece of clock work," and his Catholic contemporary Sir Kenelm Digby (1603–1665) agreed that the universe was just that. The Newtonian universe soon elevated God from a clockmaker to a master engineer and mathematician. Now the universal laws that governed the smallest portable watch also governed the movements of the earth, the sun, and all the planets.

Even after the clock ceased to be the master metaphor of the universe, it became more than ever master of daily life on this planet. The clock made it possible for Europeans to be "on time." By the late seventeenth century, when clocks were not uncommon among the literate and wealthy, the word "punctual"—which formerly had described a person who insisted upon points (from the Latin *punctus,* "points") or details of conduct—came to describe a person who was exactly observant of an appointed time. By the late eighteenth century the word "punctuality" appeared in our language to describe the habit of being in good time. "O, madam," Joseph in Sheridan's *School for Scandal* (1777) reproached, "punctuality is a species of constancy—a very unfashionable custom among ladies." The clock brought its own morality. "Punctuality"—creature of the clock—was still not listed among the dozen virtues in which Benjamin Franklin aimed to make himself perfect. By 1760, when Laurence Sterne wrote his mock-heroic *Life and Opinions of Tristram Shandy,* he opened his saga with the most modern possible interruption of his hero's conception. At the crucial moment, when Tristram's mother and father were in bed and Tristram was about to be conceived:

> "Pray, my dear," quoth my mother, "have you not forgot to wind up the clock?"
> "Good God!" cried my father. "Did ever woman, since the creation of the world, interrupt a man with such a silly question?"

Why It Happened in the West

IN Europe the clock very early became a *public* machine. Churches expected communicants to assemble regularly and repeatedly for prayers, and flourishing cities brought people together to share a life of commerce and entertainment. When clocks took their places in church steeples and town belfries, they entered on a public stage. There they proclaimed themselves to rich and poor, awakening the interest even of those who had no personal reason to mark the hours. Machines that began as public instruments gradually became some of the most widely diffused private instruments. But instruments that began their lives in private might never become diffused into the wants and needs of the whole community. The first advertisement for the clock was the clock itself, performing for new publics all over Europe.

No self-respecting European town would be without its public clock, which tolled all citizens together to defend, to celebrate, or to mourn. A community that could focus its resources in a dazzling public clock was that much more a community. The bell tolled for all and each, as the poet John Donne noted in 1623, and the tolling of the community's bells was a reminder that "I am involved in mankind."

Many communities, even before they had organized sewage disposal or a water supply, offered the town clock as a public service. In due course each citizen wanted his own private clocks—first for his household, then for his person. When more people had their private timepieces, more other people needed timepieces to fulfill their neighbors' expectations at worship, at work, and at play.

All the while, the clock was being secularized—another way of saying it was being publicized. The first European clocks, as we have seen, alerted cloistered monks for their regular prayers, but when the clock moved into the church steeple and then into the town belfry, it moved out to a secular world. This larger public soon required the clock for the whole schedule of daily life. In Europe the artificial hour, the machine-made hour, took the calculations of time out of the calendar-universe, out from the penumbra of astrology, into the bright everyday light. When steam power, electric power, and artificial illumination kept factories going around the clock, when night was assimilated into day, the artificial hour, the clock-marked hour became the constant regimen for everyone. The story of the rise of the clock in the West, then, is the story of new modes and widening arenas of publicity.

The contrast to China is dramatic and illuminating. There, circumstances conspired to prevent publicity. The first spectacular mechanical clockworks in China, as we have seen, were made not to mark the hour but to mark the calendar. And the science of the calendar—both of its making and of its meaning—was hedged in by government secrecy. Each Chinese dynasty was symbolized, served, and protected by its own new calendar. Between the first unification of the empire in the third century B.C. (c. 221) and the end of the Ch'ing, or Manchu, dynasty in 1911, about one hundred different calendars were issued, with a name identifying it with a particular dynasty or emperor. These were not required by advances in astronomy or in the technology of observation, but were needed to put the cachet of the heavens on the authority of a new emperor. Private calendar-making was punished as a kind of counterfeiting—as both a threat to the security of the emperor and an act of lèse majesté. The French Jesuit and translator of Ricci, Nicholas Trigault, reported in the early seventeenth century that the Ming emperors "forbad any to learn this Judiciall Astrology but those which by Hereditary right are thereto disigned, to prevent Innovations."

To find clues to why the mother of machines proved so unfertile there, we must recall some of the large features of life in ancient China. One of the first, most remarkable of Chinese achievements was a well-organized centralized government. As early as 221 B.C. the "Chinese Caesar," the precocious King Cheng, who had ascended the throne of Ch'in at the age of thirteen, had managed within twenty-five years to unify a half-dozen Chinese provinces into a single great empire, with a vast hierarchy of bureaucrats. He standardized the laws and the written language, established uniform weights and measures, and even fixed the length for carriage axles so they would fit into the wheel ruts.

The Chinese kings, as we have seen, had regulated the calendar, the state religion remained emphatically tied to the round of the seasons, and astronomy became "the secret science of priest-kings." Chinese farming depended on irrigation, and successful irrigation required predicting the rhythms of the monsoon rains and the melting of the snows to flood the rivers and fill the canals.

From earliest times in China an astronomical observatory was an essential part of the cosmological temple, the ruler's ritual headquarters. As the central government became stronger and better organized, Chinese astronomy, by contrast with astronomy in ancient Greece or medieval Europe, became more and more official and governmental. This meant, of course, that Chinese astronomy became increasingly bureaucratic and esoteric. There the technology of the clock was the technology of astrological indicators. Just as in the West the machinery for minting coins, for printing paper money, or for manufacturing gunpowder was tightly controlled, so in China were calendrical timepieces.

The imperial rites that survive from the era of the Chinese Caesar required the emperor to define the four cardinal points—north, south, east, and west—by observations of the polestar and the sun. The Imperial Astronomer, one of the highest-ranking hereditary officials, was expected to keep watch at night from the emperor's observatory tower. He "concerns himself with the twelve years [the sidereal revolutions of Jupiter], the twelve months, the twelve [double] hours, the ten days, and the positions of the twenty-eight stars. He distinguishes them and orders them so that he can make a general plan of the state of the heavens. He takes observations of the sun at the winter and summer solstices, and of the moon at the spring and autumn equinoxes, in order to determine the succession of the four seasons."

Another high hereditary official, the Imperial Astrologer, interpreted the message of the heavens for the destiny of the people.

> He concerns himself with the stars in the heavens, keeping a record of the changes and movements of the planets, the sun and the moon, in order to examine the movements of the terrestrial world, with the object of distinguishing good and bad fortune. He divides the territories of the nine regions of the empire in accordance with their dependence on particular celestial bodies. All the fiefs and principalities are connected with distinct stars, and from this their prosperity or misfortune can be ascertained. He makes prognostications, according to the twelve years [of the Jupiter cycle], of good and evil in the terrestrial world. From the colours of the five kinds of clouds, he determines the coming of floods or drought, abundance or famine. From the twelve winds he draws conclusions about the state of harmony of heaven and earth, and takes note of the good or bad signs which result from their accord or disaccord.

Incidentally these state astrologers produced the most remarkable continuous record of celestial phenomena before the rise of modern astronomy. The Chinese record of an eclipse in 1361 B.C. is probably the earliest verifiable eclipse reported by any people. Other Chinese records cover long periods for which we have no other accurate chronicle of celestial events. Twentieth-century radio-astronomers still use these records in their study of novae and supernovae.

While these state records survive, most of the ancient Chinese literature on astronomy has disappeared. Because astronomy was so state-oriented, so security-bound, and so secret, the old astronomy books have left few traces. By contrast, the early books on mathematics, which were used by merchants, directors of public works, and military commanders, have survived in considerable numbers. Repeated imperial edicts enforced state security for calendrical science, astronomy, and astrology. In A.D. 840, for example, when the empire had recently been disturbed by the appearance of several comets, the Emperor ordered all observers in the imperial observatory to keep their business secret. "If we hear of any intercourse between the astronomical officials or their subordinates and officials of other government departments or miscellaneous common people, it will be regarded as a violation of security regulations which should be strictly adhered to. From now onwards, therefore, the astronomical officials are on no account to mix with civil servants and common people in general. Let the Censorate look to it." The security concerns which so notoriously plagued atomic research centers at Los Alamos and Harwell in World War II had their Chinese antecedents.

The famous Heavenly Clockwork of Su Sung could not have been constructed if Su Sung had not been a high imperial official authorized to help the Emperor view the astrological destinies. This explains, too, why, within a few years, Su Sung's spectacular achievement had become only a dim legend. If Su Sung had built his clockwork not for the private garden of a Chinese emperor but for a European town hall, he would have been hailed as a heroic public benefactor. His work would have become a monument to civic pride—the object of widespread emulation.

The Emperor himself had an especially intimate need for calendrical timekeepers. For every night the Emperor in his bedchamber had to know the movements and positions of the constellations at every hour—in precisely the way Su Sung's Heavenly Clockwork made possible. In China the ages of individuals and their astrological destinies were calculated not from the hour of birth but from the hour of conception.

When Su Sung constructed his imperial clock, the Emperor had as attendants a large number of wives and concubines of various ranks. These women totaled 121 (one-third of 365, to the nearest round number), including one empress, three consorts, nine spouses, twenty-seven concubines, and eighty-one assistant concubines. Their rotation of duty, as described in the Record of the Rites of the Chou dynasty, was as follows:

> The lower-ranking [women] come first, the higher-ranking come last. The assistant concubines, eighty-one in number, share the imperial couch nine nights in groups of nine. The concubines, twenty-seven in number, are allotted three nights in groups of nine. The nine spouses and the three consorts are allotted one night to each group, and the empress also alone one night. On the fifteenth day of every month the sequence is complete, after which it repeats in reverse order.

By this arrangement, the women of highest rank would lie with the Emperor on the nights nearest to the full moon, when the Yin, or female, influence would be most potent, and so best able to match the potent Yang, or male, force of the Son of Heaven. So timely a combination, it was believed, would assure the strongest virtues in the children then conceived. The main function of the women of lower ranks was to nourish the Emperor's Yang with their Yin.

A corps of secretarial ladies kept the records of the Emperor's cohabitations with their brushes dipped in imperial vermilion. The proper order of these proceedings in the imperial bedchamber was believed essential to the larger order and well-being of the empire. In the disorderly days of the ninth century, writers lamented that the ancient tradition of "nine ordinary companions every night, and the empress for two nights at the time of the full moon" was no longer respected, with the result that "alas, nowadays, all the three thousand [palace women] compete in confusion."

The need for an accurate calendrical clock, to show the position of the heavenly bodies at each moment of the day or night, was then obvious, to ensure the best-qualified succession of emperors. The ruling houses of China did not follow the rule of primogeniture. In theory, only the sons of the empress could become emperor, but this usually left the emperor with a number of young princes from whom to choose his heir. A prudent emperor was bound to give close attention to the astrological omens at the precise moment when each prince was conceived. To record these facts accurately was the duty of the secretarial ladies with their vermilion brushes. The astronomical observations and mechanical calculations of Su Sung's Heavenly Clockwork provided the data for these records and prognostications, and so were of great political significance. But these curiosities of the imperial court had little to do with a farmer's life. The community as a whole was not expected, and did not dare, to plumb the depths of state astrology, nor to profit from the data of calendrical timepieces.

By contrast, the spread of the clock in the West came from community needs—which meant the need for publicity and portability. The crucial development, as we have seen, was the advance from the weight-driven to the spring-driven clock. Heavy weights and their accompanying pendulum rooted a clock in the site where it was first placed. But a portable spring-driven clock was versatile in its habitat. For Europeans the seafaring clock in the eighteenth century was an exploring machine—a catalyst to cartographers, travelers, merchants, botanists, and navigators, a device that encouraged sailors to go farther, helped them know where they were, and made it possible to come back again. Eventually the pocket watch, and then the wristwatch, would put a timepiece on the person of millions.

The first great Chinese clock, incarcerated in the precincts of the imperial court, was driven by a stream of water. Su Sung's escapement, the heart of his "heavenly clockwork," required a continuous stream of water, which, of course, firmly attached the machine to one site.

To confirm that there was nothing "Oriental" or "Asiatic" about the sterility of the clock in China, we have the interesting contrast of Japan. For while the Chinese remained obstinate isolationists, stubbornly suspicious of anything from outside, the Japanese combined a determination to preserve their own arts and institutions with a remarkable capacity for imitating and incorporating whatever came from abroad. Before the end of the seventeenth century the Japanese were producing their own copies of European clocks. In the next century the Japanese began to develop a clockmaking industry, turning out clocks of their own design with an adjustable "hour" plate and fixed hands. They perfected a double-escapement clock with one balance for the hours of the day and a second for the hours of the night, since the "hours" of day and night were unequal.

Until 1873 the Japanese retained the "natural" sunlight day divided into six equal hours between sunrise and sunset. Their "hour" still varied from day to day, but they succeeded in making a clock that accurately marked these unequal hours throughout the year. Since the paper walls of Japanese houses were too fragile to hold up a heavy European wall clock, they devised a "pillar clock," which hung from the timber framing of a Japanese room and marked the hours on a vertical scale. Sliding indicators on a vertical scale could be readily moved to mark the proper changing intervals of the variable hour from day to day. The Japanese retention of a system long since abandoned in Europe actually proved an incentive to their ingenuity.

The difficulty of making mainsprings delayed the manufacture of spring-driven clocks in Japan until the 1830's. Before long the Japanese were making their own elegant *inro* watches to fit into the traditional Japanese *inro,* or pillbox case, which was attached to a cord, to be worn with the pocketless Japanese costume, around the neck or tucked into the obi. Since the Japanese customarily sat on the floor, they did not develop long-case, or "grandfather," clocks.

The congestion of Japan, with its flourishing urban centers and enterprising merchants, encouraged the publicity of arts and crafts, and kept people and things in motion. Numerous ports and a network of well-traveled roads circulated all sorts of commodities. Clockmaking developed earlier in Japan than in China. Local lords, daimyos and shoguns ordered clocks for their castles, but the public taste for clocks and the opportunity for the millions to buy them did not come until the nineteenth century.

BOOK TWO

THE EARTH AND THE SEAS

There is no sea innavigable, no land uninhabitable.

—ROBERT THORNE, MERCHANT AND GEOGRAPHER (1527)

To discover the planet, mankind would have to be liberated from ancient hopes and fears, and open the gateways of experience. The largest dimensions of space, the continents and the oceans, were only slowly revealed. The West proved a vantage point, and for most of history the West would be the discoverer, the East the discovered. The first reaches from the West to another half of the planet came from laborious and lonely overland.travelers. But the full extent of the planet could be glimpsed only by organized communities of adventure on the sea, which became a highway to grand surprises.

PART FOUR

THE GEOGRAPHY OF THE IMAGINATION

Would to God your horizon may broaden every day! The people who bind themselves to systems are those who are unable to encompass the whole truth and try to catch it by the tail; a system is like the tail of truth, but truth is like a lizard; it leaves its tail in your fingers and runs away knowing full well that it will grow a new one in a twinkling.

—IVAN TURGENEV TO LEO TOLSTOY (1856)

10

The Awe of Mountains

LONG before men thought of conquering the mountains, the mountains had conquered men. Castle of the Higher Powers, the mountain long remained, in the words of Edward Whymper, the first conqueror of the Matterhorn, "an affront to man's conquest of nature." Every high mountain was idolized by people who lived in its shadow. Inspired by the Himalayas that they gazed at in awe, the people of north India imagined a still higher mountain farther north, which they called Mt. Meru. Hindus and, later, Buddhists made that mythical 84,000-mile-high mountain-above-the-mountains the dwelling place of their gods. Mt. Meru, central mountain of the universe and vertical axis of the egg-shaped cosmos, was surrounded by seven concentric mountain rings around which revolved the sun, the moon, and the planets. Between the seventh and an outer eighth ring were the continents of the earth.

On Mt. Meru, according to the sacred Hindu scriptures, "there are rivers of sweet water running in it, and beautiful golden houses inhabited by the spiritual beings, the Deva, by their singers the Gandharva, and their harlots the Apsaras." The later Buddhist tradition held "that Meru lies between four worlds in the four cardinal directions; that it is square at the bottom and round at the top; that it has the length of 80,000 yojana, one half of which rises into heaven, whilst the other half goes down into the earth. That side which is next to our world consists of blue sapphires, which is the reason why heaven appears to us blue; the other sides are of rubies, yellow and white gems. Thus Meru is the centre of the earth." The divine Himalaya —a range 1,600 miles long and 150 miles wide—were all that could be seen of the High Places. Peaks over 25,000 feet, including Everest, Kanchenjunga, Godwin Austen, Dhaulagiri, Nanga Parbat, and Gosainthan, defied human climbers even after the age of mountain climbing had arrived. They also inspired gratitude, for hidden high among them (prosaic geographers of a later age would call them the "watershed") were the secret sources of the life-giving Indus, the sacred Ganges, and the Brahmaputra.

The Japanese, too, had their Fujiyama, a goddess who dominated their landscape and never ceased to be celebrated in their art. Hokusai, the master of the popular Ukiyo-e prints, made *Thirty-six Views of Fuji* (1823–29), which showed the many faces of the sacred heights.

. In the West, the Greeks had their Olympus, rising a sudden 9,000 feet above the Aegean. Often shrouded in clouds, the veiled summit of Olympus gave gods their privacy. Only between the clouds could mortals glimpse an amphitheater of tiers of boulders where the gods sat in council. "Never is it swept by the winds nor touched by snow," wrote Homer, "a purer air surrounds it, a white clarity envelops it and the gods there taste of a happiness which lasts as long as their eternal lives." The Greeks were confident that Olympus was the highest mountain on earth. In the beginning, after Cronus had completed his creation of the world, his sons drew lots to partition his empire, and Zeus won the ethereal heights, Poseidon received the sea, and Hades was awarded the dark depths of the earth. While Hades remained alone below, Zeus allowed the other gods to share his residence on Olympus.

On the heights of Mt. Sinai the God of the Jews gave Moses the tablets of the Law.

> And on the third day in the morning, there was thunder and lightning and a dark cloud upon the mountain and a very loud trumpet blast, and all the people who were in the camp trembled. And Moses brought the people out of the camp to meet God and they assembled at the foot of the mountain. And Mount Sinai was all smoke because the Lord descended upon it in fire; and the smoke of it went up like the smoke of a kiln. And all the people trembled greatly. And as the trumpet blast grew louder and louder, Moses spoke and the Lord answered him in thunder. And Yahweh came down upon Mount Sinai to the top of the mountain. And Yahweh called Moses to the top of the mountain and Moses went up. . . .
>
> [Exodus 19:16–20]

Where there were no natural mountains, people built artificial mountains. The oldest surviving examples are the stepped pyramids—the "ziggurats" —of ancient Mesopotamia, which go back to the twenty-second century B.C. "Ziggurat" meant both the summit of a mountain and a man-made stepped tower. The vast pyramidal pile in Babylon, 295 feet square and 295 feet high, became notorious as the Tower of Babel. While the effect from a distance was that of a stepped pyramid, the ziggurat, as Herodotus described it about 460 B.C., was a pile of solid towers, each slightly smaller than the one on which it rested. "In the topmost tower there is a great temple, and in the temple is a great bed richly appointed, and beside it a golden table. No idol stands there. No one spends the night there save a woman of that country, designated by the god himself, so I was told by the Chaldeans, who are the priests of that divinity."

When the ancient ziggurats were crumbling in the fourth century, an Egyptian reported the tradition that the ziggurat "had been built by giants who wished to climb up to heaven. For this impious folly some were struck by thunderbolts; others, at God's command, were afterwards unable to recognize each other; all the rest fell headlong into the island of Crete, whither God in His wrath had hurled them." A ziggurat, according to the sacred Babylonian texts, was a "Link of Heaven and Earth."

The Tower of Babel became a symbol of man's effort to reach the heavens, to trespass on the territory of the gods. The ziggurat was said to be the earthly shape of the ladder that the patriarch Jacob, grandson of the Mesopotamian Abraham, saw. "And he dreamed, and behold a ladder set up on the earth, and the top of it reached to heaven: and behold the angels of God ascending and descending on it." All across flat Mesopotamia people felt the need for an artificial mountain to reach up to the gods, and to allow gods more easily to reach down to men. Every major city had at least one high-reaching ziggurat, probably the highest, as it was the most impressive, structure to be seen. Remains of thirty-three survived into the twentieth century. Perhaps the ziggurat was a burial mound from which the God-King Marduk could be resurrected. Or perhaps it was only a stairway from which the God could descend to the city, and on which people could ascend to make their requests.

In the Nile Valley in lower Egypt we can still see some of the most durable artificial mountains. The primeval hill, the place of the creation of life, was especially vivid to the Egyptians. Every year when the Nile flood receded, mounds of newly silted mud fertile with new life appeared above the water, and so every year Egyptians relived the story of the Creation.

The earliest Egyptian pyramid was a stepped pyramid, similar to the ziggurats of Mesopotamia. The great pyramid of Zoser (first king of the third dynasty: c. 2980 B.C.), at Sakkara in Lower Egypt, showed six steps. "A staircase to heaven is laid for him [the king] so that he may mount up to heaven thereby." The Egyptian word "to ascend" included the sign of a stepped pyramid. The later pyramids showed no steps, but took on the smooth pyramidal slope, the sacred sign of the sun-god. The God-King Pepi, the ancient Egyptians explained, "has put down this radiance as a stairway under his feet. . . . stairs to the sky are laid for him."

In Tibet, the lamas every day offered the Buddhas their own model of the earth: their little mound of rice was Mt. Meru. The Buddha instructed that his bones, after cremation, should be placed in a mound at the crossing of four highways, to symbolize the universal reign of his teachings.

During the long rule of Hinduism, countless "stupas"—artificial-mountain replicas of Mt. Meru—had symbolized the vertical axis of the egg-shaped universe. When the emperor Asoka, who reigned from c. 273 to 232 B.C, made Buddhism the religion of his vast empire, he simply transformed the Hindu stupa into a stupa for Buddhists. Two of Asoka's stupas remain

—the Great Stupa at Sanchi in central India and the Bodhnath Stupa at Katmandu in Nepal.

Like the Mesopotamian ziggurat, the Buddhist stupa was a model of the cosmos. Above a square or circular base rose the solid hemispherical dome, a replica of the dome of heaven enclosing the world-mountain that rose from earth to the sky. The world-mountain stuck up through the dome of heaven in the form of a small balcony at the summit; at the center of the dome arose a mast, the axis of the world, extending all the way up from the watery depths imagined to lie beneath.

The most impressive, largest, and most intricate of these artificial Buddhist mountains is the great stupa of Borobudur (c. eighth century A.D.) in Java. Above five walled rectangular terraces rise three round platforms bearing seventy-two small bell-shaped stupas, each containing a Buddha, and a larger solid capstone stupa surmounting all. We share the feeling of the Buddhist epic poet of Ceylon on the completion of the great stupa there: "Thus are the Buddhas incomprehensible, and incomprehensible is the nature of the Buddhas, and incomprehensible is the reward of those who have faith in the incomprehensible."

After Buddhism ebbed away from India and the Hindu faith returned, many great shrines were painted white to make more obvious their symbolic identity with the sacred snowcapped Himalaya. Hindu temples, like the Mesopotamian ziggurat, the Egyptian pyramids, and other reconstructions of the primal mountain, but unlike the Christian cathedral, were not shelters within which the faithful could gather. The artificial mountain, like the natural mountain, was an object of worship, the sacred earth at its most eminent, up which the faithful could ascend. *The builder,* who had imitated what the gods had made, was possessed of magical power.

The Hindu dynasties produced their many ornate versions of the primeval mountain—dome, spire, hexagonal or octagonal tower. The surfaces and panels, the niches and friezes of these stone monuments, bubble with images of plants, of monkeys and elephants, and of men and women in all conceivable postures. The grandest of them, the Hindu temple Kailasa ("Shiva's paradise") at Ellora, in south-central India, ingeniously used the mountain itself to make the effigy of a divine mountain. A mountain-carved-out-of-a-mountain, Kailasa was constructed by first cutting a trench into the mountain to isolate a mass of rock 276 feet long, 154 feet wide, and 100 feet high. By working from the top of the mass down, the rock cutters avoided the need for scaffolding. The product of two hundred years' labor was a worthy replica of Shiva's paradise, Mount Kailasa in the Himalaya. Hindu architects and sculptors down to their latest efforts, as at Khajraho in central India (c. 1000), never gave up their rebuilding of Mt. Meru, and spent their energy with ever greater profligacy in carving erotic images of the reunion of man and his gods. The *sikhara*, or spire, which topped the Hindu stupa also meant mountain peak.

Perhaps the most gigantic religious monument in the world is the stupa-temple complex of Angkor Wat, built by the Cambodian king Suryavarman II (1113–1150) as his sepulcher and the temple of his divinity. The stupa there, fantastically elaborated and multiplied, is a vast filigreed stepped pyramid, a sculptured mountain.

On the other side of the world simpler, starker pyramids were rising, symbols of the universal awe of mountains. In the Valley of Mexico, at Teoti-huacán, the Pyramid of the Sun rose to two-thirds the height of the Tower of Babel. On the flat peninsula of Yucatán, the Mayas set up their pyramid-temples at Uxmal and Chichén Itzá.

11

Charting Heaven and Hell

THE great obstacle to discovering the shape of the earth, the continents, and the ocean was not ignorance but the illusion of knowledge. Imagination drew in bold strokes, instantly serving hopes and fears, while knowledge advanced by slow increments and contradictory witnesses. Villagers who themselves feared to ascend the mountaintops located their departed ones on the impenetrable heavenly heights.

The heavenly bodies were conspicuous examples of disappearance and rebirth. The sun died every night and was newborn every morning, while the moon was newborn every month. Was this moon the same heavenly body that reappeared at each "rebirth"? Were the stars that were newly lit at each sunset actually the same as those extinguished every dawn? Perhaps, like them, each of us could be extinguished and yet be reborn. It is not surprising that the heavenly bodies, and especially the moon, were widely associated with the resurrection of the dead. We will illustrate these notions from ancient Greece and Rome with some reminders that such notions were not confined to the Mediterranean or the European world.

In earliest Greek antiquity, Hecate, goddess of the moon, was summoner of ghosts, queen of the infernal regions. The cold damp rays of the moon, according to a popular Eastern astrology, corrupted the flesh of the dead and so helped dislodge the soul, which then was freed from its earthly prison to reach the heavens. The ancient Syrians tried to accelerate this process by sacrifices at their tombs on the night when the moonbeams were most potent. In the Eastern Church the dates of the rituals for the dead were fixed to exploit these hopes.

"All who leave the earth go to the moon," declared an Upanishad, an ancient Hindu text, "which is swollen by their breath during the first half of the month." The Manichaean followers of the Persian sage Manes (A.D. 216?–276?) gave the moon a brilliant role in their mystic doctrines, and so compounded the doctrines of Zoroastrianism and of Christianity into an appealing new sect that tempted many early Christians, including Saint Augustine. The moon takes a crescent shape, they explained, when it is being swelled by the luminous souls that it has drawn up from the earth. The moon wanes when it has transferred these souls to the sun. Every month the boat of the moon that sails across the skies takes on a new load of souls, which it regularly passes on to the larger vessel of the sun. The crescent moon, symbol of immortality, adorned funeral monuments of the ancient Babylonians and in Celtic countries, and across Africa. In republican Rome the shoes of senators were decorated with ivory crescents, taken as a symbol of their pure spirits, since noble souls after death were transported to the heavens, where they walked on the moon.

The flight of souls to the moon was no mere metaphor. According to the Stoics, a zone of special physical qualities surrounded the moon. The soul, a burning breath, naturally rose through the air toward the fires of the sky. In the neighborhood of the moon, it found this "Porch" of ether, a substance so like the soul's own essence that the soul stayed floating there in equilibrium. Each soul was a globe of fire endowed with intelligence, and all souls together were a perpetual chorus around the luminous moon. In this case the Elysian Fields would not, as the Pythagoreans had insisted, be on the moon but in the ether surrounding the moon, into which only suitably pure souls could penetrate.

According to popular astronomy, the lowest of the seven planetary spheres was the moon, whose ether was nearest the earth's impure atmosphere. Pythagoreans and Stoics imagined souls returning to earth just after they had crossed the circle of the moon. Therefore "sublunary" (beneath the moon) came to describe everything terrestrial, mundane, or ephemeral.

Perhaps, as European folklore suggested, each person had his own star —bright or dull, according to his station and his destiny—which was illuminated at his birth and disappeared at his death. A shooting star, then, might signify some person's death. "Were there then only two stars at the time of Adam and Eve," wondered Bishop Eusebius of Alexandria in the fifth century, "and only eight after the Flood when Noah and seven other persons alone were saved in the Ark?" Everyone was born under either a lucky or an unlucky star. The Latin *astrosus* (ill-starred) meant unlucky, and today we still thank our "lucky stars."

If, as many people have thought, the departing soul becomes a bird fleeing from this earth, would not souls naturally alight on the heavenly bodies? And the multitude of the stars could be explained by the countless generations of the dead. The Milky Way, which some believed was the highway for departed souls, was such a gathering of innumerable departed spirits. Ovid recounts how Venus swooped down invisibly into the Senate and carried Caesar's soul from his bleeding body into the heavens, and how the soul took flame and flew beyond the moon to become a trailing comet. Families consoled themselves with the thought that their members departed from earth had become stars to illuminate the heavens. The Emperor Hadrian, grieving over his favorite Antinous, professed to believe that his friend had become a star which had just appeared. According to Cicero, "nearly the whole heaven is filled with mankind."

Millennia before the discovery of gravitation, the sun, that most potent of heavenly bodies, was said to govern the others, and to be somehow "the heart of the world, source of new-born souls." According to the Pythagoreans (second century B.C.) the sun was Apollo Musagetes, chorus master of the Muses, whose music was the harmony of the spheres.

People who could agree on few other facts about remote regions of the earth somehow agreed on the geography of the afterworld. Even while the shape of most of the earth's surface was still unknown, the Nether World was described in vivid detail. The practice of burying the dead in the earth made it quite natural that people should think that the dead inhabited the Nether World. A subterranean topography seemed to make that afterlife possible and even plausible. Tradition reported that the Romans, at the foundation of their city, followed an old Etruscan custom and dug a pit in the city's center so that ancestors in the Nether World could more easily communicate with the world of the living. Into this pit were thrown gifts—the first fruits of the harvest and a clod of earth from the place whence the city's settlers had come—to ease the lives of the departed and to ensure the continuity of the generations. A vertical shaft ended in a chamber with a roof curved like the heavens, which justified calling this lower-realm a world *(mundus)*. The keystone of this vault (the *lapis Manalis;* the stone of the departed spirits) was raised three times a year, on the holidays when the dead could freely return to earth.

In the beginning, life in the Nether World was simply an extension of life above. Which explains why among so many peoples the warrior was buried with his chariot, his horses, his weapons, and his wives, why his tools accompanied the craftsman to the grave, and why the housewife went with her weaving implements and cooking vessels. And so earth-life could go on beneath the earth.

In Greece there arose a sect that named itself after Orpheus, the mythical poet, whose efforts to rescue his beloved wife, Eurydice, from the underworld had made him an expert in the perils of the journey in both directions. About the sixth century B.C. these Orphic Greeks and the Etruscans who followed them developed a mythology of judgment day, an appealing eschatology that we can still see elegantly depicted on their black-figured vases.

The books of many peoples on the Descent into Hades, while varying in the cast of characters, somehow agree on the topography of the infernal regions almost as if they were describing a nearby landscape. The Greeks provided the outlines—an underground realm bounded by the river Styx and governed by Pluto and Proserpina. There were the judges Minos, Aeacus, and Rhadamanthus, the executioners Erinyes (the Furies), and a high-walled prison, Tartarus. Since there was no bridge across the Styx, all the deceased had to be ferried by Charon, a grisly old man in a dark sailor's cloak, who charged an *obolos* for the service, a coin that it became the custom to put in the mouth of the dead to ensure passage. Once across the Styx, all took a common road to the court of judgment. Such judging of the dead was, of course, familiar to the Egyptians and is commonly depicted on the tombs in the Valley of the Kings. In the Greek Nether World the judges, from whom there was no appeal and from whom nothing could be concealed, sent the wicked to the left across a river of fire to the dark tortures of Tartarus and sent the virtuous along the right-hand road toward the Elysian Fields. There were some nice problems of physics here. If, as the Stoics taught, every soul was an upward-tending burning breath, none could descend into the earth. But the Elysian Fields were relocated in the heavens above, and wicked spirits were consigned to the Inferno below.

Was the earth large enough to contain a Tartarus for all those since the beginning of time who had ever deserved its punishments? Perhaps the infernal regions should be found not under the earth but on the lower half of the terrestrial globe, in the southern hemisphere. Virgil followed the traditional geography of the Nether World when he related the descent of Aeneas into Hades. But enlightened Romans like Cicero and Seneca and Plutarch probably had ceased to believe the mythic chart of Hades. The hardheaded Pliny, for example, noted how strange it was that miners who dig deep pits and broad galleries underground had never come upon the infernal regions.

It seems that in ancient Greece and Rome the traditional topography of the Nether World was widely accepted by the populace or, at least, not actively disbelieved. We cannot be sure how many tomb inscriptions were mere metaphor. "I shall not wend mournfully to the floods of Tartarus," read the tomb of a young Roman of the Age of Augustus, announcing that he had become a celestial hero who sent this message from the ether, "I shall not cross the waters of Acheron as a shade, nor shall I propel the dusky boat with my oar; I shall not fear Charon with his face of terror, nor shall old Minos pass sentence on me; I shall not wander in the abode of gloom, nor be held prisoner on the bank of the fatal waters." Sarcophagi commonly depicted the mythic characters with their accustomed places on the map of Hades.

Although Platonism and Christianity contradicted each other on countless dogmas, both in different ways confirmed the traditional maps of heaven and of hell. When Neoplatonists in the third century revived Plato's teachings as a sacred text, they defended his vivid description of how souls lived in the bowels of the earth. Porphyry (A.D. 232?–304?), a potent antagonist of Christianity, explained that although each soul was by nature a "fiery breath" tending to rise to the heavens, yet as a soul lowered into the earthly atmosphere it tended to become damp and heavy. During a soul's life on earth, as it became encumbered with the clay of sensual life, it became still denser, until it was naturally dragged into the earthly depths. "It is true," argued Proclus (410?–485), the last of the great Greek Neoplatonists, and a still vigorous opponent of Christianity, "that the soul by force of its nature aspires to rise to the place which is its natural abode, but when passions have invaded it they weigh it down and the savage instincts which develop in it attract it to the place to which they properly belong, that is, the earth." So it was quite understandable that wicked souls should be consigned to the Nether World. Hell, then, was no mere metaphor, but a vast underground network of rivers and islands, prisons and torture chambers, irrigated by the earth's effluvia and never brightened by the sun.

During the next millennium Christianity gave new credibility and new vividness to the ancient topography of heaven and hell. Few visions were more compelling than those of the strong-willed Saint Hildegard of Bingen (1099–1179), who at the age of eight had been consigned to a nunnery with all the last rites of the dead to signify that she was buried to the world. She wrote eloquent lives of the saints, works on natural history, medicine, and the mysteries of Creation. She saw and described precisely what happened to impenitent sinners:

> I saw a well deep and broad, full of boiling pitch and sulphur, and around it were
> wasps and scorpions, who scared but did not injure the souls of those therein;
> which were the souls of those who had slain in order not to be slain.
> Near a pond of clear water I saw a great fire. In this some souls were burned

and others were girdled with snakes, and others drew in and again exhaled the fire like a breath, while malignant spirits cast lighted stones at them. And all of them beheld their punishments reflected in the water, and thereat were the more afflicted. These were the souls of those who had extinguished the substance of the human form within them, or had slain their infants.

And I saw a great swamp, over which hung a black cloud of smoke, which was issuing from it. And in the swamp there swarmed a mass of little worms. Here were the souls of those who in the world had delighted in foolish merriment.

It was not only in Saint Hildegard's visions but in many others that hell's vivid chambers of horrors became so much more interesting than the bland delights of heaven.

The most persuasive Christian geographer of heaven and hell was, of course, the greatest of Italian poets, Dante Alighieri (1265–1321). His journey to the afterworld was a pilgrimage, a return to anciently familiar scenes. The power of his *Divine Comedy* was multiplied because, unlike most of the polite literature of Europe in its day, it was written not in Latin or another scholars' language, but in Italian, a language "lowly and humble, because it is the vulgar tongue, in which even housewives hold converse." The dominant emotional experience of his life was the death in 1290 of his beloved Beatrice when he was only twenty-five, which induced him to spend most of his active life writing an epic of the afterworld where she had gone.

Dante's great work was a travel epic recounting the author's journey through the realms of the dead. One hundred cantos (14,233 lines) covered "the state of souls after death" in Dante's guided tour through Inferno, Purgatory, and Paradise. He began writing it about 1307 and was still working on it on the day he died. The last thirteen cantos of his completed work would have been lost if after his death Dante had not appeared in a dream to his son Jacopo to explain where they were hidden.

Dante translated medieval learning into a panorama of the afterlife. Virgil, whose scheme of the Nether World Dante accepts, guides him through the Inferno; Beatrice guides him through Paradise, giving way to Saint Bernard only when the presence of God is reached. His geography of the underworld is traditional. Down across the nine chasms of the underworld Virgil guides him, at each level showing the punishments of another category of the damned, until they finally reach Satan himself. Ascending a tunnel to the foot of Mount Purgatory, they climb its seven levels, each level one of the seven deadly sins, on the way to Paradise, where there are nine heavens. The tenth is where God and his angels dwell.

12

The Appeal of Symmetry

MORE appealing than knowledge itself is the feeling of knowing. And it is not surprising that human imagination has given the earth the simplest symmetrical shapes.

One of the most attractive forms for the earth was an egg. Ancient Egyptians saw the whole earth as an egg guarded at night by the moon, "a great white bird . . . like a goose brooding over her egg." The Gnostics, Christian mystics of the first and second centuries, also saw heaven and earth as a World Egg in the womb of the universe. Entwining the egg was a giant serpent, which warmed, guarded, hatched, and sometimes fed on it. "The Earth is an element placed in the middle of the world," the Venerable Bede wrote in the seventh century, "as the yolk in the middle of an egg; around it is the water, like the white surrounding the yolk; outside that is the air, like the membrane of the egg; and around all is the fire, which closes it in as the shell does."

A thousand years later the English divine Thomas Burnet (1635?–1715) combined Platonic theology, science, and Alpine-travel experience into a celebrated *Sacred Theory of the Earth* (1684) of his own. But he had to admit "that this notion of the Mundane Egg or that the World was Oviform, hath been the sence and language of all antiquity, *Latins, Greeks, Persians, Egyptians, and others.*" Burnet's "sacred theory" described the making and remaking of the earth's surface in four phases: Creation, Deluge, Conflagration, and Consummation. In the present stage, after the Deluge and preparing for the Conflagration, the sun has dried out the planet, and internal changes have been readying the whole earth for burning. After the Conflagration comes the millennium with a new heaven and a new earth; and after the millennium, when the earth will be changed into a bright star, all Scripture prophecies will be fulfilled.

No ancient Greek maps have survived, but Greek literature describes a search for symmetry. Long before the Greeks began to believe that the earth was a sphere, they were debating what other simple form the earth might have taken. Herodotus ridiculed the Homeric notion that the earth was a circular disk surrounded by the river Oceanus. It seemed obvious to him that the earth must be surrounded by a great desert. Belief in some kind of "equator"—a division of the earth into two equal parts—came even before the general belief that the earth was a sphere. The Nile and the Danube, according to Herodotus, lay symmetrically about a median line that ran through the Greek maps. A neat parallelogram was the image of the known world accepted by Aeschylus, the historian Ephorus, and other Greek writers. This "equator," which followed the longitudinal axis of the Mediterranean in Ionian maps, seemed to explain many things. It showed that Asia Minor, which lay along that axis, and hence midway between the extreme points of the summer and winter risings and settings of the sun, naturally had the ideal climate.

A square earth, too, appealed to many people. Ancient Peruvians imagined a boxlike world with a ridge-shaped roof where the great God lived. The Aztecs cast their universe in five squares—a central square and one extending from each side. Each contained one of the four cardinal points reaching out from the Middle Place, the dwelling place of the fire god Xiuhtecutli, the mother and father of the gods, who dwelt in the navel of the earth. Other peoples have seen the universe as a wheel, or even as a tetrahedron.

Grand myths and metaphors everywhere have helped make the universe seem intelligible, beautiful, and rational. The characters cast in the leading roles of Upholders of the World have been wonderfully varied. The Atlas of Greek lore, bearing the earth on his shoulders, is familiar to Europeans. In Mexico there were at least four heaven-bearing gods, of whom Quetzalcoatl was the most prominent. An ancient Hindu figure showed the hemispherical earth supported on the backs of four elephants standing on the hemispherical shell of a giant tortoise which floated on the world waters.

One of the most appealing and most universal of these protomaps of the universe was the World-Tree. The Vedic poet explained that if a little tree could lift a rock as it grew, surely a great enough tree could support the heavens. So there grew images of the Tree of Life or the Tree of Knowledge, like that in the Garden of Eden, and many peoples had their sacred tree. The Norse Eddas sang of the Cosmic Ash Yggdrasil, their World-Tree:

> The chief and most holy seat of the gods. . . . There the gods meet in council every day. It is the greatest and best of all trees. Its branches spread over the world and reach above heaven. Three roots sustain the tree and stand wide apart; one is with Asa. . . . under the second root, which extends to the Frost giants is the well of Mimir, wherein knowledge and wisdom are concealed. The third root of the Ash is in Heaven, and beneath it is the most sacred fountain of Urd. Here the gods have their doomstead.

Very early, by the fifth century B.C., Greek scholars saw that the earth was a globe. The first firm evidence is in Plato's *Phaedo*. Then serious Greek thinkers ceased thinking of the earth as a flat disk floating on the waters. The Pythagoreans and Plato based their belief on aesthetic grounds. Since a sphere is the most perfect mathematical form, the earth must of course have that shape. To argue otherwise would be to deny order in the Creation. Aristotle (384–322 B.C.) agreed for reasons of pure mathematics, and he added some physical evidence. At the center of the universe the earth would naturally become and remain a sphere. Since all falling bodies tend toward the center, the particles of earth would form a sphere as they came together from all sides. "Furthermore the sphericity of the earth is proved by the evidence of our senses, for otherwise lunar eclipses would not take such forms; for whereas in the monthly phases of the moon the segments are of all sorts—straight, gibbous, and crescent—in eclipses the dividing line is always rounded. Consequently, if the eclipse is due to the interposition of the earth, the rounded line results from its spherical shape."

In Aristotle's lifetime, mathematical geography made remarkable advances among the Greeks. They still did not have enough observed detail about the surface of the earth to draw a useful world map, but by using pure mathematics and astronomy, they arrived at some surprisingly accurate estimates. Classical writers after Aristotle, not only the great philosopher-scientists like Pliny the Elder (A.D. 23–79) and Ptolemy (A.D. 90–168) but even the popular encyclopedists, assumed and elaborated on the sphericity of the earth. This discovery was to be one of the most important legacies of classical learning to the modern world.

A spherical earth offered irresistible opportunities for the aesthetic imagination. A sphere could be symmetrically, even beautifully, subdivided in so many ways! The ancient philosopher-geographers were not slow to find them.

The first temptation was to see the sphere neatly encircled by parallel lines. If these were marked off in some regular way, might not the spaces between have a special significance? So the Greeks drew these lines around the whole sphere, dividing the earth into parallel subdivisions, which they called *climata*. These zones, unlike modern "climates," had a geographical or astronomical, not a meteorological, significance. The length of the longest day was roughly the same for all lands within a zone. *Climata* came from the Greek word *clima*, meaning "inclination," since the length of the day was always determined by the inclination of the sun as it was seen in each place. In the zone near the pole the longest day of the year lasted over 20 hours, while near the equator daylight never lasted more than 12 hours. In between were zones where the longest day lasted all the various increments.

Ancient writers disagreed over how many such zones ought to be distinguished. Some thought there were as few as three, others saw ten or more. The symmetry of such arrangements was disturbed because the zone where the longest day lasted from 14 to 15 hours would be 632 miles wide, while the zone where the longest day lasted from 19 to 20 hours would be only 173 miles wide. The most popular scheme was Pliny's division of the part of the earth well known to the Greeks and Romans (that is, to 46° north latitude) into seven parallel segments all north of the equator. He indicated three more zones for the "wilderness" farther north. Ptolemy increased the number to 21 parallel segments for the whole northern hemisphere.

Such arbitrary lines in the long run would have great significance for man's grasp of the surface of the planet. But not what ancient writers expected. The influential Strabo (64 B.C.?–A.D. 25?), for example, insisted that the torrid *climata* on both sides of the equator, where the sun stood directly overhead for half a month each year, had a characteristic flora and fauna. There, Strabo said, the parched sandy soils "produce nothing but silphium [the small terebinth tree, the source of turpentine] and some pungent fruits that are withered by the heat; for those regions have in their neighborhood no mountains against which the clouds may break and produce rain, nor indeed are they coursed by rivers; and for this reason they produce creatures with woolly hair, crumpled horns, protruding lips, and flat noses (for their extremities are contorted by the heat)." The dark complexion of the Ethiopians was said to come from the scorching sun of the tropical *climata,* and the blondness and savagery of inhabitants of the extreme north from the frigidity of the arctic *climata.*

Out of the search for *climata* and a quest for symmetry came a Ptolemaic System of the Earth. Though less widely acknowledged than the Ptolemaic System of the Heavens, which every schoolboy knows to be wrong, Ptolemy's scheme still gives us our bearings on this planet. Herodotus and other early Greeks, in their pursuit of symmetry, drew an east–west line through the Mediterranean, bifurcating their known world. This simple device, which they elaborated to fit the new spherical form that they discovered for the earth, was a crucial beginning.

Eratosthenes (276?–195 B.C.?), perhaps the greatest of ancient geographers, is known to us mostly by hearsay, and by the attacks on him from those who owed him most. Julius Caesar seems to have relied on his *Geography.* In Alexandria he served as the second librarian of the greatest library in the Western world. "A mathematician among geographers," he developed a technique for measuring the circumference of the earth which is still in use.

From travelers Eratosthenes had heard that at noon on June 21 the sun cast no shadow in a well at Syene (modern Aswan) and was thus directly overhead. He knew that the sun always cast a shadow at Alexandria. From knowledge available to him he considered Syene to be due south of Alexandria. The idea occurred to him that if he could measure the length of the shadow of the sun in Alexandria at the time when there was no shadow in Aswan, he could calculate the circumference of the earth. On June 21 he measured the shadow of an obelisk in Alexandria and by simple geometry he calculated that the sun was 7° 14′ from overhead. This is one-fiftieth of the 360° that make a full circle. This measure was remarkably accurate, for the actual difference in latitude of Aswan and Alexandria, by our best modern calculation, is 7° 14′. Thus the circumference of the earth was fifty times the distance from Syene to Alexandria. But how great was this distance? From travelers he learned that camels needed 50 days to cover the trip and that a camel traveled 100 stadia in a day. The distance from Syene to Alexandria was thus calculated at 5,000 stadia (50 × 100). He then calculated the circumference of the earth to be 250,000 stadia (50 × 5,000). We are not sure about the conversion of stadia (originally 600 Greek feet) into modern measures, but the best estimates put the Greek stadium at about 607 English feet. The Greek "stadium" from which we take our modern name was a foot-race course of precisely that length. By this calculation Eratosthenes arrived at a figure for the circumference of the earth of some 28,700 miles, which is about 15 percent too high.

It is no wonder that his measurement of angles was more accurate than his measurement of distance. For most of the history of surveying, angles have been measured with much greater precision than distances. The accuracy of Eratosthenes' figure for the circumference of the earth would not be equaled until modern times. His fruitful combination of the theory of astronomy and geometry with the evidence of everyday experience provided a model too long forgotten after his time.

More important even than Eratosthenes' final calculations was his technique for charting the earth's surface. We know this from the attacks on him by Hipparchus of Nicaea (c. 165–c. 127 B.C.), who was probably the greatest of Greek astronomers. Hipparchus discovered the precession of the equinoxes, catalogued 1,000 stars, and is generally conceded to be the inventor of trigonometry. But he was consumed by an odd personal dislike for Eratosthenes, who had died thirty years before he was born. Eratosthenes had subdivided the earth by parallel east–west lines and north–south lines, or meridians. He separated the habitable world into a Northern Division and a Southern Division by an east–west line parallel to the equator, running through the island of Rhodes and bisecting the Mediterranean. Then he added a north–south line at right angles, running through Alexandria. On Eratosthenes' map the other lines—east–west and north–south—were not laid down at regular intervals. Instead he drew his lines through ancient, familiar, and prominent places—Alexandria, Rhodes, Meroë (the capital of ancient Ethiopian kings), the Pillars of Hercules, Sicily, the Euphrates River, the mouth of the Persian Gulf, the mouth of the Indus River, the tip of the Indian peninsula. The result was an irregular network, serving human convenience by superimposing a neat grid on the earth's spherical surface.

Hipparchus took the next step. Why not mark off all *climata* lines completely around the sphere, all parallel to the equinoctial line and at *equal* intervals from the equator to the poles? Then also mark off other lines at right angles to these spaced equally at the equator. The result would be a regular grid covering the whole planet. The *climata* lines could do more than merely *describe* regions of the earth that received the sunlight at similar angles. If numbered, they would provide a simple set of coordinates for *locating* every place on earth. How easy then to tell anyone where to find any city, river, or mountain on the planet!

Eratosthenes had vaguely seen the possibilities of such a scheme. But in his day most of the places that people were interested in finding on their maps had been located only by travelers' tales and by tradition. He knew this was not good enough, but he did not possess sufficient accurate points of reference for his grid. Hipparchus went on to insist that each place be located by exact astronomical observation toward a worldwide grid of latitudes and longitudes. He not only had the right idea but saw how it could be applied in a precise, practical scheme. By using celestial phenomena common to the whole earth to locate places on the earth's surface, he set the pattern for man's cartographic mastery of this planet.

Incidentally, Hipparchus invented the mathematical vocabulary still used in modern times. Eratosthenes had divided the earth sphere into sixty parts. But Hipparchus marked off the planet's surface into 360 parts which became the "degrees" of modern geographers. He placed his meridian—or longitude—lines on the equator at intervals of about 70 miles, which remains roughly the dimension of a "degree." By combining the traditional *climata* with these, he conceived a world map based on astronomical observations of latitude and longitude.

Latitude and longitude were to the measurement of space what the mechanical clock was to the measurement of time. They signaled man's dominance over nature, discovering and marking the dimensions of experience. They substituted precise units suiting human convenience for the accidental shapes of the Creation.

A pity that Ptolemy—indisputable father of modern geography—would be indelibly identified with an obsolete astronomy! One reason why Ptolemy the geographer cuts so inconspicuous a figure in history is that we know so little about his life. A Greek Egyptian or an Egyptian Greek, he bore a name common in Alexandrine Egypt, and incidentally that of one of Alexander the Great's closest companions. Another Ptolemy became governor of Egypt on Alexander's death, then proclaimed himself king and founded the Ptolemaic dynasty, which ruled Egypt for three centuries (304–30 B.C.). But those Ptolemies were only kings, while this Ptolemy was a man of science.

Ptolemy seems to have had a great talent for improving the works of others, for fitting myriad bits of knowledge into usable generalizations. His *Almagest* on astronomy, his *Geography*, his *Tetrabiblos* on astrology, as we have seen, together with his writings on music and optics, and his chronological table of the kings of his whole known world summed up the best thinking of his day. For his geography he drew on Eratosthenes and Hipparchus. Ptolemy frequently acknowledged his debt to the prolific Strabo, the Greek historian-geographer who used tradition, myth, and his own wide travels to survey the known world.

What is most remarkable is how strong Ptolemy's influence remains two millennia after his death. The framework and the vocabulary of our maps of the world are still shaped by Ptolemy. The grid system which he adopted and improved remains the basis of all modern cartography. He was the first to popularize, and may actually have invented, the expressions for latitude and longitude. To Ptolemy, however, these words seem to have had overtones, now lost, of the "width" and "length" of the known world. In his *Geography* he gave the latitudes and longitudes for eight thousand places. He established the convention, second nature to us today, of orienting maps with the north at the top and the east at the right. Perhaps this was because the better-known places in his world were in the northern hemisphere, and on a flat map these were most convenient to study if they were in the upper right-hand corner. He marked off his world map into twenty-six regions, changing the scale to give more detail to populous areas. He established the modern scholars' distinction between *geography* (mapping the earth as a whole) and *chorography* (mapping particular places in detail). Following Hipparchus, he divided the circle and the sphere into 360 degrees, and subdivided each of these into *partes minutae primae* ("minutes") and then each of these into *partes minutae secundae* ("seconds") of the arc.

Ptolemy had the courage to face the cartographic consequences of the spherical shape of the earth. And he developed a table of chords, based on the trigonometry of Hipparchus, to define the distance between places. He devised a way of projecting the spherical earth on a plane surface, a modified spherical projection of the habitable quadrant of the earth, which still has much to be said for it. Ptolemy's errors did not come from any lack of a critical spirit. The best hypothesis, he said, was the simplest that would comprehend the facts. He cautioned us to accept only data that had been subjected to the criticism of diverse witnesses.

Ptolemy's essential weakness was his desperate lack of facts. In the long run, raw materials for a satisfactory atlas of the world would have to come from qualified observers all over the world. It is not surprising that with his limited data he fell into some crucial errors.

One of these was probably the most influential miscalculation in history. For the circumference of the earth, Ptolemy had rejected the surprisingly accurate estimate of Eratosthenes. Ptolemy calculated each degree of the earth to be only 50 miles instead of 70, and then following the Greek polymath Posidonius (c. 135–c. 51 B.C.) and Strabo, he declared the earth to be 18,000 miles around. Together with this providential underestimate, he made the mistake of stretching Asia out eastward to reach far beyond its real dimensions, for 180 degrees instead of its actual 130 degrees. On Ptolemy's maps this had the effect of grossly reducing the extent of the unknown parts of the world between the eastern tip of Asia and the western tip of Europe. How long might the European encounter with the New World have been postponed if Ptolemy had followed not Strabo but Eratosthenes? And then, if Columbus had known how large the world really was? But Columbus followed Ptolemy, than whom there then was no higher geographic authority. And he further improved his prospects by figuring a degree on the earth to be 10 percent smaller than Ptolemy's calculation.

Still, it is not only by his errors that Ptolemy must be given some credit for Columbus' exploit. By using all available facts to confirm the spherical character of the earth, and then by establishing the latitude-longitude grid on which to hang increasing knowledge, Ptolemy had prepared Europe for world exploring. Ptolemy rejected the Homeric image of a known world surrounded by uninhabitable Ocean. Instead he suggested the vastness of lands still unknown and still to be discovered, and so opened minds for knowledge. It was far more difficult to imagine the unknown than to chart the outlines of what people imagined that they knew.

Not only for Columbus, but for the Arabs and others who had put their faith in classical learning, Ptolemy remained the source, the standard, and the sovereign of world geography. If, in the millennium after Ptolemy, mariners and their royal sponsors had freely and adventurously carried on where Ptolemy had left off, the history of both the Old World and the New World might have been quite different.

13

The Prison of Christian Dogma

CHRISTIAN Europe did not carry on the work of Ptolemy. Instead the leaders of orthodox Christendom built a grand barrier against the progress of knowledge about the earth. Christian geographers in the Middle Ages spent their energies embroidering a neat, theologically appealing picture of what was already known, or was supposed to be known.

Geography had no place in the medieval catalogue of the "seven liberal arts." Somehow it fit neither into the quadrivium of mathematical disciplines (arithmetic, music, geometry, and astronomy) nor into the trivium of logical and linguistic disciplines (grammar, dialectic, and rhetoric). For a thousand years of the Middle Ages no common synonym for "geography" was in ordinary usage, and the word did not enter the English language until the mid-sixteenth century. Lacking the dignity of a proper discipline, geography was an orphan in the world of learning. The subject became a ragbag filled with odds and ends of knowledge and pseudo-knowledge, of Biblical dogma, travelers' tales, philosophers' speculations, and mythical imaginings.

It is easier to recount what happened than to explain satisfactorily how it happened or why. After the death of Ptolemy, Christianity conquered the Roman Empire and most of Europe. Then we observe a Europe-wide phenomenon of scholarly amnesia, which afflicted the continent from A.D. 300 to at least 1300. During those centuries Christian faith and dogma suppressed the useful image of the world that had been so slowly, so painfully, and so scrupulously drawn by ancient geographers. We no longer find Ptolemy's careful outlines of shores, rivers, and mountains, handily overlaid by a grid constructed on the best-known astronomical data. Instead, simple diagrams authoritatively declare the true shape of the world, though they are only pious caricatures.

We have no lack of evidence of what the medieval Christian geographers thought. More than six hundred *mappae mundi,* maps of the world, survive from the Middle Ages. They come in all sizes—some like those in copies of the seventh-century encyclopedia of Isidore of Seville, only two inches across, others like the map in Hereford Cathedral (A.D. 1275), five feet in diameter. In the days before the printing press, each of these, and the thousands of others that must have been lost, attest the willingness of individual craftsmen and patrons to invest in their special version of the world. What is most remarkable is that when all such maps were only imaginary there was so little variation in plans of the earth.

The common form of these caricatures has led them to be called "wheel-maps" or "T-O" Maps. The whole habitable earth was depicted as a circular dish (an "O"), divided by a T-shaped flow of water. East was put at the top, which was what was then meant by "orienting" a map. Above the "T" was the continent of Asia, below to the left of the vertical was the continent of Europe, and to the right was Africa. The bar dividing Europe from Africa is the Mediterranean Sea; the horizontal bar dividing Europe and Africa from Asia is the Danube and the Nile, supposed to flow in a single line. All is surrounded by the "Ocean Sea."

These were Ecumenical maps, for they aimed to show the "Ecumene," the whole inhabited world. Designed to express what orthodox Christians were expected to *believe,* they were not so much maps of knowledge as maps of Scriptural dogma. The very simplicity that offends the geographer tes-tifies to the simple clarity of Christian belief. According to Scripture, as Isidore of Seville explained, the inhabited earth had been divided among the three sons of Noah: Shem, Ham, and Japheth. Asia was named after a Queen Asia "of the posterity of Shem, and is inhabited by 27 peoples . . . Africa is derived from Afer, a descendant of Abraham [Ham], and has 30 races of 360 towns," while Europe, named after the Europa of mythol-ogy, "is inhabited by the 15 tribes of the sons of Japheth and has 120 cities."

At the center of each map was Jerusalem. "Thus saith the Lord God; This is Jerusalem: I have set it in the midst of the nations and countries that are round about her" (Ezekiel 5:5). These words of the prophet Ezekiel over-ruled any trivial earthly needs for latitude or longitude. "Navel of the world" *(umbilicus terrae)* were the words of the Vulgate, the Latin version of the Bible. Medieval Christian geographers obstinately kept the Holy City right there. New conflicts between faith and knowledge would come when explorers expanded the map eastward, then westward. Dared Christians move their Jerusalem? Or could they ignore the discoveries?

There was nothing new about putting the most sacred place at the center. As we have seen, that is where the Hindus placed their Mt. Meru, "the center of the earth." The belief in a sacred mountain, a hill of creation, with variants in Egypt, Babylonia, and elsewhere, was simply another way of saying that the most prominent place on earth had been the navel of the world. Eastern cities commonly placed themselves at the center. Babylon (*Bab-ilani*, "door of the gods") was where the gods came down to earth. In Muslim tradition the Ka'bah was the highest point on earth, and the polestar showed that Mecca was opposite the center of the sky. The capital for a perfect Chinese sovereign was where the sundial cast no shadow at noon on the day of the summer solstice. It was not at all surprising that Christian geographers, too, put their Holy City at the center, making it the place of pilgrimage, and the destination of crusades.

What was surprising was the Great Interruption. All people have wanted to believe themselves at the center. But after the accumulated advances of classical geography, it required amnesiac effort to ignore the growing mass of knowledge and retreat into a world of faith and caricature. We have already seen how the Chinese emperors produced Su Sung's Heavenly Clockwork ahead of any comparable clock in the West, then sequestered the knowledge and the technology. The Great Interruption of geography that we are about to describe was a far more remarkable act of retreat. For the advancing geographic knowledge in the West had been widespread, reaching into the cultural interstices of a varied continent.

Christian dogma and Biblical lore imposed other figments of the theological imagination on the map of the world. The maps themselves became guides to the Articles of Faith. Every episode and every place mentioned in Scripture required a location and became a tempting arena for Christian geographers. One of the most enticing of these was the Garden of Eden. In the eastern part of the world, then at the top of the map, medieval Christians commonly showed a Terrestrial Paradise with figures of Adam and Eve and the serpent all surrounded by a high wall or a mountain range. "The First place in the East is Paradise," explained Isidore of Seville (560–636), reputed to be the most learned man of the age, "a garden famous for its delights, where man can never go, for a fiery wall surrounds it and reaches to the sky. Here is the tree of life which gives immortality, here the fountain which divides into four streams that go forth and water the world." The trackless wastes that separated man from Paradise were infested with wild beasts and serpents. This orthodox view still left ample room for learned theological debate.

To fill the whole world with the rudimentary Scripture-picture, it was necessary both to embroider the Sacred Word and to ignore the real shape of the world. It is easy to forget what these medieval Christian believers enjoyed, in exchange for the scientific progress that they condemned. A harvest of imaginary delights and terrors!

Belief in Eden became a pleasure as well as a duty. In Hebrew, pious writers explained, "Eden" meant a place of delight. God had placed Eden on a height, touching the circle of the moon's orbit, so that Paradise would stay safe and dry above the waters of the Flood. Among the most popular medieval travel literature were the Journeys to Paradise. According to the *Iter ad Paradisum,* after Alexander the Great had conquered India he came upon a broad river, the Ganges, on which he embarked with five hundred men. A month later they arrived at a vast walled city where the souls of the just lived until the Last Judgment. That, of course, was the Terrestrial Paradise.

Brave monks in quest of Paradise, like voyagers to outer space in later centuries, became popular heroes. Paradise fiction became a genre of sacred literature, just as space adventure would be a form of science fiction. Adam's son, Seth, according to one popular story, brought back seeds from the Tree of Knowledge to plant in Adam's mouth after Adam had died. A tree that sprouted from this seed provided the wood for the cross on which Christ was to be crucified. Another story told how three monks set out from their monastery between the Tigris and the Euphrates to seek the place where "the earth joins the sky." Finally they reached the dark wildernesses of India, where they found dog-headed men, pygmies, and serpents, and saw the altars that had been set up by Alexander the Great to mark the outermost boundaries of his own travels. Across fantastic landscapes peopled with giants and birds that talked, the monks plodded on until, about twenty miles from the Terrestrial Paradise, they came upon the aged Saint Macarius living in a cave with two friendly lions. He delighted them with stories of the wonders of Paradise, only to turn them back with his warning that Eden could never be entered by living men.

But even on such basic matters as the location of Eden Christian geographers were not unanimous. One of the most famous mortal travelers to Paradise was the brave Irish monk Saint Brendan (484–578). Believing that Paradise was somewhere in the Atlantic Ocean, he sailed westward until after terrifying adventures he came upon a beautiful island of unsurpassed fertility. Saint Brendan confidently asserted that this was Paradise, "Promised Land of the Saints." And even those who preferred to locate their Paradise elsewhere kept "Saint Brendan's Island" on their maps and charts. The story of this heroic monk was told and retold in Latin, French, English, Saxon, Flemish, Irish, Welsh, Breton, and Scottish Gaelic. His sacred island remained plainly marked on maps for more than a thousand years, at least until 1759. And the pioneers of modern cartography and navigation dutifully tried to find its place. The classic globe-maker Martin Behaim in 1492 put Saint Brendan's Island close to the equator, west of the Canaries, while some found it nearer Ireland, and others saw it in the West Indies. Only after two centuries (1526–1721) of Portuguese expeditions searching for Saint

Brendan's Terrestrial Paradise did believing Christians finally give up the quest. They had found a better location for their Eden elsewhere.

Hardly less vivid than these delights of Eden were the menaces of Gog and Magog. Ezekiel had prophesied "against Gog, the land of Magog." "And when the thousand years are expired," the Book of Revelation had announced, "Satan shall be loosed out of his prison, And shall go out to deceive the nations which are in the four quarters of the earth, Gog and Magog, to gather them together to battle: the number of whom is as the sand of the sea." Just as Eden was usually placed at the end of the East, so Gog and Magog were generally located in the extreme north. While the existence of Gog and Magog became an article of faith, their precise northern location was long disputed, which made the source of barbarian invasion all the vaguer and all the more threatening.

One popular chronicler, Aethicus of Istria, told how Alexander the Great had driven Gog and Magog "and twenty-two nations of evil men" far northward to the shores of the Northern Ocean. There they were held back on a peninsula beyond the Caspian Gates by a wall of iron that Alexander had built with God's help—perhaps a confused reference to the Great Wall of China. Some said that the cement for this wall came from a bituminous lake at the mouth of hell. When would the frightful invasion come? And from where? The letters of the legendary Prester John were widely quoted, warning against Gog and Magog and other cannibal peoples who, in the days of Anti-Christ, would devastate all Christendom, including even the city of Rome. Roger Bacon, a medieval pioneer of science, urged the close study of geography so that by knowing the location of Gog and Magog, people could plan against the coming invasion.

Since Gog and Magog were mentioned in the Koran, learned Muslims gave special attention to this problem. The great Arab geographer Al-Idrisi (1099–1166) reported an expedition to locate the wall that held back these pagan forces of the Apocalypse. Other Muslim writers identified Gog and Magog with the ruthless marauding Vikings. Locating the people and the place of Gog and Magog became a favorite pastime of Christian geographers. Could they be found among the mysterious tribes of Central Asia? Were they perhaps the Lost Tribes of Israel? Or were they the "Goths and Magoths"? Despite these nagging doubts, the land of Gog and Magog, usually bounded by a great wall, was found clearly marked in some place or other on medieval Christian maps.

Few adventures in pursuit of myth were more seductive than the quest for the mythical kingdom of Prester John. In the twelfth century, when European Christendom was counting the years till the invasion of Gog and

Magog, the present threat of Saracens at the Holy Land stirred the quest for allies against the Muslim hordes. Tales came westward of the priest-king Prester John somewhere in the fabulous "Indies," who had already reputedly succeeded against the Muslims in his own realm. In the lands where Saint Thomas had been buried, he was said to combine shrewd military judgment, saintly piety, and the wealth of Croesus. Was he God's messenger to shift the balance of forces and so forestall a Mongol invasion?

A twelfth-century chronicle of Otto Bishop of Freising optimistically reported information to the papal court in 1145 that this Prester John was descended from the race of the Three Wise Men, and governed the lands he inherited from them with a scepter of solid emerald. "Not many years before, one John, king and priest who dwelt in the extreme Orient beyond Persia and Armenia. . . . advanced to fight for the Church at Jerusalem; but when he arrived at the Tigris and found no means of transport for his army, he turned northwards, as he had heard that the river in that quarter was frozen over in winter time. After halting on its banks for some years in expectation of a frost, he was obliged to return home."

By good fortune, about 1165, there mysteriously appeared in western Europe the verbatim text of a letter from Prester John to his friends the Byzantine emperor of Rome, Emanuel I, and the King of France promising to help them conquer the Holy Sepulcher. Scholars have never established who really wrote the letter, where or why. We do know that it was a forgery, though we do not know in what language it was originally written. "Prester John's Letter" enjoyed enormous popularity across Europe. More than one hundred Latin manuscript versions have appeared, besides numerous others in Italian, German, English, Serbian, Russian, and Hebrew.

The popularity of this ten-page manuscript booklet reminds us that, before newspapers or magazines, there was already a primitive yellow journalism to satisfy the news-hungry. Was this the Emperor out of the East who would liberate the Holy Sepulcher? Would the mysterious Prester John be a decisive new force to help Christians block the expanding Muslim Empire? Prince Henry the Navigator himself took a personal interest in locating this ally for his maritime adventures. After 1488, when the Portuguese opened a new Eastern waterway to India around the tip of Africa, there were commercial reasons to hope that Prester John was real. Two centuries later, when the Russians developed their overland trade with India, to guide themselves they sought out copies of the Russian version of Prester John's famous letter.

The sensational letter from Prester John was concocted from reports of the missionary Apostle Saint Thomas, whose body lying in India had wrought more miracles than any other saint, and who, though dead for eleven centuries, came back to preach in his Indian church every year. Tidbits were added from the romance of Alexander the Great and the adventures of Sindbad the Sailor:

> You should also know that we have birds called griffins who can easily carry an ox or a horse into their nest to feed their young. We have still another kind of birds who rule over all other fowl in the world. They are of fiery color, their wings are as sharp as razors, and they are called Yllerion. In the whole world there are but two of them. They live for sixty years, at the end of which they fly away to plunge into the sea. But first they hatch two or three eggs for forty days till the young ones come out. . . . Likewise, you would know that we have other birds called tigers who are so strong and bold that they lift and kill with ease an armored man together with his horse.
>
> Know that in one province of our country is a wilderness and that there live horned men who have but one eye in front and three or four in the back. There are also women who look similar. We have in our country still another kind of men who feed only on raw flesh of men and women and do not hesitate to die. And when one of them passes away, be it their father or mother, they gobble him up without cooking him. They hold that it is good and natural to eat human flesh and they do it for the redemption of their sins. This nation is cursed by God and it is called Gog and Magog and there are more of them than of all other peoples. With the coming of Antichrist they will spread over the whole world, for they are his friends and allies.

His realm, Prester John explained, included forty-two kings who were all "mighty and good Christians," the Great Feminie ruled by three queens and defended by one hundred thousand armed women, together with pygmies who fought an annual war against the birds, and bowmen "who from the waist up are men, but whose lower part is that of a horse." Since certain remarkable worms could live only in fire, Prester John maintained forty thousand men to keep the fires alive. In the flames these worms spun threads similar to silk, and "whenever we wish to wash them, we put them into fire whence they come clean and fresh." He described his magic mirrors, enchanted fountains, and waters from underground rivers that turned to precious stones.

The forged letter seemed to become more credible with every copying and with the passing centuries. We will never know how many true believers were seduced to search for the mythic kingdom. As late as 1573, on some of the best Dutch maps of the Great Age of Discovery we still see the empire of Prester John now moved to Abyssinia.

14

A Flat Earth Returns

"CAN any one be so foolish," asked the revered Lactantius, "the Christian Cicero," whom Constantine chose to tutor his son, "as to believe that there are men whose feet are higher than their heads, or places where things may be hanging downwards, trees growing backwards, or rain falling upwards? Where is the marvel of the hanging gardens of Babylon if we are to allow of a hanging world at the Antipodes?" Saint Augustine, Chrysostom, and others of their stature heartily agreed that the Antipodes ("anti"-"podes," a place where men's feet were opposite) could not exist.

Classic theories of the Antipodes described an impassable fiery zone surrounding the equator which separated us from an inhabited region on the other side of the globe. This raised serious doubts in the Christian mind about the sphericity of the earth. The race that lived below that torrid zone of course could not be of the race of Adam, nor among those redeemed by the dispensation of Christ. If one believed that Noah's Ark had come to rest on Mt. Ararat north of the equator, then there was no way for living creatures to have reached an Antipodes. To avoid heretical possibilities, faithful Christians preferred to believe there could be no Antipodes, or even, if necessary, that the earth was no sphere. Saint Augustine, too, was explicit and dogmatic, and his immense authority, compounded with that of Isidore, the Venerable Bede, Saint Boniface, and others, warned away rash spirits.

The ancient Greek and Roman geographers had not been troubled by such matters. But no Christian could entertain the possibility that any men were not descended from Adam or could be so cut off by tropical fires that they were unreachable by Christ's Gospel. "Yes, verily," declared Romans 10:18, "their sound went forth all over the earth, and their words unto the ends of the whole world." Neither Faith nor Scripture had any place for beings unknown to Adam or to Christ. "God forbid," wrote a tenth-century interpreter of Boethius, "that anybody think we accept the stories of antipodes, which are in every way contradictory to Christian faith." "Belief in Antipodes" became another stock charge against heretics prepared for burning. Some few compromising spirits tried to accept a spherical earth for geographic reasons, while still denying the existence of Antipodean inhabitants for theological reasons. But their number did not multiply.

It was a fanatical recent convert, Cosmas of Alexandria, who provided a full-fledged *Topographia Christiana,* which lasted these many centuries to the dismay and embarrassment of modern Christians. We do not know his real name, but he was called Cosmas on account of the fame of his geographic work, and nicknamed Indicopleustes (Indian Traveler), because he was a merchant who traveled around the Red Sea and the Indian Ocean, and had traded in Abyssinia and as far east as Ceylon. After his conversion to Christianity about A.D. 548, Cosmas became a monk and retired to a cloister on Mt. Sinai where he wrote his memoirs and his classic defense of the Christian view of the earth. This massive illustrated treatise in twelve books gives us the earliest surviving maps of Christian origin.

Cosmas rewarded the faithful with a full measure of vitriol against pagan error and a wonderfully simple diagram of the Christian universe. In his very first book he destroyed the abominable heresy of the sphericity of the earth. Then he expounded his own system, supported, of course, from Scripture, then from the Church Fathers, and finally from some non-Christian sources. What he provided was not so much a theory as a simple, clear, and attractive visual model.

When the apostle Paul in Hebrews 9:1–3, declared the first Tabernacle of Moses to be the pattern of this whole world, he conveniently provided Cosmas his plan in all necessary detail. Cosmas had no trouble translating Saint Paul's words into physical reality. The first Tabernacle "had ordinances of divine service and a worldly sanctuary; for there was a Tabernacle made; the first wherein was the candlestick, and the table and the shewbread, which is called the Sanctuary." By a "worldly" sanctuary Saint Paul meant "that it was, so to speak, a pattern of the world, wherein was also the candlestick, by this meaning the luminaries of heaven, and the table, that is, the earth, and the shew-bread, by this meaning the fruits which it produces annually." When Scripture said that the table of the Tabernacle should be two cubits long and one cubit wide, it meant that the whole flat earth was twice as long, east to west, as it was wide.

In Cosmas' appealing plan the whole earth was a vast rectangular box, most resembling a trunk with a bulging lid, the arch of heaven, above which the Creator surveyed his works. In the north was a great mountain, around which the sun moved, and whose obstructions of the sunlight explained the variant lengths of the days and the seasons. The lands of the world were, of course, symmetrical: in the East the Indians, in the South the Ethiops, in the West the Celts, and in the North the Scythians. And from Paradise flowed the four great rivers: the Indus or Ganges into India; the Nile through Ethiopia to Egypt; and the Tigris and the Euphrates that watered Mesopotamia. There was, of course, only one "face" of the earth—that which God gave to us the descendants of Adam—which made any suggestion of Antipodes both absurd and heretical.

Cosmas' work is still very much worth consulting as a wholesome tonic for any who believe there may be limits to human credulity. After Cosmas came a legion of Christian geographers each offering his own variant on the Scriptural plan. There was Orosius, the Spanish priest of the fifth century who wrote a famous encyclopedia, *Historiae adversum paganos,* where he retailed the familiar threefold division of the world into Asia, Europe, and Africa, embellished by some generalizations of his own:

> Much more land remains uncultivated and unexplored in Africa because of the heat of the sun than in Europe because of the intensity of the cold, for certainly almost all animals and plants adapt themselves more readily and easily to great cold than to great heat. There is an obvious reason why Africa, so far as contour and population are concerned, appears small in every respect (i.e., when compared with Europe and Asia). Owing to her natural location the continent has less space and owing to the bad climate she has more desert land.

Then the even more influential Christian encyclopedist Isidore Archbishop of Seville in the seventh century explained that the earth was known as *orbis terrarum* because of its roundness *(orbis)* like a wheel. "It is quite evident," he observed, "that the two parts Europe and Africa occupy half the world and that Asia alone occupies the other half. The former were made into two parts because the Great Sea called the Mediterranean enters from the Ocean between them and cuts them apart." Isidore's "wheel maps" followed the convention of the time by putting east at the top:

> Paradise is a place lying in the eastern parts, whose name is translated out of the Greek into Latin as *hortus* [i.e., garden]. It is called in the Hebrew tongue Eden, which is translated in our language as Deliciae [i.e., place of luxury or delight]. Uniting these two gives us Garden of Delight; for it is planted with every kind of wood and fruit-bearing tree having also the tree of life. There is neither cold nor heat there but a continual spring temperature.
>
> From the middle of the Garden, a spring gushes forth to water the whole grove, and, dividing up, it provides the sources of four rivers. Approach to this place was barred to man after his sin, for now it is hedged about on all sides by a sword-like flame, that is to say it is surrounded by a wall of fire that reaches almost to the sky.

Christian geographers who lacked facts to fill their landscapes found a rich resource in the ancient fantasies. While they were contemptuous of pagan science, which they considered a menace to Christian faith, their

prejudice did not include pagan myths. These were so numerous, so colorful, and so contradictory that they could serve the most dogmatic Christian
purposes. While Christian geographers feared the close calculations of
Eratosthenes, Hipparchus, and Ptolemy, they cheerfully embellished their
pious Jerusalem-centered maps with the wildest ventures of pagan imagination. Julius Solinus (fl. A.D. 250), surnamed Polyhistor, or "Teller of Varied
Tales," provided the standard source of geographic myth during all the
years of the Great Interruption, from the fourth till the fourteenth centuries. Solinus himself was probably not a Christian. Nine-tenths of his *Collec-
tanea rerum memorabilium* (Gallery of Wonderful Things), first published
about A.D. 230–240, came straight out of Pliny's *Natural History,* though
Solinus does not even mention his name. And the rest was foraged from
other classical authors. Solinus' peculiar talent, as a recent historian of
geography observes, was "to extract the dross and leave the gold." It is
doubtful if anyone else over so long a period has ever influenced geography
"so profoundly or so mischievously."

Yet Solinus' dross had wide appeal. Saint Augustine himself drew on
Solinus, as did all the other leading Christian thinkers of the Middle Ages.
The stories and fabulous images that Solinus retailed enlivened Christian
maps right down to the Age of Discovery. They became an all-encompassing network of fantasy, replacing the forgotten rational gridwork of latitude
and longitude, which had been Ptolemy's legacy. Solinus found wonders
near and far. From Italy he reported people who sacrificed to Apollo by
dancing barefoot on burning coals, pythons that grew long and fat by
feeding on the udders of milk cows, and lynxes whose urine congealed to
"the hardness of precious stone, having magnetic powers and the color of
amber." Grasshoppers and crickets in Rhegium still dared make no sound
because Hercules, irritated by their noise, had once ordered them to keep
silent. Further afield were the dog-headed Simeans of Ethiopia, ruled by a
dog-king. Along the Ethiopian coast were peoples with four eyes, while
along the Niger were ants as big as mastiffs. In Germany there was a
mule-like creature with such a long upper lip that "he cannot feed except
walking backward." Human monstrosities that were normal in remote parts
of the world included tribes who had their eight-toed feet turned backwards,
men with dogs' heads and talons for fingers who "barked for speech,"
people who had only one leg, but with a foot so large that it protected them
from the hot sun by serving as a parasol.

Perhaps the most enduring legacy of that age, so familiar that it has lost
its significance for us, was "The Mediterranean." The Romans had called
that body of water—the whole chain of inland seas between Africa, Asia,
and Europe—*mare internum* or *mare nostrum.* Solinus was one of the first

to call these the "mediterranean" seas—seas in the middle of the earth. The celebrated Isidore of Seville converted "mediterranean" into a proper name, and the authority of Isidore was not to be gainsaid!

While Christian geography in Europe was becoming this hodgepodge of fantasy and dogma, elsewhere people still advanced their knowledge of the earth, and their ability to chart earthly space. The Chinese had quite independently, without the aid of Eratosthenes, Hipparchus, or Ptolemy, devised a grid pattern that they superimposed on the irregular surface of the earth. We have seen that what the clock did for time, the rectangular map-grid did for space, providing uniform receptacles in which the endless variety of land and water, mountain and desert, could be discerned, described, discovered, and rediscovered.

While a spherical earth was the basis of the Greek cartography, a flat earth was the basis of the Chinese. By the time that Ptolemy had done his work in the West, Chinese cartographers had developed their own usable techniques of map-grids and a rich tradition of world cartography, which happily grew without the amnesiac Interruption that afflicted the West. The Greeks had developed their grid system along the lines of latitude and longitude so easily drawn around a sphere. But since it was so difficult to project a spherical surface on a flat sheet, in practice the Greek grid system of latitude and longitude was not substantially different from what it would have been if they had conceived the surface of the earth as flat.

Since the Greek grid system grew out of the requirements of a spherical shape, the Chinese rectangular grid, which made their whole cartography possible, must have had quite other origins. What were they?

From the earliest political records of the Ch'in era (221–207 B.C.), we find numerous references to maps and their uses. China, unified in 221 B.C., was both the creature and the creator of a vast bureaucracy, which had to know the features and the boundaries of its extensive regions. The Rites of Chou (1120–256 B.C.) had required the Director-General of the Masses to prepare maps of each feudal principality and register its populations. When the Chou emperor toured his realm, the Geographer-Royal was at his side explaining the topography and products of each part of the country. Under the Han dynasty (202 B.C.–A.D. 220) maps appear again and again as the indispensable apparatus of empire.

The last two millennia in China display a galaxy of cartographic talent. During the heyday of religious cartography in Europe the Chinese were marching steadily ahead toward quantitative cartography. Even before Ptolemy had done his work in Alexandria, a Chinese pioneer Chang Heng (A.D. 78–139) had "cast a network of coordinates about heaven and earth, and reckoned on the basis of it." Within two centuries the Chinese Ptolemy, Phei Hsui, appointed in A.D. 267 as Minister of Works to the first emperor of the Chin dynasty (265–420), applied these techniques to make a detailed eighteen-sheet map of China. "The great Chin dynasty has unified space in all the six directions," he explained, and it was only appropriate that map-makers should provide comprehensive maps, free of error, and drawn to scale, showing "mountains and lakes, the courses of rivers, the plateaus and plains, the slopes and marshes, the limits of the nine ancient provinces and the sixteen modern ones, . . . commanderies and fiefs, prefectures and cities . . . and lastly, inserting the roads, paths, and navigable rivers."

In the preface to his atlas, Phei Hsui gave simple directions for making a map to proper scale, with rectangular grids. "If one draws a map without having graduated divisions," he warns, "there is no means of distinguishing between what is near and what is far. . . . But . . . a true scale representation of the distances is fixed by the graduated divisions. So also the reality of the relative positions is attained by the use of paced sides of right-angled trian-gles; and the true scale of degrees and figures is reproduced by the determi-nations of high and low, angular dimensions, and curved or straight lines. Thus even if there are great obstacles in the shape of high mountains or vast lakes, huge distances or strange places, necessitating climbs and descents, retracing of steps or detours—everything can be taken into account and determined. When the principle of the rectangular grid is properly applied, then the straight and the curved, the near and the far, can conceal nothing of their form from us."

How did the Chinese develop so refined a technique for mastering the earth's irregularities? From earliest times, it seems, they had made land allotments by a scheme of coordinates. From the times of the Ch'in dynasty, the emperor's maps had been drawn on silk. The terms *(ching* and *wei)* which Phei Hsui used for the coordinates on his maps were the same words that had long been used for the warp and the weft in the weaving of textiles. Was the idea of a rectangular grid on a map suggested by finding that a place on a silken map could be located by following a warp and a weft thread to their meeting place? Or was the origin possibly in the diviner's board of Han times that used a system of coordinates to represent the whole cosmos? Or was it connected, somehow, with the form of the early Chinese chessboard, which located pieces by their coordinates? Whatever the origins, the result is plain enough: a well-developed and widely used system of rectangular grids. In 801, during the T'ang dynasty (618–907), the emperor's cartogra-pher completed a grid map of the whole empire on a scale of 1 inch to 100 *li* (one-third of a mile: 33⅓ miles), which measured 30 feet long and 33 feet high. Maps became so popular that they were found even in the imperial baths.

Cartographers found ways to link geographical with celestial coordinates, and without interruption they continued to elaborate their grid system. While Christian cosmographers were fantasizing their maps to fit Scriptural texts, Chinese map-makers were making unprecedented advances.

More and more the grid pattern imposed for human convenience dominates their design. By the Sung era (960–1279), they are regularly placing north at the top. The Mongol unification of Asia under Genghis Khan and Kublai Khan in the thirteenth century brought masses of new geographical information to imperial map-makers. With the passage of time, the grid becomes more rather than less prominent on the face of Chinese maps. A whole new cartography developed, called the Mongolian style, in which the map itself became simply a grid, with no effort to delineate the earth's erratic shape, but with the names of places and tribes inserted in their proper squares.

By the mid-twelfth century, even before the general revival of Ptolemy in Europe, when the Arab geographer Al-Idrisi made his world map in 1150 for Roger II, Norman king of Sicily, he too used a grid scheme which, like that on Chinese maps, makes no allowance for a curved earth. Perhaps, as Joseph Needham suggests, the long tradition of grid maps of China had reached the Arabs in Sicily through the Arab colony at Canton and through the increasing number of Arab travelers to the East. And so perhaps the Chinese played a part in ending the Great Interruption—setting European geographers once again on the path of knowledge, rediscovering the quantitative tools that were the heritage from Greece and Rome.

PART FIVE

PATHS TO THE EAST

From the East, light. (Ex Oriente, lux.)

—LATIN PROVERB

Too far East is West.

—ENGLISH PROVERB

15

Pilgrims and Crusaders

THE same faith that had fantasized the landscape and imprisoned Christians in dogmatic geography would lure pilgrims and crusaders from Europe on paths of discovery to the East. The Star of Bethlehem, which drew the Three Kings, guided countless faithful in later centuries to their Holy Land. Pilgrimage became a Christian institution and paths of faith would become paths of discovery.

Within a century after the death of Jesus a few intrepid believers were journeying to Jerusalem for penance, for thanksgiving, or simply to walk the earth where their Savior had walked. After Emperor Constantine became a Christian, his mother, Empress Helena, went to Jerusalem in 327, turned archaeologist, found the Mount of Calvary, collected supposed pieces of the True Cross, and even uncovered the Holy Sepulcher, where Jesus reputedly had been buried. There Constantine himself built the first Church of the Holy Sepulcher. The learned Saint Jerome settled in 386 in a monastery in Bethlehem provided by the noble Roman lady Saint Paula where he instructed pilgrims after they had visited the holy places. By the early fifth century there were two hundred monasteries and pilgrim hospices near Jerusalem. Saint Augustine and other Church Fathers warned that the Christian tourist to the Holy Land might be distracted from his journey to the Heavenly City. Still the stream of pilgrims grew, aided by countless handy guides and a chain of hospitable lodgings all along the way.

The pilgrim, blessed by his priest before setting out, carrying staff and mussel shell, wearing his broad flat-crowned hat and bearing the badge of his destination, became a picturesque figure in the medieval panorama. The Latin *peregrinatio* came to mean any wandering, and *peregrinus* for pilgrim, became a synonym for stranger. But the pilgrim, properly speaking, was someone, whatever his regular occupation, on the way to a sacred destination and the "palmer," so called from the palm branches brought back from Rome, was a religious vagrant who might spend his whole life wandering from one holy place to another.

The decline of the western Roman Empire, along with the rise of pirates, Vandals and others, made the pilgrim's lot difficult and dangerous. The extensive Arab conquests around the Mediterranean, the rise of Islam, and the increase of Muslim pilgrims congested the Christian pilgrim routes and brought a bitter contest for Jerusalem. It was from the rock at the Temple Mount, site of the Temple of Solomon, that Muhammad had ascended into heaven. Some Muslim traditions made Jerusalem, and not Mecca, the center and the Navel of the Earth, "the highest of all countries and the nearest to heaven." When the caliph Omar, riding on a white camel, entered a surrendered Jerusalem only six years after the death of Muhammad, he opened a millennial battle for the holy places.

The great age of Christian pilgrimage had begun in the tenth century. Muslims were generally tolerant, if contemptuous, of these passionate "unbelievers." But as the distant Holy Land became less accessible, pious Christians found the balm of pilgrimage nearer home. They produced a hybrid literature of history, sociology, myth, and folklore. In the popular *Guide de Pèlerin* one could read of the pilgrim who asked a woman in Villeneuve for a piece of the loaf of bread she was baking under hot ashes in her oven. She refused, and when she went for her loaf, she found only a round stone. Other pilgrims en route in Poitiers searched a whole street before they could find anyone to lodge them. That very night all houses in the street burned—except the one that had given them hospitality. Popular epics like the *Chansons de Geste* celebrated heroic pilgrims.

Santiago de Compostela in northwestern Spain had become holy because of the miraculous discovery there, about 810, of the body of Saint James, who was known to have been executed in Jerusalem and presumably had been buried there. Charlemagne was reputed to be among the first of the pilgrims who flocked from all Europe. When the advancing Moors set about conquering Spain for Islam, the cult arose of Saint James the Moor-slayer, *Santiago Matamoros*.

The great magnet for pilgrims in Europe was Rome, "threshold of the Apostles." The Venerable Bede (673–735) reported the trek to Rome of Britons "noble and simple, layfolk and clergy, men and women alike" who yearned to spend some of their "earthly pilgrimage" in the vicinity of holy places, "hoping thereby to merit a warmer welcome from the saints in heaven." From which, perhaps, was derived the English verb "to roam." By 727 King Ina of Wessex had founded a special hostel there for Saxon pilgrims. The stream from Britain and elsewhere would broaden after the failure of crusaders to recapture Jerusalem.

Meanwhile the monks of Cluny organized to assist pilgrims going eastward. Able-bodied young Scandinavians journeyed all the way from Iceland, Norway, or Denmark, spent a few years serving in the Emperor's famous Varangian Guard in Constantinople, and then went on to Jerusalem before returning northward with their military earnings. A Danish prince, to expiate a murder, set out in the mid-eleventh century but never reached the Holy Place because he went barefoot for his sins, and died of exposure in the mountains on the way.

Then at the height of the pilgrims' enthusiasm Alp Arslan (1029–1072), the Seljuk sultan of Persia, whose Turkic-speaking people had spread across Asia to eastern Siberia, swept westward, defeated the forces of Byzantine Christians at Manzikert in 1071, and occupied most of Asia Minor, including the pathways to the holy places. Christian pilgrims and all Eastern Christianity faced new perils.

At the same time forces farther west, a new commercial life, and a growing population were swelling the pilgrim tide. The Normans, descendants of the Norsemen who had swept in the tenth century into "Normandy" on the northern coast of France, were converted to Christianity, and sent their conquering force in all directions. William the Conqueror led them north to England in 1066. They roamed the Mediterranean, overran southern Italy and by 1130 had set up the Kingdom of Sicily, where Christians, Jews, and Arabs exchanged knowledge, arts, and ideas.

When Urban II became pope in 1088, his Church was in dire need of reform—rotten with the buying and selling of pardons and church offices, and split by the claims of an anti-Pope. A muckraking reformer, he used his organizing talents and his eloquence to cleanse and to heal. Alexius Comnenus, the Eastern emperor, seeing the capital of his Byzantium threatened by militant Islam, sent envoys to Urban appealing for military aid. The energetic Urban saw this as an opportunity to unite the Churches of East and West while the holy places would be redeemed.

To a historic Council at Clermont in south-central France he summoned French bishops and representatives of the faith from across Europe. When the Council met on November 18, 1095, it became a mass meeting, too large for the cathedral, and moved to a field outside the eastern gates of the city. There in the open air the Pope stirred the crowd with an eloquence we can still savor from the report by Robert the Monk, who was there:

> Jerusalem is the navel of the world, a land which is more fruitful than any other, a land which is like another paradise of delights. This is the land which the Redeemer of mankind illuminated by his coming, adorned by his life, consecrated by his passion, redeemed by his death, and sealed by his burial. This royal city, situated in the middle of the world, is now held captive by his enemies and is made a servant, by those who know not God, for the ceremonies of the heathen. It looks and hopes for freedom; it begs unceasingly that you will come to its aid. It looks for help from you, especially, because God has bestowed glory in arms upon you more than on any other nation. Undertake this journey, therefore, for the remission of your sins, with the assurance of "glory which cannot fade" in the kingdom of heaven.

To rescue the holy places, all Christians were to go east as soon as the harvest was taken in the next summer, but not later than the Feast of the Assumption, August 15, 1096. God would be their guide, the white cross their symbol, and their battle cry *"Deus le volt!"* (God wills it!). Back home their belongings would be under the protection of the Church.

With this call to arms Pope Urban II marshaled the forces of Christian Europe to transform pilgrims into crusaders. As the medieval Latinist put it, while the pilgrimage was only a *passagium parvum,* a little journey by an individual, the crusade was to be a *passagium generale,* a communal or mass pilgrimage. People on the move could not help becoming discoverers. But, for the most part, they did not find what they went to seek, and they found much they had not imagined.

The Crusades would be one of the most miscellaneous, most unruly movements in history. A portent of things to come was Peter the Hermit. He was appropriately miscalled the Hermit because he usually wore a hermit's cape, but he was by no means a hermit, for he loved crowds and knew how to move them. Peter set up his own corps of recruiting agents and began gathering his motley pilgrim army in the county of Berry in central France. By the time he reached Cologne in western Germany on Holy Saturday, April 12, 1096, some fifteen thousand pilgrims of all ages, sexes, shapes, and sizes had joined his party. "All the West and all the barbarian tribes from beyond the Adriatic as far as the Pillars of Hercules," a Byzantine princess, Anna Comnena, fearfully reported, "were moving in a body through Europe towards Asia, bringing whole families with them."

The arrival of Peter's horde at Constantinople brought new troubles. There they joined forces with Walter the Penniless, and moved on toward the Holy City, plundering as they went. One group under Rainald, an Italian nobleman, sacked Christian villages en route, tortured their inhabitants, and were even reported to roast Christian babies on spits over open fires. The Byzantine emperor Alexius I tried to persuade adventuring knights to submit to his rule, but the more ambitious of them conquered and pillaged to establish new kingdoms of their own. These Christian forces defeated the Turks in several battles and entered Jerusalem in triumph in July 1099, so bringing to an end what came to be called the First Crusade.

Jerusalem was speedily organized into a new Latin Kingdom. This was only the beginning of the hectic movement that lasted two centuries to make secure the way of the pilgrims. But in one sense, too, it was the end of the Crusades, for it was the last successful expedition to redeem the holy places. Later "Crusades" turned out to be only expeditions to help the Christians already established in the East. After the fall of Jerusalem to the Turkish Saladin in 1187, the more accessible holy destinations in the West attracted Christian pilgrims more than ever.

For the faithful in Britain the most sacred place was Canterbury. It was in Canterbury Cathedral, where the second Saint Augustine (d. 604) had been the first archbishop, that Thomas à Becket championed the church against King Henry II, and was martyred on December 29, 1170. King Henry II himself marked the pilgrim path when he journeyed there to do public penance. "As soon as he neared the city," Roger of Hoveden, a contemporary chronicler, reported, "in sight of the Cathedral in which lay the body of the blessed martyr, he dismounted from his horse, and having taken off his shoes, with bare feet and dressed in woollen garments, he walked three miles to the tomb of the blessed martyr, with such meekness and repentance that it really may be held to have been the work of Him who looketh down on the earth and maketh it to tremble." Chaucer gave the shrine of Saint Thomas at Canterbury literary immortality by his description of thirty-one varieties of pilgrims.

> Thanne longen folk to goon on pilgrimages,
> And palmeres for to seeken straunge strondes,
> To ferne halwes, couthe in sondry londes;
> And specially from every shires ende
> Of Engelond to Canterbury they wende . . .

After the Crusades came to an end, pilgrimage still remained a living force in European Christendom, and for many Rome replaced Jerusalem. In 1300, in the spirit of Urban II, Pope Boniface VIII proclaimed the first Jubilee Year, offered special indulgences to the faithful who visited Rome and attracted more than twenty thousand pilgrims. Jubilee Years with their attendant indulgences for pilgrims to Rome continued every fifty years till the Pope in 1470 reduced the interval to twenty-five years.

In Islam, from its very beginning, pilgrimage was a holy duty. Every good Muslim was, and is, obliged, if he can afford it and can support his family in his absence, to visit Mecca at least once. During the *hajj* from the seventh to the tenth month of the Muslim year, the pilgrim wears two seamless white garments, symbols of equality before God. He neither shaves nor cuts his hair or nails during the ceremony, he must circumambulate the Ka'bah seven times, and follows certain other rites around Mecca before returning home. Forever after, the returned pilgrim is honored by the title of *hajji*.

Mecca had been a place of pilgrimage for idolatrous Arabs centuries before Muhammad. They went for their annual festival to welcome the renewed year, to light bonfires to persuade the sun to rise, and to work charms to prevent drought. Mecca never ceased to be the Muslim pilgrim's goal and in Western languages became a synonym for any goal of pilgrimage. In the late twentieth century, the *hajj* was so popular that some Muslim countries annually limited the number of outgoing pilgrims to avoid foreign-exchange problems. By 1965 some one and a half million pilgrims were visiting Mecca each year, about half from outside Arabia.

Ibn Battuta (1304–1374), the greatest Muslim traveler of the Middle Ages, at the age of twenty-one left his home in Tangier at the northwestern tip of Africa as a pilgrim "swayed by an overmastering impulse . . . and a desire long-cherished to visit these illustrious sanctuaries." His popular chronicle of lifelong travels made him a kind of Muslim Marco Polo, celebrated as "the traveler of Islam." Despite his rule "never to travel any road a second time," he made four pilgrimages to Mecca. Altogether he covered some seventy-five thousand miles, probably more than any other recorded traveler before his day. He visited every Muslim country and neighboring lands, served as Muslim judge, or *cadi,* in Islamic communities as far away as Delhi, the Maldive Islands, and Ceylon, and became an envoy from sultans to Chinese unbelievers. Still, his extensive travels were no enticement into the unknown, but a kind of encyclopedia of Muslim life and customs in various climates and terrains. He showed how much a devout Muslim man of curiosity and energy could discover of the world if he was willing to move around, to risk bandits, pirates, the Black Death, and the whims of despotic sultans. So he acquired a Muslim liberal education, but his imagination did not range far beyond Islam and his learning was inhibited by his faith.

In the heart of Asia, too, faithful crowds seeking the balm of their own holy places, en route began to discover their world. No one knows exactly how, why, or when ancient Benares on the Ganges—some called it the most ancient city in the world—first became sacred, but by the seventh century the city held a hundred temples to Siva. In the eleventh century, the Muslim Alberuni reported how Hindus venerated Benares, "their anchorites wander to it and stay there for ever, as the dwellers of the Ka'ba stay for ever in Mecca. . . . that their reward after death should be the better for it. They say that a murderer is held responsible for his crime and punished with a punishment due to his guilt, except in case he enters the city of Benares, where he obtains pardon."

The Buddhists, too, taught that the Deer Park at Sarnath where the Buddha (c. 500 B.C.) had preached his first sermon was a rung on the ladder to heaven. The north Indian Emperor Asoka, who converted to Buddhism about 200 B.C., led pilgrimages to all the Buddhist sacred places. As he visited them he repaired old shrines, *stupas,* and built new ones. Wherever he went, he erected commemorative stone pillars, many still standing. From remote corners of Asia, men and women, noble and peasant, scholar and illiterate, followed Asoka's footsteps. From the Chinese imperial capital of Sian on the Wei River in central China the Buddhist enthusiast Fah Hian, about A.D. 400, traversed deserts and mountain ranges to visit the Buddhist shrines in north India, then crossed the peninsula to enjoy the sanctity of the tooth of Buddha in Ceylon.

India became a land of sacred places. According to the Buddha, "all mountains, all rivers, holy lakes, places of pilgrimage, the dwellings of *rsis,* cow pens, and temples of the gods are places which destroy sin." The cults of local spirits and countless local priesthoods multiplied till one traveler to Kashmir observed that there was "not a space as large as a grain of sesame seed without a place of pilgrimage."

The Lutheran Reformers' declaration of faith, the Augsburg Confession (1530), condemned pilgrimages—along with set fasts, the worshipping of saints, and the counting of rosaries—as "childish and needless works." But in retrospect, the Crusades of embattled pilgrims proved to be great awakeners. They were both a symptom and a cause of a new vitality, a new curiosity, a new openness and mobility in the life of the West. Countless novelties came with and from the Crusades, by-products of people on the move. The Crusader states in the Eastern Mediterranean developed trade with the Muslim world. Italian banks flourished as they financed kings and popes, and lent funds to pious travelers. Returning crusaders brought back tales of Oriental splendor, along with tastes for damasks, silks, perfumes, and spices which gave Venice the exotic charm still visible in the Doges' Palace and the Piazza San Marco.

Yet the failure of the Crusades was in many ways a blessing for Christendom and a catalyst for European discovery of the Eastern world. The great organizing international institution of Islam remained the pilgrimage, what Ibn Battuta called "the annual congress of the Muslim world," which continued to meet at Mecca, a familiar Arab-Muslim stronghold. But there was no counterpart place of Again-and-Again, no accessible site of obligatory return, for all Christians. With no prospect of recapturing Jerusalem and the pathways there, Western Christendom turned to missions. Pilgrimage gathered the faithful, but missions reached out to the stranger even in unknown lands. The history of expanding Christianity was a history of missions.

Of course, as we have seen, missions were an ancient institution for all world-reaching religions. King Asoka had sent Buddhist missionaries abroad in the second century B.C., and they were found across China in succeeding centuries. But missions played a greater role in Christianity than in any other world religion. As early as the second century, a missionary college was founded in Alexandria, and another in Constantinople in 404. Saint Patrick, Saint Augustine of Canterbury, and Saint Boniface who carried the gospel to Ireland, England, and Germany were missionaries, and lesser missionaries spread the word across the continent. In the high mountains of Switzerland, in the Rhine Valley, in the Swedish forests, and in frigid Russia, monasteries became centers of civilization and of Christian preaching. By the time Europe was substantially Christianized in the seventh century, and the papacy was organizing and supporting professional missionaries and monks across the continent, Muhammad came with his competing faith. Militant Islam forced Christian retreat from the Near East, northern Africa, and the Iberian peninsula, and created its own empire where Christianity was sometimes tolerated but could not advance. Islam did not allow the propaganda of other faiths, and punished apostasy with death. But Nestorian Christian missions had extraordinary success in China beyond the reach of Islam. By the mid-seventh century they had translated the Scriptures into Chinese for the imperial library and were declared a tolerated religion by imperial edict.

After the Crusades the crusading energies were channeled into missions. Missionary monks had accompanied crusaders on their way east, and when the Crusades came to an end, zealous Franciscan and Dominican friars, as we shall see, continued to reach out to the little-known farther East. The popes took a renewed interest in missions to distant parts. They issued papal bulls protecting and supporting the relentless missionary friars, and sent envoys to Mongol Khans and Chinese emperors to open their way. The friars would be the vanguard for Europe's overland discovery of Asia.

Of course Islam, beginning with the Prophet himself, was a vigorously proselytizing religion and every Muslim was declared to be a missionary. But Muslim missions never were so well organized or so widely diffused as those of Christianity. While Islam had its mullahs, these were teachers rather than a priesthood, and there was no Islamic institution like the missionary friars of Christendom. Nor were there Muslim missionary societies until the late nineteenth century. The *jihad,* the religious duty to spread Islam by waging war, was long the primary authorized way of extending the empire of the Prophet. The all-embracing character of the Muslim faith, which saw no distinction between the realm of Caesar and the realm of God, made the reach of the faith coterminous with the reach of the sword. While Muslim swordsmen conquered for the faith, Christian missionaries were willing to explore tentatively on the frontiers of empire, hoping to bring the good news to even a few more souls.

In Islam the pilgrim remained a devotee of the Faith fulfilling the prescribed ritual journey to a known holy destination. In the parlance of modern Christendom, a "pilgrim" was seldom on his way to Jerusalem, but every person "sojourning in the flesh" was passing through this earth to a mysterious state of future bliss. So, in familiar American usage, were the Pilgrim Fathers. "They knew they were but pilgrims," wrote William Bradford in 1630, "and looked not much on those things, but lift up their eyes to the heavens, their dearest countrey." The Christian "Pilgrim's Progress" was toward no earthly destination. In the West, the "crusade," too, ceased to be a battle against unbelievers and acquired more exploratory overtones, as when Thomas Jefferson urged his good friend to "preach a crusade against ignorance."

16

How the Mongols Opened the Way

THE land-faring pioneers of Europe's first Age of Discovery who went eastward in the mid-thirteenth century needed resources quite different from those of the later, the seafaring, age. Columbus would have to raise a large sum of money, find ships, enlist and organize a crew, secure supplies, keep the crew happy and unmutinous, and navigate a trackless ocean. Quite other talents were required of the earlier overland traveler. He could go with one or two companions along main-traveled roads—though the roads had not been frequented by Europeans before them. They could live off the land, finding food and drink along the way. While they did not need to be fund-raisers or master organizers, they had to be adaptable and affable. Columbus' men became mutinous when the voyage stretched a few weeks beyond what they had expected, but the overland pioneers could extend their journey as long as necessary, by another month, another year, even another decade. While the seafarers went over long stretches of cultural emptiness, and at sea news usually meant trouble, the landfarers—merchants or missionaries—could practice their vocation along the way, learning as they went. If the solitary landfarer took ship for some stages of his trip, he was a passenger. The vessel was usually commanded and supplied by someone of the region. The land-faring pioneer would be both more lonely and less lonely than his counterpart on the sea. For if he lacked the companionship and support of fellow countrymen, like those who went with Columbus on the *Santa María,* yet he had the opportunity for many novel, warm, and casual associations on his days and nights along the way.

The perils of the sea were quite the same everywhere—wind and wave and storm, loss of bearings—but the perils of the land were as varied as the landscape, and helped make the journey interesting and suspenseful in surprising ways. Were robbers lurking in this inn? Could you digest the local food? Should you wear your own or a native costume? Would you be allowed entry in this city gate? Could you crash the barriers of an unknown language to explain your wants and show that your mission was harmless?

Overland travel was not an adventurous communal leap, but a laborious, individual trek. From that age came our English word "travel"—originally the same as "travail," meaning labor, especially of a painful or oppressive nature—an accurate description of what it meant to go long distances overland. A few pioneers took up this travail and opened the way from Europe to Cathay.

While Europeans were still plunged in the darkness of dogmatic geography, they had long been entertained by legends of the mysterious East. A few men and women enjoyed the exotic luxuries from that other end of the earth—sleek silk from China and sparkling diamonds from Golconda. In rooms draped with costly carpets from Persia, they feasted on dishes spiced from Ceylon and Java, and passed the hours with ebony chessmen from Siam.

Yet the merchants of Venice, Genoa, or Pisa who prospered by selling these exotic Eastern commodities had themselves, of course, never seen India or China. Their Eastern contact was in the Levant ports in the eastern Mediterranean. Their precious stock had been brought by one of two main routes. One, the fabled Silk Road, was an all-land route from eastern China through central Asia, by way of Samarkand and Baghdad, finally reaching the coastal cities of the Black Sea or the eastern Mediterranean. The other came through the South China Sea, the Indian Ocean, and the Arabian Sea, either up the Persian Gulf to Basra or up the Red Sea to Suez. To reach the European market these goods would still have to go overland, across Persia and Syria or else through Egypt. On either of these routes, Frankish and Italian merchants found their way blocked as soon as they tried to advance eastward from the Mediterranean ports. Muslims gladly traded with them at Alexandria or even in Aleppo or Damascus, but the Muslim Turks would not allow Europeans to advance a step farther. This was the Iron Curtain of the late Middle Ages.

Then for a single century, from about 1250 to about 1350, that curtain was lifted, and there was direct human contact between Europe and China. During this interlude the bolder and more enterprising Italian merchants no longer had to wait until their exotic goods reached Aleppo, Damascus, or Alexandria. Now they themselves took caravans across the Silk Road to the cities of India and China, where they could hear Christian missionaries, Frankish and Italian friars, saying mass. What might have been the beginning of a continuous mutual enrichment, a widening and enlivening of the visions both of East and of West, proved to be only a brief and tempting curtain-raising, an adventurous episode, after which the curtain again came down with a thud. This would prove to be another kind of interruption, an interlude of light in the darkness that for most of modern history blanketed both the Eastward and the Westward vision. Decades would pass before the discovery of the ocean would make it possible for Europeans once again to touch the coastal fringes of India and southeast Asia. Centuries would elapse before Europeans were again permitted to visit the ports of China. Central Asia would remain long unvisited, and inland China, after an entr'acte of only two centuries, would remain inhospitable or hostile to visitors from the West.

It was not the march of Christian soldiers, nor the maneuvers of European statesmen, that lifted the curtain. Like many other world-awakening events, it was a by-product. If credit must be given for opening the way to Cathay, it must go, surprisingly, to a people of the same stock as those Turks who so long blocked the way for Europeans—to a Mongol people from central Asia, the Tartars (or Tatars). A threat to Europe in the Middle Ages, they have been much maligned. Featured in our European historical pantheon as reckless destroyers, their very name has become an English synonym for barbarian. The word "horde," which has come to mean a disorderly swarm, came from the Turkish *ordū*, which simply meant a "camp." Their reputation has been fixed by European writers who knew, or who had heard of, the horrors of the first Tartar onslaughts on the West. But few of these writers had ever seen a Tartar, and they knew nothing of the remarkable solid achievements of the Tartar Khans.

The Mongol empires were land empires, twice the size of the Roman Empire at its greatest extent. Genghis Khan and his hordes came down from Mongolia to Peking in 1214. In the half-century after, they took nearly all eastern Asia, then turned westward across Russia, even into Poland and Hungary. When Kublai Khan came to the Mongol throne in 1259, his empire reached from the Yellow River in China to the shores of the Danube in eastern Europe and from Siberia to the Persian Gulf. The Mongol Khans, from Genghis Khan through his sons and grandsons—Batu Khan, Mangu Khan, Kublai Khan, and Hulagu—were as able a dynasty as ever ruled a great empire. They showed a combination of military genius, personal courage, administrative versatility, and cultural tolerance unequaled by any European line of hereditary rulers. They deserve a higher place and a different place than they have been given by the Western historian.

Without the peculiar talents and special achievements of these Mongol rulers and their people, the way to Cathay would probably not have been opened when it was. When would there then have been a path for Marco Polo? Without Marco Polo and the others who stirred the European imagination with impatience to reach Cathay, would there have been a Christopher Columbus?

In 1241 a swarm of Tartar horsemen ravaged Poland and Hungary, defeating an army of Poles and Germans at the Battle of Lignitz in Silesia, while another of their armies defeated the Hungarians. Terror struck Europe. In the North Sea, even the courageous fishermen of Gothland and Friesland were frightened into staying away from their usual herring grounds. The learned Holy Roman emperor Frederick II (1194–1250), a patron of science and literature, who had led the successful Sixth Crusade (1228–29), had actually captured Jerusalem and then made a ten-year truce with the sultan of Egypt, now feared that the Tartar flood would overwhelm Christendom. He called on King Henry III of England and others to unite against this new "Scourge of God," in the hope "that these Tartars would be driven finally down to their Tartarus" *(ad sua Tartara Tartari detrudentur)*. Pope Gregory IX proclaimed a new crusade, this time against the Tartars. But because of the bad blood between the Pope and Frederick II, who had already been twice excommunicated, the King of Hungary's plea for help was answered only by words. After all, Europe was saved by an Act of God, when the Tartar hordes at the height of their successes received word that their great Khan Okkodai was dead in Asia and that they must hasten home.

Despite the alarms of Christian rulers, and the Tartar massacres of Poles and Hungarians, the Tartars would prove powerful allies against the Muslims and the Turks who blocked the eastward path. For the Tartars, after succeeding in their campaigns against the "Assassins," or Ismailians, on the southern shores of the Caspian, went on to overcome the Caliph of Baghdad and Syria. The conquering Tartar general in Persia had actually sent his embassy to Saint Louis, King Louis IX of France, who was then at Cyprus on a Crusade, offering an alliance and asking for collaboration. If Christian kings and the Pope himself had been willing to join in such an alliance, they might have shared the glory and the profit of the conquest of the Muslim Turks, and eventually have accomplished the aims of the Christian Crusades with pagan help. But instead of postponing conversion until after a worldly victory, they determined to ally only with fellow Christians, and so spent themselves on futile efforts to convert the Khans before joining them as allies. This mistake of judgment decisively shaped the future of much of Asia. The power of Islam was then in retreat. If Christian leaders had only been willing first to become comrades in arms against a common enemy, Pope Innocent IV and the Christian powers might soon enough have made them comrades in faith.

Western Christendom vainly awaited the sudden conversion of the

Khans. Meanwhile the Europeans would unintentionally profit from the religious vagueness, the indifference and tolerance of the Tartars. After the Tartars extinguished the Caliphate of Baghdad and took possession of Syria, Persia, and the lands that had become an Iron Curtain, the path was suddenly opened for the European traveler. The way of thinking of the Tartar Khans was most alien to the medieval Christian West. In 1251 at the court of Mangu Khan in the Tartar capital of Karakorum far north of the Great Wall, the Franciscan friar William of Rubruck was surprised to find priests from all over and from all religions—Catholics, Nestorians, Armenians, Manichaeans, Buddhists, and Muslims—peacefully debating, vying for the support of the Khan. The Khans also believed in free commercial intercourse among nations. They made merchants welcome by lowering tolls and taxes, by protecting caravans, by guarding roads against bandits.

The "barbarian" Tartars, who did not care enough for any dogma to persecute in its name, opened the pathway from the Christian West. The Tartar conquest of Persia brought the usual Mongol policy of low customs, well-policed roads, and free passage for everybody—and so opened the road to India. The Tartar conquest of Russia opened the road to Cathay. The great overland Silk Road traversing Asia, though heavily traveled for centuries, was not frequented by Europeans until the years of Tartar conquest. Egyptian roads, still in the hands of Muslims, remained forbidden to Europeans, and goods passing there were so heavily taxed by the Mamluk sultans that Indian merchandise trebled in cost by the time it reached an Italian merchant.

17

Missionary Diplomats

BY the mid-thirteenth century the hopes to convert the Tartars and their Khans to Christianity were nourished by recent events. For the Tartar conquest of the Muslim Turks had made Tartars the unwitting allies of Western Christendom. These contagious hopes made militant Christians eager to confuse Genghis Khan with Prester John. Rumors that the Great Khan himself had been converted in Cathay seemed substantiated by the report that the wives and the mother of the Great Khan had taken up

Christianity, and the fact that considerable numbers of Nestorian Christians spread over the Tartar realm were allowed to practice their religion freely.

Franciscan friars became geographic pioneers. "Just at the time when God sent forth into the eastern parts of the world the Tartars to slay and to be slain," a pious chronicler noted, "he also sent forth in the west his faithful and blessed servants Dominic and Francis, to enlighten, instruct, and build up in the Faith." The energetic Pope Innocent IV, soon after his election in 1243, organized Christendom against the new threat of Tartar invasion. He called a church council at Lyons in 1245 for "finding a remedy for the Tartars and other spurners of the faith and persecutors of the people of Christ." Recalling recent Tartar atrocities in Poland, Russia, and Hungary and the rising Tartar flood, the council desperately exhorted faithful Christians to block every road over which the invaders might come, by digging ditches, building walls, and erecting other barriers. The Church itself would contribute to the cost of these defenses, and also would enforce contribution by all Christians in the neighborhood.

At the same time the Pope decided to try to stop the trouble at its source by sending an emissary to convert the Great Khan, Kuyuk Khan himself, at his capital in northern Mongolia. Undaunted by the fact that no European had gone to the Tartar capital and returned to tell the tale, Innocent IV dispatched his emissary on April 16, 1245, even before the council had met. His providential choice was a Franciscan friar, John of Pian de Carpine (1180?–1252), who had actually been a companion and disciple of the great Saint Francis of Assisi (1182–1226). Born near Perugia, just a few miles from Assisi, he was then in charge of the Franciscan order at Cologne. Friar John proved the perfect man for the assignment. His thirty-page report of his two-year journey remains, despite its brevity, the best description of Tartar customs in the Middle Ages. Another hardy Franciscan, Friar Benedict the Pole, became his companion to Mongolia and back.

These two pioneer Franciscan friars trekking across eastern Europe and central Asia suffered the howling winds and numbing cold of the high steppes, the deep snow of the Altai mountain passes, and the heat of the Gobi Desert.

Thence then, by the grace of God having been saved from the enemies of the Cross of Christ, we came to Kiew, which is the metropolis of Ruscia. And when we came there we took counsel . . . as to our route. They told us that if we took into Tartary the horses which we had, they would all die, for the snows were deep, and they did not know how to dig out the grass from under the snow like Tartar horses, nor could anything else be found (on the way) for them to eat, for the Tartars had neither straw nor hay nor fodder. So, on their advice, we decided to leave our horses there . . . I was ill to the point of death; but I had myself carried along in a cart in the intense cold through the deep snow, so as not to interfere with the affairs of Christendom.

Friar John, while not concealing his mission, actually cajoled his unwilling hosts along the way into supplying guides and fresh mounts to hasten their journey. From Batu's camp on the Volga, a journey of three and a half months brought the friars to the court of the Great Khan, Kuyuk Khan, at Karakorum in the heart of Mongolia. When the two Franciscans arrived there in mid-August two thousand Tartar chiefs had met to elect and enthrone their new emperor, in a tent "resting on pillars covered with gold plates, fastened with gold nails." Kuyuk's first audience as newly enthroned Khan enacted European fantasies of the fabled East. "They asked us if we wished to make any presents; but we had already used up nearly everything we had, so we had nothing at all to give him. It was while here that on a hill some distance from the tent there were more than five hundred carts, all full of gold and silver and silken gowns, all of which was divided up between the Emperor and the chiefs; and the various chiefs divided their shares among their men as they saw fit." The Franciscans then were offered the opportunity to deliver their message from the Pope declaring his wish that all Christians be friends of the Tartars and that the Tartars be mighty with God in Heaven. But to achieve this the Tartars must embrace the faith of the Lord Jesus Christ. Saddened that the Tartars could have slain so many Christians when the Christians had done nothing to injure the Tartars, the Pope urged them to repent what they had done and write him what they intended to do in all these matters.

The Great Khan obliged by giving Friar John two letters to the Lord Pope. Unfortunately these contained nothing of substance, for the Khan was not persuaded to embrace Jesus Christ. Even so, Friar John was not discouraged, because Christians employed in the Khan's household told him that the Khan was about to become a Christian. When Kuyuk Khan proposed sending his own ambassadors back to the Pope with the two Franciscans, Friar John demurred. "We feared they would see the dissensions and wars among us, and that it would encourage them to march against us." On November 13, 1246, Kuyuk Khan gave Friar John permission to leave.

"We were travelling the whole winter," Friar John reported, "resting most of the time in the snow in the desert, save when in the open plain where there were no trees we could scrape a bare place with our feet; and often when the wind drifted in we would find, on waking, our bodies all covered with snow." Reaching Kiev in early June, they were met with rejoicing as if they had risen from the dead. Similar jubilation greeted them all across Europe. In the fall of 1247, a year after leaving Karakorum, Friar John delivered the Khan's letter to the Lord Pope Innocent IV, and gave him their report in person.

This was not the end of Friar John of Pian de Carpine's role in the meeting of East and West. Louis IX, king of France, was about to set out for Cyprus on the first stage of his promised Crusade (the Seventh Crusade, 1248–54). To persuade Louis that he could serve Christendom better by remaining in France to protect Innocent IV against the Tartars and against the Hohenstaufen "arch-fiend" Frederick II, the Pope sent the two worldly-wise Franciscan friars to Paris. In that mission they failed. But another remarkable Franciscan friar, William of Rubruck, a native of French Flanders, who had the confidence of King Louis, was stirred by their fresh account of their Mongol adventure. When King Louis went on his Crusade, he took along this Friar William. Soon after Louis IX arrived on Cyprus in September 1248, a man who represented himself as an emissary from the Great Khan brought a message of good cheer. He reported that the Great Khan was eager for an alliance against Islam. It was on the day of the Epiphany three years before, this emissary reported, that Kuyuk Khan himself, following his mother's conversion, had become a Christian. All the leading Tartar princes had followed his example and the Tartar people were now eager to join hands against the Saracen enemy.

The credulous King Louis speedily dispatched as his ambassador the Dominican Friar Andrew of Longumeau, who knew Arabic and had earlier visited the camp of Batu. After another remarkable overland journey, Andrew finally reached the court of the Great Khan, where his mission came to a melodramatic anticlimax. He expected to be embraced by Kuyuk Khan as a companion in the Faith and so open a grand alliance. But Kuyuk Khan had died, and the empire was now in the hands of the Regent Mother Ogul Gaimish, who was surely no Christian. She dismissed him like a tribute bearer, with insolent letters for his sovereign.

The overland homeward journey lasted a year. Friar Andrew's mission brought back reports that the Tartars, who originally came from the far end of a great sandy desert that began at the eastern end of the world, had escaped long ago from the wall of mountains (the Great Wall of China?) which held back Gog and Magog. They recounted the conversion of Kuyuk's grandfather Genghis Khan to Christianity after a vision in which God promised him dominion over Prester John. They described the piles of bleached human bones in the path of the Tartar conquests, and eight hundred Christian chapels mounted on carts, which they had seen in a single camp. They brought cheering rumors of a Mongol chief, Sartach, son of Batu, whom they reported to be a Christian.

King Louis was in the Holy Land when he received this optimistic report. With him once again was William of Rubruck, who by the standards of his day was not unprepared for a long expedition to the Khan. He knew some Arabic, had an aptitude for languages and could manage for himself in the Tartar language. King Louis gave him a Bible and a small sum for expenses, together with letters to Sartach and the Great Khan. Queen Margaret gave him an illuminated psalter and some church vestments. He also carried his own prayer book, favorite devotional works, and, for some unexplained reason, a rare Arabic manuscript. To avoid the indignity of another rebuff, King Louis purposely did not name the Friar as his ambassador. This Friar William, accompanied by another friar, Bartholomew of Cremona, a dipsomaniac guide-interpreter, and two servants, left Constantinople on May 7, 1253, took ship across the Black Sea to the Crimea, then went overland across the Don. When they finally reached Sartach, their reputed friend, he indignantly denied being a Christian and "he mocked the Christians." Pushing on across the Volga, the corpulent friar suffered hunger, frostbitten toes, alpine winds, and desert heat before finally arriving at the camp of the Great Khan, Mangu, in the heart of Mongolia on December 27, 1253. The Khan, out of "compassion," gave him permission to remain for two months, till "the great cold" would be over.

At court there were a considerable number of heretical Christians, Nestorians, who were bringing Christianity into bad repute. Nor was the missionary Friar William encouraged by the tolerant spirit of Mangu Khan himself, at his last audience:

> He began confiding to me his creed: "We Mongols," he said, "believe that there is only one God, by whom we live and by whom we die, and for whom we have an upright heart." Then I said: "May it be so, for without His grace this cannot be." . . . Then he added: "But as God gives us the different fingers of the hand, so he gives to men divers ways. God gives you the Scriptures, and you Christians keep them not. You do not find (in them, for example) that one should find fault with another, do you?" "No, my lord," I said; "but I told you from the first that I did not want to wrangle with anyone." . . . "I do not say it," he said, "for you. God gave you therefore the Scriptures, and you do not keep them; He gave us diviners, we do what they tell us, and we live in peace."

"If," Friar William recorded with regret, "I had had the power to work by signs and wonders like Moses, perhaps he would have humbled himself."

Like John of Pian de Carpine, he refused to take Tartar ambassadors back with him. But he did carry the Khan's letters to King Louis. Though the friar took a different way home, surviving the hardships of hunger, thirst, cold and heat, by mid-June 1255 he was back in Cyprus. King Louis had returned to France, and the provincial of the Franciscan Order refused to allow Friar William to follow. The friar was ordered to the Franciscan house in Acre and instructed to send his report from there. If Friar William could have reported in person to King Louis, he might not have reported so eloquently to us.

Friar William of Rubruck brought to Europe a treasure of facts about the opposite side of the earth. He described the course of the Don and the Volga, and showed that the Caspian was not a gulf but a lake. For the first time in European literature, he observed that Cathay was the same as the land of the "Seres," which the Romans had described as the source of silk. "The Cathayans write with a brush such as painters paint with, and they make in one figure the several letters containing a whole word." In this first Western reference to Chinese writing, he shows an understanding that eluded others for centuries. The religious rituals of the lamas, the northern Buddhist monks, and their prayer formula—*Om mani padme, hum*—are precisely noted.

When Friar William finally was given leave to go to Paris, luckily he encountered the pioneer English scientist and fellow Franciscan Roger Bacon (c. 1220–1292). The notorious Friar Bacon, "Doctor Mirabilis," was suspected of necromancy and heresy by the chiefs of his Franciscan order, and had been confined in Paris, where his superiors could watch him. Bacon studied William of Rubruck's account of his journey, and then embodied Friar William's findings in his own *Opus majus,* the encyclopedia he prepared for Pope Clement IV (1268). Through Bacon's work, Friar William's discoveries reached the Christian West. Friar William's own Franciscan order long ignored him in their annals, and not until 1600 was even a portion of his delightful and informative travel book published by Richard Hakluyt.

William of Rubruck, on the very last page of his reflections, wrote a swan song of the overland pioneers. A half-century before, in 1201, the Doge of Venice had contracted with the Council of the Fourth Crusade to carry the crusading army to the Holy Land by sea for the astronomical fee of £180,000. Against this extravagance the friar contrasted the frugality of travel overland. Should not the Army of the Church better pass by land—from Cologne to Constantinople and thence to the Holy Land?

In times past valiant men passed through these countries, and succeeded, though they had most powerful adversaries, whom God has since removed from the earth. Nor should we, if we followed this road, be exposed to the dangers of the sea or to the mercies of the sailor men, and the price which would have to be given for a fleet would be enough for the expenses of the land journey. I state it with confidence, that if your peasants—I speak not of the princes and noblemen— would but travel like the Tartar princes, and be content with like provisions, they would conquer the whole world.

18

The Discovery of Asia

MARCO POLO excelled all other known Christian travelers in his experience, in his product, and in his influence. The Franciscans went to Mongolia and back in less than three years, and stayed in their roles as missionary-diplomats. Marco Polo's journey lasted twenty-four years. He reached farther than his predecessors, beyond Mongolia to the heart of Cathay. He traversed the whole of China all the way to the Ocean, and he played a variety of roles, becoming the confidant of Kublai Khan and governor of a great Chinese city. He was at home in the language, and immersed himself in the daily life and culture of Cathay. For generations of Europe, his copious, vivid, and factual account of Eastern ways was the discovery of Asia.

Venice at the time was a great center for commerce in the Mediterranean and beyond. Marco Polo was just fifteen years of age in 1269, when his father, Nicolò, and his uncle Maffeo Polo returned to Venice from their nine-year journey to the East. Another of Marco's uncles, also named Marco Polo, had trading houses in Constantinople and at Soldaia in the Crimea, where Nicolò and Maffeo had joined him in 1260 in his trading ventures. Marco Polo opens his own book with an account of these earlier travels in which he had no part. Nicolò and Maffeo laid in a stock of jewels at Constantinople which they took by sea to Soldaia, then north and east along the Volga to the splendid court of Barka Khan, son of Genghis Khan. Barka Khan not only treated them courteously and with honor but, what was more to the point, bought their whole stock of jewels, as Marco Polo observes, "causing the Brothers to receive at least twice its value."

When a war between Barka Khan and a rival Tartar prince cut off the Polo brothers' return to Constantinople, they decided to take their trading ventures farther eastward. A seventeen-day journey across the desert took them to Bokhara, where they fell in with some Tartar envoys who were en route to the court of the Great Khan, Kublai Khan. These envoys persuaded the Polos that Kublai Khan, who had never before seen any Latins, intensely desired to see them, and would treat them with great honor and liberality. The envoys promised to guard them on the way. The Polo brothers took up this invitation, and after a full year's journey, "seeing many marvels of divers and sundry kinds," arrived at the court of Kublai Khan. The Great Khan, every bit as friendly as had been promised, proved to be a man of wide-ranging curiosity and alert intelligence, eager to learn everything about the West.

Finally he asked the two brothers to be his envoys to the Pope, requesting one hundred missionaries educated in all the Seven Arts to teach his people about Christianity and Western science. He also wanted some oil from the lamp at the Holy Sepulcher in Jerusalem. When Nicolò and Maffeo departed, they carried the Emperor's Tablet of Gold, his certificate of safe passage, ordering everybody en route to supply their needs. Arriving at Acre in April 1269, the two brothers learned that the Pope had died and his successor had not yet been named. They returned to Venice to await the result. When the new pope, Gregory X, was finally named, he did not offer the requested hundred missionaries, but instead only assigned two Dominican friars to accompany the Polos.

In 1271, when Nicolò and Maffeo Polo set out from Venice on their return journey to Kublai Khan, they took with them Nicolò's seventeen-year-old son, Marco, who was destined to make their trip historic. At Lajazzo on the eastern Mediterranean the two Dominicans left in panic. The three Polos, now alone, proceeded to Baghdad, then on to Ormuz at the mouth of the Persian Gulf, where they might have taken ship for a long journey through the Sea of India. Instead they chose to go north and east overland through the Persian Desert of Kerman to the frigid mountains of Badakhshan, noted for its rubies and lapis lazuli, and its fine horses. There used to be horses here, Marco tells us, which were "directly descended from Alexander's horse Bucephalus out of mares that had conceived from him and they were all born like him with a horn on the forehead." There they stayed a year to allow Marco to recover from an illness by breathing the pure mountain air.

Then up still higher, across a land of glaciers, with many peaks over twenty thousand feet—Pamir, which the natives accurately called "The Roof of the World." "Wild game of every sort abounds. There are great quantities of wild sheep of huge size [now known as *ovis Poli*, though William of Rubruck had noted them before]. Their horns grow to as much as six palms in length and are never less than three or four. From these horns the shepherds make big bowls from which they feed, and also fences to keep in their flocks." "No birds fly here because of the height and the cold. And I assure you that, because of this great cold, fire is not so bright here nor of the same colour as elsewhere, and food does not cook well." They then took the old southern caravan route through northern Kashmir, where no European would be seen again till the nineteenth century, then eastward to the edge of the Gobi Desert.

The party rested at Lop, a town at the western edge of the desert, where travelers usually took supplies to strengthen them against the terror of the crossing.

> Beasts and birds there are none, because they find nothing to eat. But I assure you that one thing is found here, and that a very strange one, which I will relate to you.
>
> The truth is this. When a man is riding by night through this desert and something happens to make him loiter and lose touch with his companions, by dropping asleep or for some other reason, and afterwards he wants to rejoin them, then he hears spirits talking in such a way that they seem to be his companions. Sometimes, indeed, they even hail him by name. Often these voices make him stray from the path, so that he never finds it again. And in this way many travellers have been lost and have perished.

Across the desert they entered Tangut, in extreme northwestern China, traversed the Mongolian steppes and arrived in the court of the Great Khan after a trek of three and a half years.

Kublai Khan received the Venetians with great honor. Sensing the talents of the twenty-one-year-old Marco, the Khan at once enlisted him in his service, and sent him on an embassy to a country six months away. When we read Marco Polo's travels today, we all reap the fruits of the voracious curiosity of that thirteenth-century Mongol emperor.

> Now he had taken note on several occasions that when the Prince's ambassadors returned from different parts of the world, they were able to tell him about nothing except the business on which they had gone, and that the Prince in consequence held them for no better than fools and dolts, and would say: "I had far liever hearken about the strange things, and the manners of the different countries you have seen, than hearing of the affairs of strange countries." Mark therefore, as he went and returned, took great pains to learn about all kinds of different matters in the countries which he visited, in order to be able to tell about them to the Great Khan. . . . Thereafter Messer Marco abode in the Khan's employment some seventeen years, continually going and coming, hither and thither, on the missions that were entrusted to him. . . . And, as he knew all the sovereign's ways, like a sensible man he always took much pains to gather knowledge of anything that would be likely to interest him, and then on his return to Court he would relate everything in regular order, and thus the Emperor came to hold him in great love and favour. . . . And thus it came about that Messer Marco Polo had knowledge of, or had actually visited, a greater number of the different countries of the World than any other man; the more that he was always giving his mind to get knowledge, and to spy out and enquire into everything in order to have matter to relate to the Lord.

It seemed, the Khan would exclaim, that only Marco Polo had learned to use his eyes!

We do not know how Nicolò and Maffeo Polo spent their time at the court of the Khan, except that, at the end of the seventeen years, they had "acquired great wealth in jewels and gold." Every year Kublai Khan became more reluctant to lose Marco's services. But in 1292 an escort was required for a Tartar princess who was to become the bride of the Ilkhan of Persia. Envoys of the Ilkhan had already failed in their efforts to deliver the seventeen-year-old bride overland. Returned to the court of Kublai Khan, they hoped to secure sea passage. Just then Marco had come back from an assignment that had taken him on a long sea voyage to India. The Persian envoys, who knew the seafaring reputation of Venetians, persuaded Kublai Khan to allow the Polos to accompany them and the bride by sea. The Khan outfitted fourteen ships, with an entourage of six hundred persons and supplies for two years. After a treacherous sea voyage through the South China Sea to Sumatra and through the Sea of India from which only eighteen of the six hundred survived, the Tartar princess was safely delivered to the Persian court. She had become so attached to the Venetians that she wept at the parting.

The Polos, returning home overland by way of Tabriz in northern Persia, Trebizond on the south coast of the Black Sea, to Constantinople, finally reached Venice in the winter of 1295, after their absence of twenty-four years. The Polo family had long since given them up for dead. A plausible tradition reports that when these three shabby strangers appeared, looking more like Tartars than Venetians, their noble relatives would have nothing to do with them. But the relatives' memories were quickly jogged when the unkempt wanderers ripped open the seams of their sordid garments and produced their secret treasure—a shower of rubies, diamonds, and emeralds. The returned travelers were affectionately embraced, and then entertained at a luxurious banquet, where music and jollity were mixed with exotic reminiscence.

Those were years of bitter rivalry between Venice and Genoa for the Mediterranean seafaring trade. On September 6, 1298, a climactic sea battle between Venice and Genoa at Curzola off the Dalmatian coast left the Genoese victors, with seven thousand prisoners. Among these was a "gentleman commander" of a Venetian galley, Marco Polo. Brought back in chains to a prison in Genoa, he became friendly with another prisoner, relic of an earlier Genoese victory over the Pisans. This Rustichello happened to be a writer of romances who already had a considerable reputation for his retelling of the tales of King Arthur and his Round Table. Not a literary genius, Rustichello still was master of his genre, industrious and persuasive. In Marco Polo's reminiscences he saw the raw material for a new kind of romance—"A Description of the World"—and he persuaded the Venetian to cooperate. Marco Polo must somehow have managed to secure his notes from home. Then, profiting from his enforced leisure and from their confinement together, the Venetian dictated a copious account of his travels to Rustichello, who wrote it all down.

If either Marco Polo or Rustichello had not fought in the wars against Genoa, we might have no record of Marco Polo's travels and might not even have heard his name. Luckily, Rustichello was a writer congenial to the great Venetian traveler, and he knew the makings of a romance to charm the world for seven hundred years. Of course he could not restrain himself from occasionally embellishing Marco Polo's facts with his own fancies. Some of the more colorful episodes are adapted from earlier writings by Rustichello or others. For example, the extravagant praise that Kublai Khan lavished on the young Marco when he first arrived at the court recalls what King Arthur said, according to Rustichello's own romance, when he received the young Tristan at Camelot. This was not the first or the last time a writer made the reputation of an adventurer. The formula, "as told to," which nowadays appears less often than it rightly should on title pages of books, has a surprisingly respectable history. Why did the energetic Venetian, who was literate in several languages, who must have written much to please Kublai Khan and extensive detailed notes for his own use, not write down for himself his personal adventures? Perhaps if promptly on his return to the commercial city of Venice in 1295 he had been tempted by a publisher's contract, he might have written his own book. But two centuries would pass before a publishing trade flourished.

Other great medieval travelers—Friar Odoric of Pordenone, Nicolo de' Conti, and Ibn Battuta—and the noted French chronicler and biographer of Saint Louis, Jean de Joinville (1224?–1317?), also dictated their books. The rewards of money or celebrity were not yet dangled before so many, nor was literacy required to get or to hold political power. The opening sentence of the prologue of Marco Polo's book exhorts: "Emperors and kings, dukes and marquises, counts, knights, and townsfolk, and all people who wish to know the various races of men and the peculiarities of the various regions of the world, take this book, take this book *and have it read to you.*"

Rustichello wrote Marco Polo's book in French, which in Western Europe in those days was current among the laity just as Latin was among the clergy. Before long it was translated into most European languages, and numerous manuscripts survive. Never before or since has a single book brought so much authentic new information, or so widened the vistas for a continent.

19

The Land Curtain Comes Down

IN the decades of the overland pioneers there was a flourishing, but small-scale and specialized, trade between Europe and easternmost Asia. Scores of European merchants must have reached those far parts, but not many merchants besides the Polos left a firsthand report of their travels. A vivid if not always reliable record of communities of Europeans in Chinese cities was kept for us by a remarkable succession of intrepid Franciscan friars. One of the most enterprising was the Italian Franciscan friar John of Montecorvino. Dispatched by Pope Nicholas IV in 1289, he arrived in Peking in 1295, where he "presented the letter of our lord the Pope, and invited him [the Great Khan] to adopt the Catholic Faith of our Lord Jesus Christ, but he had grown too old in idolatry. However he bestows many kindnesses upon the Christians, and these two years past I am abiding with him." In the Peking cathedral with a bell tower and three bells that he built just across the street from the Khan's palace, he baptized, according to his own count, some six thousand townspeople. There he organized and trained a choir of a hundred and fifty boys. "His Majesty the Emperor moreover delights much to hear them chanting. I have the bells rung at all the canonical hours, and with my congregation of babes and sucklings I perform divine service, and the chaunting we do by ear because I have no service book with the notes." Friar John was appointed Archbishop of Cambaluc (Peking) in 1307 and within a few years received three suffragan bishops to assist him.

Another Franciscan friar, Odoric of Pordenone, dictated to a fellow friar one of the most picturesque and copious records of life in China, where he lived for three years, before returning to Padua by way of central Asia in 1330. He noted many items not reported by Marco Polo—the custom of fishing with cormorants, the habit of letting fingernails grow long, and the tradition of compressing women's feet. By the time Friar John Marignolli, an aristocratic Florentine, reached Peking, in 1342, he observed that the Archbishop of Peking had a residence appropriate to his high station and all the Christian clergy "had their subsistence from the Emperor's table in the most honorable manner." In the seaport city of Zayton (Tsinkiang) he found three great Franciscan churches and a bathhouse for the use of European merchants. The greater part of Marignolli's recollections, however, is given over to a detailed description of Paradise, and the Garden of Eden—in Ceylon—with its delightful mountains, fountains, and rivers.

About 1340 the alert Francesco Balducci Pegolotti, agent for a Florentine banking family, the Bardi, prepared a helpful handbook for merchant-travelers that leaves us clues to a flourishing trade. His mercantile Baedeker gives much information that an overland merchant-traveler would need: the distances between places, and the local dangers; weights and measures; prices and rates of exchange; customs regulations; practical hints on customs regulations, on what to eat, and what not to eat, and where to sleep.

In the first place, you must let your beard grow long and not shave. And at Tana you should furnish yourself with a dragoman. And you must not try to save money in the matter of dragomen by taking a bad one instead of a good one. For the additional wages of the good one will not cost you so much as you will save by having him. And besides the dragoman it will be well to take at least two good men servants who are acquainted with the Cumanian tongue. And if the merchant likes to take a woman with him from Tana, he can do so; if he does not like to take one there is no obligation, only if he does take one he will be kept much more comfortably than if he does not take one. Howbeit, if he do take one, it will be well that she be acquainted with the Cumanian tongue as well as the men. . . .

Whatever silver the merchants may carry with them as far as Cathay the lord of Cathay will take from them and put into his treasury. And to merchants who thus bring silver they give that paper money of theirs in exchange. This is of yellow paper, stamped with the seal of the lord aforesaid. And this money is called; and with this money you can readily buy silk and other merchandize that you have a desire to buy. And all the people of the country are bound to receive it. And yet you shall not pay a higher price for your goods because your money is of paper. . . .

(And don't forget that if you treat the custom-house officers with respect, and make them something of a present in goods or money, as well as their clerks and dragomen, they will behave with great civility, and always be ready to appraise your wares below their real value.)

These days of lively overland traffic between the ends of the earth were not to last long. John of Montecorvino was to be both the first and, for centuries, the last effective archbishop of Peking. The successor whom Pope John appointed in 1333 appears never to have reached his destination. The land paths to the East that were so abruptly opened in the mid-thirteenth century were closed no less abruptly only one century later.

The strength and unity of the vast Mongol Empire had opened, had kept open, and had protected the European overland passage to India and China. During these years, sometimes called the Mongol Century, when some Europeans came eastward, some Chinese also went westward. Homeward-bound Westerners and traveling Chinese carried with them playing cards, porcelain, textiles, motifs in art, styles in furniture, which shaped the every-day life of the European upper classes. A few items—paper money, printing, and gunpowder—were world-shaking. These novelties reached the Middle East directly, and then arrived in Europe indirectly and interstitially, through the Arabs and others. Such momentous notions were carried by a very few.

The empire that they had conquered on horseback, the Mongols discov-ered, could not be governed on horseback. They needed an elaborate ad-ministration for their vast empire. Within China, where they were aliens and invaders, it was never easy to keep their subject peoples under control. The Mongols put themselves or other foreigners like Marco Polo into the high government positions. Meanwhile the Chinese, with an ancient literary tradition, a developed technology, and a fastidious ceremonialism, found many reasons to condemn their barbarian conquerors. The Mongols, from the dry grasslands of the north, had never formed the habit of bathing. "They smell so heavily that one cannot approach them," a Chinese traveler to Mongolia reported. "They wash themselves in urine." Marco Polo was awed by the ruthlessness and hardiness of the Mongol soldiers, who drank mare's milk, carried almost no baggage, and were "of all men in the world the best able to endure exertion and hardship and the least costly to main-tain and therefore the best adapted for conquering territory and overthrow-ing kingdoms." Those Mongol soldiers he found in Cathay had become degenerate and dissolute, and he observed the restlessness of the native Chinese as he traveled around the country. All the ways of the Mongol rulers, and not the least their tolerance of foreign religions, irritated tradi-tional Confucians.

By the mid-fourteenth century famine in the north and disastrous flood-ing of the Yellow River multiplied problems for the ruling Mongols. There were outbreaks of rebellion all over the country. Toghon Temür Khan (1320–1370), the last of the Mongol emperors, a man of Caligulan dissolute-ness, came to his insecure throne in 1333. He took ten close friends into the "palace of deep clarity" in Peking, where they adapted the secret exercises of Tibetan Buddist tantra into ceremonial sexual orgies. Women were sum-moned from all over the empire to join in functions that were supposed to prolong life by strengthening men with women's powers. "All who found most pleasure in intercourse with men," current rumor recounted, "were chosen and taken to the palace. After a few days they were allowed out. The families of the common people were glad to receive gold and silver. The nobles were secretly pleased and said: 'How can one resist, if the ruler wishes to choose them?' "

But the people did resist. The climax came in 1368, when Hung Wu (Chu Yüan-chang, 1328–1398), a self-made man of great talents, emerged as the leader of Chinese rebellion to found the Ming dynasty. The natives had ingeniously organized rebellion under the very noses of the Mongols. In these last years of Mongol rule, folk history recalls, the nervous Khans placed an informer in nearly every family, and forbade people to gather in groups. The Chinese were not allowed to carry arms, which meant that only one family in ten was permitted to possess a carving knife. But somehow the Mongols had forgotten to suppress the Chinese custom, at the coming of the full moon, of exchanging little round full-moon cakes, decorated with pictures of the moon hare and which, like a fortune cookie, carried a piece of paper inside. The wily rebels, we are told, used these innocent-looking moon cakes for their messengers. Inside were instructions for the Chinese to rise and massacre the Mongols at the time of the full moon in August 1368.

The dissolute Toghon Temür Khan, instead of staying to defend his patrimony, fled with the Empress and his concubines—first to the fabled summer palace at Shangtu, the famed Xanadu, then to Karakorum, the original Mongol capital, where he met his death. Meanwhile Mongol princes and generals were fighting among themselves, and the Mongol Empire was coming apart. In the very year of the Peking uprising the great Tamerlane, with his headquarters far to the west in Samarkand, was consummating the first stage of his own plan for world conquest. Tamerlane's empire, vast by any other standard, was only a southwesterly quarter of the realm of Genghis Khan. But Tamerlane stood astride the land paths to the East.

Dismemberment of the Mongol Empire interrupted the paths of safe passage that Pegolotti had described only a few decades before. Tamerlane did keep the passage open within his realm where Europeans could go as far as Tabriz, in Persia. Samarkand, where Tamerlane's power stopped, briefly was the Athens of Asia. Soon after his death in 1405, Samarkand, the once busy way station on the Silk Road, lay in ruins, an Asian ghost town. Tamerlane's empire was only a memory.

Land passage to Cathay for Europeans was now closed. Even news of Cathay became a scarce commodity in Europe. The Pope himself, whose continent-wide communications network excelled all others, could not get word of the happenings in Peking. He went on appointing archbishops or bishops of Peking long after there was no hope that they could reach their see. Even if a European did reach the borders of Cathay, he would not be admitted. The new rulers of China, with memories of foreign tyranny still fresh, resumed their ancient isolationism.

After the disruption of that first Mongol Empire the European West lost touch with the farthest East. Only a trickle of travelers' reports comes to us from the era of Tamerlane and of the second Mongol Empire. These sources remind us how few, how limited, and how insignificant had become the European overland contacts with Asia.

A Madrid nobleman, Ruy González de Clavijo, and two companions, was sent by King Henry III of Castile in 1403 to solicit the alliance of Tamerlane against the Turks. This delegation went by ship to Trebizond on the far corner of the Black Sea, then overland as far as Samarkand. They witnessed the splendor of Tamerlane's capital with its communities of captive craftsmen—silk weavers, potters, glassworkers, armorers, silversmiths—gathered from the cities of conquest. There they heard rumors that it was a six-month journey to Cambaluc, but that was not for them. Before Clavijo could return from Samarkand, Tamerlane was dead, the subject princes were rising in revolt, and anarchy once again descended on the overland routes to the West. The Spanish party had to meander homeward to avoid robbers and escape the countless battlefields that had sprung up all across the remnants of Tamerlane's realm.

The last well-recorded European in the East in this overland age was no bold adventurer, no missionary, no diplomat or merchant, but an involuntary traveler. Hans Schiltberger, a Bavarian of good family, was only fifteen years old when he was captured at the Battle of Nicopolis (1396), where he had been fighting in the Crusade against the Turks led by King Sigismund of Hungary. For the next thirty-two years Schiltberger was a slave of the Ottoman sultan Bajazet, and then of the Tartar Tamerlane and his successor. He served as a runner and did other menial jobs to keep himself alive until he could make the dash for freedom. Carried along on the ebb and flow of battle, he was handed from one side to another, meanwhile getting a slave's-eye view of life among Turks and Tartars. Like Clavijo, Schiltberger never reached farther east than Samarkand. His account is spiced by episodes like the victory secured by Tartar Amazons, giant female warriors, led by a vengeful princess. He reports a Cairo with twelve thousand streets and twelve thousand houses in every street. But his book of travels, his *Reisebuch,* dictated after his return to his ancestral property in Bavaria, also gives us, along with these flights of fancy, a rare firsthand account of medieval life among the lowly.

The frustrations of Clavijo and Schiltberger attested the end of the heroic age of overland travel. Instead of vivid eyewitness accounts of life in the capital of Cathay by Europeans honored by the Great Khan, Europeans now had to depend on rumor, on the casual reports of captives and slaves for word from the fabled East.

PART SIX

DOUBLING THE WORLD

Enough for us that the hidden half of the globe is brought to light, and the Portuguese daily go farther and farther beyond the equator. Thus shores unknown will soon become accessible; for one in emulation of another sets forth in labours and mighty perils.

—PETER MARTYR (1493)

20

Ptolemy Revived and Revised

THE blocking of the land paths proved a godsend. Driven by new incentives to go to sea, Europeans would discover waterways to everywhere. The science of cartography first flourished on the sea. There the needs of working mariners shifted the interest of geographers and map-makers from the wholesale to the retail. Christian geography had become a cosmic enterprise, more interested in everyplace than in anyplace, more concerned with faith than with facts. Cosmos-makers confirmed Scripture with their graphics, but these were no use to a sea captain delivering a cargo of olive oil from Naples to Alexandria.

The mariner who did not find much help in Cosmas Indicopleustes' neat box of the universe needed to know the precise location of rocks and shoals outside the ports that fed Athens or Rome, or how to find the clear way between the small islands of the Adriatic. During the Great Interruption in European geographic knowledge, mariners going about their business accumulated bits of information about the Mediterranean that would smooth their paths, make their passage safer, quicker, more sure. Collecting information on a scale and in a form they could use, they accumulated a stock of knowledge that had nothing to do with the speculations of philosophers, theologians, and cosmos-concocters. Caring not at all about the grand shape of the Ecumene, about the precise location of Eden or from what direction to expect the invasion of Gog and Magog at the end of the world, they recorded countless details of the seacoasts to guide themselves and those who followed. As early as the fifth century B.C. some mariners in the Mediterranean were noting down their experiences of landmarks, coastal features, and other useful miscellaneous facts. Such a record was called a periplus (for "sailing around"), and we might call it a coast pilot.

The oldest of these ancient peripli to survive was made by Scylax in the service of Darius the Great, Persian emperor of the sixth century B.C. His detailed sailing directions describe the perils and passages of the Mediterranean—the best way to get from the Eastern point, the Canopic mouth of the Nile in Egypt, to the Pillars of Hercules (Gibraltar), and many shorter voyages, always indicating how many days' sailing each voyage required with a favorable wind and fair weather. "This whole coasting from the Pillars of Hercules to Cerne Island takes twelve days. The parts beyond the isle of Cerne are no longer navigable because of shoals, mud, and sea-weed. This sea-weed has the width of a palm, and is sharp toward the points so as to prick." Scylax's critical faculties leave him when he moves away from the sea. There, luckily, misstatements and exaggerations will not wreck a vessel or delay passage into port. "These Ethiopians are the tallest of all people we know, greater than four cubits; some are even five cubits [seven and a half to ten feet]; and they wear a beard and long hair and are the most handsome of all people. And he rules them who happens to be the tallest." Of course, written records—whether of fact or fiction—were useful only to mariners who could read. It would be many centuries in the future before mariners were literate. Until then there was no large market for a written text. Yet it was hard to provide a useful picture of the seacoast because the arts of cartography remained primitive. The shortest and safest passage from one port to another, besides being a mariner's trade secret, was a valuable state secret, for it spelled the commercial opportunity that could enrich a city or promote an empire.

It is not surprising, then, that manuscript coast pilots were few. From the whole period of the Great Interruption, the fourth until the fourteenth century, no mariner's charts survive. In that age of widespread illiteracy, sailors passed on their traditional knowledge by word of mouth. From about 1300, however, we do find Mediterranean sea charts, offering the kind of useful detail found in the ancient peripli. While the ancient guides were written texts telling about the coasting and sailing conditions, the later guides are charts. These coast charts of the Mediterranean, according to historians of cartography, are "the first true maps" because they "first laid down any considerable part of the earth surface from close, continuous, and what we may call scientific observation." They came to be known by their Italian name, *portolano,* or harbor guide. They could have been called "handy guides," for that is what they were. Being portable, they could be tested and corrected on the spot against the actual experience of each place.

Despite—or perhaps because of—their humble, utilitarian origins, the portolanos were the source of some of the most reliable information to be found in the grand printed atlases down to the middle of the sixteenth century. The pioneer-masters of modern cartography, Mercator and Ortelius, who opened a new era in the mapping of the planet, found little that was useful in all the speculations of Christian theologian-cosmographers. But they gratefully incorporated the piecemeal everyday findings of working mariners. As late as 1595, the world's leading seafaring merchants, the Dutch mariners, were being guided by the coastal outlines, hints, and cautions of mariners who had compiled sea charts two centuries before.

Like Scylax, these later observant professionals somehow lost their critical faculties when they went inland. Coastal guides either left the interior blank or sprinkled it with myth and rumor. It was on the edge of the sea, where the outlines of the earth were tested by everyday experience, that the living truths of modern cartography were born.

There were other reasons, too, why the sea was the nursery of scientific, precise, usable charts of the earth. Christian theologians, whose business it was to know about the universe and man's destiny, naturally put the Garden of Eden at the head of their maps. Scripture, in the apocryphal Book of II Esdras 6:42, declared "six parts hast thou dried up." The earth therefore must be six-sevenths covered by land, and only one-seventh by water, and the seas would be only a minor marginal element in their scheme. "Facts" about the *land* were drawn mainly from such literary, and usually sacred, sources. In the Middle Ages, and all the long ages before printing, those manuscript sources accumulated and glossed one another.

Slow to change, the sacred written sources acquired credibility by repetition. Sea charts, however, were tested not by literature but by experience. No amount of theology would persuade a mariner that the rocks his ship foundered on were not real. The outlines of the seacoast, marked off by hard experience, could not be modified or ignored by what was written in Isidore of Seville or even in Saint Augustine. The farther man went out to sea, the less the opportunity or the temptation to be seduced by literary sources. For the sea had no memory. While the topography of the land remained servile to the written word, to rumor, myth, and tradition, the seascape remained a realm of freedom, freedom to learn from experience, to be guided by fact, and to increase knowledge.

To reach Asia by water from the Mediterranean countries meant, of course, leaving the closed for the open sea. Mediterranean voyages were mostly coastwise sailing, which meant relying on personal experience of those particular places—local winds and currents, familiar landmarks, well-known offshore islands and the distinctive silhouette of a neighboring mountain. Beyond the Pillars of Hercules lay new problems. When Portuguese sailors advanced southward down the coast of Africa, they left behind all familiar landmarks. The farther down they went, the farther away they were from the reassuring details of the portolanos. There was no accumulated experience and there were no handy guides.

The Mediterranean was nowhere more than five hundred miles from its southern to its closest northern coast, which meant a difference of only some seven degrees of latitude. Mediterranean pilots therefore seldom worried about their latitude, especially since the modes of defining latitude were still so crude. But the African continent stretched from 38 degrees north latitude to 38 degrees south latitude, one-fifth around the globe. When a seacoast was so little known, when local inhabitants were so hostile, and when the offshore hazards were so uncharted, latitude was the best and sometimes the only way of defining a ship's position. Prudent mariners then would have to learn how to mark latitude. At first they could estimate latitude by the altitude of the polestar. But as they reached farther south the polestar sank, and they had to use tables of declination with a sea astrolabe or a quadrant or cross-staff as they observed the midday altitude of the sun. These techniques, crucial for navigation over long distances and in unfamiliar waters, developed in the late fifteenth century when the Portuguese were pressing down the coast of Africa. By the early sixteenth century, marine charts began to show scales of latitude, and gradually numerous places on the African coast had their latitude defined.

Such aids to navigation promoted sailing southward and northward. But, as we have seen, the definition of longitude, to mark and measure east–west distances, would be much more complicated. Mariners continued to rely on "dead reckoning." This meant estimating position without astronomical observation, by calculating or guessing the course and distance traveled from a previously determined position. As we have seen, it was not until the eighteenth century that a seafaring clock made it possible for mariners to define their longitude with a precision sufficient to enable them to use longitude to guide their return to where they had been, and to guide those who wanted to follow them. Besides all these problems, leaving the Mediterranean obviously carried the risk of being driven off course into the trackless open ocean.

The schematic Christian T-O map was little use to Europeans seeking an eastward sea passage to the Indies. European sovereigns and other investors financing long sea voyages had to abandon the theologian's for the mariner's point of view. No longer would Jerusalem remain the center, the Garden of Eden must be relegated to another world, and in their place appeared the geometry of latitude and longitude.

Here entered—or, rather, reentered—the great Ptolemy. It was just about this time, when the land curtain thudded down across the European land paths to the East, that the geography of Ptolemy was revived to refresh and reform the thinking of European Christians. If there is a connection between these events, we do not know it, but the coincidence was pregnant for the future of the world.

Even before the revival of Ptolemy, by the time Marco Polo had returned to Venice the results of the portolano charts, the observations of careful mariners, were being garnered into larger maps and atlases and for posterity. The most impressive of these that survive was the so-called Catalan Atlas of 1375, made for the King of Aragon by Cresques le Juif, Abraham Cresques, a Jew of Palma on the island of Majorca, map-maker and instrument-maker to the King. His services along with those of other Jews of skill and learning were a reward that the kings of Aragon received for their

policy of toleration, which created a whole Jewish school of cartography at
Majorca. When King Charles V of France sent to the King of Aragon for
a copy of his best world map, he received this one, which, luckily, now
survives in the Bibliothèque Nationale in Paris.

As the persecution of the Jews resumed in Aragon late in the fourteenth
century, Abraham's son, Jehuda, who was carrying on his father's work,
was forced to emigrate. Accepting the invitation of Prince Henry the Navigator, he took refuge in Portugal, where he helped the Portuguese prepare
the maps and charts for their grand overseas adventures. It was no accident
that Jews played a leading role in the liberation of Europeans from the
slavery of Christian geography. Driven from place to place, they helped
make cartography, still the special preserve of princes and high bureaucrats,
into an international science, offering facts equally valid in lands of all
faiths. Marginal both to Christians and to Muslims, the Jews became teachers and emissaries bringing Arab learning into the Christian world.

The Catalan Atlas aimed to provide a "mappamundi, that is to say, image
of the world and of the regions which are on the earth and of the various
kinds of people who inhabit it." It expressed the dominant interests of
European mariners of the Age of the Land that was nearing its end. The
east–west stretch that was the center of their world was depicted on twelve
leaves mounted on boards to fold like a screen. It did not show northern
Europe, northern Asia, or southern Africa, but did show the Orient and the
little that was known of the Western Ocean. By contrast with Christian
maps, it was a triumph of empiricism. It showed what could be learned by
adding up the experiences of countless individuals, including Arabic seamen
and the very latest European world travelers. Of course, the map-makers
had to start with something and they inevitably started with the familiar
T-O circular forms. Jerusalem is still near the center, and there too are the
tribes of Gog and Magog held back by the "Caspian" mountains, and other
relics of the orthodox landscape. But this is essentially a portolano atlas,
which means that the seacoasts of the Black Sea, of the Mediterranean, and
of Western Europe are delineated from the "normal" portolano, from the
countless sketches of these coasts made by active seamen and recorded on
their coastal guides. Abraham Cresques also drew on the reports brought
back by recent travelers to Asia.

We know that Cresques's patrons, King Peter IV of Aragon and his son,
went to special trouble to secure manuscripts of Marco Polo's "Description
of the World," of Friar Odoric of Pordenone's travels, and of the travels
of the apocryphal "Sir John Mandeville," to facilitate the cartographer's
work. As a result Cresques's atlas offered, at last, a recognizable version of
the Asian continent. The least inaccurate part is that stretch through the
heart of the continent across which the Polos traveled and on which Marco

Polo's book had provided a useful gloss. The southeastern peninsula of Asia is entirely omitted, but now for the first time in the West, India itself is properly represented as a great peninsula.

This Catalan Atlas, however primitive it may seem to the modern eye, was a masterpiece of a burgeoning empirical spirit. Much of the legendary data that had populated maps for all the Christian centuries was omitted. The cartographer's greatest act of self-control was to leave parts of the earth blank, and in the portolano spirit the Catalan Atlas leaves vast regions of the north and the south undescribed. Southern Africa, long a favorite habitat for anthropophagi and mythical monsters, is now left vacant to await facts that would satisfy a hardheaded sea captain.

For most of history the human mind has abhorred a vacuum, and preferred myths and factitious facts to the label "Terra Incognita." How could people, especially the "learned," be awakened to a willing confession of ignorance? The portolano atlases were a beginning.

The discovery and plotting of the earth could not, however, be accomplished by the unaided empirical spirit. Here Ptolemy's large *a priori* aesthetic concepts were essential. Like the portolano makers, Ptolemy had left behind the Homeric notion of a primordial, all-surrounding Oceanus encircling the earth and the sea. He admitted that there might be unknown lands beyond the bounds of his known mapped world. But he had something else to add. The portolano charts were projectionless. Although they appear to have been based on measurements and careful calculation, the manuscript copyists never made any two of them precisely alike. For they had no regular, constant, and universal system of coordinates—nothing like latitude and longitude. The geometric pattern characteristic of the portolano charts was a scheme of "wind roses," each a center for radiating lines. The number of these on any particular chart depended on the size of the sheet on which it was drawn. There was usually a central point of radiation about which eight or sixteen other focal points are found in a circle. The decorative roses showed the orientation of the winds, and also could help the mariner by numerous radiating lines, one of which might correspond to the course of a particular voyage. The portolano charts did place north at the top, but they did not provide any net of parallels and meridians. Not until the sixteenth century did the sea charts show latitudes, nor was there need to define latitude while sailing in enclosed waters and by dead reckoning.

Ptolemy's great contribution was the scientific, quantitative spirit. His scheme of latitude and longitude, unlike the decorative wind rose, was uniform and universal. Any two maps properly made according to his prescription would be precisely alike. The coordinates he provided did not depend on the size of the sheet or the particular area mapped. In the first book of his geography where he offers his instructions on map-making, he explores the problem of transposing a spherical surface, the earth, onto a flat surface of a sheet of parchment. There he explains the need to indicate parallels of latitude and meridians of longitude. He describes the difficult process of making a modified spherical projection of the habitable quadrant of the earth, and also tells how to make a geometrically precise simple conic projection "for the sake of those who, through laziness, are drawn to that earlier method." Unlike the Christian cosmos-makers who started from, and never departed from, their dogmatic oversimplifications, Ptolemy champions both a holistic and a mathematical approach to the surface of the planet. This he explains in the definition of geography which opens his work:

> Geography is a representation in picture of the whole known world together with the phenomena which are contained therein.
> It differs from Chorography in that Chorography . . . treats more fully the particulars of . . . the smallest conceivable localities, such as harbors, farms, villages, river courses, and such like.
> Geography looks at the position rather than the quality, noting the relation of distances everywhere, and emulating the art of painting only in some of its major descriptions. Chorography needs an artist, and no one presents it right unless he is an artist. Geography does not call for the same requirements, as any one, by means of lines and plain notations can fix positions and draw general outlines. Moreover Chorography does not have need of mathematics, which is an important part of Geography. In Geography one must contemplate the extent of the entire earth, as well as its shape, and its position under the heavens, in order that one may rightly state what are the peculiarities and proportions of the part with which one is dealing, and under what parallel of the celestial sphere it is located . . . the length of its days and nights, the stars which are fixed overhead, the stars which move above the horizon, and the stars which never rise above the horizon at all. . . .
> It is the great and the exquisite accomplishment of mathematics to show all these things to the human intelligence. . . .

The revival of Ptolemy, then, would mean the awakening, or the reawakening, of the empirical spirit. Now men would use their experience to measure the whole earth, to mark off the known from the unknown, and to designate newfound places for return. The rediscovery of Ptolemy was a signal event in the revival of learning that marked the Renaissance, a prologue to the modern world.

Manuscripts of Ptolemy's geography, in the Greek language, come to us from the early thirteenth century. But since the ability to read Greek, even among the learned in Europe, was rare, knowledge of Ptolemy's work could not spread until it was translated into Latin. In 1400 a copy of Ptolemy in Greek was brought from Constantinople to Florence by Palla Strozzi (1373–

1462), one of the family who used the wealth acquired in commerce to become patrons of learning. There the geography was translated from Greek into Latin by the celebrated Manuel Chrysoloras (1355–1415) and his pupils. By the early fifteenth century numerous Latin manuscripts of Ptolemy's geography were circulating in Western Europe. More than forty from this period have survived. Some of these manuscripts were accompanied by what purported to be Ptolemy's maps, usually twenty-seven in number. The earliest printed version of this Latin translation (Vicenza, 1475) produced only the texts. Modern scholars are puzzled over what became of Ptolemy's work during the Great Interruption. Where were his text and his maps during the millennium between Ptolemy's death and the revival of his work? It now seems likely that only the first theoretical book of the *Geography* survives substantially as Ptolemy wrote it. The remaining books, including the lists of cities located by his system and the maps, appear to have been compiled over the centuries by Byzantine and Arabic scholars and fathered on his eminent name.

As Ptolemy was revived, more and more manuscript copies were made of the "Ptolemy" text and the maps, and the work as a whole acquired a nonpareil reputation. Not only the technique of map-making, described in Book One, but the whole text and all the maps that came with it were accepted as gospel, with the double authenticity of a newly discovered classic. While Ptolemy's *theory* of map-making could not be faulted, the maps that had become attached to his *Geography* contained some crucial errors which would shape the future of world exploration. For example, Ptolemy's gross underestimate of the circumference of the earth and his gross overestimate of the eastward extent of Asia combined to make Asia seem much closer to Europe, by way of the Western Ocean, than it really was. These "facts" from Ptolemy eventually lured Columbus westward. Meanwhile other errors on "Ptolemy's" classic world map would have proved that it was impossible to reach India and China by sailing eastward around Africa. The part of Africa below the equator, labeled "Terra Incognita," was shown as an enormous southern continent reaching all the way around till it joined with the upper Asian continent on the northeast. The effect was to make the whole Indian Ocean and the China Sea into a vast enclosed lake, and so, of course, to make any sea passage to Asia southward and eastward inconceivable.

Before European seafarers could respond to the challenge offered them by the closing of the land passages to Asia, this south African part of Ptolemy's world map would have to be revised. In fact, the very meaning of "Ocean" would have to be changed. Until that time Europeans made a sharp distinction between the Ocean and a sea *(mare)*. There was, in fact, only *one* Ocean. In Greek mythology it was *Oceanus*, the great circular stream that was supposed to encompass the disk of the earth. Hence, in English, until about 1650, the Great Outer Sea of Boundless Extent was commonly called the Ocean Sea, from *mare oceanum,* and was opposed to the Mediterranean inland sea or the other inland seas.

Christian map-makers in the Middle Ages generally followed the ancient Greek legends and depicted the habitable parts of the world as surrounded by the Ocean Sea. "Let the waters under the heaven be gathered together into one place," declared the Book of Genesis (1:9), "and let the dry land appear: and it was so." While Christians disagreed over the features of the all-surrounding Ocean, there was general agreement that even if that Ocean should somehow prove to be navigable, it ought not to be penetrated. Somewhere beyond or outside it lay Paradise, which no living man should or could reach.

In those days the Ocean led nowhere; in the next centuries people would see that it led everywhere. Only gradually, in the course of the fifteenth century, did "Ocean" come to have this revolutionary modern significance. Till then the Atlantic Ocean was not commonly listed among the "seas" of the earth. The sea route to India would have to be opened in men's minds, and on their maps, before it could be traversed in ships. And so it was. The openings in the mind preceded and made possible the struggle for the openings on the sea. In this respect the finding of a sea route to the Indies was quite different from the finding of America, which occurred on the earth before it had happened in men's minds.

By the mid-fifteenth century some world maps made in Europe showed Africa as a freestanding peninsula and the Indian Ocean as an open sea, which could be entered by water around the African continent en route to India and China. These openings of the mind and of the map occurred decades before we know of any European actually rounding the Cape into the newfound Indian Ocean. For example, on the beautiful and famous "planisphere" (1459) of Fra Mauro. This projection of the whole sphere of the earth onto a flat circle was the last of the great medieval maps. In a sense, Fra Mauro's is also one of the first modern maps. For he now shows the ocean no longer as the forbidden road to nowhere, but as a sea highway to the Indies. He pays his respects to Ptolemy, but explains that to follow the Master's scheme of latitude and longitude he must change some of Ptolemy's maps to add places unknown in Ptolemy's time. In this way he justifies his filling in some areas that Ptolemy had labeled "Terra Incognita."

This opening of the ocean—in which Fra Mauro's map was not the first —was not yet proven by mariners' experience. It still was largely speculative, based on rumors and reports that came through overland travelers. The principal source for this crucial modernizing of Ptolemy's maps was probably a lone Venetian merchant-adventurer. Even after the breakup of the Mongol Empire when the route straight eastward from Syria across Asia was no longer protected for Europeans, the merchants of Venice would not give up their Eastern trade. They tried to keep up a prosperous commerce in goods from Asia by controlling the paths that ran southeastward, overland through Egypt, then through the Red Sea and the Gulf of Aden and across the Arabian Sea. One of these latter-day Venetian merchants was Nicolo de' Conti, who traveled for twenty-five years after leaving Venice in 1419. His far-ranging adventures took him across the Arabian Desert, in search of precious stones down the west coast of India to the tip of the peninsula and to the burial place of Saint Thomas near Madras, into the cinnamon forests of Ceylon, and onto the island of Sumatra, which he noted for its gold, its camphor, its pepper, and its cannibalism, then to Burma to observe the tattooed people and the elephants, rhinoceroses and pythons, to many-domed Pegu, and even as far as Java. During these travels he married an Indian wife, who bore him four children. While returning to Venice, Conti stopped in the Holy Land, where he met a Spanish traveler who solemnly recorded Conti's tall tales of his imagined adventures at the court of Prester John.

Such tales are all we might have known of Conti's travels. But during these decades in the East, Conti had renounced Christianity. On his return to Venice in 1444, Pope Eugenius IV therefore ordered him to do penance by dictating his full story to the papal secretary Poggio Bracciolini. The product was one of the best European accounts of south Asia in those years after the stream of overland merchant-travelers had ceased, and before the seafaring travelers had yet arrived. While Conti's other observations may have been more picturesque, none were more influential than his speculations on the possibility of reaching the Oriental spice islands by sailing around Africa. Mid-fifteenth-century makers of *mappae mundi* grasped optimistically at Conti's account. Using this latest information, they boldly revised Ptolemy to open the ocean to India.

Even after some of the most beautiful and most reputable world maps were showing a sea path around Africa to the Indies, the older Ptolemaic pictures of Africa continued to circulate. During the great age of seafaring discovery that was about to come, the old maps of Ptolemy would still remain standard. The newest atlases on their title pages claimed to be prepared "after the original maps of Ptolemy." Publishers relied on the fame of Ptolemy, as later dictionary makers used the name of Webster, to make their product seem authentic.

The rise of printing, as we shall see, would change not only the content, but the currency and the uses of geographic knowledge. The effects were not entirely progressive. With the rise of the printed image, of wood and metal engraving, it was no accident that it was metalworkers, goldsmiths, and goldsmith-painters in central Germany and the Rhineland who turned to copperplate printing. The heavy investment of atlas publishers in their plates, together with the overwhelming reputation of "Ptolemy," combined to keep the older maps in circulation, and not always as mere historical facsimiles. Even after those plates had been made obsolete by recent geographic discoveries, they were used, sometimes alongside newer maps which contradicted them. People who had difficulty accustoming themselves to the notion that it was possible to sail around Africa into an open Indian Ocean could continue to solace themselves with the familiar picture in Ptolemy. For the representation of a peninsular Africa and an open Indian Ocean by Fra Mauro and a few others by the mid-fifteenth century was still crude and seemed fanciful. Long after the Portuguese had rounded the Cape and reached India by water, until as late as 1570, the "best" atlases still offered reissues of Ptolemy's obsolete maps. Many printed atlases circulated before the end of the fifteenth century, when map-making had already begun to be highly profitable, but it was not until 1508 that a printed map gave a reasonably precise portrait of the southward extension of Africa.

The opening of the Indian Ocean was the Europeans' first world-shaking, world-shaping revision of Ptolemy. The centuries after the closing of the traders' land routes to the East would revise Ptolemy in countless other ways. Ptolemy's world, which had terminated at 63 degrees north latitude, about halfway up the Scandinavian peninsula, would have to be extended toward both the north and the northwest. And, of course, a whole New World between Europe and Asia would eventually be added. The scientific spirit of Ptolemy, his admissions of his ignorance, his plea for latitude and longitude, cheered on cartographers and mariners.

21

Portuguese Sea Pioneers

AMONG those most encouraged were the seamen of Portugal, who had an assignment from geography for their role in history. On the westernmost edge of the Iberian peninsula, the nation attained its modern borders very early, in the mid-thirteenth century. Portugal had no window on the Mediterranean—the "Sea-in-the-Midst-of-the-Land"—but was blessed by long navigable rivers and deep harbors opening oceanward. Cities grew up on the shores of waters that flowed into the Atlantic. The Portuguese people, then, naturally faced outward, away from the classic centers of European civilization, westward toward the unfathomed ocean, and southward toward a continent that for Europeans was also unfathomed.

An organized long-term enterprise of discovery, the Portuguese achievement was more modern, more revolutionary than the more widely celebrated exploits of Columbus. For Columbus pursued a course suggested by ancient and medieval sources, the best information of his day, and if he had succeeded in his avowed purpose, he would have confirmed them. There was no uncertainty in his mind either about the landscape en route to Asia or about the direction to be taken. Only the sea was unknown. Columbus' courage was in taking a direct sea passage to "known" lands in a known direction but without knowing precisely how long the passage might be.

By contrast, the Portuguese voyages around Africa and, it was hoped, to India, were based on risky speculative notions, rumors, and suggestions. Unknown lands would have to be skirted, used as supply bases for food and water en route. The journey would go where Christian geography threatened mortal dangers, far below the equator. Portuguese discoveries, then, required a progressive, systematic, step-by-step national program for advances through the unknown. Columbus' Enterprise of the Indies was a bold stroke, the significance of which would not be known for decades. The Portuguese voyagers were on a century-and-a-half enterprise, the actual meaning of which was imagined long in advance, the accomplishment of which was known immediately. Columbus' greatest achievement was something he never even imagined, a by-product of his purposes, a consequence of unexpected facts. The Portuguese achievement was the product of a clear purpose, which required heavy national support. Here was a grand prototype of modern exploration.

Long-term planning was possible only because the Portuguese had undertaken a collaborative national adventure. Earlier national epics of European people sang the courage and the exploits of some particular hero, a Ulysses or an Aeneas or a Beowulf. The Portuguese epic of seafaring could not sing as Virgil did "Of Arms and *the Man.*" Now the hero had become plural. "This is the story," Camoens begins his *Lusiads* (appropriately named after the *sons* of Lusus, companion of Bacchus and mythical first settler of Portugal), "of *heroes* who, leaving their native Portugal behind them, opened a way to Ceylon, and further, across seas no man had ever sailed before." The dimensions of life had broadened and become more public and more popular. While ancient lays celebrated a god-hero, modern lays would celebrate heroic peoples.

The adventure, too, was more plural and far wider. Water passages were no longer merely familiar well-marked roads within a closed sea, a Mediterranean. The new highways led across the open seas, and they led everywhere.

Separated only by a tiny strait from Africa, the Portuguese were remarkably free of racial prejudice or provincialism. Their ancestors were Celts, Iberians, and Englishmen. They intermarried with Africans and Asians. Portugal became a small proto-America, a place for the mixing of people —Christians, Jews, and Muslims. A Muslim occupation had left its mark on institutions. The divers physical, mental, temperamental, traditional, aesthetic, and literary resources enriched one another, providing the varied energies and the motley knowledge required to reach out to the open ocean and come back home again.

The ability to come home again was essential if a people were to enrich, embellish, and enlighten themselves from far-off places. In a later age this would be called *feedback*. It was crucial to the discoverer, and helps explain why going to sea, why the opening of the oceans, would mark a grand epoch for humankind. In one after another human enterprise, the act without the feedback was of little consequence. The capacity to enjoy and profit from feedback was a prime human power. Seafaring ventures, and even their one-way success, were themselves of small consequence and left little record in history. Getting there was not enough. The internourishment of the peoples of the earth required the ability to get back, to return to the voyaging source and transform the stay-at-homes by the commodities and the knowledge that the voyagers had found over there. Fourth-century coins made in Carthage have been found in the Azores, and ancient Roman coins seem to have been left in Venezuela by vagrant wind-driven vessels. Vikings from Norway and Iceland appear to have touched North America from time to time in the Middle Ages. In 1291 the Vivaldi brothers from Genoa set out to round Africa by sea, but they disappeared. It is possible, too, that in pre-Columbian times a Chinese or Japanese junk may have been driven off course all the way to the shores of America. But these acts and accidents that produced no feedback spoke only to the wind.

For most countries of Western Europe the fifteenth century—the epoch of the Hundred Years War and the Wars of the Roses—was a time of civil strife and/or fears of invasion. The Turks, who captured Constantinople in 1453, menaced the whole Levant and the Balkans. Spain, the only country that shared some of Portugal's peninsular advantage (though diluted by the competition of her prosperous Mediterranean ports), was torn by civil strife that for most of the century kept her near anarchy. Portugal, in sharp contrast to all these others, was a united kingdom for the whole of the fifteenth century and was hardly touched by civil disturbance.

Still, to exploit its many advantages, Portugal needed a leader—someone to draw people together, to organize resources, to point the direction. Without such a leader all other advantages would have been nothing. Prince Henry the Navigator was a curious combination of a bold heroic mind and an outreaching imagination, with an ascetic stay-at-home temperament. Frigid to individuals, he was passionate for grand ideals. His talent of obstinacy and his power to organize proved essential for the first great enterprise of modern discovery.

In the perspective of history, it is not so surprising that the pioneer of modern exploration never himself went out on an exploring expedition. The great medieval adventure in Europe—crusading—called for risk of life and limb against the infidel. Modern exploration had to be an adventure of the mind, a thrust of someone's imagination, before it became a worldwide adventure of seafaring. The great modern adventure—exploring—first had to be undertaken in the brain. The pioneer explorer was one lonely man thinking.

The traits of personality that made this lone adventure possible were not all attractive. Henry the Navigator compared himself with Saint Louis, but he was a much less engaging person. He lived like a monk, his biographers observe, and it is said that he died a virgin. At his death he was found to be wearing a hair shirt. All his life he was torn between crusading and exploring. His father, King John I, whose alternative sobriquets were the Bastard or the Great, founder of the Aviz dynasty, had seized the Portuguese throne in 1385. At the decisive battle of Aljubarrota, with the aid of English archers, John defeated the King of Castile and so secured the independence and the unity of Portugal. King John cemented his English alliance by marrying the devout and strong-willed Philippa of Lancaster, daughter of John of Gaunt, but he still kept his mistress in the palace to which she came as queen. "She found the Court a sink of immorality," a pious and optimistic latter-day Portuguese historian observes, "she left it as chaste as a nunnery." And she bore the King six sons, the third of whom, Henry, was born in 1394.

To celebrate his formal treaty of friendship with Castile in 1411, King John followed the chivalric custom of the age by planning a tournament, to last a full year. Knights were to be invited from all over Europe, and the jousts would give the King's three eldest sons who had just reached manhood the opportunity to earn their knighthoods by public acts of chivalry. But the three princes, reinforced by the King's treasurer, dissuaded King John from this expensive panoply. They urged him instead to offer them opportunity for deeds of Christian valor by launching a Crusade against Ceuta, a Muslim stronghold and trading center on the African side opposite Gibraltar. There, too, the King could atone for the Christian bloodshed in his earlier campaigns by "washing his hands in the blood of the infidel." Young Prince Henry helped plan this expedition, which, in a number of unexpected ways, was to shape his life.

Prince Henry, still only nineteen, was assigned the task of building a fleet up north in Oporto. After two years' preparation, the Crusade against Ceuta was launched in an aura of miracles and omens. A monk near Oporto had a vision of the Virgin Mary handing a glittering sword to King John. There was an eclipse of the sun. Then Queen Philippa, after a long and ill-advised religious fast, fell mortally ill. Summoning the King and her three eldest sons, she gave each a fragment of the True Cross to wear in holy battle. To each prince she also gave a knightly sword, and with her expiring breath she blessed the expedition against Ceuta. A papal bull, solicited for the occasion, offered all the spiritual benefits of a Crusade to those who died in this effort.

The Portuguese armada stormed the fortress at Ceuta on August 24, 1415, in a one-sided battle. Well armed and armored, and supported by a contingent of English archers, they overwhelmed the Muslims, who were reduced to hurling rocks. Within a day the Portuguese crusaders had taken the Infidel stronghold and provided Prince Henry his moment of glory. Only eight Portuguese had been killed, while the city streets were piled with Muslim bodies. By afternoon the army had begun sacking the city, and the spiritual rewards of killing infidels were supplemented by more worldly treasure. This occasion gave Prince Henry his first dazzling glimpse of the wealth that lay hidden in Africa. For the loot in Ceuta was the freight delivered by the caravans that had been arriving there from Saharan Africa in the south and from the Indies in the east. In addition to the prosaic essentials of life—wheat, rice, and salt—the Portuguese found exotic stores of pepper, cinnamon, cloves, ginger, and other spices. Ceutan houses were hung with rich tapestries and carpeted with Oriental rugs. All in addition to the usual booty of gold and silver and jewels.

The Portuguese left a small garrison and the rest went home. When Prince Henry was sent back to Ceuta against renewed Muslim attack, he spent several months learning about the African caravan trade. Under the Muslims, Ceuta had been alive with some twenty-four thousand shops selling the gold, silver, copper, brass, silks, and spices, all brought by caravan. Now that Ceuta had become a Christian city, the caravans no longer arrived. The Portuguese possessed a profitless city of the dead. Either they must come to terms with surrounding infidel tribes or they must conquer the hinterland.

Prince Henry gathered information about the interior lands from which the treasures of Ceuta had come. He heard tales of a curious trade, "the silent trade," designed for peoples who did not know each other's language.

The Muslim caravans that went southward from Morocco across the Atlas Mountains arrived after twenty days at the shores of the Senegal River. There the Moroccan traders laid out separate piles of salt, of beads from Ceutan coral, and cheap manufactured goods. Then they retreated out of sight. The local tribesmen, who lived in the strip mines where they dug their gold, came to the shore and put a heap of gold beside each pile of Moroccan goods. Then they, in turn, went out of view, leaving the Moroccan traders either to take the gold offered for a particular pile or to reduce the pile of their merchandise to suit the offered price in gold. Once again the Moroccan traders withdrew, and the process went on. By this system of commercial etiquette the Moroccans collected their gold. Tales of the bizarre process fired Prince Henry's hopes. But still the crusader, he organized a Portuguese fleet and declared his intention to capture Gibraltar from the infidels. When King John forbade that expedition after it was already underway, Prince Henry returned home sulking. Instead of joining the court in Lisbon where he would have shared the burdens of royal government, he went far southward through the Algarve to Portugal's Land's End, Cape Saint Vincent, the southwestern tip of Europe.

Ancient geographers had given a mystic significance to that extremity of land, the borderland of the watery unknown. "Sacred Promontory" *(Promentorium Sacrum)* was what Marinus and Ptolemy had christened it. The Portuguese, translating this into *Sagres,* made it the name of the nearby village. The visitor to Portugal today can see a lighthouse on the ruins of the fortress that Prince Henry made his headquarters for forty years. There Prince Henry initiated, organized, and commanded expeditions on the frontier of mystery. In the first modern enterprise of exploring, from that spot he sent out an unbroken series of voyages into the unknown. Today's visitor to the harsh inhospitable cliffs of Sagres senses the appeal that place must have had for an ascetic prince who wanted to separate himself from the formalities of an effete court.

At Sagres Prince Henry became the Navigator. There he applied the zeal and energy of the crusader to the modern exploring enterprise. Prince Henry's court was a primitive Research and Development Laboratory. In the crusader's world the known was dogma and the unknown was unknowable. But in the explorer's world the unknown was simply the not-yet-discovered. And all the trivia of everyday experience could become signposts.

Prince Henry had learned his destiny from the astrologers. It was fixed in the stars, the contemporary chronicler Gomes Eanes de Zurara reported, that "this prince was bound to engage in great and noble conquests, and above all was he bound to attempt the discovery of things which were hidden from other men, and secret." From newfound distant lands he would bring merchandise to enrich Portuguese commerce. Incidentally he would gather useful facts about the extent of Muslim power, and hoped to find new Christian allies, perhaps Prester John himself, against the infidel. Along the way he would, of course, convert countless souls to Jesus Christ.

Prince Henry, for all these reasons, made Sagres a center for cartography, for navigation, and for shipbuilding. He knew that the unknown could be discovered only by clearly marking the boundaries of the known. This meant, of course, junking the caricatures drawn by Christian geographers and replacing them with cautious, piecemeal maps. And this required an incremental approach.

In the spirit of the portolanos, the coast pilots, he accumulated the bits of many mariners' experiences to fill out an unknown coast. The Jews, wherever they were, had long since become powerful cultural ambassadors and cosmopolitanizers. The Catalan Jew from Majorca, Jehuda Cresques, son of the cartographer Abraham Cresques, whom we have already met, was brought to Sagres, where he supervised the piecing together of the geographic facts brought back by Prince Henry's seafaring explorers.

Prince Henry encouraged, then required, his mariners to keep accurate logbooks and charts, and to note for their successors everything they saw of the coast. Until that time, as a letter (October 22, 1443) of King Alfonso V complained, navigators' records had been haphazard and slovenly, not "marked on sailing charts or mappe mondes except as it pleased the men who made them." Now Prince Henry ordered all details to be marked accurately on navigation charts, which were to be brought back to Sagres, so that cartography could become a cumulative science. To Sagres came sailors, travelers, and savants from all over, each adding some new fragment of fact or some new avenue to facts. Besides Jews, there were Muslims and Arabs, Italians from Genoa and Venice, Germans and Scandinavians, and, as exploring advanced, tribesmen from the west coast of Africa. At Sagres, too, were manuscript records of the great travelers that Prince Henry's brother Pedro had collected during his grand tour (1419–28) of the European courts. In Venice Pedro had received a copy of Marco Polo's travels along with a map "which had all the parts of the earth described, whereby Prince Henry was much furthered."

With these facts came the latest navigating instruments and newest navigating techniques. The mariner's compass was already well known, but its use was still haunted by superstitious fears of its occult power, believed to be akin to necromancy. Only a century before, tricks such as those performed with the lodestone had got Roger Bacon in trouble. At Sagres the compass, like other instruments, was tested only by whether it helped the mariner reach out farther, and then find his way home.

When Prince Henry's mariners went farther south than Europeans had ever gone before, they faced new problems of defining latitude, which could best be done now by measuring the altitude of the sun at noon. In place of the delicate, costly, and complicated astrolabe, Prince Henry's men used the simpler cross-staff—a handy graduated stick with a moving crosspiece which could be aligned with the horizon and the sun to measure the angle of the sun's elevation. The cosmopolitan community at Sagres helped to make the quadrant, the new mathematical tables and other novel instruments, which became part of Prince Henry's exploring equipment.

At Sagres and at the nearby port of Lagos, experiments in shipbuilding produced a new type of ship without which Prince Henry's exploring expeditions and the great seafaring adventures of the next century would not have been possible. The caravel was a ship specially designed to bring explorers back. The familiar heavy, square-rigged *barca* or the still larger Venetian carrack was suited for sailing with the wind. These worked well enough within the Mediterranean, where the size of a trading vessel was a measure of its profit, and by 1450 there were Venetian square-riggers of six hundred tons or more. A larger ship meant a bigger profit from more cargo.

A discovery ship had its own special problems. It was not a cargo-vessel, it had to go long distances in unfamiliar waters and had to be able, if necessary, to sail into the wind. An exploring ship was no good unless it could get there *and back*. Its important cargo was *news*, which could be carried in a small parcel, even in the mind of one man, but which was definitely a return product. While discovery ships did not need to be big, they had to be maneuverable, and adept at the return. Feedback was the explorer's commodity. Seafarers were naturally tempted to sail out with the wind, which obviously meant having to come back against the wind. The vessels best suited for profitable trading in the closed Mediterranean did not serve the explorer on the unknown open sea.

Prince Henry's caravel was specially designed for these explorer's needs. He found some clues in the *caravos*, ships used by Arabs since ancient times off the Egyptian and Tunisian coasts, modeled on the still more ancient fishing vessels that the Greeks had made of rushes and hide. These dhows, rigged with "lateen," slanting and triangular sails, carried Arab crews of as many as thirty, in addition to seventy horses. A similar smaller, even more maneuverable vessel, called the *caravela* (*-ela* = diminutive) was in use on the Douro River in northern Portugal. Prince Henry's shipbuilders produced the famous caravel, which combined some of the cargo-carrying features of the Arab *caravos* with the maneuverability of the Douro River *caravelas*.

These remarkable little vessels were large enough to hold an explorer's supplies for a small crew of about twenty, who usually slept on deck but in bad weather went below. The caravel displaced about fifty tons, was about seventy feet in length and about twenty-five feet in the beam, and carried two or three lateen sails. "The best ships that sailed the seas" was what Alvise da Cadamosto (1432?–1511), the experienced Venetian mariner, called the caravels in 1456 after his African voyage in a caravel organized by Prince Henry. The caravel became the discoverer's standard ship. Columbus' three ships—the *Santa María,* the *Pinta,* and the *Niña*—were all of caravel design, and the *Santa María* was only one-fifth as big as the large Venetian square-riggers of his day. The caravel proved that bigger was not always better.

Prince Henry's African enterprises would show that the caravel had a crucial, and unprecedented, capacity to get back home. Its shallow draft qualified it to explore inshore waters and also made it easier to beach the vessel for careening or for repair. In seafaring terms, the ability to return meant the ability to sail into the wind, at which the caravel excelled. While the old square-rigged seafaring *barca* could not sail closer than 67 degrees from into the wind, the lateen-rigged caravel could sail at 55 degrees. This meant that the distance that required the *barca* to tack five times would require only three tacks from the caravel. This saving of about one-third in time and distance could cut off many weeks at sea. Seamen who knew they were going out in a vessel specially designed for their sure and speedy return were more cheerful, more confident, and more willing to risk longer outward voyages.

Under Prince Henry's stimulus, Lagos, a few miles along the coast from Sagres, became a center for caravel-building. Oak for keels came from Alentejo, bordering on the Algarve. Pine for the hulls grew along Portugal's Atlantic seaboard, where it was protected by law. The cluster pines also produced resin to waterproof the rigging and to calk the seams of the hull. Around Lagos there soon developed flourishing crafts of sail-making and rope-making.

While Prince Henry at Sagres did not actually build a modern research institute, he did bring together all the essential ingredients. He collected the books and the charts, the sea captains, pilots, and mariners, the map-makers, instrument-makers, and compass-makers, the shipbuilders and carpenters, and other craftsmen, to plan voyages, to assess the findings, and to prepare expeditions ever farther into the unknown. The work Prince Henry started would never end.

22

Beyond the Threatening Cape

UNLIKE Columbus, who would aim straight for the Indies, Prince Henry the Navigator had a larger, a vaguer, and more modern destination—true to his horoscope. "The noble spirit of this Prince," the admiring reporter Gomes Eanes de Zurara explained, "was ever urging him both to begin and to carry out very great deeds. . . . he had also a wish to know the land that lay beyond the isles of Canary and that Cape called Bojador, for that up to his time, neither by writings, nor by the memory of man, was known with any certainty the nature of the land beyond that Cape. . . . it seemed to him that if he or some other lord did not endeavour to gain that knowledge, no mariners or merchants would ever dare to attempt it, for it is clear that none of them ever trouble themselves to sail to a place where there is not a sure and certain hope of profit."

We have no evidence that Prince Henry had in mind the specific purpose of opening a sea-way around Africa to India. What beckoned him was the unknown, which lay west and southwest into the Sea of Darkness and southward along the uncharted coast of Africa. The Atlantic islands—the Azores (one-third of the way across the Atlantic Ocean!), the Madeiras, and the Canaries—had probably been discovered by Genoese sailors in the mid-fourteenth century. Prince Henry's efforts in that direction were less an enterprise of discovery than of colonization and development. But when his people landed in Madeira (*madeira* means wood) in 1420 and set about clearing the thick woods, they set a fire that raged for seven years. Although they never planned it that way, the potash left from the consumed wood would prove a perfect fertilizer for vineyards of the Malmsey grapes imported from Crete to replace those forests. The justly famous "Madeira" wine was the lasting product. Yet, as his stars foretold, Prince Henry was by nature and by preference not a colonizer but a discoverer.

When we look at a modern map of Africa, we look long and need a magnifying glass before we can find Cape Bojador (Portuguese for "Bulging Cape"), on the west coast, just south of the Canary Islands. Some thousand miles north of the continent's greatest westward bulge we see a tiny bump on the coastal outline, a "bulge" so slight that it is almost imperceptible on maps of the full continent. The sandy barrier there is so low that it can be seen only when one comes close, where there are treacherous reefs and unmanageable currents. Cape Bojador was no worse than a score of other barriers that skillful Portuguese sailors had passed and survived. But this particular Cape Bojador they had made their *ne plus ultra*. You dare not go beyond!

When we see the enormous risky promontories, the Cape of Good Hope or Cape Horn, that European seafarers would manage to round within the next century, we must recognize Bojador as something quite else. It was a barrier in the mind, the very prototype of primitive obstacles to the explorer. The eloquent Zurara tells us "why ships had not hitherto dared to pass beyond Cape Bojador."

> And to say the truth this was not from cowardice or want of good will, but from the novelty of the thing and the wide-spread and ancient rumour about this Cape, that had been cherished by the mariners of Spain from generation to generation. . . . For certainly it cannot be presumed that among so many noble men who did such great and lofty deeds for the glory of their memory, there had not been one to dare this deed. But being satisfied of the peril, and seeing no hope of honour or profit, they left off the attempt. For, said the mariners, this much is clear, that beyond this Cape there is no race of men nor place of inhabitants . . . and the sea so shallow that a whole league from land it is only a fathom deep, while the currents are so terrible that no ship having once passed the Cape, will ever be able to return. . . . these mariners of ours . . . [were] threatened not only by fear but by its shadow, whose great deceit was the cause of very great expenses.

At home in Sagres Prince Henry knew that he could not conquer the physical barrier unless he first conquered the barrier of fear.

He would never reach farther into the unknown unless he could persuade his seamen to go beyond Cape Bojador. Between 1424 and 1434 Prince Henry sent out fifteen expeditions to round the inconsequential but threatening cape. Each returned with some excuse for not going where none had gone before. At the legendary cape the sea bounced with cascades of ominous red sands that crumbled from the overhanging cliffs, while shoals of sardines swimming in the shallows roiled the waters between whirlpools. There was no sign of life along the desert coast. Was this not the very image of the world's end?

When Gil Eannes reported back to Prince Henry in 1433 that Cape Bojador was in fact impassable, the Prince was not satisfied. Would his Portuguese pilots be as timid as those Mediterranean or Flemish sailors who plied only the familiar ways? Surely this Gil Eannes, a squire whom he knew well in his own household, was made of bolder stuff. The Prince sent him back in 1434 with renewed promise of reward for yet another try. This time, as Eannes approached the cape he steered westward, risking the unknown perils of the ocean rather than the known perils of the cape. Then he turned south and discovered that the cape was already behind him. Landing on the African shore, he found it desolate, but by no means the gates of hell. "And as he purposed," Zurara reported, "so he performed—for in that voyage he doubled the Cape, despising all danger, and found the lands beyond quite contrary to what he, like others, had expected. And although the matter was a small one in itself, yet on account of its daring it was reckoned great."

Having broken the barrier of fear "and the shadow of fear," Prince Henry was on his way. Year after year he dispatched expeditions, each reaching a bit farther into the unknown. In 1435, when he sent out Eannes once again, this time with Afonso Baldaya, the royal cupbearer, they reached another fifty leagues down the coast. There they saw footprints of men and camels, but still did not encounter the people. In 1436, when Baldaya went out again, with orders to bring back an inhabitant for the Prince to interview at Sagres, he reached what seemed to be the mouth of a huge river, which he hoped would be the Senegal of "the silent trade" in gold. They called it the Rio de Ouro, even though it was only a large inlet and not a river, for the Senegal actually lay another five hundred miles farther south.

The relentless step-by-step exploration of the west African coast proceeded year by year, although commercial rewards were meager. In 1441, from Prince Henry's household went Nuno Tristão and Antão Gonçalves, reaching another two hundred fifty miles farther to Cape Branco (Blanco) where they took two natives captive. In 1444 from that area Eannes brought back the first human cargo—two hundred Africans to be sold as slaves in Lagos. Zurara's eyewitness account of this first European episode in the African slave trade was a painful glimpse of the miseries to come. "Mothers would clasp their infants in their arms, and throw themselves on the ground to cover them with their bodies, disregarding any injury to their own persons, so that they could prevent their children from being separated from them."

But Zurara insisted that "they were treated with kindness, and no difference was made between them and the free-born servants of Portugal." They were taught trades, he said, were converted to Christianity, and eventually intermarried with the Portuguese.

The arrival of this human merchandise from Africa, we are told, caused a change in the public attitude toward Prince Henry. Many had criticized the Prince for wasting the public substance in his frolics of exploration. "Then those who had been foremost in complaint grew quiet, and with soft voices praised what they had so loudly and publicly decried." "And so they were forced to turn their blame into public praise; for they said it was plain the Infant was another Alexander; and their covetousness now began to wax greater." Everybody wanted a share of this promising Guinea trade.

When Dinis Dias rounded Cape Verde, the western tip of Africa, in 1445, the most barren coast had been passed, and the prosperous Portuguese trade with west Africa soon engaged twenty-five caravels every year. By 1457 Alvise da Cadamosto—a Venetian precursor of the Italian sea captains like Columbus, Vespucci, and the Cabots who served foreign princes—advancing down the coast for Prince Henry had accidentally discovered the Cape Verde Islands and then went up the Senegal and Gambia rivers sixty miles from the sea. This Cadamosto proved to be one of the most observant as well as one of the boldest of Prince Henry's explorers. By his engaging accounts of curious tribal customs, of tropical vegetation, elephants, and hippopotami, he enticed others to follow.

At the time of Prince Henry's death in Sagres in 1460 the discovery of the west African coast had only begun, but it was well begun. The barrier of groundless fear had been breached in what became the first continuous organized enterprise into the unknown. Prince Henry therefore is properly celebrated as the founder of continuous discovery. For him each new step into the unknown was a further invitation.

Prince Henry's death caused only a brief hiatus in the exploring enterprise. Then in 1469 King Alfonso V, Prince Henry's nephew, in financial difficulty, found a way to make discovery into a profitable business. In an agreement quite unlike any we have heard of before between sovereign and vassal, Fernão Gomes, a wealthy citizen of Lisbon, committed himself to discover at least one hundred farther leagues, about three hundred miles, of the African coast each year for the next five years. In return, Gomes obtained a monopoly of the Guinea trade, from which the King received a share. The rest of the story has the inevitability of a steadily rising curtain. Discovery of the whole west African coast by the Portuguese now was a question no longer of *whether* but of *when*.

The supposed Portuguese policy of secrecy poses tantalizing problems for the historian because the policy itself seems to have been kept secret. When we chronicle Portuguese advances into the hitherto unknown, we must wonder whether any particular Portuguese voyage was unrecorded because of this "policy of secrecy" or simply because it was never made. Portuguese historians have been understandably tempted to treat the absence of a record of pre-Columbian voyages to America as a kind of evidence that such voyages really were made. The Portuguese did have some diplomatic and compelling reasons to advertise their discoveries in America. In Africa, however, they had every reason to conceal both the knowledge they had gained of the actual shape of the coast and word of the treasure they were extracting there. The records that remain of these earliest Portuguese discoveries in Africa are probably only minimal notices of their exploits.

The Gomes contract, we know, produced an impressive annual series of African discoveries—around Cape Palmas at the continent's southwestern tip, into the Bight of Benin, the island of Fernando Po at the eastern tip of the Guinea coast and then down southward across the equator. It had taken Prince Henry's sailors thirty years to cover a length of coast that Gomes, under his contract, covered in five. When Gomes' contract expired, the King gave the trading rights to his own son, John, who became King John II in 1481, opening the next great age of Portuguese seafaring.

King John II had some advantages that Prince Henry had lacked. The royal treasury was now enriched by the feedback of imports from the west African coast. Cargoes of pepper, ivory, gold, and slaves had already become so substantial that they gave their names to the parts of the continent that faced the Gulf of Guinea. For centuries these would be called the Grain Coast (Guinea pepper was known as "Grains of Paradise"), the Ivory Coast, the Gold Coast, and the Slave Coast. King John protected Portuguese settlements by building Fort Elmina, "the mine," in the heart of the Gold Coast. He supported land expeditions into the interior, to the back-country of Sierra Leone and even as far as Timbuktoo. And he pushed on down the coast.

As we have noted, when mariners advanced below the equator they could no longer see the North Star, and so had to find another way to determine their latitude. To solve this problem King John, like Prince Henry, collected experts from everywhere, and he set up a commission headed by two learned Jewish astrologer-mathematicians—a Portuguese dividend from the persecutions across the border in Spain. In 1492, when the Spanish inquisitor-general Torquemada gave Jews three months to convert to Christianity or leave the country, the brilliant Abraham Zacuto left the University of Salamanca and was welcomed to Portugal by King John II. Zacuto's disciple at Salamanca, Joseph Vizinho, had already accepted the King's invitation ten years before, and in 1485 had been sent out on a voyage to develop and apply the technique of determining latitude by the height of the sun at midday. He was to accomplish this by recording the declination of the sun along the whole Guinea coast. The most advanced work for finding position at sea by the declination of the sun, as would be necessary in sailing below the equator, was the *Almanach Perpetuum* which Zacuto had written in Hebrew nearly twenty years before. After Vizinho translated these tables into Latin, they guided Portuguese discoverers for a half-century.

Meanwhile King John, carrying on Prince Henry's work, kept sending his discovery voyages farther down the west African coast. Diogo Cão reached the mouth of the Congo (1480–84), and began the custom of setting up stone markers *(padroes),* surmounted by a cross, as proof of first discovery and tokens of Christian faith.

These advances down the coast brought new rumors of the famous but still unseen Prester John. While Prince Henry's first objective was to move into the unknown, another objective, his chronicler Zurara reported, was "to know if there were in those parts any Christian princes, in whom the charity and the love of Christ was so ingrained that they would aid him against those enemies of the faith." This conjectural potential ally must have been Prester John, whose "letter," as we have seen, had been circulating in Europe for two centuries. By this time the locale of the legendary priest-king had been transferred from "furthest Asia" to Ethiopia. Whenever one of Prince Henry's voyages found another great river—the Senegal, the Gambia, the Niger—debouching on the west coast, he found new hope that this at last might be the "Western Nile" that would lead to Prester John's Ethiopian kingdom. When King John II's men reached Benin at the eastern tip of the Gulf of Guinea, he received the interesting report that the kings of Benin had sent gifts to a king called Ogané whose realm was twelve months' journey inland to the east, and who returned gifts inscribed with small crosses. Prince Henry had already unsuccessfully tried to make contact with Prester John by journeys inland, and King John II's own envoys to Jerusalem had also failed to find him.

By 1487 King John II had organized a two-pronged grand strategy for reaching the long-sought Christian ally. He would send one expedition southeastward overland and another by sea around the African coast. If there really should be a sea passage to India, a Christian ally was more desirable and more necessary than ever, not only for crusading but also to serve as a way station and supply base for future trading enterprises.

The overland expedition that left Santarém on May 7, 1487, like others before it, was characteristically small, consisting of only two men. After a considerable search, the King had chosen Pero da Covilhã (1460?–1545?) and Affonso de Paiva for the dangerous assignment. Covilhã, a married man with children, in his late twenties, had already proven himself bold and versatile. He had spent much of his life abroad, had taken part in ambushes in the streets of Seville, had served as the King's secret agent at the court of Ferdinand and Isabella, and had undertaken diplomatic missions to the Barbary States in north Africa. Covilhã's mission to Tlemcen, famous as "the Granada of Africa," and then to Fez, accustomed him to the Muslim ways, which did not vary much from Morocco to Calicut, and so equipped him to travel through Islam without rousing suspicion. While in those days knowledge of Arabic was not unusual in Portugal, his contemporaries praised Covilhã as "a man who knew all the languages which may be spoken by Christians, Moors, or the heathen." An attractive man of courage and decision, he also possessed the required powers of observation and of memory. Of his companion, Paiva, we know only that he was a gentleman of the court, and that he, too, spoke Spanish and Arabic.

Covilhã and Paiva had an audience with the King and were then briefed by the King's chaplain, his physicians and his geographers, in a session shrouded in secrecy. From the plans presented in Portugal some time before by Christopher Columbus, these experts had drawn information that was expected to be useful. A Florentine banker in Lisbon gave the travelers a letter of credit, which they drew on for expenses as they moved eastward through Spain and Italy. At Barcelona they took ship for Naples, then sailed to Rhodes. There, as they took off into Muslim territory, the knowledgeable Knights Hospitalers of Saint John of Jerusalem warned that they would now be no more than "Christian dogs." The prospering agents of Venice and Genoa whom they would encounter would want no Portuguese competition. They were urged therefore to dress and act the part of Muslim merchants, ostensibly dealing in a cargo of honey. In this guise they reached Alexandria, where they both nearly died of fever, then on to Cairo and Aden at the mouth of the Red Sea.

There they separated. Paiva was to make his way directly to Ethiopia and Prester John, while Covilhã would go on to India. Paiva disappeared, but Covilhã finally reached Calicut and Goa on the southwestern shores of India, where he witnessed the prosperous trade in Arabian horses, spices, fine cottons, and precious stones. In February 1489 Covilhã took ship westward to Ormuz at the entrance to the Persian Gulf, then to the east African port of Sofala opposite Madagascar and back north to Cairo. Having completed his mission of assessing the European trade with India, he was eager to return home. But there in Cairo he encountered two Jewish emissaries from King John II who carried a letter instructing Covilhã, if he had not already done so, to proceed at once to the realm of Prester John to collect information and promote an alliance.

Unable to disobey his sovereign, Covilhã took up the mission, meanwhile sending to King John a momentous letter with all he had learned about Arab seafaring and the commerce of India. In 1493, after a side trip to Mecca, six years after his departure from Portugal, he finally arrived in Ethiopia. In this Realm of Prester John, actually ruled by Alexander "Lion of the Tribe of Judah, and King of Kings," he became a Portuguese Marco Polo, so useful at court that the King would not let him leave. Convinced that he would never return home, Covilhã married an Ethiopian wife who bore him several children.

Meanwhile, Covilhã's letter, which has not survived and which is known only secondhand, would have a powerful influence on the future of Portugal and of Asia. For it appears to have informed King John II, from reports Covilhã had heard on the African coast, "that his [the king's] caravels, which carried on trade in Guinea, navigating from land to land, and seeking the coast of this island [Madagascar] and of Sofala, could easily penetrate into these Eastern seas and come to make the coast of Calicut, for there was sea everywhere."

23

To India and Back

THE other prong of King John II's discovery enterprise, in the modern seafaring mode, was a long-planned, carefully organized undertaking, with large capital investment and a numerous crew. For commander he chose Bartholomeu Dias, who had superintended the royal warehouses in Lisbon and had taken a caravel down the African coast. Dias' expedition comprised two caravels of fifty tons each and a store ship, never before added to a discovery trip, which would enable the others to range farther, to stay at sea longer, and go far out to sea. Dias carried with him six Africans who had been on earlier Portuguese voyages. Well fed and dressed European style, they were deposited at several places along the coast, together with specimens of gold, silver, spices, and other products of Africa so that, in the way of "the silent trade," they could show the natives what goods the Portuguese wanted. Having landed the last of these African emissaries, Dias' ships ran into a storm that became a gale. Running before the northerly wind with close-reefed sails in a rough sea for thirteen days, the ships were driven far from shore, then southward into the open sea. The crew, who had just suffered tropical heat at the equator, were panicked. "And as the ships were tiny, and the seas colder and not such as they were in the land of Guinea . . . they gave themselves up for dead." After the storm, Dias steered east with all sails hoisted, but for several days he still sighted no land. Turning north for 150 leagues, he suddenly viewed high mountains. On February 3, 1488, he anchored in Mossel Bay, about two hundred thirty miles east of what is now Cape Town. The providential storm, achieving what no planning could yet accomplish, had driven him around the southern tip of Africa. When the men landed, the natives tried to drive them off with stones, Dias himself killed one of them with an arrow from his cross-bow, and that was the end of their encounter. He followed the coast, which now plainly ran to the northeast, another three hundred miles to the mouth of the Great Fish River and into Algoa Bay.

Dias wanted to go on into the Indian Ocean, and so fulfill the hope of many centuries, but the crew would have none of it. "Weary, and terrified by the great seas through which they had passed, all with one voice began to murmur, and demand that they proceed no farther." Provisions were low, and could be replaced only by hastening back to the supply ship left far behind. Was it not enough for one voyage to bring back news that Africa really could be rounded by sea? After a meeting of his captains, who all signed a sworn document declaring their decision to turn back, Dias agreed. As the ships turned about, Dias passed the stone marker they had set up to note their coming, "with as much sorrow and feeling as though he were taking his last leave of a son condemned to exile forever, recalling with what danger to his own person and to all his men they had come such a long distance with this sole aim, and then God would not grant it to him to reach his goal."

En route home, they returned to the supply ship, which they had left nine months before with nine men aboard. Only three were still alive, and one of these "was so overcome with joy at seeing his companions that he died of a sudden, being very weak through illness." The worm-eaten supply ship was unloaded and burned, and the two caravels made their way back to Portugal in December 1488, sixteen months and seventeen days after their departure.

When Dias' weatherbeaten caravels came sailing into Lisbon harbor, there awaiting them was the still-obscure Christopher Columbus. The outcome of Dias' voyage was of immediate personal interest to him. For Columbus was then in Lisbon making another effort to persuade King John II to support his own seaborne expedition to the Indies by sailing westward across the Atlantic. In 1484, when Columbus had first come for that purpose, the King had referred the project to a commission of experts, who turned him down, probably because they thought he had grossly underestimated the sea distance westward to the Indies. But Columbus had then impressed the King with his "industry and good talent" and now had come back to renew his request. Dias' moment of triumph was a season of disappointment for Columbus. For the eastward sea route to the Indies around Africa was now feasible and Columbus' project was superfluous. Columbus noted in the margin of his copy of Pierre d'Ailly's *Imago mundi* that he was present when Dias gave to the King the epoch-making report. Columbus would have to seek support from a nation that had not yet found its own way around Africa.

Dias was never properly rewarded by his sovereign, and he became the forgotten man of Portugal's Age of Discovery. He did supervise the building of ships for Vasco da Gama, but was not included in Gama's climactic voyage to India. Only his death, in 1500 while playing an inconspicuous part in Cabral's thirteen-ship fleet off the coast of Brazil, was appropriate. A hurricane sank four of Cabral's ships, one of which was commanded by Dias, "casting them into the abyss of that great ocean sea . . . human bodies as food for the fish of those waters, which bodies we can believe were the first, since they were navigating in unknown regions."

A quick follow-up to Dias' discovery might have been expected. But the next step was delayed—by domestic problems in Portugal, by a disrupted succession to the crown, and especially by the running dispute that kept Portugal on the brink of war with Spain. Ironically, the discoveries of Columbus himself proved to be the principal cause of these troubles, which postponed for a full decade the sequel to Dias' rounding of the Cape.

When King John II received word of Columbus' discovery of new islands in the Atlantic, he announced, in March 1493, that these new lands, because of their proximity to the Azores and for other reasons, rightly belonged to Portugal. The ensuing disputes between King John of Portugal and King Ferdinand of Castile, and their rivalry for the support of the Pope, who presumably had the power to assign to Catholic kings the worldly governance of all newly discovered parts of the earth, resulted in the famous Treaty of Tordesillas (June 7, 1494). Spain and Portugal both acquiesced in a papal line which was to run north and south at 370 leagues (about 1,200 nautical miles) west of the Cape Verde Islands. Lands to the west of the line would belong to Spain, those eastward to Portugal. This agreement did evade war for the moment, and remains one of the most celebrated treaties in European history. But it was so full of ambiguities that nobody knows whether it really went into effect. From which Cape Verde island should the line be measured? Precisely how long was a league? And centuries would pass before the technology existed to draw the required precise line of longitude. In any event, in addition to securing Portugal's claim to Brazil, whose existence then may or may not have been known, the treaty did affirm Portugal's right to the eastward sea path to the Indies.

On coming to the Portuguese throne in 1495, the bold twenty-six-year-old King Manuel I was nicknamed "The Fortunate" because he inherited so many grand enterprises. He set in motion a scheme to follow up Dias' discoveries with a new voyage of discovery that would take the sea route all the way to India, would open the way for trade, and possibly too for conquest. The young King's cautious advisers warned against the enterprise. How could so small a country succeed in conquest at so great a distance? And would not this enterprise excite the enmity of all the great powers—the Spanish, the Genoese, the Venetians, and, of course, the Muslims—whose commercial interests would be threatened? The King, overruling objections, chose for leader of the expedition a gentleman of his household, Vasco da Gama (c. 1460–1524), son of a minor official from the south coast. Gama had proven himself to be both a sailor and a diplomat. As King Manuel foresaw, the expertise of a sailor, though perhaps enough for coasting down the sparsely inhabited west African seaboard, would not suffice for dealing with sophisticated Indian potentates. Events proved Vasco da Gama to be brilliantly qualified. Though ruthless and of violent temper, he would show the courage, the firmness, and the broad vision required for dealing with humble seamen and arrogant sultans.

After two years of preparation, Gama's fleet of four vessels—two square-rigged ships of shallow draft, each about 100 tons, a lateen-rigged caravel of about 50 tons, and a store ship of some 200 tons—sailed out of Lisbon harbor on July 8, 1497. The ships carried provisions for three years. They were well supplied, too, with maps, with astronomical instruments and tables of declination prepared by Zacuto, and they carried carved stone pillars to mark Portuguese claims. There was of course a priest and the customary number of convicts who, being considered expendable, could be used whenever there was mortal risk. Altogether the crew numbered some 170.

Columbus' dazzling fame, at least from the American perspective, has blinded us to other achievements of seafaring discovery as great or even greater in that first Age of the Sea. The immediate effects of Vasco da Gama's voyage were incomparably more fulfilling than those of Columbus. Columbus promised the fabled cities of Japan and India, but he reached only uncertain savage shores. When after decades his enterprise finally paid off, it was in the most unexpected ways. Gama proposed to reach the trading capitals of India and to initiate a profitable trade—which he did. He promised to circumvent the Asian trade monopolies of the Muslims of the Levant and of the merchants of Genoa and Venice—which he did.

Columbus took the initiative, promised a gold mine, and only found a wilderness. The initiative for Vasco da Gama's voyage was not his own but that of his King. Not for qualities of character but for the magnitude of seafaring achievement, Vasco da Gama must overshadow Columbus. Columbus' first voyage went due westward before a fair wind, twenty-six hundred miles from Gomera in the Canaries to the Bahamas, remaining at sea for thirty-six days. Gama's course, which required subtle navigation, took him in a wide circle nearly all the way across the south Atlantic, then against opposite currents and contrary winds. He made the dangerous decision not to hug the African coast but to swing around through the mid-Atlantic from the Cape Verde Islands to the Cape of Good Hope, a distance of more than thirty-seven hundred miles, before reaching the Bay of St. Helena just above the present Cape Town, remaining at sea for ninety-three days. From there, his skill in navigation, in managing his crew, and in dealing with the hostile Muslim rulers at Mozambique, Mombasa, and Malindi, finally took him and his fleet across the Arabian Sea and the Indian Ocean, to arrive at Calicut, his intended destination on the south-western coast of the Indian peninsula on May 22, 1498. Until then, there had been no seafaring achievement of equal scope.

Unfortunately, Gama, unlike Columbus, did not leave us his own records. But luckily a member of Gama's crew did keep a journal, which offers vivid glimpses of the variety of problems conquered en route. The perils of nature and the sea somehow seemed the least threatening, for the sea in those remote parts was empty of human enemies, and nature did not dissemble. But as he crept up the east coast of Africa, where no European ship had been before, and where there were no useful maps, Gama had to use every device to secure an Arab pilot who would guide him across the vast Arabian Sea. At one place after another, at Mozambique and at Mombasa, the pilot he found or had assigned to him by the local ruler proved ignorant or treacherous. Finally at Malindi he secured an Arab pilot able to guide his fleet the twenty-three days across the Arabian Sea to Calicut.

The first greeting to Gama's fleet the morning after arrival at Calicut, on the southwestern coast of India, recorded in his crew-member's vivid journal, proved that the King of Portugal had acted none too soon.

On the following day these same boats came again alongside, when the captain-major [Gama] sent one of the convicts to Calecut, and those with whom he went took him to two Moors from Tunis, who could speak Castilian and Genoese. The first greeting that he received was in these words: "May the Devil take thee! What brought you hither?" They asked what he sought so far away from home, and he told them that we came in search of Christians and of spices. They said: "Why does not the King of Castile, the King of France, or the Signoria of Venice send hither?" He said that the King of Portugal would not consent to their doing so, and they said he did the right thing. After this conversation they took him to their lodgings and gave him wheaten bread and honey. When he had eaten he returned to the ships, accompanied by one of the Moors, who was no sooner on board, than he said these words: "A lucky venture, a lucky venture! Plenty of rubies, plenty of emeralds! You owe great thanks to God, for having brought you to a country holding such riches!" We were greatly astonished to hear his talk for we never expected to hear our language spoken so far away from Portugal.

The shrewd Gama spent three months in palaver with the King, or *Samuri*, of Calicut. He tried to persuade the local ruler that the Portuguese were mainly in search of the Christian kings said to rule in those parts, "not because they sought gold or silver, for of this they had such abundance that they needed not what was to be found in this country." But the Samuri of Calicut was insulted that Gama had not brought him costly presents, and spurned the cheap trading goods that might have served well enough on the Guinea coast. Gama tried to explain that his ships had come "merely to make discoveries. . . . The King then asked what it was he had come to discover: stones or men? If he came to discover men, as he said, why had he brought nothing?"

Gama's fleet left Calicut in late August 1498, "greatly rejoicing at our good fortune in having made so great a discovery. . . . having agreed that, inasmuch that we had discovered the country we had come in search of, as also spices and precious stones, and it appeared impossible to establish cordial relations with the people, it would be as well to take our departure." After contrary winds, obstruction by the Muslim rulers en route, and the curse of scurvy, which decimated the crew, finally two of Gama's four ships, the square-rigged *San Gabriel* and the caravel *Berrio,* made a triumphal entry into Lisbon in mid-September 1499. Of the crew of 170 who went out, only 55 lived to return.

Not many heroes of discovery have the good luck themselves to enjoy the fruits of their discovery. Vasco da Gama was one. His voyage, which finally proved a feasible sea route between West and East, changed the course of both Western and Eastern history. In February 1502 he set out again from Lisbon, this time with a Portuguese squadron to make Calicut into a Portuguese colony. Arriving off the Malabar coast, when he sighted a large dhow, the *Meri,* carrying Muslim pilgrims home from Mecca, he demanded all the treasure on board. When the owners were slow to deliver, the result was recorded by one of his crew. "We took a Mecca ship on board of which were 380 men and many women and children, and we took from it fully 12,000 ducats, and goods worth at least another 10,000. And we burned the ship and all the people on board with gunpowder, on the first day of October." On October 30, Gama, now off Calicut, ordered the Samuri to surrender, and demanded the expulsion of every Muslim from the city. When the Samuri temporized and sent envoys to negotiate peace, Gama replied without ambiguity. He seized a number of traders and fishermen whom he picked up casually in the harbor. He hanged them at once, then cut up their bodies, and tossed hands, feet, and heads into a boat, which he sent ashore with a message in Arabic suggesting that the Samuri use these pieces of his people to make himself a curry. When Gama departed for Lisbon with his cargo of treasure, he left behind in Indian waters five ships commanded by his mother's brother, the first permanent naval force stationed by Europeans in Asiatic waters.

The next sequence of steps to establish an empire in India seemed as obvious as the stages down the west African coast. The first Portuguese viceroy of India, Francisco de Almeida, destroyed the Muslim fleet in 1509. Afonso de Albuquerque, the next Portuguese governor of India, conquered Ormuz, the gateway to the Persian Gulf, in 1507, made Goa capital of the Portuguese possessions in 1510, took Malacca in 1511, and then opened sea trade with Siam, the Moluccas, or Spice Islands, and China. The Portuguese now ruled the Indian Ocean.

The consequences reached around the world. Much Italian splendor had been based on the wealth of the East pouring through Venice and Genoa. Now the traffic in Asiatic treasure—spices, drugs, gems, and silks—would no longer come through the Persian Gulf, the Red Sea and the Levant, but on Portuguese ships around the Cape of Good Hope to Atlantic-facing Europe. The Egyptian sultans had been able to keep the price of pepper high by keeping the consignments down to about two hundred and ten tons per year. So quickly was the effect of the Portuguese sea route felt that by 1503 the price of pepper in Lisbon was only one-fifth of what it was in Venice. The Egyptian-Venetian trade was destroyed. The wealth of Asia, the fabled treasures of the Orient, were flowing west. The new Age of the Sea moved the entrepôts of commerce and civilization from the coasts of a finite body, the closed Mediterranean, the "Sea-in-the-Midst-of-the-Land," to the shores of the open Atlantic and the boundless world-reaching Oceans.

24

Why Not the Arabs?

IF, as it turned out, Africa was a peninsula, if there really was an open sea passage from the Atlantic Ocean to the Indian Ocean, as we are accustomed to say, then obviously there also was a sea passage from the Indian Ocean to the Atlantic Ocean. The Arabs who lived around the western and north-western borders of the Indian Ocean were at least as far advanced in the seafaring sciences—in astronomy, geography, mathematics, and the arts of navigation—as their European contemporaries. Why didn't the Arabs take the sea passage westward?

One answer could be the same as that given by a Boston lady of an old New England family who was asked why she never traveled. "Why should I?" she replied. "I'm already there!" When Vasco da Gama finally arrived on the Malabar coast, as we have seen, he was greeted by Arabs from Tunis. These were members of a considerable Arab community, merchants and shipowners, who already dominated foreign commerce at Calicut. Long before a continuous sea passage was found between West and East, Arabs of North Africa and the Middle East were firmly rooted in India.

Taboos of caste, it seems, prevented Hindus from freely joining in overseas commerce. Some were forbidden by their religion from passing over salt water. Meanwhile, the astonishing expansion of Islam in the generations after Muhammad carried the Muslim empire across the Indus River and into India before the mid-eighth century. Arab traders swarmed into the cities of the Malabar coast.

Muslims from anywhere were remarkably at home wherever they went in Allah's commonwealth. As we have seen, Ibn Battuta, the Marco Polo of the Arab world who had been born in Tangier, during his extensive travels served comfortably as a judge in Delhi and in the Maldive Islands and was sent by an Indian sultan as ambassador to China. The Calicut that Gama saw included a prospering Arab quarter. Arab-operated warehouses and shops were found all over the city, and the Arab community was judged by its own cadis. Hindu rulers remained tolerant of the religion of the merchants who kept their city's commerce flourishing. Many a Hindu family hoped their daughter would become the wife of a rich Arab merchant. It was hardly surprising that the Arabs of Calicut did not welcome the Portuguese intruders.

Seafaring in the Indian Ocean had flourished long before the birth of the Prophet Muhammad. At first the trip from Egypt and the Red Sea to India was made by following along the coast. The sailing traffic was enormously increased when the monsoons were discovered and put to use. A characteristic feature of the Indian Ocean, the monsoon (from Arabic *mausim,* meaning season) is a wind system that reverses direction seasonally. From a planetary perspective, it is a product of the special relations of land, sea, and atmosphere—a result of the differences in heating or cooling of land masses in contrast to that of the oceans. In India and in Southeast Asia, the monsoon blows in one direction at one season and contrariwise at another, and so offers convenient motive power both for going eastward in the Indian Ocean and for returning. *The Periplus of the Erythraean [Indian] Sea* (c. A.D. 80) credited a Greek pilot named Hippalus, who steered ships a century before under the late Ptolemaic kings of Egypt, as "the pilot who by observing the location of the ports and the conditions of the sea, first discovered how to lay his course straight across the ocean." After Hippalus showed how to use the southwest monsoon that blows across the Indian Ocean from June to October, to carry ships from the Red Sea to the shores of India, that wind came to be called the Hippalus.

Under the Roman Empire of Augustus the thriving sea trade between the Red Sea and India reached a hundred and twenty ships a year. Pliny complained, during the reign of Nero, that the empire was being drained of its currency in exchange for the baubles of India. The large hoards of Roman coins found in India prove how widespread that trade must have been.

Arab merchants were a familiar sight in India long before the landward expansion of Islam, but after the Prophet Muhammad the crusading motive was added to the commercial. In the mid-fourteenth century Ibn Battuta noted that Arab merchants were being carried from the Malabar coast to China in Chinese ships. In Canton as early as the ninth century there had been a Muslim community with its own cadi, and we have early records of Muslims as far north as Korea.

From the European perspective we have formed the stereotype that the Arabs have never been enthusiastic or successful seafarers. And the story of the Arabs in the Mediterranean does lend some substance to this notion. Caliph Omar I (581–644), who organized Muslim power and carried on the great landward expansion of the Muslim Empire into Persia and into Egypt, was wary of the sea. His Governor of Syria asked permission to raid Cyprus. "The isles of the Levant," the Governor argued, "are close to the Syrian shores; you might almost hear the barking of the dogs and the cackling of the hens; give me leave to attack them." Omar sought the advice of his wisest general. "The sea is a boundless expanse," General 'Amribn-al-As warned, "whereon great ships look tiny specks; nought but the heavens above and the waters beneath; when calm, the sailor's heart is broken; when tempestuous, his senses reel. Trust it little, fear it much. Man at sea is an insect on a splinter, now engulfed, now scared to death." When Omar forbade the excursion, he expressed the traditional Arab distrust of the sea. In Arabic, you "rode a ship" *(rakaba markab)* as you rode a camel, and when Muslims reached the shore around the Arabian peninsula, they saw the sea as a desert to be crossed en route to raid or to trade. There the northern Arabs rarely felt at home. The seafaring adventures of the ancient Arabs in the Mediterranean were sallies of commerce or of piracy, then not sharply distinguished from each other. They did not build a seafaring empire.

But even in the Mediterranean the Arabs were forced to go to sea. After a Byzantine fleet retook Alexandria (A.D. 645), it was plain that the Muslim

Empire could not do without a navy. Alexandria became their maritime center, a new headquarters for naval training and for building ships with timber brought from Syria. By 655 the Arab fleet at Dhat al-Sawari defeated a Byzantine force of five hundred vessels. According to Arab tradition, the Arabs would have preferred to fight these enemies, too, on land, but the Byzantines preferred the sea. The Arabs, however, maneuvered the encounter into a kind of land battle conducted on shipboard. While Byzantine and Arab vessels were locked together, the Arabs slaughtered their enemies with swords and arrows.

The Arab-Muslim Empire spread landward around the Mediterranean. The Iberian peninsula, where the land of Europe came down to meet the land of Africa, was the part of the west European mainland that came under the Muslim sway. Historians, stirred especially by Henri Pirenne, still debate whether the Mediterranean ever really became a Muslim lake. It was the land-based strength of the Arabs who controlled both ends of the Mediterranean, whether or not they dominated the traffic inside, that shaped the future of seafaring in and from Europe.

With the minor exceptions of the islands of Cyprus, Crete, and Sicily, it was not necessary for the Arabs to cross a sea to pass from one part of their empire to another. If the Arabs of the north, those who settled and expanded around the southern shores of the Mediterranean, had been more like the Romans, more adept and more at home on the sea, less fearful of wide expanses of water, the later history and even the religion of Europe might have been quite different. Alexandria might have become a Muslim Venice. But instead, that once great metropolis, which in its earlier heyday contained a population of 600,000, had only 100,000 in the late ninth century. The caliphs of the ninth and tenth centuries allowed the city to decay. The famous Pharos lighthouse, marking Alexandria's harbor, which had been one of the Seven Wonders of the ancient world, became a ruin. Then even its ruins were appropriately destroyed by an earthquake in the fourteenth century. Arabic thought and Arabic literature looked landward.

But in the Mediterranean, empires were repeatedly won and lost on the water. There the ship was the sword of empire builders. During the centuries when Allah's empire was retreating in the West, the Indian Ocean, an area of nature's turmoil, remained remarkably peaceful. That was where Arab maritime prowess developed freely. The brilliant embodiment of that prowess, Ibn Majid, son and grandson of eminent Arab navigators, who called himself "The Lion of the Sea in Fury," achieved fame as the man who knew most about navigation in the dread Red Sea and in the Indian Ocean. He became a Muslim patron saint of seamen, in whose memory orthodox mariners would recite the first chapter of the Koran, the Fatiha, before venturing out on dangerous waters. Author of thirty-eight works in prose and poetry, he covered every maritime topic of his day. Most useful to Arab navigators was his *Kitab al-Fawa'id,* or Nautical Directory (1490), a compendium of everything then known of nautical science, which included information to guide seamen through the Red Sea and the Indian Ocean. Even today, for some areas, his work is said to be unequaled.

A divine providence must have been watching over Vasco da Gama on his first voyage. By an astonishing coincidence, when on reaching Malindi he finally secured a competent and trustworthy Arab pilot to steer his fleet across the Indian Ocean, it was this very same Ibn Majid. The Portuguese captain did not know how lucky he was. Nor could Ibn Majid have realized, as they sailed into Calicut harbor, that they were enacting one of the majestic ironies in history. The great Arab master of navigation was unwittingly guiding the great European sea captain to a success that meant the defeat of Arab navigation in the Indian Ocean. Later Arab historians have tried to explain away Ibn Majid's role by saying he must have been drunk to confide to Vasco da Gama the information that would guide him safely to his Indian destination.

Once admitted, the Portuguese and their European successors were not to be expelled from that ocean. In the later nineteenth century the Suez Canal would make the passage to India easier than ever for European sailors. By the mid-twentieth century Arabs still sailing from Kuwait and Aden to East Africa and to India seem to have forgotten much that Ibn Majid knew, for they were once again clinging to the coast.

Long before Prince Henry the Navigator had even begun to reach down the west coast of Africa, the Arabs knew the east coast of Africa all the way down to Sofala, opposite the island of Madagascar, and less than a thousand miles north of the Cape. There, at the Mozambique Channel, they found *their* Cape Bojador. Dare not to go beyond! The Koran twice declared that God had separated "the two seas" by a barrier that man could not overcome. Scholars explained that these two enclosed bodies of water were the Mediterranean Sea and the Indian Ocean, including the Red Sea. But the Prophet had also said, "Seek knowledge, even in China." Somehow, during the later Middle Ages, the Arab scholars were less securely imprisoned by their Muslim faith than the European scholars were by their Christian dogmas. Arab scholars proved willing to criticize and even to alter some revered classical texts, including Ptolemy's *Geography.*

The distinctive broadener of Muslim vistas was, as we have seen, the pilgrimage—the duty of every Muslim, man or woman, wherever living, to visit Mecca once before dying. We must recall the narrow orbit at that time of the life of a Scottish or Norwegian or French peasant, who might never journey beyond the nearest fair. But while the pilgrim tradition focused Arab-Muslim travel, it did not encourage exploratory seafaring.

Yet Arab geography flourished. And while medieval European cosmographers reposed in dogmatic slumber, Arab geographers were at home in the works of Ptolemy, which the West kept buried for a thousand years. The Arabs were even beginning to revise Ptolemy, suggesting that the Indian Ocean was not Ptolemy's closed sea but that it actually flowed into the Atlantic. One of the most influential of these pioneer Arab geographers was the versatile Al-Biruni (973–1050?), one of the greatest Muslim scientists of the Middle Ages. He combined precise observation with omnivorous curiosity, and even before he was seventeen he had made an improved device for determining latitude. Though born in central Asia, Al-Biruni became entangled in the far-ranging political intrigues of the rival dynasties of Persia, Turkey, and Iraq. "After I had barely settled down for a few years," he wrote, "I was permitted by the Lord of Time to go back home, but I was compelled to participate in worldly affairs, which excited the envy of fools, but which made the wise pity me." He expressed some of the more advanced Arab speculations about the shape of Africa.

> The Southern Sea commences at China and flows along the shores of India towards the country of the Zendj [Zanzibar]. . . . Navigators have not passed this limit, the reason being that the sea on the north-east penetrates into the land . . . while on the south-west, as if by way of compensation, the continent projects into the sea. . . . Beyond this point, the sea penetrates between the mountains and valleys which alternate with one another. The water is continually set in motion by the ebb and flow of the tide, the waves for ever surging to and fro, so that ships are broken in pieces. This is why the sea is not navigated. But this does not prevent the Southern Sea from communicating with the Ocean through a gap in the mountains along the south coast [of Africa]. One has certain proofs of this communication although no one has been able to confirm it by sight. It is because of this inter-communication that the habitable part of the world has been placed in the centre of a vast area environed on all sides by the sea.

Since this controversial version of the linking of the oceans was known to Ibn Majid, who apparently accepted it, he could not have been unduly surprised to meet Vasco da Gama's fleet at Malindi. Ibn Majid himself was pleased to be able to point out that Al-Biruni's notions and his own had now been proven by the Portuguese, "the experienced ones." Since these "Franks" (a name in the East for all Europeans) had entered the Indian Ocean through *al-madkhal* (the place of entry), the perilous channel between the island of Madagascar and the African coast of Mozambique, Ibn Majid christened that sea path "the passage of the Franks."

For the Arabs, the Mozambique Channel, like Cape Bojador, had accumulated legendary menaces with the passing centuries. *The Arabian Nights* had embroidered the real perils with terrifying tales of a gigantic bird, the ruc or gryphon. Marco Polo reported what his reliable sources told him of Madagascar:

> You must know that this Island lies so far south that ships cannot go further south or visit other Islands in that direction, except this one, and that other of which we have to tell you, called Zanghibar [Zanzibar]. This is because the sea-current runs so strong towards the south that the ships which should attempt it never would get back again. . . .
>
> 'Tis said that in those other Islands to the south, which the ships are unable to visit because this strong current prevents their return, is found the bird *Gryphon,* which appears there at certain seasons. The description given of it is however entirely different from what our stories and pictures make it. For persons who had been there and had seen it told Messer Marco Polo that it was for all the world like an eagle, but one indeed of enormous size; so big in fact that its wings covered an extent of 30 paces, and its quills were 12 paces long, and thick in proportion. And it is so strong that it will seize an elephant in its talons and carry him high into the air, and drop him so that he is smashed to pieces; having so killed him the bird gryphon swoops down on him and eats him at leisure. The people of those isles call the bird *Ruc,* and it has no other name. So I wot not if this be the real gryphon, or if there be another manner of bird as great. But this I can tell you for certain, that they are not half lion and half bird as our stories do relate; but enormous as they be they are fashioned just like an eagle.

The Great Khan, he added, had received the gift of a Ruc's feather "which was stated to measure 90 spans, whilst the quill part was two palms in circumference, a marvellous object!" Oddly enough, the name of the rook in chess, though its modern guise gives no clue, appears originally to have come from the name of this bird.

The technology of Arab shipbuilding in the Indian Ocean in that era before the Portuguese arrived was a curious combination of strengths and weaknesses. The lateen sail, which the Arabs brought to the Mediterranean, by its adeptness at sailing into the wind had made the Portuguese ventures possible. The Arabs also pioneered in developing the stern rudder, which made any ship more maneuverable. They were skilled at using the stars for navigation. "He it is," said the Koran, "who hath appointed for you the stars that ye guide yourselves thereby in the darkness of land and sea; we have made the signs distinct for a people who have knowledge."

For reasons still unclear, instead of using nails in their ships, Arabs stitched together planks with cords made from coconut husks. Ships held together in this fashion would not long survive the buffeting of winds or the scraping of rocks. Why did they build their ships in this way? There was a widespread legend that magnetic rocks, or lodestones, in the sea would pull out iron fastenings and so drag apart ships held together by nails. And the high cost and scarcity of nails must also have had something to do with it. Once a distinctive style was adopted, naturally conservative seamen made it into a firm tradition.

Certain special features of the Arabian peninsula, homeland of the Arabs and of Islam, remind us of the difficulties faced by their seamen. Arabia possesses almost none of the naval stores—neither wood nor resin nor iron nor textiles—required for shipbuilding. To a man of the sea, the geography of Arabia could hardly have been more uncongenial. There were no navigable rivers, few good harbors, nor any populous or hospitable hinterland. Coral reefs surrounding the coasts produced wrecks to encourage pirates, from whom there was no convenient refuge. There was no easy source of fresh water. And the menacing northerly winds came down without respite the year round.

All these features of Arab lands and Arab civilization help us understand why they were not inclined to take their sailing ships around Africa and up the west coast to Europe. The best explanation, perhaps, is the most obvious. Why organize a continuous venture into the unknown? The modern organized exploring enterprise initiated by Prince Henry the Navigator was without known precedent. Seamen, eminently practical men, had habitually embarked because they had a cargo or passengers for a particular destination. Or they arrived to pick up a particular cargo somewhere. The seaman, like the landman, generally does not go in quest of the unknown, nor to confirm some concept of the earth or the oceans, but, as E. G. R. Taylor explains, "goes to sea like a man to his office, along a set route for a set purpose, his livelihood." Just as, on land, people dreaded mountaintops and preferred familiar paths, so too there were familiar paths on the sea.

The Arabs in the Indian Ocean were "already there." Both in the East and in the West. Why would they have wanted to go by sea to Portugal or northern Europe? Muslims were already just across the Straits of Gibraltar from the Christian world. Their domain already included the rich tropical variety of plants and animals and minerals and incense-laden ways. What the Arab world had to gain from the Europeans had already been tested and tried in the Iberian peninsula. Their encounters with the crusaders in the Middle East seemed to promise only a vast reservoir of infidels in need of conversion.

25

The Chinese Reach Out

WHEN Prince Henry the Navigator was sending his ships inching down the west coast of Africa, on the other side of the planet Chinese navigators possessed a navy unparalleled in numbers, in skills, and in technology. Their grand fleet had already sailed beyond the China Sea and around the Indian Ocean, reaching down the east coast of Africa to the very tip of the Dark Continent. But while the exploits of Prince Henry's ships were a prologue to seafaring voyages that would discover a whole New World and circumnavigate the globe, the grander Chinese expeditions of the same era were a dead end. They prefaced the catastrophic withdrawal of the Chinese into their own borders, with consequences we still see today.

What made the Great Withdrawal of 1433 so dramatic was that the Chinese seafaring outreach had been so spectacular. The hero, whose name became a byword for Chinese seafaring power, the designer and commander of the most remarkable of these wide-ranging ventures, was Chêng Ho, the Admiral of the Triple Treasure, commonly known as the Three-Jewel Eunuch, perhaps from the three precious elements of Buddhism—the Buddha, the Dharma, and the Sangha—or from the gems he gave as gifts and received as tribute. The fact that Chêng Ho was a eunuch, as we shall see, helps explain how he managed to develop these grand adventures, and also why they were so abruptly concluded.

In the West, *castrati* are known to history not for their political influence but mainly for their vocal peculiarities. In addition to removing the power to procreate, the castrating operation retards the deepening of the voice, and leaves the eunuch a soprano. From Constantinople the practice spread of using eunuchs in choirs. In the eighteenth century Handel's operas featured *castrati,* who then began to dominate the opera scene, sometimes requiring composers to write in parts especially for them. Until the early nineteenth century *castrati* sang in the papal choir in Rome. The Italian practice of castrating boys to prepare them to become adult male sopranos did not end till the reign of Pope Leo XIII in the late nineteenth century.

Religious dogma led some pious men to emasculate themselves to avoid sexual sin or temptation. "There be eunuchs," preached the Gospel according to St. Matthew (19:12), "which have made themselves eunuchs for the kingdom of heaven's sake. He that is able to receive it, let him receive it." The Church Father Origen (A.D. 185?–254) followed this advice, and after him there grew up a whole sect who castrated themselves the better to enter the kingdom of heaven. Such a sect persisted into twentieth-century Russia.

Wherever eunuchs have been able to exercise political influence, it has been a symptom both of the secluded position of women and of despotic government. The surgical act that deprived the eunuch of his sexual powers qualified him to be "keeper of the couch" ("eunuch" is from Greek "bed-watcher"). Where the monarch maintained a harem of wives and concubines, only members of his immediate family were permitted to reside within the palace precincts. The eunuchs detailed to attend on the women of the harem, being no menace to the purity of the imperial line or to the chastity of the royal consorts, were an exception. They became a privileged class. Knowledge of the daily habits and personal tastes of the emperors gave eunuchs a peculiar opportunity to anticipate the monarch's whims. In the arbitrary governments of the East this meant an opportunity to seize power. The power of eunuchs was conspicuous under the Byzantine emperors, when the brilliant Justinian I, codifier and preserver of Roman law, made the eunuch Narses (478?–573?) one of his generals. This choice was justified when Narses led the Byzantine armies in Italy to drive out the Goths, Alamanni, and Franks (553). The Ottoman sultans also put their eunuchs in commanding positions. The institutionalized influence of eunuchs over their royal masters became so familiar in Egypt that the term "eunuch" came to be used for any officer of the court, whether castrated or not. At times people were tempted to call their government a "Eunarchy."

Chinese imperial institutions proved peculiarly favorable to the power of eunuchs. The rigid court etiquette, as early as Han times in the reign (126–144) of Han Shun Ti, confined the emperor to his palace and the palace garden, like that where the Heavenly Clockwork would be displayed. On the rare occasions when he left those confines the emperor's advance men cleared the road of the populace and he was protected from the public gaze. Even the ministers of state could not converse with him familiarly. They saw him only at formal audiences when they were expected to address him through other officials also "below the steps" who stood nearer to the emperor's high throne. The term of respectful address to the emperor in Chinese *(Chieh Hsia),* the equivalent for the Western term "Your Majesty," meant "from below the steps." By contrast, the favored imperial eunuchs living inside the palace had daily converse with the emperor. Ministers of

state could only offer formal reports and written memorials, but the eunuchs could whisper in the emperor's ear.

If the emperor had grown up outside the palace and came to the throne
only as an adult, the eunuchs were less likely to exert political power. But
again and again in later Chinese history the heir, born in the palace, grew
up under constant tutelage of the eunuchs. When such an emperor, still a
child, succeeded to the throne, the imperial eunuchs would control the
child-emperor's decisions or those of the empress-regent. These eunuchs,
who first became influential under the late Han emperors, were usually
drawn from the lowest levels of society. Having no future outside the palace,
they had no reason not to merit their reputation for being mercenary and
unscrupulous. They collected bribes, distributed honors, and meted out the
punishments of the torture chamber.

But gradually a new scholarly class, disciples and interpreters of Confucius, also recruited from the poorer classes, became organized into a civil
service. The lines between the pro- and anti-eunuch forces were now clearly
drawn. The scholar-bureaucrats feared, envied, and despised the eunuchs
who excelled them in power, even without being able to recite a single
passage from the Confucian classics. The military class, led by generals who
had risen by their competence, had reasons of their own to feel contempt
for these effeminate imperial confidants of the bedchamber who had never
fought a battle. The scholars and the generals somehow never managed to
combine against those advisers who lived where adversaries could not penetrate.

One of these strategically placed eunuchs was Chêng Ho. For reasons
that are themselves a part of our story, we know far less about Chêng Ho
than about Henry the Navigator, Vasco da Gama, Columbus, Vespucci, or
Magellan, who were his European counterparts. We do know that he was
born a Muslim, but we know little else, except that he probably was a person
of low birth who came from the south-China province of Yunnan.

The setting for Chêng Ho's exploits on the sea had been prepared a
century earlier, when the last Mongol emperor was driven from Peking by
the upstart "Chinese Napoleon." Chu Yüan-chang, the clever son of a poor
farm laborer, was born in the eastern province of Anhwei, where his whole
family died in an epidemic when he was only seventeen. He entered a
Buddhist monastery, but when he was twenty-five, he took off his saffron
robes and went out to lead his province against the Mongol intruders. After
thirteen years of struggle his forces finally occupied Peking in 1368. Still only
forty, he proclaimed himself the first emperor of the new Ming dynasty.
While keeping his own capital in Nanking, the Southern Capital, he sent
officials from the north to administer the south and officials from the south
to administer the north, hoping in that way to draw the nation together.

During his thirty-year rule he managed to consolidate the nation, which had long been split by Mongol domination of the north.

The Emperor Chu Yüan-chang never ceased to resent reference to his lowly origins or his early years as a Buddhist monk. When two unlucky Confucian scholars in a congratulatory message made the mistake of using the word "birth" *(sheng)* which might have been misconstrued to be a pun for the word "monk" *(seng),* they were sentenced to death. No religious crusader, he surrounded himself with Buddhist monks at the same time that he promoted Confucian academies and Confucian rituals.

As he grew older he suspected plots everywhere. It became a capital crime even to petition against any of his policies, and when he thought he noticed a rebellious spirit in Nanking he executed fifteen thousand people at one swoop.

The person who filled the post of prime minister, though appointed by the emperor, was usually promoted upward from the bureaucracy. Since the prime ministers had come from outside the palace circle, often having risen from the common people, they reached their position by academic attainments and personal competence. The prime minister thus was a wholesome check on the whims of the emperor and on the influence of the palace clique. But Chu Yüan-chang consolidated his personal rule by putting an end to the post of prime minister. When he suspected his prime minister of treason, he summarily abolished the post, providing that "anyone who dares to petition for its reestablishment will be ordered to perish immediately together with the rest of his family." This inevitably increased the power of the others who had the Emperor's ear, who of course were the eunuchs.

At the same time that this first Ming emperor took steps to lessen the powers of the bureaucracy he inaugurated other policies which hardened into hatred the suspicions that the regular civil servants had long felt against the eunuchs. In a peculiarly contemptuous gesture, he defied the ancient Chinese tradition, reinforced by Confucian teachings, that a gentleman or scholar should never be humiliated. If a scholar failed in his duty he could be ordered to his death, and was expected to commit suicide, but he was not to be publicly degraded. But the parvenu Chu Yüan-chang seemed to enjoy humiliating his intellectual superiors. At his court he made a public ritual of the flogging of high officials who seemed independent-minded or insufficiently sycophantic. All the court were required to be present in their ceremonial robes to watch their colleague stripped and beaten to death by scores of blows with a wooden rod. The Emperor's defenders argued that this practice actually did reduce bribery among the bureaucracy. It was the eunuchs, with their intimate access to the Emperor, who controlled and assigned these ritual floggings.

After the thirty-year rule of Chu Yüan-chang, and the brief reign of his reformist son who championed the Confucian way, a palace revolt was engineered by Yung Lo (1359–1424), the Emperor's uncle, with the assistance of the court eunuchs. Just as Kublai Khan had tried to build a Chinese empire for the Mongols, so Yung Lo now set out to encompass the Mongol Empire for the Chinese. In 1409 he boldly moved his headquarters northward from Nanking, the Southern Capital, to Peking, the Northern Capital, on the very borders of Mongol power, defiantly close to the Great Wall. He reshaped Peking into a rectangle-dominated imperial capital with a "Violet-purple Forbidden City," the abode of the Emperor, at its very center, embellished by a splendid ensemble of palaces, terraces, artificial lakes and hills, gardens, and vistas of shrubs and flowers brought from the far corners of the empire.

The megalomaniac Yung Lo soon decided to send out naval expeditions with messages of his grandeur into all the surrounding seas. For the command he chose Chêng Ho. These expeditions (1405–33), the vastest until then seen on our planet, enlisted some thirty-seven thousand in their crews, in flotillas of as many as three hundred and seventeen ships. Vessels ranged in size from the largest, the Treasure Ship carrying nine masts, 444 feet long with a beam of 180 feet, down through the ranks of Horse Ship, Supply Ship, Billet Ship, to the smallest, the Combat Ship, which carried five masts and measured 180 feet by 68 feet. Ibn Battuta, a century earlier, and Nicolo de' Conti, who was a passenger on a Chinese ship about this time, were both astonished that these vessels were so much larger than any they had seen in the West.

Westerners also noted the remarkable construction that prevented water in one part of the hull from flooding the whole ship. Bulkheads, a series of upright partitions dividing the ship's hold into compartments to prevent spread of leakage or fire, though then novel to Europe were an old story in China. They were probably suggested by the septa, the transverse membranes of the bamboo. Already in ancient pre-Han China, this design gave the strength and resiliency that made possible the multistoried ships which dazzled visitors from abroad with their high overhanging stern gallery, from which was suspended a gargantuan rudder with a blade of 450 square feet. These were only a few of the remarkable features of Chêng Ho's navy. Of course he used the compass, and perhaps other directional instruments, along with elaborate navigational charts showing detailed compass bearings. Though the Chinese had long been using the grid for charting the land, Chêng Ho's maps yield no evidence that they used latitude and longitude at sea.

Chêng Ho took his navy—we must not call it an armada, for it was not designed for battle—to nearly every inhabited land bordering the China Sea and the Indian Ocean. For at least five hundred years, since the glorious renaissance in the T'ang dynasty, the Chinese had been trading overseas with the Islamic world. To their own maps they had added the Nile, the Sudan, Zanzibar, and even some south Mediterranean places. Perhaps this knowledge came indirectly from Arab traders, but recent finds of T'ang and Sung coins and porcelain all along the African coast from Somalia to Zanzibar suggest that the Chinese themselves were there. Chêng Ho's expeditions were so well equipped with Chinese speaking the languages of these places that they must have had a long experience of dealing overseas.

Seven expeditions reached farther and farther west. The first, which set out in 1405, visited Java and Sumatra, then Ceylon and Calicut. The following expeditions reached Siam, made Malacca a headquarters for visiting the East Indies, then went on to Bengal, to the Maldive Islands and as far west as the Persian sultanate of Ormuz at the entrance to the Persian Gulf. Pacific squadrons visited Ryukyu and Brunei, while others went farther westward from Ormuz to Aden at the mouth of the Red Sea, then southwestward down the African coast to Mogadishu in Somaliland, to Malindi north of Mombasa, and to the Zanzibar coast. The adventuring sixth expedition (1421–22), within only two years, visited thirty-six states stretching the full width of the Indian Ocean from Borneo to Zanzibar. It was an ill omen for the grand enterprise when Chêng Ho's patron, the emperor Yung Lo, died in 1424. His successor supported the anti-maritime party by stopping the voyage planned for that year.

Chêng Ho's voyages then became pawns of the imperial succession. After the anti-maritime emperor's short reign his successor, a maritime enthusiast, supported the seventh and most expansive of all the voyages. Carrying 27,550 officers and men, this two-year expedition ranging farther than any of its predecessors, on its return in 1433 had established diplomatic or tributary relations with twenty realms and sultanates from Java on the east through the Nicobar Islands all the way to Mecca on the northwest and far down the east coast of Africa. Now these distant peoples, who for a thousand years had known small Chinese junks in their waters, were overwhelmed by many-storied ships, vaster than any seen before or any that the Portuguese would bring their way. They must have been puzzled that so potent a navy should pretend to have no warlike mission.

The purpose of these Grand Treasure Fleets is difficult for the Western mind to grasp. Chêng Ho's interests and aims were as far apart as the poles from those of the European fleets of the Age of Discovery. The Portuguese voyaging down the west African coast and around the Cape to India hoped to increase their nation's wealth, to secure the staples and the luxuries of the East, and to convert the heathen to Christianity. For his trading goods, as we have seen, Vasco da Gama brought bolts of striped cloth, washbasins, strings of beads, and lumps of sugar—which made the Samuri of Calicut laugh in contempt. The "goods" the Portuguese extracted included slaves by the thousands, who from Angola alone before the mid-seventeenth century numbered one and a third million. With ships heavily armed for battle, they were uninhibited in using terror. We have seen how Vasco da Gama cut up the bodies of casually captured traders and fishermen, and sent a basketful of their hands, feet and heads to the Samuri of Calicut simply to persuade him into a quick surrender. Once in power, the Portuguese governed their India in the same spirit. When Viceroy Almeida was suspicious of a messenger who came under a safe-conduct to see him, he tore out the messenger's eyes. Viceroy Albuquerque subdued the peoples along the Arabian coast by cutting off the noses of their women and the hands of their men. Portuguese ships sailing into remote harbors for the first time would display the corpses of recent captives hanging from the yardarms to show that they meant business.

Chêng Ho's navy came from another world. The purpose of his vast, costly, and far-ranging expeditions was not to collect treasure or trade or convert or conquer or gather scientific information. Few naval expeditions in recent history have had any other purpose. Chinese chroniclers repeated the report that Chêng Ho was first sent out to track down Emperor Yung Lo's nephew, whose throne Yung Lo had usurped, who had fled from Nanking and was said to be wandering abroad. But other, larger motives developed as the expeditions proceeded.

The voyages became an institution in themselves, designed to display the splendor and power of the new Ming dynasty. And the voyages proved that ritualized and nonviolent techniques of persuasion could extract tribute from remote states. The Chinese would not establish their own permanent bases within the tributary states, but instead hoped to make "the whole world" into voluntary admirers of the one and only center of civilization.

With this in mind, the Chinese navy dared not loot the states that it visited. Chêng Ho would not seek slaves or gold or silver or spices. Nothing would suggest that the Chinese needed what other nations had. While peoples of Asia would be struck by the Portuguese power to seize, the Chinese would impress by their power to give. They would unwittingly dramatize the Christian axiom that it was nobler to give than to receive. Instead of shoddy trinkets and childish gewgaws, they offered treasures of the finest craftsmanship. European expeditions to Asia revealed how desperately Europeans wanted the peculiar products of the East, but the prodigal gestures of Chinese expeditions would show how content the Chinese were with what they already had. This "tribute" system, which then dominated Chinese relations with other Asian states, was bafflingly different from any to which the Western mind has become accustomed. A state bringing tribute to China was not submitting to a conqueror. Rather, it was acknowledging that China, by definition the *only* truly civilized state, was beyond the need for assistance. Tributes therefore were less economic than symbolic. A tributary state declared its willingness to enjoy the benevolence of Chinese culture, and in return China demonstrated "the generosity and abundance of the Central Kingdom." No wonder the Chinese found it hard to imagine a community of sovereign nations! Only China was truly sovereign, for only China was worthy of sovereignty. The corrosive consequences of this frame of mind persisted into the twentieth century.

During the days of Chêng Ho the Chinese practiced what they preached, with costly consequences. The lopsided logic of the tributary system required China to pay out more than China received. Every new tributary state worsened the imbalance of Chinese trade. The accidents of history that cast Chinese public relations in this curious frame help explain why Chinese communication with the outside world was stultified for centuries to come. Meanwhile, the tributary system became a blind for the burgeoning commercial demands of other nations. Foreign sovereigns were not reticent to receive "gifts" from the Chinese emperor which were really bribes to encourage them to give peacefully what might have been taken violently. The Chinese government became the cat's-paw for foreign powers. Over the centuries the weakening Chinese government continued to receive foreign traders in the flattering guise of "tributaries." But in the time of Chêng Ho the Chinese Emperor managed, at least for a while, to give substance to his assertion that the Central Kingdom needed nothing from anybody and had nothing to learn from anybody.

Just as the Chinese were not traders or conquerors, so they were not crusaders. The Portuguese brought to Asia a peculiarly Western intolerance along with a ruthless determination to convert the heathen. Muslims, Buddhists, Hindus, and heretic Christians became targets of their proselytizing and their persecution. In 1560, when the Portuguese actually established the Holy Inquisition in Goa, they inaugurated a pious reign of terror enforced by the logic of the torture chamber.

The Chinese had quite another view of religion, a tradition of live-and-let-live. Tolerance is too weak a word for their complaisant pluralism. Chêng Ho's forces were not only unwilling to persecute for God's sake, but wherever they went they actually spent their resources to support whatever religions the people there professed.

Visible witness of this generous spirit survives in the town of Galle on the southwestern coast of Ceylon on the Indian Ocean. There an upright stone slab, inscribed in three languages—Chinese, Tamil, and Persian—and dated

1409, records a visit by Chêng Ho's fleet. The Chinese version is translated as follows:

> His Imperial Majesty, Emperor of the Great Ming, has despatched the Grand Eunuchs Chêng Ho, Wang Chhing-Lien and others, to set forth his utterances before the Lord Buddha, the World-Honoured One, as herein follows.
>
> Deeply do we reverence Thee, Merciful and Honoured One, of Bright perfection, wide-embracing, whose Way and virtue passes all understanding, whose Law pervades all human relations, and the years of whose great *kalpa* rival the river-sands in number; Thou whose controlling influence ennobles and converts, inspiring acts of love and giving intelligent insight (into the nature of this vale of tears); Thou whose mysterious response is limitless! The temples and monasteries of Ceylon's mountainous isle, lying in the southern ocean far, are imbued and enlightened by Thy miraculously responsive power.
>
> Of late we have despatched missions to announce our Mandate to foreign nations, and during their journeys over the oceans they have been favoured with the blessing of Thy beneficent protection. They have escaped disaster or misfortune, journeying in safety to and fro, ever guided by Thy great virtue.
>
> Wherefore according to the Rites we bestow offerings in recompense, and do now reverently present before the Lord Buddha, the World-Honoured One, oblations of gold and silver, gold-embroidered jewelled banners of variegated silk, incense-burners and flower-vases, silks of many colours in lining and exterior, lamps and candles, with other gifts, in order to manifest the high honour of the Lord Buddha. May His light shine upon the donors.

To prove that Chêng Ho's tribute had substance, the inscription then itemizes the presents he offered to the Lord Buddha: 1,000 pieces of gold, 5,000 pieces of silver, 100 rolls of silk, 2,500 catties (1 catty equaled 1⅓ pounds) of perfumed oil, and various gilded and lacquered bronze ecclesiastical ornaments.

Since this same slab also bore inscriptions in Tamil, the language of south India and Ceylon, and in Persian, historians long assumed that, as on the Rosetta Stone, all versions carried the same message. The surprising facts proved this was only a provincial Western assumption. For the Tamil version, in its turn, used phrases no less extravagant than those lavished on the Lord Buddha, to proclaim the Chinese emperor's adoration of the god Tenavarai-nayanaŕ, an incarnation of the Hindu god Vishnu, and then in turn the Persian version similarly glorified Allah and the Muslim saints. Each of these indiscriminate paeans appended a list of the lavish presents for whatever God they named. And all sets of presents were quite the same as those offered to the Lord Buddha. Chêng Ho's fleet, bringing this stele with them from China, came well equipped to smother with equally extravagant generosity each of the three competing religions of Ceylon.

26

An Empire without Wants

THE navigating powers in the West would never be satisfied with mere ritual recognition. From earliest times they went in search of whatever they lacked. For the perfumes of Araby, the silks of China, and the spices of India, ships of the ancient Roman Empire plied the Indian Ocean. The best Roman cookbooks required pepper for nearly every recipe. As the satirist Persius (A.D. 34–62) observed:

> The greedy merchants, led by lucre, run
> To the parch'd Indies and the rising sun;
> From thence hot Pepper and rich Drugs they bear,
> Bart'ring for Spices their Italian ware . . .

Roman coins were scattered over Asia and the gold treasures of the Han dynasty found their way to Rome.

By the end of the fifteenth century, when the Portuguese led the way to Asian seas, pepper was no longer a luxury table condiment but a staple of the European kitchen. The need for pepper was a by-product of the European system of animal husbandry. Without a satisfactory winter fodder, which would not be developed yet for several centuries, European farmers could keep over the winter only the few animals needed for draft and for reproduction. Meat from the others, which had to be killed, was generally preserved by "salting"—a process that required large quantities of pepper, in addition to salt, to inhibit the unpalatable effects of the salt itself.

In the early nineteenth century, when the British Empire's silver was being drained East to pay for silk, tea, and lacquerware, the British East India Company shrewdly introduced opium, which they could import to China from India and elsewhere, as a new medium of exchange. While solving their foreign-exchange problem, they set the stage for the Opium War (1839–42), which brought the decisive foreign occupation of China. But back in the age of Chêng Ho in the early Ming dynasty, the Chinese had no such wants. Characteristic European products like woolen cloth and wine also held little charm for them. The proclamations that Chêng Ho carried to foreign potentates boasted that the rest of the world had little to give except their awe and their friendly alliance.

It was not asceticism but complacency that stultified Chinese exploration. Even while they made it a crime to reach out for foreign products, they expressed supreme confidence in their natural immunity to enlarged desires. A seventeenth-century Chinese treatise on navigation boasted:

> Coming into contact with barbarian peoples you have nothing more to fear than touching the left horn of a snail. The only things one should really be anxious about are the means of mastery of the waves of the sea—and, worst of all dangers, the minds of those avid for profit and greedy of gain.

For long centuries the Chinese obstinately resisted alien desires, a contagion from the West. When the first British diplomatic representative, Lord Macartney, arrived in Peking to open trade with China, the response of the Manchu emperor was hardly encouraging. "There is nothing we lack," the Emperor observed in 1793, "as your principal envoy and others have themselves observed. We have never set much store on strange or indigenous objects, nor do we need any more of your country's manufactures."

Since by definition the Central Kingdom would not be interested in other countries' staple products, Chinese interest would have to focus on rarities and curiosities. These did not shape the economy but brightened the imperial cabinet and the imperial zoo. During the reign (A.D. 8–23) of Emperor Wang Mang the capital was delighted when a tributary state brought a live rhinoceros. Chinese tales from Madagascar elaborated reports of the enormous *p'êng* bird, or gryphon, which Marco Polo said could consume an elephant. And now, in the wake of Chêng Ho's expeditions, lions, tigers, oryxes, nilgais, zebras, and ostriches were delivered by foreign ambassadors to the imperial zoo.

The spirit of Chêng Ho's expeditions was nowhere better summarized than in the reception accorded one of the most spectacular of these imported zoological curiosities. On September 20, 1414, the tribute that arrived from Bengal for the Emperor was an animal never before seen in China, a giraffe. No other foreign product—animal, vegetable, or mineral—had made so great a stir. The immediate response of the court was not to exclaim over the wonders of a country where such strange beasts grew. Instead the occasion exercised the Chinese capacities to see the world in the mirror of China. The giraffe brought an orgy of self-congratulation, ingeniously nourished by all the rich resources of Chinese folklore, religion, poetry, and chauvinism.

The giraffe, as we all know, is an animal of extraordinary and charming design. To Western eyes, even in the sixteenth century, such an animal lent romance to its native habitat. But this was not the view of the Ming dynasty Chinese, the Board of Rites, the Imperial Academy, and other courtiers. Some convenient coincidences confirmed their interesting conclusions. In the language of Somali, the east African country where giraffes come from, the giraffe is called *girin*. To the Chinese this word sounded very like *k'i-lin* (modern pronunciation, *ch'i-lin*), their name for a legendary animal resembling the Western unicorn. The graceful, pure white unicorn symbolizing virginity, and often associated with the Virgin Mary, was featured on hunting scenes in tapestries of the late Middle Ages and the Renaissance.

In Chinese folklore the unicorn had a wider, even a cosmic significance. The appearance of a *k'i-lin* was more than a good omen. It showed Heaven's favor and proved the virtue of the ruling emperor. Under a perfect regime the cosmic forces would manifest their surplus energy in the creation of extraordinary beings like dragons and *k'i-lin*'s with wonderful beneficent powers. There seemed to be a striking resemblance between the reputed form of a *k'i-lin,* which had "the body of a deer and the tail of an ox," ate only herbs, and harmed no living being, and everything that was known of the giraffe. When Chêng Ho and the other eunuchs saw the giraffe, they had no doubt that it must be a *k'i-lin.* What a grand opportunity to flatter the Emperor! When they learned that the giraffe was not a native of Bengal, whose tributary ruler had sent it, they managed to persuade the African country of Malindi, the native habitat of giraffes, to become a tributary. The King of Malindi sent another giraffe, which arrived the following year. It was not the appeal of slaves, of gold, or of silver, but the charm of the giraffe that drew Chêng Ho's later expeditions to Malindi and the far reaches of the eastern coast of Africa.

In the year before the giraffe arrived at court, there had been a number of other favorable omens. When the first giraffe arrived at court from Bengal, the eunuchs and other court flatterers insisted on making the giraffe a proof and a symbol of the Emperor's cosmic perfection, but the Emperor, with affected humility, brushed off their sycophancy. Declining to receive a Memorial of Congratulations from the Board of Rites, he observed that "even without *K'i-lins* there is nothing that hinders good government." But he finally gave in, and when the second giraffe arrived from Malindi, the Emperor in full panoply greeted the happy omen at the city gate. There he modestly acknowledged after all that the *k'i-lin* did certify the "abundant virtue" of the Emperor's father and the devotion of his ministers, and signaled that they continued to "cling to virtue."

So wonderful an event gave the Imperial Academy an opportunity to express their adulation. One of their addresses reads as follows:

> Respectfully I consider that Your Majesty succeeded to the Emperor T'ai-tsu's Grand Heritage and that Your virtue transforms [the world] and causes the Three Luminaries to follow their regular course and all living souls to perform their duty. Consequently a Tsou-yü [vegetarian tiger] has appeared, Wonderful Ears of grain are produced, Sweet Dew has descended, the Yellow River has been Clear and Savoury Springs have gushed forth. All the creatures that spell good fortune arrive. In the 9th month of the year *chia-wu* of the Yung-lo period (1414) a K'i-lin came from the country of Bengal and was formally presented as tribute to the Court. The ministers and the people all gathered to gaze at it and their joy knows no end. I, Your servant, have heard that, when a Sage possesses the virtue of the utmost benevolence so that he illuminates the darkest places, then a K'i-lin appears. This shows that Your Majesty's virtue equals that of Heaven; its merciful blessings have spread far and wide so that its harmonious vapours have emanated a K'i-lin, as an endless bliss to the state for a myriad myriad years. I, Your servant, joining the throng, behold respectfully this omen of good fortune and kneeling down a hundred times and knocking my head on the ground I present a hymn of praise as follows.
>
> Oh how glorious is the Sacred Emperor who excels both in literary and military virtues.
>
> Who has succeeded to the Precious Throne and has accomplished Perfect Order and imitated the Ancients!

After a lengthy paean to the Emperor's perfection, there follows a paean to the giraffe:

> Truly was produced a K'i-lin whose shape was high 15 feet,
> With the body of a deer and the tail of an ox, and a fleshy boneless horn,
> With luminous spots like a red cloud or a purple mist.
> Its hoofs do not tread on living beings and in its wanderings it carefully selects its ground,
> It walks in stately fashion and in its motion it observes a rhythm,
> Its harmonious voice sounds like a bell or a musical tube.
> Gentle is this animal that in all antiquity has been seen but once,
> The manifestation of its divine spirit rises up to Heaven's abode.

The world's curiosities had become mere symptoms of China's virtue. So was revealed a Chinese Wall of the Mind against the lessons of the rest of the planet. The Emperor Yung Lo, celebrated by the appearance of the *k'i-lin,* it was said, received more missions from foreign countries than any other emperor in Chinese history. But the Chinese had developed a traditional immunity to the world experience. Confucian teachings would accommodate and sequester the most astonishing novelties.

No less remarkable than Chêng Ho's far-flung naval enterprises themselves was the suddenness with which they came to an end. Had this Chinese Columbus been followed by a procession of Chinese Vespuccis, Balboas, Magellans, Cabots, Corteses, and Pizarros, then the history of the world might have been quite different. But Chêng Ho had no successor, and Chinese naval activity abroad came to a sudden halt. Energies formerly spent sending out expeditions were all at once spent enforcing withdrawal. The Europeans' race for colonies and their quest for *terra incognita* had no counterpart in modern Chinese history. And the exploring spirit remained alien to China.

Chinese reclusiveness was an old story. The Great Wall, which goes back to antiquity, the third century B.C., was given its present form in the Ming dynasty, in the age of Chêng Ho. There is nothing like it in scale or in chronological continuity anywhere else in the world. The spirit of the Great Wall was expressed in countless other ways. One was the Great Withdrawal, when the emperor forbade his subjects to go abroad. Chinese seen outside their country were there illegally, and indiscreet travelers were punished with beheading. Chêng Ho's grand seventh voyage was China's last. His return home in 1433 marked an end to his country's organized seafaring adventures. An imperial edict in that year, and others that followed (1449, 1452) imposed increasingly savage punishments on Chinese who ventured abroad.

Of course there were practical reasons for the Great Withdrawal. Adding and maintaining tributary states was costly. For, as we have seen, the burden of the Chinese tribute system lay mostly on the "receiver." Impressing so many countries at such a distance was a heavy expense with meager economic return. Was so costly an ego trip for the nation, for the Emperor, and for his eunuchs really necessary? If China really was the all-perfect Center, was not such expensive reassurance superfluous?

Opposition to Chêng Ho's exploits was only another skirmish in the age-old battle of the Confucian bureaucrats against the court eunuchs. The centralized bureaucracy, dominated by scholars in the Confucian tradition, had been one of the most precocious of Chinese achievements. The bureaucrats sensibly argued that the imperial treasure be spent on water-conservation projects to help farmers, on granary projects to forestall famine, or on canals to improve internal communication, and not on pompous and reckless maritime adventures. What had these brought in, besides a few precious gems, and useless curiosities like rhinoceroses and giraffes?

Some incidental advantages did come from opening communications with all the countries around the Indian Ocean and the China Sea. But the Chinese balance of trade remained adverse, and when a drastic currency

depreciation brought paper money down to 0.1 percent of face value, overseas trade relations could be kept up only by exporting gold and silver.
Meanwhile, the thousand-mile-long Grand Canal reaching from Tientsin in
the north to Hangchow in the south, begun two thousand years earlier, was
perfected into a full-capacity, all-season seaway. Canal shipping displaced
the seafaring vessels formerly needed to carry food from one part of the
country to another, and maritime transport of grain was abolished.

At the same time, threats from Mongols and Tartars on the northwest
frontiers required heavy military expenditures. The 1,500-mile-long Great
Wall had to be repaired, and it was rebuilt into its present shape. Within
fifteen years after Chêng Ho's return from his last voyage, the same Chinese
emperor who had suppressed the Grand Treasure Fleets would be captured
by Mongol and Tartar armies. By 1474 the main fleet of 400 warships had
declined to 140. Shipyards disintegrated, sailors deserted, and shipwrights,
fearing to become accomplices in the crime of seafaring, were hard to find.
The ban on foreign maritime ventures was extended to include coastal
shipping. Within a few years "there was not an inch of planking on the sea."
Within a century—the century of Henry the Navigator, when European
conquistadores and circumnavigators were reaching across the oceans and
around the world—the Chinese were perfecting laws and organizing officials to suppress all seafaring. By 1500 it had been made a capital offense
even to build a seagoing junk with more than two masts. In 1525 coastal
officials were ordered to destroy all such ships and to arrest mariners who
continued to sail in them. In 1551 the crime of espionage was redefined to
include all who went to sea in multiple-masted ships, even if they went only
to trade. The party of anti-maritime bureaucrats had triumphed. China
turned back on herself.

The Chinese had long since developed their own version of the *oikou-
mene,* the habitable world, which put them at the center. They were their
own Jerusalem. Since the Ming emperors were the Sons of Heaven, they
were by definition supreme rulers and superiors of all other people on earth.
While other peoples would exclude foreigners because they were not of their
tribe, the Chinese automatically incorporated the rest of the world into
theirs, in the role of barbarian acolytes. How natural that all non-Chinese
peoples should pay tribute! And how obvious that the Chinese had little to
gain overseas! Commerce abroad was beneath the needs of the Sons of
Heaven.

When Europeans were sailing out with enthusiasm and high hopes, land-
bound China was sealing her borders. Within her physical and intellectual
Great Wall, she avoided encounter with the unexpected. The unit of Chinese geographic description had long been the *kuo,* or "country," an inhabited land under an established government. And only such a government

could be a tributary to the Sons of Heaven. The Chinese therefore showed little interest in lands uninhabited or out of reach. Confucian orthodoxy since the second century had confirmed their inward emphasis. Why should Confucian scholars concern themselves with the mere physical layout of the outer world? The sphericity of the earth interested them less as a phenomenon of geography than as a fact of astronomy. The Greek notion of five bands of *climata* extending around the globe and companion doctrines characterizing the plants and animals that grew in each zone were not congenial. Instead, they described the cultural features of all parts of the globe by their relationship to the one Central Kingdom, and felt no impulse to find seaways to exotic lands or to quest for *terra incognita*. Fully equipped with the technology, the intelligence, and the national resources to become discoverers, the Chinese doomed themselves to be the discovered.

PART SEVEN

THE AMERICAN SURPRISE

*A man of genius makes no mistakes. His errors are volitional
and are the portals of discovery.*

—JAMES JOYCE, *Ulysses* (1922)

27

The Wandering Vikings

WE have seen how suddenly and willfully the Chinese withdrew from the threshold of the world and turned back on themselves. The Chinese had to make an effort, and even establish a policy, to withdraw. But peoples who were not organized or equipped to go out and discover the world on the sea were never confronted with such a choice. This was the situation of most of Europe during the Middle Ages. In the era of the great seafaring adventures of the Vikings (c. 780-1070), the rest of Christian Europe was least outgoing. The Muslim Empire, bottling up the Mediterranean, had reached its farthest extent, from the Pyrenees across Gibraltar all around the Maghreb in northwest Africa and through the Middle East to the shores of the Indus River. Within western Europe the movements of merchants, pilgrims, invaders, and bandits were mostly land-bound.

With lightning suddenness in the late eighth century, seaborne swoops of Northmen shocked the peoples around the Baltic and the North Sea. Centuries before, these Northmen of Germanic speech had settled in Europe's great northern peninsula and its surrounding islands and were now gradually being distinguished into Danes, Swedes, and Norwegians. They swept down on peaceful settlements without warning or provocation, and left bewilderment and terror. Intermittent movements of Scandinavian peoples out into Europe had been going on for a thousand years. Now their movements became a plague of murder, rape, and robbery.

We call them "Vikings," a word of richly ambiguous origins. In Old Norse and Icelandic, when *víking* meant a pirate raid or the practice of marauding for loot, *víkingr* came to mean a pirate or raider. This word, in turn, is said to come from the Old Norse *vík,* which means a creek, inlet, or bay, where the marauding Vikings lurked. The word is also related to *wīc* or *wīcing,* the Old English word for camp or temporary settlement, since the Vikings lived in temporary camps while they were going about their business. This word came also to mean a fighter or a soldier. Perhaps "Viking" is related, too, to the word for a man of the town *(wīc,* Latin *vicus),* which later came to mean a seafarer or trader. And then it may be connected with the Old Norse verb *víkja,* "to move speedily," describing them as fast movers, men who receded into the distance, or who took a long turn away from home.

The unlucky targets of the first Viking raids were western Europe's least fortified deposits of treasure, that is to say, the monasteries and churches.

In Christian Europe the treasures collected and deposited there needed little protection from the pious, or even the impious, local laity. Robbing a church was an especially heinous crime. Still, rulers like Charlemagne's grandfather Charles Martel and Aethelbald of Mercia, who would not rob a particular church, could despoil churches wholesale by expropriating their lands and diminishing clerical privileges. The sacred treasuries, which the neighboring Christian worshippers dared not violate, seemed specially reserved for heathen marauders.

When the Vikings discovered this providential opportunity, they did not hesitate. Isolated monasteries were their first convenient victims. Lonely islands off the Atlantic coast of Ireland, where monks felt safe from worldly temptations, stood naked before them. Those solitary high stone towers now rising in open fields near monastery sites which the tourist to Ireland sees today were the monks' answer. In these chimneylike structures, reaching 150 feet in the air, they found temporary refuge. At the first word of a Viking attack the monks clambered up a ladder to the lowest entrance, fifteen feet above ground, then pulled the ladder inside. Thus they could avoid instant massacre, but of course they were not equipped for a siege. Into those towers they crammed their sacred silver vessels, croziers, and jeweled reliquaries till the marauders had gone away. But the Vikings learned to wait them out, and then demanded "protection" money for a promised respite.

The earliest recorded Viking raid in the west attacked the defenseless island of Lindisfarne (3 by 2 miles) off the northeast coast of Northumberland. Holy Island, it was called, because there in 635 Saint Aidan, a famous Irish monk, first bishop of Lindisfarne, had founded a monastery, and Saint Cuthbert had retreated to a hermit's cell on nearby Farne Island in 676. The miracles performed by Saint Cuthbert's body made these islands a favorite destination for pilgrims. The Anglo-Saxon Chronicle records that early in 793 after ominous flashes of lightning, flying dragons, and bitter famine, in June a storm of heathens suddenly appeared from the sea. These Norwegian Vikings pillaged church and monastery, slaughtered monks, then robbed and burned the buildings. Naturally, the attack was explained as the wrath of God for the people's sins. Why otherwise would God have allowed the desecration of Saint Cuthbert's church?

But the sinfulness must have been widespread, for the next century witnessed a plague of Viking raids all along the Baltic and the North Sea, to Scotland, northern England, Ireland, and the Isle of Man, and even to the remote Orkneys, Shetlands, and Hebrides. For three centuries Viking raiders haunted western Europe. The bellicose Charlemagne himself felt menaced. Once while he was dining in a maritime town, the chroniclers record, pirate Northmen came to rob the port, "and their flight was so rapid that they withdrew themselves not only from the swords, but even from the eyes of those who wished to catch them." His face bathed in tears, Charle-magne remained long looking out the window toward the east, whence the marauders had come. He was overwhelmed with the sorrows that he feared his posterity would suffer from these Northmen.

Throughout the ninth century the Northmen struck fear into all who lived within range of Viking ships—in the ports, up the rivers, on islands or peninsulas. By the tenth century this plague had become so regular that the English found it prudent to institutionalize their payoff with a form of regular blackmail called "Danegeld." Viking raiders fully deserved their reputation for sensuality and recklessness. In 1012, for example, when they feasted with the captive Archbishop of Canterbury, whom they had care-fully guarded in the hope of selling his life for a high ransom, the festivities got out of hand and the Archbishop was pelted to death with the bones of animals eaten at the banquet. Then there was one particularly gentle Viking who was nicknamed "the children's man," because he would not impale captive children on the point of his lance "as was the custom among his companions." No wonder, then, that the churches of northern and western Europe amended their litany with the petition, "From the fury of the Northmen, good Lord, deliver us."

For the hit-and-run robber the sea was the best avenue. From the sea he could strike his victim without warning, then carrying heavy booty make a quick getaway with least danger of pursuit. When raiders came across the land, news of their coming usually preceded them, giving victims time to hide their treasure and disappear. But the empty sea was the Vikings' advance moat. And on the sea where the paths lay in all directions how could the victim know where to follow?

Not until the mid-eighth century had the Vikings perfected their ships for piratical enterprise. "It is nearly 350 years that we and our fathers have inhabited this most lovely land," the English scholar Alcuin wrote in 793, the year of the raid on Lindisfarne, "and never before has such a terror appeared in Britain as we have now suffered from a pagan race, nor was it thought that such an inroad from the sea could be made." In their own homeland the Vikings already had a long seafaring experience around the fjords of the Norwegian coast, the sandy shores of the Danish peninsula, and up the rivers of Sweden. The marine architecture of piracy came out of that experience. The beautiful ships, called knorrs, they were making soon after 800, like the one recently unearthed at Gökstad, were wonder-fully adapted to the needs of raiders. A Viking leader commanded a ship some 76 feet long from stem to stern, 17 feet broad in the beam, about 6 feet from the bottom of the keel to the gunwale amidships. The keel, more than 57 feet long, was a single oak timber, which, with other features, made the ship remarkably flexible. Clinker-built of sixteen strakes of varying thickness, she was caulked by tarred animal hair or wool. While fitted for oars as an auxiliary form of power, this was basically a sailing ship, which could be tented at night to provide sleeping quarters for her crew of thirty-five. Yet, with a full load of ten tons she had a draft of only 3 feet, and each additional ton of ballast increased the draft by only one inch.

Such vessels seemed made to order for Viking raiders. The shallow draft let them go up the rivers so that instead of arriving at the deepwater port where the victims might expect them, they could slip up behind on sandy shores. The steerboard (from which comes our word "starboard"), a side rudder fastened to the starboard quarter, was well suited to these maneuvers. When William of Normandy invaded England in 1066, as the Bayeux tapestry shows us, his ships of classic Viking design could quickly lower their masts and speedily land their horses. Without ships so well suited for unannounced arrival and for quick departure, there might have been few Viking raids. In the eleventh century, when Viking trade in bulky commodities increased, deeper ships were needed to carry heavier loads. Though no longer fit for marauding, these ships delivered corn, timber, cloth, fish, and building stone to the harbors of established centers of commerce.

Gradually the raiders became settlers. Instead of returning home every fall to winter in frigid Scandinavia, they found it convenient to transform their base camps along the target coasts into villages where they remained to launch attacks again the following spring. Norsemen and Northmen became "Normans," and so gave their name to Normandy. In 911 the Frankish king, Charles the Simple, offered upper Normandy, the area around Rouen, as a fief to Rollo the Ganger (860?–931?), a Viking chief who had brought his invaders twenty years before. Legend reports that when Rollo went through the feudal ritual of homage, he willingly placed his hands between those of the King, which "neither his father, nor his grandfather, nor his great-grandfather before him had ever done for any man," but when he was asked to kneel and kiss the King's foot, he exclaimed, "No, by God!" The lieutenant to whom he assigned this humiliating act then did his duty with such aggressive awkwardness that the King was knocked over. Only a century and a half after Rollo forcibly brought Northmen into France, William Duke of Normandy took his Normans across the Channel to invade the British Isles.

Everywhere the Normans went across Europe they showed a genius for adapting. In France and Germany they fitted into the feudal hierarchy. In England they catalyzed a unifying nation. And they helped consolidate Kievan Russia by their trade along the waterways. In Sicily they played a very different role. Finding a polyglot community of many religions— Muslims, Christians, Jews; speaking Arabic, Greek, or Italian—they became mediators. Under the tolerant auspices of the Norman king Roger II, the brilliant court at Palermo became a bustling crossroads of southern Europe, a place for the interbreeding of ideas and the arts. It was a Norman, Tancred (1078?–1112), who led the First Crusade, captured Jerusalem (celebrated in Tasso's *Jerusalem Delivered*) and then established still another kind of Norman kingdom in Syria.

Though adept at migration, at the assimilation of peoples and the cementing of nations, the Normans had no talent or appetite for exploration. Viking ships were not well suited for long voyages, nor for colonizing on the far side of a broad ocean. Their cargo capacity was not enough to feed a numerous crew or passengers for weeks at sea. With a crew of about 35, the ninth-century Gökstad ship had a cargo capacity of only about 10 tons, by contrast with Columbus' *Santa María* (crew, about 40; cargo capacity, about 100 tons) or the Pilgrims' *Mayflower* (passengers and crew, about 100; cargo capacity, about 180 tons). While the Viking mode of flexible construction, requiring a keel hewn from a single oak timber, set limits on length, the Viking ship of the era of the pirate raids was wonderfully seaworthy. This was demonstrated in 1893, when Captain Magnus Andersen made a 28-day passage (April 30–May 27) from Bergen to Newfoundland through a stormy sea in an exact replica of the Gökstad ship. To provide a keel long enough for Andersen's vessel, a Canadian oak had to be brought to Norway.

While some of these Viking ships were carrying Norsemen on hit-and-run raids for the treasures of churches and monasteries, others carried Norse settlers in similar ships to search for land. These were well designed to ride the stormy North Sea to nearby islands where land could be had for the taking. The Norse had already acquired a legendary reputation as a "fast-breeding" people. Tradition, based on no statistics, reported that the Scandinavian countries were "like a mighty hive, which, by the vigor of propagation and health of climate, growing too full of people, threw out some new swarm at certain periods of time, that took wing, and sought out some new abode, expelling or subduing the old inhabitants, and seating themselves in their rooms." Western observers were so impressed by this "swarming" of the Scandinavian peoples that they guessed there must be some peculiar institution, perhaps polygamy, to explain why the population grew so fast. We do know that some of their leaders were remarkably prolific. Harold Fairhair (850?–933), who made a strong kingdom in Norway and forced many lesser chieftains to leave the country, had nine sons who reached manhood. His son and successor, Eric Bloodaxe, had eight sons. It is no wonder they wanted more land.

Some of these sons, with the deposed chieftains and still others, went out to settle islands of the North Sea and the neighboring Atlantic. Their colonies in the Orkneys, the Shetlands, the Faeroes, the Hebrides, and Iceland became bywords of remoteness, marginalia of European history.

Meanwhile the Swedes, whose rivers and bays faced eastward into the Baltic, made their own way up and down the rivers of Russia—the Dvina, the Dnieper, the Volga—trading with Muslims and dominating the life of Kiev and Novgorod. The word "Russia" itself seems to derive from the Old Norse *Rothsmenn,* meaning seafarers, from *rōthr,* "to row." From our later perspective, we too easily forget that it was not only the silks and spices and gems of the Orient and the tropics which nourished worldwide trade. The Muslims traded for the slaves whom the Swedes captured in the forests of northern Russia. Some of the special products of the Arctic north, including walrus tusks, then still the main source of ivory in Europe, and furs, were much in demand by southern and eastern traders. For seven hundred years, from the time of Emperor Constantine to the Crusades, the Scandinavians were the chief agents of European expansion toward the south, toward the east—and toward the west.

28

Dead End in Vinland

THE Vikings going west were restless, vagrant island-hoppers. A look at the map in those far-northern latitudes just below the Arctic Circle, where later history has given us little reason to look with interest, helps us understand the Norsemen's way west. In the waters between the Arctic Circle and 60 degrees north latitude all across the Ocean from Bergen to American shores, land is encountered about every five hundred miles. How unlike the vast unrelieved open waters in the southern latitudes, the domain of Columbus and Vespucci! By about 700 they had reached the Faeroe Islands some 200 miles north of Scotland, and then in 770 went on and began settling Iceland. From the Hebrides off the northwest coast of Scotland the Norse went on to Ireland, where they founded Dublin in 841.

When the learned men of Christian Europe heard that Vikings had settled in Iceland, they called that land Thule. This name that Polybius and Ptolemy had applied to the most northern land on earth became Ultima Thule to describe the furthest goal of any human endeavor. But the Ultima Thule of learned Europe was no Ultima Thule for the unlettered Vikings, who stumbled on from one Ultima Thule to another, in defiance of classical literature and Christian dogma. By about 930 it seems that most of Iceland's habitable land was fully occupied. The "tyranny" of King Harold Fairhair, who was unifying Norway, drove out many petty chiefs. The familiar land shortages and the vague hope for better things produced another "swarming," which soon filled Iceland itself to overflowing. And famine forty years later increased the pressure for Norse Icelanders to move on.

The movement to settle Greenland, the Vikings' next Ultima Thule, was led by a habitual criminal named Eric the Red. In 982, when Eric had been outlawed for manslaughter from his native Norway, he fled to Iceland, where he settled at Haukadal in the west. Outlawed for more killings, he moved on to a town on Breida Fjord, on a peninsula reaching out from Iceland's west coast. And when he was outlawed again for more killings and sentenced to three years' banishment, he decided to go still farther west. This time he sailed out toward a rumored land which had been reportedly sighted a half-century earlier by a storm-driven Norwegian sailor called Gunnbjörn.

When Eric had sailed only about five hundred miles westward, he discovered to his delight that the rumors had substance. Finding a vast subcontinent, he sailed down Greenland's east coast to Cape Farewell. From there Eric reached up its western coast where he found inviting grassy slopes, impressive fjords, and headlands reminiscent of his native Norway. During the three years of his banishment Eric and his crew laid out sites for farms and homesteads. The land was rich in game animals, bears, foxes, and caribou. The sea was full of fish and offered large sea mammals, seal and walrus. Even the air provided a harvest—clouds of birds, all innocent of the hunter, and ready for the catching. Best of all, there was no sign of earlier human habitation.

Hoping that a flattering name would help him recruit settlers, Eric christened this new country Greenland. When his three years' banishment had expired, he returned to congested Iceland to gather settlers. In 986 he sailed from Iceland again, this time with an emigrant fleet of twenty-five ships carrying men and women and domestic animals. Only fourteen ships survived the stormy passage, which brought to this first Viking settlement in Greenland about four hundred fifty immigrants. Some of Eric's party stayed at an inlet just west of the southern tip. Others went farther north along the western coast. Although the climate, like that of Iceland, was not well suited to agriculture, people raised cows, horses, sheep, pigs, and goats, and could live well enough off their butter, milk, meat, and cheese. Excavation of the remains of Eric's own farm has revealed a surprisingly spacious and comfortable establishment with thick walls of stone and turf against freezing wind and heavy snow.

Again and again we witness this familiar scenario of the seagoing Viking vagrants on the lookout for any likely landing. Finding a place by accident, they would then remain as settlers. The Vikings had discovered Iceland when a certain Swede, Gardar Svavarsson, pressed by his mother, a seeress, and seeking his wife's inheritance in the Hebrides, sailed out from Scandinavia and was driven off course. He accidentally encountered eastern Iceland. Later, as we have seen, it was another storm-driven sailor who came upon some unexpected lands to the south and west of Iceland, and when Eric the Red pursued that report he ran into Greenland. The Viking encounter with Vinland would be one more version of the same plot.

The story of the accidental Viking settlements in America begins with Bjarni Herjolfsson, who owned a ship that traded between Norway and Iceland. In the summer of 986 he took a full cargo to Iceland intending to follow his usual practice of spending the winter there with his father Heriulf. This time, dismayed to find that Heriulf had sold his estates in Iceland and had already left for Greenland with Eric the Red's party, Bjarni and his crew decided to follow. They must have known that passage to the southwestern fjords of Greenland was dangerous. Of course they had never made the journey before, and they had no chart or compass. They could not have been surprised when they were fogged in and lost their bearings. When finally they sighted a land "level and covered with woods," Bjarni knew that this could not be Greenland. Following the coast, they first found more "flat and wooded country," and farther north all they saw was glacier-covered mountains. Being a practical and incurious man, who had gone looking for his father in Greenland, Bjarni was simply puzzled and troubled. He refused to let his crew go ashore, turned his prow out to sea, and after four days' sailing arrived at Herjolfsnes, the exact place he had been seeking on the southwestern tip of Greenland.

The Greenlanders' sagas preserve the record of Bjarni's glimpse of that unsought unidentified land to the west, which would prove to be America. For the next fifteen years there is no record that any other Greenlanders found out what the storm-driven Bjarni had seen. Yet when Greenlanders climbed those high mountains behind their settlements and looked westward out to sea, on the far horizon they could see what looked like land, or at least the kinds of clouds they usually found above land.

Leif Ericsson, according to the sagas, a "big and strong, of striking appearance, shrewd, and in every respect a temperate, fair-dealing man," who had come to Greenland with his father Eric the Red, bought Bjarni's ship. In 1001 he gathered a crew of thirty-five, and set out for the land Bjarni had sighted but did not have the courage or the curiosity to explore. Leif had offered to put his father Eric in charge, but Eric's horse stumbled and threw him as they were riding down to the ship, which confirmed his hunch that the trip was not in his destiny—and "that of all their family Leif would still command the best luck."

Leif and his crew, sailing due west, "lighted on that land first which Bjarni and his people had lighted on last. The background was all glaciers, and right up to the glaciers from the sea as it were a single slab of rock. The land impressed them as barren and useless." This was Baffin Island, just north of Hudson Strait, and they called it Helluland, or Flat-stone Land. Coasting southeast, they next saw a flat country, now known as Labrador, that was covered with forest, which they called Markland, or Woodland. Farther on they found an attractive wintering place, which they called Vinland, or Wineland, after its plentiful grapes. But the word in the sagas, crudely translated as "wild grapes," probably meant "wineberry," the wild red currant, the gooseberry or the mountain cranberry, which does grow abundantly that far north. Their campsite has now been unearthed on Newfoundland's northeastern tip at a place called L'Anse aux Meadows.

Finding the land unexpectedly attractive, Leif's party brought their ship up a river and into the lake from which it flowed. "They cast anchor, carried their skin sleeping-bags [a Viking invention] off board, and built themselves booths [huts]. Later they decided to winter there and build a big house." The Greenland these men had just come from was no Eden, which helps explain their enthusiasm reported in the sagas:

> There was no lack of salmon there in river or lake, and salmon bigger than they had ever seen before. The nature of the land was so choice, it seemed to them that none of the cattle would require fodder for the winter. No frost came during the winter, and the grass was hardly withered. Day and night were of a more equal length there than in Greenland or Iceland. On the shortest day of winter the sun was visible in the middle of the afternoon as well as at breakfast time.

The next summer, Leif and his party returned to Greenland. When Leif's father, Eric the Red, died, and the family responsibilities fell on Leif, he would have to stay near home. Leif then lent his ship to his brother Thorvald who wanted to see the Vinland which Leif had praised so extravagantly. Thorvald and his crew of thirty had no difficulty finding the very place of Leif's encampment. They spent the summer exploring the coast, then wintered at Leifsbudir (Leif's Hut). The next summer, when they first encountered the natives, eight natives were killed, Thorvald himself received a fatal arrow wound, and Thorvald's party returned to Greenland.

The Vikings had not yet tried to plant a permanent colony. This next step was undertaken by one of Leif Ericsson's in-laws, an Icelander named

Thorfinn Karlsefni, who had sailed to Greenland with a trading cargo. Viking Vinland seems somehow to have become a family affair, which proved a fatal weakness. "Thorfinn Karlsefni was a very well-to-do man," the sagas report, "and spent the winter at Brattahlid with Leif Ericsson. It did not take him long to set his heart on Gudrid [the attractive widow of Leif's brother]; he asked for her hand, and she left it to Leif to answer for her. So now she was betrothed to him and their wedding took place that winter." But Gudrid was no stay-at-home and she knew what she wanted. When Karlsefni finally gave in to her urging to lead a colony to Vinland, he organized an expedition of three ships carrying some two hundred and fifty men and women, together with all sorts of livestock. They arrived in Vinland before autumn, and settled at a comfortable bay not far from the old site of Leif Ericsson's encampment. That winter was so mild they could let their cattle graze outdoors. Then, the sagas tell us, came omens of trouble.

> After that first winter came summer. It was now they made acquaintance with the Skraelings [natives: Indians or Eskimos], when a big body of men appeared out of the forest there. Their cattle were close by; the bull began to bellow and bawl his head off, which so frightened the Skraelings that they ran off with their packs, which were of grey furs and sables and skins of all kinds, and headed for Karlsefni's house, hoping to get inside there, but Karlsefni had all the doors guarded. Neither party could understand the other's language. Then the Skraelings unslung their bales, untied them, and proffered their wares, and above all wanted weapons in exchange. Karlsefni, though, forbade them the sale of weapons. And now he hit on this idea; he told the women to carry out milk to them, and the moment they saw the milk that was the one thing they wanted to buy, nothing else. So that was what came of the Skraelings' trading: they carried away what they bought in their bellies, while Karlsefni and his comrades kept their bales and their furs. And with that they went away.

When the Skraelings returned, the Vikings had no plan either for trading with them or for subduing them. The Skraelings happened to covet pieces of the Vikings' red cloth, for which they would exchange their best unblemished skins and gray fur pelts. "When the cloth began to run short they [the Vikings] cut it up so that it was no broader than a fingerbreadth, but the Skraelings gave just as much for it, or more."

Then one day there came a sudden attack by "a great multitude of Skraeling boats." The Skraelings stormed the Viking camp, swinging their battle staves "anti-sunwise" (there was not yet any "anti-clockwise"), yelling and showering missiles from their war slings. What most terrified the brave Vikings was the Skraelings' primitive buzz bomb. When the Skraelings launched this ball-shaped object (probably a blown-up moose bladder), fearless Freydis, Leif Ericsson's sister, daughter of Eric the Red, came out-of-doors and saw how the Vikings had taken to their heels. "Why are you running from wretches like these?" she cried. "Such gallant lads as you, I thought for sure you would have knocked them on the head like cattle. Why, if I had a weapon, I think I could put up a better fight than any of you!"

> They might as well not have heard her. Freydis was anxious to keep up with them, but was rather slow because of her pregnancy. She was moving after them into the forest when the Skraelings attacked her. She found a dead man in her path, Thorbrand Snorrason—he had a flat stone sticking out of his head. His Sword lay beside him; she picked it up and prepared to defend herself with it. The Skraelings were now making for her. She pulled out her breasts from under her shift and slapped the sword on them, at which the Skraelings took fright, and ran off to their boats and rowed away. Karlsefni's men came up to her, praising her courage.

The Skraeling threat, reinforced by a creature they described as a hopping "uniped" who shot poisoned arrows, drove these Vikings out of Newfoundland and back to Greenland. There Freydis organized what proved to be the final Viking expedition to Vinland. Arriving slightly ahead of Freydis, two Icelandic brothers, Helgi and Finnbogi, had promptly occupied Leif's house. When she arrived in Vinland, the brothers explained that they expected to share the house with Freydis' crew. But she dispossessed them and taunted her husband with cowardice.

> He could not endure this baiting of hers. He ordered his men to turn out immediately and take their weapons, which they did, and crossed straightway to the brothers' house and marched in on the sleeping men, seized them and bound them, and led them outside, each man as he was bound. And Freydis had each man killed as he came out.
> Now all the men were killed, but the women were left, and no one would kill them.
> "Hand me an axe," said Freydis.
> Which was done, and she turned upon the five women they had there, and left them dead.

Freydis seized the brothers' possessions, which she distributed among her crew to persuade them not to reveal her crimes.

Early the next spring in the brothers' ship Freydis and her party sailed back to Greenland. They gave out the story that the brothers had decided to remain behind. Leif tortured three of her crew to learn the truth. He still did not have the heart to punish his own sister, but he laid a curse on her and all her offspring, which seems to have had some effect. By the year 1020 the Viking settlements, the first recorded European settlements in America, had left history and entered the domain of the archaeologists.

After 1200 the Greenland climate grew colder, glaciers moved southward, the timberline moved lower down the mountains. As the sea temperatures sank, the ice drifting off the north coast of Iceland began to envelop Cape Farewell at Greenland's tip, and so shut off the Viking settlements up the western coast. Expanding polar ice imperiled the familiar seaways to Greenland from Iceland and Norway. Meanwhile, too, Greenland's special products began to lose their markets. Furs now poured out of northern Russia, woolens in larger quantities and of better quality were coming from England and the Netherlands, and walrus tusks, Greenland's most distinctive product, were considered an inferior kind of ivory once French craftsmen came to know the elephant ivory trickling in from Africa and farther east. Now fewer ships came to the remote Greenland ports, and the regular Greenland carrier from Norway made its last run in 1369. The Greenland trade was a royal Norwegian monopoly belonging to the town of Bergen. When Bergen suffered the Black Death in 1349, one in three of the Norwegian population died. The harbor town was burned and sacked, and Greenland lost its lifeline to the motherland. After the Black Death also decimated Greenland, the Eskimos, Greenland's Skraelings, advanced southward to take possession of Viking settlements. Pious Christian chroniclers imagined that Greenland was being punished for having abandoned the true faith, though there was not a shred of evidence for apostasy. Before the end of the fourteenth century the Viking settlements in Greenland, like those in Vinland, were only a memory.

The Vikings were probably the first European settlers in America, which is far from saying that they "discovered" America. Their settlement across the stormy ocean was an act of physical but not of spiritual courage. What they did in America did not change their own or anybody else's view of the world. Was there ever before so long a voyage (L'Anse aux Meadows is a full forty-five hundred miles as the crow flies from Bergen!) that made so little difference? There was practically no feedback from the Vinland voyages. What is most remarkable is not that the Vikings actually reached America, but that they reached America and even settled there for a while, without *discovering* America.

Their America was no new encounter. In fact, these voyages were hardly an encounter at all. The word "encounter" (Latin *in,* in; *contra,* opposite, against) means a coming up against, a meeting that conflicts with the familiar or the already known. The Viking westward movement across the North Atlantic from Bergen remained in the same climatic zone. From

Bergen to Iceland a sailor went only a few degrees of latitude to the north; the Viking settlements in Greenland were on the same latitude as Bergen; and southward to Vinland differed only some 10 degrees. The Vinland climate proved somewhat milder, but to the Greenlander the Vinland flora and fauna were not exotic. The sagas described the Skraelings prosaically. "They were small, ill favoured men, and had ugly hair on their heads." "They had big eyes and were broad in the cheeks." The two Vinland curiosities were the Skraelings' moose-bladder buzz bomb and an occasional hopping "uniped."

Between the two cultures that met in Vinland there was little difference. The Vikings, it turned out, had not the technology, the will, or the man-power to dominate or enslave the Skraelings. Nor had they the materials, the will, or the organization to develop a Viking-Skraeling trade. The appeal of their red cloth to the Skraelings was purely accidental. If the Vikings, like their Spanish and Portuguese successors in America, had possessed firearms, they might have scared away the Skraelings and dug in. With ships larger than their 76-foot knorr the Vikings might have brought more set-tlers, and then might have had the supporting manpower. But their most important contribution to ship design, the steerboard, their kind of rudder on the right side, was hard to use on larger vessels.

The Vikings had neither chart nor compass when they reached America. Their usual technique of navigation depended on intimate acquaintance with the seas to be traversed. It was not much help in remote parts, and could hardly be used at all in strange latitudes. For longer voyages, al-though they had not yet invented the idea of latitude, they used a kind of "latitude sailing" in which the sailor put himself on the latitude of his destination, then simply did what he could to stay on that latitude. For example, a Norseman going from Bergen to Iceland would sail up along the coast of Norway until he reached a point where the angular height of the North Star above the horizon and the declination of the sun at noon were what he knew them to be at their destination in Iceland. Of course this amounted to marking latitude, but the Vikings had not begun to think of it in that way. At sea they used any crude device—a notched stick would do, but the length of an arm, a hand, or a thumb would also serve—to keep on a course where the observed angles remained uniform. Of course the Vikings often missed their destination, and that was how they first encoun-tered Iceland, Greenland, and Vinland.

This primitive form of "latitude sailing" had to be supplemented by familiarity with the waters sailed in. A Viking sea captain could not rely exclusively on observing the North Star and the sun, for in northern seas these were often obscured by cloud or fog. He needed to know the birds, the fish, the currents, the driftwood, the weeds in the sea, the color of the water, the iceblink (the yellowish glare in the sky above an ice field), the clouds, and the wind. Floki, the great Viking sailor in the ninth century, found Iceland by sending up a raven from his ship to lead him there. Viking sailors had an uncanny feel for the neighborhood where their ancestors had sailed so often. It was only by island-hopping that they actually reached the shores of America. The seafaring distance from Greenland to North America was only half the distance between Iceland and Greenland or between Norway and Iceland which they had long since been traversing.

29

The Power of the Winds

APART from the directions of the rising and setting of the sun, which, of course, varied with place and season, the most noticeable directions that could help a sailor were those of the winds. At least as early as the first century B.C. the Chinese were writing about their "wind-seasons." They developed elaborate classifications into the twenty-four seasonal winds, and used kites to test their different behavior. It is not surprising that very early the Chinese made weather cocks or vanes, probably the pioneer of all the other pointer-reading devices which later served the natural sciences. Ancient Greeks were so accustomed to use the names of the winds to indicate the directions from which they came that for them "wind" became a synonym for direction. The puffed cheeks and exhaled strong breath of the symbolic winds are not mere decoration but the main direction-markers on early maps. The Spanish sailors on Columbus' crew thought of direction not as degrees of compass bearings but as *los vientos,* the winds. Portuguese sailors continued to call their compass card a *rosa dos ventos,* a wind rose. When the religious brotherhood of pilots commissioned the Madonna for their chapel in Cordova, it was no accident that she was *Nuestra Señora del Buen Aire,* "Our Lady of the Fair Wind." Before the age of the magnetic compass, sailors all over Europe identified "direction" with the wind.

The winds, the energy that carried men across the seas, were of course a subject of great interest, prolific of romantic myth and scientific speculation. The leading source of doctrine was that popular Roman man of letters,

Lucius Annaeus Seneca (c. 4 B.C.–A.D. 65), who was tutor to the young Nero, and for a while dominated Nero's court until he was forced to commit suicide at Nero's command. "Air flowing one way" was Seneca's widely repeated definition of wind. Some Christian mystics like Hildegard of Bingen said it was the winds that moved the firmament from east to west, and somehow kept the other forces of the planet in order. Without the winds' constant movement, the fires of the south, the waters of the west, and the dark shadows of the north might cover the whole earth. God's "wings," the four winds, kept the elements separated and in proper balance. Just as the soul's breath held man's body together, so the breath of the winds held together the firmament and prevented its corruption. Like the soul, the winds, too, were invisible, both sharing the mystery of God.

Elaborate theories, like that proposed in the twelfth century by William of Conches, gave the winds a primary role in the making of climates, the moving of oceans, the stirring of earthquakes. Boreas from the north blew cold, Auster from the south blew hot. One of the most influential medieval encyclopedias, published in the year of Columbus' first voyage, by a Franciscan friar, Bartholomew the Englishman (fl. 1230–1250), popularized a wind-oriented anthropology. "The north wind dries and cools the land, yet owing to its clearness it is clear and subtle," hence its coolness closes the pores of the body, which then better retains heat. As a consequence "men in the north are tall in stature and fair in body." The hot moist south wind naturally has the contrary effect. "Therefore the men of the southern lands are different from those of the north in stature and appearance. They are not so bold nor so wrathful."

One delightful legend, retailed by Gervase of Tilbury, told of a valley in the ancient Kingdom of Arles, sometimes called Burgundy, which remained sterile for ages because it was so enclosed by mountains that no winds could enter. Finally Caesarius, a benevolent archbishop in the time of Charlemagne, determined to help the people there. Seeing what they needed most, he simply filled his glove with sea breezes which he released in the valley. This produced a wind known as the *pontianum*, which suddenly transformed the valley-desert into a fertile haven.

European makers of maps and sailing charts in the Middle Ages adopted the classical names for the winds. Ancient Greek sailors named the four principal wind directions and marked four other points in between. The elegant octagonal Tower of the Winds in Athens (second century B.C.) shows the visitor today the vivid symbolism attached to each of the eight winds. Where, as among the Germanic peoples, the winds were less regular, only the four cardinal wind points were given names. Laymen were still inclined to describe the four quarters of the sky in relation to the daily pathways of the sun.

The Arab world had a peculiar advantage in the quest for absolute direction, because Islam required that their mosques face Mecca. Only by finding geographical coordinates could they be certain that they were properly oriented toward a distant place. Even during the Great Interruption, when Christian Europe was imprisoned in its dogmatic theological geography, mathematically inclined Muslim scientists were using astrology as a proto-astronomy to improve on Ptolemy's figures for latitude and longitude.

After the Great Interruption, Christian Europe used the magnetic compass to open a new world of direction naming and direction finding. No longer were the directions only local and relative, marked by the breezes in a particular place. The magnetic compass suddenly enabled the sailor to find an absolute direction anywhere on the globe without complicated astronomical calculations. Using his magnetic compass, Columbus could take his bearings direct for Cipangu and stay on the same latitude, without the aid of celestial navigation.

After the introduction of the magnetic direction-finding needle in the twelfth century the wind rose (the card marked off with the four, eight, or twelve "winds") was gradually displaced by the magnetic compass rose, with its sharper definition of sixteen or thirty-two directional points. It took some time before the old "compass" and the new one were combined. At first the sailor's "compass card" was circular, inscribed with the pattern of the wind rose, and kept lying flat on a table. Beside it in a dish a bare magnetic needle was left afloat on a piece of cork or straw. The pilot could then turn the card as the magnetic needle indicated. In the fourteenth century at last someone had the idea of attaching the needle beneath the card so they would both float together and the magnet would keep the compass card pointed in the right direction.

The magnetic compass was, of course, a catalyst of exploration, a new lure into the unknown. Sailors, leaving behind their crude sketch maps, rough diagrams of familiar places, now could take along true maps, which oriented them to the whole world. The magnetic poles, a characteristic feature of this planet, are not the same as the geographic poles around which the earth rotates. The reason for the location of the magnetic poles remains a mystery, and the earth's magnetic field, historians of paleomagnetism tell us, has reversed its polarity many times in the geologic past.

Still, in practice the compass provided a worldwide absolute for space comparable to that which the mechanical clock and the uniform hour provided for time. Both these epochal discoveries occurred in Europe in the same centuries. From the very nature of our spherical spinning planet, the marking of time and the marking of space were inseparable. When you moved any great distance from your home out into the uncharted great oceans, you could not know precisely *where* you were unless you had a way of finding precisely *when* you were.

Locating yourself on the whole planet meant finding your place on the grid of latitudes and longitudes. Ptolemy had made a beginning, but then came the thousand years of obscurantist Christian geography in the European Middle Ages. For a new era of exploration the magnetic compass was needed. The apparatus and the skills of celestial navigation would not come for another two centuries after Columbus. Meanwhile, the wonderfully simple and inexpensive magnetic compass gave sailors a new confidence that they would be able to find their way back. Anyone could make one and even the illiterate could use one. And with the compass sailors would move ahead more willingly into the unknown.

The application of the magnetized needle to navigation occurred in China about A.D. 1000. But the first notice of the compass in European writings does not appear until two centuries later, in the works of Alexander Neckam (1157–1217), an English monk lecturing at the University of Paris. We do not know how the compass came to Europe or, as is more likely, when and how and by whom it was independently invented there. As late as the seventeenth century the magnetic compasses used by European surveyors and astronomers—unlike the compasses used by sailors—were marked to "point" toward the south. The Chinese needles had been marked in this way for many centuries. Perhaps, as Joseph Needham hints, this is a clue that the magnetic compass was first transmitted westward overland from China and later adapted by European sailors, who made it "point" north.

Wherever people noted the remarkable powers of the lodestone—the stone that "leads"—they were tempted to associate it with dark necromantic forces. In China, for example, the powers of the lodestone were probably first used as an aid to the arts of fortune-telling. Originally the game of chess seems to have been a technique of divination by witnessing the outcome of the battle between the universal forces of Yin and Yang. In early Chinese chess the Great Bear, or Northern Dipper, was represented in a spoon that was spun about. The spoon came to be made of magnetite after its magical properties were discovered, and so served as a divining device when it was rotated according to the complicated rules of the game.

Saint Augustine recounts his amazement when he saw that magnetite could not only attract iron but actually give iron the power to attract other iron, creating a chain held together by an unseen force. No wonder that in the West, too, magnets were part of the magician's equipment. Roger Bacon, the most eminent of medieval European necromancer-scientists, played a leading role in the histories and legends of the compass. Even the origin of the name "magnetite," from which came "magnet" and "magnetism," is mysterious. Although a common mineral in many parts of the world, it probably takes its name from Magnesia in ancient Thessaly along the Aegean Sea. There popular legend tells of a shepherd named Magnes whose iron-bound staff and shoes with iron nails stuck to the ground, where he discovered the magic mineral. Under the pillow of an unfaithful wife, a piece of lodestone could make her confess her sins. This mineral was said to be so potent that a small piece could cure all sorts of ailments and even act as a contraceptive. But it was also believed that the effects of magnetite could be counteracted by the breath of onion or garlic. Therefore sailors were not to be given these foods on shipboard for fear of demagnetizing the compass needles.

Since the inexplicable power of a magnetized needle to "find" the north smacked of black magic, common seamen were wary of its powers. For many decades the prudent sea captain consulted his compass secretly. All this makes it doubly difficult for us to find the origins and trace the history of the compass in Europe. But it also helps explain the origins of the binnacle, the "little house," or box, where the compass was kept. During the centuries when the compass was still considered an occult instrument, the pilot would doubtless have wanted to keep his magic needle out of public view. The pilot's "navigating house," or "binnacle," must have become a "little house" for the compass. After the compass had lost its occult flavor and become every sailor's everyday tool, it came out onto the open deck. Still, in Columbus' day, a pilot who used the magnetic compass might be accused of trafficking with Satan. Some pious compass-guided navigators replied that since the magnetic needle was floated by being pierced at right angles through a straw, and so formed a Holy Cross, it could not possibly be Satan's instrument. At Sagres, Prince Henry the Navigator combated such superstitions by accustoming his pilots to the everyday use of the compass. By Columbus' time, the magnetic compass had become so indispensable that, for security, the captain himself would carry extra magnetized needles. Magellan took thirty-five needles on his flagship to replace the one under the circular compass card if it lost north. Sometimes weak needles were remagnetized by a precious piece of lodestone guarded by the captain.

Just as the clock freed mankind from the daily need to measure time by the sun and the stars, the compass oriented mankind anew in space, and so extended the times and seasons of seafaring. "When the mariners cannot see the sun clearly in murky weather, or at night, and cannot tell which way their prow is tending," Alexander Neckam wrote about 1180, "they put a needle above a magnet which revolves until its point looks north and then stands still." The compass thus began as a guide across familiar seas, a mariner's aid in bad weather or whenever he could not find his direction by the sun.

In the Mediterranean on overcast days, without a compass even experienced sailors risked losing their way. For that reason among others, as late as the thirteenth century seafaring trade between distant points in the Mediterranean was cut off in winter. The copious records of Italian cities show that vessels were expected to stay in port during the "bad weather," which lasted from October to March. A fleet that plied to the Levant to keep Venetian merchants stocked with merchandise from the farther East could make only one round trip each year. One fleet would leave Venice at Easter and return in September, before the season of overcast skies set in. Another, the "winter fleet," would leave Venice in August, to arrive at its destination before bad weather, but then had to spend the winter months overseas, returning to Venice the following May. For all practical purposes, during half the year these trading fleets were immobilized.

By the fourteenth century the compass had come to the Mediterranean and seafaring commerce was enlivened. A Venetian fleet no longer confined to port by overcast weather could make two round trips to the Levant each year.

Prevailing winds in the Mediterranean were such that there were advantages to sailing with the winds of the cloudy months. During the months of clear skies, May to October, ships returning to Venice from Egypt faced prevailing north and northwesterly winds, and so had to take a roundabout course to Cyprus and then work west. But during the "bad weather" months a following wind eased their direct passage. The compass broke the traditions of millennia by opening the Mediterranean seas in winter. Again, the mastery of time and of space was one.

In the Indian Ocean, by contrast, the monsoon winds were so regular, as they shifted with the seasons, that they served as a kind of compass. Pilots took their direction from the winds themselves. Nor was there any problem of cloudy overcast under the clear tropical skies. Sailors blessed by a wind-compass hardly needed another.

For quite other reasons, sailors in the North Sea and the Baltic were slower to feel a need for the magnetic compass. Much of their sailing was in shallow waters where sailors had long since been finding their course by feeling their way along the bottom. Off northwest Europe the vast extension of the continental shelf, so different from the deep basins of the Mediterranean, made shallow seas. There the tides were strong and wide-ranging, and knowing depth was a matter of life and death. "In this sea," Fra Mauro explained on his classic map (1459), "they do not navigate by compass and chart but by soundings." Their soundings traced the shape and character of the floor of the sea by "lead and line." A lead weight covered with tallow was lowered to the sea floor to find the depth and also to bring up a sample of the sand or mud found there. Experienced northern pilots became familiar with the floor of their sea. After the compass arrived, pilots off the north European coasts still felt more secure when they could combine the new device with their old, reliable lead and line. This is how, in the mid-fifteenth century, the oldest surviving English book of sailing directions instructed sailors plying from Cape Finisterre on the northwest tip of Spain to England:

> When you come out of Spain, and when you are at Cape Finisterre set your course northeast. When you reckon you are two thirds of the way across if you are bound for the Severn you should go north by east until you come into soundings. If you then find 100 fathoms deep or 90, then go north until you sound again [and find] at 72 fathoms fair gray sand. And that is the ridge that lieth between Cape Clear [Ireland] and [the] Scilly [Islands]. Then go north until you come into soundings of ooze, and then set your course East North East or else East and by North.

The English pilot books, with their emphasis on tides, depths, and grounds, were strikingly different from the Italian pilot books of the same era, which emphasized distance. Sailors in ancient times, as Herodotus tells us, also tried to navigate by lead and line, but most of the Mediterranean was too deep. Immediately offshore they found themselves "out of soundings." They could not feel their way along the sea floor.

Naturally, it was the Mediterranean sailors who most welcomed the magnetic compass. By the early fifteenth century the portolanos for the Mediterranean were much improved and simplified. Charts that once had shown a complicated zigzag of tacks now marked the course by a single compass bearing. The portolanos for the Mediterranean had already acquired a remarkable new accuracy, even while the Atlantic and North Sea coasts were still vague. The compass, adding new precision to the ancient techniques of dead reckoning, had become the prime or even the only essential instrument of navigation.

Without that characteristically Mediterranean incentive to use the compass, Columbus might never have had the one instrument he needed to get him to "the Indies" and back. The "north-pointing" needle, an obstinate Europe eventually learned, was really pointing to unimagined new worlds. "The Load-stone," Samuel Purchas (1575?–1626) observed only a century after the death of Columbus, "was the Lead-stone, the very Seed and ingendring stone of Discoverie, whose soever Joviall Braine first conceived that Minerva."

30

"The Enterprise of the Indies"

"THAT noble and powerful city by the sea," Genoa, where Columbus spent the first twenty-two years of his life, had long struggled against Venice for maritime dominance of the eastern Mediterranean. The Venetian Marco Polo had dictated his travels from a Genoese prison. In Columbus' youth Genoa was a flourishing center of shipbuilding and seafaring enterprise whose map-makers dominated the market for portolano charts in the western Mediterranean. They were even making maps of the parts of the African coasts newly discovered by the Portuguese disciples of Henry the Navigator. It was in Genoa that Columbus very likely began to learn the arts of map-making, which he and his brother later practiced in Lisbon. While Genoa would remain a birthplace and nursery of individual explorers like Columbus (1451?–1506) and John Cabot (1450–1498), the grand seafaring enterprises required larger resources, a vaster hinterland, and, when the Muslims held so much of the eastern Mediterranean, a more western exposure.

In 1476, when Columbus was serving on a Flemish vessel in a Genoese convoy escorting a cargo through the Straits of Gibraltar and up to northern Europe, his ship was attacked and sunk by a French armada. Luckily this occurred near Lagos, off the coast of Portugal only a few miles from where Prince Henry the Navigator had made his headquarters. The twenty-five-year-old Columbus used one of the floating long oars as a life raft and propelled himself ashore.

In those years there could have been no happier or more providential landing for an ambitious young mariner. The friendly people of Lagos dried and fed Christopher, then sent him to join his younger brother Bartholomew in Lisbon. Henry the Navigator had made Portugal the exploring center of Europe, and perhaps of the world. By 1476 those exploits were paying off in rich cargoes of Negro slaves, ivory, malagueta pepper, and gold dust. The rewards of seafaring were visible all around. The Columbus brothers, Christopher and Bartholomew, went into the newly flourishing business of making and selling mariner's charts. In Lisbon they could update the old charts by adding the latest information brought back by the adventuring Portuguese ships. With sharper knowledge of new seacoasts arriving every month, the chart-maker had to be a kind of marine journalist.

When Columbus and his brother set up shop in Lisbon, Portuguese ships were still inching down the west coast of Africa and had only reached the Gulf of Guinea. But the full shape of Africa, which Ptolemy had hitched around to join southeast Asia, making a closed sea of the Indian Ocean, had not yet been traced by mariners. In late 1484, when Columbus offered what he called his "Enterprise of the Indies" to King John II of Portugal, it still seemed that a westward sea passage might be not only the shorter but perhaps the only maritime route to the Indies.

A full decade earlier a westward sea passage to the Indies seems to have been under consideration by King John's predecessor, Alfonso V. He had sought the expert opinion of a famous Florentine physician-astrologer-cosmographer Paolo dal Pozzo Toscanelli (1397–1482), who, in a letter dated June 25, 1474, proposed "a shorter way of going by sea to the lands of spices, than that which you are making by Guinea." Basing his argument mainly on Marco Polo's report of the vast eastward extent of Asia and the location of "the noble island of Cipangu," or Japan—"most fertile in gold, pearls and precious stones, and they cover the temples and the royal residences with solid gold"—supposedly some fifteen hundred miles off the Chinese coast, Toscanelli confidently urged a trial of the westward passage. "Thus by the unknown ways there are no great spaces of the sea to be passed." Toscanelli, himself one of the more advanced cartographers of his age, had actually drawn a nautical map of the Atlantic Ocean, a copy of which he sent along with his letter to Lisbon.

In late 1481 or early 1482, when Columbus heard of this letter, he wrote to Toscanelli in great excitement and asked for more information. In return he received an encouraging letter along with another chart, which he eventually carried with him on his trip to prove that Toscanelli was right.

While Columbus had already been convinced, he now became passionate for this grand untested opportunity. Those who could finance him were harder to persuade. To convince investors to put their treasure in so novel an enterprise, he had to be at home in the writings of travelers, cosmographers, theologians, and philosophers. For, as we have already seen, geography as a separate discipline was in neither the trivium nor the quadrivium, and still had no place in the medieval Christian gamut of knowledge. Genoese, the tongue to which Columbus was born, was a spoken dialect, not a written language, and so was no help in his effort to document his Enterprise of the Indies. But Italian, which was a written language, and might have helped him, was a language that Columbus could not speak or write. He had no formal schooling where he might have learned Italian, and when by his own efforts he became literate, he wrote Castilian, which was then the favorite language of the educated classes on the Iberian peninsula, including Portugal. When Columbus wrote Castilian, he used Portuguese spelling, which suggests that he first spoke Portuguese. He may have written Portuguese too, but we have nothing in that language from his hand. He did somehow learn to read Latin, which was essential for his efforts to persuade the learned.

In 1484 Columbus made his first formal presentation of the Enterprise of the Indies to King John. At first the King was much taken by the enthusiasm of the engaging young Genoese. Columbus, having "read a good deal in Marco Polo . . . reached the conception that over this Western Ocean Sea one could sail to this Isle Cypango and other unknown lands," and for this purpose he asked the King to man and equip three caravels. But the King "gave him small credit," finding Columbus to be "a big talker and boastful in setting forth his accomplishments, and full of fancy and imagination with his Isle Cypango."

Despite his own doubts, the King was enough persuaded by the fast-talking Columbus to refer the project to a committee of experts. This group, which included an eminent cleric and two Jewish physicians respected for their knowledge of celestial navigation, turned down Columbus. Contrary to vulgar legend, their rejection was not based on any disagreement about the shape of the earth. Educated Europeans by this time had no doubt about the earth's sphericity. But the committee seems to have been troubled by Columbus' gross underestimate of the sailing distance westward to Asia. And, in the end, their misgivings proved better founded than were Columbus' hopes.

Of course, Europeans had no notion that there might be a land barrier between Europe and Asia in the form of two vast continents. At the very most, some of them suspected that in the Western Ocean there might be islands like Antillia, the mythical Island of the Seven Cities, and possibly others, which could serve as way stations. Columbus' optimistic calculations indicated that the direct westward sea voyage from the Canary Islands to Japan would be only twenty-four hundred nautical miles. A tempting prospect! And by no means beyond the capacity of Portuguese vessels in that day. The furthest Portuguese venture down the west coast of Africa, Diogo Cão's discovery of the Congo River in that very year, 1484, was more than five thousand nautical miles from Lisbon. And there still was no sign of where, or whether, the African continent really could be rounded by ships en route to the Indies. If Portuguese ships could go out and return safely to a destination five thousand miles away, past treacherous shoals and hostile natives, surely they could reach out a mere half that distance due west across the friendly open ocean.

King John II's committee did not allow themselves to be persuaded by Columbus' will to believe. Still, in 1485, presumably with the committee's expert concurrence, the King authorized two Portuguese, Fernão Dulmo and João Estreito, to try to discover the island of Antillia in the Western Ocean. This expedition was to be at their own expense, and they would be hereditary captains of any land they discovered. But they promised that after they had sailed west for forty days they would return home whether or not they had found any islands. We know no more of that ill-fated expedition than that they set sail in 1487. And that, unlike Columbus, they made the mistake of setting out from the Azores, in the high latitudes where the strong westerly winds made their expedition nearly impossible. Seeking Antillia for forty days was one thing, to go all the way to farthest Asia, quite another. The King's committee of experts was, of course, much nearer the truth than was the enthusiastic Columbus. The actual air-line distance from the Canaries to Japan is ten thousand six hundred nautical miles, and their estimates were probably in that neighborhood. They dared not encourage their king to invest in so speculative an enterprise.

Fourteen eighty-five proved a bad year for Columbus in more ways than one. In that year his wife died, and with his five-year-old son, Diego, he left the country where he had spent most of his adult life. He moved on to Spain, hoping there for better luck in promoting his monomaniacal project.

Columbus' successful enterprise would be almost as much a feat of salesmanship as of seamanship. Helped by his brother Bartholomew, he spent most of the next seven years peddling the Enterprise of the Indies in the courts of western Europe. In Spain he first awakened the interest of the Count of Medina Celi, a wealthy Cádiz shipowner. Celi might have financed the three caravels for Columbus' voyage if the Queen had not refused her assent. Such an expedition, if it went out at all, should certainly be a royal enterprise. She let a year pass before she would receive Columbus in audience. Then she, too, appointed a commission, under Hernando de Talavera, her confessor, to hear Columbus' proposal in detail, and to make recommendations.

Columbus now suffered through dreary years of academic and bureaucratic palaver by Queen Isabella and her Spanish minions. Meanwhile the commission proved its academic qualifications by neither approving nor rejecting the project. The professors learnedly debated the width of the Western Ocean and kept Columbus on the string with the pittance of a tiny monthly royal grant.

As the Spanish negotiations dragged on he recalled that King John II of Portugal had been personally friendly to him back in 1484–85, and so he decided to go back to Lisbon and try again. From Seville Columbus wrote the King of Portugal telling of his hopes. But Columbus had left Portugal in dire financial straits with many unpaid bills behind. He dared not return to Lisbon unless the King guaranteed him safe-conduct and freedom from arrest for his debts. The King agreed, praising Columbus' "industry and good talent," and urged him, "our particular friend," to come ahead. The King's renewed interest was doubtless due to the fact that the voyage of Dulmo and Estreito to Antillia had been abortive. Also there was still no word from Bartholomeu Dias, who had been gone seven months seeking the eastward sea passage to India, in the twentieth Portuguese attempt in that direction.

As it turned out, Columbus could not have chosen a worse moment. For, as we have seen, when Christopher and his brother Bartholomew arrived in 1488, they were just in time to be on the dock to see Bartholomeu Dias and his three caravels triumphantly sail up the Tagus with the good word that he had rounded the Cape of Good Hope and found that there really was an open eastward sea passage to India. Dias' success and all that it promised naturally killed King John's interest in Columbus. If the eastward sea passage was open and clear, why speculate in the other direction?

The Columbus brothers desperately hoped that this very Portuguese success to the east would stimulate the interest of rivals in a competing project in the other direction. Bartholomew, it seems, went to England, where he tried unsuccessfully to awaken the interest of King Henry VII, then went on to France where he solicited King Charles VIII. The French king himself was at first not receptive, but Bartholomew, encouraged by the friendly support of the King's elder sister, stayed on in France and was there supporting himself as a map-maker when news finally came of Columbus' great discovery.

Meanwhile Christopher returned from Lisbon to Seville, where he found Ferdinand and Isabella still vacillating. In disgust, he was actually en route to take ship for France to help Bartholomew persuade King Charles VIII when Queen Isabella, urged on by her keeper of the privy purse, suddenly decided to stake Columbus. Columbus' advocate had pointed out that support of his enterprise would cost no more than a week's royal entertainment of a visiting dignitary. Perhaps Isabella was persuaded by the fact that Columbus had shown his intention to offer the bargain enterprise to her rival sovereign next-door. She would pledge her crown jewels if needed to finance the trip. Fortunately, this proved unnecessary.

In her melodramatic last-moment decision, the Queen sent a messenger to catch Columbus before he embarked for France. Not till April 1492, eight years after Columbus had made his first offer to the king of Portugal, were the contracts, the so-called Capitulations, between Columbus and the Spanish sovereigns finally signed. The years of persuasion and promotion were at an end. Now Columbus' element would be the sea, where personal charms would not help, for there were no friends at Neptune's court.

Columbus had spent years collecting evidence and "expert witnesses" for the feasibility of a westward voyage to reach the Indies. The project, though surely not harebrained, was without doubt speculative. Yet its feasibility depended on two simple propositions that were not at all unorthodox.

The first, a Christian cartographic dogma, was that the surface of the earth was covered mostly by land. "Six parts hast thou dried up," declared the prophet Esdras (II Esd. 6:42). Among the orthodox it had become axiomatic that the surface of our planet was six-sevenths dry land and only one-seventh water. God's rationale for this seemed obvious, since He had set man above all the rest of Creation. "Nature could not have made so disorderly a composition of the globe," affirmed João de Barros, the Portuguese historian, who gives us our best account of Columbus' efforts to sell the king of Portugal on his project, "as to give the element of water preponderance over the land, destined for life and the creation of souls." If all the oceans together amounted to only one-seventh the surface of the earth and, as all the learned believed, the earth was a sphere, then there was not much sea available to separate Spain to the westward from the Indies, the Western Ocean could not be extensive, and Columbus' enterprise was feasible. *Q.E.D.*

The second proposition concerned the eastward extent of the land mass of Asia and the size of the whole earth. Obviously, the more extensive and eastward-reaching Asia was imagined to be, the narrower became the sea passage that Columbus proposed to make. On this subject the opinions of the most respected authorities varied wildly. Agreeing that the earth was a sphere encompassing all the way round 360 degrees of longitude, they expressed their estimates in the number of degrees of longitude between Cape St. Vincent in Portugal and the east coast of China. These estimates ranged from 116 degrees (in the Catalan Atlas of 1375) or 125 degrees (Fra Mauro, 1459) or 177 degrees (Ptolemy, A.D. 150), to maximum estimates of 225 degrees (Marinus of Tyre, A.D. 100) or 234 degrees (Martin Behaim, 1492). We now know that the correct figure is 131 degrees.

For a sailor going west, the practical meaning of any of these estimates in nautical miles depended on still another, even more important matter of opinion—the circumference of the earth. The length of one degree of longitude at the equator, 1/360 of the circuit of the earth, plainly varied with the size one assigned to the whole planet. Here, too, the most respected authorities widely, though not quite so wildly, disagreed. Their estimates of the circumference varied by some 25 percent, from the Catalan Atlas figure of about 20,000 miles to Fra Mauro's figure of about 24,000. Translating these figures into the length of a degree at the equator, they produced estimates that ranged from about 56 miles to about 66 miles. The correct figure is 69 miles.

Now it is easy to see that whether or not you thought Columbus' Enterprise of the Indies was feasible depended on which combination of figures you selected. If you believed that the land mass of Eurasia extended eastward from Cape St. Vincent to the coast of China only for some 116 degrees of longitude, that left the vast reach of 244 degrees for the westward oceanic distance between Portugal and China. And that made a sea voyage of 14,000 miles! We cannot be surprised that Columbus chose another set of figures.

We know a great deal about what Columbus read. For clues to Columbus' reaction to what he read, we have at least 2,125 of his "postils," his manuscript comments in the margins of his own copies of authoritative books. These books discussed the question of the eastward extent of Eurasia, the breadth of the Western Ocean, and the size of the earth. Of the books that Columbus owned, we have his copy of Plutarch's *Lives,* his copy of Ptolemy's *Geography* (1479) with no writing in it by Columbus, except his signature, and three other geographical works, all copiously annotated in Columbus' own hand.

Most extensively annotated was the *Imago mundi,* a world geography written about 1410, before the widespread revival of Ptolemy in Christian Europe. The author, the French theologian-astrologer Pierre d'Ailly, documented Columbus' hopes on the crucial questions of the eastward extent of Asia and the width of the Western Ocean. Columbus appears to have kept the *Imago mundi* by him for years, underlining passages in a variety of pens and inks, adding comments, summarizing points in the text, drawing an index finger to emphasize a sentence, with comments also in the hand of his brother Bartholomew. D'Ailly served Columbus well, not only because he adopted Marinus of Tyre's long stretch (225 degrees) for the eastward extent of Eurasia, but also because he made the Western Ocean conveniently narrow. More than that, d'Ailly explicitly refuted Ptolemy, whose shorter estimates of about 177 degrees for Eurasia made him a powerful witness against Columbus. "The length of the [Eurasian] land toward the Orient," the *Imago mundi* declared, "is much greater than Ptolemy admits . . . because the length of the habitable Earth on the side of the Orient is more than half the circuit of the globe. For, according to the philosophers and Pliny, the ocean which stretches between the extremity of further Spain (that is, Morocco) and the eastern edge of India is of no great width. *For it is evident that this sea is navigable in a very few days if the wind be fair* [Columbus' heavy underlining], whence it follows that the sea is not so great that it can cover three quarters of the globe, as certain people figure it."

Another heavily annotated work in Columbus' library—the *Historia rerum ubique gestarum* (1477) of Aeneas Sylvius (Pope Pius II, or Piccolomini)—collected scraps of enticing information about China, drawn from Marco Polo, Odoric of Pordenone, and others, with special emphasis on the Great Khan and the Emperor of China, along with tales of Amazons and anthropophagi. Then, of course, there was Marco Polo's Travels, which Columbus owned and extensively marked, and which had provided the basis for all the later estimates of the vast eastward extent of China.

The magic of the East that captured Columbus was concocted from the eloquent reminiscences of Marco Polo, the extravagant imaginings of Sir John Mandeville and others who had been inspired by them, the myths of Asian treasure, and the fables of fantastic animals and peculiar peoples. And by the despair over the Christians' failure to dislodge the infidel from the Holy Sepulcher which now diverted missionary efforts toward the heathens in Asia. Columbus must have been persuaded, too, by the axiom attributed to Aristotle, that one could cross from Spain to the Indies in a few days. And by the oft-repeated prophecy of Seneca, "An age will come after many years when the Ocean will loose the chain of things, and a huge land lie revealed; when Tiphys will disclose new worlds and Thule no more be the Ultimate."

31

Fair Winds, Soft Words, and Luck

CHRISTOPHER COLUMBUS' single-minded devotion to his Enterprise of the Indies and all the treasure of Ferdinand and Isabella would have counted for naught if Columbus had not had fair winds and if he had not known how to harness the winds to carry him there *and back*. The long-past Age of Sail has taken away with it the wonder we should feel for Columbus' mastery of the winds. Columbus was, of course, grossly mistaken about the continents. He did not really know the lands, but he did know the sea, which in his time especially meant to know the winds.

When, at the age of forty-one, Columbus won his opportunity to try his great enterprise, he already had a wide-ranging seafaring experience behind him. Under the Portuguese flag he had sailed from above the Arctic Circle nearly to the equator, and from the Aegean westward to the outer Azores. One trip had been in a vessel engaged in trading wool and dried fish and wine between the far-north regions of Iceland and Ireland, the Azores, and Lisbon. Then for a while Columbus had lived in Porto Santo, in the Madeira Islands, where one of his sons was born. From there he had sailed, and once actually commanded, voyages to São Jorge da Mina, the flourishing Portuguese trading post on the Gold Coast of the Gulf of Guinea. His varied experience in sailing northern latitudes and encountering all the perils of the sea would now at last be turned to a single grand purpose.

Columbus might have set out from Cádiz, the main Spanish seaport on the Atlantic, but Cádiz was crowded on his appointed day, for it had been designated as the principal point of embarkation for departing Jews. His day of departure, August 2, 1492, had also been fixed by their Most Catholic Majesties, Ferdinand and Isabella, as the deadline for the expulsion of all Jews from Spain. Any who remained thereafter were to be executed unless they embraced the Christian faith. Thousands whose only crime was faith in their God were on that very day being crammed into the holds of ships that crowded the narrow Rio Saltés. Out of the Gulf of Cádiz they would be shipped to a well-known, unfriendly old Christian world. Some would seek refuge in the Netherlands, others in the more tolerant world of Islam.

In the early morning of August 3 the very same tide that carried the hapless Jews to an old world of persecutions carried Columbus' three vessels from Palos de la Frontera near the mouth of the Río Tinto toward their unwitting discovery of a new refuge for the persecuted.

Columbus' own journal recorded that his trip was ordered only after the realm had been cleansed of Jews. His Catholic sovereigns now dispatched him to the idolaters of India on another high Christian mission—"their conversion to our Holy Faith, and ordained that I should not go by land (the usual way) to the Orient, but by the route of the Occident, by which no one to this day knows for sure that anyone has gone." As we have seen, Columbus' was not the first voyage from the Iberian peninsula to try the way westward into the Atlantic. Dulmo and Estreito, who had set out in 1487 to find the fabled island of Antillia, had the bad judgment to set sail directly west from the Azores in the high latitudes, and had never been heard from since. They had not come to terms with the wind.

Instead of setting his course due west from Spain, Columbus first sailed southward to the Canaries and so prudently avoided the strong westerly winds of the North Atlantic. At the Canaries, after this useful week-long "shake-down cruise," he turned due west, exploiting the advantages at that season of the northeasterly trade winds, which would carry him direct to his destination. An incidental advantage of this route from Columbus' point of view was that the Canary Islands happened to be on the same parallel of latitude as Cipangu (Japan), the particular destination that he had chosen from his reading of Marco Polo. He could run directly "westing," following his latitude until he reached his desired point in the Indies. It was on the parallel of the Canaries that Orient and Occident were said to come closest together, for there the islands of Japan, according to Marco Polo, lay fully fifteen hundred miles off the east coast of China.

Having set that course, Columbus found clear sailing. With a following wind, Columbus' ships scudded ahead. The strong following wind was so constant that Columbus' crew even began to fear that in those regions they might never find the westerly wind needed to take them home. They must actually have felt some relief, then, on September 19, when Columbus hove the deep-sea lead and still found no bottom at two hundred fathoms, that they ran temporarily into an area of variable winds. On October 5 the restless crew was encouraged by flocks of birds flying their way. Then, thirty-three days out, at two o'clock on the morning of October 12, a lookout on the *Pinta* claimed the announced 5,000-maravedi bonus, for he was the first to shout *"Tierra! Tierra!"*

For the homeward journey Columbus planned to go north, above the "horse latitudes," to the neighborhood of 35 degrees, where he would meet the westerly trades. While Columbus' plan for using the winds was quite correct, the homeward passage was troubled by storms.

In those days it was not for nothing that mariners were called "sailors," men of the sail. They were expected to fill their sails with the winds to take them where they wanted. "The wind bloweth where it listeth" was Saint John's way of describing the mystery of that world. "And thou hearest the sound thereof, but canst not tell whence it cometh, and whither it goeth: so is every one that is born of the Spirit" (John 3:8). For the skillful mariner, the Columbus, the winds were the mystery he had to master just as the steamer's captain must master the machinery of his ocean liner. Modern racing yachtsmen agree that a sailing vessel today, after all that has been learned in the last five centuries, could not do better than follow Columbus' route.

Was his course the product of solid knowledge of the winds or the dictate of a faultless seaman's instinct? Even before he set out, he already had personal experience of the ways of the winds in all the different latitudes where his voyage to the Indies would take him, and so he was well equipped to find the best course out and back. Acolytes of the sea (like Samuel Eliot Morison, whose *Admiral of the Ocean Sea* sings the mystique of seamanship) prefer to credit Columbus' intuition.

The "discovery" of America has overshadowed Columbus' other discoveries, which the passing of the Age of Sail has made it hard for us to appreciate. Even on his first voyage, as another seafaring historian, George E. Nunn, reminds us, Columbus really made *three* momentous discoveries. Besides finding land that Europeans had not found before, he discovered both the best westward sea passage from Europe to North America and the best eastward passage back. Columbus discovered the pathways necessary for ships whose source of energy was the wind. Although he might not have known where he really was going or where he had finally arrived, he was a knowledgeable master of the winds, which would make it possible for others to follow him.

Of course, he also had to manage his men, and keeping up the morale of a crew sailing into the unknown was no easy matter. Mutiny threatened on more than one occasion during the 33-day outward voyage. Passage to the Indies had to be accomplished before the patience of the crew wore thin. At the outset Columbus promised his men that they would find land when they had gone 750 leagues, or about 2,250 statute miles, west of the Canaries. They needed to be reassured that they were not going to some point of no return.

Columbus was not above using devious, even deceptive, techniques to keep his crew in good spirits and devoted to the common purpose. He did not forget his crew's concern for getting home, and in good time. To be sure that he would not discourage the men, he falsified his daily journal of the voyage. In noting down his estimates of distance traveled, "he decided to reckon less than he made, so that if the voyage were long the people would not be frightened and dismayed." For example, on September 25, Columbus himself believed that they had sailed a full 21 leagues, "but the people were told that 13 was the distance made good; for it was always feigned to them that the distances were less, so that the voyage might not appear so long." As it turned out, Columbus was deceiving them less than he intended. He did not realize that his own habit was to overestimate distance. The result was that the "phony" reckoning that he gave out to the crew turned out to be closer to the facts than his own "true" journal.

There were more than a few tricky moments during the outward passage. On September 21–23, for example, when they sailed into the Sargasso Sea, the vast oval area of the mid-South Atlantic which was blanketed with bright green and yellow sargassum gulfweed, the crew, who had never seen anything like this before, took alarm. Fearing their ships would be "frozen" in sargassum, they demanded that the captain change course to find open water. Columbus plowed straight ahead. Still today, however, sailors keep alive the superstitious fear of getting stuck in the Sargasso Sea.

Even the calm weather, lack of rain, and smooth seas became a cause for grumbling. If there was no rain, how to replenish the supply of fresh water on the salt ocean? If Columbus was to carry them endlessly westward, as some of the crew feared, perhaps their only hope of seeing their families again would be to toss him overboard. Columbus turned aside complaints with "soft words" and visions of the Indies' treasure which all would share. But he also reminded them of the dire consequences for the whole crew should they return to Spain without him. On his first outward voyage

Columbus had that most precious additional ingredient, luck. The weather was beyond compare, so (in his own words) "that the savor of the mornings was a great delight." "The weather was like April in Andalusia, the only thing wanting was to hear the nightingales."

Not the least remarkable, though the least celebrated, of Columbus' achievements was his ability on his later voyages to return to the lands he had first so accidentally and unwittingly encountered. Doubly remarkable, too, because Columbus' navigating techniques were so rudimentary. In Columbus' time, celestial navigation was quite undeveloped. He was unable to use even so elementary an instrument as the astrolabe. Despite romanticized illustrations to the contrary, he probably never saw a cross-staff. With his simple quadrant he was unable to do any useful sighting until he had been ashore at Jamaica for a full year. Only many years after Columbus' death did celestial navigation become normal equipment for the professional European pilot.

To set his direction and find his bearings at sea, Columbus depended on dead reckoning. This was less a scientific technique than a practical skill. He used the magnetic compass to fix direction, then estimated distance by guessing the speed at which they were sailing, by watching the bubbles, gulfweed or some other object float by. His estimates were crude, because not until the sixteenth century did someone invent the chip log to measure speed.

Dead reckoning served well enough for going from one known location to another, where the landscape, shoals, and currents were familiar. But it gave you no bearings into the unknown. Columbus, we must recall, thought he was traveling to a known destination.

32

Paradise Found and Lost

ON shipboard off the Azores in mid-February 1493, returning from his first voyage, Columbus wrote his own report of what he thought, and wanted others to think, that he had accomplished. Since it would have been disrespectful for him to address Ferdinand and Isabella directly, he reported to them in a "letter" addressed to Santangel, the crown official who had persuaded Queen Isabella, at the very last moment, to support Columbus' Enterprise of the Indies. Columbus' letter, written in Spanish, was printed in Barcelona about April 1, 1493, then translated into Latin dated April 29, and again printed in Rome in May as an eight-page pamphlet entitled *De Insulis Inuentis*. Frequently and speedily reprinted, by the standards of its day it became a best seller. At Rome there were three further editions in 1493, and six different editions printed at Paris, Basel, and Antwerp in 1493–94. By mid-June 1493 the Latin letter had been translated into a 68-stanza poem (printed in Rome and twice in Florence in 1493) in Tuscan, the dialect of Florence.

Northern Europe only slowly received news of Columbus' exploit. The famous Nuremberg Chronicle, an illustrated world history from the Creation to the present (printed on July 12, 1493), made no mention of Columbus' voyage. Not until late March 1496 do we find word of Columbus in England, and the first German translation of Columbus' letter was printed at Strasbourg in 1497.

What was the news that Columbus brought? The first illustrated Latin edition of his report (Basel, 1493) carried crude woodcuts that had already been used in earlier Swiss books that had no connection with Columbus, the Indies, or the New World. One woodcut purported to show the landing of Columbus in the Indies, in a forty-oared Mediterranean galley; another, supposed to represent the Bahama Islands, could have depicted any south-European seaside village.

Columbus, having convinced himself that a trip across the Western Ocean would take him to the Indies, now set about convincing a wider audience. He had a heavy vested interest in his destination actually being the Indies. In this first public announcement of his momentous voyage, Columbus was careful not to mention disasters or near disasters—the loss of the flagship, *Santa María,* the insubordination of Martín Alonso Pinzón, the commander of the *Pinta,* or the mutinous spirit of the crew. Following the national-security regulations of his day, he omitted information on the courses taken or the precise distance covered in order to prevent competitors from following where he had led. While Columbus conceded that he had not actually seen the Great Khan or the court of gold-rich Cipangu, he detailed numerous clues reinforcing his belief that he was just off the coast of China. The resplendent Great Khan, he was confident, would be found just a little farther on, doubtless on the next voyage.

Although Columbus was a hardheaded observer of the winds and waves, on the crucial question of where he had arrived he remained the slave of his hopes. He was determined to find signs everywhere that he had reached the fringes of Asia. Botany, still a vague wilderness whose images were not yet standardized by printing, was his happy hunting ground. From the moment when he first touched the north coast of Cuba on his first voyage, he had no trouble finding the Asiatic flora. A shrub that smelled like cinnamon he eagerly called cinnamon and so made it a hint of untold spice treasures. The aromatic West Indies gumbo-limbo, he insisted, must be an Asiatic form of the mastic tree of the Mediterranean that yielded resin. A small inedible nut, the *nogal de pais,* he hastily mistook for the coconut described by Marco Polo. The ship's surgeon examined some roots that the men had dug up and obligingly pronounced them the valuable medicinal Chinese rhubarb, a strong cathartic drug. Actually it was only the common garden rhubarb that we now use for pies and tarts, *Rheum rhaponticum,* not the pharmacist's *Rheum officinale.* But so many false scents somehow seemed to add up to the authentic odor of the Orient.

In the mind of Columbus such clues quickly clinched the thesis that had secured support for his Enterprise of the Indies. Typical of his frame of mind and his exploring techniques was his first expedition into Cuba. On October 28, 1492, Columbus' caravels entered Bahía Bariay, a beautiful harbor in the Oriente province of Cuba. There the captive San Salvador natives whom he had brought along as interpreters interviewed the local Indians and told Columbus that there was gold at Cubanacan (meaning mid-Cuba) only a short trek inland. Columbus eagerly assumed that they had meant to say "El Gran Can," the Great Khan of China, and at once sent an embassy to meet that Oriental potentate. An Arabic-speaking scholar whom he had brought along for just such missions was put in charge, accompanied by an able seaman who years before had once encountered an African king in Guinea, and so was supposed to know how to deal with exotic royalty. They took along diplomatic paraphernalia—their Latin passport, a letter of credence from their Catholic Majesties to his Chinese Majesty, and a rich gift for the Khan—along with glass beads and trinkets to buy food en route. Led on by visions of Cambaluc, which had been named by Marco Polo as the Mongol capital of China where the Khan held his splendid court, they hiked up the valley of the Cocayuguin River. What they found were some fifty palm-thatched huts. The local cacique feasted them as messengers from the sky and the people kissed their feet. But they got no word of the Great Khan.

On the trail back to the harbor, Columbus' two ambassadors did have one epochal encounter. They met a walking party of Taino Indians—"many people who were going to their villages, with a firebrand in the hand, and herbs to drink the smoke thereof, as they are accustomed." The long cigar that they carried would be relighted at every stop by small boys carrying along firebrands, then passed around for each member of the party to take a few drags through his nostrils. After the relaxing interval the Tainos resumed their journey. This was the first recorded European encounter with tobacco. Obsessed by visions of China's gold, Columbus' embassy saw only a primitive custom. Some years later, when the Spaniards had colonized the New World and learned to enjoy tobacco themselves, they introduced it to Europe, Asia, and Africa, where it was to become a source of wealth, delight, and dismay.

Meanwhile, Columbus remained back at the harbor, working up the figures from his dead reckoning to confirm his conviction that Cuba really was Marco Polo's Province of Mangi. He occupied his spare moments gathering the botanical specimens that he believed could not be found anywhere except in Asia.

The conspicuously pious Christian nomenclature that Columbus affixed on the lands he first visited—San Salvador, Navidad, Santa María de Guadalupe, S. M. de Monserrate, S. M. la Antigua, S. M. la Redonda, San Martín, San Jorge, Santa Anastasia, San Cristóbal, Santa Cruz, Santa Ursula y las xi mil Virgenes, San Juan Bautista—bore witness to his proverbial piety. It was his divinely appointed errand to enlarge the realm of the True Faith with the souls of pagan millions. His confidence as God's messenger had given him strength to bear years of ridicule, to risk mutiny, and would continue to shape his views of world geography.

Columbus' first voyage had many features of a Caribbean cruise, for he mainly enjoyed the sights and sounds and curiosities which he could witness from the coast, with only occasional short excursions inland. He had speedily coursed through the Bahamas, then skirted the northern coasts of eastern Cuba and of Hispaniola. Just three months after he first sighted the land of "the Indies," the island of San Salvador, his caravels set sail on January 16, 1493, from Samaná Bay on the eastern end of the island of Hispaniola to return home.

After only so brief a journey to the peripheral islands, with so little experience of the interior and such ambiguous clues to the Oriental character of the country, Columbus remained undaunted in his faith. His report revealed no doubt that he had reached "the Indies." And he generalized with all the confidence of the quickie tourist. The natives, he asserted, were "so ingenuous and free with all they have, that no one would believe it who has not seen it; of anything that they possess, if it be asked of them, they never say no; on the contrary, they invite you to share it and show as much love as if their hearts went with it, and they are content with whatever trifle be given them, whether it be a thing of value or of petty worth." "In all these islands, I saw no great diversity in the appearance of the people or in their manners and language, but they all understand one another, which is a very singular thing, on account of which I hope that their Highnesses will determine upon their conversion to our holy faith, towards which they are much inclined." The place he chose for La Villa de Navidad was "in the best district for the gold mines and for every trade both with this continent and with that over there belonging to the Gran Can [Grand Khan], where there will be great trade and profit." To their Catholic Highnesses he promised "as much gold as they want if their Highnesses will render me a little help; besides spices and cotton, as much as their Highnesses shall command; and gum mastic, as much as they shall order shipped . . . and aloe wood, as much as they shall order shipped, and slaves, as many as they shall order, who will be idolators. And I believe that I have found rhubarb and cinnamon, and I shall find a thousand other things of value, which the people whom I have left there will have discovered, for I have not delayed anywhere, provided the wind allowed me to sail. . . ."

During the next twelve years Columbus undertook three more voyages to "the Indies." They were called voyages of discovery, but more precisely they should have been called voyages of confirmation. For someone less committed they might have produced tantalizing puzzles, planting seeds of doubt. When these successive voyages still failed to connect with the Great Khan or to discover Oriental splendors, it became harder to persuade others back home. Although Columbus was ingenious at inventing new strategies of explanation, as his explanations became increasingly farfetched he once again became a butt of ridicule, a casualty of his own faith.

Only six months after returning from his first voyage, Columbus set out again. This time his expedition was on a far grander scale. Instead of three small caravels he had an armada of seventeen vessels and at least twelve hundred men (still no women), now including six priests to oversee the work of conversion, numerous officials to enforce order and keep books, colonists hoping to make their fortunes in the Indies, and, of course, the crews. While the first voyage could have been called merely exploratory, this second voyage was planned to make the exploration pay off. Columbus was now commissioned to set up a trading post in Hispaniola. He was under greater pressure than ever to prove that he had found the fabled treasure trove of the Indies. This time Columbus' feat of seamanship was even more impressive. While crossing the ocean, he managed to keep the seventeen ships together, and, as Samuel Eliot Morison boasts, "Columbus hit the Lesser Antilles at the exact spot recommended by sailing directions for the next four centuries!" His actual discoveries were also significant, for he found the Lesser Antilles, Jamaica, and Puerto Rico, explored the south coast of Cuba, and established the first permanent European settlement on this side of the Atlantic. Still, for Columbus this was not enough. He demanded the shores of Asia.

On this second trip, as Columbus proceeded through countless small islands of the Lesser Antilles he was encouraged by recalling Sir John Mandeville's observation that there were five thousand islands in the Indies.

By the time he reached the southern tip of Cuba he was convinced that he had reached the mainland of Asia. As he coasted Cuba westward from the Gulf of Guacanayabo he was confident that he was following the shores of Marco Polo's Mangi in southern China. When he reached Bahía Cortés, a point where the coast turns sharply toward the south, he knew that he was at the starting point of the eastern shores of the Golden Chersonese (the Malay Peninsula). If he had not yet found the sea passage that Marco Polo said would lead him to the Indian Ocean, he had found the peninsula at the end of which the sea passage would surely appear. But at this point his caravels were leaking, his ships' rigging tattered, his supplies low, and his crew showing signs of mutiny. Columbus decided to turn back. A pity. If he had gone only fifty miles farther, he could have discovered that Cuba was an island.

To protect himself against accusations of timidity or cowardice and to "confirm" his geographic ideas, from the officers and crew of all the three ships which he had detached for this exploring sally he extracted sworn depositions. This was not an unprecedented procedure, as Columbus must have known. Only six years earlier, in 1488, Columbus had been at Lisbon when Dias returned and had to justify his reversing course at the critical moment when he had clear sailing to India. As we have seen, Bartholomeu Dias had taken this same precaution to prove that it was his crew that forced him to return. But while Dias' crew had been asked only to certify Dias' courage and seamanship, Columbus' crew was asked also to certify Columbus' geography. The deposition that all were required to sign declared that the coast they had sailed three hundred and thirty-five leagues from east to west was longer than any island they had seen, that therefore they were certain that this coast must be part of a continent, which was obviously Asia, and that if they had sailed farther, they would have encountered "civilized people of intelligence who know the world." Columbus threatened to prove it by continuing the voyage until they had circumnavigated the globe. As an additional argument, Columbus explained that anyone who refused to sign would be fined ten thousand maravedis and have his tongue cut out. If the obstinate sailor happened to be a boy, he would take a hundred lashes on his bare back.

The return of Columbus' ships to Spain in March 1496 was anything but a triumph. He was warmly received at court, but the discovery of islands of the Indies in the Western Ocean no longer created a sensation. Like a second moon landing, Columbus' feat had somehow been minimized by showing that it could be repeated. Except among a few men of learning, word of this voyage was received with indifference. One reason surely was that the commercial payoff for the large investment had been so negligible. Some of his closest collaborators were beginning to doubt that Columbus'

"Indies" really were Asia. Juan de la Cosa, who had been master of the *Santa María* on Columbus' first voyage and was with him too on the second voyage, had actually signed the required "Cuba no island" deposition. But when La Cosa made his famous world map in 1500, he showed Cuba as an island. Dubious European map-makers for many years showed two Cubas —one an island, the other conforming to the shape of Marco Polo's Mangi as the mainland of south China.

The multiplying doubts of others hardened Columbus' obstinacy. "The Admiral . . . named the first coast he touched [in Cuba] Alpha and Omega," reported the chronicler Peter Martyr in 1501, "because he thought that there our East ended when the sun set in that island, and our West began when the sun rose. . . . He expected to arrive in the part of the world underneath us near the Golden Chersonese, which is situated to the east of Persia. He thought, as a matter of fact, that of the twelve hours of the sun's course of which we are ignorant he would have lost only two."

Now it was only with difficulty, and after two full years of promotional activity, that he collected a fleet of six ships for a third voyage to depart on May 30, 1498. New rumors and reports already suggested that a great land mass, perhaps not Asia, might lie somewhere to the west of the islands Columbus had found. But Columbus was not impressed. Instead he was more eager than ever speedily to find the sea passage around the Golden Chersonese into the Indian Ocean and so vindicate his hopes. On this third trip he encountered some geographic conundrums, which led him into fantasies which would discredit him among the best cartographers of his age. The faith of medieval Christian geographers was still very much alive in the mind of Christopher Columbus.

On this voyage his first "discovery" was the island he christened Trinidad in honor of the Holy Trinity. Then he happened into the Gulf of Paria, the bay formed by the delta of the great Orinoco River. Till then, of course, it had been an article of his faith that there could be no land mass down that way. But how explain this great freshwater sea, and the breadth of the freshwater rivers that poured into it? Was there, after all, a land mass not found in Ptolemy, which gathered this great flood of fresh water? "I believe that this is a very great continent, until today unknown," Columbus recorded with surprise in his journal. "And reason aids me greatly because of that so great river and fresh-water sea, and next, the saying of Esdras . . . that the six parts of the world are of dry land, and one of water. . . . Which book of Esdras St. Ambrose approved in his *Examenon* and so St. Augustine in the passage *morietur filius meus Christus,* as Francisco de Mayrones alleges; and further I am supported by the sayings of many Carib Indians whom I took at other times, who said that to the south of them was mainland. . . . and . . . that in it there was much gold. . . . and if this be a continent, it is a marvelous thing, and will be so among all the wise, since so great a river flows that it makes a fresh-water sea of 48 leagues."

Columbus the zealot was not unprepared for some grand new revelation about the shape of the planet. "God made me the messenger of the new heaven and the new earth of which he spoke in the Apocalypse by St. John, after having spoken of it by the mouth of Isaiah; and he showed me the spot where to find it." This revelation required revision of the orthodox dogmas of the earth:

> I have always read that the world comprising the land and the water was spherical, and the recorded experiences of Ptolemy and all others have proved this by the eclipses of the moon and other observations made from east to west, as well as by the elevation of the pole from north to south. But I have now seen so much irregularity that I have come to another conclusion respecting the earth, namely that it is not round as they describe, but of the form of a pear . . . or like a round ball, upon one part of which is a prominence like a woman's nipple, this protrusion being the highest and nearest the sky, situated under the equinoctial line, and at the eastern extremity of this sea where the land and the islands end. . . . [The ancients] had no certain knowledge respecting this hemisphere, but merely vague suppositions, for no one has ever gone or been sent to investigate the matter until now that Your Highnesses have sent me to explore both the sea and the land.

Here, finally, was the actual earthly location of the Scriptural landscape which medieval Christian cosmographers had so long featured at the top of their maps.

> I am convinced that it is the spot of the earthly paradise whither no one can go but by God's permission. . . . I do not suppose that the earthly paradise is in the form of a rugged mountain, as the descriptions of it have made it appear, but that it is on the summit of the spot which I have described. . . . I think also that the water I have described [i.e., of the Orinoco] may proceed from it, though it be far off, and that stopping at the place which I have just left, it forms this lake [the Gulf of Paria]. There are great indications of this being the terrestrial paradise, for its site coincides with the opinion of holy and wise theologians, and, moreover, the older evidences agree with the supposition, for I have never either read or heard of fresh water coming in so large a quantity in close conjunction with the water of the sea. The idea is also corroborated by the blandness of the temperature; and if the water of which I speak does not proceed from the earthly paradise, it seems to be a still greater wonder, for I do not believe that there is any river in the world so large or so deep.

Columbus' location of the Terrestrial Paradise on this unexpected mainland to the south was no random fantasy but the only rational explanation to reconcile the existence of some vast source of fresh water with his Christian doctrine, with his Ptolemaic geography, with the Asiatic identity of Cuba, and the certainty of a direct sea passage around the Golden Chersonese to the Indian Ocean.

To understand Columbus' problem, and why he sought refuge in his Terrestrial Paradise, we must remind ourselves of the Ptolemaic-Christian view of the lands of the earth. All the habitable parts of the earth were believed to be parts of a single land mass, the Island of the Earth, or *Orbis Terrarum,* including Europe, Asia, and Africa, surrounded by a comparatively small amount of water. The Book of Esdras argued the unity of all land that was not covered by the sea. Another vast land mass like the Americas separated by oceans on two sides from the Island of the Earth would not fit into that picture at all. Such a possibility suggested more water than there could possibly be (*ex hypothesi* on only one-seventh of the planet's surface). Moreover, lands of that sort would block the fulfillment of Columbus' great hope—a straight westward passage to India.

Another serious objection to new continents, from Columbus' orthodox point of view, was that, as we have seen, Christian doctrine refused to admit the possibility of habitable lands below the equator. Detached continents not continuous with the three-continent Island of the Earth, although suggested by pagan authors, were explicitly denied by the Church Fathers. Confronted with the possibility of a vast land mass where Christian doctrine insisted it could not be, the pious Columbus fitted it into his faith as the Terrestrial Paradise. In the best Christian sense that would be, in Columbus' phrase, *orbis alterius* or *otro mundo*—another world or another "island of earth."

Still, Columbus could round out his argument, confirm his Christian faith, and fulfill his obsessive reach for the Indies only by finding the sea passage around Marco Polo's Golden Chersonese. With this specific purpose, the same for which ten years earlier he had undertaken his first voyage, Columbus set out on his fourth and last voyage. With four caravels he left Seville on April 3, 1502. Somewhere between Cuba, which he still believed to be China, and the Terrestrial Paradise to the south he determined to find the strait through which Marco Polo had sailed from China into the Indian Ocean. This time he carried a letter of introduction from their Catholic Majesties to Vasco da Gama, whom he expected to meet in India. Of course, the Pacific Ocean, still unknown to Europe, did not figure in anybody's calculations.

A fair wind carried his four caravels across the Atlantic, from the Canaries to Martinique, in only twenty-one days. Columbus, who was fifty-one years old when he set out, called this fourth voyage *El Alto Viaje,* the High Voyage.

Still not discovering Cuba to be an island, he sailed southwestward from Cuba until he touched the Atlantic littoral of the present Republic of

Honduras. He then followed the coast eastward and southward, ever seeking the opening around this imagined Golden Chersonese to the Indian Ocean. He continued to find reassuring clues to the Asiatic character of this land in the form of botanical specimens and rumors of gold mines like those described by Marco Polo. After several disillusions—for example, when he explored Bahía Almirante near the Panama–Costa Rica border—he concluded that there was no sea passage in this area.

Instead of giving up his Asiatic hypothesis, Columbus seems to have concluded that there were actually two Asiatic peninsulas of the Golden Chersonese, one much longer than had been supposed. If only he had gone far enough southward along the coast, he still insisted, he would eventually have found his way around into the Indian Ocean. Perhaps, after all, the Gulf of Paria was not part of a separate *Orbis Terrarum,* but was merely an extension of this part of Asia. Columbus died believing that while he had incidentally found some Asiatic islands and peninsulas that had not yet appeared on the maps, all along he had been following the east coast of Asia.

33

Naming the Unknown

It was appropriate that the name America should be affixed on the New World in a manner casual and accidental, since the European encounter with this new world had been so unintentional. While the name and the person of Christopher Columbus were to be celebrated throughout the Americas, and his birthday would become a holiday, Amerigo Vespucci has been scarcely recognized and surely has not become a folk hero. "In this whole hemisphere," an eminent Latin American historian complains, "from Alaska to Tierra del Fuego, not one statue has been erected to him." This pioneer of the Age of the Sea, who deserves fame as an opener of the modern mind, has been caught in the cross fire of chauvinists, pedants, and ignorant but enthusiastic men of letters. "Strange . . . that broad America must wear the name of a thief," shouted the American pundit Ralph Waldo Emerson, with his eloquent indifference to facts, "Amerigo Vespucci, the pickle-dealer at Seville . . . whose highest naval rank was boatswain's mate in an expedition that never sailed, managed in this lying world to supplant

Columbus and baptize half of the earth with his own dishonest name." In these rotundities there was not a word of truth. More accurate was the early-eighteenth-century inscription by fellow citizens of Florence on the Vespucci family mansion, which described him as "a noble Florentine, who by the discovery of America rendered his own and his country's name illustrious; the Amplifier of the World."

Amerigo Vespucci was born of an influential family in Florence in 1454 in the seedtime and on the seed ground of the Italian Renaissance. There he spent the first thirty-eight years of his life, there he acquired the voracious curiosity and intellectual ambitions which governed his life. When Vasari went to study with Michelangelo in Florence, he stayed with Amerigo's uncle, who also played host to the poet Ludovico Ariosto. The Vespucci family was on friendly terms with Botticelli and Piero di Cosimo. Leonardo da Vinci so admired the face of Amerigo's grandfather that he followed him about the streets to sharpen in his mind the features that he later drew in a unique crayon portrait. Ghirlandaio painted the Vespucci family portrait, including Amerigo, in his fresco at the Church of the Ognissanti. When still a young man, Amerigo entered the service of the Medici family to help manage their wide-ranging affairs. Like his patron Lorenzo the Magnificent, Amerigo read widely, collected books and maps, and developed a special interest in cosmography and astronomy. Amerigo was sent to Spain in 1492 to look after Medici business interests. In Seville he became an outfitter of ships on his own account, and as he saw and learned more of seafaring adventure, his activities shifted from merchandising to exploring.

By 1499 Amerigo Vespucci's commercial and geographic interests had combined to draw him decisively into this new calling. It was obvious by then that the future of Spanish trade to the Orient would have to lie through the Western Ocean. The Portuguese had preempted the route around Africa, but Columbus had shown that lands could be reached by sailing west. Vespucci would try to fulfill Columbus' hopes of reaching Asia. The third voyage of Columbus, which had confirmed his wild fantasies about the Terrestrial Paradise, still revealed no passage to India. "It was my intention," Vespucci explained, "to see whether I could turn a headland that Ptolemy called the Cape of Catigara, which connects with the Sinus Magnus." Catigara, shown on Ptolemy's maps as the southeast tip of the Asian continent, was described by Marco Polo as the point around which Chinese treasure poured en route to the Sinus Magnus and the Sinus Gangeticus, the two great bays of the Indian Ocean. Since Ptolemy had located Catigara at eight and a half degrees south of the equator, that was where Vespucci would try to find the passage that had eluded Columbus.

In command of two ships, Vespucci joined the expedition headed by

Alonso de Ojeda which departed Cádiz on May 18, 1499. The expedition sighted land south of where Columbus had arrived on his third voyage. When Ojeda's other ships went northward for the treasures of the "Coast of Pearls," Vespucci coursed southeastward, groping for the passage around Catigara. "After we had sailed about four hundred leagues continually along one coast, we concluded that this was mainland; that the said mainland is at the extreme limits of Asia to the eastward and at its beginning to the westward." Vespucci was still ready to continue the search, but teredos (shipworms) had eaten the hulls of his ships, his provisions were low, and the wind and currents were against him. Reluctantly he made his way back to Spain.

Soon after returning to Seville, he made a "new decision to take another turn at discovery." "In due course," he wrote to a Florentine friend, "I hope to bring back very great news and discover the island of Tabrobana [Ceylon], which is between the Indian Ocean and the Gulf or Sea of the Ganges." The report of his first voyage, which Vespucci wrote for this Florentine friend and patron, revealed new worlds of thought and feeling. When Vespucci, like Columbus, reached across the Ocean, he, too, was thinking in Ptolemy's world. But now he spoke in a new voice.

> It appears to me, most excellent Lorenzo, that by this voyage of mine the opinion of the majority of the philosophers is confuted, who assert that no one can live in the Torrid Zone because of the great heat, for in this voyage I found it to be the contrary. The air is fresher and more temperate in this region, and so many people are living in it that their numbers are greater than those who live outside of it. Rationally, let it be said in a whisper, experience is certainly worth more than theory.

He refused to package random details into wholesale generalization. "Sailing along the coast, we discovered each day an endless number of people with various languages." "Very desirous of being the author who should identify the polar star of the other hemisphere, I lost many a night's sleep in contemplation of the motion of the stars around the South Pole, in order to record which of them had the least motion and was nearest to the pole." Instead of lines from a Church Father, he quoted Dante's verses, from *Purgatorio,* Book I, on what might be a view of the Antarctic pole.

The problem of determining longitude, crucial in westward voyages across the ocean, had long puzzled Vespucci. On the track of a new way of solving the problem, he had brought with him astronomical tables of the moon and planets. During twenty days of enforced leisure, August 17 to September 5, 1499, when his crew was recovering from a battle with the Indians, he turned to this question.

As to longitude, I declare that I found so much difficulty in determining it that I was put to great pains to ascertain the east-west distance I had covered. The final result of my labors was that I found nothing better to do than to watch for and take observations at night of the conjunction of one planet with another, and especially of the conjunction of the moon with the other planets, because the moon is swifter than any other planet. . . .

After I had made experiments many nights, one night, the twenty-third of August, 1499, there was a conjunction of the moon with Mars, which according to the almanac [for the city of Ferrara] was to occur at midnight or a half hour before. I found that when the moon rose an hour and a half after sunset, the planet had passed that position in the east.

Using this data, Vespucci calculated how far west he had come. His astronomical method could eventually yield results far more accurate than the dead reckoning used by Columbus and others at the time, but, for lack of precision instruments, was not yet practical. Even so, during his calculations of the length of a degree, he improved the current figure and produced an estimate of the earth's equatorial circumference which was the most accurate until his time—only fifty miles short of the actual dimensions.

When Vespucci set out on his next voyage, which would offer him the occasion to announce his doubts about Ptolemy, to break with the hallowed traditions of cosmography and declare a new world, he went under another flag. Now he sailed not for Ferdinand and Isabella of Spain but for King Manuel I of Portugal.

Vespucci's easy change of flags recalls the remarkable collaboration and mutual restraint between the two great competing seafaring powers, Spain and Portugal, in this pioneer age of seafaring discoveries. For more than a quarter-century after the first voyage of Columbus, Spain and Portugal remained peaceful and even cooperative in their separate efforts to discover the new world in the Western Ocean. The intermarriages of the heirs and sovereigns of Portugal with those of Castile and of Aragon were not the whole story. Competitors, they somehow became fellow seekers. In advance they made rules for sharing a new world of unknown dimensions and unknowable resources. In advance Spain and Portugal divided the whole un-Christian world between them.

What made that agreement possible and gave force to their agreement was their common acceptance of an outside authority, the pope, who, without an army or navy, exercised enormous spiritual power. Respect for papal authority was all the more remarkable at this time, for when Columbus made his first voyage the seat of Saint Peter was occupied by a notoriously dissolute Borgia, Alexander VI (1431?–1503; pope, 1492–1503). As priest and cardinal, he had enjoyed many mistresses and fathered numerous children. Born near Valencia in Aragon, he had secured his election to the papacy through bribery and the intervention of Ferdinand and Isabella.

The Christian community in Europe had long recognized the pope's right to allot temporal sovereignty over any lands not already claimed by a Christian ruler. "The Pope who is the vicar of Jesus Christ," declared the thirteenth-century decretals of the Church, "has power not only over Christians, but also over all infidels. . . . For all, the faithful and infidels alike, are the sheep of Christ through the creation, though they may not be of the flock of the Church." Even before Columbus, the kings of Portugal had endorsed this papal power by securing papal bulls to confirm their rights to the African coast all the way to the Kingdom of Prester John, "as far as the Indians who are said to worship Christ." Then Columbus' voyage and the surprising islands of the "Indies" had stirred up new possibilities, which the Spanish sovereigns were alert to notice.

In mid-April 1493, within a month after Columbus returned from his voyage, Columbus' letter describing his exploit was known in Rome, and extracts from it were actually included in a papal bull on the subject of these new lands which Alexander VI issued on May 3. "Of all the pontiffs who have ever reigned," Machiavelli had certified this Alexander VI as the one who "best showed how a Pope might prevail both by money and by force."

Even while conniving with Spain's enemies, in a series of four bulls the devious Pope gave to Spain all newfound lands in the newly explored "Indies." In these bulls he drew the famous demarcation line running from the North to the South Pole "one hundred leagues towards the west and south from any of the islands commonly known as the Azores and Cape Verdes." All lands discovered west of that line and not already possessed by a Christian prince would belong to Spain.

It seems that this line was proposed by Columbus himself. Its rationale was defended by some arrant pseudoscientific nonsense. Columbus asserted that just beyond that 100-league mark the climate suddenly changed "as if you had put a hill below the horizon," the temperature became mild "and no change winter or summer," the sea suddenly became full of weeds. "Up to the Canaries and 100 leagues beyond, or in the region of the Azores, many are the lice that breed; but from there on they all commence to die, so that upon raising the first islands [of the Indies] there be no man that breedeth or seeth one." In his fourth bull, the Pope actually transferred the eastern routes to the Indies and lands there discovered to Spain.

King John II of Portugal, who had the advantage of a superior navy, would not stand by while the Pope gave away his empire. With his navy behind him, he negotiated with Ferdinand and Isabella to avoid the consequences of the Pope's declarations. At Tordesillas in northern Spain, on June 7, 1494, an epoch-making treaty moved the line of demarcation farther west, to the meridian three hundred seventy leagues west of the Cape Verde Islands.

Both Spain and Portugal showed remarkable goodwill in their efforts to obey the terms of their treaty, even though the technology of that time could not yet precisely locate the prescribed meridian. One lasting result of this agreement was the Portuguese settlement with the Portuguese language in Brazil, and the prevalence of the Spanish people and language elsewhere in South America. This complaisance of the principal powers competing for seafaring empire would last only as long as Spain and Portugal, both deferring to papal sovereignty, were dominant. After the Protestant Reformation, when European rulers disagreed over the sources of spiritual sovereignty, such pacific agreements were difficult or impossible. When the English, the Dutch, and others joined in a free-for-all, spheres of discovery would be defined only by the competing powers of armies and navies.

Vespucci's first voyage under the Spanish flag had suggested to him that in order to reach the passage to the Indies around Ptolemy's "Strait of Catigara" he would have to follow the coastline eastward and then southward, through areas decisively in the Portuguese realm. It was not surprising, then, that on his next voyage to "the Indies" Vespucci went under the auspices not of Spain but of Portugal. Other considerations may have led to his changing flags. Since the Spanish sovereigns did not yet know that the eastern point of South America was actually on the Portuguese side of the agreed line, perhaps they deliberately cut the Florentine Vespucci out of their expeditionary force, preferring to see this potential empire explored by their own nationals.

On May 13, 1501, nearly a decade after Columbus' first crossing, Amerigo Vespucci, commanding three caravels, departed Lisbon on the momentous sixteen-month voyage that would reap the harvest which Columbus had prepared. Delayed by the doldrums, Vespucci's passage "across the ocean wastes in search of new land" required sixty-four days. "We arrived at a new land which, for many reasons that are enumerated in what follows, we observed to be a continent."

Vespucci had followed the South American coast for about eight hundred leagues, about twenty-four hundred English miles, "always in the direction of southwest one-quarter west," which took him well down into Patagonia, near the present San Julián, only some four hundred miles north of the southern tip of Tierra del Fuego. When Vespucci returned to Lisbon in September 1502, he wrote again to his Florentine friend and patron.

We coursed so far in those seas that we entered the Torrid Zone and passed south of the equinoctial line and the Tropic of Capricorn, until the South Pole stood above my horizon at fifty degrees, which was my latitude from the equator. We navigated in the Southern Hemisphere for nine months and twenty-seven days [from about August 1 to about May 27], never seeing the Arctic Pole or even Ursa Major and Minor; but opposite them many very bright and beautiful constellations were disclosed to me which always remain invisible in this northern Hemisphere. There I noted the wonderful order of their motions and their magnitudes, measuring the diameters of their circuits and mapping out their relative positions with geometrical figures. . . . I was on the side of the antipodes; my navigation extended through one-quarter of the world. . . .

The inhabitants there were numerous, yet the infinite variety of trees, the sweet-smelling fruits and flowers and display of brilliantly plumed birds stimulated "fancies" of the Terrestrial Paradise. "What should I tell of the multitude of wild animals, the abundance of pumas, of panthers, of wild cats, not like those of Spain, but of the antipodes; of so many wolves, red deer, monkeys, and felines, marmosets of many kinds, and many large snakes?" Vespucci was led to the heretical conclusion that "so many species could not have entered Noah's ark."

With omnivorous curiosity and the studied elegance of a Florentine man of the Renaissance, Vespucci described the natives' faces and figures, their marriage customs, childbirth practices, religion, diet, and domestic architecture. Since these people used only bows and arrows, darts and stones, all their blows were, in Petrarch's phrase, "committed to the wind." Flaunting no pious hopes for conversion of the natives, only once did Vespucci cite a Christian author. "The natives told us of gold and other metals and many miracle-working drugs, but I am one of those followers of Saint Thomas, who are slow to believe. Time will reveal everything." And no word any more of Ptolemy!

Despite all these fascinating New World novelties, the wish for a westward passage to India still dominated. The unexpected continent continued to seem less a resource for new hopes than an obstacle to the old. Vespucci, too, appeared less interested in exploring this Fourth Part of the World than in finding passage through it to the proven treasure trove in the true Asian Indies. In a month after he returned to Lisbon from his momentous voyage he changed flags once again, and moved back to Seville. Vespucci's voyages, and his work in revising the map of the western Atlantic, had persuaded him that Ptolemy's Strait of Catigara would not be found on this unexpected Fourth Continent. He had followed the whole coastline that belonged to Portugal without finding an opening, and he knew therefore that if there was a passage through to India, it would have to be farther west, on the Spanish side of the agreed line of demarcation. At that time, too, when Portugal was already amassing treasure from her monopoly of the eastward sea trade to India, the Spanish sovereigns were making an orga-

nized effort to improve Spanish seamanship to find a better westward passage. Foreign scholars were welcomed, the University of Salamanca was
re-endowed, and Queen Isabella herself had taken to collecting printed
books, a new resource of knowledge.

The Spanish sovereigns welcomed Vespucci and immediately assigned
him the task of provisioning caravels for an expedition to sail "westward,
north of the equator, to seek for discovery of a strait not found by Columbus." Vespucci's eminence was attested in 1508, when Queen Joanna of
Castile, who had succeeded to the throne of Isabella, named Amerigo
Vespucci to the newly created post of "pilot-major of Spain." He was to
found a school for pilots and was given exclusive authority to examine and
license "all the pilots of our kingdoms and lordships who voyage hereafter
to the said lands of our Indies, discovered or to be discovered." Returning
pilots were ordered to report to him all their findings, which would keep
Spanish maps up to date. Against the resistance of illiterate, practical-
minded pilots, Vespucci tried to popularize his complex method for finding
longitude. He made plans for another voyage of his own in ships covered
with lead, which protected their hulls against the teredos, to "go west to
find the lands which the Portuguese found by sailing east." But still suffering from the malaria which he had contracted on his last voyage, and for
which there was no known cure, Amerigo Vespucci died in 1512.

It is not surprising that the newness of the New World, with all its unimagined opportunities, did not take Europe by storm. Booksellers and map-
makers had a vested interest in the supposed accuracy of items on their
shelves and in the wood blocks and plates from which these were made. The
most respectable maps and globes and planispheres had left no room at all
for a Fourth Continent. The vocabulary of papal bulls and the administrative forms of government departments all encouraged people to stay in
linguistic ruts. Since Columbus had "discovered" those lands in the "Indies," it seemed prudent as well as convenient to keep thinking of the new
overseas empire in that way, and not to give it the legally perilous implications which a newfangled name might suggest. The Spanish government
continued to convene its Council of the Indies, promulgated its Laws of the
Indies, and indelibly christened the New World natives as Indians. In
Spanish, histories of the New World multiplied as histories of the "Indies."
Even if the New World should turn out not to be part of the Asian continent, it was still safer for the moment to think of it as an outpost of Asia.

But some, excited by Vespucci's voyages, enjoyed the tonic notion of an
unexpected part of the earth. The christening of their New World was done
not ceremoniously by sovereigns or by a prestigious convocation of the
learned but casually and informally in a place that Vespucci himself never

visited and had probably never heard of. Vespucci never named the continent after himself, although he was often accused of that conceit. It was
Alexander von Humboldt (1769–1859), the great German explorer and naturalist, who, by his own account, "earned a modest merit by having proved
that Amerigo Vespucci had no part in the naming of the New Continent,
but that the name America originated in a hidden spot of the Vosges
Mountains."

The christening was the work of Martin Waldseemüller (1470?–1518), an
obscure clergyman, who had studied at the University of Freiburg. Waldseemüller was a man of broad interests with a poetic feeling for words, and a
passion for geography. When he became canon of the little town of Saint-Dié,
built around a monastery founded in the seventh century by Saint Deodatus
in the Vosges Mountains in the Duchy of Lorraine in northeastern France, he
found the post congenial. The ruling duke of Lorraine, Renaud II of Vaude-
mon, wanting to cultivate the arts, had organized a provincial learned society, a kind of salon, and Waldseemüller became a member of this *Gymnase
Vosgien.* A wealthy member of the group, Canon Walter Ludd, indulged his
vanity by setting up a printing press in 1500 to publish his own works and
incidentally those of other members of the society.

Waldseemüller's idiosyncrasy was a penchant for making up names.
Wanting an impressive Latin pseudonym for himself, he combined the
Greek word for "wood," the Latin word for "lake," and the Greek word
for "mill" to make "Hylacomylus" for use in his learned publications.
Translated back into the German vernacular this became his family name,
Waldseemüller. Led by Waldseemüller, the little group had ambitious plans
to print a new edition of Ptolemy's geography as the first item from their
press. Then one of them reported seeing a printed copy of a French letter
entitled "Four Voyages," in which

> . . . a great man, of brave courage, yet small experience, Americus Vespucius, has
> first related without exaggeration of a people living toward the south, almost
> under the antarctic pole. There are people in that place . . . who go about entirely
> naked, and who not only (as do certain people in India) offer to their king the
> heads of their enemies whom they have killed, but who themselves feed eagerly
> on the flesh of their conquered foes. The book itself of Americus Vespucius has
> by chance fallen in our way, and we have read it hastily and have compared
> almost the whole of it with the Ptolemy, the maps of which you know we are at
> this time engaged in examining with great care, and we have thus been induced
> to compose, upon the subject of this region of a newly discovered world, a little
> work not only poetic but geographical in its character.

The Saint-Dié group suddenly dropped their grand plan for an edition of
Ptolemy. Instead, they produced a little volume of 103 pages called *Cosmo-*

graphiae Introductio, which summarized the traditional principles of cosmography, including definitions of axes and climata, the divisions of the earth, the winds, and the distances from place to place. It also offered something sensationally new, an account of a *fourth* part of the world revealed in the voyages of Amerigo Vespucci. In a summary chapter Waldseemüller casually observed:

> Now, these parts of the earth [Europe, Africa, Asia] have been more extensively explored and a fourth part has been discovered by Amerigo Vespucci (as will be described in what follows). Inasmuch as both Europe and Asia received their names from women, I see no reason why any one should justly object to calling this part Amerige [from Greek "ge" meaning "land of"], i.e., the land of Amerigo, or America, after Amerigo, its discoverer, a man of great ability.

In two other places in his text Waldseemüller reinforced this suggestion. To accompany these items, as the third part of the *Cosmographiae,* he printed an impressive large map from twelve wood blocks made in Strasbourg. Each sheet measured eighteen by twenty-four and a half inches, and when pasted together, the map covered about thirty-six square feet. At the top, Waldseemüller emphasized his new message by two dominating portraits: Claudius Ptolemaeus faced east, and Americus Vespucius faced west. In this astonishing cartographic prophecy the South American continent on which "America" was inscribed showed a contour remarkably similar to its actual shape. On the inset map the two Americas were actually connected. Farther west appeared a whole new ocean, broader than the Atlantic, separating the New World from Asia.

Whatever those brave and famous explorers may have done, it took the obscure Martin Waldseemüller to put America on the map. This first book from the press of Saint-Dié, in April 1507, was so popular that a second edition was published in August. In the next year Waldseemüller boasted to his partner that their map was known and commended throughout the world. Soon he announced that they had sold one thousand copies.

The printing press could disseminate, but it could not retrieve. To his annoyance, Waldseemüller himself learned the fantastic, irreversible reach of this new technology. When Waldseemüller changed his mind and decided that after all Amerigo Vespucci should not be credited as the true discoverer of the New World, it was too late. On all the three later maps that he published showing the New World, he had deleted the name "America." But the printed messages advertising America were already diffused into a thousand places and could not be recalled, and "America" became indelibly imprinted on maps of the world. So appealing was the name, which Waldseemüller himself had applied only to the southern continent, that when Gerardus Mercator published his large map of the world in 1538, he doubled its application. Mercator's map showed both a "North America" *(Americae pars septentrionalis)* and a "South America" *(Americae pars meridionalis).*

The printing press, still only a half-century old, revealed its unprecedented power to diffuse information—and misinformation. The widened audience, a new reading market created by and for the new technology, was already shaping the printed product to its tastes. The Columbus Letter of 1493, though reprinted numerous times, had nothing like the appeal of the sensational account of Vespucci's adventures, the *Mundus Novus* of 1502. While readers were, of course, curious about how Columbus had reached the fabled well-documented "Indies," they were more interested in this surprising "Fourth Part of the World." During the quarter-century after the first published account of Vespucci's voyages, publications about Vespucci's travels exceeded those of Columbus threefold. In those years, of all the works we can find printed in Europe describing the New World discoveries, about one-half dealt with Amerigo Vespucci. Vast audiences were now being equipped to receive messages of new worlds.

PART EIGHT

SEA PATHS TO EVERYWHERE

An age will come after many years when the Ocean will loose the chain of things, and a huge land lie revealed; when Tiphys will disclose new worlds and Thule no longer be the Ultimate.

—SENECA, *Medea*

And if there had been more of the world,
They would have reached it.

—CAMOENS, *The Lusiads,* VII, 14

34

A World of Oceans

WITHIN a few decades the European world concept would be transformed. The dominant Island of the Earth, a connected body of *land* comprising six-sevenths of the surface, was displaced by a dominant Ocean of the Earth, a connected body of *water* comprising two-thirds of the surface. Never before had the arena of human experience been so suddenly or so drastically revised. And the earth became more than ever explorable.

We can witness this discovery of the ocean in the well-chronicled feats of two hero-leaders—Balboa and Magellan—men of contrasting temperaments and talents, both from the Iberian peninsula.

Vasco Núñez de Balboa (1474–1517), an adventurer born of obscure parents in an inland village of southwestern Spain, went to sea at the age of twenty-five, but he was destined to do his historic work on land. In 1500 he joined an expedition to explore the Spanish Main, and remained to become a planter in Santo Domingo. Plainly not suited to a sedentary life, he accumulated debts and to flee his creditors he stowed away on a ship on its way to Spanish settlements on the east coast of the Gulf of Darien, where the Isthmus of Panama joins the South American mainland. The Spanish settlers there had been decimated by starvation and the poisoned arrows of Indians. When the newly arrived commander, Martín Fernández de Enciso (1470?–1528), a prosperous and scholarly bachelor of laws, proved unequal to the task of organizing a new colony, the upstart Balboa seized command. He moved to a better site where food was available and where the Indians had no poisoned arrows, christening it Santa María de l'Antigua del Darién, now known as Darien. Diego, Columbus' son, then governing this region from a capital in Santo Domingo, authorized Balboa to keep command, but Enciso and the other original officers resisted and Balboa shipped his rivals back to Spain. Then Balboa conciliated the local Indians by helping Comaco, the cacique, fight his wars, and marrying one of Comaco's daughters.

Comaco thanked his new allies with a gift of four thousand ounces of gold. But as the Spaniards were weighing it out to separate the crown's share, "brabbling and contention arose." A son of the cacique became so disgusted by this spectacle that he knocked over the scales, scattered the gold on the ground and delivered a concise sermon against their greed. At the same time, according to the contemporary chronicler Peter Martyr, he volunteered a gem of geographical information worth more than all the gold of the Indies:

What is the matter, you Christian men, that you so greatly esteeme so little portion of gold more than your owne quietnesse . . . If your hunger of gold, bee so insatiable, that onely for the desire you have thereto, you disquiet so many nations, . . . I will shewe you a region flowing with golde, where you may satisfie your raveninge appetites. . . . When you are passing over these mountaines (poynting with his finger towarde the south mountaines) . . . you shall see another sea, where they sayle with shippes as bigge as yours, using both sayles and ores as you doe, although the men be naked as we are.

At once the alert Balboa selected one hundred and ninety of his own men and several hundred native guides and porters and set out to follow this lead across the mountainous Isthmus of Panama. Careful to conciliate the Indians who might have threatened his rear, he employed them as "guides and bearers who went on ahead and opened the trail. They passed through inaccessible defiles inhabited by ferocious beasts, and they climbed steep mountains."

The dark recesses of the tropical rain forest were like nothing they had known before. Later explorers found that Balboa's route still taxed them to the limits of courage and endurance. In the mid-nineteenth century a French explorer reported that he could not see the sky for eleven days, while a German botanical expedition, trying to struggle through, lost every man. To cross the numerous swamps and lakes, Balboa's men had to strip naked and carry their clothes on their heads, risking poisonous snakes and the arrows of unknown tribes. When blocked by the primitive Quarequas, who were armed only with bows and arrows and two-handed wooden swords, Balboa's men took them apart "like butchers cutting up beef and mutton for market. Six hundred including the cacique, were thus slain like brute beasts." As Peter Martyr reported:

Vasco discovered that the village of Quarequa was stained by the foulest vice. The king's brother and a number of other courtiers were dressed as women, and according to the accounts of the neighbours shared the same passion. Vasco ordered forty of them to be torn to pieces by dogs. The Spaniards commonly used their dogs in fighting against these naked people, and the dogs threw themselves upon them as though they were wild boars or timid deer. The Spaniards found these animals as ready to share their dangers as did the people of Colophon or Bastabara, who trained cohorts of dogs for war; for the dogs were always in the lead and never shirked a fight.

After twenty-five days of "many adventures and great privation" the cordillera was finally crossed.

On September 25, 1513, their Quarequa guide pointed them toward a nearby peak. Vasco commanded his men to halt while he ascended, and from the top he glimpsed an ocean in the distance. "Kneeling upon the ground, he raised his hands to heaven and saluted the south sea; according to his account, he gave thanks to God and all the saints for having reserved this glory for him, an ordinary man, devoid alike of experience and authority." He then beckoned to his men to join him on the peak, where they all knelt and gave thanks together. "Behold the much-desired ocean!" he exclaimed. "Behold! all ye men, who have shared such efforts, behold the country of which the sons of Comogre and other natives have told us such wonders!" He heaped up stones to form an altar, while his men carved the name of their king on tree trunks around the slopes. In accordance with Spanish custom, the notary they had brought along drew up a sworn statement signed first by Balboa himself, then by all the others.

Another four days' trek took them to the shore of this newly discovered ocean. In a climactic gesture, Balboa, wearing his armor and carrying his unsheathed sword, waded into the surf, raised the banner of Castile, and for his Catholic sovereigns formally took possession of this Mar del Sur, the Southern Sea. Balboa called it the "Southern" Sea for an obvious reason. The Isthmus of Darien, which he had just crossed, runs east and west. Starting from the Caribbean, he had come southward and in that direction he first sighted the Pacific. He took possession, too, "of all that sea and the countries bordering on it" by a brief ceremonial ride in dugout canoes borrowed from the local Indians.

This was the height of Balboa's fortunes. Incidentally, on his way back across the Isthmus, friendly or frightened Indian tribes presented him two hundred forty select pearls and four pounds of lesser quality with six hundred fourteen pesos in gold. News of his discovery did not reach Spain in time to overshadow Enciso's disastrous reports of Balboa's coup. To displace Balboa as governor came Pedrárias Dávila, whose only qualification was his marriage to a lady-in-waiting to Queen Isabella. With twenty ships and fifteen hundred men, Pedrárias initiated a program for enslaving the natives which had the immediate effect, according to Balboa himself, of changing sheeplike Indians into "fierce lions." Meanwhile Balboa, planning to explore the shores of the Southern Sea, had transported shipbuilding materials across the Isthmus. In 1517 he had nearly completed four ships when Pedrárias' men, including a certain Francisco Pizarro, came to arrest Balboa and take him back across the Isthmus to Darien. There Pedrárias falsely accused Balboa of planning to abandon his Spanish sovereigns and set himself up as emperor of Peru. Before Balboa's supporters could consolidate their defense, Balboa and four companions were beheaded in the public square, and their bodies were thrown to the vultures.

Spanish adventurers had now firmly located themselves in the West Indies. But they continued to believe these were only outposts toward Asia. Was it not logical to extend that domain still farther west to the valuable Spice Islands?

The Treaty of Tordesillas, as we have seen, had fixed a line of demarcation 370 leagues west of the Azores and Cape Verde Islands and established the New World boundary at 46 degrees west longitude, cutting through the bulge of South America. Since the pope's dominion covered the whole globe, this meridian line extended through both the poles and went all the way around the planet on the other side. This same line therefore served also to separate the domains of Spain from those of Portugal on the Asian half of the planet. There the meridian of 134 degrees east longitude became the dividing line between the two great powers. But scientific instruments could not yet mark off precisely this line of longitude. In practice the division meant that Portugal would be sovereign over all pagan or still-undiscovered areas eastward from the western boundary of Brazil across the Atlantic, Africa, and the Indian Ocean to the East Indies, while Spain would rule from the western boundary of Brazil westward across the Pacific to the East Indies.

Nobody yet knew what intervened between this new Fourth Part of the World and Asia. Spanish hopes were still high that Ptolemy, Marco Polo, and Columbus had been correct in extending the Asian continent so far eastward. Perhaps it was only a short hop, maybe along a chain of still to be discovered Asiatic islands, from America to the East Indies. The Spanish Emperor Charles V naturally hoped that the Spice Islands would prove to be on the eastern, Spanish side of the dividing line when it was extended on that Asian half of the globe. Why not send out an expedition to mark off that line and then assert Spanish claims? Here was Magellan's opportunity.

The mountainous northern part of Portugal where Ferdinand Magellan (1480?–1521) had been born of a noble family was known among his countrymen as a place that experienced "nine months of winter and three months of hell." From this toughening climate he went to the mild life of the court of Queen Leonor, consort of King John II, where he was raised as a page. At the age of twenty-five he joined the fleet of Francisco de Almeida, the first viceroy of Portuguese India (1505–1509), then served with Afonso de Albuquerque, founder of the Portuguese empire in Asia, and explored the Spice Islands, the Moluccas, where he personally assessed the treasure to be captured there. By the time Magellan returned to Portugal in 1512 he had reached the rank of captain and he was raised to *fidalgo escudeiro,* a higher rank of nobility. Wounded while fighting with Portuguese forces against the

Moors in North Africa, he was lamed for life. When he was accused of trading with the enemy, he lost the favor of King Manuel and so ended his Portuguese career.

Magellan publicly disowned his loyalty to Portugal, and left for the Spanish court of Emperor Charles V. With him he brought an old friend, Rui Faleiro, a megalomaniac mathematician-astrologer who mistakenly thought he had solved the problem of determining longitude, who had a great reputation as a cosmographer, and who passionately advocated the southwestern seaway to Asia. To promote his grandiose westward expedition halfway around the world to the Indies, Magellan played his cards shrewdly. He married the daughter of an influential Portuguese émigré who controlled the Spanish voyages to the Indies, and then secured the enthusiastic approval of Juan Rodríguez de Fonseca (1451–1524), the powerful bishop of Burgos, organizer of the Council of the Indies, who had been a chief enemy of Columbus. And for funds he cultivated a representative of the international banking firm of Fuggers who had a grudge against the King of Portugal. On March 22, 1518, Charles V announced his support for Magellan's expedition. The familiar objective was to reach the Spice Islands by sailing westward. This time the plan was more precise—to find a strait at the extreme tip of South America. Magellan and Faleiro were to receive one-twentieth of the profits, and they and their heirs would be given the government of all lands discovered, with the title of Adelantados.

The Portuguese tried in vain to stop Magellan's voyage. But after a year and a half spent outfitting the expedition, the determined Magellan departed on September 20, 1519. For his round-the-world voyage he set out with five barely seaworthy ships varying in burden from seventy-five to one hundred twenty-five tons. They were heavily armed and well supplied with trading goods, which included, in addition to the usual hawkbells and brass bracelets, five hundred looking glasses, bolts of velvet, and two thousand pounds of quicksilver—all chosen to appeal to the sophisticated princes of Asia. The crew of about two hundred and fifty included Portuguese, Italians, Frenchmen, Greeks, and one Englishman, since it was not easy to find Spaniards for so dangerous a voyage under the command of a foreign adventurer. Magellan's old friend the pretentious Faleiro at the last minute decided not to go along because his horoscope showed that he would not survive the trip.

On risky voyages at that time it was customary for the captain to submit crucial decisions to a council including all the ship's officers and even the crew. For Magellan, who did not like to share his decisions, this was sure to cause trouble. It seems, too, that from the very beginning the three Spanish captains on his staff had plans to do him in.

Among the lucky coincidences of Magellan's voyage none was more fortunate for us than the presence on board of Antonio Pigafetta (1491–1534?), an Italian gentleman-adventurer from Vicenza, a Knight of Rhodes, who had come along for the ride. Personable, with a voracious appetite for facts and a boundless admiration for Magellan, he kept a detailed journal from which he wrote his *Primo Viaggio Intorno al Mondo,* the most vivid of the eyewitness accounts of the great voyages of that age. Pigafetta's affability and his talent for languages often led Magellan to send him ashore to conciliate the natives. Again and again he showed his genius for survival, and happily he was one of the eighteen who completed the round-the-world voyage.

Magellan's feat, by any measure—moral, intellectual, or physical— would excel even that of Gama or Columbus or Vespucci. He would face rougher seas, negotiate more treacherous passages, and find his way across a broader ocean. He commanded a more mutinous crew, yet managed his more difficult command firmly and humanely. He bore more excruciating pangs of palate and stomach. "Among the other virtues which he possessed," observed Pigafetta, "he was always the most constant in greatest adversity." "He endured hunger better than all the rest and more accurately than any man in the world, he understood dead reckoning and celestial navigation. No other had so much talent, nor the ardor to learn how to go around the world which he almost did."

Two months' sailing brought his ships from the Canary Islands to the eastern tip of Brazil, where they followed the coast southwestward, on the alert for the opening to take them into Balboa's South Sea. Again and again they tried what looked like promising sea passages to the west, at the Rio de Janeiro and farther south at the Gulf of San Matías, hoping that earlier explorers had been wrong. All these were dead ends. By the time they reached Port San Julián, midway on the coast of Patagonia, it was late March and the southern winter was setting in. Magellan made the crucial decision to wait there, to go on short rations, and weather wind and cold until the arrival of spring. When the men began to grumble, demanding they return north to winter in the tropics, he said he would die before turning back.

Magellan faced his two great tests, one of command and the other of seamanship, even before he entered the Pacific Ocean. In Port San Julián, when three of his ships—the *Concepción,* the *San Antonio,* and the *Victoria* —mutinied during the night, Magellan found support only from his own ship, the *Trinidad,* and from the *Santiago,* a vessel of seventy-five tons, the smallest of the five. The mutineers had three ships against his two. Magellan dared not allow them to take their ships and go home. On an empire-building voyage, every ship and every man was needed. Knowing that he had many supporters on the *Victoria,* Magellan sent on board a boatload of loyal men, ostensibly to parley over the terms of return. Following instructions, his emissaries managed to kill the mutinous captain, and then persuaded the vacillating crew to return to duty. With his three loyal ships he blocked the mouth of the bay. When the *San Antonio* tried to escape, it was overcome, and then the *Concepción*, the only remaining rebel ship, surrendered. In his final accounting, Magellan executed only one man, a ringleader who had murdered a loyal officer. And he marooned the leader of the conspiracy along with a priest who had helped organize the revolt. While he sentenced some other mutineers to death, he later pardoned them all.

During the remaining winter at Port San Julián, the *Santiago* was wrecked while exploring the coast, and her crew had to make a painful overland march to return to the other ships in port. In territory they thought uninhabited they encountered a living Patagonian, and so substantiated the legend of a race of giants. "So tall was this man that we came up to the level of his waistbelt," reported Pigafetta. "He was well enough made and had a broad face, painted red. . . . His hair was short and coloured white, and he was dressed in skins."

In late August 1520 Magellan's remaining four ships moved farther south to the mouth of the Santa Cruz River, where they stayed until October, when the southern spring appeared. Now Magellan faced his second great test, a trial of seamanship with few equals outside the *Odyssey*. Magellan had to find a passage to lead him through a continent of unknown breadth. Then, however tortuous and meandering it might be, he had to thread his way. How could he be confident that any entrance was not an opening to a dead end? How could he know that he was not losing himself ever deeper in the heart of the continent?

On October 21, only four days' sailing beyond the Santa Cruz River, once again they "saw an opening like unto a bay," as they rounded Cape Virgins just beyond 52 degrees south latitude. This time would the bay open to the precious strait? The crew thought it could not be, for it seemed "closed on all sides." They naïvely imagined that the strait must be a simple open passageway, like the Strait of Gibraltar, that you could see through. But Magellan was somehow prepared to find a "well-hidden strait." Perhaps, as Pigafetta observed, Magellan had actually seen "in the treasury of the King of Portugal" a secret map that showed a devious passage.

Yet the maps or globes like those by Martin Behaim or by Johan Schöner that he was most likely to have seen showed the southernmost tip of "America" separated at about 45 degrees south by a narrow but straight and open seaway from a conjectured great Antarctic continent which stretched around the world. Such maps and the map in Magellan's head were still substantially based on Ptolemy, whose picture had been modified only by inserting the new continent as several large islands still of uncertain dimensions in the Western Ocean. Ptolemy's Zipangu, or Japan, his Cape of Catigara and the Golden Chersonese, or Malay Peninsula, would still be there. The sought-for hidden strait would lead from the Western Ocean into Ptolemy's Great Gulf, between the Cape of Catigara and the Golden Chersonese, and there the Spice Islands would be found. Then, following Ptolemy, the sea to the west of the newfound American "islands" would be narrow and heavily sprinkled with large islands. In that case Japan would be separated from America only by a narrow channel. Most encouraging for Magellan, this view of the planet placed the round-the-world extension of the Line of Demarcation far west of the Spice Islands, which left those desirable treasures of empire well within the Spanish zone.

To say that the strait through the American land barrier was "well hidden" turned out to be the understatement of the centuries. The Strait of Magellan—the narrowest, most devious, most circuitous of all the straits connecting two great bodies of water—was a wonderful ironic prop for a seafaring melodrama. This meandering narrow maze debouched unexpectedly into the most open, most vast of all the seas. We must view on a modern map the tortuous passage, the angular disorder of small islands, the countless unexpected slots of water, to grasp the full measure of expertise, the persistence, courage—and luck—required to find the way. While the entrance at Cape Virgins on the Atlantic side goes through mild and pleasant country with low grassy banks, the exit at Cape Pillar on the Pacific is a gargantuan fjord between craggy ice-capped mountains. To sail the three hundred thirty-four miles between the oceans took Magellan thirty-eight days. Drake's sixteen days would be the sixteenth-century record, others would take more than three months, many would simply give up.

Only Magellan's iron courage against the elements and his deft mastery of men kept him going. He had suffered another unpleasant surprise, in addition to the mutiny. At Port San Julián, the last stop on the Patagonian coast before Cape Virgins, when they unloaded the ships to prepare for careening, he received a shock. The suppliers back in Seville had cheated him by actually loading provisions not for the full specified year and a half but for only six months, and then had falsified the records. Were they in the pay of Portuguese saboteurs? The crew now tried to replenish supplies by catching fish and seabirds and wild llamas, but these were not enough. The barren shores about Port San Julián offered no wood and hardly any fresh water, and the local natives, not being seafarers, were no use as guides.

Having lost the *Santiago* at Port San Julián, Magellan entered the strait with four ships. Groping his way at first, he sent off the biggest of his ships, the *San Antonio* (120 tons), to investigate one of the possible seaward openings, which proved to be a cul-de-sac. Unable to sight the ship, he retraced his course over at least two hundred fifty miles looking vainly for some trace. Unknown to Magellan, the pilot of the *San Antonio*, Esteban Gómez, who hated the captain-general for not having given him a command, mutinied, put his captain in irons, and piloted the ship back to Spain. The astrologer on Magellan's own ship, when asked about the whereabouts of the *San Antonio*, plotted the stars, consulted his books and gave the captain-general a precise circumstantial account of what actually proved to have happened.

It is remarkable that there were no more mutinies and the remaining three ships were kept together. Some of the narrows were less than two miles wide. The passage so meandered, with countless misleading bays and rivers, that not until the very end was there any view of open sea. When Magellan sensed that they might be near the end of the strait he sent a well-equipped boat out ahead. "The men returned within three days," Pigafetta notes, "and reported that they had seen the cape and the open sea. The captain-general wept for joy, and called that cape, Cape Dezeado [Desire], for we had been desiring it for a long time."

Bizarre winds, the williwaws, blasted the western half of the strait. "These were compressed gales of wind," Captain Joshua Slocum observed in 1900, "that Boreas handed down over the hills in chunks. A full-blown williwaw will throw a ship, even without sail set, over on her beam end." Having threaded a maze, having survived encroaching land and rocks, Magellan was now cast out to a vast watery emptiness. For more than a hundred days Magellan and his crew suffered trials of a seemingly endless, boundless, landless world of water.

On the best available evidence, he expected the crossing of these waters, familiar as Ptolemy's "Great Gulf," to take only a few weeks. There was still no known way of precisely marking longitude, and without that it was impossible to know the exact distance between any two points around the earth. All the authoritative estimates that Magellan had seen or could have known about, Samuel Eliot Morison concludes, were at least 80 percent short of the actual extent of the ocean. Even a century after Magellan, "reliable" maps still underestimated that dimension by 40 percent. For Magellan, the extent of the Pacific Ocean was an excruciating surprise! Of course, it would also be his greatest and his most unwilling discovery.

They knew now that they had only one-third of their planned provisions, for a trip three times longer than they expected. Let Pigafetta, who was there, tell the story:

Wednesday, November 28, 1520, we debouched from the strait, engulfing ourselves in the Pacific Sea. We were three months and twenty days without getting
any kind of fresh food. We ate biscuit, which was no longer biscuit, but powder
of biscuits swarming with worms, for they had eaten the good. It stank strongly
of the urine of rats. We drank yellow water that had been putrid for many days.
We also ate some ox hides that covered the top of the mainyard to prevent the
yard from chafing the shrouds, and which had become exceedingly hard because
of the sun, rain, and wind. We left them in the sea for four or five days, and then
placed them for a few moments on top of the embers, and so ate them; and often
we ate sawdust from boards. Rats were sold for one-half ducado [about $1.16 in
gold] apiece, and even then we could not get them. The gums of both the lower
and the upper teeth of some of our men swelled, so that they could not eat under
any circumstances and therefore died. Nineteen men died from that sickness, and
the [Patagonian] giant together with an Indian from the country of Verzin.

But they were lucky in their weather. During the whole three months and
twenty days during which they sailed about twelve thousand miles through
open ocean, they had not a single storm. Misled by this one experience, they
named it the Pacific.

Neither did they encounter any land during all those weeks "except two
desert islets, where we found nothing but birds and trees, for which we
called them Ysolle Infortunate [i.e., the Unfortunate Isles]. . . . We found
no anchorage, [but] near them saw many sharks. . . . Had not God and His
blessed mother given us so good weather we would all have died of hunger
in that exceeding vast sea. Of a verity I believe no such voyage will ever be
made [again]."

Had Magellan not been a master of the winds, he would never have made
it across the Pacific. Leaving the straits, he did not go directly northwest
to reach his desired Spice Islands, but first sailed north along the west coast
of South America. His purpose must have been to catch the prevailing
northeasterly trade winds there that would carry him not to the Moluccas,
where the Portuguese were rumored to be in control, but to other spice
islands still open for Spanish taking. Whatever his motive then, the course
he chose is the one still recommended by United States Government Pilot
Charts for sailing from Cape Horn to Honolulu in that season.

On March 6, 1521, at last they anchored for refreshment and supplies at
Guam. There they were greeted by good-natured but greedy natives, who
swarmed over all three ships, on deck and below, promptly removing everything movable—crockery, lines, belaying pins and even the longboats.
Magellan christened these the Islas de Ladrones or Isles of Thieves, now
known as the Marianas. They stayed for only three days to pick up rice,
fruit, and fresh water. In another week they were on the east coast of

Samar island in the Philippines, in the neighborhood of Leyte Gulf, which would be the scene some four centuries later of the greatest naval battle in history.

In the regions Magellan was now approaching, where the Chinese, Portuguese, and others were engaged in a busily competitive seafaring commerce, the high stakes went to the shrewd merchant and the canny diplomat. Magellan's life, just salvaged from the worst elements that nature could muster, would be forfeited by a single act of imprudence. The king of the island of Cebu, feigning conversion to Christianity, persuaded Magellan to become his ally—"to fight and burn the houses of Mactan to make the King of Mactan kiss the hands of the King of Cebu, and because he did not send him a bushel of rice and a goat as tribute." Magellan's officers and crew begged him not to go, "but he, like a good shepherd refused to abandon his flock." There, on the beach of the tiny island of Mactan, on April 27, 1521, Magellan was repeatedly wounded by the poisoned arrows, spear thrusts, and scimitar cuts of Mactan tribal warriors. He fell face down on the sand.

Magellan could have retreated more speedily and saved himself, but he chose to cover the retirement of his men. "Thus they killed our mirror, our light, our comfort and our true guide," Pigafetta lamented. "When they wounded him, he turned back many times to see whether we were all in the boats. Then, seeing him dead, we wounded made the best of our way to the boats, which were already pulling away. But for him, not one of us in the boats would have been saved, for while he was fighting the rest retired."

In a sense, Magellan did complete his circumnavigation. For on his earlier voyages for the Portuguese, in coming around Africa to these islands he had probably already been farther east than Cebu.

The expedition was not abandoned. The *Concepción* had become so unseaworthy that it had to be burned. The *Trinidad,* also judged incapable of the journey back to Spain by the western route, made an unsuccessful effort to cross the Pacific to Panama, and returned to the East Indies. The semi-seaworthy *Victoria,* under Juan Sebastián del Cano, took the western route around the Cape of Good Hope. To the already familiar trials of hunger, thirst, and scurvy now was added the hostility of the Portuguese, who imprisoned nearly half of Del Cano's crew when they put in at the Cape Verde Islands in the Atlantic. On September 8, 1522, only twelve days less than three years from the day of their departure, a feeble remnant of the original two hundred fifty, eighteen shipworn men, arrived at Seville. The next day, to fulfill their penitential vows all eighteen walked barefoot, wearing only their shirts, each carrying a lighted candle, the mile from the waterfront to the Cathedral shrine of Santa María de l'Antigua.

35

The Reign of Secrecy

WHEN the Portuguese pilot Péro d'Alemquer, who had sailed with Dias and Gama, returned home, he boasted at court that he knew how to take *any* ship, not only a caravel, to the coast of Guinea and back. King John II publicly rebuked him, then took him aside to explain privately that he simply wanted to discourage foreign interlopers from profiting by the Portuguese experience. Prince Henry the Navigator and his successors did everything in their power to establish and preserve a monopoly over the commerce with their newly discovered coasts of Africa. This meant not letting out the word about where the places were and how to reach them. When King Manuel developed his plans for a pepper monopoly in 1504, he ordered that all the navigating information be kept secret. "It is impossible to get a chart of the voyage," an Italian agent complained after Cabral's return from India, "because the King has decreed the death penalty for anyone sending one abroad."

This policy was not easy to enforce, since the Portuguese kings had to rely on foreigners like Vespucci to do their work of discovery. In 1481 the Portuguese Cortes petitioned King John II to exclude foreigners, especially Genoese and Florentines, from settling in the country, because they habitually stole the royal "secrets as to Africa and the islands." Yet a few years later the young Genoese, Christopher Columbus, went on his voyage to help the Portuguese build their fort at São Jorge da Mina on the Guinea coast. And it was a Fleming, Fernão Dulmo, whom King John II sent with Estreito even before Columbus to seek islands in the Western Ocean.

Still, the Portuguese conspiracy of silence was effective—at least for a while. Until the middle of the sixteenth century other countries seeking information about the Portuguese seafaring commerce to Asia had to rely on scraps gathered from ancient writers, casual overland travelers, occasional turncoat sailors, and spies. It was despite this policy that maps of Asia leaked out to the rest of Europe.

The Spanish, trying to enforce a similar policy, kept their official charts in a lockbox secured with two locks and two keys, one held by the pilot-major (Amerigo Vespucci was the first), the other by the cosmographer-major. Fearing that their official maps would be deliberately corrupted or would not include the latest authentic information, in 1508 the government created a master chart, the *Padron Real,* to be supervised by a commission of the ablest pilots. But all these precautions were not enough. The Venetian-born Sebastian Cabot (1476?–1557), while serving as pilot-major to Emperor Charles V, tried to sell "The Secret of the Strait" both to Venice and to England.

Fear of stimulating domestic competitors prevented these successful exploring nations from exploiting the full patriotic benefits of the epic adventures sponsored by their governments. Outside Spain and Portugal, as we have seen, reports of Vespucci's voyages were the most widely printed of all the voyages to the New World in the thirty-five years after Columbus first sailed west. Sixty editions of Vespucci appeared all over Europe in Latin and in the rising vernacular languages, including even the Czech. But during all these years no edition appeared in either Spain or Portugal. This curious fact suggests that the rulers on the Iberian peninsula did not want to threaten their government monopoly by stirring up the interest of private competitors even among their own people.

An interesting aside: just as secrecy breeds monopoly, so does monopoly breed secrecy. A curious parallel to this experience had occurred on the other side of the earth not many years before. As we have seen, after the seafaring exploits of the eunuch Chêng Ho took the ships of China all over the East, in 1433 the empire retreated into itself and forbade later expeditions. Then in 1480 another Chinese eunuch who had risen to great power wished to start a maritime expedition against Annam. But the high officials of the Chinese War Office destroyed the records of the earlier expeditions to prevent his carrying on the forbidden work.

Even for a voyage so inspiring of national pride as Sir Francis Drake's voyage around the world (1577–80) the authentic original record oddly disappeared. On his return to England, Drake and his cousin presented to Queen Elizabeth their own illustrated log. This top-security document with so much information useful to foreign competitors must have been kept locked in a safe place, but it has never turned up. There seems to have been an embargo on other accounts of the great voyage. How else explain that so grand an adventure was not reported in print for more than a decade? In 1589, when Richard Hakluyt published his famous compendium of notable *Voiages and Discoveries of the English Nation made by sea or over land to the most remote and farthest distant quarters of the earth* there still was no account of Drake's circumnavigation. But the ban appears to have been lifted within the next decade, when extra pages were bound into the volume to recount Drake's famous voyage.

Secrecy did create problems both for recruiting the crews and for keeping up morale on long voyages to uncertain destinations. Captains seeking crews to sail unexplored waters were wary of scaring off the men, and then at sea feared that the dangerous facts would incite mutiny. Drake did not reveal his full purpose in advance to his crew and only gave his officer-colleagues on board the bare information necessary to conduct the ship to the next port.

Expanding empires become obsessed with keeping secrets. In the Roman Empire, Suetonius reports, maps of the world were exclusively for government use, and it was a crime for a private person to possess one. Perhaps this helps us understand why no original Ptolemy maps survive and the earliest Ptolemy manuscripts date from the thirteenth century.

The secrecy of the seafaring powers in the great Age of Discovery has itself been made the basis of extravagant claims. Portuguese historians, determined to suggest that Portuguese voyagers really "discovered" America before their Spanish rivals, argue that such voyages naturally would not have been recorded. "But the only evidence of a Portuguese policy of secrecy with regard to the discovery of America," Samuel Eliot Morison concludes, "is lack of evidence of a Portuguese discovery of America!" We have clues to a conspiracy of silence in the meagerness of surviving fifteenth- and sixteenth-century charts, maps, and chronicles of Portuguese seafaring. How secretive did the Portuguese wish to be about their policy of secrecy? Rulers in the past have not been ignorant of the maxim that "a secret may sometimes be best kept by keeping the secret of its being kept a secret." Historians, like sailors, have been both plagued and intrigued by the efforts of empire builders to bury their secrets. In most ages the national archives have become a literary cemetery, where historical remains are preserved and revered only after they are no longer useful or dangerous.

It was from a quite unexpected quarter that the policy of secrecy would be defeated. Not by spies or treacherous pilots-major like Sebastian Cabot. But by a new technology that created a new kind of merchandise. After the arrival of the printing press geographic knowledge could be conveniently packaged and profitably sold.

Of course there had long been a trade in the sea charts that sailors made their living by. The hand-copied portolanos had taken a form in the thirteenth century that served practical-minded Mediterranean seamen, and by the fourteenth century chart-makers were operating prosperous establishments. Until the mid-fifteenth century in Europe these chart-makers were the only active professional cartographers. Their charts tended to be quite uniform, despite the fact that each was hand-drawn and the product of several specialized craftsmen. But secrecy and monopoly produced a black

market offering shoddy work and counterfeits pretending to be stolen originals. As maritime trade to Asia and the West Indies became more competitive, there was a new premium on scraps of geographic information—clues
to secret watering places, snug harbors, or shorter passageways.

Private trading companies prepared their own "secret" atlases. The
Dutch East India Company, for example, employing the best cartographers
in the Netherlands, put together for the exclusive use of the company some
one hundred eighty maps, charts, and views showing the best routes around
Africa, to India, to China, and to Japan. Such a collection, long suspected
to exist, did not come to light until years later in the library of Prince
Eugene of Savoy in Vienna. Official government charts, generally speaking,
did not become available to the public until what they contained was
already common knowledge.

The rediscovery of Ptolemy's *Geography* and the maps found with the
Byzantine manuscripts, more perhaps than any other single event, brought
into being the professional *map*-maker. While sea charts served the everyday needs of sailors, maps had a grander purpose. In addition to being
decorative, maps helped stay-at-home scholars, priests, and merchants find
their bearings in the whole world. The map-makers' insecure profession, as
we have seen, had no basis in the medieval trivium and quadrivium. Ptolemy now provided them a sacred text to make their work serious and
respectable. He had comprehended and depicted the whole world, and
opened the way to mathematical cartography. And once the world was
marked by latitudes and longitudes, every place could be located on a
scheme that the whole world could use.

Gutenberg's Bible, printed with movable type, came off the press in
Mainz in 1454. Despite clerical wariness of any machine-made version of
Holy Scripture, the infant printing industry received its main support from
churches. In Europe by 1480 printing presses were found in iii towns, and
by 1500 the number exceeded 238. These presses offered books not commonly found in churches—ancient classics like Aristotle, Plutarch, Cicero,
Caesar, and Aesop's fables, along with Boccaccio's love stories. When these
works could be bought in the marketplace, there was a new incentive for
people to learn to read.

A full half-century before Gutenberg, engravers of wood and metal had
been "printing" illustrations for handwritten books. Goldsmiths and silversmiths developed a technique of inking and transferring their decorative
designs to paper, at first for their own records and then for sale. Maps, books
of maps, and travelers' tales offering subjects for lively fanciful illustrations
could be enjoyed by people who were not accustomed to reading lengthy
texts. In an age of growing curiosity about sea voyages to exotic places and
"discoveries" of all sorts, these proved wonderfully marketable.

What good fortune that Ptolemy's *Geography* was there to be printed! It had everything required to make a beautiful book and a salable product, and incidentally spread abroad the authentic version of the planet. Even before 1501, during the era of "incunabula" (from Latin *cunae*) when printing was said to be still in its cradle, the presses turned out seven folio editions of Ptolemy's *Geography*. In the next century there were at least thirty-three. His book became canonical. Until 1570, for more than a century after the first printed edition of Ptolemy, European books of geography, maps, and atlases offered only slight variations on his themes and images. Ptolemy's name on the title page made a book respectable, much as Webster's name later served American dictionaries. Printing presses in those days before copyright made it easy for Ptolemy's ideas to be widely diffused at the very time when many of them were being proved wrong. For example, even after Dias and Gama had demonstrated that ships could sail around Africa and that the Indian Ocean was an open sea, Ptolemy's maps continued to show the Indian Ocean as a land-bound lake, an Asian Mediterranean. Sometimes in the very same volumes that depicted the exploits of Dias and Gama.

In this way the power of print was emphatically conservative. *Ars artium omnium conservatrix* (The art preservative of all arts) were the words inscribed in the mid-sixteenth century on the house of Laurens Janszoon Coster (d. 1441), whom some Dutch scholars credit with the invention of printing. And the press had a new power to conserve obsolete ideas.

36

Knowledge Becomes Merchandise

THE press also had a daemonic power to open the world and diffuse knowledge of discoveries in convenient packages. Hundreds and thousands of printed maps went abroad at random. Merely by its power to multiply the product, the printing press would be a champion of freedom, providing myriad unstoppable channels for dangerous facts and ideas, sending out countless items which could not be traced or withdrawn. Once the printing press had done its work, there was no force on earth, no law or edict, that could retrieve the message. A later printed work, as Waldseemüller discovered to his dismay, might contradict the earlier, but could never erase or abolish it. Book burners and book censors and Index promulgators would always fight a losing battle.

Unlike a manuscript, which required only pen and ink and paper and the skills of the copyist, a printed book required heavy capital investment. In addition to larger quantities of ink and paper for the multiplied product, there had to be the fonts of type and a press. The preparation of a wood block or a copperplate for a printed map was costly. Book printers and map printers were investing in the future. They would not lightly abandon their product, even if the ideas had gone out of fashion or the maps had been revised by new discoveries. Whatever they had once produced they had to try to sell. Works with a proven market were apt to have the patina of the centuries, which had nothing to do with their truth. The year 1530 saw three reprintings of the apocryphal Sir John Mandeville's alleged *Travels,* which many thought had been confirmed by Columbus. Copper was costly and the life of copperplates was often a good deal longer than that of the "truths" inscribed on them. Merchandisers of maps had a vested interest in obsolescent information. Europe's map-making headquarters went where the technology was most developed. After 1550 when the best maps began to be put on copperplates instead of wood blocks, the center of European map-making moved to the Netherlands, where there were the best line engravers.

Sailors, too, were naturally conservative, slow to accept new ideas. Even after Mercator provided his convenient new projection, which made it possible to plot the rhumb lines of navigation as straight lines, it was nearly two centuries before seamen gave up their old ways. Of course they were reluctant to accept a new continent or a new ocean. Meanwhile portolanos, the sailing charts, only slowly went into print. Even into the seventeenth century, European pilots remained suspicious of printed charts, preferring those that were hand-drawn because they were familiar. Perhaps also these actually were more reliable since they were easily revised and so more likely to be up to date.

Despite the reluctance of sailors, by the standards of that day map-making soon became big business. As we have seen, less than twenty years after Gutenberg's Bible, the first printed edition of Ptolemy's voluminous *Geography* appeared, and numerous others followed. After 1500, maps came off the presses regularly and in profusion. Henricus Martellus, who made the updated version of Ptolemy on which Columbus had relied, worked with a Francesco Rosselli of Florence, who was the first map printer and dealer known to specialize in the business. In Waldseemüller's product we have seen how influential even a small press in a remote place could be by 1507.

Gerardus Mercator (1512–1594) was the most original and the most influential of those who grasped the opportunity. Christian geographers who set Jerusalem at the center of their world view had been more concerned to guide the faithful to salvation than to help sailors reach their next port or to direct explorers across oceans to strange continents. Mercator transformed the world of maps for the new secular age. Cosmography became geography, and the convenience of merchants, military men, and seamen was served not just by coastal charts but by new views of the whole planet.

Mercator's epochal service to sailors was the "Mercator projection." Sailors had found it difficult to lay out their courses on a chart because their charts made no allowance for the sphericity of the earth. On the spherical earth the meridians converged to a point at the poles. How could a segment of this sphere be put on a flat piece of paper so that a sailor could lay out his compass course by a straight line? Mercator found a way. He imagined the lines of longitude to be like cuts in the rind of an orange, then peeled off segments of the skin and laid them down next to each other on a table. Treating these segments as if they were elastic, he stretched the narrow points, spreading them to make each segment into a rectangle touching the next one from top to bottom. The whole rind of the sphere, which represented the surface of the earth, thus became a single large rectangle, with the meridians of longitude parallel to each other from North Pole to South Pole. By careful stretching, the shapes on the surface could be kept, though their dimensions were enlarged. This was the Mercator projection, on which the spherical surface of the whole earth was now conveniently presented as a flat rectangle subdivided into a grid by the *parallel* lines of latitude and of longitude. Then with simple drawing instruments a navigator could mark off his constant compass bearing as a straight line which would cross all meridians at the same angle. In the late twentieth century, deep-sea navigators still do more than 90 percent of their work on Mercator's projection.

A man of action and enterprise, Mercator had the advantage of the best academic training. Born in Flanders, he studied philosophy and theology at the University of Louvain, then turned to mathematics and astronomy and incidentally learned the arts of engraving, instrument-making and surveying. His first work, in 1537, was a small-scaie map of Palestine. Then he spent three years, doing all the work from surveying to drafting and engraving, for his *Most Exact Description of Flanders (Exactissima Flandriae Descriptio),* which was so much better than anything before that it secured him a commission to do a terrestrial globe for Emperor Charles V. When Mercator delivered the globe in 1541, the Emperor ordered from Mercator a set of drafting and surveying instruments, including a sundial, to use on military campaigns.

Louvain, where Mercator lived and worked, was a hotbed of fanaticism and persecution. Only by good luck did he escape the fires of the *auto da fé*. The Catholic regent Mary, queen dowager of Hungary, who then happened to rule Flanders, ordered that all heretics be executed, "care being only taken that the provinces were not entirely depopulated." In 1544 Mercator was caught in a roundup of suspected Lutherans. Of the forty-two alleged heretics arrested with Mercator, two were burned at the stake, two were buried alive, and one was beheaded. While all unrepentant heretics were supposed to go to the stake, the humane regent Mary dictated that any who recanted be spared that torture. Instead, the men would be put to the sword and the women buried alive. Mercator was imprisoned for some months, but the efforts of his parish priest finally secured his release.

Mercator moved to a friendlier climate in 1552, when he was invited to be professor of cosmography in a new university at the Prussian town of Duisburg on the Rhine. But the professorship did not materialize, and he became cosmographer to the Duke of Cleves, settling permanently in Duisburg. There he published the first modern maps of Europe and of Britain and, in 1569, the epoch-making first world map on the projection that he had invented.

Mercator's projection followed Ptolemy's grid of latitudes and longitudes, which he gave a new utility for navigators. His first map of the world (1538), the first on which both a "North America" and a "South America" appear, still showed the heavy influence of Ptolemy. But he was no servile disciple. On his great map of Europe (1554) the Mediterranean ceased to be elongated in the traditional Ptolemaic style, and instead appeared as only 52 degrees long, which is much closer to its actual dimensions. He also set a new standard for map engraving and established the italic style for map lettering.

Incidentally, Mercator left us the most authentic surviving edition of Ptolemy's maps. The numerous earlier editions of Ptolemy had casually incorporated each editor's own "improvements." By defining what Ptolemy himself really had depicted, Mercator could plainly show how that picture had to be corrected. Mercator expressed a strikingly modern sense of history when his 1578 edition offered, unadulterated and unimproved, twenty-seven of Ptolemy's own maps, along with a more accurate version of the text of Ptolemy's *Geography*.

In his pioneer historical work of four hundred and fifty folio pages, he corrected the dates of historic events by the contemporary references to solar and lunar eclipses. *Mercator's Chronology . . . from the beginning of the world up to the year 1568, done from eclipses and astronomical observations (Chronologia, hoc est temporum demonstratio . . . ab initio mundi usque ad annum domini 1568, ex eclipsibus et observationibus astronomicis)* compared the dating of events by the different systems of the Assyrians, Persians, Greeks, and Romans.

Mercator's enterprising young friend Abraham Ortelius (1527–1598) had never been to a university, but he richly possessed the talents of the businessman. Ortelius, too, was haunted by the Inquisition. While the southern provinces of the Netherlands where Ortelius was born into a Catholic family remained predominantly Catholic, Calvinism was growing in the northern Netherlands. Philip II, king of Spain and conquering ruler of the Netherlands, continued the fanatical policies of Isabella. The Duke of Alva's Council of Blood, which began its work in 1567, could summon anyone to answer charges of heresy, which also meant disloyalty to the Spanish crown, and failure to appear before the council brought confiscation of all the accused's property. Printers and publishers were always suspect because the printed word was a notorious channel of heresy.

Who could tell what the Inquisition might find heterodox or obscene? Simply selling the engraved portrait of a suspected heretic like Erasmus was itself a grave offense. Any large map, covered with decorative motifs, coats of arms, and political and ecclesiastical boundaries, was a happy hunting ground for the Inquisitor. In those years it took courage to print or publish anything in Antwerp.

Unlike Mercator, Ortelius had come to cartography not through mathematics and astronomy but from handling maps as merchandise. At twenty he was already illuminating maps and had been admitted to the proper guild. To support his mother and two sisters after his father died, he became a dealer. He bought maps, which his sisters mounted on linen, then he would color them to sell in Frankfurt or at some other fair. As his business grew he made regular circuits through the British Isles, Germany, Italy, and France, buying the maps locally produced and selling his own illuminated product. In this way he collected the best current maps from all over Europe, which he brought back to his Antwerp headquarters.

In those turbulent times Antwerp merchants urgently needed reliable up-to-date maps reporting the latest results of the religious and dynastic wars. For without these they could not plan the shortest, least risky routes for their merchandise. One of the more enterprising of them, Aegidius Hooftman, had prospered by keeping himself well informed, and collecting in his office stocks of the best current charts and maps of all shapes and sizes. The large maps could not be used without being unrolled. But the small print on the handy city plans made place-names barely legible. Finding this map miscellany a nuisance, Hooftman and another merchant friend persuaded Ortelius to seek for them the most reliable maps in a uniform size. Every map chosen would have to be printed on a single "sheet" of paper, about twenty-eight by twenty-four inches, which was the largest size that papermakers turned out in those days. Then thirty of these could be bound together like a book, in a format convenient to store and handy to use.

When Ortelius did this for Hooftman, he unwittingly produced a new kind of book, the first modern geographical atlas. It seemed such a good idea that he put together more books of this kind for the general market. With the help of his friend Mercator, he collected the best maps, had large-scale maps reduced to his standard size, and secured the collaboration of Christophe Plantin, another friend, whose Antwerp press was doing some of the best work in Europe. *The Picture of the World (Theatrum Orbis Terrarum),* the first modern atlas, came off Plantin's Antwerp press on May 20, 1570, after ten years of work. Considerably larger than Hooftman's volume, it now contained fifty-three copperplate maps, along with a descriptive text. A novel feature was the editor's list acknowledging by name all eighty-seven of the authors of maps consulted or copied. He heralded a new age of incrementalism, when any individual could add piecemeal to the stock of knowledge. Cartographers would no longer have to father their works on Ptolemy in order to make them respectable.

Ortelius' atlas was an immediate commercial success. A second edition was required in three months, and then the Latin text was translated into Dutch, German, French, Spanish, Italian, and English. There were twenty-eight editions by the time of Ortelius' death in 1598, and by 1612 there were forty-one. Fame and fortune came to Ortelius himself, who advised the leading geographers of the age as he traveled around Europe. After being certified for his Catholic orthodoxy, he was appointed geographer to King Philip II of Spain.

Fan mail poured in. "Ortelius, you, the everlasting ornament of your country, your race, and the universe," one reader effused, "you have been educated by Minerva. . . . By the wisdom she has imparted to you, you unfold the secrets of nature, and declare how this stupendous frame of the world has been adorned with countless towns and cities by the hand and labour of men, and by the command of kings. . . . Hence all extol your Theatrum to the skies and wish you well for it." Mercator himself praised "the care and elegance with which you have embellished the labours of the authors, and the faithfulness with which you bring out the geographical truth, which is so corrupted by mapmakers." At long last, Mercator testified, Ortelius had brought together the latest and best information about the whole earth in one convenient portable volume, and at a reasonable price. Ortelius kept his work up to date by adding maps sent him by agents and admirers.

Ortelius' title page for the first time displayed *four* symbolic human figures, one for each of the continents, which now included America. The title pages of the editions of Ptolemy had, of course, shown only three: one each for Europe, Asia, and Africa. The general arrangement in the book was familiar—first a map of the world called *Typus Orbis Terrarum,* then a map of each of the known continents, followed by special maps for countries and regions. Not yet entirely liberated from Ptolemy or from folklore, Ortelius still showed Ptolemy's legendary southern continent reaching out from the South Pole and always, of course, the realm of the irrepressible Prester John. Still, Ortelius did much to free map-makers, and all literate Europeans, from the grosser errors in Ptolemy. Ortelius wrote to Mercator in Germany with the news that Sir Francis Drake had been sent out on an expedition, to which Mercator replied that the English had also sent out a Captain Arthur Pitt to explore the northern coast of Asia. The atlas was making the pursuit of knowledge more than ever a cooperative enterprise.

These pioneer cartographers, map printers and dealers brought the discoveries of Columbus and Vespucci, Balboa and Magellan, to the people, whose lives their discoveries would transform. Before the printing press there were two great traditions of cartography in Europe. Cosmographers produced grand works to ornament palaces and libraries while chart-makers furnished pilots with the portolanos they needed at sea. Now a new format, the atlas, in many sizes and prices, could inform all who wanted to learn.

Mercator himself had plans for a three-volume atlas including the best maps of the whole world. He managed to publish two parts before his death in 1594, and the work was finally completed in 1595 by his son Rumold, under the old-fashioned, flamboyant title that Mercator had chosen: *Atlas, or cosmographical meditations upon the creation of the universe, and the universe as created (Atlas sive Cosmographicae meditationes de fabrica mundi et fabricati figura).* Within a few years thirty-one editions were published in folio. Although Ortelius had already provided an atlas, this was the first time that the word "Atlas" was used in print to describe such a work.

Just as the portable clock made the world's time accessible to everybody, so when atlases became portable, millions could share a view of the world's space. In the early eighteenth century, the geographer to the French King Louis XV complained in his introduction to the *Pocket Atlas, for the use of Travelers and Officers (Atlas de Poche, à l'usage des voyageurs et des officiers,* Amsterdam, 1734–38) that the full-size folio atlases were "priced so that many scholars are unable to afford them." "Because of their grandeur . . . they are, so to speak, nailed up in the book-case, usually adorned with a very appropriate binding . . . one shows them in one's library rather

as a decorative ornament than as a useable tool . . . and I know individuals
who have never profited from the money these atlases have cost them."
Once the world atlas in folio had proven itself, inexpensive portable atlases
began to appear. Mercator's large atlas was published in smaller format as
the *Atlas Minor* in at least twenty-seven editions, which included one in
Turkish. Ortelius' *Theatrum* appeared before long in several languages in
more than thirty pocket-size *Epitomes*. Interested Europeans could now
carry around in their pocket the latest version of the earth.

37

The Ardors of Negative Discovery

THE same natural conservatism which made sailors slow to give up hand-
drawn charts for printed ones, or to embrace the possibility of new continents, would make them reluctant to abandon their time-honored illusions.
Perhaps the most appealing and certainly the most long-lived of these was
the belief in a great Southern Continent. It was embellished precisely because it could not yet be disproved, and it answered the universal love of
symmetry. The ancient Greeks, knowing that the earth was a sphere and
that there was a large land mass north of the equator, believed that to
balance it there had to be a similar land mass in the south. Then Pomponius
Mela, in the oldest surviving geographical work in Latin, about A.D. 43,
made this Southern Continent so large that Ceylon was its northern tip.
Maps that purported to follow Ptolemy continued to show a vast Antarctic
continent inscribed "Unknown Land according to Ptolemy." In the late
fifteenth century this mythic continent solidly attached to Africa made the
Indian Ocean a great lake, which could never be reached from Europe by
sea.

When Dias rounded the Cape of Good Hope and proved that there was
a water passage into the Indian Ocean, the Southern Continent had to
shrink in that part of the globe. And when Magellan finally threaded his
way through the straits that bear his name into the Pacific, map-makers still
believed that Tierra del Fuego, to the south, was the north shore of the
mythic continent. On the first modern atlas, Ortelius' *Theatrum Orbis
Terrarum*, the whole South Polar area was covered by *Terra Australis*

nondum cognita (the Southern Continent not yet discovered). Seventeenth-century European maps continued to show this continent, somewhat vaguely defined but reaching northward toward the equator. After the Dutch navigator Willem Schouten (c. 1580–1625) rounded Cape Horn in 1616, map-makers once again had to move the northern extent of the Southern Continent to unexplored territory.

European explorers of the Pacific never ceased to be tantalized by Marco Polo's description of a far-southern Eldorado which he called Lokach, which he said the Great Khan would have conquered if it had been more accessible, for there "gold is so plentiful that no one who did not see it could believe it." As the Americas were gradually delineated, and the outlines of Africa and Asia became sharper, Western map-makers exercised their imaginations to fill the empty Antarctic spaces on the globe.

European discoveries of some lands Down-Under only pushed the desirable continent farther southward. In 1642 Abel Tasman (1603–1659), perhaps the greatest of the Dutch navigators, was commissioned by Anton van Diemen, governor-general of the Dutch East Indies, to go exploring to the "Great South Land" (Australia), which had already been touched at its northern and western coasts. He was to discover the "remaining unknown part of the terrestrial globe" which would "comprise well-populated districts in favourable climates and under propitious skies" and so "in the eventual discovery of so large a portion of the world . . . be rewarded with certain fruits of material profit and immortal fame." On this and his next voyage Tasman circumnavigated Australia, proving that it, too, was not part of the mythical Southern Continent.

In the next century a Scottish geographer working for the British East India Company made this conjectural Great South Land his obsession and produced the most copious and detailed argument yet offered. The dyspeptic Alexander Dalrymple (1737–1808) had made a profession of charting the seas and their currents, and would become the first Hydrographer to the Navy, in 1795. From youth, his heroes were Columbus and Magellan, whom he hoped to rival by discovering his own continent. His *Account of the Discoveries made in the South Pacifick Ocean, Previous to 1764* (1767), arguing "as well from the analogy of nature, as from the deduction of past discoveries," described the vast extent of the Southern Continent, which "was wanting on the South of the Equator to counterpoise the land to the North, and to maintain the equilibrium necessary for the Earth's motion." From the equator to 50 degrees north latitude there were roughly equal surfaces of land and of water, but looking southward from the equator, the lands so far discovered were barely one-eighth the surface of the water. The erratic winds noted by past explorers in the southernmost Pacific were symptoms of large bodies of land nearby. He confidently concluded that a vast continent must lie down there, and that nearly all uncharted areas from the equator to 50 degrees south must be land, "a greater extent than the whole civilized part of Asia, from Turkey eastward to the extremity of China." This would be an ample substitute for the restive American colonies, which still had only two million inhabitants, for the new continent might one day hold fifty million people, and "the scraps from this table would be sufficient to maintain the power, dominion, and sovereignty of Britain by employing all its manufacturers and ships."

It happened that a transit of the planet Venus across the sun was calculated to occur on June 3, 1769. From observing this phenomenon (not to occur again for another century) at widely separated points on the earth would come more accurate figures for the earth's distance from the sun and improved data for celestial navigation. The Royal Society in London therefore was planning an expedition to Tahiti. The government saw it as a cover for a new effort to sail down into the unexplored southernmost rim of the Pacific, seeking the boundaries of the fabled Great South Land. If conceivably the land did not exist, the voyage might dispel the myth once and for all.

Alexander Dalrymple, who considered himself the leading living authority on the uncharted continent, hoped to command the expedition. Although still only in his thirties, he was a competent mathematician, already a Fellow of the Royal Society, and a scion of the influential Scottish family that held the earldom of Stair. His older brother, Lord Hailes, was a prominent judge and a friend of Dr. Johnson. Moreover, a risky two-year journey into uncharted waters among "savage" peoples seemed no plum for sedentary scholars or ambitious sea captains in an age of buccaneering.

Unfortunately for Dalrymple, the British Navy had recently benefited from the sweeping reforms of Lord Anson (1697–1762), who could take much of the credit for recent British naval victories. His historic four-year round-the-world privateering voyage had taken him into the Pacific, where he captured a Spanish treasure which sold for £400,000. He had set new professional standards for the naval commands, and the appointment of well-connected aristocrats was no longer so easy. Dalrymple had proven himself by temperament and by physique ill suited to such an exacting assignment. He had been dismissed from the British East India Company for his tactless handling of their relations in the Pacific Islands, and, besides, he was suffering from a bad case of the gout. Lord Hawke at the Admiralty was willing to allow him to go along as a civilian observer. But for the command there had to be an officer of the navy. Dalrymple withdrew in disgust.

Hawke's shrewd choice, offending people of title, wealth, and learning, was a little-known noncommissioned officer named James Cook (1728–

1779). The bright son of a migrant farm laborer from Scotland who had settled in Yorkshire, Cook had only a rudimentary formal education in reading, writing, and arithmetic at a dame school. Working in a general store, he became acquainted with sailors and shipowners plying the east coast. At eighteen he was taken on as a sea apprentice by a local shipowner who ran a fleet of sturdy coal carriers—"collier-barks"—in the rough North Sea. For nine years he worked the crudely charted lee shores in unpredictable winds. In his spare hours he studied mathematics, for which he had a natural aptitude, he became a skilled navigator and soon was the mate of a collier. He might have had a secure career on private vessels in the North Sea, but he preferred adventure, and when he was offered his first collier command he refused and, instead, volunteered in 1755 as an able seaman in the Royal Navy. Tall and vigorous, he attracted attention by his commanding presence, his affability, and his skill in uncharted waters. During the Seven Years War Cook advanced through the noncommissioned ranks. His expert survey of the difficult passages of the St. Lawrence helped make possible the capture of Quebec and the final victory.

After the War he went back to Newfoundland, where for five years he commanded a schooner surveying the coast, spending the winters in England perfecting his charts. In Newfoundland when he observed an eclipse of the sun in 1766, he broke precedent by volunteering his calculations to the Royal Society in London.

It is not surprising that the Admiralty named Cook to command the Tahiti expedition. Though still only a noncommissioned officer, he had proven himself in combat and on rough seas, he was a competent surveyor of treacherous coasts, and he had registered both his competence and his curiosity as an observer of astronomical phenomena. The choice of Cook, as it turned out, was also the choice of the ship, for on his advice the Admiralty commissioned a squat cat-built Whitby collier of precisely the kind on which Cook had served his apprenticeship in the North Sea. This was a four-year-old stout vessel of 368 tons, 98 feet long with a beam of 29 feet, of a design noted for roominess and sturdiness rather than for the grace of its lines. Fastidious seamen are still offended by its lack of decoration, for it did not even carry a figurehead in an age when such ornaments were universal. Hardly impressive for the Royal Navy's expedition halfway around the world!

In May 1768 James Cook was promoted to the commissioned rank of lieutenant. The collier was christened the *Endeavour,* given a sheathing of wood filled with nails against the tropical teredos, and victualed for eighteen months. Linnaeus' correspondent John Ellis reported to him that "No people ever went to sea better fitted out for the purpose of Natural History, or more elegantly." The considerable contributions of the voyage to botany and zoology were due to Joseph Banks (1743–1820), the most prominent English patron of natural history, who was destined to become president of the Royal Society (1778), and who used his wealth in various ways to promote the new community of scientists. He later developed the Royal Botanic Gardens at Kew and sent botanical explorers across the world. Daniel Solander, the principal naturalist, whom Banks brought along, was a disciple of the Swedish naturalist Linnaeus. The Royal Society supplied the scientific instruments needed at Tahiti, but the ship still carried no chronometer. Although, as we have seen, the prize for a seagoing clock to find longitude had been awarded in 1765 to John Harrison, the Admiralty had put none on board. This meant that to find longitude, Cook, aided by an astronomer from the Greenwich Observatory, had to try to fix their position by complicated lunar calculations. Assisted by the charts and reports of earlier explorers, which Cook had been careful to take along, their calculations were remarkably accurate.

The *Endeavour* sailed from Plymouth on August 26, 1768, with its full complement of ninety-four persons, crowded by the last-minute demands of Joseph Banks' "Suite comprising eight persons with their Baggage." These included, in addition to Solander, a personal clerk, two botanist-draftsmen, two footmen from the Banks estates, and two black servants—with their artists' equipment, fishing nets, trinkets for "savages," chemicals and containers for preserving specimens, and two large greyhounds. Sailing southwestward in fair weather, to Madeira, then to Rio de Janeiro and around Cape Horn, Cook arrived in Tahiti on April 10, 1769, in plenty of time to prepare for the observations on June 3. When the astronomical observations were completed, Cook took up his secret larger assignment, to seek out the Great Southern Continent, and possibly prove that it did not exist.

To succeed in negative discovery—to prove that some mythical entity really did *not* exist—was far more exacting and more exhausting than to succeed in finding a known objective. The westward sea passage from Europe to Asia, which Columbus sought, was a path to a known goal. Going westward on the latitude of Japan, he remained confident that he had reached his destination. When he was proven wrong, it was only because an unexpected continent stood in the way, and in the long run Columbus really did open a roundabout westward sea passage to Asia. So long as the existence and precise location of the Great South Land were legendary, the explorer had to scour all conceivable places, and in fact would have to circumnavigate the globe, before he dared assert that it would never be discovered.

Captain James Cook was suited to be the world's greatest negative discoverer—by his restless energy, his organizing ability, his vast knowledge of charts and the sea, and his persistence in trying possibilities that others had not had the courage or the vigor to pursue. This grand elusive enterprise began with his departure from Tahiti. Before him, explorers in that area had usually sailed west and west-northwest with the favoring winds, but Cook reached south and southwest seeking the supposed continent down to 40° south latitude. When to that point he found no land, he made westward, where he encountered New Zealand and spent six months circumnavigating and charting the 2,400 miles of coasts of both the north and the south island. Incidentally he proved that these really were islands, and not continuous with any Southern Continent. This was a first step, but only a small one in proving that Dalrymple's clues were misleading.

Cook's orders had given him a choice of returning home either eastward, as he had come, or westward around the Cape of Good Hope. By the end of March 1770, when the southern summer was at its end, to sail eastward on the Antarctic latitudes was looking for trouble. So he decided to go westward, to explore the east coast of New Holland (Australia), then sail up to the East Indies and homeward around the Cape of Good Hope. While this deprived him of an opportunity on this trip to provide more facts concerning the great Southern Continent, it would enrich science in unexpected ways. On the southeast coast of Australia they found Stingray Harbor, but Banks, Solander, and the artists were so delighted there by the countless specimens unknown in Europe that they renamed it Botany Bay. And so it remains a vivid reminder in the South Pacific of how the naturalists' quest was enriching the European's vision of the whole world.

Their next discovery was less welcome. The Great Barrier Reef, off the northeast coast of Australia, is the largest structure ever built up by living creatures. Stretching some 1,250 miles at a distance from 10 to 100 miles offshore, it covers 80,000 square miles, with multicolored coral polyps and the remains of coralline algae, at least three hundred fifty coral species built up over twenty-five million years. In the twentieth century it has become so popular as a tourist attraction that its survival is threatened. Until Cook's first voyage, the reef was not known to Europeans.

The old charts called this a dangerous coast, warning of banks or shoals, yet Cook was managing to thread his way through. Making a coastal survey as he went, he had to stay within sight of the shore. He did not know it, but in June he was sailing inside the Great Barrier Reef. Before retiring on one moonlit tropic night, he was reassured to learn from his leadsman that the soundings showed seventeen fathoms, more than a hundred feet. Then suddenly a screeching of the keel on the coral, and the *Endeavour* was perched on a reef. Cook, who had gone to his cot, was quickly on deck, still "in his drawers." Water was pouring into the hold, where it soon measured four feet. The crew heaved the anchors to try to pull the ship free, then tossed overboard some fifty tons of ballast including several precious can-non. Even the gentlemen manned the pumps to keep the bark from founder-ing, and if this ship had not been a stout Whitby collier it doubtless would have been lost.

By a combination of luck, pluck, and skill, and the good fortune of a rising tide, the bark was finally floated off the reef. But something had to be done about the holes in the keel if she was not to sink before reaching shore. A crewman remembered that once when he had been wrecked off Virginia he had seen his ship saved by "fothering." Cook decided to try it, which meant passing a sail in the water under the ship's bottom. On the sail were sewn bits of wool and oakum, covered with rope ends and dung from the ship's livestock, and when the sail was pulled around the keel by lines, it was hoped that flotsam might be sucked into the seams. Luckily, a large slice of coral from the reef had plugged the biggest hole. This worked well enough to keep the *Endeavour* afloat until it could be taken into the mouth of a nearby river where she was beached, and a month was lost while she was repaired. Meanwhile Cook and his crew learned more of survival in the tropics, eating kangaroo, birds, turtles, clams, and fish. Food was scarce, but Cook insisted on equal shares for all ranks.

His perilous passage from there confirmed that Australia was separated from New Guinea in the north. Cook went on to Batavia, on Java in the East Indies, then rounding the Cape of Good Hope, he returned to England on July 12, 1771, a month less than three years after his departure.

Cook's talent for realistic assessment of his own achievements was rare among the great seafarers. "Altho' the discoveries made in this voyage are not great, yet I flatter my self," he reported to the Admiralty, "that they are such as may merit the attention of their Lordships, and altho' I have faild in discovering the so much talk'd of southern continent (which perhaps do not exist) and which I my self had much at heart, yet I am confident that no part of the failure of such discovery can be laid to my charge. . . . Had we been so fortunate not to have run ashore much more would have been done in the latter part of the voyage than what was, but as it is I presume this voyage will be found as compleat as any before made to the South Seas, on the same account." On his return he was presented to King George III, and promoted to the rank of commander.

A great deal of the interest in Cook's first voyage came from the rich haul of specimens that the naturalists had collected, and Banks now joined in urging another voyage of discovery. Unfortunately, Banks could not be accommodated this time, for he enlarged his "Suite" from eight to fifteen, including not only Solander and a portrait painter but several draftsmen, additional servants, and a horn player. For this voyage, Banks wanted a large East Indiaman, but Cook stayed with his trusted Whitby colliers, which could not meet Banks' extravagant demands. In a huff, Banks made off with his entourage to Iceland.

Cook secured two newly built Whitby colliers—the *Resolution*, of 462 tons, and the *Adventure*, of 340 tons—both well fitted and competently manned. For his naturalist he engaged Johann Reinhold Forster (1729–1798), a pedantic German scholar of wide reputation who had published works on natural history, and his son, Georg, to be aided by the Swede Anders Sparrman, another Linnaeus disciple. In addition, there was one astronomer on each ship from the Board of Longitude. And something new —four chronometers to help find longitude, of which only one, built on John Harrison's prize-winning model, would prove satisfactory. Cook called it "our never-failing guide" and "trusty friend."

Cook's plan this time was wholly directed at solving the problem of the Great South Land. For this purpose the voyage had to be a full circumnavigation of the earth at the southernmost possible latitude. His last trip had come into the Pacific by way of Cape Horn. This time he proposed to try the other way, down the Atlantic past the Cape of Good Hope, then, in the farthest southern latitude he could manage, proceed eastward all around the South Polar regions of the globe. If there really was a southern continent reaching up into inhabitable zones, he could not possibly miss it. The plan required that Cook be at the Cape of Good Hope by early October "when you would have the whole summer before you and . . . might, with the prevailing westerly winds, run to the eastward in as high a latitude as you please and, if you met with no lands, would have time enough to get round Cape Horne before the summer was too far spent." After traversing the whole southernmost rim of the Pacific in the Antarctic regions "if you met with no lands," there would still be time to go "to the northward and after visiting some of the islands already discovered . . . proceed with the trade wind back to the westward" in search of other still conjectural islands. "Thus," it concluded, "the discoveries in the South Sea would be complete." The instructions for this voyage, drawn up with Cook's advice, took account of both possibilities—that the continent did, or did not, exist. If Cook found any part of the mythic continent, he was to survey it, claim it for Britain, and distribute medals among the inhabitants. New islands too were to be surveyed and claimed. But in any case officers and crew were to maintain strict secrecy about the voyage, and all logs and journals were to be confiscated before the ships' return.

The voyage on which Cook's two Whitby collier-barks departed from Plymouth on July 13, 1772, would be one of the greatest—as it was undoubtedly one of the longest—sailing-ship discovery voyages in history. He would sail more than seventy thousand miles. But it was also novel in other ways. Never before in modern times had there been so long a voyage with one focused inquiring purpose. Not to seek an Eldorado, nor find gold or silver or precious gems, nor capture slaves. Earlier expeditions, like Cook's first voyage, had aimed to make astronomical observations at a distant place. Dias, Gama, Columbus, Magellan, and Drake sought the safest or the shortest passage to a desired destination, or hoped to find strategic locations for capturing foreign treasure. Now Cook went, in a modern skeptical spirit, mainly to answer a question. Was the rumored Southern Continent really there?

Incidentally, this question carried Cook into some of the most inhospitable areas of the globe, and revealed a seascape the like of which had never been seen. For the Antarctic was dangerously different from the Arctic. Cook would unveil a new scene of perpetual mountainous ice beyond the belief of temperate Europe, in an area where extravagant *a priori* speculation had flourished for centuries. One medieval work, *De vegetabilibus* (erroneously ascribed to Aristotle in the Middle Ages), actually argued that because the sun shone continuously for half a year at the poles, and then never sank far below the horizon, no plants or animals could survive there, since they would be regularly burned up by the sun.

But there really was a four-month Antarctic summer, and Cook hastened to make the most of it. The *Resolution* and the *Adventure* left Cape Town on November 23, 1772, went south and within two weeks were within the Antarctic Circle (66° 33′ south latitude). Arctic waters were, of course, not unknown to European sailors, who had been seeking a Northwest Passage for two centuries—ever since the time of Jacques Cartier, Martin Frobisher, and Henry Hudson. Those North Polar regions were a vast frozen ocean surrounded by land. A ship that could avoid or manage the ice pack would somehow make its way. On the fringes of the pole, well within the Arctic Circle, were large areas inhabited year round by Lapps, Greenlanders, and Eskimos. What Vilhjalmur Stefansson called the Friendly Arctic was teeming with edible fauna—duck, geese, salmon, crab, and copious fish. But in the Antarctic the few animals were not good eating, and there were no Eskimos. A travesty of the romanticized Great South Land, the Antarctic was a frigid continent girded by icebergs, some the size of mountains, others smaller, and called "growlers," all tossed and churned by gusty winds and unpredictable heavy seas. Cook, though trained for the unexpected, had no reason to expect what he found.

Arriving in the Antarctic summer in January, Cook and his men were overwhelmed by the blue-white beauty of the iceberg-alps which they saw ahead. They kept on southward until they could not move for the pack ice. Impenetrable icebergs were crunching and grumbling and tumbling about them. Luckily, they avoided the crush, but when they encountered a strong gale and a heavy sea, they dared not go on into the mist. At one point Cook came within seventy-five miles of the Antarctic continent, but he could not see it, and it was hopeless to try to chart the coast—if there was one. Then Cook desperately turned north out of the ice and carried on eastward. His two ships were separated in the fog, but rendezvoused according to plan in Dusky Bay in southwest New Zealand to pass the southern winter. For their second polar season, pursuing the circumnavigation, they proceeded eastward and southward sometimes below the Antarctic Circle. Nothing changed. On January 30, 1774, when they touched the farthest south Cook would ever reach, ice pack blocked the way, the mist revealed only icebergs in the distance, and he recorded in his journal:

In this field we counted Ninety Seven Ice Hills or Mountains, many of them vastly large. . . . I will not say it was impossible anywhere to get in among this Ice, but I will assert that the bare attempting of it would be a very dangerous enterprise and what I believe no man in my situation would have thought of. I whose ambition leads me not only farther than any other man has been before me, but as far as I think it is possible for man to go, was not sorry at meeting with this interruption, as it in some measure relieved us from the dangers and hardships, inseparable with the Navigation of the southern Polar regions. Since therefore we could not proceed one Inch farther South, no other reason need be assigned for our Tacking and stretching back to the North, being at that time in the Latitude of 71° 10′ South, Longitude 106° 54′ W.

He spent the next winter exploring the South Pacific, where he charted the Easter Islands and Tonga, and discovered New Caledonia before making his way again eastward in the high southern latitudes. En route to the Cape of Good Hope in the Atlantic, he discovered the South Sandwich Islands and South Georgia. He was back in England on July 30, 1775, three years and seventeen days after his departure.

Again in his journal he summed up his achievements:

I have now made the circuit of the Southern Ocean in a high Latitude and traversed it in such a manner as to leave not the least room for the Possibility of there being a continent, unless near the Pole and out of the reach of Navigation; by twice visiting the Pacific Tropical Sea, I had not only settled the situation of some old discoveries but made there many new ones and left, I conceive, very little more to be done even in that part. Thus I flater my self that the intention of the Voyage has in every respect been fully Answered, the Southern Hemisphere sufficiently explored and a final end put to the searching after a Southern Continent, which has at times ingrossed the attention of some of the Maritime Powers for near two Centuries past and the Geographers of all ages.

The great Negative Discoverer would release energies for searches not yet discovered to be fruitless.

The British Admiralty had still another, more focused assignment for Cook on the frontiers of myth, hope, and geography. Was there really a Northwest Passage? The quest for a northern seaway from the Atlantic to the Pacific had enticed voyagers ever since the discovery of America. Cook's exploits in the myth-ridden Pacific suggested to the Royal Society that he was the man to answer the question from the Pacific side. Within less than a year after his return from his second voyage, Cook was off on this quest for a passage that might (or might not) be there. The *Resolution* was refitted and another Whitby collier, the *Discovery,* provided, and Cook was on his way eastward around the Cape of Good Hope, across the Indian Ocean, through Cook Strait between the two islands of New Zealand, to the north-west coast of America. His search along the coast through the Bering Sea, to the southern borders of ice in the Arctic Ocean proved fruitless. There was no Northwest Passage—at least none usable by sailing ships. Returning from his arduous quest for a rest in Hawaii, this man who had braved Antarctic ice, coral reefs, and tropical storms, and had managed crews on voyages lasting years, met his end in an unheroic scuffle sadly reminiscent of the death of Magellan in the Philippines just two hundred and fifty years before. The Polynesians, with whom Cook had done much to establish friendly relations, had an exasperating passion for anything that could be pried loose from the ships, especially anything of iron. They had even invented a way by diving under the ship and using a flint fixed to a stick to pry out the long nails that held the sheathing to the ship's bottom. When one of Cook's large boats was stolen, he could stand it no more. He went ashore with an armed guard to recover the boat or secure a hostage. The enraged Hawaiians attacked him with knives and clubs and held him under the water to drown.

Cook's signal recognition in his day came not from his navigating exploits but from what he did to improve the health and save the lives of men at sea. He did more than any other explorer in the sailing days of long ocean voyages to cure the curse of seamen—scurvy. The lethargy and anemia, bleeding gums, loosened teeth, stiffness of the joints, and slow wound heal-ing were vividly described by Samuel Taylor Coleridge in "The Rime of the Ancient Mariner." On Vasco da Gama's voyage around the Cape of Good Hope, scurvy is said to have taken a hundred of his hundred and seventy men. When James Lind (1716–1794), a Scottish naval surgeon, demonstrated that citrus fruit could prevent and cure the disease, and published his findings in 1753, he did attract the notice of Lord Anson, whose reforms of naval personnel would help make Cook's career possible. But the Admiralty delayed so long acting on his findings that this has become the sociologists' classic case of bureaucratic apathy.

Apparently Cook himself never knew Lind's work, but he had heard of the use of citrus fruits and other possible preventives of scurvy. And he went to the trouble of testing new fruits and grasses. He enforced cleanliness on board by regularly inspecting the men's hands, and he punished the dirty-handed by stopping their daily grog. But he was no martinet, and used flogging only sparingly. The result of his experiments with oranges, lemons, and their juices, along with sauerkraut, and miscellaneous items like the onions of Madeira, the wild celery and "scurvy grass" of Tierra del Fuego was quite remarkable. On his first voyage he lost men by accidents and other ailments, but appears not to have lost a single one from scurvy, and his record on the second voyage was just as impressive. On his return from his second voyage, Cook was elected a Fellow of the Royal Society in late February 1776, and then received their highest award, the Copley Medal, for his methods of preserving the health of his men on long voyages. In his journal he told his parable of leadership:

The Sour Krout the Men at first would not eate untill I put in practice a Method I never once knew to fail with seamen, and this was to have some of it dress'd every Day for the Cabbin Table, and permitted all the Officers without exception to make use of it and left it to the option of the Men either to take as much as they pleased or none at all; but this practice was not continued above a week before I found it necessary to put every one on board to an Allowance, for such are the Tempers and disposissions of Seamen in general that whatever you give them out of the Common way, altho it be ever so much for their good yet it will not go down with them and you will hear nothing but murmurings gainest the man that first invented it; but the Moment they see their Superiors set a Value upon it, it becomes the finest stuff in the World and the inventer a damn'd honest fellow.

BOOK THREE

NATURE

*The investigation of nature is an infinite pasture-ground,
where all may graze, and where the more bite, the longer
the grass grows, the sweeter is its flavor, and the more it
nourishes.*

—THOMAS HENRY HUXLEY (1871)

The discovery of nature, of the ways of planets, and plants
and animals, required first the conquest of common sense.
Science would advance, not by authenticating everyday
experience but by grasping paradox, adventuring into the
unknown. Novel instruments, telescopes and microscopes
among others, would offer disturbing new perspectives. In
parliaments of science—communities of knowledge, not in
learned languages but in the vernacular—amateurs could
challenge professionals and professionals could challenge
one another. The public became a witness and a patron.
Novelty came to be prized. Nature itself had a history, and
in the eons of the planet's extended past appeared countless creatures no longer on earth. Here were new incentives to ransack the world for undiscovered species and
seek clues to the mystery of an ever-changing nature.

PART NINE

SEEING
THE
INVISIBLE

Where the telescope ends, the microscope begins. Which of the two has the grander view?

—VICTOR HUGO, *Les Misérables* (1862)

38

Into "the Mists of Paradox"

NOTHING could be more obvious than that the earth is stable and unmoving, and that we are the center of the universe. Modern Western science takes its beginning from the denial of this commonsense axiom. This denial, the birth and the prototype of science's sovereign paradoxes, would be our invitation to an infinite invisible world. Just as Knowledge was what led Adam and Eve to discover their nakedness and put on their clothing, so the guilty knowledge of this simple paradox—that the earth was not as central or as immobile as it seemed—would lead man to discover the nakedness of his senses. Common sense, the foundation of everyday life, could no longer serve for the governance of the world. When "scientific" knowledge, the sophisticated product of complicated instruments and subtle calculations, provided unimpeachable truths, things were no longer what they seemed.

Ancient cosmologies used picturesque and persuasive myths to embellish the verdicts of common sense, and to describe how the heavenly bodies moved. On the walls of the tombs of the Egyptian pharaohs in the Valley of the Kings we can find colorful caricatures of how above the Earth the God of Air supported the dome of heaven. There, too, we observe how every day the sun-god Ra rode his boat through the sky. Every night, riding another boat through the waters beneath the earth, he arrived back where he began his daily journey again. This mythic view, as we have seen, did not prevent the Egyptians from developing the most precise calendar of the solar year that was known for millennia. To the ordinary Egyptian such myths made sense. They did not contradict what he saw every day and every night with his naked eye.

The Greeks developed the notion that the earth was a sphere on which man lived while the heavens above were a rotating spherical dome that held the stars and moved them about. The spherical nature of the earth, as we have seen, was demonstrated by such commonsense experience as the disappearance of departing ships below the horizon. The spherical nature of the heavens was also confirmed by everybody's naked-eye experience, day and night. Outside that dome of stars, according to the Greeks, there was nothing, not space, nor even emptiness. Inside the sphere of the stars, the sun went around the earth in its daily and yearly courses. Plato described the creation of this two-sphere universe with his usual mythic felicity. "Wherefore he made the world in the form of a globe, round as from a lathe, having its extremes in every direction equidistant from the center, the most perfect and the most like itself of all figures; for he considered that the like is infinitely fairer than the unlike."

In his work *On the Heavens* Aristotle elaborated this commonsense vision into an attractive dogma. "Ether," transparent and weightless, was the pure material of the heavens and of the concentric nesting heavenly spheres that carried the stars and the planets. Though some of his disciples disagreed, Aristotle said that these ethereal shells numbered precisely fifty-five. The varying distance of each planet from the earth was explained by the movements of each planet from the innermost to the outermost edges of its own special sphere. For many centuries the speculations of leading Western astronomers, astrologers, and cosmologists were only modifications of this picture.

To understand the paradoxical beginnings of modern science, we must recall that this beautiful symmetrical scheme, much ridiculed in the modern classroom, actually served very well for both astronomer and layman. It described the heavens precisely as they looked and fitted the observations and calculations made with the naked eye. The scheme's simplicity, symmetry, and common sense made it seem to confirm countless axioms of philosophy, theology, and religion. And it actually performed some functions of a scientific explanation. For it fitted the available facts, was a reasonably satisfactory device for prediction, and harmonized with the accepted view of the rest of nature. In addition, it aided the astronomer's memory with a convenient coherent model, replacing the list of miscellaneous facts then known about the heavens. More than that, while this much maligned geocentric, or "Ptolemaic," scheme provided the layman with a clear picture to carry around in his head, it helped the astronomer reach out to the unknown. Even for the adventurous sailor and the navigator it served well enough, as Columbus proved. The modern advance to Copernicus' heliocentric system would be hard to imagine if the geocentric system had not been there available for revision. Copernicus would not change the shape of the system, he simply changed the location of the bodies.

Of course the traditional geocentric system of Aristotle and Ptolemy and so many others over centuries had its own weaknesses. For example, the system did not explain the irregularities observed in the motions of the planets. But the layman hardly noticed these irregularities, and anyway they seemed adequately described by the supposed movement of each planet within its own special ethereal sphere. Astronomers were adept at explaining away what seemed only minor problems by a variety of complicated epicycles, deferents, equants, and eccentrics, which gave them a heavy vested interest in the whole scheme. The more copious this peripheral literature became, the more difficult it became to retreat to fundamentals.

If the central scheme was not correct, surely so many learned men would not have bothered to offer their many subtle corrections.

Why did Nicolaus Copernicus (1473–1543) go to so much trouble to displace a system that was amply supported by everyday experience, by tradition, and by authority? The more we become at home in the Age of Copernicus, the more we can see that those who would remain unpersuaded by Copernicus were simply being sensible. The available evidence did not require a revision of the scheme. Decades would pass before astronomers and mathematicians could gather new facts and find new instruments, a century or more before laymen would be persuaded against their common sense. To be sure, despite all the arcane modifications that astronomers and mathematicians devised, the old scheme did not quite fit all known facts. But neither would Copernicus' own simplification.

It seems that Copernicus was animated not by the force of facts but by an aesthetic, metaphysical concern. He imagined how much more beautiful another scheme could be. Copernicus possessed an extraordinarily playful mind and a bold imagination. But there was nothing extraordinary about his career. Although he never took holy orders, he led his whole working life comfortably in the bosom of the Church. In fact, it was the Church that made it possible for him to pursue his wide-ranging intellectual and artistic interests. He was born in 1473 in the busy commercial town of Thorn on the banks of the Vistula River in northern Poland. When he was only ten, his father, a prosperous wholesale merchant and city official, died. His uncle and guardian, who became bishop of Ermeland, a see in northern Poland, arranged for Nicolaus to be taken care of by Mother Church. In the Bishop's cathedral seat, the city of Frauenburg, his nephew Nicolaus was appointed canon at the age of twenty-four, and this post remained his worldly support until the day he died.

As an astronomer Copernicus was a mere amateur. He did not make his living by astronomy or by any application of astronomy. By our standards, at least, he was wonderfully versatile, which put him in the mainstream of the High Renaissance. He was born when Leonardo da Vinci (1452–1519) was in full career, and Michelangelo (1475–1564) was his contemporary. He began by studying mathematics at the University of Cracow, where he picked up enough skill at painting to be able to leave us a competent self-portrait. After receiving his convenient appointment as canon of Frauenburg he quickly took leave for an extended journey to Italy to study canon law at Bologna and Ferrara, to study medicine at Padua, and incidentally to hear some lectures in astronomy. Back in Frauenburg, he served as the Bishop's personal physician until his uncle's death in 1512. In those turbulent times his post as canon was no sinecure. He had to keep the accounts, to see that the chapter's political interests were protected, and serve as commissary of the whole diocese. Along the way he provided the Polish provincial Diet of Graudenz with a scheme for improving its currency. Copernicus developed his heliocentric theory as an avocation, and only the enthusiasm of friends and disciples induced him to have it published.

Copernicus was well aware that his system seemed to violate common sense. For that very reason his friends had "urged and even importuned" him to publish the work. "They insisted that, though my theory of the Earth's movement might at first seem strange, yet it would appear admirable and acceptable when the publication of my elucidatory comments should dispel the mists of paradox."

Copernicus' own first comprehensive sketch of his system, *Commentariolus* or "Sketch of his Hypotheses for the Heavenly Motions," was not printed during his lifetime. Only a few handwritten copies were circulated among his friends. Oddly enough, the first description to the world of Copernicus' revolutionary system was not by Copernicus himself but by a brilliant and erratic twenty-five-year-old disciple. This young Austrian, born Georg Joachim (1514–1574), had taken the name Rheticus to avoid bearing the stigma of his father, a town physician who had been beheaded for sorcery. Rheticus arrived in Frauenburg in the summer of 1539 to meet Copernicus and learn more about his new cosmology, still not available in print. He had just received his M.A. from the University of Wittenberg for a thesis which proved that the Roman law did not forbid astrological predictions, because like medical predictions they were based on observable physical causes. Rheticus was obviously a young man of some courage and considerable powers of persuasion. Although Copernicus had repeatedly refused to give in to requests that he himself publish his stirring new ideas, he now granted his young visitor permission to do the job for him.

Within a few months, by late September that same year, Rheticus had written his *First Report (Narratio Prima)* of Copernicus' system, in the form of a letter to his former teacher, which was printed in Danzig early in 1540. For Copernicus the advantages of such a trial balloon were obvious. If the reception was favorable, he could confidently publish his own amplified account. If not, he could leave it alone or modify his own statement. Copernicus' doubts were dispelled when the demand for Rheticus' *First Report* required a second edition in 1541. He then turned to revising for publication the manuscript of his own great work, which he had nearly completed a full decade before. Copernicus assigned to Rheticus the task of seeing the epochal book through the press. When, at the last moment, Rheticus, for personal reasons, could not finish the job, he unfortunately handed it over to an acquaintance Andreas Osiander (1498–1552). This militant, quarrelsome and Machiavellian Lutheran theologian believed that divine revelation was the sole source of truth, and, as we shall see, was determined to do all he could to cast Copernicus' suggestions in the image of his own orthodoxy. Copernicus, who lay dying in Frauenburg far from the site of publication, would be powerless to intervene.

The revolutionary suggestion of Copernicus was that the earth itself moved. If the earth moved around the sun, then the sun and not the earth was the center of the universe. Might not the whole scheme of the heavens become suddenly simpler if the sun, instead of the earth, were imagined at the center?

Copernicus' objective was not to devise a new system of physics, much less a new scientific method. His single revision—a moving earth no longer in the center—leaves the large features of the Ptolemaic system untouched. He stays with the doctrine of spheres, which was crucial to the Ptolemaic system, and avoids the debated question whether the celestial spheres were imaginary or real. He does not say whether the "spheres" (orbes) in which the planets and, according to his system, the earth too revolve are only a convenient geometric device for describing how they move, or whether each "sphere" really is a thick shell made of some ethereal transparent material. For Copernicus, orbis simply meant sphere, and he plainly keeps the traditional concept of spheres in his own system. The title of the climactic book in which he finally summed up his theory, De Revolutionibus Orbium Caelestium, does not refer to planets, but means "Concerning the Revolutions of the Heavenly Spheres." On another crucial question, whether the universe is finite or infinite, Copernicus again decisively refuses to commit himself. He leaves this matter "to the discussion of the natural philosophers."

Just as Columbus relied on Ptolemy and the other traditional texts whose suggestions he thought had not been energetically enough pursued, so, too, Copernicus found his clues among the ancients. First of all, from Pythagoreanism, the influential doctrine of the followers of Pythagoras of Samos, a Greek philosopher and mathematician of the sixth century B.C. None of Pythagoras' own work has survived, but the ideas fathered on him by his followers would be among the most potent in modern history. Pure knowledge, the Pythagoreans argued, was the purification (catharsis) of the soul. This meant rising above the data of the human senses. The pure essential reality, they said, was found only in the realm of numbers. The simple, wonderful proportion of numbers would explain the harmonies of music which were the beauty of the ear. For that reason they introduced the musical terminology of the octave, the fifth, the fourth, expressed as 2:1, 3:1, and 4:3.

For astronomy the Pythagorean adoration of numbers carried an overwhelming message. "They say that the things themselves are Numbers," was Aristotle's succinct summary in his *Metaphysics,* "and do not place the objects of mathematics between Forms and sensible things." "Since, again, they saw that the modifications and the ratios of the musical scales were expressible in numbers—since, then, all other things seemed in their whole nature to be modelled on numbers, and numbers seemed to be the first things in the whole of nature, they supposed the elements of numbers to be the elements of all things, and the whole heaven to be a musical scale and a number. . . . and the whole arrangement of the heavens they collected and fitted into their scheme; and if there was a gap anywhere, they readily made additions so as to make their whole theory coherent." In Copernicus' time Pythagoreans still believed that the only way to truth was by mathematics.

The other fertile source for Copernicus' ideas and for the pragmatic foundations of modern science was just as surprising—Plato and his mystic followers, the Neoplatonists. Although Copernicus was to be the unconscious prophet of the scientific belief in the sovereignty of the senses, his godfather was Plato, who believed that all the data of the senses were mere unsubstantial shadows. Plato's "real" world was a world of ideal forms, and from his point of view geometry was more real than physics. Over the entrance to Plato's Academy, we are told, stood the warning, "Let no one destitute of geometry enter my doors."

Plato's Neoplatonist followers, too, built their whole view of the world on an ideal mathematics. Numbers offered the best human vision of God and the world-soul. "All mathematical species . . . have a primary subsistence in the soul," observed Proclus (A.D. 410?–485), the last and greatest of the Greek exponents of Neoplatonism, "so that, before sensible numbers, there are to be found in her inmost recesses, self-moving numbers . . . ideal proportions of harmony previous to concordant sounds; and invisible orbs, prior to the bodies which revolve in a circle. . . . we must follow the doctrine of Timaeus, who derives the origin, and consummates the fabric of the soul, from mathematical forms, and reposes in her nature the causes of everything which exists."

Neoplatonism, reborn in the Renaissance—the age into which Copernicus was born—took up battle against the frigid, prosaic spirit of the scholastics. Aristotle's hardheaded commonsense approach had been reinforced by the finding of new Aristotelian texts in the twelfth century. Against this the Neoplatonists championed poetry and the free-flying imagination. When Copernicus studied at Bologna, his teacher was Domenico Maria de Novara, an enthusiastic Neoplatonist who attacked the Ptolemaic system. Surely the heavenly scheme must be too simple to need all that pedantic apparatus of epicycles, deferents, equants, etc., etc. The astronomers must somehow have missed the essential charm of celestial numbers.

In Copernicus' own Preface to his *De Revolutionibus,* he spoke with the voice of his teacher and put himself squarely in the ranks of the Neoplatonists. The Ptolemaic system for explaining the motions of the planets, he said, required "many admissions which seem to violate the first principle of uniformity in motion. Nor have they been able thereby to discern or deduce the principal thing—namely the shape of the Universe and the unchangeable symmetry of its parts." Copernicus believed that his system actually accorded better than the older geocentric system with the way the universe *ought* to be. He believed he was describing the actual truths of an essentially mathematical universe.

Heavenly motions must be perfect circles. In Copernicus' time, all this reminds us, astronomy was still a branch of mathematics—in E. A. Burtt's phrase "the geometry of the heavens." Following Pythagorean and Neoplatonic doctrine, this carried implications too for mathematics itself, which, instead of being the deductive study of abstract constructs, purported to describe the actual world. It would be some time before this notion changed. Meanwhile, this proved to be a fruitful confusion, luring astronomers and others through the gateway to modern science.

Copernicus had some authorities, and some appealing assumptions, but he did not yet have the evidence to support his hunches. In this, too, he was like Columbus, who thought the westward voyage to the Indies worth trying even though direct evidence was still lacking, and though Gama had done well enough by going eastward. Similarly the Ptolemaic system had for centuries provided a usable calendar. The scheme that Copernicus now proposed, for all its aesthetic appeal, fit the observed facts no better. Nor could he predict the position of the planets with anything like the demonstrated accuracy of the older system.

How seriously did Copernicus take his own proposals? Did he think that he had finally solved the central problems of astronomy? Or was he merely offering a tentative suggestion for others to explore? The first printed edition of Copernicus' great work, the *De Revolutionibus* (1543), which reached him only on his deathbed, carried a lengthy unsigned introduction which seemed to answer this question beyond doubt.

> Since the novelty of the hypotheses of this work has already been widely reported, I have no doubt that some learned men have taken serious offense because the book declares that the earth moves, and that the sun is at rest in the center of the universe; these men undoubtedly believe that the liberal arts, established long ago upon a correct basis, should not be thrown into confusion. But if they are willing to examine the matter closely, they will find that the author of this work has done nothing blameworthy. For it is the duty of an astronomer

to compose the history of the celestial motions through careful and skillful observation. Then turning to the causes of these motions or hypotheses about them, he must conceive and devise, since he cannot in any way attain to the true causes, such hypotheses as, being assumed, enable the motions to be calculated correctly from the principles of geometry, for the future as well as for the past. The present author has performed both these duties excellently. For these hypotheses need not be true nor even probable; if they provide a calculus consistent with the observations, that alone is sufficient. . . . So far as hypotheses are concerned, let no one expect anything certain from astronomy, which cannot furnish it, lest he accept as the truth ideas conceived for another purpose, and depart from this study a greater fool than when he entered it. Farewell.

Only later was it discovered that this introduction was not written by Copernicus at all. In the cause of Lutheran orthodoxy, the unscrupulous Andreas Osiander had secretly suppressed Copernicus' own introduction and substituted this unsigned one, which he concocted himself. It was the great Johannes Kepler (1571–1630) who identified the anonymous author and defended Copernicus from Osiander's "most absurd fiction," the slander on his scientific integrity. Osiander had thought he was defending Copernicus, but his exercise in timidity proved superfluous. By the time the *De Revolutionibus* was spread abroad, Copernicus himself was dead and safely beyond reach of retribution by any church on earth. "He thought that his hypotheses were true," the outraged Kepler insisted, "no less than did those ancient astronomers. . . . He did not merely think so, but he proves that they are true. . . . Therefore Copernicus is not composing a myth but is giving earnest expression to paradoxes, that is, philosophizing, which is what you required in an astronomer."

Copernicus himself was no party to the theological compliance to which Osiander had tried to commit him. But Kepler, always an enthusiast, appears to have become more of a Copernican than Copernicus himself. Copernicus seemed to have realized that he had only pushed the door ajar. He enjoyed giving his contemporaries a glimpse of what might be in store for them. This itself required courage. He was not yet ready for a bold exploration of his New World. He did not and could not yet realize how new was the New World he had opened. For, again like Columbus, he was still relying heavily on ancient maps.

Copernicus described his system as "hypotheses." And in the language of the Ptolemaic Age a "hypothesis" was more than a mere experimental notion. It was, rather, the principle or the fundamental proposition (his synonyms were *principium* or *assumptio*) on which a whole system was based. This meant, according to Copernicus, that his propositions had two essential qualities. First, they must "save the appearances" *(apparentias salvare)*, which meant that the conclusions drawn from them must agree with the actual observations. Some interesting ambiguities in this simple phrase would surface within the next century, when the telescope supplied "appearances" not visible to the naked eye. In 1543 "saving the appearances" still seemed a self-evident and self-defining criterion. Yet, fitting what the eye saw was not enough. A second requirement was that a scientific proposition had to fit with and confirm the basic *a priori* notions accepted as the axioms of physics. For example, it must not be inconsistent with the axiom that all the motions of heavenly bodies are circular and that every such motion is uniform. While, according to Copernicus, the Ptolemaic system fitted well enough with observed appearances, it did not adequately provide for the required uniformity and circularity. A "true" system by Copernicus' standards would not merely satisfy the eye *(apparentias salvare)* but it would also have to please the mind.

If Copernicus had fears that his astronomic system would put him down as a heretic, these proved quite unfounded, not only during his lifetime but for a half-century after his death. His friends high in the Church, including a cardinal and a bishop, had long been urging him to publish his *De Revolutionibus*. He actually dedicated his great work to Pope Paul III, whose own mathematical education, he hoped, would arouse his special interest.

The prophets of Protestantism—Luther (1483–1546), Melanchthon (1497–1560), and Calvin (1509–1564)—all close contemporaries of Copernicus, carried a strong fundamentalist, anti-intellectual message. "An upstart astrologer," was Luther's epithet for Copernicus in his *Table Talk* in 1539. "This fool wishes to reverse the entire science of astronomy; but sacred Scripture tells us that Joshua commanded the sun to stand still, and not the earth." "Now, it is in want of honesty and decency," Luther's disciple Melanchthon added a few years after Copernicus' death, "to assert such notions publicly, and the example is pernicious. It is the part of a good mind to accept the truth as revealed by God and to acquiesce in it." Calvin seems never to have heard of Copernicus, but his fundamentalist bias made him and his disciples plainly unsympathetic. The Osiander who naïvely tried to cover Copernicus' theological flank with his forged apologetic introduction was a well-known Lutheran preacher with a Protestant notion of orthodoxy. This helps explain, too, why the *De Revolutionibus* was not published as we might have expected, in Wittenberg, where Rheticus was professor at the university. For Wittenberg, where Luther nailed his 95 Theses on the door of All Saints Church, had become the headquarters for the preachings of Luther and Melanchthon.

The Catholic Church took a more sophisticated and more tolerant view of speculations in secular science. After the fourteenth century the Church had not officially proclaimed any orthodox cosmology. Perhaps the follies and frustrations of Christian geography and the stirring secular revelations of the new seafaring age had something to do with this. But, whatever the reasons for this openness, Copernicus' *De Revolutionibus* was actually read in some of the best Catholic universities. The Church had survived many a secular novelty. Wiser heads continued to hope that the eternal truths of revelation and divine reason could be kept safely isolated from the shifting explanations of the practical world. It was decades after Copernicus' death before this separation had ceased to be possible.

In astronomy, more than in other sciences, there was a simple public test of any system. A perfect theory of the heavens would regularly and accurately forecast the dates of the summer and the winter solstice, the arrival of summer and winter. By Copernicus' time the discrepancy of the calendar offered public evidence that the generally accepted theory of the heavens was not quite right. When Julius Caesar drew on the Egyptian calendar to reform the Roman calendar in 45 B.C., as we have seen, he introduced the system of three years of 365 days followed by a leap year of 366 days. This produced a year of 365¼ days, which still proved to be some 11 minutes and 14 seconds longer than the actual solar cycle. Over the centuries, the accumulation of this error, like that of a clock that runs too slow, had produced a noticeable dislocation of the calendar. As a result, when Copernicus lived, the vernal equinox, which traditionally marks the beginning of spring in the northern hemisphere, had moved back from March 21 to March 11. Farmers could no longer rely on their calendar for the planting and the gathering of their crops, merchants could not depend on the calendar in signing contracts for the delivery of seasonal products.

Copernicus himself had used this disorder in the calendar as a reason for trying some alternative to the Ptolemaic system. "The mathematicians," he declared in his Preface to the *De Revolutionibus,* "are so unsure of the movements of the Sun and Moon, that they cannot even explain or observe the constant length of the seasonal year." Surely, Copernicus argued, there must be something the matter with a theory that had produced this calendar.

Meanwhile the Renaissance city-states and a seafaring commerce that reached around the world had brought new needs for a calendar that was precise and reliable. It is not surprising that Renaissance popes undertook calendar reform. But when they asked Copernicus to help with the project, he said the time was not yet ripe. While the old geocentric Ptolemaic system could not produce a calendar of the required accuracy, the evidence was not yet available to prove that his own heliocentric system would work any better. With the facts available at the time, as historians of astronomy remind us, Copernicus' revised scheme actually would not work as well.

Even so, Copernicus' notions were co-opted into the service of the Church to help Pope Gregory XIII produce the reformed calendar, which we still use. During the next half-century the only direct public application of Copernicus' theories was for this very practical purpose. Yet this "proof" of the truth of Copernicus' system was not by Copernicus himself and was presented in such a way as not to seem an endorsement of any risky cosmological shift.

The work on the calendar was done by another enthusiastic disciple of Copernicus with a genius and a passion for astronomical computation. At the age of twenty-five Erasmus Reinhold (1511–1553) was appointed professor of astronomy *(mathematum superiorum)* at the University of Wittenberg in 1536 by Luther's redoubtable lieutenant Philip Melanchthon. In the 1540's, when printing had made textbooks inexpensive enough to come into general use in universities, Reinhold produced popular versions of the standard works expounding the Ptolemaic system and the solid heavenly spheres. His colleague Rheticus, then also a professor at Wittenberg, had brought back an enthusiastic account of Copernicus. In Reinhold this awakened "a lively expectancy" and a hope that Copernicus would "restore astronomy." When the *De Revolutionibus* appeared, Reinhold began annotating his copy and was stimulated to prepare a set of astronomical tables fuller than any then available. After seven years of labor on "this huge and disagreeable task" (in Kepler's phrase), Reinhold finally published his calculations in 1551.

Reinhold's *Prutenic Tables,* named after his patron the Duke of Prussia, were so far superior to anything else at the time that they soon became the standard astronomical tables in Europe. To revise the older tables, Reinhold had freely used observations that Copernicus had included in his work. Of course he did not realize that Copernicus' notions of the positions and the movements of the planets, which he assumed to be combinations of simple circles, were far from the fact. Nevertheless Reinhold's product was an improvement and came into general use. While he generally acknowledged his debt to Copernicus, he did not even refer to the heliocentric system. The conjectural new arrangements of sun and planets seemed merely a means toward a better table of numbers, and were themselves of no special interest. When Pope Gregory XIII made his new calendar in 1582, he relied in turn on Reinhold's tables. Their superior accuracy seems to have been a bizarre historical coincidence, testimony to Reinhold's intuition rather than the truth of Copernicus' system.

39

The Witness of the Naked Eye

THE powers of the naked eye to observe and interpret the heavens were pushed to their limit by a tireless Danish astronomer born just three years after Copernicus died. Tycho Brahe (1546–1601) was the eldest son of a wealthy Danish nobleman who encouraged him to develop the gentlemanly breadth of interests and the sybaritic tastes which would make his name a byword among cultivated Europeans of his day. At the Lutheran University of Copenhagen he was introduced to all seven of the liberal arts—the trivium (grammar, rhetoric, and logic) and the quadrivium (geometry, astronomy, arithmetic, and music). There he ingested a heavy dose of Aristotle and was introduced to Ptolemy's system of the heavens. Of course he studied astrology, an "interdisciplinary" study combining astronomy and medicine, which made astronomers seem useful in everyday affairs. Then he went to Leipzig to round out his education by studying law.

A bookish education had not stifled Tycho's precocious passion for observing the heavens. Observational astronomy of course was not in the university curriculum. When Tycho had not yet reached his fourteenth birthday he was astonished and delighted that a predicted solar eclipse actually took place on the scheduled day. It seemed to him "something divine that men could know the motions of the stars so accurately that they could long before foretell their places and relative positions."

But since Tycho's family preferred that he follow more conventional studies, he had to pursue his passion in secret. At Leipzig, where they hired a private tutor to keep him on track, he obliged them by pursuing legal studies during the day. Then at night when the stars came out and his tutor slept, he pursued his real interest. He saved his allowance money to buy more astronomical tables, and taught himself the constellations with a miniature celestial globe, no larger than his fist, which he kept hidden from his tutor.

As if to confirm the astrological dogmas of that day, a lucky conjunction of the planets set Tycho on his career. In August 1563, during a widely anticipated conjunction of Saturn and Jupiter, Tycho, not yet seventeen, seized the opportunity to begin his own astronomical observations. His only instrument was a pair of ordinary draftsman's compasses. He held the center close to his eye, pointed each leg of the compass to one of the planets, and then placed the compass on a piece of paper on which he had drawn a circle divided into 360 degrees and half-degrees. He recorded his first observation, of many thousands to come, on August 17, 1563. On August 24 he found Saturn and Jupiter so close together that there was no noticeable interval between them. To his surprise, the old Alphonsine tables were a whole month off in their prediction, and even Reinhold's improved Prutenic tables were several days off.

The next year Tycho added to his equipment a simple cross-staff of the kind familiar in his time. The cross-staff was nothing but a light graduated rod about a yard long on which another graduated rod of about half its length was arranged to slide, so that there was always a right angle between them. When the observer looked through sights fixed at the ends of both rods, by sliding the shorter rod till both bodies were seen, he could measure the angular distances. While his tutor slept Tycho secretly practiced using his cross-staff. Finding it too crude to give the correct angles, he wanted to buy a better one. He dared not ask for the money, but instead devised his own table of corrections to compensate for the crudity of his instrument. From this effort "by that Phoenix of astronomers, Tycho, first conceived and determined on in the year 1564" the great Kepler dated the "restoration of astronomy."

Tycho himself was a phenomenon, not only the paragon of the observational astronomer but one of the more bizarre personalities of the age. As a twenty-year-old student at the University of Rostock, while attending a dance at a professor's house he quarreled with another student over which of them was the better mathematician. When the issue was settled in a duel "in perfect darkness" at seven o'clock on the night of December 29, 1566, Tycho lost a piece of his nose. He repaired the defect by devising an ingenious replacement of gold and silver. Tycho's nose became only one of the many impressive features of his extravagant establishment on the Danish island of Hven which King Frederick II had given him for his observatory. (When Tycho's tomb was opened on the anniversary of his death in 1901, a green stain was found on his skull at the nasal opening, indicating that his nosepiece must have been adulterated with copper.)

Tycho's observations added more to the stock of astronomical facts than those of anyone before. Most were collected during his twenty years on the 2,000-acre island of Hven in the sound between Denmark and South Sweden. King Frederick II had granted him all the rents from the island's tenants. Drawing also on his private fortune, Tycho built a grand scientific establishment. He called it the Heavenly Castle (Uraniborg), but it might have been called the Heavenly City, for it was a whole community devoted to studying the heavens. In addition to workshops for the artisans who built the instruments, there were a chemical laboratory, a paper mill, and a printing press, a mill for grinding corn and preparing hides, sixty fish ponds, flower gardens and herbaries, an arboretum with trees of some three hundred species, a windmill and pumping apparatus to supply running water —all for the care and feeding of astronomers.

This pioneer scientific "think tank" would have been the envy of a twentieth-century scientist. The palatial observatory included an elegant library with a five-foot celestial globe, studies, conference rooms, and bedrooms for scholars and their assistants. In a Castle of the Stars (Stjerneborg), a smaller separate observatory nearby, there were additional instruments with portraits of famous astronomers ancient and modern, climaxing, of course, in a portrait of the great Tycho Brahe himself. All Tycho's instruments were simple devices for directing and repeating observations by the naked eye. But they were the best available at the time, and he improved them by making the instruments larger, by providing more precisely graduated scales, and by making it easier to rotate them in the vertical or horizontal planes. At the same time he invented ways to fix them in place so that successive observations could be made from the same point.

He even tried clepsydras, hoping they might be more accurate than his mechanical clocks. His most celebrated instrument was a giant "mural quadrant" with a radius of six feet which, because of the large scale of the graduations on its arc, increased the accuracy of its measurements. Tycho made his observations with scrupulous regularity, repeating them, combining them, and trying always to allow for the imperfection of his instruments. As a result he reduced his margin of error to a fraction of a minute of an arc, and provided the sharpest precision achieved by anyone before the telescope.

Assisted by numerous students and fellow workers, in his *Progymnasmata* (1602) he catalogued the positions of 777 fixed stars. To help others assess the limits of his accuracy he included descriptions and diagrams of his observational methods and the instruments he used. Tycho's copious work soon displaced Ptolemy's classic catalogue. He eventually included an additional 223 stars, to a grand total of 1,000.

Even before arriving at Hven, Tycho had discovered a new star in the constellation Cassiopeia which showed it to be supralunar, and so required that old theories about the celestial spheres be revised. At Hven his observations were unexcelled, but the theory he devised to explain them was not of comparable quality. He was, Matthew Arnold might have said, "wandering between two worlds, one dead, the other powerless to be born." His own system of the heavens attested to both the inadequacy of the old Ptolemaic geocentric scheme and the insufficiency of evidence for the new Copernican heliocentric scheme.

Tycho would not abandon his belief in a motionless earth at the center of the universe. He was too much attached to Aristotelian physics with its heavy immobile earth. If the earth actually rotated, Tycho remarked, then a cannonball fired in the direction of the earth's rotation should go farther than if it were fired in the opposite direction. But this did not happen. And then there was the clear argument from Scripture, for the Book of Joshua declared that the sun stopped in the heavens.

Yet, seeing how a heliocentric system might simplify the world picture, Tycho devised his own compromise with its own kind of simplicity. He left an unmoving earth at the center while the sun continued, as in the Ptolemaic system, to go around the earth. But in Tycho's new scheme, the other planets rotated around the Sun following the sun's movements about the earth.

On his deathbed Tycho Brahe bequeathed the voluminous records of his observations to a younger, more liberated, more erratic mind. He begged Johannes Kepler to translate them into improved astronomical tables. And he expressed his hope that Kepler would use the tables to prove the Tychonic (*not* the Copernican!) theory.

While Copernicus had dared to change the relations of the heavenly bodies, he had not dared to alter the perfect circularity of their movements, or the circular shape of the whole system. Kepler took this next step. Seeking a more subtle mathematical symmetry in the orbits of the heavenly bodies and the relations between their distances and their periods, he dared to abandon the Aristotelian circular perfection of celestial movements. In retrospect we can see that incidentally he made the Copernican system more plausible by accommodating all the observed movements to simple empirical laws expressed in mathematical form.

The modern era in astronomy is commonly dated from Kepler's enunciation of his laws of planetary motion. But Kepler had come to his inquiries convinced in advance for theological and metaphysical reasons, that it was the scientist's duty to display a harmony that was surely there. He went in quest not of the ways of nature but of the harmonies.

Johannes Kepler (1571–1630) was born in Württemberg in South Germany, on the Lutheran-Catholic battleground. There the Thirty Years War (1618–48) would decimate the population, devastate agriculture, ruin commerce, and torture the peasantry. Kepler's Lutheran family was disgraced when his ne'er-do-well father went to fight as a mercenary against the Protestant uprising in the Netherlands. Until he was twenty-two, Kepler prepared himself for the ministry. He refused tempting offers of money and position which would have taken him into the Catholic camp. From first to last he remained a passionate Lutheran Christian, every day seeing evidences of the divine design.

> Yesterday, when weary with writing, I was called to supper, and a salad I had asked for was set before me. "It seems then," I said, "if pewter dishes, leaves of lettuce, grains of salt, drops of water, vinegar, oil and slices of eggs had been flying about in the air from all eternity, it might at last happen by chance that there would come a salad." "Yes," responded my lovely, "but not so nice as this one of mine."

Savoring the delights of the heavenly salad, he went in search of God's recipe.

If Kepler's family had been wealthy, he might never have turned to astronomy. Theology was his first love, and only reluctantly did he give up preparation for the ministry at Tübingen to make his living as a mathematics teacher in a little town in south Austria. He supplemented his income by issuing astrological calendars which predicted the weather, the fates of princes, the uprising of peasants, and the dangers of invasion by the Turks. Astrology remained Kepler's source of income when all else failed. Prognosticating, he said, was at least better than begging.

"I wanted to become a theologian," he explained in 1595 to the Tübingen professor Michael Maestlin, who had first introduced him to Copernican astronomy, "for a long time I was restless. Now, however, behold how through my effort God is being celebrated in astronomy." His first book, to which he was referring, the *Mysterium Cosmographicum* (1596), was a *tour de force* of mathematical mysticism, and set the course for his lifework. He recounted how, certain that there was mathematic beauty in the relative sizes of the planets and their orbits, he went in search.

> Almost the whole summer was lost with this agonizing labor. At last on a quite trifling occasion I came nearer the truth. I believe Divine Providence intervened so that by chance I obtained what I could never obtain by my own efforts. I believe this all the more because I have constantly prayed to God that I might succeed if what Copernicus had said was true. Thus it happened 19 July 1595, as I was showing my class how the great conjunctions [of Saturn and Jupiter] occur successively eight zodiacal signs later, and how they gradually pass from one trine to another, that I inscribed within a circle many triangles, or quasi-triangles such that the end of one was the beginning of the next. In this manner a smaller circle was outlined by the points where the lines of the triangles crossed each other.

When he compared these two circles, the inner one corresponded to Jupiter, the outer circle to Saturn. Was this his clue?

Kepler suddenly recalled a remarkable coincidence: in geometry there were *five* types of regular polyhedrons and, besides the earth, there were only *five* other planets.

> And then . . . it struck me: why have plane figures among three-dimensional orbits? Behold, reader, the invention and whole substance of this little book! In memory of that event, I am writing down for you the sentence in the words from the moment of conception: The earth's orbit is the measure of all things; circumscribe around it a dodecahedron, and the circle containing this will be Mars; circumscribe around Mars a tetrahedron, and the circle containing this will be Jupiter; circumscribe around Jupiter a cube, and the circle containing this will be Saturn. Now inscribe within the earth an icosahedron, and the circle contained in it will be Venus; inscribe within Venus an octahedron, and the circle contained in it will be Mercury. You now have the reason for the number of planets. . . .
>
> This was the occasion and success of my labors. And how intense was my pleasure from this discovery can never be expressed in words. I no longer regretted the time wasted. Day and night I was consumed by the computing, to see whether this idea would agree with the Copernican orbits, or if my joy would be carried away by the wind. Within a few days everything worked, and I watched as one body after another fit precisely into its place among the planets.

His geometric fantasy actually worked. If we allow space for the eccentricity of planetary paths, and overlook a little problem about Mercury, all the planets do fit, within a margin of only 5 percent, into Kepler's neat scheme.

Whatever we may think of Kepler's "method," his product was impressive. To him it seemed to justify his switch from theology to astronomy. Oddly enough, this book by the twenty-five-year-old Kepler a full half-century after Copernicus' *De Revolutionibus* was the first outspoken defense of the new system after Copernicus himself.

In Copernicus' system, which had displaced the earth by putting the sun at the center, the sun still performed merely the optical function of illuminating the whole universe of planets. But the sun did not *cause* their motions. Kepler took a giant step forward when he saw the sun as a center of force. He noted that the more distant a planet was from the sun, the longer was its period of revolution. For this fact medieval astronomers had offered only mystic or animistic explanations. For example, the Stoics, whom Kepler had studied in Julius Caesar Scaliger's classic textbook, believed that each planet possessed its own *mens*—spirit or intelligence— to guide it through the heavens. The prevalent medieval view which attached each planet to its own transparent sphere also declared the spheres to be moved by celestial intelligence.

When Kepler tried to account for the decrease in the linear velocity of a planet as its distance from the sun increased, at first he, too, imagined that each planet had its own "moving spirit" *(anima motrix)*.

> We therefore have to establish one of the following two facts: either the *animae motrices* [of the planets] are feebler as they are more distant from the Sun or there is only one *anima motrix* in the center of all the orbits, that is, in the Sun, which impels a body more violently as it is nearer, but which becomes ineffective in the case of the more distant bodies, on account of the distance and the attendant weakening of its power.

Kepler himself later momentously added that his whole system of celestial physics made perfect sense "if the word soul *(anima)* is replaced by force *(vis),*" So he boldly pointed the way from an organic to a mechanical explanation of the universe. "Spirits" and "celestial intelligences" would be replaced by forces.

Kepler, if we would believe him, was not impelled by any desire for a mechanistic interpretation. Quite the contrary. For Kepler, the Copernican system as expounded by Copernicus was actually not spiritual enough. He preferred to see the immovable sun, source of Light and Power and Enlightenment, as God the Father, while the immovable fixed stars beyond the planets stood for the Son. The sun's moving force which pervaded the space in between was the Holy Ghost. On this sacred foundation Kepler built his theory of forces in the universe.

While Kepler's description of these forces as a kind of magnetic emanation was off the mark, his intuition was prophetic. When the English physician William Gilbert (1544–1603) published his epoch-making work on magnetism in 1600, Kepler thought that at last he saw the force that accounted for celestial movements. Might it not be possible, Kepler asked, "to show that the celestial machine is not so much a divine organism but rather a clockwork . . . inasmuch as all the variety of motions are carried out by means of a single very simple magnetic force of the body, just as in a clock all motions arise from a very simple weight."

From Copernicus' visionary theory, from the voluminous data gathered by Tycho Brahe, "without whose observation books everything that I [Kepler] have brought into the clearest light would have remained in darkness," and shaped by Kepler's own mystic mathematic passion, Kepler formulated the three laws of planetary motion that made him a pioneer of a science that would lead to modern physics.

His ecstasy at discovering the third law of planetary motion recalls other great religious prophets:

Now, since the dawn eight months ago, since the broad daylight three months ago, and since a few days ago, when the full sun illuminated my wonderful speculations, nothing holds me back. I yield freely to the sacred frenzy; I dare frankly to confess that I have stolen the golden vessels of the Egyptians to build a tabernacle for my God far from the bounds of Egypt. If you pardon me, I shall rejoice; if you reproach me, I shall endure. The die is cast, and I am writing the book—to be read either now or by posterity, it matters not. It can wait a century for a reader, as God himself has waited six thousand years for a witness.

40

A Vision Troubled and Surprised

THE leap from naked-eye observation to instrument-aided vision would be one of the great advances in the history of the planet. But nobody set out to invent a telescope. One of the deepest and most widespread of human prejudices was faith in the unaided, unmediated human senses.

We do not know who invented eyeglasses, how or where. Everything we know suggests that they were invented by chance, and by laymen who were not learned in optics. Perhaps an aged glassmaker making glass disks for leaded windows tested a disk by looking through it and found to his delight that he could see much better. We can suspect that the inventor was not an academic, for professors delight in boasting of their inventions, and before the thirteenth century we have no record by any such self-styled inventor. The Italian word *lente* (English "lens," from "lentil," the edible seed) or *lente di vetro* (glass lentil) first used to describe the invention is conspicuously unacademic. It is not the sort of word that a learned professor would use to describe the application of his optical theories. From the first recorded uses of eyeglasses before 1300 until the invention of the telescope nearly three hundred years later, lenses were ignored by learned scholars. For this there were plenty of reasons. Very little was known about the theory of light refraction. Unluckily, the few inquiring physicists, instead of studying refraction by simple curved surfaces, were seduced once again by their love of perfect forms, circles and spheres. They began by studying refraction in a complete glass sphere, which involved the most complex aberrations, and actually led them nowhere.

In seeking the effects of lenses, natural philosophers were blocked by their theories of light and vision. From early times, the speculations of European philosophers had been dominated by thinking about *how people see* rather than by the nature of light as a physical phenomenon. The ancient Greeks thought of vision as the active process of a living human eye, rather than the passive recording of physical impressions from outside. Euclid's theory of perspective made the eye, and not the seen object, the point of origin of the lines of vision. Plato and the Pythagoreans described the process of seeing as emanations from the eye which somehow encompassed the object seen. Ptolemy shared this approach. Democritus and the atomists, on the contrary, suggested that emissions from the seen object somehow entered the eye and produced images. Galen, the arbiter of European anatomy, raised the commonsense objection that large images, like those of mountains, could not possibly squeeze through the tiny pupil of the eye. Nor could the atomists explain how a single object could produce enough emissions to reach all the many people who might be seeing it at one time. Galen developed a compromise theory, which he tried to relate to the physiology of the eye. During the Middle Ages, Christian Europe was still dominated by the notion of the "active" eye whose visual experience depended on the soul within, which meant that the eye itself was not merely an optical instrument, nor was light a phenomenon in physics.

And there were religious obstacles to studying optics or making instruments to aid the naked eye. "Ye are the light of the world," Jesus declared in the Sermon on the Mount (St. Matthew 5:14); "God is light," declared John, "and in him is no darkness at all" (1 John 1:5). On the very first day of Creation, students of Scripture noted, "God said, Let there be light: and there was light" (Genesis 1:3). Sun, moon, and stars were not created till the fourth day! Tampering with light, or treating light as a merely physical phenomenon was like inquiring into the chemistry of the Eucharist.

Theology was reinforced by folklore and common sense. Why had eyes been given to men if not to know the true shape, size, and color of objects in the external world? Then were not mirrors, prisms, and lenses devices for making visual lies? And man-made instruments for multiplying, deflecting, enlarging or reducing, and doubling or inverting visual images were means for distorting the truth. Devout Christians and honest philosophers would have nothing to do with such trickery.

Still, some practical people went ahead. They were glad to put eyeglasses on their noses, simply because it helped them see. The first use of eyeglasses seems to have been to correct presbyopia, or farsightedness, the defect of vision that comes in advancing age with the hardening of the eye's crystalline lens when the eye cannot focus sharply on nearby objects. Early in the fourteenth century the inventory of a Florentine bishop's estate actually included "one pair of eyeglasses framed in gilded silver." In Venice, by 1300, the making of eyeglasses was so common that a law had to be passed against eyeglass-makers who deceived customers by pretending that they were giving real crystal when they were only giving them glass. "To my annoyance," Petrarch (1304–1374) complained in his autobiographical *Letter to Posterity,* "when I was over sixty years of age . . . I had to seek the help of eyeglasses." Kepler himself wore eyeglasses. By the mid-fourteenth century eminent Europeans had their portraits painted wearing eyeglasses. It is hard to know the full story because craftsmen who discovered how to make eyeglasses had good commercial reasons not to publish their secrets or inform their competitors.

"We are certain," Galileo wrote in 1623, "the first inventor of the telescope was a simple spectacle-maker who, handling by chance different forms of glasses, looked, also by chance, through two of them, one convex and the other concave, held at different distances from the eye; saw and noted the unexpected result; and thus found the instrument." This lucky combination of lenses would probably have occurred in several glassmakers' shops at about the same time. The most likely story puts the crucial episode in the shop of an obscure Dutch spectacle-maker named Hans Lippershey, in Middelburg about 1600. Two children who happened into Lippershey's shop, we are told, were playing with his lenses. They put two lenses together and when they looked through both at the same time toward a distant weathervane on the town church, it was wonderfully magnified. Lippershey looked for himself and then began making telescopes.

This Lippershey was reputed to be an "illiterate mechanick." But he was not so illiterate that he did not know how to profit from his good luck. On October 2, 1608, the States General, governing body of the Netherlands, received a petition from him,

> . . . spectacle-maker, inventor of an instrument for seeing at a distance, as was proved to the States, praying that the said instrument might be kept secret, and that a privilege for thirty years might be granted to him, by which everybody might be prohibited from imitating these instruments, or else grant to him an annual pension, in order to enable him to make these instruments for the utility of this country alone, without selling any to foreign kings and princes. It was resolved, that some of the Assembly do form a committee, which shall communicate with the petitioner about his said invention, and enquire of him whether it would not be possible to improve upon it, so as to enable one to look through it with both eyes. . . .

With the Netherlands battling for independence against the well-financed armies of Philip II of Spain, this was the psychological moment to sell a new military device. Prince Maurice of Nassau, the brilliant commander of the forces of independence and patron of science, would appreciate the battlefield uses of "an instrument for seeing at a distance." After testing Lippershey's device from a tower on the Prince's palace, a committee declared it "likely to be of utility to the State."

It was Lippershey's bad luck that at the same moment other Netherlanders were claiming the honor and the profit of the telescope. One of these, a James Metius of Alkmaar, declared that he had already made a telescope as good as Lippershey's, that he knew the secrets of glassmaking, and with the States General's support could make a far better one. When the authorities would not take up his offer at once, the eccentric Metius refused to let anyone see his telescope, and at his death he had his tools destroyed to prevent anyone from claiming his honors. As word of the invention got around, it was tempting to claim the invention for yourself or for your father. One of the more brazen of these retroactive inventors was a certain Zacharias Jansen (1588–1631?), another spectacle-maker of Middelburg. He had prospered by counterfeiting Spanish copper coins in order to harass the enemy, and then used his skills at home to counterfeit gold and silver coins. For this he was convicted and sentenced to be boiled in oil. His son later testified under oath that his father's idea for a telescope had been stolen by his fellow townsman Lippershey (at a time when his father would have been only two years old).

In the confusion, the States General turned down Lippershey's petition. They gave no claimant either credit or cash for the new device. Meanwhile, the telescope was becoming known. In 1608 the French ambassador at The Hague secured a telescope for King Henry IV, and the very next year telescopes were being sold in Paris. In 1609 a telescope was seen at the Frankfurt fair. Under the name of "Dutch trunks," "perspectives," or "cylinders," they appeared in Milan, Venice, and Padua, and before the year's end were actually being made in London.

Still, prudent people were reluctant to allow the firsthand evidence of their sight to be overruled by some dubious novel device. To persuade "natural philosophers" to look through Galileo's instrument was not easy. They had so many learned reasons to distrust what they had not seen with their naked eyes. The eminent Aristotelian Cesare Cremonini refused to waste his time looking through Galileo's contraption just to see what "no one but Galileo has seen . . . and besides, looking through those spectacles gives me a headache." "Galileo Galilei, Paduan mathematician, came to us at Bologna," reported another hostile colleague, "bringing his telescope with which he saw four feigned planets. I never slept on the twenty-fourth or twenty-fifth of April, day or night, but I tested this instrument of Galileo's in a thousand ways, both on things here below and on those above. Below, it works wonderfully; in the sky it deceives one, as some fixed stars are seen double. I have as witnesses most excellent men and noble doctors . . . and all have admitted the instrument to deceive. Galileo fell speechless, and on the twenty-sixth . . . departed sadly." At first the famous Father Clavius, professor of mathematics at the Collegio Romano, laughing at Galileo's pretended four satellites of Jupiter, said he, too, could show them if he were only given time "first to build them into some glasses."

Galileo himself would look at an object through his telescope, then go up to the object to be sure he had not been deceived. By May 24, 1610, he declared that he had tested his telescope "a hundred thousand times on a hundred thousand stars and other objects." A year later he was still testing. "Over a period of two years now, I have tested my instrument (or rather dozens of my instruments) by hundreds and thousands of experiments involving thousands and thousands of objects, near and far, large and small, bright and dark; hence I do not see how it can enter the mind of anyone that I have simple-mindedly remained deceived in my observations."

Simpleminded, indeed! Galileo was an early crusader for the paradoxes of science against the tyranny of common sense. The grand message of the telescope was not what it showed of the earthly objects that Galileo could go and verify in person and with his naked eye, but rather the infinity of "other objects" that could not be personally examined, or not seen at all by the naked eye.

Telescopic vistas troubled people long before they were fully persuaded. In 1611 John Donne (1572?–1631), the English poet, noted that Copernican ideas, which "may very well be true," were "creeping into every man's mind," and he gave classic expression to the modern malaise:

> And new Philosophy calls all in doubt,
> The Element of fire is quite put out;
> The Sun is lost, and th'earth, and no mans wit
> Can well direct him where to looke for it.
> And freely men confesse that this world's spent,
> When in the Planets, and the Firmament
> They seeke so many new; then see that this
> Is crumbled out againe to his Atomies.
> 'Tis all in peeces, all cohaerence gone;
> All just supply, and all Relation. . . .
> And in these Constellations then arise
> New starres, and old doe vanish from our eyes . . .

In 1619, when Donne visited the Continent, he took the trouble to visit Kepler in the remote Austrian town of Linz.

John Milton (1608–1674) also was discomfited by the new cosmology—

and uncertain what it might mean. When barely thirty, he called on the blind Galileo in Arcetri near Florence where the astronomer had been confined by papal order. Then, in his *Areopagitica* (1644), published two years after Galileo's death, Milton described Galileo as a heroic victim. "This was it which had damped the glory of Italian wits . . . nothing had been there written now these many years but flattery and fustian. There it was that I found and visited the famous Galileo grown old, a prisoner of the Inquisition, for thinking in Astronomy otherwise than the Franciscan and Dominican licensers thought." Yet, two decades later, when Milton wrote *Paradise Lost,* to "justify the ways of God to men" he himself stayed close to the traditional Ptolemaic-Christian cosmology. He actually described the two different world systems but would not explicitly choose between them. He did reveal his choice, however, when he located his epic in the Biblical cosmos. His story made sense only with a Heaven above and a Hell beneath a motionless earth which God had created especially for Man. On Milton's Ptolemaic stage, Satan climbs the stairway to the heavens, descends to the sun, and from there is directed to the earth. A full century after Copernicus' *De Revolutionibus,* Milton was still unable or unwilling to revise his thought to match the newly discovered universe.

A series of coincidences brought Galileo Galilei (1564–1642) and the telescope together. These had nothing to do with anyone's desire to revise the Ptolemaic cosmos, to advance astronomy, or to learn the shape of the universe. The immediate motives were in the maritime and military ambitions of the Venetian Republic and the experimental spirit inspired by its commercial enterprises.

Within a month after Lippershey applied to Prince Maurice, word of his telescope reached Venice. The man who first heard the news was Paolo Sarpi (1552–1623), a versatile friar with a passion for science. As state theologian to the Venetian Senate and principal counsel in their running quarrel with the papacy, he was expected to keep informed on events abroad. Sarpi himself had been excommunicated by Paul V, and was the target of an assassination plot. He was a friend of the ingenious instrument-maker Galileo, whose invention of a new calculating device he had lately defended against the claims of a malicious Milanese plagiarist. At this time Galileo had already served for fifteen years as professor of mathematics at the university in nearby Padua, a post awarded by the Venetian Senate. He had frequently visited shops of the Venetian arsenal, and in Padua he operated a small shop of his own. There he made surveying instruments, compasses, and other mathematical apparatus. The income from the shop supplemented his meager honorarium as professor, provided dowry for his sisters, and supported his brothers and his aged mother. By now Galileo had a secure reputation as an instrument-maker.

When a stranger arrived in Venice to sell the Senate a telescope, the matter was referred to Sarpi. Though convinced that a telescope could be useful to a rising maritime power, he was confident that their own Galileo could make a better one, and so advised the Senate to refuse the stranger's offer.

Sarpi's confidence in the local instrument-maker would quickly be justified. In July 1609 Galileo himself, who happened to be in Venice, had heard rumors that there was such an instrument as a telescope, and at the same time heard that a foreigner had arrived in Padua with one of them. To satisfy his curiosity, he immediately returned to Padua, only to find that the mysterious foreigner had already left for Venice. Having learned how the foreigner's telescope had been made, Galileo at once went about making one for himself. Before the end of August Galileo returned to Venice where he astonished the Senate and gratified Sarpi with a nine-power telescope, three times as powerful as the one offered by the stranger. Galileo continued to improve the instrument until at the end of 1609 he had produced a telescope of thirty power. This was the practicable limit for any telescope of the design then in use—a plano-convex objective and a plano-concave eyepiece—and it came to be known as the Galilean telescope.

With a grand gesture, instead of trying to sell the device, Galileo made a gift of it to the Venetian Senate in a ceremony on August 25, 1609. In return the Senate offered Galileo a renewal for life of his professorship, which was to expire the next year, and an increase of his annual salary from 520 florins to 1,000 florins. This bargain produced resentment among his envious fellow academics, who would plague him the rest of his life. They objected that since others had invented the telescope, the most Galileo was entitled to was a good price for his instrument.

With no special insight into the science of optics, Galileo, a deft instrument-maker, had made his device by trial and error. But if Galileo had been merely a practical man, the telescope would not have been such a troublemaker. Other nations would have shared the Venetian Senate's enthusiasm for a new device that served commerce and warfare by making distant objects seem close. For some reason, however, Galileo would not leave it at that. Early in January 1610 he did what now seems most obvious—he turned his telescope toward the skies. Today this would require neither courage nor imagination, but in Galileo's day it was quite otherwise. Who would dare use a toy to penetrate the majesty of the celestial spheres? To spy out the shape of God's Heaven was superfluous, presumptuous, and might prove blasphemous. Galileo was no better than a theological Peeping Tom.

A half-century had passed since Copernicus' *De Revolutionibus* (1543) proposed a moving earth and a heliocentric universe—with no publicly disturbing consequences. Copernicus' theory, we must recall, was not an astronomical discovery, nor was it based on new observations. "Mathematics are for mathematicians!" was the cautious Copernicus' warning. During the decades after Copernicus' death, his complicated demonstrations and his aesthetic-philosophical speculations had not reached the laity, nor had they unduly ruffled the theologians.

The prevailing attitude toward the Copernican speculations was expressed, only a decade before Galileo made his telescope, by the polymath Jean Bodin (1530–1596), who had a reputation for being open-minded:

> No one in his senses, or imbued with the slightest knowledge of physics, will ever think that the earth, heavy and unwieldy from its own weight and mass, staggers up and down around its own center and that of the sun; for at the slightest jar of the earth, we would see cities and fortresses, towns and mountains thrown down. A certain courtier Aulicus, when some astrologer in court was upholding Copernicus' idea before Duke Albert of Prussia, turning to the servant who was pouring the Falernian, said: "Take care that the flagon is not spilled." For if the earth were to be moved, neither an arrow shot straight up, nor a stone dropped from the top of a tower would fall perpendicularly, but either ahead or behind. . . . Lastly, all things on finding places suitable to their natures, remain there, as Aristotle writes. Since therefore the earth has been allotted a place fitting its nature, it cannot be whirled around by other motion than its own.

In 1597 Galileo himself was actually supporting the Ptolemaic system in a series of lectures at Padua, and the *Cosmography* he was then writing revealed no doubts about the traditional cosmos. Still, in that year he did write to a former colleague from Pisa defending the Copernican hypothesis against unwarranted criticism. When he received Kepler's first book, the *Mysterium Cosmographicum,* with its defense of the Copernican system, he acknowledged it to Kepler sympathetically. What appealed to him in Kepler's version of Copernicus was not its astronomy but its consistency with Galileo's own special theory of the earth's tides. And when Kepler urged him to speak out for the new world view, Galileo refused.

What Galileo viewed through his telescope when he first turned it on the heavens so amazed him that he promptly published a description of what he saw. In March 1610 *The Starry Messenger (Sidereus Nuncius),* a mere pamphlet of twenty-four pages, astonished and troubled the learned world. Galileo ecstatically reported the "most beautiful and delightful sight. . . . matters of great interest for all observers of natural phenomena . . . first, from their natural excellence; secondly, from their absolute novelty; and lastly, also on account of the instrument by the aid of which they have been presented to my apprehension." Until then, all the fixed stars seen "without artificial powers of sight" could actually be counted. Now the telescope "set distinctly before the eyes other stars in myriads which have never been seen before, and which surpass the old, previously known, stars in number more than ten times." Now the diameter of the moon appeared "about thirty times larger, its surface about nine hundred times, and its solid mass nearly 27,000 times larger than when it is viewed only with the naked eye: and consequently any one may know with the certainty that is due to the use of our senses, that the Moon certainly does not possess a smooth and polished surface, but one rough and uneven, and just like the face of the Earth itself, is everywhere full of vast protuberances, deep chasms, and sinuosities."

Next, the telescope settled disputes about the Galaxy or Milky Way: "All the disputes which have tormented philosophers through so many ages are exploded at once by the irrefragable evidence of our eyes, and we are freed from wordy disputes upon this subject, for the Galaxy is nothing else but a mass of innumerable stars planted together in clusters. Upon whatever part of it you direct the telescope straightway a vast crowd of stars presents itself to view. . . ."

"But that which will excite the greatest astonishment by far," he announced, "and which indeed especially moved me to call the attention of all astronomers and philosophers, is this, namely, that I have discovered four planets, neither known nor observed by any one of the astronomers before my time, which have their orbits around a certain bright star." These were actually the four satellites of Jupiter.

Each of his simple observations shook another pillar of the Aristotelian-Ptolemaic universe. Now, with his very own eyes, Galileo had seen fixed stars beyond his capacity to count them (Was the Universe infinite?). He had seen that the moon was no more perfect in shape than the earth itself (Was there perhaps no difference, after all, between the substance of celestial bodies and that of the Earth?). The Milky Way then proved to be simply a mass of countless stars (Was there nothing, after all, to the Aristotelian theory of celestial exhalations? Were heavenly processes not essentially different from those on Earth?). While these brief and casual observations began to remove the traditional obstacles of dogma, yet none of them would actually confirm Copernicus.

Still, for Galileo, what he saw was enough to convert him. In this little pamphlet he dared announce his sympathy for the Copernican system. Though Kepler had been unable to persuade him, he was now persuaded by the telescope. To him the four newly discovered satellites circulating around Jupiter seemed his most important discoveries, for they were the most obvious evidence that the earth might not be unique in the universe. How many other planets had satellites of their own? And it proved that a body like the Earth with another body circulating around it could itself circulate around still another. So Galileo concluded:

> . . . we have a notable and splendid argument to remove the scruples of those who can tolerate the revolution of the planets around the Sun in the Copernican system, yet are so disturbed by the motion of one Moon about the Earth, while both accomplish an orbit of a year's length about the Sun, that they consider that this theory of the universe must be upset as impossible: for now we have not one planet only revolving about another, while both traverse a vast orbit about the Sun, but *our sense of sight* [italics added] presents to us four satellites circling about Jupiter, like the Moon about the Earth, while the whole system travels over a mighty orbit about the Sun in the space of twelve years.

These startling discoveries speedily promoted Galileo's career. But the envy of his rivals in Padua and Venice seems to have had some effect, for the Venetian Senate did not live up to their generous promises. Galileo looked elsewhere for an academic sinecure where he could pursue his new astronomical interest. With this in view he christened the four newly discovered moons of Jupiter the "Medicean planets" after the family of Grand Duke Cosimo II de' Medici of Florence. And he sent the Grand Duke an "exquisite" telescope.

These compliments quickly produced the desired effect. From the Grand Duke came a gold chain and a medal, and in June 1610 a letter appointing Galileo "Chief Mathematician of the University of Pisa and Philosopher of the Grand Duke, without obligation to teach and reside at the University or in the city of Pisa, and with a salary of one thousand Florentine scudi per annum." Florence would remain his academic base for the rest of his life.

Kepler, whose faith had anticipated the facts, rejoiced that Galileo had at last "put aside" his doubts, and wrote two little books in his support. Meanwhile Galileo continued his telescopic observations, which produced still more clues to the plausibility of the Copernican system. He noted the oval appearance of Saturn. And the phases of Venus, which had not been seen by the naked eye, now added to the probability that Venus revolved around the sun. These observations began to provide direct evidence of a heliocentric system.

Galileo was invited to Rome, where he enjoyed a quite unexpected triumph. Arriving on April 1, 1611, he was promptly received in audience by the Borghese Pope Paul V who showed him rare deference by refusing to let him remain kneeling. The Jesuit fathers held a special meeting in the Collegio Romano where the encomium was entitled "The Starry Messenger of the Roman College." Galileo persuaded some of the Church authorities to look through his telescope. They enjoyed what they saw, but still did not accept Galileo's interpretation.

On the night of April 14, 1611, a banquet in Galileo's honor was held on a grand hillside estate outside Rome's St. Pancratius Gate, by the pioneer scientific society, the Academy of the Lynxes (Accademia dei Lincei). The academy's crest showed a bold lynx, with upturned eyes, tearing at the entrails of Cerberus, the three-headed dog of darkness that guarded the gates of the underworld—Truth battling against Ignorance. "The guests were several theologians, philosophers, mathematicians, and others," it was reported. "After Galileo showed them the satellites of Jupiter, together with a number of other celestial marvels, he let them see with his instruments the benediction gallery of St. John Lateran, with the letters of the inscription of Sixtus V appearing very distinctly. And yet . . . the distance was three miles."

On this occasion Galileo's instrument was christened. The name was announced by the noble host, Federico Cesi, hereditary marquis of Monticelli and Duke of Acquasparta, but the word "telescope" was actually devised by a Greek poet-theologian who happened to be present, and so began the custom of giving to instruments of modern science names borrowed from ancient Greece.

41

Caught in the Cross Fire

GALILEO himself, back in Florence, began marshaling arguments for the simultaneous truth of the Bible and the Copernican theory. Anxious to preserve his orthodoxy, he offered an ingenious explanation of the apparent disagreements between the words of Scripture and the facts of Nature. There is only one Truth, he said, but it is communicated in two forms—the language of the Bible and the language of Nature. Both are God's languages. In Scripture, God wisely spoke the vernacular, while in Nature He spoke a more recondite tongue. As Galileo explained:

> . . . both the Holy Scriptures and nature proceed from the Divine Word, the former as the saying of the Holy Spirit and the latter as the most observant executrix of God's orders. And since it is needful in the Scriptures, in order to accommodate these to the understanding of ordinary people, to say many things

which appear different (as to the meaning of the words) from absolute truth, while on the other hand nature is inexorable and immutable and cares nothing about having her hidden reasons explained to the understanding of man, so long as she never oversteps the bounds of the laws imposed on her, it seems that none of the physical effects that are either placed before our eyes by sensible experience or are the conclusions of necessary demonstrations should ever on any account be placed in doubt by passages of the Scriptures which seem to have a different verbal import. . . . two truths can never contradict one another. . . .

Meanwhile, in Rome the Jesuit fathers were not so easily satisfied. Prodded by the Pope's champion, the brilliant and belligerent Cardinal Robert Bellarmine (1542–1621), they sniffed heresy. And Bellarmine, the master of theological polemics and Aristotelian orthodoxy, had common sense on his side. He reminded his tempted brothers that Saint Augustine himself had argued that the literal meaning of Scripture should always be taken as correct, unless the contrary was "strictly demonstrated." Since man's everyday experience "tells him plainly that the earth is standing still," and, from the nature of the case, the rotation of the earth and its revolutions around the sun could not be "strictly demonstrated," the literal Scriptures must be defended. King Solomon's observation that the sun "returns to its place" must mean precisely what he said.

Galileo made the mistake of going to Rome to defend himself. This trip, which would provide the grist for historians' mills for centuries to come, produced no good for Galileo. In Galileo's notorious trial by the Inquisition seventeen years later the charges against him would rest on what had or had not been said during his audience with Cardinal Bellarmine and Pope Paul V back in 1616. Had he actually been ordered to desist from teaching Copernican doctrine? What really had been said? If he had not gone back to Rome, the Church might have had to give Galileo a very different sort of trial. During this visit Galileo did not succeed in persuading the Church authorities that they should approve his teaching that the earth moved. That notion was expressly condemned, but Galileo himself was not personally condemned, nor were his books forbidden.

Some of the most sophisticated modern philosophers of science—for example, the French physicist Pierre Duhem (1861–1916) and the English philosopher Sir Karl Popper—still argue that in the light of modern positivism Cardinal Bellarmine was actually closer to the truth than was Galileo. They say that Galileo had not explained what really happens while Bellarmine did recognize that the Copernican theory merely "saves the appearances."

In 1624 Galileo went to Rome again to pay his respects to the newly enthroned Pope Urban VIII. He made a futile request for papal permission, despite the prohibition of 1616, to publish an evenhanded book comparing the Ptolemaic and the Copernican doctrines. When Galileo returned to Florence, he spent the next six years on his *Dialogue on the Two Chief World Systems.* Though no polemic for the Copernican system, this was a persuasive exposition of the new cosmos. In the tradition of Plato, Galileo allowed the arguments for and against the Copernican system to emerge from conversation among three friends: a Florentine nobleman who believed the Copernican system, an imaginary Aristotelian champion of the geocentric theory, and an open-minded Venetian aristocrat for whose benefit they laid out their arguments.

If it was Galileo's purpose, as has been said, "to bamboozle the censors," he did not succeed. Just as *The Starry Messenger* had announced a telescopic universe of infinite space, so the *Dialogue on the Two Chief World Systems* brought unmistakable news of a heliocentric universe. Could our earth really be only another "planet"—one more wanderer about the sun?

The Copernican doctrine had lain semidormant for a half-century after Copernicus. Without the telescope the heliocentric theory might long have remained an interesting but unpersuasive hypothesis. Now the telescope made all the difference. What he *saw* persuaded Galileo of the truth of what he had read. And he was not alone. Until the telescope, the defenders of Christian orthodoxy felt no need to ban Copernican ideas. But this new device, which spoke directly to the senses, short-circuited the priests' appellate jurisdiction over the heavens. Astronomy was transformed from a preserve of arcane theories in learned language into a public experience.

When Galileo, a man of forty-five years, first looked through his telescope, he had already challenged the Aristotelians. If he ever actually performed his famous experiments from the Tower of Pisa, it was probably to discredit them. But now suddenly he was thrust into the stormy world of cosmological controversy. He did not shrink from its challenges. Of a naturally belligerent temperament, he welcomed the chance as far as orthodoxy would allow to carry a new gospel. With this enticing new device, his biographer Ludovico Geymonat recounts, he undertook a dual campaign of persuasion: to interest literate laymen in this new way of looking at the universe, and to persuade the Church to accept the inevitable.

The reception of Galileo's *Dialogue,* published in Florence on February 21, 1632, encouraged him to think that his campaign of publicity was succeeding. In Europe most scientific works were still written in Latin, but to reach a lay audience Galileo had offered this work in Italian. By midsummer he was flooded with fan letters. "There appear new theories and noble observations which you have reduced to such simplicity that even I, of a different occupation, am certain I shall be able to understand at least some parts of it." "You have succeeded with the public to a point at which no one else has arrived." "Frankly, who in Italy cared about the Copernican system? But you have given it life, and, what really counts, have laid bare the breasts of Nature." A few caught the larger import. "I see how much stronger your argument is than that of Copernicus, though his is fundamental. . . . These novelties about old truths, new worlds, new stars, new systems, new nations, and so on, are the commencement of a new age."

Causes that had nothing to do with astronomy conspired to defeat his hopes of converting the Church of Rome. Galileo would be caught in the cross fire between Catholics and Protestants. The rising attacks of Protestantism made it necessary that Pope Urban VIII respond by showing the determination of the Church of Rome to preserve the purity of ancient Christian dogmas. Protestants must have no monopoly on fundamentalism. To put down an old favorite of his like Galileo would dramatize the Pope's apostolic zeal. And some typographic trivia helped confirm the irascible Pope's impulses. The Florentine press of Landini, which printed Galileo's *Dialogue*, marked the book with its usual colophon, three fish. Was this perhaps, Galileo's enemies asked, a libelous reference to the three nephews of dubious competence whom the Pope had promoted in the Church hierarchy? And, they added, was not the conservative defender of geocentrism with the uncomplimentary name, Simplicio, in Galileo's *Dialogue* meant to be a caricature of Pope Urban VIII himself?

The story of Galileo's brutal trial by the Inquisition is familiar. When the Pope's summons reached Galileo in Florence, he was in bed, gravely ill. Medical certificates declared that his removal to Rome might prove fatal. Nevertheless the Pope threatened to remove Galileo in chains if he did not come of his own accord. The Grand Duke of Florence provided a litter, and Galileo was carried all the way to Rome in the wintry February of 1633. The trial focused on technicalities, on what Galileo had or had not been told by Cardinal Bellarmine back in 1616, on how clearly he had known the papal disapproval of Copernican doctrines. To ensure the truthfulness of his testimony, Galileo was threatened with torture, although it was never actually applied. The Pope's verdict, recorded on June 16, chose the most humiliating of all the alternatives. The Pope might merely have prohibited the *Dialogue* until it was "corrected" or might have condemned Galileo to private penance and house arrest. Instead, the *Dialogue* was totally forbidden, Galileo was to make public and formal abjuration, and also to be imprisoned for an indefinite period. On the morning of Wednesday, June 22, Galileo knelt before the tribunal and obediently declared:

I, Galileo, son of the late Vincenzio Galilei, Florentine, aged seventy years, arraigned personally before this tribunal and kneeling before you, Most eminent and Lord Cardinals Inquisitors-General against heretical pravity throughout the entire Christian Commonwealth, having before my eyes and touching with my hands the Holy Gospels, swear that I have always believed, do believe, and with God's help will in the future believe all that is held, preached and taught by the Holy Catholic and Apostolic Church. But, whereas, after an injunction had been lawfully intimated to me by this Holy Office to the effect that I must altogether abandon the false opinion that the sun is the center of the world and immobile, and that the earth is not the center of the world and moves, and that I must not hold, defend, or teach, in any way, verbally or in writing, the said false doctrine, and after it had been notified to me that the said doctrine was contrary to Holy Scripture, I wrote and printed a book in which I treated this new doctrine already condemned and brought forth arguments in its favor without presenting any solution for them, I have been judged to be vehemently suspected of heresy, that is, of having held and believed that the sun is the center of the world and immobile and that the earth is not the center and moves.

Therefore, desiring to remove from the minds of Your Eminences, and of all faithful Christians, this vehement suspicion rightly conceived against me, with sincere heart and unpretended faith I abjure, curse, and detest the aforesaid errors and heresies and also every other error, error and sect whatever, contrary to the Holy Church, and I swear that in the future I will never again say or assert verbally or in writing, anything that might cause a similar suspicion toward me; further, should I know any heretic or person suspected of heresy, I will denounce him to this Holy Office or to the Inquisitor or Ordinary of the place where I may be. . . .

Without impugning the decision of the court, Galileo asked the judges, while enforcing their just verdict, "to take into consideration my pitiable state of bodily indisposition, to which, at the age of seventy years, I have been reduced by ten months of constant mental anxiety and the fatigue of a long and toilsome journey at the most inclement season."

Confined in a secluded house in Arcetri outside Florence, he was allowed no visitors except with the permission of the Pope's delegate. Soon after his return to Florence, the death of his beloved daughter, his lone solace, plunged him into deep gloom. He seemed to lose interest in everything. But his ebullient curiosity could not be repressed. Within four years he produced a book, on "two new sciences"—one concerned with mechanics and the other with the strength of materials. This, too, was written in Italian and cast into a dialogue between Salvati, Sagredo, and Simplicio. Since the Inquisition had forbidden his books, the work had to be smuggled out of the country and published by the Elzevirs at Leyden. This, Galileo's last book, laid the foundation on which Huygens and Newton could develop the science of dynamics and, eventually, a theory of universal gravitation.

During his last four years Galileo was blind, perhaps from hours spent peering at the sun through his telescope. It was during these years that John Milton visited him, and found additional inspiration (besides his own blindness) for *Samson Agonistes*. Eventually the Pope allowed him the companionship of a young scholar, Vincenzo Viviani, who reported Galileo's death on January 8, 1642, a month before his seventy-eighth birthday. "With philosophic and Christian firmness he rendered up his soul to its Creator, sending it, as he liked to believe, to enjoy and to watch from a closer vantage point those eternal and immutable marvels which he, by means of a fragile device, had brought closer to our mortal eyes with such eagerness and impatience."

42

New Worlds Within

THE microscope was a product of the same age that made the telescope. But while Copernicus and Galileo have become popular heroes, prophets of modernity, Hooke and Leeuwenhoek, their counterparts in the microscopic world, have been relegated to the pantheon of the specialized sciences. Copernicus and Galileo played leading roles in the much publicized battle between "science" and "religion," Hooke and Leeuwenhoek did not.

We do not know who invented the microscope. The leading candidate is Zacharias Jansen, an obscure spectacle-maker of Middelburg. We do know that the microscope, like eyeglasses and the telescope, was in use long before the principles of optics were understood, and it was probably as accidental as the telescope. It could hardly have been devised by anyone eager to peer into a microscopic world still unimagined. Soon after the first telescopes were made, people simply tried using them to enlarge objects nearby. In the beginning the same Italian word, *occhialino,* or the Latin *perspicillum* served for both telescope and microscope. Galileo himself tried using his telescope as a microscope. "With this tube," Galileo reported to a visitor in Florence in November 1614, "I have seen flies which look as big as lambs, and have learned that they are covered over with hair and have very pointed nails by means of which they keep themselves up and walk on glass, although hanging feet upwards, by inserting the point of their nails in the pores of the glass." He discovered to his dismay that while a telescope focused on the stars needed to be only two feet long, to magnify small objects nearby required a tube two or three times that length.

As early as 1625 a member of the Academy of the Lynxes, the physician-naturalist John Faber (1574–1629), had a name for the new device. "The optical tube . . . it has pleased me to call, after the model of telescope, a microscope, because it permits a view of minute things."

The same suspicions that made Galileo's critics unwilling to look through his telescope, and then reluctant to believe what they saw, also cursed the microscope. The telescope was obviously useful in battle, but there were no battles yet where the microscope could help. In the absence of a science of optics, sensible people were especially wary of "optical illusions" *(deceptiones visus)*. This medieval distrust of all optical devices was the great obstacle to a science of optics. As we have seen, it was believed that any device standing between the senses and the object to be sensed could only mislead the God-given faculties. And to a certain extent the crude microscopes in those days confirmed their suspicions. Chromatic and spherical aberration still produced fuzzy images.

In 1665 Robert Hooke (1635–1703) published his *Micrographia,* an enticing miscellany expounding his theory of light and color, and his theories of combustion and respiration, along with a description of the microscope with its uses. But the widespread suspicions of optical illusions would plague Hooke. At first the "new world" he claimed to see through his lenses was the butt of ridicule—for example, in Thomas Shadwell's popular farce *The Virtuoso* (1676).

What Galileo's *Sidereus Nuncius* had done for the telescope and its heavenly vistas, Hooke's *Micrographia* now did for the microscope. Just as Galileo did not invent the telescope, neither did Hooke invent the microscope. But what he described seeing in his compound microscope awakened learned Europe to the wonderful world within. Fifty-seven amazing illustrations drawn by Hooke himself revealed for the first time the eye of a fly, the shape of a bee's stinging organ, the anatomy of a flea and a louse, the structure of feathers, and the plantlike form of molds. When he discovered the honeycomb structure of cork, he said it was made of "cells." Frequently reprinted, Hooke's illustrations remained in textbooks into the nineteenth century.

Just as the telescope had brought together the earth and the most distant heavenly bodies into a single scheme of thought, now microscopic vistas revealed a minuscule world surprisingly like that seen on a large scale every day. In his *Historia Insectorum Generalis* Jan Swammerdam (1637–1680) showed that insects, like the "higher" animals, possessed an intricate anatomy and did not reproduce by spontaneous generation. In his microscope he saw insects developing as man did, by epigenesis, the gradual development of one organ after another. Still, belief in other forms of spontaneous generation survived. As we shall see, it was not until Louis Pasteur's brilliant experiments with fermentation in the nineteenth century, and his practical application of his ideas for the preservation of milk, that the dogma ceased to be scientifically respectable.

The microscope opened dark continents never before entered and in many ways easy to explore. The great sea voyages had required large capital, organizing genius, talents of leadership, and the charisma of a Prince Henry or a Columbus, a Magellan or a Gama. Astronomic exploring required the coordinated observations of people in many places. But a lone man anywhere with a microscope could venture for the first time where there were neither experienced navigators nor skilled pilots.

Antoni van Leeuwenhoek (1632–1723) with his microscope pioneered this new science of other-worldly exploration. In Delft, where he was born, his father made baskets to pack the famous delftware for the world market. Antoni himself made a good living by selling silk, wool, cotton, buttons, and ribbons to the city's comfortable burghers, and had a substantial income as head of the City Council, inspector of weights and measures, and court surveyor. He was a close friend of Jan Vermeer, and on the painter's death was appointed trustee of Vermeer's bankrupt estate. He never attended a university, and during his whole ninety years he left the Netherlands only twice, journeying once to Antwerp and once to England.

Leeuwenhoek did not know Latin and could write only in the vernacular Dutch of his native Delft. But the modern instrument-aided experience of the senses was trans-lingual. No longer did one need Hebrew, Greek, Latin, or Arabic to join the community of scientists.

That age of bitter commercial rivalry between the Dutch and the British for the treasures of the East Indies saw their vigorous collaboration in science. Even while the guns of British and Dutch admirals were firing at each other, British and Dutch scientists were cordially exchanging information and sharing new scientific vistas. An international community of science was growing. In 1668 the *Philosophical Transactions* of the Royal Society of London published an extract from an Italian learned journal telling how an Italian lens-maker, Eustachio Divini (1610–1685), using a microscope, had discovered "an animal lesser than any of those seen hitherto." Five years later, in the heat of the Anglo-Dutch naval wars, Henry Oldenburg (who was born in Germany, educated at the University of Utrecht, and was now in London publishing the *Philosophical Transactions*) received a letter from the Dutch anatomist, Regnier de Graaf (1641–1673):

> That it may be the more evident to you that the humanities and science are not yet banished from among us by the clash of arms, I am writing to tell you that a certain most ingenious person here, named Leewenhoeck, has devised microscopes which far surpass those which we have hitherto seen, manufactured by Eustachio Divini and others. The enclosed letter from him, wherein he de-

scribes certain things which he has observed more accurately than previous authors, will afford you a sample of his work; and if it please you, and you would test the skill of this most diligent man and give him encouragement, then pray send him a letter containing your suggestions, and proposing to him more difficult problems of the same kind.

With this "encouragement," Leeuwenhoek was drawn into a community of science where he enjoyed fifty years of communication with a world of unseen colleagues.

Careful drapers like Leeuwenhoek were in the habit of using a low-power magnifying glass to inspect the quality of cloth. His first microscope was a small lens, ground by hand from a glass globule and clamped between two perforated metal plates, through which an object could be viewed. Attached was an adjustable device for holding the specimen. All his work would be done with "simple" microscopes using only a single-lens system. Leeuwenhoek ground some five hundred fifty lenses, of which the best had a linear magnifying power of 500 and a resolving power of one-millionth of a meter. In the traditions of alchemy, of instrument-making, and of cartography, Leeuwenhoek was secretive. What visitors to his shop saw, he declared, was nothing compared with what he himself had seen with the superior lenses that he was not at liberty to show them. His fellow townsmen called him a magician, but this did not please him. He remained wary of the eager visitor from abroad who, he said, "was much rather inclined to deck himself out with my feathers, than to offer me a helping hand."

The Royal Society encouraged Leeuwenhoek to report his findings in one hundred ninety letters. Since he had no systematic program of research, a letter was his perfect format for reporting his unexpected glimpses of the innards of anything and everything. Some of his first chance observations proved to be his most startling. If Galileo was so excited by distinguishing stars in the Milky Way, and four new satellites of the planet Jupiter, how much more exciting to discover a universe in every drop of water!

Once Leeuwenhoek had a microscope he began looking for something to do with it. In September 1674, out of curiosity, he filled a glass vial with some greenish cloudy water, which the country folk called "honey-dew," from a marshy lake two miles outside of Delft, and under his magnifying glass he found "very many small animalcules." He then turned his microscope on a drop of pepper water:

    I now saw very plainly that these were little eels, or worms, lying all huddled up together and wriggling; just as if you saw, with the naked eye, a whole tubful of very little eels and water, with the eels a-squirming among one another: and the whole water seemed to be alive with these multifarious animalcules. This was for me, among all the marvels that I have discovered in nature, the most marvellous of all; and I must say, for my part, that no more pleasant sight has ever yet come before my eye than these many thousands of living creatures, seen all alive in a little drop of water, moving among one another, each several creature having its own proper motion: . . .

In his famous Letter 18 to the Royal Society (October 9, 1678), he concluded that "these little animals were, to my eye, more than ten thousand times smaller than the animalcule which Swammerdam has portrayed, and called by the name of Water-flea, or Water-louse, which you can see alive and moving in water with the bare eye."

Like Balboa speculating on the extent of his great Southern Ocean, or Galileo delighting in the new infinity of the stars, so Leeuwenhoek lux-uriated in the minuteness of these tiny creatures and their infinitely vast populations. He put into a slender glass tube an amount of water as big as a millet seed, marked off thirty divisions on the tube "and I then bring it before my microscope, by means of two silver or copper springs, which I have attached thereto . . . to be able to push it up or down." The visitor to his shop at the time was amazed. "Now supposing that this Gentleman really saw 1000 animalcules in a particle of water 1/30th of the bigness of a millet-seed, that would be 30000 living creatures in a quantity of water as big as a millet-seed, and consequently 2730000 living creatures in one drop of water." Yet, Leeuwenhoek added, there were much smaller crea-tures not revealed to the visitor, "but which I could see by means of other glasses and a different method (which I keep for myself alone)."

It is no wonder that those who read these reports were troubled by doubts. Some accused him "of seeing more with his imagination than with his magnifying glasses." To persuade the Royal Society, he produced signed testimonials of eyewitnesses, not fellow scientists but simply respectable citizens, notaries public, the pastor of the English congregation in Delft, and others. Each called himself a *testis oculatus*, who had seen the little animals with his very own eyes.

Having discovered the world of bacteria, Leeuwenhoek went on to dig-nify its inhabitants. Contradicting Aristotle's dogmas about the "lower animals," he declared that each of the animalcules had its full complement of the bodily organs needed for the life it led. Therefore there was no reason to believe that the small animals, insects and intestinal worms, would arise spontaneously out of filth, dung, dirt, and decaying organic matter. Rather, as the Bible hinted, each reproduced after its kind and was the offspring of a predecessor of the same species.

When Leeuwenhoek sent the report of his microscopic observations of human semen to the Royal Society, he discreetly apologized. "And if your Lordship should consider that these observations may disgust or scandalize the learned, I earnestly beg your Lordship to regard them as private and to publish or destroy them, as your Lordship thinks fit." Some years earlier, William Harvey, in his *De Generatione* (1651), had described the egg as the sole source of new life. The prevailing notion was that semen provided nothing more than fertilizing "vapors." When Leeuwenhoek, who equated motility with life, saw the lively spermatozoa swimming about, he went to the other extreme and gave them the dominant role in creating new life.

An irrepressible explorer, he went down many dead ends—explaining pepper's pungent taste by its spiny microscopic texture, and human growth by the "preformation" of organs in the sperm. But he also opened vistas into microbiology, embryology, histology, entomology, botany, and crystallography. His well-earned election as a Fellow of the Royal Society of London (February 8, 1680) delighted him beyond measure. It signaled a new world of scientists, international and unacademic, where knowledge would be advanced not only by the traditional custodians of knowledge. Mere "mechanicks," amateurs, would come into their own.

43

Galileo in China

DURING the Middle Ages great advances in optical theory and the understanding of the eye had come from the Arab physicians and natural philosophers. Al-Kindi (813–873), sometimes called the first Arab philosopher, developed the notion of rectilinear rays traveling from the illuminated object to the eye. The pioneer experimenter was Alhazen (Ibn al-Haytham; 965–1039), who carried further the idea, not yet accepted by Christian philosophers, that vision was the product of an agent wholly external to the seeing eye. He went further to develop the notion that the rectilinear rays emanated from every point on an illuminated surface. He experimented with the problem of dazzle, noted the persistence of images on the retina, and began to treat the eye as a piece of optical machinery. The Arab scientists were in the mainstream of optical science.

Nowhere on the long list of Chinese priorities do we find either the telescope or the microscope. But the Chinese had mastered the techniques of making mirrors as early as the seventh century B.C. Very early they made burning mirrors and curved mirrors, they were adept at glass technology at least from the fifth century B.C., and they were actually wearing eyeglasses in the fifteenth century. The camera obscura was their plaything by the eleventh century. The Mohist classic of Chinese physics as early as the fourth century B.C. had elaborated a theory of optics which anticipated many of the more sophisticated notions of post-Renaissance Europe. The Chinese, perhaps because they were not confused by belief in a "soul," were not blocked by notions of optical emissions from the eye. Instead they studied the activity of light rays coming from the object.

As we have seen, the Chinese observed and recorded celestial phenomena with industry and precision. Yet when Father Ricci arrived in China, he immediately noted the backward state of their astronomy. He remarked that they had counted four hundred more stars than those counted in the West, but only because they included the fainter stars. "And yet with all this, the Chinese astronomers take no pains whatever to reduce the phenomena of celestial bodies to the discipline of mathematics. . . . they center their whole attention on that phase of astronomy which our scientists term astrology, which may be accounted for by the fact that they believe that everything happening on this terrestrial globe of ours depends upon the stars. . . . The founder of the family which at present regulates the study of astrology prohibited anyone from indulging in the study of this science unless he were chosen for it by hereditary right. The prohibition was founded upon fear, lest he who should acquire a knowledge of the stars might become capable of disrupting the order of the empire and seek an opportunity to do so." Among the false notions of the Chinese, Ricci noted that "they do not believe in crystalline celestial spheres." Lacking the Greek adoration of the circle as the most perfect of all geometrical figures, they did not have the motives of Euclid and Plato to confine the movements of the planets and the rotations of the fixed stars in that ideal design.

From Peking Father Ricci wrote to his Roman superiors on May 12, 1605, asking that a competent astronomer be sent out to join him. "These globes, clocks, spheres, astrolabes, and so forth, which I have made and the use of which I teach, have gained for me the reputation of being the greatest mathematician in the world. . . . if the mathematician of whom I spoke came here, we could readily translate our tables into Chinese characters and rectify their year. This would give us great face, would open wider the gates of China, and would enable us to live more securely and freely." Ricci was writing even before Galileo had made his startling observations.

When finally word reached the East of Galileo's triumphal reception by the learned Jesuits in Rome, it strengthened the determination of the Jesuits in China to impress the Chinese with their astronomical prowess. It happened that a Father John Schreck, a former student of Galileo at Padua and a member of the Academy of the Lynxes, was with the Jesuit mission in Peking. Schreck had been at the famous party for Galileo in Rome and recalled that one of the guests had refused to look through Galileo's telescope, lest he be forced to believe some unwelcome facts. Already before the end of 1612 when a Jesuit missionary in India had heard of Galileo's discoveries, he begged for a telescope, or at least for instructions on how to make one. In 1615, a Jesuit Father in Peking added to his little Chinese book on astronomy a last page describing the telescope. It had taken five years, in those days not unduly long, for the message of Galileo's *Starry Messenger* to reach from Rome to Peking.

When Galileo himself refused to help the missionaries with astronomical data, they turned to Kepler, who did lend a hand. The Jesuit General in Rome finally sent out several competent mathematicians, including a convinced Copernican, to strengthen the Peking mission. Father Schall, who had been at the convocation of the Collegio Romano which honored Galileo in May 1611, had not forgotten the Galilean message. Now stationed in Peking, in 1626 he produced an illustrated book with complete instructions on how to build a telescope. His preface extolled the eye, which guides "from the visible to the invisible" and which received new power from the telescope. In 1634 a telescope made under Jesuit direction was ceremoniously presented to the Emperor.

Suspicions had already been raised in the Emperor's court that a device so useful in astrology might have a subversive purpose. The machine was speciously justified as "only an instrument for reaching where other instruments do not reach." "If there should break out unexpectedly a military revolution," a Chinese scholar had explained, ". . . one can look at, from a distance, the place of the enemy, the encampments, the men, the horses, whether armed more or less, and to know thus whether one is ready or not, whether it is fitting to attack or to defend oneself, and also whether it is fitting to discharge the cannon. Nothing is more useful than this instrument."

The Jesuits in China could not yet know of Galileo's 1633 trial and condemnation. When they learned of it, their enthusiasm for the telescope was unabated, but they ceased to champion the Copernican theory of a heliocentric universe and a mobile earth. We have seen how Galileo himself had acquiesced in the papal condemnation. At Galileo's death in 1642 the learned community had not yet been converted to the Copernican doctrine. The attitudes of individual Jesuits were clouded by personal animosities, envy, and jealousy. Some of Galileo's friends thought he had "ruined himself by being so much in love with his own genius, and by having no respect for others." And a leading Jesuit mathematician, Christopher Schreiner (1575–1650), accused Galileo of stealing from him the credit for first observing sunspots.

The upshot of all this was that when the telescope did reach China, it was not an effective propagandist for the heliocentric system. Recently Jesuit fathers have tried to justify the missionaries' public retreat from heliocentrism. Since Chinese traditional science had put the earth at the center of the universe, they say, to have insisted on the heliocentric system would have created unnecessary antipathy to the Jesuits and so would have discredited the Christian faith they were there to propagate. They now suggest that the shift to a Copernican universe required social conditions not yet found in China. By 1635 the telescope was actually being used there to direct artillery in battle. Within a decade after the *Sidereus Nuncius,* according to Needham, two Chinese "optician virtuosi" were making optical apparatus, including perhaps compound microscopes and magic lanterns. Before Galileo's death a few Chinese scholars had phoneticized the name of a barbarian astronomer, *Chia-li-lê-lo.*

Elsewhere in East Asia, diffusion of the telescope, as we might expect, was only through the few official channels. A Korean ambassador en route to Peking in 1631 encountered a Portuguese Jesuit missionary, Father John Rodriguez, who had been taking refuge in Macao. When the ambassador expressed an interest in astronomy and the improvement of the calendar, Father Rodriguez supplied two books on astronomy which included descriptions of Galileo's discoveries and then actually gave him a telescope. This was called the "mirror of a thousand *li,*" or of 400 kilometers, since it could supposedly see that far.

How the telescope then migrated from Korea across the narrow sea to Japan we do not know. But we do know that by 1638, even before Galileo's death, there was a telescope at Nagasaki, the single permitted entry point for foreigners, where it helped warn the wary Japanese of unwelcome visitors. On the southeast side of the city an official with a telescope was stationed in the Observatory for Foreigners overlooking the harbor. His duty was to note the arrival of foreign ships, and then send a boat with a black flag to warn the Office of Voyages. Within a half-century the device was being turned to other purposes. An illustration of Ihara Saikaku's novel, *The Man Who Spent His Life in Love* (1682), shows the nine-year-old hero perched on a roof where he can train his telescope on a maidservant in her bath.

The Copernican and Galilean ideas eventually reached Japan through Chinese books printed by the Jesuit missionaries of Peking. Among those whom they probably influenced was the "Japanese Newton," Seki Kowa (1642?–1708?), who invented his own brand of calculus. Nagasaki remained a port of entry for these foreign ideas. By the end of the eighteenth century the Copernican theory had been accepted by numerous Japanese astronomers,. and although "disbelievers are by far the majority," it was being popularized in books by reputable scholars. In Nagasaki, Dutch merchants in place of Jesuit missionaries were advance agents of Western science. Although the Copernican ideas reached Japan late, when they did arrive they met less obstinate resistance than they had in Europe, for in the early nineteenth century the prestige of Western science gave the doctrine special appeal.

The Asian noncrusading, pluralist view of the world was finally producing some scientific dividends. In Japan there was no religious opposition to the Copernican theory. Japanese beliefs, G. B. Sansom reminds us, were "neither anthropocentric nor geocentric and were consequently not thought to be endangered by a theory that made the earth a satellite and diminished the importance of man." Before long the Japanese honored the Copernican theory by claiming it as their very own invention. Several scientists professed to have discovered the theory earlier and quite independent of Europe. Traditional Japanese scholars began to explain that the sun, at the center of a heliocentric universe, was really their ancient god Ame-no-minaki-nushi-no-kami, "the god who rules the center of the heavens," and so Copernicanism had been their orthodox faith all along.

PART TEN

INSIDE
OURSELVES

*Experience does not ever err, it is only your judgment that errs
in promising itself results which are not caused by your experiments.*

—LEONARDO DA VINCI (C. 1510)

44

A Mad Prophet Points the Way

IN sixteenth-century Europe, common sense and folk wisdom, like that which stood between man and the stars, obstructed his vision of himself, and his exploration of the human body. Yet, unlike astronomy, human anatomy was a subject on which no one could avoid some firsthand acquaintance. In Europe the knowledge of the human body had been codified and put in the custody of a powerful, exclusive, and respected profession. Stored in learned languages (Greek, Latin, Arabic, and Hebrew), such knowledge was the preserve of monopolists who called themselves Doctors of Physick. Handling the body, for treatment or dissection, was the province of another group more akin to butchers and sometimes called barber-surgeons.

Not until about 1300 were human bodies dissected for teaching and learning anatomy. In those days, dissecting a cadaver was an especially unpleasant business. Since there was no refrigeration, it was necessary to dissect the most perishable parts first—beginning with the abdominal cavity, then the thorax, and finally the head and the extremities. A dissection, known as an "anatomy," went on hastily and continuously for four days and nights and was usually conducted out of doors. Illustrations in the earliest printed textbooks of anatomy show a professor of physick, the physician, impeccably dressed in hat and gown, seated high on his throne-like chair, the *cathedra,* while a barber-surgeon standing on the turf below handles the entrails of a body extended on a wooden bench and a demonstrator holds a pointer to indicate parts of the body. In the physician's hands we see a book, probably Galen or Avicenna, from which he reads at an antiseptic distance.

Doctors of Physick kept their secrets locked in languages their patients could not read. It is not surprising that they enjoyed the prestige of learning and the awe of the occult. Aristocrats of the academic world, custodians of the means of life and death, they remained invulnerable to attacks from laymen. Rather than pay their high fees or risk painful and drastic treatment, people commonly consulted the nearest apothecary, who was little more than a spice merchant or grocer.

That world of medicine was a world of separations—of books from bodies, of knowledge from experience, and of learned healers from those who most needed healing. Yet it was all those separations that made the dignity of an awesome profession.

At the end of the fifteenth century, any physician who had labored to learn the academic languages and had become the disciple of some eminent professor of medicine had a heavy vested interest in the traditional lore and the accepted dogmas. "Strive to preserve your health," Leonardo da Vinci warned, "and in this you will the better succeed in proportion as you keep clear of the physicians, for their drugs are a kind of alchemy concerning which there are no fewer books than there are medicines." To attack this citadel demanded a willingness to defy the canons of respectability, to uproot oneself from the university community and from the guild. Such a venture required as much passion as knowledge, and more daring than prudence. To open the way, a man needed the knowledge of a professional and yet must not be committed to the profession. He should be in the physician's world but not of it.

The path to modern medicine obviously could not be opened by a tame professor of high eminence. What was required was a vagrant and a visionary, a man of mystic recklessness. The man who dared point the way would have to use the vernacular, and not speak but shriek.

Paracelsus (1493–1541) was suspect in his day, and never lost his reputation as a charlatan. His faith in God drove him to a new vision of man and of the healing arts. Just as Kepler's belief in the divine symmetry of the universe confirmed his faith in a Copernican system of the heavens, so faith in the divine order in the human body inspired Paracelsus.

"Paracelsus," the nickname by which he lived in history, is itself a mystery. Perhaps it meant that he classed himself with the great Roman medical authority Celsus, perhaps simply that he wrote *para*doxical works that contradicted the common opinions of his profession. His real name was Theophrastus Philippus Aureolus Bombastus von Hohenheim. He was really not the origin of the English word "bombast," but might well have been. In eastern Switzerland where he was born, his father was a physician of illegitimate birth, and his mother a bondswoman of the Benedictine abbey of Einsiedeln. His mother died when he was nine, and his father moved to a mining village in Carinthia in Austria, where he grew up. His education was casual and erratic, picked up from his father or from churchmen versed in medicine and occult lore. He probably never received a diploma as doctor of medicine. He never settled down, and in the course of his wanderings he worked in the Fugger mines in the Tyrol and served as surgeon for the Venetian armies in Denmark and in Sweden. He ventured even to the island of Rhodes and farther east.

For a while he prospered in Strasbourg as a practicing physician. Then it was his good luck to be called to Basel to consult on the critical illness of the eminent Johann Froben (1460–1527), who had founded one of the most influential early presses and published the first printed New Testament in Greek. Froben's recovery was credited to Paracelsus. Living with Froben at the time was the great Erasmus (1466–1536), whom he also treated. They were both so impressed by the young Paracelsus' good sense that in 1527 they secured for him the post of municipal physician and had him appointed professor at the university. But the professors ostracized him because he had refused to take the Hippocratic oath, and was not even a certified doctor of medicine.

The thirty-three-year-old Paracelsus combined the arrogance of a self-made man with the articulateness of a self-appointed spokesman for God. Sponsored by the leading publicist of humanism, he seized this lucky opportunity in Basel to blast the medical establishment. At the same time he issued his own belligerent manifesto for the healing arts which he hoped would take the place of the traditional Hippocratic oath. Just as Luther ten years earlier had appealed to the primitive Church, so Paracelsus appealed over the heads of the bishops and cardinals of medicine to the pristine principles of medicine. He showed that he meant business by throwing a copy of Galen's works and of Avicenna's revered *Canon* into a student bonfire on St. John's Day, June 24, 1527. And he boldly announced that his courses in medicine would be based on his own experience with patients.

He further outraged the professors of physick when, instead of using Latin, he lectured in the local Swiss dialect of German called Schweizerdeutsch. Incidentally, this also violated the professional Hippocratic oath in which the proper physician swore to guard his professional knowledge, presumably to protect the laity from incompetent practitioners. "Give not that which is holy unto the dogs," reads the Gospel according to St. Matthew (7:6), "neither cast ye your pearls before swine, lest they trample them under their feet, and turn again and rend you."

The learned doctors turned on Paracelsus. When Froben, his strongest supporter, died suddenly in October 1527, all his enemies—the professors, the apothecaries whom he had attacked for their large profits and little knowledge, and even the students who enjoyed ridiculing his passion—joined forces. Paracelsus' fortunes collapsed after he lost his lawsuit to collect an exorbitant medical fee from a high ecclesiastic. The dignitary, acutely ill with an abdominal disorder, had actually promised the large sum to Paracelsus if he was cured. Then, when Paracelsus cured him simply with a few laudanum pellets, the churchman refused to pay the fee. The judge found against Paracelsus, and when Paracelsus denounced the judge he was forced to leave Basel. Paracelsus' stormy two-year stint in Basel was his last regular employment. Never again was he attached to an institution. Paracelsus became an academic picaro, a medical Don Quixote. In 1529 he stayed in Nuremberg long enough to attack the common treatment of syphilis by poisonous doses of mercury and by guaiacum, a drug derived from a New World hardwood tree, and supposed to be intended by God to cure the disease that came from there. He browbeat the local clergy and the medical faculty, which disputed his right to publish. Then he put on "beggar's garb" for visits to Innsbruck and the Tyrol to study miners' diseases. His wanderings took him through Augsburg and Ulm, across Bavaria and Bohemia. By 1538 he had returned to Villach, the town where his father had died four years earlier.

Though his own health had been broken by poverty, exposure, and the tribulations of a wandering life, he was still trying to practice medicine. His shrill truculence increased with the years. When the learned doctors forced him to move on, he went finally to Salzburg, where he died, aged forty-eight, on September 24, 1541. He was buried there in the almshouse of St. Sebastian, honored with a flattering epitaph that concluded: "Here is buried Philippus Theophrastus, distinguished Doctor of Medicine, who with wonderful art cured dire wounds, leprosy, gout, dropsy, and other contagious diseases of the body, and wished his goods to be distributed to the poor."

The opposition of the doctors prevented most of his writings from being published in his lifetime. But within a few decades of his death, the printing presses had diffused his ideas beyond the academic reach of the doctors. And he became a romantic hero, celebrated by Christopher Marlowe, Goethe, Robert Browning, and Schnitzler, and by the music of Berlioz.

Paracelsus' original concept of disease, despite—or because of—its mystic source, would provide axioms for modern medicine. The prevailing view of disease in medieval Europe was inherited from classical authors, elaborated by the Doctors of Physick. Disease, they said, was the upset of the balance of "humors" in the body. Medical theory was only part of their general theory of human nature. Within each person there were four "cardinal humors" ("humor" from Latin *umor,* meaning fluid or moisture)— blood, phlegm, choler, and melancholy (or black choler). Health consisted of the proper balance of these four humors, and disease came from an excess or an insufficiency of one or another of them. Each person's "temperament" was his unique balance of the four cardinal humors, hence some people were "sanguine," others "phlegmatic," "choleric," or "melancholic."

It followed that there were as many different diseases as there were different individuals, because disease was a disorder of a person's unique humoral relations. Since there was no one norm for the body's temperature, Francis Bacon could observe that among learned men there were "persons of all temperatures." "It is evident," Sir Walter Raleigh wrote in 1618, ". . . that men differ very much in the temperature of their bodies." What was a fever for one might be normal for another. Before the invention of the clinical thermometer and even for some time afterwards, bodily "temperature" was only a synonym for "temperament."

The all-encompassing theory of humors was at once a physiology, a pathology, and a psychology. Ben Jonson's *Every Man in his Humour* (1598), in which Shakespeare himself acted, elaborated a comedy around the "humours" of a jealous husband. Robert Burton's *Anatomy of Melancholy* (1621), which Sir William Osler called "the greatest medical treatise written by a layman," was a comprehensive study of another kind of humoral disorder. It became an English classic because it touched every subject of human interest. Disease Burton defined as "an affection of the body contrary to nature." Since a disease was a disordering of all the elements in the body, then the cures for diseases would have to treat the body as a whole. The humoral lore taught physicians how to discover the unique "natural" balance of humors in each person, and then how to restore that balance in the whole body by such treatments as sweating, purging, bloodletting or inducing vomiting.

Paracelsus championed a radically different theory based on a radically different notion of disease, with far-reaching consequences for medical science. The cause of a disease, Paracelsus insisted, was not the maladjustment of the bodily humors within a person, but some specific cause outside the body. "Humors" and "temperaments" he ridiculed as figments of the learned imagination. But he was also impatient with the few pioneer anatomists who were trying to put medicine on a more solid footing. When God ordered the whole universe, according to Paracelsus, He provided a remedy for every disorder. The causes of disease were mainly minerals and poisons borne from the stars in the atmosphere. This insight Paracelsus cast in his own revised language of astrology. When he pointed outside the body, when he insisted on the uniformity of causes and the specificity of diseases, he was pointing the way to modern medicine. If his arguments were not right, his insights and his hunches were.

Paracelsus' faith led him to believe that there were no incurable diseases, only ignorant physicians. "For God it is, who commanded: Thou shalt love thy neighbour as thyself and thou shalt love God above all things. If, now, thou wilt love God, thou must also love His works. If thou wilt love thy neighbour, thou must not say: For thee there is no help. But thou must say: I cannot do it and I understand it not. This truth shields thee from the curse that descends on the false. So take heed what is told thee; the rest shall be sought after until the art is found from which good works proceed." The physician must always search for new remedies, never limiting his prescriptions to those allowed by Galen.

Academic doctors had generally confined their prescriptions to herbal remedies, which, being organic, were believed for that reason to be appropriate for the human body. Hence botany was a regular subject in the medical curriculum, and for centuries the prejudices that inhibited medicine also narrowed the study of botany. The plant kingdom became the realm of herbal remedies. Myths from everywhere—from Egypt, Sumeria, China, and Greece—told how herbs had been made from the flesh of the gods, and how the gods then instructed men in their use. The "herbal," a medico-botanical genre, was one of the first best sellers among early printed books. Handsomely illustrated herbals found a ready market among well-to-do physicians and prosperous merchants. The ancient works on botany that had the greatest influence in the European Middle Ages were not philosophical treatises on the nature of plants, like those of Theophrastus, but practical guides to medicinal uses. The standard work on botany, and the foundation of pharmacology for fifteen hundred years, was *De materia medica* by Dioscorides, a first-century Greek who served as physician in the armies of Nero.

Medicine and botany had become Siamese twins. It seemed that neither could advance without the other. But Paracelsus prophesied their separation. Why should not doctors use *all* the resources that God had created —mineral as well as vegetable and animal, inorganic as well as organic— to cure the body's ills? "Each disease has its own physic." Who dared say that minerals and metals should not heal? In a few cases, as in the use of mercury against syphilis, the doctors had reluctantly tried inorganic remedies. The objection that inorganic materials were "poison" because they were foreign to the body was quite silly, Paracelsus observed, since "every food and every drink, if taken beyond its dose, is poison."

Paracelsus' faith also led him back to the popular doctrine of "signatures," which held that the shape or color of an herb might suggest the organ it was designed to cure. For example, the orchid might be intended for ills of the testicles, or a yellow plant for diseases of the liver. Try "the hair of the dog that bit you." Unlike his more respectable colleagues, Paracelsus respected folk remedies.

In his time there was no proper science of chemistry, and the study of minerals and metals was dominated by the alchemists' quest for the "philosopher's stone," to transmute other elements into gold. Paracelsus assigned the alchemist a new task: to make minerals and metals into medicines. He hoped to divert the alchemists from the search for wealth to the search for health.

While the comfortable Doctors of Physick were pronouncing on the humoral balance of their wealthy patients, Paracelsus made his pioneer study of occupational diseases. Paracelsus knew the life of the miner, for when he was only nine his father had moved to the mining town of Villach in southern Austria and as a young man he worked in the iron smelters in Schwaz in the Tyrol. His later wanderings in Denmark, Sweden, and Hungary, and in the Inn Valley took him into the mines. He finally went back to Villach to manage the Fugger metallurgical works. During all these years he had noted the working conditions of miners and metal workers, he observed their special ailments, and he experimented with remedies. *On the Miners' Sickness and other Miners' Diseases (Von der Bergsucht und andern Bergkrankheiten),* like other Paracelsus books, was not published during his lifetime. It was put into print in 1567, a quarter-century after his death, and it bore fruit in the next centuries.

Miners' sickness, he explained, was a disease of the lungs and also produced stomach ulcers. These came from the air that the miner breathed and from minerals taken in through the lungs or the skin. He distinguished between acute and chronic poisoning and noted the differences between the disorders caused by arsenic, by antimony, or by the alkalis. In a special section on mercury poisoning he accurately noted its symptoms—shivering, gastrointestinal disorders, putrefaction of the mouth, blackening of the teeth. His therapy for mercury poisoning was based on the assumption that since mercury collects in certain parts of the body, the physician must make openings through which the mercury can escape. This was done by applying a corrosive plaster to make an ulcer, or by baths, a therapy still used.

"Because so much lies in the knowledge of natural things which man himself cannot fathom, God has created the physician . . . ," Paracelsus observed. "And in the same manner as the devil is driven out of man, the poisonous diseases are expelled by means of such physic, just as evil expels evil and good retains good. . . ." He challenged the Doctors of Physick to match the successes of folk medicine. "Doctors of medicine," he warned, "should consider better what they plainly see, that for instance an unlettered peasant heals more than all of them with all their books and red gowns. And if those gentlemen in their red caps were to hear what was the cause, they would sit in a sack full of ashes as they did in Nineveh."

45

The Tyranny of Galen

FOR fifteen hundred years the main source of European physicians' knowledge about the human body was not the body itself. Instead they relied on the works of an ancient Greek physician. "Knowledge" was the barrier to knowledge. The classic source became a revered obstacle.

Of all the ancient writers on science, except for Aristotle and Ptolemy, none was more influential than Galen (c. 130–200). Born of Greek parents during the reign of the emperor Hadrian in Pergamum in Asia Minor, he began studying medicine at the age of fifteen. After working under professors of medicine in Smyrna, Corinth, and Alexandria, at the age of twenty-eight he returned to his native Pergamum to become physician to the gladiators. In an age when cadavers for dissection were tabooed, he profited from his opportunity to learn from what he saw inside the gladiators' wounds. When he moved to Rome he cured some eminent patients, he gave public lectures on medicine, and finally became court physician to the Stoic philosopher-emperor Marcus Aurelius (121–180) and his son Commodus. Galen, one of the most prolific writers of antiquity, was said to have produced five hundred treatises in Greek—on anatomy, physiology, rhetoric, grammar, drama, and philosophy. More than a hundred of these works survive, including a treatise on the order of his own writings, and in their modern edition they fill twenty thick volumes.

Although Galen's works were prolix, their sheer volume, preserved by lucky coincidences for a distant posterity, overwhelmed his competitors. He collected and organized the medical lore of the more ancient physicians who had preceded him. But he was no mere compiler. He produced his own philosophy of medical procedures. "I do not know how it happened," Galen himself boasted, "miraculously, or by divine inspiration, or in a frenzy or whatever you may call it, but from my very youth I despised the opinion of the multitude and longed for truth and knowledge, believing that there was for man no possession more noble or divine." He noted, too, that his medical colleagues, who prospered by serving Rome's wealthy and powerful, actually criticized him "for pursuing truth beyond moderation." Even he conceded that he would not have been able to succeed as a physician if he had not "called on the mighty in the morning and dined with them in the evening." In his own way he anticipated Paracelsus' scorn for possessions, and for the money-hungry physician. For, he said, he needed no more than two garments, two house slaves, and two sets of utensils.

According to Galen, since knowledge was cumulative, the progressive physician must learn from Hippocrates and all the other greats who had gone before. The conspicuous progress of medicine, he said, was like the impressive improvement of Roman roads over the centuries. The ancients had marked off the paths and made the first cuts through the wilderness, then each later generation built dikes and bridges, and added stone surfacing. "There is thus no cause for wonder if we, while acknowledging that Hippocrates discovered the therapeutic method, have ourselves undertaken the present work." Galen urged his colleagues, while learning from experience, to focus on the useful knowledge that would cure patients. He made a special study of the pulse, and showed that the arteries did not carry air, as others believed, but carried blood. He was said to be brilliant in diagnosis, and he even wrote a treatise on malingering.

His most influential work, which would come to some seven hundred printed pages, was *On the Usefulness of the Parts of the Body*. There he describes each limb and organ and explains how it is designed to serve its particular purposes. For example, in his opening Book on "The Hand":

> Thus man is the most intelligent of the animals and so, also, hands are the instruments most suitable for an intelligent animal. For it is not because he has hands that he is the most intelligent, as Anaxagoras says, but because he is the most intelligent that he has hands, as Aristotle says, judging most correctly. Indeed, not by his hands, but by his reason has man been instructed in the arts. Hands are an instrument, as the lyre is the instrument of the musician, and tongs of the smith. . . . every soul has through its very essence certain faculties, but without the aid of instruments is helpless to accomplish what it is by Nature disposed to accomplish.

Even while leaning on Aristotle, Galen urges his reader to be wary of pedantic medicine. "If anyone wishes to observe the works of Nature, he should put his trust not in books on anatomy but in his own eyes and either come to me, or consult one of my associates, or alone by himself industriously practise exercises in dissection; but so long as he only reads, he will be more likely to believe all the earlier anatomists because there are many of them." By his own lights Galen was an experimental physician, constantly appealing to experience.

In a familiar irony of history, as Galen's books became sacred texts his own spirit was forgotten. For centuries "Galenism" would be the physicians' dominant dogma. Just as the writings of Aristotle became the basis of scholastic philosophy, so Galen's much more voluminous works founded scholastic medicine. Since he had written in Greek, his first influence was in Alexandria and Constantinople, the eastern remnants of the Roman Empire, and among the Muslim neighbors.

Physicians established a Galenic canon of sixteen works which were declared to be the most authoritative. Such a selection violated Galen's doctrine, for he had insisted that his disciples should first study his writings on method. When the Arab world assimilated Greek science, they translated Galen into Arabic and they too made him their model physician. Even his autobiography became a pattern for the biographies of Arabic scientists. By the tenth century, the title "Galen of Islam" was the highest honor that Muslim physicians could pay to Avicenna (980–1037) or any other medical great.

In the Arab world the Galenic texts became corrupted and were combined with Arabic texts. Galen sometimes suffered in competition with Rhazes, Avicenna, Averroës, and Maimonides, who had dared to write their own critiques of Galen. Still Galen remained the unifier of medieval medicine, and physicians called themselves members of "Galen's family."

Some of Galen's Greek works had probably been translated into Latin by the sixth century, and then with the rise of Arab-Muslim power in the Mediterranean, and their occupation of Spain and Sicily, the texts of Galen finally reached western Europe. There, too, by the eleventh century Galenism was becoming rigidified. Aristotelianism continued to be dominated by the words of Aristotle. But Galenism was compounded of the original Galen and the Byzantine and Arabic texts, with the commentaries through which he reached the West. At the very time when European Christians were traveling across the Mediterranean to crusade against Muslim infidels, and were burning heretics and Jews in town squares, Christian physicians in Europe were daily curing bodily ills by the wisdom of modern Muslim and Jewish doctors. Already there were signs that modern science would respect no national or confessional boundaries. On the Canterbury pilgrimage Chaucer's Doctor of Medicine "well knew" both the Greek and the Arab doctors—not only Aesculapius, Hippocrates and Galen, but also Rhazes, Avicenna, and Averroës.

The Renaissance, which we credit with the birth of modern science, had some curiously contradictory and unheralded consequences. Few of Galen's works of science had been known in Europe before the fourteenth century. His most important work on anatomy was not translated and so not generally available in the West before the Renaissance, with its revival of the Greek classics. The first printed Latin translation of substantial works of Galen was made in 1476. From the press founded by Aldus Manutius in Venice came the first printed Greek edition of Galen (1525). Few other products of the Aldine Press had a more far-reaching effect. European physicians now for the first time had their own copies of the texts of their revered master in his original language. The printers who merchandised these texts by the thousands reinforced the Galenic orthodoxy. The product of their collaboration was not medical science, not experiment but pedantry.

The medical faculty of Paris bought the Aldine edition of Galen's works the year after they were published. Jacobus Sylvius, the leading professor of anatomy there, taught that Galen was always right. His study of medicine, therefore, was a search for what Galen really meant, and for him "anatomy" was a branch of classical philology. He and other Galenists believed that the most important contribution to a better knowledge of the human body would be a more accurate Latin rendering of the purest Greek text of Galen. Medical debates came to resemble the theologians' quibbles over the meaning of words in Scripture. Leading anatomists went to any length to defend Galen. Sylvius, for example, shared a popular view that if a dissected body did not show all the features described by Galen's text, it was because the human body had actually changed, and because, in the passing centuries, the human species had declined from the ideal form seen by Galen.

Even the most up-to-date Renaissance professors of medicine sought their image of the human body in the mirror of antiquity. The refurbishing of Galen was only a polishing of that mirror. For example, Thomas Linacre (1460?–1524), physician to Henry VIII, an M.D. from Padua and founder of the Royal College of Physicians (1518), reinforced his medical reputation by translating six of Galen's works from Greek into Latin.

But most of what Galen was describing was what he had never seen! The great authority on human anatomy, whose word was gospel for fifteen hundred years, had himself probably made studies of human bodies but never dissected a cadaver. According to him, on only two occasions could he study the whole bony structure of the human body. Once he had the opportunity to study a skeleton that had been cleansed of flesh by scavenging birds, and on the other occasion a skeleton had been washed clean in a river.

Since Roman custom in his time forbade dissection of the human body, Galen had performed all his anatomies on monkeys for external anatomy and on pigs for internal. He then projected what he found onto his anatomy of the human body. He made no secret of this, and wrote nostalgically of those excellent earlier times when dissection had been allowed. In the influential works in which he purported to describe human anatomy, he tacitly assumed that what he found in "the other animals most closely resembling man" would also be found in man.

Generations of physicians who made Galen their source of anatomical knowledge accepted supinely, even enthusiastically, this crucial flaw in Galen's materials. It eased their task and justified their following his example. "Because the structure of the internal parts of the human body was almost wholly unknown," a twelfth-century anatomy text from Salerno explained, "the ancient physicians, and especially Galen, undertook to display the positions of the internal organs by the dissection of brutes. Although some animals, such as monkeys, are found to resemble ourselves in external form, there are none so like us internally as the pig, and for this reason we are about to conduct an anatomy upon this animal."

Christianity had a curiously varied influence on the rise of anatomy. The Christian belief in an immortal soul, and contempt for the body as mere dross that was sloughed off at death, did not encourage a passionate interest in human anatomy. At the same time this separation of the physical body from the soul, which was the essential immortal person, in the long run made it easier than it had been in Egypt or in Rome to allow the dissection of cadavers.

Medieval Islam was never reconciled to the dissection of the human body. From the eighth to the thirteenth centuries the anatomical knowledge of the learned Muslim doctors was merely (in historian C. D. O'Malley's phrase) "Galen in Moslem dress." When the best Muslim physicians corrected Galen's anatomy it was not methodically or by their own dissections, but by accident or good luck. For example, an eminent Arab physician traveling through Egypt in the early thirteenth century luckily happened on a pile of human skeletons that had accumulated during a recent pestilence, and by examining these he was able to correct Galen's faulty description of the human mandible.

Galen's efforts to describe the human body by analogy had seduced him into so many errors that some critics in the next generation derided Galen as the exponent of "ape anatomy." And the climax of Paracelsus' career came at the very moment of publication of the definitive Aldine edition of Galen's works in Greek. But to expose Galen's errors the prophetic enthusiasm of a Paracelsus was not enough.

Even in the age of Galen, an acute and determined observer like Leonardo da Vinci (1452–1519) could describe what he could see for himself. Leonardo intended to write a treatise on anatomy, along with others on painting, architecture, and mechanics. He never published any of them, but after his death a work on painting and another on the movement and measurement of water were compiled from his notes. If he had completed his anatomical treatise, and if it had been published, medical science might have progressed more speedily. But Leonardo seldom finished anything. An unkind providence left unfinished two of his most important paintings, the Sforza monument and the mural of the Battle of Anghiari.

After his death the five thousand pages of his manuscript notes were widely dispersed as collectors' items. Nearly every page revealed the cosmic miscellany of his mind, the indiscriminate reach of his curiosity. A single page, for example, taking off from his interest in curves, shows an exercise in the geometry of curves, a drawing of curly hair, grasses curling around an arum lily, sketches of trees, curvesome clouds, rippling waves of water, a prancing horse, and the design of a screw press.

Leonardo exercised his ingenuity to make these scatterings still more illegible and cryptic. He invented his own shorthand and spelling, he combined and divided words according to a system of his own, and employed no punctuation at all. Further to mystify posterity, he wrote the characters backwards and with his left hand, so that reading them requires a mirror. Not until the late nineteenth century did Leonardo's notes reach a wide scholarly audience.

Then finally Leonardo was revealed as a pioneer anatomist. "The eye," he wrote, "the window of the soul, is the chief means whereby the understanding can most fully and abundantly appreciate the infinite works of Nature; and the ear is second." It is not surprising that Leonardo's sensitive eyes and nose found something repulsive in a cadaver. Yet to him every mark and vein and pimple of the actual world was sacrosanct. To deny anything one saw was sacrilege. "Experience does not ever err, it is only your judgment that errs in promising itself results which are not caused by your experiments." So Leonardo was slow to translate observed facts into universal "principles," in such matters as the circulation of the blood.

The anatomy that we garner from the thousands of sheets of his cryptic miscellany reveals that Leonardo saw and recorded what others before him had not seen. If he had managed to put his views together and had not been distracted by his universal interests, he might well have become the successor to Galen. Leonardo secretly bypassed Galen and read the body itself. The parts of the body, he said, had to be shown from all directions. His unpublished views of the human skeleton were drawn from back, front, and side. He urged systematic and repeated dissections. "You will need three [dissections] in order to have a complete knowledge of the veins and arteries, destroying all the rest with very great care; three others for a knowledge of the membranes, three for the nerves, muscles and ligaments, three for the bones and cartilages. . . . Three must also be devoted to the female body, and in this there is a great mystery by reason of the womb and its fetus." He explored the anatomy of the eye by constructing a glass model of the eye and the lens so that he could look through it to confirm his theory that the optic nerve carried the visual impressions. His belief that the body was a machine led him to make remarkably accurate pictures of the muscles and their operation on the bones. He pioneered in drawing the coils of the small and the large intestines, and was probably the first to sketch the appendix. In detail he showed that the atria of the heart are contracting chambers that propel blood into the ventricles. He took casts of parts of the body, and prepared his specimens by injecting wax.

Yet, despite his consummate art, his industry, and his unexcelled powers of observation, Leonardo added only to his own knowledge, and little or nothing to the anatomical knowledge of his time. Nor were his own observations enriched as they might have been. For, as we shall see, the public forum of printed matter has a way of improving the product, and Leonardo's work remained private.

46

From Animals to Man

ANDREAS VESALIUS (1514–1564), no universal genius, was not distracted from his main subject. He was born just inside the city walls of Brussels within sight of the hill where condemned criminals were tortured and executed. As a child he must often have seen the bodies that were left hanging to be picked clean by the birds. His father was apothecary to the emperor Charles V, and the family was well known in the medical profession. Unlike Paracelsus, Vesalius had the best medical education offered in his time. He matriculated at the University of Louvain in 1530, then went on to the University of Paris, where he studied under Professor Sylvius, the celebrated champion of Galen. When war broke out between France and the Holy Roman Empire, Vesalius, an enemy alien, was obliged to leave Paris. Returning to Louvain, he received his Bachelor of Medicine degree in 1537, then proceeded to Padua, the most esteemed medical school in Europe. There he submitted himself to two days of examination and received his degree of Doctor of Medicine *magna cum laude.* He must have been well versed in the conventional learning, for at the age of twenty-three, only two days after passing his examination, he was installed in the university's chair of surgery and anatomy.

When Vesalius took up his professorship, he gave surgery and anatomy a new significance. For he no longer considered his primary duty to be the expounding of Galen's texts. In conducting his required "anatomy" (from Greek for "cutting up") he departed from custom. Unlike the professors before him, Vesalius did not stay seated high in his professorial cathedra while a barber-surgeon with bloody hands pulled organs out of the cadaver. Instead, Vesalius himself handled the body and dissected the organs. To help his students, he prepared some new teaching aids in the form of four large anatomical charts, detailed enough to show the student the body structure when there was no cadaver at hand. Each part was labeled with its technical name. An accompanying glossary and index listed the names of the parts in Greek, Latin, Arabic, and Hebrew.

To use charts at all was a great novelty. During the European Middle

Ages anatomical drawings had been rare. In the sixteenth century, when texts of Galen were rediscovered, scrupulously edited, newly translated and printed, they were still not provided with anatomical drawings. Some of the leading anatomy professors, including Vesalius' respected teacher Sylvius, actually campaigned against the use of figures and diagrams. Students should be reading the authentic text instead!

Vesalius' *Six Anatomical Tables* (*Tabulae Anatomicae Sex,* Venice, 1538) was the first effort to give comprehensive visual form to the teachings of Galen. If there had not been a printing press at the time, Vesalius might not have been tempted to publish the charts he prepared for his students. But when one of his charts was plagiarized and it seemed that the others might be, Vesalius had them all published. Three were drawings of a skeleton by Titian's Dutch pupil, John Stephen of Calcar, "from the three standard aspects" known to art students in the Middle Ages. The other three "tables" were radically original in concept: Vesalius' own drawings of the veins, the arteries, and the nervous system. Their novelty was not so much in what they showed as in their way of showing. And with these anatomical "tables" Vesalius invented the graphic method in anatomy. Today it seems surprising that so obvious a teaching aid ever had to be invented. Yet on second thought it is not quite so surprising. For centuries, even while the training of doctors in the best medical schools in Europe had included some anatomy, the opportunities to see inside a human body had been few and far between.

Not only the "humoral" dogmas, which were Paracelsus' special target, but also the widely practiced medical astrology ignored anatomic details. The popular diagrams of "zodiac man" simply showed a connection between each bodily part as a whole and its corresponding sign of the zodiac to indicate the best and the worst seasons for certain remedies. The English word "influenza" is a relic of this linkage. When Vesalius studied medicine, learned doctors were still using the word (borrowed from the Italian with the same senses as the English "influence") to describe the medical effects of an unfortunate astral "influence." At first it meant any sudden visitation of an epidemic disease and was a synonym for epidemic until the present use for a specific disease began to appear in the eighteenth century.

After his *Six Anatomical Tables* Vesalius still had a long way to go, for his tables, following Galen, again and again silently made the leap from animal to human anatomy. For example, they showed a *rete mirabile,* a "marvelous network," at the base of the human brain where, according to Galen, man's "vital spirit" was transformed into "animal spirit." But while found in hoofed animals, this "network" does not exist in man. The "great blood vessels" (superior and inferior venae cavae) seen in Vesalius were also peculiar to the ungulates. His drawings of the shape of the heart, the branches of the aortic arch, the location of the kidneys, and the shape of the liver, depicted, as in Galen's text, not a man but an ape.

Only on rare and gruesome occasions was the interior of the human body actually examined. For example, Emperor Frederick II (1194–1250), celebrated across Europe for his versatility, wanted to satisfy his curiosity about the processes of human digestion. A chronicler reported how he "provided an excellent dinner for two men after which he ordered one to go to sleep and the other to go hunting. The following evening he ordered their stomachs to be emptied in his presence in order to see who had better digested the meal, and the surgeons decided that he who slept had had the better digestion." Incidentally, in 1238 the Emperor commanded the medical school of Salerno to conduct a public dissection once every five years.

A macabre opportunity to study the human skeleton came from time to time during the Crusades when the bodies of those who had died en route were dismembered and boiled so their bones could be conveniently shipped back for burial in the crusader's home soil. This custom was so widespread that it occasioned a bull (1299) by Pope Boniface VIII prohibiting the practice. Although many individual churchmen opposed dissection of the human body, it seems that the Pope himself never announced his opposition in principle. During the fourteenth century, human dissection became more familiar in medical faculties, and when Pope Alexander V died suddenly in Bologna in 1410 a postmortem was conducted there upon his body.

Still, the acts of dissection somehow seemed unnatural and contrary to God's will. To "anatomize" also described the delivery of a baby by Caesarean section. Occasionally a postmortem would be ordered by the court to settle whether the wounds on the deceased had actually been the cause of death.

When the health of the community was at stake, autopsies might be tolerated or even required. After the Black Death of 1348, the Public Health Department in Padua ordered that whenever a person died of unknown causes, the body could not be buried without the certificate of a doctor who had examined the body and found no signs of the plague. To discover the enlarged lymph nodes, which were diagnostic of the plague, required dissection of the body, and Paduan medical students learned from these specimens.

Autopsies on the bodies of prominent persons whose death had caused widespread concern would sometimes add bits of medical knowledge. Vesalius reports his experience during a visit to Brussels in 1536:

When I had returned from my visit to France I was invited by the physician of the Countess of Egmont to attend the autopsy of an eighteen-year-old girl of noble birth who, because of an enduring paleness of complexion and difficulty in breathing—although otherwise of agreeable appearance—was thought by her uncle to have been poisoned. Since the dissection had been undertaken by a thoroughly unskilled barber I could not keep my hand from the work, although except for two crude dissections lasting three days, which I had seen at school in Paris, I had never been present at one.

From the constriction of the thorax by a corset the girl had been accustomed to wear so that her waist might appear long and willowy, I judged that the complaint lay in a compression of the torso around the hypochondria [under the ribs] and lungs. Although she had suffered from an ailment of the lungs, yet the astonishing compression of the organs in the hypochondria appeared to be the cause of her ailment, even though we found nothing that would indicate strangulation of the uterus except some swelling of the ovaries. After the attendant women had left to shed their corsets as quickly as possible and the rest of the spectators had departed, in company with the physician I dissected the girl's uterus for the sake of the hymen. The hymen, however, was not entirely whole but had not quite disappeared, as I have found is usually the case in female cadavers in which one can barely find the place where it had been. It looked as if the girl had ripped the hymen with her fingers either for some frivolous reason or according to Rhazes's prescription against strangulation of the uterus without the intervention of a man.

Since the bodies of executed criminals were the main source of cadavers, female cadavers were especially scarce, which added still another obstacle to discovering the processes of procreation and gestation.

Only gradually did anatomy cease to mean the occasional opening of a body to answer a specific question and come to mean the systematic study of the body. A compendium of anatomy by Mondino de Luzzi of Bologna in 1316, which added to Galen some suggestions from Arabic authorities, dominated the teaching of Galenic anatomy for two hundred years. The order of Mondino's exposition still followed the urgencies of the time by describing first the organs in the abdominal cavity which were the most perishable and hence dissected first, then going on to the bones, the spinal column, and the extremities. Mondino repeated the old errors borrowed from animal anatomy, and he added nothing visually.

As we have seen, there were many practical obstacles to the sustained scrutiny of the insides of the human body. Lack of refrigeration made it necessary for an anatomy to be conducted hastily before the body decomposed, and even in the best universities dissections were performed only annually or biennially. During the four days and poorly illuminated nights of these rare performances the bleary-eyed crowd of medical students hardly had time or inclination to ask questions, to reflect, or to look again. Vesalius himself described

> that detestable procedure by which usually some conduct the dissection of the human body and others present the account of its parts, the latter like jackdaws aloft in their high chair, with egregious arrogance croaking things they have never investigated but merely committed to memory from the books of others, or read what has already been described. The former are so ignorant of languages that they are unable to explain their dissections to the spectators and muddle what ought to be displayed according to the instructions of the physician who haughtily governs the ship from a manual since he has never applied his hand to the dissection of the body. Thus everything is wrongly taught in the schools, and days are wasted in ridiculous questions so that in such confusion less is presented to the spectators than a butcher in his stall could teach a physician.

For centuries in Europe the only bodies legally available for dissection continued to be those of executed criminals, which seldom arrived intact. In England hanging was usual, but some persons of high station were allowed the privilege of being decapitated. In the Republic of Venice and elsewhere on the Continent, beheading was more common. Mondino's textbook explained that an "anatomy" began by laying out "the body of one who has died from decapitation or hanging." This fact inevitably skewed the student's view—for example, distorting what could be seen of internal phenomena, such as the circulation of the blood. Yet even these cadavers were only rarely available. Of all the public executions in Padua from 1562 to 1621 only one body was delivered for dissection. In this case the body of the young man who had been hanged for murder was tied to a horse's tail and so dragged from the Piazza della Signoria to the medical school. Seldom were there opportunities to anatomize bodies that had not been mutilated in execution.

Resourceful professors seized every opportunity to pick up bits and pieces of human bodies, with the most unappetizing consequences. Vesalius' eminent teacher Jacobus Sylvius, one of his students reported, had sources of his own.

> I have seen him bring in his sleeve, because he lived all his life without a servant, sometimes a thigh or sometimes the arm of some one hanged, in order to dissect and anatomize it. They stank so strongly and offensively that some of his auditors would readily have thrown up if they had dared; but the cantankerous fellow with his Picard head would have been so violently incensed, threatening not to return for a week, that everyone kept silent.

Vesalius seized every opportunity, legal or illegal, to garner specimens. He recorded one of these escapades in 1536:

Because of the outbreak of war I returned from Paris to Louvain where, while out walking with that celebrated physician and mathematician Gemma Frisius and looking for bones where the executed criminals are usually placed along the country roads—to the advantage of the students—I came upon a dried cadaver similar to that of the robber Galen mentions having seen. As I suspect the birds had freed that one of flesh, so they had cleaned this one, which had been partially burned and roasted over a fire of straw and then bound to a stake. Consequently the bones were entirely bare and held together only by the ligaments so that merely the origins and insertions of the muscles had been preserved. . . . Observing the body to be dry and nowhere moist or rotten, I took advantage of this unexpected but welcome opportunity and, with the help of Gemma, I climbed the stake and pulled the femur away from the hipbone. Upon my tugging, the scapulae with the arms and hands also came away, although the fingers of one hand and both patellae as well as one foot were missing. After I had surreptitiously brought the legs and arms home in successive trips—leaving the head and trunk behind—I allowed myself to be shut out of the city in the evening so that I might obtain the thorax, which was held securely by a chain. So great was my desire to possess those bones that in the middle of the night, alone and in the midst of all those corpses, I climbed the stake with considerable effort and did not hesitate to snatch away that which I so desired. When I had pulled down the bones I carried them some distance away and concealed them until the following day when I was able to fetch them home bit by bit through another gate of the city.

Finally, by such expedients as these he was able to piece together a whole skeleton at Louvain. To avoid incriminating himself, he managed to convince people that he had brought it with him from Paris. Luckily the burgomaster of Louvain, who had taken an interest in anatomy, "was so favorably disposed toward the studies of the candidates in medicine that he was willing to grant whatever body was sought from him."

Later, at Padua, Vesalius interested a judge of the criminal court in his research, who not only offered him the bodies of executed criminals, but was thoughtful enough to delay executions so that the bodies would be fresh when Vesalius was ready to dissect them. There were even rumors that medical students were stealing respectable bodies from their graves and then after dissection throwing the pieces in the river or tossing them to the dogs. As a result, in 1597 a Padua ordinance required a public funeral for the parts of every dissected body. It seemed there never would be enough cadavers to satisfy medical students. As late as the eighteenth century in England the grave-robbing profession flourished, and "one who makes open profession of dealing in dead bodies" was known as a "resurrection man." Dickens' Jerry Cruncher, in *A Tale of Two Cities,* introduces us to this profitable occupation.

While teaching from Galen's text, Vesalius had noted so many instances where Galen described what was not found in the human body that he soon realized that Galen's ostensibly "human" anatomy was really only a compendium of statements about animals in general. "I gave careful consideration," Vesalius noted as a revelation in 1539, "to the possibility that anatomical dissection might be used to check speculation." He then determined to produce a whole new anatomy text based entirely on his own observations of the *human* body. For his public anatomy at Bologna in 1540 Vesalius had put together two skeletons—one of an ape, the other of a man—to show that the appendage that Galen described as extending from the vertebra to the hip appeared only in the ape. He found this discrepancy so significant that he specially illustrated it in his *Fabrica.* At his anatomical demonstrations Vesalius insisted that students see, feel, and decide for themselves. To students who asked whether the arteries really followed the movement of the heart, Vesalius would reply, "I do not want to give my opinion, please do feel yourselves with your own hands and trust them."

His anatomical studies culminated in the book that brought him fame as it sped across Europe. Vesalius' *Structure of the Human Body (De Humanis Corporis Fabrica),* commonly called the *Fabrica,* a handsome printed folio of 663 pages, appeared in August 1543, the very year of Copernicus' *De Revolutionibus.* Destined to be for anatomy what Copernicus' work was for astronomy, it would have justified the work of a lifetime but it was completed sometime between his twenty-sixth and twenty-eighth year.

Since he was determined to show with the utmost precision only what he had confirmed by his own eyes and hands, he knew that the scientific value of his product would depend on the quality of his illustrations. So he sought out and then supervised the best artists to make the drawings. The most skilled Venetian wood-block cutters were engaged for reproductions. A talented draftsman, he did some figures himself. The rest were by artists of the school of Titian, probably by the same John Stephen of Calcar who had done the sketches for the *Six Anatomical Tables.*

Leonardo da Vinci had warned against the delusive precision of verbal anatomy texts. "And you who think to reveal the figure of man in words," he warned in the privacy of his notebook, "with his limbs arranged in all their different attitudes, banish the idea from you, for the more minute your description the more you will confuse the mind of the reader and the more you will lead him away from the knowledge of the thing described. It is necessary for you to represent and describe." The moment was auspicious for a Vesalius to free anatomy from its literary shackles. Renaissance artists like Leonardo were announcing a new realism on the walls of palaces and churches. When Leonardo enumerated the qualities of the successful anatomist, he listed patience, perseverance, a "love for such things," and the courage for "living in the night hours in the company of those corpses, quartered and flayed and horrible to see." But now to his list he significantly added "the skill in drawing . . . combined with knowledge of perspective." Leonardo boasted in his notebooks that he himself had dissected "more than ten human bodies," and that he would combine what he learned from all of them into a single drawing. Naturally, there were striking resemblances between Leonardo's drawings and some in Vesalius' *Fabrica.* But there is no solid evidence that Vesalius ever saw any of Leonardo's drawings. The new techniques of perspective now helped all skilled artists depict the same original.

Selecting the right printer for his work was crucial, and Vesalius knew it. It might have seemed natural for a professor at Padua in the prosperous Venetian Republic to have his work printed in the capital city. Venice, "Queen of the Adriatic," from the first days of printing had been a headquarters for great printing houses. Early in the sixteenth century the printer's craft had already reached a climax in the elegant products of Aldus Manutius. But by the 1540's there were legal complications. The university commissioners in Padua had to approve books before they were submitted to the Venetian Council of Ten for their imprimatur. The reputable Venetian firm of Giunta had made a business of publishing improved texts of Galen's *Opera omnia* and had actually enlisted Vesalius as an editor in their most recent definitive edition. But in Venice standards of printing had declined. The risks of transport across the steep and slippery alpine passes were notorious, but to ensure the quality of the printed *Fabrica,* the scrupulous Vesalius still chose to send his heavy manuscript and the numerous wood blocks for illustrations over the mountains to Oporinus in Basel. He would not be disappointed.

For Vesalius' confidence in his "dearest friend Oporinus" there were good reasons. Johannes Oporinus, the son of an artist, had worked at the celebrated Froben press, and also had been professor of Latin and Greek. As student and secretary to the formidable Paracelsus in Basel, he had accompanied Paracelsus briefly on his prophetic wanderings. Oporinus was willing to take risks. He even dared to publish Theodor Bibliander's Latin translation of the Koran, for which he served a term in jail. Now, he was so confident that he could meet Vesalius' exacting standard that at the beginning of the book he actually printed Vesalius' instructions to the printer so that the reader himself could judge. To oversee the production, Vesalius himself had gone to Basel.

The famous illustrated title page of Vesalius' sumptuous *Fabrica* displayed the crowded scene of a "public anatomy," such as was required by the statutes of the University of Padua. The professor himself, Vesalius, touches the exposed abdominal organs of an opened female cadaver. To emphasize that this is a *human* anatomy, a human skeleton is seated just above the corpse, while a naked male figure looks on from the side. In the foreground, seated disconsolately under the dissecting table, are two barbers, who in the pre-Vesalian era would themselves have been performing the dissection. Now they sharpen the professor's razors. In the left-hand corner is a pet monkey, in the right-hand corner a dog, neither a subject of the professor's concern.

Vesalius' title *De Humanis Corporis Fabrica* suggested that he was interested in both the *structure* (on analogy of "fabric") and the *working* (on analogy of the French *fabrique* or German *Fabrik,* meaning factory) of the human body. For his students Vesalius would emphasize the internal structure of the body by using a piece of charcoal to mark on the flesh of the cadaver the shape of the skeleton within. Now at last, in the *Fabrica,* he departs from the usual order of dissection and description that had been determined by the rate of putrefaction. Instead, he begins with the bones —the basic structure of the body—and then proceeds to the muscles, the vascular system, the nervous system, the abdominal organs, the thorax and the heart, and finally to the brain.

Vesalius seems on the whole to have fulfilled his high hopes in the sections on bones, muscles, heart, and brain, which were more than half the work. What he shows and tells in those sections he had verified personally. We cannot be surprised that the twenty-eight-year-old Vesalius proved unable to survey the whole human anatomy exclusively from his own observation. He repeats Galen's error of showing perforations within the ventricular system of the heart. The rest of his work still follows the traditional scheme of Galen, but no section fails to show revisions of Galen's flagrant errors. For example, in the skeleton he demonstrates in the mandible, the sternum, and the humerus how Galen had projected animal structures onto man. He exposes Galen's mistake of describing multiple lobes in the human liver on the analogy of monkeys, dogs, and sheep, and he depicts the human liver as a single mass. Now he corrects an error that he had perpetuated in his earlier *Six Tables* and shows that the *rete mirabile* does not exist in man at all.

Within a scant half-century, Vesalian anatomy prevailed in European medical schools. The study of anatomy in the West would never be the same again. What Vesalius said about the heart or the brain was unimportant compared with the path he opened for future students to learn about all organs of the body. Discrediting Galen was not enough. There had to be a new enthusiasm for repeated comparative dissection. In no other way could the doctor be sure that he was not describing anomalies.

The customs of the time and continuing prejudice against dissection still made problems. The *Fabrica* gave unashamed accounts of the professor's body-snatching and advertised gruesome techniques for avoiding detection:

> The handsome mistress of a certain monk of San Antonio here [in Padua] died suddenly as though from strangulation of the uterus or some quickly devastating ailment and was snatched from her tomb by the Paduan students and carried off for public dissection. By their remarkable industry they flayed the whole skin from the cadaver lest it be recognized by the monk who, with the relatives of his mistress, had complained to the municipal judge that the body had been stolen from its tomb.

He also reports how, in order to satisfy his curiosity about the fluid in the pericardium, he made sure to be present when a criminal was quartered alive. Then he quickly carried off for study "the still-pulsating heart with the lung and the rest of the viscera." Rumor reported that his greed for specimens sometimes led him to dissect bodies before they were dead.

As he learned from more dissections Vesalius revised his own work. The second edition of the *Fabrica,* twelve years later, carried crucial corrections. While he still evaded the explosive theological question of whether the heart was the site of the soul, he conciliated critics by receiving their corrections sympathetically. When the great Gabriello Fallopio (1523–1562) published a respectful critique of the *Fabrica,* Vesalius went to the trouble of writing a detailed reply, which accepted some of the corrections. His respected teacher Jacobus Sylvius blasted Vesalius for irreverence toward the infallible Galen. But luckily, Vesalius' successors in the influential chair of anatomy at Padua were his own disciples, who answered his call for a fully *human* anatomy.

After the *Fabrica* was published, the young Vesalius impetuously abandoned the study of anatomy for the practice of medicine, and secured appointment as physician to the court of Emperor Charles V. The medical practice to which he was now committed proved quite specialized. Since lechery and gluttony were the prevalent sins at court, Vesalius found himself preoccupied with "the Gallic disease, gastrointestinal disorders, and chronic ailments, which are the usual complaints of my patients." He lived for another twenty years, but he had already done his work.

47

Unseen Currents Within

FOR fourteen centuries Galen dominated European physiology as well as anatomy. His persuasive account of the life process started from the three "souls," or *pneuma,* that Plato said governed the body. The rational in the brain ruled sensation and motion, the irascible in the heart controlled the passions, and the concupiscible in the liver gave nutrition. After being inhaled, air was transformed into *pneuma* by the lungs, and the life process transformed one kind of *pneuma* into another. The liver elaborated "chyle" from the alimentary tract into venous blood carrying the "natural spirit," which ebbed and flowed in the veins with a kind of tidal motion. Some of this natural spirit entered the left ventricle of the heart where it became a higher form of *pneuma,* the "vital spirit." Then the vital spirit was carried to the base of the brain where, in the *rete mirabile,* the blood was transformed into a still higher form of *pneuma,* the "animal spirit." This highest form of *pneuma* was diffused through the body by the nerves, which Galen supposed to be hollow.

Each aspect of the soul possessed its own special "faculty," corresponding to its *pneuma*-producing power. "So long as we are ignorant of the true essence of the cause which is operating," Galen explained, "we call it a *faculty.* Thus we say that there exists in the veins a blood-making faculty, as also a digestive faculty in the stomach, a pulsatile faculty in the heart, and in each of the other parts a special faculty corresponding to the function or activity of that part."

This, in brief, was the grand structure of Galen's physiology, which was essentially a *pneuma*-tology. He explained everything, yet no one could accuse him of pretending to more knowledge than he really had, for he freely admitted the elusive character of all the elements in his system. Appealing to the indefinable to explain the inexplicable, his vocabulary provided an arena of debate for philologically-minded Doctors of Physick.

At the heart of Galen's system was a special theory about the human heart. For the *innate heat,* which, according to Hippocrates and Aristotle, pervaded the whole body and distinguished the living from the dead, had its source in the heart. Nourished by the *pneuma,* the heart naturally was the hottest organ, a kind of furnace that would have been consumed by its own heat if it had not been conveniently cooled by air from the lungs. The heat that came with life itself was therefore *innate,* the hallmark of the soul.

Since the heart was plainly the citadel of Galenic physiology, before doctors could discard their "spirits" and their *pneuma* someone would have to provide another persuasive account of how the heart functioned. This would be accomplished by William Harvey (1578–1657). Born near Folkestone, England, into a family of substance, he would have every advantage that an aspiring physician could have desired. After attending King's School in Canterbury, he went on to Gonville and Caius College at Cambridge.

The college had become a unique center of medical education since it was refounded by John Caius (pronounced "Keys"; 1510–1573), a man of dynamic energy who had championed medical professionalism in an earlier generation. As a student at Padua, Caius actually lived with the great Vesalius, who was still teaching anatomy there. But he remained a devotee of Galen. "Except for certain trivial matters," Caius insisted, "nothing was overlooked by him, and all those things that recent authors consider important could have been learned solely from Galen." As president of the College of Physicians in London, Caius asserted the power of the college to license physicians and to banish quacks. In order to raise the level of medical education he persuaded the judges, following their similar grant to the United Company of Barber-Surgeons in 1540, to provide every year the bodies of four executed criminals for dissection, two of which were to go to his college at Cambridge. While Caius was physician to Edward VI, Mary, and Elizabeth he made a fortune, with which he rebuilt his old Cambridge college as "Gonville and Caius" and funded the first scholarships in the university for the study of medicine.

When the fifteen-year-old Harvey came to Gonville and Caius in 1593, he had been awarded a scholarship in medicine to support him for six years. Then, in 1599, following Caius' own example, Harvey went to Padua, where he won the confidence of his fellow students, and became the representative of the "English Nation" on the university council. Of course, the lectures were in Latin, which Harvey could read and speak. Student life was turbulent without being intellectually exciting. Harvey usually wore a weapon and was "too apt to draw out his dagger upon every slight occasion." But luckily one stimulating professor pointed the way for Harvey's life in medicine.

The famous Fabricius ab Aquapendente (1533–1619), who had once treated Galileo as his patient, was an indefatigable researcher, but still a devotee of Galen. When a group of students rebelled against his mocking manner, he managed to conciliate them with a precious cadaver for their very own dissection. The anatomical theater that Fabricius constructed in 1595 made it possible for the first time to perform teaching anatomies indoors. Five flights of wooden stairs wound up to six circular galleries above a narrow pit. From all these galleries students could lean on the balustrades as they peered down through the darkness at a table in the center where students held up candelabras to illuminate the cadaver as it was dissected. This provided three hundred students at one time with a clear view. The scarcity of cadavers and the infrequency of anatomies made this a signal advance in medical education. Here Harvey witnessed Fabricius' theatrical anatomies. For a while Harvey lived with Fabricius in his country house, with a pleasure garden attached, just outside Padua.

About 1574, long before Harvey came to Padua, in the course of his dissections Fabricius had noticed that the veins of the human limbs contained tiny valves that allowed the blood to flow in one direction only. He noted that such valves were not found in the large veins of the trunk of the body which took blood directly to the vital organs. Fabricius aptly fitted these new facts into Galen's old theories of the centrifugal movement of blood outward to feed the viscera:

> My theory is that Nature has formed them [the valves] to delay the blood to some extent, and to prevent the whole mass of it from flooding into the feet, or hands and fingers, and collecting there. Two evils are thus avoided, namely, under-nutrition of the upper parts of the limbs, and a permanently swollen condition of the hands and feet. Valves were made, therefore, to ensure a really fair distribution of the blood for the nutrition of the various parts. . . .

The memory of these wonderful valves, which Fabricius demonstrated to the young Harvey at Padua, would remain vivid in Harvey's mind to trouble and stimulate him.

When Harvey returned to England, he married the daughter of the doctor who had been physician to Queen Elizabeth, became a Fellow of the College of Physicians and acquired a prosperous, aristocratic medical practice. At the same time he regularly gave the lectures on surgery at the college from 1615 to 1656. And he served as the royal physician, first to James I and then to Charles I, in times when it was politically risky to befriend the king. Harvey's circle included the philosopher-scientist Francis Bacon, the Rosicrucian Robert Fludd, the lawyer John Selden, and Thomas Hobbes, and his interests covered the universe.

Galen had diffused the vital processes into separate organs, each satisfying a particular bodily need. In Galen the blood played no unifying role, for the unity of life processes was in the collaboration of the several "spirits" or *pneuma*. The blood, concocted in the liver, was only another specialized vehicle carrying a nourishing cargo out to certain organs. Harvey went in search of a unifying vital phenomenon. The success of his quest he revealed in his *De Motu Cordis et Sanguinis in Animalibus* (On the Motion of the Heart and of Blood in Animals), a poorly printed tract of seventy-two pages published in 1628.

When we read Harvey's little book today, we are still impressed by its cogency. Step by step he leads us to his conclusion that the heart propels the blood, and that the movement of the blood is circular throughout the whole body. First he marshals the available facts about the arteries, veins, and the heart, their structure and operation. All along the way his observations are "gauged from dissection of living animals."

When Harvey began to study the heart, doctors were not yet agreed on whether the heart was at work when it expanded, which seemed to coincide with the expansion of the veins, or when it contracted. He starts with a rudimentary description of how the heart works.

> In the first place, then, in the hearts of all animals still surviving after the chest has been opened and the capsule immediately investing the heart has been divided, one can see the heart alternating between movement and rest, moving at one time, devoid of movement at another. . . . Muscles in active movement gain in strength, contract, change from soft to hard, rise up and thicken; and similarly the heart. . . .
>
> At one and the same time, therefore, the following events take place, namely, the contraction of the heart, the beat of the apex [of the heart] (felt outside through its striking against the chest), the thickening of the heart walls, and the forcible expulsion of the contained blood by the constriction of the ventricles.
>
> Thus the exact opposite to the commonly accepted views is seen. The general belief is that the ventricles are being distended and the heart being filled with blood at the time when the apex is striking the chest and one can feel its beat from the outside. The contrary is, however, correct, namely, that the heart empties during its contraction. Hence the heart movement which is commonly thought to be its diastole is in fact its systole. And likewise its essential movement is not diastole [expansion] but systole [contraction]; and the heart does not gain in strength in diastole but in systole—then indeed is when it contracts, moves, and becomes stronger.

Harvey goes on to describe the movement of the arteries, how they expand when the heart contracts and pumps blood into them. "An idea of this generalized pulsation of the arteries consequent upon the expulsion of blood into them from the left ventricle can be given by blowing into a glove, and producing simultaneous increase in volume of all its fingers." "Hence," he explains, "the pulse which we feel in the arteries is nothing but the inthrust of blood into them from the heart."

Harvey then traces the blood as it leaves the right chamber of the heart. From the right ventricle the blood is carried through the lungs on its way into the left auricle, and from there it is expelled through the left ventricle. This implied another new notion—the "lesser," or pulmonary, circulation of the blood, the circulation of the blood through the lungs.

This notion, which proved essential to Harvey's larger system, had already been expounded by Realdo Colombo (1510–1559), who was no deep student of Galen but a bold experimentalist, and successor to Vesalius at Padua. The Italian physician and botanist Andrea Cesalpino (1519–1603) had described the cardiac valves and the pulmonary vessels connected to the heart. It also happened that the Spanish virtuoso, Michael Servetus, whom Calvin burned at the stake for heresy in 1553, had incidentally described the pulmonary circulation of the blood in his most culpable theological tract, *Christianismi Restitutio* (1553), of which only a few copies survive. And as early as the thirteenth century the Arab physician Ibn al-Nafis seems also to have had this idea.

It was Colombo who supplied Harvey with the essential facts. Two crucial sets of his observations were missing pieces in Harvey's cardiovascular puzzle. The first was the fact, which had not been known to Vesalius, that the blood passed from the right ventricle of the heart into the left by way of the lungs. The second was the accurate description of the workings of the heart, and the proper meaning of systole and diastole. Colombo insisted that the heart did its work when it contracted, in systole. He listed the rhythm of the heart among "things which are most beautiful to behold. You will find out that while the heart is dilating the arteries are constricted, and again, while the heart is constricting that the arteries are dilated." This simple fact, as Harvey himself noted, gave him a clue he needed on how to use vivisections, rescuing him in "the task so truly arduous and full of difficulties, that I was almost tempted to think, with Fracastorius, that the motion of the heart was only to be comprehended by God."

When Harvey brought together Colombo's insights into the heart's pumping action with Fabricius' descriptions of valves in the veins which permitted the flow in only one direction, he began to see the light. The heart was not a furnace but a pump, and the blood flowed out to nourish the organs. But still other facts were needed to prove the *circularity* of the blood's movement. Harvey had to take the momentous step from the mere *circulation* of the blood—which even Galen had suggested after his fashion —to the *circularity* of the movement, which became the foundation concept of modern physiology. The reasoning that made this step possible was

momentous in a larger sense. It opened the way from qualities to quantities
—from the ancient world of "humors" and vital spirits to the modern world
of thermometers and sphygmomanometers, electrocardiograms, and countless other measuring machines.

Having described the paths of the blood into and out of the heart and the
function of the heart in constantly propelling the flow, Harvey had posed
the essential question. He had found an Amazon within and the force that
kept the current flowing. But he had not yet charted the full course of the
rivers and rivulets of the blood. "The remaining matters, however," Harvey
explained in his crucial Chapter 8, "(namely, the amount and source of the
blood which so crosses through from the veins to the arteries), though well
worthy of consideration, are so novel and hitherto unmentioned that, in
speaking of them, I not only fear that I may suffer from the ill-will of a few,
but dread lest all men turn against me. To such an extent is it virtually
second nature for all to follow accepted usage and teaching which, since its
first implanting, has become deep-rooted; to such extent are men swayed
by a pardonable respect for the ancient authors."

Here he posed a novel *quantitative* question, "How much blood passes
from the veins into the arteries?"—to which he was determined to give a
quantitative answer. "I also considered the symmetry and size of the ventricles of the heart and of the vessels which enter and leave them (since Nature,
who does nothing purposelessly, would not purposelessly have given these
vessels such large size)." By opening the arteries of living animals he hoped
to find his answer. He examined "how much blood was transmitted and in
how short a time." "*So great a quantitie* cannot be furnished from those
things we eat, and . . . it is far greater than is convenient for the nutrition
of the parts [italics added]." If the filling of the bloodstream was being
constantly replenished only by the juice from newly ingested food, the result
would be both the quick emptying of all the arteries and the bursting of the
arteries by the excessive inrush of the blood.

What was the answer? Within the bodily system there was no explanation, "unless the blood somehow flowed back again from the arteries into
the veins and returned to the right ventricle of the heart. In consequence,
I began privately to consider if it had a movement, as it were, in a circle."

This explanation was beautiful in its simplicity. Having in his own mind
confirmed his hypothesis against all the objections that he himself could
raise, Harvey then tried to persuade his colleagues by marshaling ancient
authorities in his support. He quoted at length the sovereign Galen—"that
divine man, that Father of Physicians"—to support his own view of the
relation of arteries and veins to the lungs. He repeatedly and respectfully
cited Aristotle, whose anatomical ideas had been somewhat eclipsed since
the dissemination of printed texts of Galen in the sixteenth century.

Harvey had long felt a kinship with Aristotle's way of looking at living processes. For Aristotle, too, saw life as the single process of the whole living organism—not as something that occurred when "spirits," or *pneuma,* were added to bodily organs. Aristotle's view of the unity of the life process was an incentive to this quest, and finally a justification of Harvey's conclusions. Harvey explained in Chapter 8, where he first expounded the circular movement of the blood:

> We have as much right to call this movement of the blood circular as Aristotle had to say that the air and rain emulate the circular movement of the heavenly bodies. The moist earth, he wrote, is warmed by the sun and gives off vapours which condense as they are carried up aloft and in their condensed form fall again as rain and remoisten the earth, so producing successions of fresh life from it. In similar fashion the circular movement of the sun, that is to say, its approach and recession, give rise to storms and atmospheric phenomena. . . .
>
> This organ deserves to be styled the starting point of life and the sun of our microcosm just as much as the sun deserves to be styled the heart of the world.

We are naturally tempted to seek a connection between Harvey's belief in the circular movements of the blood, with the heart at the center, and Copernicus' heliocentric theory, with planets going around the sun at the center. There is no evidence to support this seductive surmise. Galileo was teaching at Padua when Harvey was a student, but so far as we know, none of his students was a physician. And anyway, in his lectures at that time, Galileo was still faithfully expounding the Ptolemaic system.

Harvey repeatedly insisted that what he described was only simple fact, neither the application nor the embroidering of a philosophy. "I do not profess to learn and teach Anatomy from the axioms of the Philosophers," he explained in the introduction to his *De Motu,* "but from Dissections and the Fabrick of Nature." And near the end of his life he recalled, "I would say with Fabricius, 'Let all reasoning be silent when experience gainsays its conclusion.' The too familiar vice of the present age is to obtrude as manifest truths, mere fancies, born of conjecture and superficial reasoning, altogether unsupported by the testimony of sense."

Still there was a gap in Harvey's circle which he could not bridge. The large quantities of blood were always being speedily propelled from the heart into the arteries, then to the veins, and so back to the heart. But the whole system would not work unless blood was constantly being transmitted from arteries to veins.

Harvey finally had no answer as to how that happened. Yet his faith in the large simple circle of the blood made him confident that the last crucial link in the pathway must be there. He never could find the connecting passageways ("anastomoses," the doctors later would call them), but he expressed his fervent belief that the connection was actually accomplished by some still undiscovered "admirable artifices." Though Harvey occasionally used a magnifying lens, he had no microscope, and that would be required to discover the capillaries. Ultimately, he had to rest his theory on his faith that Nature had not failed to complete the circle.

48

From Qualities to Quantities

THE classic criticism of Harvey's work from the orthodox Galenists was by his contemporary Professor Caspar Hofmann, a noted professor of medicine at the University of Altdorf, near Nuremberg. Speaking for reputable physicians, Hofmann accused Harvey of being reckless of his profession's reputation when he "doffed the habit of the anatomist" and suddenly played the mathematician. "Of a truth, you do not use your eyes or command them to be used, but instead rely on reasoning and calculation, reckoning at carefully selected moments how many pounds of blood, how many ounces, how many drachms have to be transferred from the heart into the arteries in the space of one little half-hour. Truly, Harvey, you are pursuing a fact which cannot be investigated, a thing which is incalculable, inexplicable, unknowable." Harvey's pettifogging quantitative approach, according to Hofmann, had misdirected the whole debate. The proper argument, Hofmann insisted, concerned the grand purposeful scheme of Nature:

I. You would seem to accuse Nature of stupidity in that she went so far astray in a work of almost prime importance, the making and distribution of nutriment! And this being once admitted, what amount of confusion will not follow in all the other works which are dependent on the blood!

II. For that very reason you would seem to condemn the universally accepted maxim concerning Nature, the which you praise with your own words, namely that she is neither deficient in those things which are necessary nor indeed redundant in those which are superfluous etc.

In spite of the Galenists' outrage, Harvey attracted some respectable attention to "pettifogging" questions of quantities. Harvey was not alone.

Others all over Europe were now beginning to speak the language of machines, parsing experience by novel grammars of measurements. Familiar experience was transformed. Nothing was more remarkable than the new way of thinking about heat and cold. Hot and cold, dry and moist, were distinctions obvious to the touch. According to the ancient Greeks, these qualities combined to make the earth, air, fire, and water of which the whole world was made. Just as today we treat odors or tastes as different *kinds* rather than different quantities, so it then was, as we have seen, with temperature. In the English language, for example, before the seventeenth century the word "temperature" (from the verb "to temper," to mix, combine, or keep in due proportion), as we have seen, had a number of meanings, none of which was absolute or quantitative.

So long as medicine was ruled by the Galenic theory of humors, there could be no quantitative way of comparing the internal conditions of bodies against some external norm. The appropriate mixture of humors in a person produced health, disturbance of that mixture produced illness.

The most conspicuous differences of heat and cold were those of climate and weather. The notion that there might be a scale of warmth seems first to have been applied to the weather. It suited the Ptolemaic zones around the earth. Some notion of a scale of temperature, in the modern sense of degrees of heat, appeared even before there was an instrument for its measurement. Galen himself had suggested that four "degrees of heat and cold" might be measured in both directions from a neutral point defined by the mixture of equal quantities of ice and boiling water. He left his definition vague, and, of course, he believed that the heart was the hottest organ of the body.

Until there was a way of measuring the body temperature on a universal scale, it was natural to believe that body temperatures varied in different parts of the world. People who lived in the tropics would have a warmer body temperature than those in colder climates. The earliest European book we know on medical mathematics (*De Logistica Medica,* by Johannis Hasler of Berne, 1578) poses as its very first problem: "To find the natural degree of temperature of each man, as determined by his age, the time of year, the elevation of the pole [that is, the latitude] and other influences." The author provided a chart indicating how much heat or cold could be expected in a person living in a certain latitude so that the physician could adjust the "temperature" of his medicines accordingly.

There were "thermo*scopes,*" devices that indicated a change of temperature, long before there were "thermo*meters,*" which measured that change on a scale. Ancient scientists—Philo of Byzantium (second century B.C.) and Hero of Alexandria (first century A.D.)—showed how heat made water rise, and they suggested an experimental "fountain that drips in the sun."

Though Galileo was probably not the first, we do know that he made a device to measure changes in the temperature of the air. The first recorded use (1633) of the word "thermometer" in English describes it as "an instrument to measure the degrees of heat and cold in the aire."

Variations in temperature were, of course, noted long before variations in barometric pressure, which had to await the discovery that air had weight. Meanwhile the scaling of all such instruments remained confused. In seventeenth-century England the discovery that changes in the air would make a liquid go up and down in a tube produced a "weather glass," which soon became a staple of glassmakers and instrument-makers. In his *Novum Organum* (1620), Francis Bacon described how to make such an instrument.

The unanswerable question of who made the "first" air thermometer confronts us with a menagerie of pseudoscientists, quacks, and mystics. A friend of Harvey's, the amazing Rosicrucian Dr. Robert Fludd (1574–1637) about 1626 modestly disclaimed credit for being the thermometer's first inventor because he had received his philosophical principles from Moses, "figured or framed out by the finger of God." He did boast that he had recovered the idea for the thermometer "in a manuscript of five hundred years antiquity at the least." He himself then "made use of it for demonstration's cause." Even before there was a practical device to measure changing temperature by the rise and fall of liquid in a closed tube, natural philosophers were prepared to harness the thermal movement of the liquid for less prosaic purposes. Salomon de Caus, engineer and architect to the elector palatine, Frederick, at Heidelberg, had a scheme in 1615 to use the phenomenon for a perpetual-motion machine. And on the same basis an enterprising Hollander, Cornelis Drebbel, who had been trained as an engraver, patented (1598) "a watch or time-piece, which may be used for fifty, sixty, yea, one hundred years without being wound up or having anything done to it, as long as the wheels and other works are not worn out." In due course the changes in barometric pressure would be harnessed into elegant and precise modern "atmospheric" clocks.

But Galen's own dogmas could somehow tempt an inventive spirit into the new world of measurement. Just as Columbus went on the course marked by Ptolemy, so Santorio Santorio would follow the paths of Galen. In fact, he believed he had found quantitative techniques that could verify Galen and make the classic scheme even more useful. According to the Galenic classification of diseases, for each person there was a continuous scale of disorders, ranging from precisely the right mixture of humors ("eucrasia") all the way to the worst possible mixture ("dyscrasia"), which caused death. The mathematically minded Santorio calculated that all possible mixtures of the humors came to about eighty thousand, which meant that there must be that many possible "diseases." Before the end of his life Santorio's interest in measuring and counting would lead him far beyond Galen.

Santorio (1561–1636) had the advantage of being born into a well-to-do aristocratic family on an island in the Republic of Venice, where world-ranging commerce, civic pride, and the struggle against papal orthodoxy kept ideas bubbling. Respectable Venetian citizens tried experiments and offered notions which in the dominion of Rome might have required a bold rebellious spirit. His father, a wealthy nobleman, was a chief of ordnance for the Venetian Republic, and his mother was an heiress of noble birth. As was then fashionable, they gave their eldest son his family name for his given name. At the age of fourteen young Santorio Santorio entered the University of Padua, where, according to custom, he first studied philosophy, and then received his medical degree in 1582, when he was twenty-one. He traveled to Croatia, where he served as physician to a noble family. On the Adriatic coast he used the opportunity to test his wind gauge and his device for measuring water currents.

When he returned in 1599 to establish a medical practice in Venice, he enjoyed the effervescent company of artists, physicians, alchemists, and mystics, which included such men as Galileo, Paolo Sarpi, Fabricius, and Giambattista della Porta. The Venetian Republic had been championed against the papacy by the versatile and vigorous Venetian prelate Fra Paolo Sarpi, and when Sarpi was the victim of an assassination attempt, Santorio had his lucky chance. Sarpi had been left for dead, but Santorio and Fabricius treated his wounds so successfully that he recovered, which made Sarpi a prestigious champion of experiment and inquiry in Venice.

Santorio himself believed, quite correctly, that he had invented a new branch of medicine, which he called "Static Medicine," from the Latin *staticus* and the Greek word for the art of weighing. Santorio's *Ars de Medicina Statica* (1612), published in Venice, spread his fame across Europe, not only in Latin but also in English, Italian, French, and German. The Latin version went through twenty-eight editions, and the second edition (1615) was reprinted, with commentaries, at least forty times. Within a century, leading physicians classed it with Harvey's book on the circulation of the blood as one of the two foundations of modern scientific medicine. "No other invention in Medicine," the pioneer zoologist Martin Lister (1639–1712), the leading English physician of the day, declared in the English edition (1676), "except the one of the Circulation of the Blood, is comparable to this." And the great Dutch physician Hermann Boerhaave (1668–1738) proclaimed Santorio's the "most perfect" of all medical books.

The ancient physicians were Santorio's point of departure and he firmly founded his work on theirs. In his early works he aimed "to combat the errors in the art of medicine" by using his own experience to correct the medical works of Hippocrates, Galen, Aristotle, and Avicenna. When he sent *Static Medicine* to his friend Galileo in 1615, in the accompanying letter he explained his two principles. "The first, enunciated by Hippocrates, that medicine is addition and subtraction, adding that which is deficient and taking away that which is superfluous: the second principle is experiment." The ingenious Santorio was confident that he could advance the science of humors into a new quantitative era by his instruments for measuring the phenomena and the qualities inside the human body. He unwittingly invented an arsenal that would eventually conquer the Galenic citadel of humors and qualitatives. The thermoscope, which Galileo and others had used to note changes in the heat of the surrounding air, Santorio adapted to measure thermal changes within the body. The old air thermoscopes consisted of a leaden or glass bulb filled with liquid and attached to a tube in which the liquid visibly rose and fell as the air surrounding the bulb became hot or cold. Santorio modified the device to measure the temperature of a human body. "The patient grasps the bulb," Santorio explained, "or breathes upon it into a hood, or takes the bulb into his mouth, so that we can tell if the patient be better or worse, so as not to be led astray in knowledge of prognosis or cure."

True to his Galenic humoral theory of health and disease, Santorio did not provide any absolute scale of temperature. Anyway, that would have been irrelevant, since the balance of humors in each individual was unique. By the dogmas of Hippocratic medicine, when any "sign" in the body deviated from that particular body's norm, it was a symptom of "disease." Santorio transformed the thermoscope into a thermometer by adding a scale divided into equal units between the temperature of snow and the temperature of a candle flame. This was not to establish a "normal" temperature for all human bodies, but rather to check the variation of each individual's temperature between his body heat when he was in health and when he was diseased. The more the deviation from that individual's norm, the worse the prognosis.

How long should the patient grasp the bulb, or breathe into the hood, or take the bulb into his mouth, to provide a proper measure of his temperature? Santorio instructed: "For ten beats of the pulsiloge." It is not surprising that a friend of Galileo's should have contrived this timing device. Portable clocks were in their infancy, minute hands and second hands were still unknown. And as we have seen, when the young Galileo watched the swinging of the lamp in the cathedral in Pisa, he had reputedly timed it by the beating of his own pulse. Now, in an ingenious reverse application of the principle, Santorio found that a pendulum could be used to time the period of the pulse.

To make such a device, all the physician needed was a string with a weight tied to the end. The physician would shorten or lengthen the string until the period of the pendulum exactly corresponded to the beating of the patient's pulse. Then the length of the string would express quantitatively the pulse rate of that patient. The device was improved by winding the string around a drum and affixing a pointer to the axle of the drum to indicate pulse rate on a dial. The obvious analogy to a clock (horologe) led the device to be called a "pulsiloge."

When Santorio found a knowledge of atmospheric humidity useful in treating diseases, he invented a simple hygrometer. A cord was stretched horizontally on a wall, and from its center a ball was suspended. Increasing moisture in the air tightened the cord and caused the ball to be lifted. The amount of the lift was registered against a vertical scale inscribed on the wall.

The health of a body, the proper balance of its humors, according to Hippocrates and Galen, was an equilibrium between the living body and everything outside. Disease therefore was an imbalance between what the body received and consumed and what the body rejected or extruded. Santorio set himself the task of studying this equilibrium. It proved to be both difficult and unpleasant, for it meant measuring everything that came into or out of his body. For this purpose he constructed his "static chair," which became known as the Sanctorian weighing chair. To a specially designed, delicately calibrated scale (or steelyard) he suspended a chair in which he sat to weigh himself before and after eating, sleeping, exercising, and sexual activity. He weighed the food he ate and also his excreta, and he noted all the variations.

Santorio was founding a modern science of metabolism, the study of the transformations that were the life process. He was so successful in his efforts to use measurements to prove Galen's theories that he ended by destroying the whole Galenic scheme. In Galen's system, hot and cold, dry and moist —the four elementary humors—were different qualities. They were not only objectively real, they were the only important realities for human health and disease. The distinctions between them were absolute. But when heat and cold were measured on the scale of a thermometer, when moist and dry were measured on the scale of a hygrometer, then each of the four qualities became more or less of something else. In the modern physical sciences, therefore, "hot" and "cold" would be merely secondary, subjective qualities, sensed in a particular body under certain circumstances. By transforming the Galenic humors into quantities, Santorio had struck a death blow at ancient medicine.

But Santorio's "Static Medicine" did not stop here. It opened a whole new arena for a world where life processes would be explored and explained by *quantities.* What Santorio's scrupulous observations showed was that when he weighed his food and then weighed all his excreta, the weight of his excreta was a great deal less than the weight of his food. At the same time he found that his own body's weight was a great deal less than was accounted for by all his excreta, including feces, urine, and visible perspiration. There must be some other process disposing of what he ingested. What was it?

Santorio's answer was "insensible perspiration." In his day "perspiration" still carried its original Latin meaning (from *per + spirare = "*to breathe through")—to evaporate, breathe out, or exhale. The basic, food-consuming process in the body was something still unexplained. Santorio began to chart out the dimensions of what needed explaining. When to "perspiration" Santorio added the word "insensible" *(perspiratio insensibilis),* in his time he seemed quite redundant, but he did emphasize that he was not describing any of the visible excretions.

With the enthusiasm of a pioneer, he insisted that the phenomenon he described was quantitatively the most significant of all body processes. As he explained in the Aphorisms of his work on Static Medicine:

Aphorism IV. Insensible Perspiration alone, discharges more than all the servile Evacuations together.

Aphorism V. Insensible Perspiration is either made by the Pores of the Body, which is all over perspirable, and covers the Skin like a Net; or it is performed by Respiration through the mouth, which usually in the Space of one Day, amounts to about the Quantity of half a Pound, as may plainly be made appear by breathing upon a Glass.

Aphorism VI. If eight Pounds of Meat and Drink are taken in one Day, the Quantity that usually goes off by Insensible Perspiration in that time, is five Pounds. . . .

Aphorism LIX. Sixteen Ounces of Urine is generally evacuated in the Space of one Night; four Ounces by Stool and forty Ounces and upwards by Perspiration.

Aphorism LX. There is as much carried off by Insensible Perspiration in the Space of a natural Day, as by Stool in the Course of five Days. Any physician who failed to take account of this process "will only deceive his Patient and never cure him."

The ancient Greek physicians had believed that not only the lungs but the whole body breathed in and out. Galen explained that the purpose of breathing was to cool the flame in the heart, and to produce the natural, animal, and vital spirits that kept the organism alive and growing. Sweat, he said, was a sign of excess fluid in the whole body. The body's health required the proper openness of all the body's openings, especially the skin pores so that the "vapours" from body processes could escape. "Perspiration" was the name for these vapours. Not until the late nineteenth century did it specifically denote droplets of sweat. Since at the time so little was known of the structure of the skin, it was hard to explain how sweat passed out of the body. That puzzle would be solved only after Nicolaus Steno (1638–1686) and Marcello Malpighi studied the skin through a microscope.

This long-recognized process of insensible perspiration was what Santorio was finally putting into quantitative terms. "It is a new and unheard of thing in Medicine," Santorio boasted, "that anyone should be able to arrive at an exact measurement of Insensible Perspiration. Nor has anyone, either Philosopher or Physician, dared to attack this part of Medical Inquiry. Indeed, I am first to make the Trial, and, unless I am mistaken, I have by Reasoning, and the Experience of thirty Years, brought this Branch of Science to Perfection." The science had only begun—and the thermometer, the pulsiloge, and Santorio's own static chair were carrying physicians into new unknowns.

For years Santorio ate and worked and slept in his weighing chair. He was adept at devising other simple tools like his "trocar" (a triangular surgical syringe for extracting bladder stones), as well as complicated contraptions like his bathing bed, in which a patient could be drenched with cold water or warm water to lower or raise his temperature while the room stayed dry. His fellow physicians elected him president of the Venetian College of Physicians, and during the disastrous 1630 epidemic the Venetian Senate put him in charge of measures to combat the plague.

Santorio's mind remained a hodgepodge of the new and the old. While his attacks on astrology stirred the hostility of his colleagues, he supported the Copernican system, agreed with Galileo on astronomy and mechanics, and with Kepler on optics. But he did not understand the large significance of Harvey's discoveries. His extravagant claims for "Static Medicine" as a new technique of Galenic medicine were, of course, ill founded. But his quantitative method, which was his pride and his delight, would make Galen obsolete.

49

"The Microscope of Nature"

MODERN anatomy, as we have seen, progressed when Vesalius and others insisted on studying the human body by dissecting the *human* body. Yet, within a few decades, some unlikely comparisons would reveal the human body in surprising ways. Harvey found clues to the circulation of the blood by his experiments on chickens, frogs, toads, serpents, and fishes. But the circle of Harvey's circulation was not yet complete, and would be completed only by some ingenious observations of "lower" animals, by a new *comparative* anatomy. The range of these comparisons would be far wider, far bolder, and more outlandish than anything Galen had dared.

The hero of this story, Marcello Malpighi (1628–1694), was a great scientist whose work had no dogmatic unity. He was one of the first of a new breed of explorers who defined their mission neither by the doctrine of their master nor by the subject that they studied. They were no longer "Aristotelians" or "Galenists." Their eponym, their mechanical godparent, was some device that extended their senses and widened their vistas. What gave his researches coherence was a new instrument. Malpighi was to be a "microscopist," and his science was "microscopy," a word first noted in English in Pepys' Diary in 1664. His scientific career was held together not by what he was trying to confirm or to prove, but by the vehicle which carried him on his voyages of observation.

Usually called the founder of microscopic anatomy, Malpighi was one of the first of these new-style explorers, diverting attention from the cosmos to the increment, from the universe to the fact. Malpighi's writings might have been called "Voyages with the Microscope," for his work was the miscellaneous journal of a traveler into a world invisible to the naked eye. Vesalius had discovered the large outlines of the Continent of Man, Harvey had discovered the Mississippi. Now Malpighi would describe the topography, the inlets, the rivulets and tiny islets within. No wonder that his work had little theoretical coherence! On such subtly involuted territory the delights of discovery were everywhere.

Two looks through Galileo's telescope, Malpighi exclaimed, had revealed more of the heavens than was seen in all the millennia before. When a critic attacked Malpighi for wasting his time on microscopic minutiae and contrasted Galen's wholesome focus on the large visible shapes, Malpighi was ready with a reply. He noted that Galen, too, had reported the smallest shapes that he could see. "I am no astrologer," Malpighi observed, "so that I can be sure precisely what Galen would have said, but I should think it probable that he would have sung a hymn to God to thank Him for having revealed so many parts, even the smallest, that he had not known about."

Unfortunately, we do not know much about the particular instrument that Malpighi looked through. We do know that he frequently used a microscope with a single lens which he called a "flea glass," and sometimes one with two lenses. He considered his microscopes the essential tools of his research, and in 1684 when a fire consumed his house and all his microscopes in Bologna, he was inconsolable. Trying to repair the loss, the Royal Society in London ordered specially ground lenses to be sent to him, and several scientifically minded noblemen added their own microscopes to the gift.

Malpighi had been amply trained in the professional tradition, and his liberation from dogmatic medicine would take time. Born near Bologna in 1628 into a wealthy family, he graduated from the University as Doctor of Medicine and Philosophy in 1653. There he lectured on logic, then went on as professor of theoretical medicine to the University of Pisa, where he met a professor of mathematics twenty years his senior who would be the great influence in his life. Giovanni Alfonso Borelli (1608–1679) had been born in Naples but studied at the University of Pisa, where Galileo also had been professor of mathematics. Had it not been for Malpighi, the talented Borelli might have remained only a respectable disciple of Galileo and Kepler, who had traced the motions of the moons of Jupiter. Had it not been for Borelli, Malpighi might have remained only another expounder of "theoretical medicine." Borelli was by temperament and training a physicist and a mathematician. "What progress I have made in philosophizing," Malpighi recalled, "stems from him. On the other hand, dissecting living animals at his home and observing their parts, I worked hard to satisfy his very keen curiosity." Malpighi focused Borelli's interest on the subtle movements of living creatures.

After a brilliant and stormy career in the Accademia del Cimento (Academy for Experiment) in Florence Borelli left Tuscany, and became a vagrant member of the new community of European science. The scientific academies focused his interests, fertilized his activities, and provided an eager audience. Borelli became a founder of "iatrophysics" (Greek *iatro* means physician), the application of physics to medicine. He took the physical principles that he had been applying to the motions of fluids, and the eruption of Mt. Etna in 1669, and turned them inward on the human body. In 1675 Borelli joined the new Accademia Reale that Queen Christina of Sweden, after her melodramatic conversion to Catholicism, had set up in Rome. Hoping to be elected to the Royal Academy of Sciences in Paris recently founded by Louis XIV, he offered as credentials his two copious manuscript volumes *On the Movement of Animals,* but since he had only one copy, he dared not entrust it to the uncertain mail from Rome to Paris. It was finally printed in Rome after Borelli's death.

In this work Borelli showed that movements of the human body were like those of all other physical bodies. When a man's arm lifted a weight, the work was accomplished according to Archimedes' familiar principles: the bone was the lever, moved at its shorter segment by the pulling force of the muscle. Movements of the limbs in lifting, walking, running, jumping, and skating also followed the laws of physics. Borelli showed how the very same laws governed the wings of birds, the fins of fishes, and the legs of insects. Having explained the body's "external" motions in his first volume, he proceeded in his second volume to apply these same physical laws to the movements of muscles and the heart, the circulation of the blood, and the process of respiration.

Meanwhile Malpighi was focusing his microscope on the internal organs to discover the subtlety of their structure. As a young medical student at the University of Bologna, he had been powerfully impressed by Harvey's work, which he said signaled "the growing new knowledge of anatomy." He believed that when Harvey explained the function of the heart and the blood, he had given a wonderfully new coherence to all human physiology, and that his experimental technique, the tightness of his logic, and his exclusion of other possible explorations were persuasive. But in Malpighi's time the acceptance of Harvey's doctrine was far from unanimous. Doctors were still bandying about competing doctrines like that of Cesalpino, who had asserted that blood went through the arteries to the outer parts of the body when the animal was awake and returned to the innermost parts through the veins while it slept. Since Harvey, few others had come up with theories about the motion of the blood, "except perhaps those who are so addicted to one sect that there is no hope of their recovering from Galen."

Harvey, Malpighi said, had clearly demonstrated that the blood circulates through the body many times a day. Yet there remained a crucial missing link in Harvey's theory. If so much blood passed through the heart so speedily and the body produced blood so slowly, the blood itself must surely be in a process of renewal and recirculation. The same blood must be in constant movement from the arteries into the veins to keep the life stream ever flowing. It was easy enough for a skilled anatomist to trace the arteries or the veins. But what was their connection? Until that mystery, which had troubled Harvey himself, was solved, there remained room to doubt Harvey.

Malpighi located the mystery in the lungs. And there he would solve the mystery by new techniques of comparative anatomy. In 1661 he announced his discoveries in two brief letters from Bologna to his old friend in Pisa, Giovanni Borelli. These were quickly published in Bologna as a book *On the Lungs* and became a pioneer work in modern medicine.

In the orthodox anatomy of Galen the lungs were supposed to be fleshy viscera, the sources of a hot-humid temperament and a sanguine nature. Malpighi wondered whether that really was their structure. "Since Nature is wont to place in the imperfect the rudiments of the perfect, we reach the light by degrees." By dissecting "lower" creatures and observing through his microscope, he hoped to find new clues to the anatomy of man. Whether by shrewd calculation, scientific intuition, or good luck, Malpighi happened on a place where the missing link in the circulation of the blood could be plainly seen. In his letters to Borelli, Malpighi recalled:

I have sacrificed almost the whole race of frogs, something that did not come to pass even in Homer's savage battle between the frogs and the mice. It was in the dissection of frogs, undertaken with the help of my distinguished colleague, Carol Fracassati, in order to arrive at greater certainty about the membranous substance of the lungs, that I chanced to see such wonderful things that I could with greater aptness than Homer exclaim with him, "Mine eyes beheld a certain great work." Indeed, things show up much more clearly in frogs, because in them this membranous substance has a simple structure and the vessels and almost all the rest of it are transparent, permitting a view of deeper structures.

Now microscopical observation reveals things even more wonderful, for while the heart is still beating . . . the motion of the blood in opposite directions may be observed in the vessels, so that the circulation is clearly revealed. . . .

Since my eye was powerless to see any farther in the living animal, I consequently believed that the blood issues into an empty space from which it is collected by a gaping vessel. Yet a dried frog's lung made me doubt this, for its smallest parts (vessels, as I later learned) had by chance remained blood-red, and with a sharper lens I saw not spots resembling shagreen but small vessels so interconnected as to form rings, and so great is the divarication of these vessels proceeding from a vein on one side and an artery on the other that the condition of a vessel is no longer retained and a rete [network] appears, formed from the branches of the two vessels. This observation I could confirm in the turtle's lung, which is equally membranous and diaphanous.

From this I could clearly see that the blood is divided and flows through tortuous vessels and that it is not poured out into spaces, but is always driven through tubules and distributed by the manifold bendings of the vessels. . . .

To help others verify his findings, Malpighi gave instructions on how to prepare and mount a specimen of the frog's lung on a plate of crystal, how to illuminate it, and then how to observe it either with a "flea glass" of one lens or with a microscope having two lenses.

Even after the frog's pulse had stopped, one could still see the course of the blood. The conclusions for human anatomy and the structure of the human lung were plain.

From analogy, therefore, and from the simplicity Nature employs in all her works, we may conclude . . . that the network I once believed to be nervous is really a vessel intermingled with the vesicles and sinuses and carrying the mass of blood to them or away from them. And although in the lungs of perfect animals a vessel sometimes appears to end and gape in the middle of the network of rings, it is nevertheless probable that, as occurs in the cellules of frogs and turtles, it has minute vessels that are propagated farther in the form of a network, though these elude even the keenest sight because of their small size.

Malpighi had discovered the capillaries. Incidentally he revealed the structure and function of the lungs, opening the way to understand the process of respiration.

By ingenuity, patience, careful laboratory technique, an eagerness to seek analogies, and the industry to accumulate evidence, Malpighi had elaborated a bold new comparative anatomy. What for Galen had been a source of error became for him a resource of knowledge. This new comparative anatomy used what Malpighi called the "Microscope of Nature."

Malpighi showed how endless were these microscopic vistas. In the tongue he noted the budlike taste organs, or papillae, and began to describe their function. He revealed the structure of the glands. He pioneered the anatomy of the brain by observing the distribution of gray matter and the subtle structures of the cerebrum and cerebellum. He discovered the pigmentary layer of the skin. The twentieth-century medical student finds the name of Malpighi also attached to parts of the kidney and the spleen, which he was the first to describe. Finally, he advanced embryology by his ingenious microscopic observations of the development of the chick in the egg. Malpighi went willingly wherever the microscope beckoned, even into the world of the "lower" animals and insects, which Aristotle had not even credited with a full complement of organs. His classic study of the silkworm provided the first detailed treatise on the anatomy of an invertebrate. The silkworm also helped him understand the processes of respiration by the intricate system of tracheae spread throughout its body. With his microscope he compared the cells and the vesicular system of plants to the tracheal system of insects, and so founded phytotomy, the anatomy of plants.

Despite his bias against theory, Malpighi was stimulated to some grand hypotheses about all life processes. What he saw in the texture of wood, in the tracheae of insects, in the lungs of frogs and of men suggested that the more "perfect" is an organism, the smaller proportionately are its respiratory organs. While plants have their respiratory organs spread all over their surface, and the tracheae of insects are spread throughout their body, and fishes have large, extensive gills, man and the other higher animals are served by a comparatively small pair of lungs.

In his century, Malpighi observed, the study of insects, fishes and "the unelaborated outlines of animals" had discovered "much more than was achieved by previous ages, which limited their investigations to the bodies of perfect animals only." The higher animals, he warned,

> enveloped in their own shadows, remain in obscurity; hence it is necessary to study them through the analogues provided by simple animals. I was therefore attracted to the investigation of insects; but this too has its difficulties. So, in the end, I turned to the investigation of plants, so that by an extensive study of this kingdom I might find a way to return to early studies, beginning with vegetant nature. But perhaps not even this will be enough, since the yet simpler kingdom of minerals and elements should take precedence. At this point the undertaking becomes immense, and absolutely out of all proportion to my strength.

Malpighi's strength would be tested, too, by the envy and malice of his medical colleagues, whom he was anxious to refute in advance. "Casting off the dark fog of verbal philosophy and vulgar medicine, which inculcate names alone," Malpighi had set out to test the theories of Galenists and Aristotelians by "sensory criteria." Following the example of Galileo's *Dialogues,* he put his own ideas in the mouth of a "mechanist surgeon," ostensibly refuted by a Galenist and also by a neutral interlocutor. In other ways, too, the example of Galileo offered intriguing parallels. The know-nothings had refused to look through Galileo's telescope or refused to believe what they saw. The "flea glass" was so handy that it was harder for them to refuse to look through it, but they, again, objected that the microscope distorted natural shapes, added colors that were not there, and was a device for counterfeiting reality. Such attacks came from the most respectable sources, and even, to Malpighi's special pain, from his own students.

In 1689, in the awesome presence of ecclesiastical dignitaries, a full-scale formal indictment was pronounced on Malpighi at Rome in the Library of the Friars Servants of the Blessed Mary. Four theses, devised and defended by one of his own students, condemned Malpighi's rash work and pronounced it all useless. First, since "Almighty God has prepared a wonderful domicile in the body for the most noble human soul . . . it is our firm opinion that the anatomy of the exceedingly small, internal conformation of the viscera, which has been extolled in these very times, is of use to no physician." So much for the microscope! Second, to assert "that the humors are separated . . . solely by means of a structure that behaves like a sieve . . . is absolutely untrue." So much for the capillaries and the lungs! Third, although "the anatomy of insects and plants, arrived at by the exquisite resolution of the parts composing them, is certainly an outstanding labor of our times . . . . a knowledge of the marvelous conformation of these entities will not advance the art of curing the sick." So much for comparative anatomy! Fourth, and finally, the *only* anatomy of the human body that is useful is to "learn the difference between the diagnostic and prognostic signs and symptoms, and the positions of the organic parts, by means of which the names of diseases and their periods and their outcomes are known." In with verbal medicine, out with experiment!

The great sadness and frustration of Malpighi's life was not the carping of the dogmatists and the know-nothings, though some of them had been his students. Malpighi had addressed his momentous work to his old friend and colleague Borelli. But in this competitive world of scientific pioneering, in the battle for priority, even the longest and deepest intellectual companionships would be strained to the breaking point. Malpighi angrily broke off his long and fertile correspondence with Borelli in 1668 for the bizarre reason that an acquaintance of Borelli had published a tract (which Malpighi suspected Borelli had had a hand in) disagreeing with Malpighi's new explanation of the function of the dermal papillae. The eloquence of friends did not succeed in reconciling them. Borelli and his old friend Malpighi became locked in an acrimonious quibble over which of them had first noted the "spiral structure" of the fibers of a macerated beef heart. Malpighi claimed that he had seen them first and then pointed them out to Borelli. "I first happened to see this marvellous structure," Borelli counterclaimed, "at Pisa in 1657 while the most eminent Malpighi was standing by." In 1681, when Borelli's great work *On the Movement of Animals,* which Malpighi had helped inspire, finally appeared, though Borelli was now dead Malpighi churlishly attacked the book as a shameless effort by an old friend to "invalidate" his work.

When Malpighi died, his own colleagues gave him a more charitable verdict. In 1697, the Transactions of the Royal Society published an obitu-

ary letter from the Professor of Anatomy at Rome. "The Incomparable
Malpighi," the professor wrote, "who naturally applied himself only to
serious Studies, which he seldom interrupted, and that against his will, to
take some Recreation; had employed all his time to discover New Worlds
by Anatomies, and to refute (in imitation of Great Men) by his Vertue and
great Learning, the Calumnies of the Envious." Appropriately, the obituary
carried a detailed account of the autopsy on Malpighi's body—a deformity
of the right kidney, a heart that was "bigger than ordinary," and the broken
blood vessel that caused the stroke that brought on his death. It concluded
that few, if any, men of the day had so richly contributed to the "Commonwealth of Learning."

PART ELEVEN

SCIENCE
GOES PUBLIC

As to science itself, it can only grow.

—GALILEO, *Dialogue* (1632)

50

A Parliament of Scientists

"TRUTHS," Descartes observed, "are more likely to have been discovered by one man than by a nation." The generations that produced Galileo, Vesalius, Harvey, and Malpighi needed new forums of science bringing together the truths discovered by individuals for mutual enrichment and for other discoverers everywhere. The communities of science became parliaments of scientists conducted in the vernacular languages. What was offered did not have to be part of a grand scheme of meaning. It was enough that it be "interesting," unusual, or novel. The edges were fuzzied between science and technology, between the professional and the amateur. Out of a new mechanics for information exchange came a new incremental concept of science.

The parliaments of scientists would need a new kind of scientific statesman or politician with a knack for stimulating, cajoling, and conciliating. A friend of the great and the ambitious, yet he could not be a rival for their fame. He had to be at home in the main vernacular languages, since by the sixteenth and seventeenth centuries few men of science spoke any language but their native tongue, and many leading scientists were no longer writing their works in Latin.

A very model of this new Man of Science was Marin Mersenne (1588–1648). Born into a laboring family in northwestern France, after attending a Jesuit college and studying theology at the Sorbonne, he joined the recently founded Franciscan order of Minims, which was even stricter than other Franciscans in its rule of humility, penance, and poverty. Mersenne entered the Minims monastery in Paris near the Place des Vosges where, except for short trips, he lived till his death. His personal charm made his monastery a center of scientific life for Paris, and he helped make Paris the intellectual center of Europe. In the beginning Brother Marin seems to have been as much concerned to preserve religion as to advance science. According to his own estimate, there were fifty thousand atheists in Paris alone. Against these "atheists, magicians, deists, and suchlike" he saw the new discoveries of science confirming the truths of religion. There in the house of the Minims, Mersenne gathered some of the liveliest, most inquisitive minds of his day, and not only from France. His regular conferences included Pierre Gassendi (friend of Galileo and Kepler), the Descartes (father and son), and many others. Mersenne's correspondence reached from London to Tunisia, Syria, and Constantinople, bringing together the latest notions and discoveries of Huygens, van Helmont, Hobbes, and Torricelli. It was there in Mersenne's cell that Pascal first met Descartes. Mersenne's gentleness and generosity made him the perfect go-between in a community of irascible and vitriolic pundits, among whom he was said to have no enemies—except that impossible English mystic Robert Fludd. No prima donna himself, he secured their confidence and they sought his advice.

Mersenne drew far-separated thinkers into his net of international correspondence. To the Italian disciples of Galileo he explained that Galileo had really not been condemned for heresy. While he brought out a French version of unpublished works of Galileo, he was still wary of endorsing the new astronomy. An English friend requested "any new observations, magnetical, optical, mechanical, musical, mathematical" that Mersenne might have from Italy or from Paris, and at the same time reported that he in turn would soon be sending a short treatise on the Roman system of measurement and another on the pyramids of Egypt. He also promised to keep Mersenne informed of an Irishman's invention rumored "to write in such a way that the message may be read at once in all languages." Mersenne sent out word of Parisian experiments with telescopes and a new definition of the problem of the cycloid, reported on the chemistry of tin, and spread news of a "sensitive plant" from the West Indies. When foreign intellectuals visited Paris, they called on Mersenne, joined a conference on some subject of interest, then returned to Rome, Altdorf, London, or Amsterdam, where they remained members of Mersenne's network. Mersenne expressed his own credo in a miscellany of "Questions—Theological, Physical, Moral, and Mathematical."

Mersenne's spirit was abroad in varied incarnations. An informal group of literary men was incorporated by Cardinal Richelieu into the French Academy in 1635. A quite different academy was organized by the wealthy Parisian Henri-Louis Habert de Montmor, and its members gathered in his mansion to air their scientific concerns. "The purposes of the conferences," the Montmor Academy's constitution of 1657 declared, "shall not be the vain exercise of the mind on useless subtleties, but the company shall set before itself always the clearer knowledge of the works of God, and the improvement of the conveniences of life, in the Arts and Science which seek to establish them."

Mersenne had developed an especially active exchange with England, importing English books and supplying French books for English scientists. There he inspired another, more formal parliament of scientists. The man who brought it all together was the little-celebrated Henry Oldenburg (1617?–1677), who was not among the great scientific minds of his generation, but who had a talent for organizing and inspiring those who were.

Born in the busy city of Bremen, son of a professor of medicine and philosophy, he studied Latin, Greek, and Hebrew, received the degree of Master of Theology, then went on to the University of Utrecht. During the next dozen years, as tutor to young English noblemen he visited France, Italy, Switzerland, and Germany, becoming fluent in French, Italian, and English, in addition to his native German.

Sent to England to persuade Oliver Cromwell to allow Bremen to continue her trade during the Anglo-Dutch Wars, he actually secured Cromwell's aid in negotiating Bremen's independence of Sweden, and kept open the city's thriving commerce. Meanwhile, still in his mid-thirties, Oldenburg had come to know the leading English thinkers, including John Milton, Thomas Hobbes (1588–1679), and most important, Robert Boyle (1627–1691). "You have indeed learnt to speak our language," Milton wrote, "more accurately and fluently than any other foreigner I have ever known." Oldenburg made his way not through the power of ideas but through the fluency and personal charm that would be essential for scientific diplomacy. Robert Boyle's sister, Lady Ranelagh, attracted by the knowledgeable young Oldenburg, engaged him to tutor her son, and when Oldenburg accompanied Richard Jones to Oxford in 1656, he met the scientists gathered around Boyle, including John Wilkins, the versatile mathematician-astronomer, and the others who would become the nucleus of the Royal Society.

Oldenburg was dazzled by his glimpse of the new science. "I have begun to enter into companionship with some few men who bend their minds to the more solid studies, rather than to others, and are disgusted with Scholastic Theology and Nominalist Philosophy. They are followers of nature itself, and of truth, and moreover they judge that the world has not grown so old, nor our age so feeble, that nothing memorable can again be brought forth." Boyle himself had already christened this informal companionship of scientific enthusiasts wherever they were as the Invisible College.

Fresh with this enthusiasm, in 1657 Oldenburg took his young charge—now no mere Richard Jones but Lord Ranelagh—on a tour of the Continent. Ranelagh's rank made them welcome in the salons of French scientists and amateurs. Oldenburg's visit to Paris at this time was providential. The "invisible colleges" were flourishing, and the spirit of Mersenne, who had died ten years before, was still very much alive. Oldenburg brought Ranelagh along to meetings of the Montmor Academy, where they joined in the conversations on everything under the sun. "Each of the members of the company is obliged to treat of a certain topic, either physical, medical, or mechanical. Among these topics are several very fine and remarkable, such as The Source of the Variety of Popular Opinions, The Explanation of the Opinions of Descartes, The Insufficiency of Movement and Figure to explain the Phenomena of Nature (undertaken to be proven by an Aristotelian). Then, of the Brain, of Nutrition, of the Use of the Liver and Spleen, of Memory, of Fire, of the Influence of the Stars, If the Fixed Stars are Suns, If the Earth is animated, of the Generation of Gold, If our Knowledge Springs from the Senses, and several others, which I do not at this moment remember." But Oldenburg noted that "the French naturalists are more discursive than active or experimental. In the meantime the Italian proverb is true: *Le parole sono femine, li fatti maschii* [Words are feminine, facts masculine]."

When Oldenburg returned to England, he had visions there of a more "masculine" community of science. Arriving in London just in time to witness the entry of Charles II, he found hope that a renewal of science would come with the restoration of order and monarchy. Under the King's patronage a group of English men of science met in Gresham College on November 28, 1660, to found a new academy for the advancement of the sciences. "It is composed of extremely learned men," Oldenburg recounted, "remarkably well versed in mathematics and experimental science." The president of the new society was the versatile John Wilkins, just named Dean of York, whom Oldenburg had come to know during his brief stay in Oxford, and Robert Boyle was one of the leaders. Though not at the founders' meeting, Oldenburg was on the first list of those proposed for membership in December, and then duly admitted. Early in February 1661 he was named to a committee "for considering of proper questions to be enquired of in the remotest parts of the world."

When Charles II chartered the Gresham College group as the Royal Society in 1662, he gave Oldenburg the opportunity of his life. "This Curious German," a knowledgeable French visitor reported, "having well improved himself by his Travels, and pursuant to the Advice of Montaigne, rubbed his Brains against those of other People, was upon his Return into England entertained as a Person of great Merit, and so made Secretary to the Royal Society." Technically John Wilkins was "first secretary" and Oldenburg was only "second secretary," but he played the leading role until his death. While others happily provided the scientific observations Oldenburg organized them all into a uniquely productive new parliament of scientists.

The company, no longer confined to eminent and respectable residents of one capital, became an "invisible college." To be heard at the Royal Society in London, it was no longer necessary to attend a meeting. John Beale could write from Herefordshire in the west of England describing problems of orchards, advising the best way to prepare cider, and offering his harebrained cure-alls for farmers' ills. Nathaniel Fairfax from Suffolk reported on some people who ate spiders and toads. But the list also included John Flamsteed, who wrote from Derbyshire on astronomy, and

Martin Lister, who wrote from York on biology. Of course there were frequent communications from Boyle and Newton.

Oldenburg's wide acquaintance and his knowledge of languages paid off. The current of correspondence widened, and, along with the books sent in, letters provided subjects for the society's weekly meetings. In 1668 Oldenburg reported that his job as Secretary was to ensure the performance of the experimental tasks recommended, to write all letters abroad, and carry on regular correspondence with at least thirty foreign scientists, taking "much pain in inquiring after and then satisfying forrain demands about philosophicall matters."

By this time letters were already a familiar form of communication among scientists. In Paris, for example, scientists would put their ideas in a letter to a friend, pay to have it printed, and then send out hundreds of copies. To keep posted on new inventions and discoveries, they wanted correspondents in other centers of learning. Few could manage this on their own, and those who could ran heavy risks. In an age of continual warfare a latent ambiguity or a thoughtless phrase could land a natural philosopher in jail for treason, when all he wanted was further observations of the rings of Saturn, news of experiments in blood transfusion, or the description of an exotic insect. In 1667 Oldenburg himself was suddenly imprisoned in the Tower of London for a few ill-considered words in a scientific communication which the Secretary of State resented as a criticism of his conduct of the Anglo-Dutch War.

A letter had obvious advantages over a book. While works of science were often large tomes easy to stop for censorship, the novel observations in a letter could slip in unnoticed or be delivered with the "ordinary post." There still was no regular "parcel post," but even in the seventeenth century the "ordinary post" might go once a week between London, Paris, and Amsterdam. Yet it was heavily dependent on weather and on political conditions, it was erratic, costly, and went only to nearby destinations. The enterprising Oldenburg developed a more extensive and more reliable service. As agents he enrolled young members of the staff of British embassies, who would post their reports through diplomatic channels to a code address concocted from his name, "Grubendol, London." Once there, in the office of the Secretary of State, they would be forwarded to Oldenburg, who in return obliged the Secretary of State by supplying any political news that might happen to be included.

When Oldenburg became Secretary of the Royal Society, the rudimentary British postal service was still very much an organ of national security, serving as an agent of censorship as well as of counterespionage. All unlicensed carriers had been suppressed. An Act of 1711 would describe postal charges as taxes, to help pay for Britain's interminable wars. Not until nearly the end of the eighteenth century were mounted "postboys" supplanted by the famous mail coaches. Meanwhile Oldenburg used all available means to open channels of scientific communication from London to the nation and the world.

The letter, which remained for centuries the most "swift, certain, and cheap" vehicle of long-distance communication, also expressed a new attitude toward science and new hopes for technology. A letter was suited to communicate a fact or a small cluster of facts. It signaled an incremental rather than a cosmic approach to experience. The printed scientific "paper" or "article," which was simply a later version of the letter, would be the typical format in which modern science was accumulated and communicated. This form, and the attitude that led scientists to be engrossed with it, declared the appearance of the experimental scientist, in place of the "natural philosopher." The letter was an ideal vehicle for the increasing numbers of men dispersed over Europe who no longer expected to storm the citadel of truth, but hoped to advance knowledge piece by piece.

Even without the instructions of the society, Oldenburg wrote to anyone who he suspected had or was able to find a bit of novel scientific information. Sometimes he prodded the society to instruct him to start an official correspondence. For example, he initiated an exchange with Johannes Hevelius (1611–1687), whose notes on a solar eclipse, seen from the observatory he had built with the profits of his brewery, along with his map of the surface of the moon, were published by the society. From this English connection Hevelius received the special lenses he wanted for his observation, and his telescope design reached across Europe. The reports Oldenburg received from French physicians kept English doctors up to date on the acrimonious French debate over blood transfusion.

Letters came in all the main European languages. The amateur Leeuwenhoek, who did not know Latin, wrote in his native Dutch. Oldenburg would summarize such communications or translate them into English, from which, in turn, the French might translate them for their own publications. Ignorance of Latin no longer excluded an inventive Delft draper or anyone else from the community of scientists.

Yet for the world of science the rise of the vernacular languages was a mixed blessing. It set up new barriers. So long as Latin was the universal language of European science, as it had been through the sixteenth century, a printer of works in Latin could expect a wide sale even of costly technical or heavily illustrated books. The spread of literacy and the rise of vernacular languages, which came with the printing press, proportionately reduced the market for books in Latin. The new markets were narrowly regional. Even in Italy the scientific community would not read a book in Latin if there was one in Italian. They "love every whit as well to read books in Italian,"

Oldenburg reported to Boyle in 1665, "as the English doe to read them in English." Of course, this widened the opportunity for public education and opened an audience for works of popular science. But it also made new problems for men of science. The standard Latin vocabulary, which still survives in our nomenclature of botany and zoology, was increasingly confused by colloquialisms. While Latin had once sufficed for the serious scholar of European science, he now had to read a half-dozen vernaculars. And he still would be much less certain of what was being said. As national communities of literacy were created in the languages of the marketplace, the international community of learning was dissolved, or at least attenuated. Gradually mathematics and universal measurements would provide a new laboratory language. But mathematics dealt only in quantities.

These multiplying vernaculars created a special need for a network of correspondence. No longer was it enough to be in touch with Venice or Paris, or the other centers of Latin publication. Now there was always the extra problem and the cost of translation. Oldenburg tried to leap the new language barrier by promoting translations into French and English. He also tried to reach the remnant of the old ecumenical audience by arranging translations (for example, of some of Boyle's English works) into Latin.

In Oldenburg's day most English Fellows of the Royal Society still read Latin. Newton wrote Latin as well as English, but few were at home in any vernacular except their own. Robert Hooke was said not to believe anything that was written in French. And French scientists were generally ignorant of English. German was only beginning to be elaborated into a learned language. All this made the brief communication, the letter, especially convenient, economic, and useful. When a communicant wanted to spread news of his latest observation or invention, no considerable investment was required either by him or by a printer, as would have been the case with a book. Here, too, was a possible escape from some of the political and religious controls over the conspicuous treatise.

The enterprising Oldenburg, combining the fragmentary informal character of the letter with the outreach of the printed word, invented the profession of the scientific journalist. A new literary genre, scientific journalism would convey some of the most momentous news of modern times.

At first Oldenburg was not paid. Then in December 1666 the Council of the Royal Society voted him forty pounds for all his work during the previous four years, and two years later he was given an annual salary of forty pounds and the help of an amanuensis. Meanwhile, Oldenburg conceived the idea of collecting and publishing the correspondence, which was all considered to be his property. On March 6, 1665, he opened a new era in science when he published the first issue of *Philosophical Transactions: giving some Accompt of the Present Undertakings, Studies, and Labours, of the Ingenious in many Considerable Parts of the World.* The *Journal des Sçavans,* which had appeared in Paris two months earlier, is sometimes awarded priority as a scientific periodical, but it was given over to book reviews and literary matters. When its Jesuit enemies forced it to become bland, it was discontinued in 1668.

From its very beginnings, Oldenburg's *Phil. Trans.* (as it was familiarly known) had a grand purpose. As he declared in his Introduction to Number 1:

> Whereas there is nothing more necessary for promoting the improvement of philosophical Matters, than the communicating to such, as apply their Studies and Endeavours that way, such things as are discovered or put in practice by others; It is therefore thought fit to employ the *Press,* as the most proper way to gratifie those, whose engagement in such Studies, and delight in the advancement of Learning and profitable Discoveries, doth entitle them to the knowledge of what this Kingdom, or other parts of the World, so, from time to time, afford, as well of the Progress of the Studies, Labors and attempts of the Curious and Learned in things of this kind, as of their complete Discoveries and Performances: To the end, that such Productions being clearly and truly communicated, desires after solide and useful knowledge may be further entertained, ingenious Endeavours and Undertakings cherished, and invited and encouraged to search, try, and find out new things, impart their knowledge to one another, and contribute what they can to the Grand Design of improving Natural knowledge, and perfecting all Philosophical Arts, and Sciences. All for the Glory of God, the Honor and Advantage of these Kingdoms, and the Universal Good of Mankind.

The publication of this first scientific journal was interrupted only twice during Oldenburg's lifetime—once briefly by the plague when the issues came from Oxford instead of London, and then when Oldenburg was put in the Tower of London for his few ill-considered words.

While the *Philosophical Transactions* realized Oldenburg's hopes in ways he could never have imagined, the money rewards were meager. Each monthly issue of about twenty pages printed in twelve hundred copies returned little more than the cost. The enterprise, as Oldenburg's dedication to the society showed, was very much his own, and not until the mid-eighteenth century was publication officially assumed by the society. The *Phil. Trans.* became a model for modern scientific publications. "If all the books in the world except the *Philosophical Transactions* were destroyed," Thomas Henry Huxley observed in 1866, "it is safe to say that the foundations of physical science would remain unshaken, and that the vast intellectual progress of the last two centuries would be largely, though incompletely, recorded."

In retrospect it is easy to forget that the Royal Society was a company of pioneers. When science was still securely tied to religion, novelty had the stigma of heterodoxy. In the early years the defense of the Royal Society was less a catalogue of its useful work than an effort to prove that the work of the society was really innocent. When Bishop Sprat published his copious *History of the Royal Society* (1667), he devoted a third of the book to prove "that the promoting of Experiments, according to this idea, cannot injure the Virtue, or Wisdom of Mens minds, or their former Arts, and Mechanical Practices; or their establish'd wayes of life: Yet the perfect innocence of this design, has not been able to free it from the Cavill of the Idle, and the Malicious; nor from the jealousies of the particular Professions, and Ranks of Men, I am now in the Last place to remove; and to shew that there is no Foundation for them."

In the long run the defenders of novelty would win. One of the most eloquent, the maverick English clergyman Joseph Glanvill (1636–1680), whose own writings included a defense of the preexistence of the soul and a treatise on the menace of witchcraft, boasted in 1668 that this "great ferment of useful and generous knowledge, makes a bank of all useful knowledge, and makes possible the mutual assistance that the practical and theoretical part of physics affords each other. . . . the Royal Society . . . has done more than philosophy of a notional way since Aristotle opened shop."

51

From Experience to Experiment

THE Royal Society's motto, *Nullius in Verba,* has been best translated, "Take nobody's word for it; see for yourself." The new currency of knowledge was the product of a special form of experience, to be known as experiment. While the older language of science aimed at meaning and certainty, the new language aimed at precision.

The intention of the Royal Society, Bishop Sprat explained, was "not the Artifice of Words, but a bare knowledge of things." At that moment in British history, the voluble Puritans, despite their professed aim at a "plain style," had given eloquence a bad name. To many their lengthy florid sermons and their parliamentary bombast seemed fuel for civil disorder.

Their "superfluity of talking" had led Bishop Sprat and other respectable Fellows to declare "that eloquence ought to be banish'd out of all civil Societies, as a thing fatal to Peace and good Manners." To reform ways of speaking would refresh ways of thinking.

The Royal Society, hoping to accomplish this, therefore "exacted from all their members, a close, naked, natural way of speaking; positive expressions; clear senses; a native easiness: bringing all things as near the Mathematical plainness, as they can: and preferring the language of Artizans, Countrymen, and Merchants, before that, of Wits, or Scholars." The "Universal Temper" of the British, Sprat boasted—"our climate, the air, the influence of the heaven, the composition of the English blood; as well as the embraces of the Ocean"—all tended "to render our Country, a Land of Experimental Knowledge."

It was not enough that the language of science be simple. It had to be precise—and, if possible, international. Sprat was on track when he called for "Mathematical plainness." The difference of language would be a clue to the difference between Experience and Experiment. Experience was always personal, and never precisely repeatable. The travels of Marco Polo, the voyages of Columbus and Magellan, were experiences to be recounted, and enjoyed in the reading and hearing. In the new world of "Experimental Knowledge" this was not good enough. To be an experiment an experience had to be repeatable.

The Fellows of the Society, with Sprat, were committed even when word of an experiment came from a remote place to bring it "within their own Touch and Sight." They laid it down "as their Fundamental Law, that whenever they could possibly get to handle the subject, the Experiment was still perform'd by some of the Members themselves. The want of this exactness, has very much diminish'd the credit of former Naturalists." Earlier collections, the miscellaneous experiences of naturalists, had been random, often frivolous, and sometimes purposely misleading. Now cast in the rigorous form of experiment, experience could be coordinated, confirmed, and added piecemeal to the stock of knowledge. To make experience into experiment for scientists everywhere, there had to be a universal language of calculation and of measurement.

Mathematics would be the Latin of the modern scientific world and, like Latin, would overleap vernacular barriers. The common measurements since ancient times had been the product of usage in the local market. They came from the bodily measurements which were available everywhere. The "digit" was the width of a finger, the "palm" was the width of four fingers, the "cubit" was the distance between the elbow and the tip of the middle finger, the "pace" was one step, and the "fathom" the distance between outstretched arms. By such "rules of thumb" it had been possible to build a Great Pyramid, with a disparity in the length of the sides of only one part in four thousand.

In England the early development of a strong central government tended to produce common standards. The early Tudors established a "furlong" (i.e., a "furrow-long") as 220 yards. Then Queen Elizabeth I decreed that the traditional Roman mile of 5,200 feet should instead be 5,280 feet, precisely eight furlongs, and therefore more convenient for everyday use. Even so, the variety of common units caused daily inconvenience and was an incentive for fraud. After Saxon times the "pound" was usual for weight and for money, but there were at least three different pounds in general use. Weight could be measured also by the clove, the stone, the hundredweight, or the sack, and for capacity there were the pottle, the gallon, the bushel, the firkin, the stake, and the cartload. Every trade had its own vocabulary. For apothecaries there were minims and drams, for seamen fathoms, knots, and cable lengths. A gallon of wine was not the same measure as a gallon of ale. A bushel of wheat was sold rounded or "heaped," but corn was leveled.

Elsewhere in Europe the practice was not much simpler. A dictionary of local units of weights and measures used in France before the Revolution comes to two hundred printed pages. The chaos and the local variety everywhere expressed the variety of needs.

"Next to the inconvenience of speaking different languages," James Madison observed in 1785, "is that of using different and arbitrary weights and measures." An international language of mathematics that would be useful to scientists confirming one another's experiments would have to provide a convenient way of expressing and dividing the smallest fractional units. The hero of this effort was a phenomenal late-blooming Belgian tradesman, Simon Stevin (1548–1620). Born in Bruges, the natural son of two wealthy citizens, he did not enter the University of Leyden till his mid-thirties. He was celebrated in his lifetime for his "Sailing Chariot," an amphibious boat, in which Prince Maurice and his party of twenty-eight sped along the seashore near Scheveningen "flying in two hours . . . to Petten at a distance of fourteen Dutch miles." One of the passengers on this historic excursion, Hugo Grotius (1583–1645), founder of modern international law, solemnly memorialized the adventure in a Latin poem, *Iter currus veliferi.* The chariot attained literary immortality in Sterne's *Tristram Shandy,* and guests of the House of Orange enjoyed riding it until the very end of the eighteenth century.

Stevin's other works were more practical. His *Table of Interest Rates* (1582), published in Antwerp by Christophe Plantin (1520?–1589), who had worked with Ortelius on his *Picture of the World* and who was famous for his eight-volume Polyglot Bible, marked a new era in banking. Before then there had been interest tables, but like the maps of the best trade routes, these were kept secret by bankers, who guarded them as valuable capital equipment. Now Plantin's handsomely printed tables could be bought on the open market, with rules for computing simple and compound interest along with tables for speedily figuring discounts and annuities.

When Stevin was appointed tutor to Maurice of Nassau (1567–1625), Prince of Orange, the military genius of the age, he prepared his revolutionary textbook, *The Art of Fortification* (1594), which replaced the designs made against the bow and arrow with new angles to defy firearms. The versatile Stevin wrote a book on astronomy (1608) supporting Copernicus (even before Galileo), a treatise on perspective, handbooks on mechanics, texts on navigation and how to determine longitude, an improved scheme for steering a ship along a loxodrome to keep its bearings, and a book on the theory of musical tuning with the scale of "equal temperament," a design for a mechanically driven spit to roast meat which used his own theory of the parallelogram of forces, and a manual advising citizens on how to survive periods of civil disorder. His motto was "What seems a wonder is not really a wonder."

But his greatest invention was so simple that we can scarcely believe that it ever had to be invented. In a 36-page booklet, *The Tenth* (1585), which Plantin published in Leyden, Stevin offered the decimal system. The English translation in 1608 first introduced the word "decimal" into our language. Earlier systems for handling fractions had been cumbersome. Stevin's solution was to treat all fractional units as integers. Take, for example, the quantity 4 and 29/100. Why not, Stevin asked, simply treat this as 429 items of the unit 1/100? Simply reduce the size of the designated unit to the *smallest* quantity under consideration, and then treat both the integer and the fraction as multiples of that. Experimenters now could deal exclusively in whole numbers.

For everyday use Stevin showed how his decimal system would simplify the problems of merchants and their customers, of bankers and borrowers. Decimals could be used too for weights and measures and coinage, even for divisions of time and degrees of the arc of a circle. Stevin showed the advantages of "tenth" numbers for surveying, for measuring cloth and wine casks, for the work of astronomers and mintmasters. And he explained the advantage of grouping soldiers in units of 10 or 100 or 1,000.

Stevin did not think of a decimal point. Instead he suggested that after his "unit of commencement," the integer, each succeeding digit be distinguished with a sign (1, 2, 3, etc.) above or beside it, signaling that the units were tenths, hundredths, thousandths, etc. For others it was an easy step from superscripts to the decimal point. John Napier (1550–1617), the Scottish mathematician and an inventor of logarithms, by inserting a "decimal point," assimilated the whole system into the Hindu-Arabic positional scheme, and so made decimal fractions still more appealing for daily use.

The enthusiastic Stevin urged that all sorts of computation, even the degrees of an arc and the units of time, be revised into his decimal system. But the sexagesimal system, rooted in earliest antiquity and sanctified by the perfect circle and the heavenly movements, could not be displaced— from astronomy, from the circle, or from the units of time so closely related to them.

When Galileo saw the connection between the period and the length of a pendulum, he opened the way to use time as a basis for a uniform measure of space. When Christiaan Huygens, as we have seen, invented the pendulum clock, he began to accomplish this. In the long run the search for a common measure of *time* would advance the quest for other universal units, and in this sense, too, the clock would be the mother of machines. For some reason or other, Gabriel Mouton (1618–1694), a priest in Lyons who never left his native city, became obsessed by the quest. Studying the period of a pendulum, he found, to his astonishment, that the length of a pendulum with a frequency of one beat per second would vary with the latitude. He then suggested that these variations could be used to calculate the length of a degree of the terrestrial meridian. A fraction of time, or one "minute" of a degree, could become a universal unit of length.

The effort to use the pendulum, along with a simplified and comprehensive decimal system, to define a universal unit of measure would eventually bear fruit. In April 1790 Talleyrand (1754–1838) called on the National Assembly of the French Revolution to devise a national (he hoped it would become international) system of weights and measures based on the precise length of a pendulum that beat one second at 45 degrees latitude in the very middle of France. To make the measurements and calculations required for this purpose, the assembly decreed:

> The King shall also beg His Majesty of Britain to request the English Parliament to concur with the National Assembly in the determination of a natural unit of measures and weights; and . . . under the auspices of the two nations the Commissioners of the Academy of Sciences of Paris shall unite with an equal number chosen by the Royal Society of London . . . to deduce an invariable standard for all the measures and all the weights.

Luckily, the French Academy did not wait for the Royal Society to join them, for the British never did. Instead, the French went ahead on their own, recommending that the new units be based on decimals, and that the basic unit should be one ten-millionth of the length of a quadrant of the earth's meridian (i.e., of the length of an arc between the equator and the North Pole). Soon this unit would be christened a "meter," from the Greek word for measure, and from it all other metric units would be derived. A cube one meter on each side would be the measure of capacity, and the cube filled with water would be the unit of mass. There was a natural constant base for the whole system in the one-second pendulum, which could be made to serve for every kind of quantity, all expressed in multiples of ten.

Thomas Jefferson (1743–1826), too, was eager for projects to bring humankind together by science. The federal Constitution (Article 1; section 8) had given the Congress of the new United States power "to fix the Standards of Weights and Measures," and at great personal pain Jefferson produced his *Report . . . on the Subject of Establishing a Uniformity in the Weights, Measures and Coins of the U. S.* (1790). Jefferson did not see Talleyrand's proposal until he had already published his own, which started from the need for uniform weights and measures to unify a nation. For the calculations he consulted his friend David Rittenhouse, the leading American mathematician. And Jefferson led the way toward a decimal monetary system. Enforcing a new scheme of weights and measures, he lamented, would be more difficult.

For the measurement of length Jefferson sought some universal standard, preferably a unit found in nature. But he was troubled by the fact that changes in temperature affected the length of objects, and so he proposed to base his system on time and motion. The rate of the rotation of the earth on its axis was presumably uniform and was equally accessible everywhere. Following the traditions of Stevin, Galileo, and Huygens, Jefferson fixed on a pendulum. "Let the standard of measure," he reported, "then be a uniform cylindrical rod of iron, of such length as, in latitude 45°, in the level of the ocean, and in a cellar, or other place, the temperature of which does not vary through the year, shall perform its vibration in small and equal arcs, in one second of mean time." At first he had named the latitude of 38 degrees (through the heart of Virginia), but when Talleyrand proposed 45 degrees (through the heart of France) ostensibly because it was precisely halfway between the equator and the poles, Jefferson gave in.

When modern science was taking shape in Europe, the great instrument-making countries were also the great science-advancing countries. Britain, France, the Netherlands, Germany, and Italy, where the theory-building scientists were nurtured, were also where the best instruments were crafted. Modern scientific instruments were transforming the old Aristotelian world of qualities into a new Baconian world of quantities. Mersenne insisted that the goal of the natural philosopher should be precision. Newton's epoch-making work, which we misleadingly call the *Principia,* bore the full title *Philosophiae Naturalis Principia Mathematica,* or the *Mathematical Principles of Natural Philosophy.* When science became mathematical, when measurement became the test of scientific truths, then those who made the instruments of measurement became first-class citizens in the republic of science and the scientific community was vastly enlarged.

The new instruments also transformed unique experiences into repeatable experiments. In Europe a scientific instrument-making industry, which, of course, included clockmaking, developed in the seventeenth century. By the eighteenth century, as we have seen, scientific and mathematical instruments were among the significant exports of England and the Netherlands.

Instruments that began as tools of observation became the tools of measurement and then the apparatus of experiment. The astrolabe, an ancient instrument of astronomers and navigators for observing the altitudes and positions of heavenly bodies, was improved into a refined measuring device by the Portuguese mathematician-cosmographer Pedro Nunes (1502–1578). Finding that the traditional instrument could not precisely measure small portions of an arc, he invented a simple attachment. The "nonius" (named after Nunes) consisted of forty-four concentric circles, each marked with equal divisions extending out to its quadrant. The outermost circle had 89 divisions, the innermost only 46. Each circle had one less division than the circle outside it, and one more than that inside. By reading the scale on the circle most closely approximating the position of the sight, it was possible to measure fractions of a degree of arc.

A French military engineer, Pierre Vernier (1584–1638), while helping his father survey for a map of Franche-Comté, found that the nonius was not precise enough for his purpose. So he devised the improvement that makes his name familiar in machine shops around the world. His idea was simply to replace the inner concentric circles, which had been marked off on the static face of the instrument, with a mobile concentric segment, which could be rotated to find a line that precisely matched the line of sight. This improvement was crucial because at the time engraving technology was not refined enough to mark off in legible fashion all the lines required by the nonius. Vernier dispensed with most of these lines, for the central disk could be rotated to the required position. This, the first "vernier," was adapted to the caliper and other instruments, and improved the technology of surveying and navigating in the next centuries.

Galileo himself had first attained fame as a telescope maker. Advances in lens grinding and polishing, the invention of achromatic lenses, and mechanical methods for dividing limbs and graduated rules all made it worthwhile to use telescopic sights for new purposes. It was not long before micrometers were installed in telescopes to measure the diameters of the planets and the stars. "I have either found out, or stumbled on . . . a most certain and easy way," William Gascoigne (1612–1644), the self-made English astronomer, recorded modestly, "whereby the distance between any the least stars, visible only by a perspective glass, may be readily given, I suppose to a second [of a degree of arc]; affording the diminutions and augmentations of the planets strangely precise." Gascoigne reported how the "All Disposer" had caused a spider obligingly to weave a thread in an open case while he was experimenting with a view of the sun and so had given him the idea.

The micrometer for telescopes then was improved by the use of hairs in the sights, and other devices, and all these were applied to the microscope. Leeuwenhoek's achievement was not simply to *see* microscopic objects, but actually to *measure* them. His letters to the Royal Society reported that a coarse sand grain had a diameter of 1/30 of an inch, a fine sand grain about 1/80 or 1/100 of an inch. He observed that twenty hairs from his wig would fit into 1/30 of an inch, which suggests to modern experts that his wig must have been made from the hair of an Angora goat. The eye of a louse, he wrote, measured between 1/250 and 1/400 of an inch. One of the red globules of human blood, he noted, was 25,000 times smaller than a fine sand grain, and "the complete globules that make our blood red are so small that a hundred of them laid length-wise would not make up the axis of a coarse sand-grain of 1/3,000th of an inch."

52

"God Said, Let Newton Be!"

THE first popular hero of modern science was Isaac Newton (1642–1727). Before him, of course, there were others known across Europe for their mastery, real or imagined, of the forces of nature. Aristotle was the approved classic source. But when Roger Bacon (c. 1220–1292), the most celebrated European scientist of the Middle Ages, sought "to work out the natures and properties of things"—which included studying light and the rainbow and describing a process for making gunpowder—he was accused of black magic. He failed to persuade Pope Clement IV to admit experimental sciences to the university curriculum, he had to write his scientific treatises in secrecy, and was imprisoned for "suspected novelties." The legendary Dr. Faustus, fashioned after a real magician-charlatan of the sixteenth century, dramatized the perils of intruding into nature's secrets, and became a literary stereotype. In the unforgettable lines of Christopher Marlowe and Goethe, he gratified audiences with the spectacle of his damnation.

But Newton, whose vision of nature's processes was more grandiose and more penetrating than Bacon's or Faustus', was publicly acclaimed and apotheosized. While earlier experimentalists were assumed to be in league with the Devil, Newton was placed at the right hand of God. Unlike Galileo, his greatest predecessor, Newton was swimming with the scientific currents of his time. He probably exercised greater influence over scientific thought than any secular figure since Aristotle. There would not be another such hero until Einstein. Although Newton's works are difficult or impossible for the layman to grasp, in his time he was understood well enough to be made a demigod. When Queen Anne knighted him at Trinity College, Cambridge, in 1705, he was the first person ever to be so honored in England for scientific achievements. This was only a small measure of his glamour as the Galahad of the Scientific Quest.

In Newton converged and climaxed the forces advancing science. His age, as we have seen, was already going "the mathematical way." New parliaments of science, for the first time, were exposing observations and discoveries for discussion, endorsement, correction, and diffusion. For a quarter-century as president of the Royal Society in London he made it an unprecedented center of publicity and of power for science.

Yet if a novelist had planned it so, the circumstances of his birth in 1642 and his youth could hardly have been better designed to feed Newton's feelings of insecurity. His father was a small farmer, a "yeoman" who could not sign his name. His ancestors on his father's side may have been of even lower station. He was a weakly infant. At birth, it was said, he could have fitted into a quart mug, and it was doubtful whether he would survive. His father died three months before Isaac was born, and when he was only three his mother married and moved away to live with a well-to-do clergyman in the neighborhood, leaving the infant Isaac in the charge of his maternal grandmother in a lonely farmhouse. He so resented his mother's second marriage that at the age of twenty he could still recall "threatning my father and mother Smith to burne them and the house over them." When he was eleven, his mother, on the death of her second husband, returned to his household with her three young children. She withdrew him from school, hoping to make him into a farmer, but he was inept at farm chores. Encouraged by the local schoolteacher and a clergyman uncle, he went back to the schoolroom, where he acquired a good grounding in Latin but very little mathematics. At nineteen, older than the other undergraduates, he entered Trinity College, Cambridge, as a "subsizar," a poor scholar working his way through. Despite all his worldly honors he never lost the insecurity of those years. Very early he began calling himself a "gentleman" and claiming a family connection with lords and ladies. He would always overvalue the honors of the court and the dignity of inherited position. And he remained, in public at least, a scrupulous and loyal Anglican.

Newton received his Bachelor of Arts degree in the early summer of 1665 just when the university was being closed on account of the plague, and he retreated to his home in Lincolnshire for about two years. When the university reopened and he returned to Cambridge in 1667, he was elected a fellow of Trinity College, and two years later, at the age of twenty-six, was named Lucasian Professor of Mathematics. When Newton went to Cambridge, the physics of Aristotle, based on the distinctions of qualities, was being displaced by a new "mechanical" philosophy of which Descartes (1596–1650) was the most famous exponent. Descartes described the physical world as consisting of invisible particles of matter in motion in the ether. Everything in nature, he said, could be explained by the mechanical interaction of these particles. According to Descartes's mechanistic view of the world, there was no difference, except in intricacy, between the operation of a human body, of a tree, or of a clock. Elaborated in various theories of atomism, Descartes's ideas dominated the new physical thinking in Europe. Everything in nature was to be explained by these minute invisible particles in motion and interaction. To Newton the prevalent philosophy seemed to depend on "things that are not demonstrable" and so were no more than "hypotheses." The physics or "natural philosophy" of the age when Newton came to Cambridge was replete with elaborations of Descartes's notions into "corpuscles," "atoms," and "vortices."

Reacting against these pretentious suppositions, Newton determined to stay on the strait path of mathematics. He believed that although he might seem now to explain less, in the long run his experimental philosophy would surely explain more. The versatile Descartes, also, had a genius for mathematics, he invented analytic geometry and made other advances in algebra and geometry. But he soared on to his expansive theories of sensation and of physiology, and he pretended even to have unraveled the secret of human reproduction. Equipped with his mechanistic dogma, Descartes would not admit any secrets of Nature to be beyond his reach. Although, as we shall see, Newton was by temperament no more modest than Descartes, he almost always managed to keep his scientific efforts channeled into the search for physical laws expressed in mathematical form.

As an undergraduate and during his two years of retreat in the Plague Years, Newton drew the main outlines for his experimental approach to nature. Even before his twenty-sixth year, when he would become a fellow of Trinity College, he had discovered the binomial theorem, and was well on the way to formulating the calculus. His "experimental philosophy" was a kind of self-discipline. In his often quoted disclaimer Newton was not being merely sententious. "I do not know what I may appear to the world; but to myself I seem to have been only like a boy playing on the seashore, and diverting myself in now and then finding a smoother pebble or a prettier shell than ordinary, whilst the great ocean of truth lay all undiscovered before me."

The essence of his new method was revealed in his very first significant experiments, his work with light and color. This, as historian Henry Guerlac has shown, proved to be a perfect parable of Newton's "experimental philosophy." For of all natural phenomena light was the most awesome in its temptations to romance, metaphor, and theology and most unlikely to be confined in the discipline of numbers. Yet this was precisely what the young Newton would manage. Just after he had received his bachelor's degree, as he reported to Henry Oldenburg:

> in the beginning of the year 1666 (at which time I applied myself to the grinding of Optick glasses of other figures than Spherical) I procured me a Triangular glass-Prisme, to try therewith the celebrated Phaenomena of colours. And in order thereto having darkened my chamber, and made a small hole in my window shuts, to let in a convenient quantity of the Sun light, I placed my Prisme at its entrance, that it might be thereby refracted to the opposite wall. It was at first a very pleasing divertisement, to view the vivid and intense colours produced thereby; but after a while applying myself to consider them more circumspectly, I became surprised to see them in an oblong form; which according to the received laws of Refraction I expected should have been circular. . . .

To explain this phenomenon, he devised what he called his *experimentum crucis*. Through a small hole he directed a part of the oblong spectrum— a ray of a single color—toward a second prism. He found that the light refracted from the second prism was not further dispersed, but remained a single color. From this he concluded simply "that Light consists of Rays differently refrangible, which . . . were, according to their degrees of refrangibility, transmitted towards diverse parts of the wall." Which meant that "Light itself is a Heterogeneous mixture of differently refrangible Rays." There was an exact correlation, he noted, between color and "degree of refrangibility"—the least refrangible being red and the most refrangible being a deep violet. In this way he disposed of the ancient commonsense notion that colors were modifications of white light. He then confirmed his surprising suggestion that all colors were the components of white by using a biconcave lens to bring the rays of the complete spectrum to a common focus. The colors disappeared altogether when they joined to produce white light. By these elegantly simple experiments, Newton had reduced the "qualitative" differences of color to quantitative differences. Or, as he put it, "to the same degree of Refrangibility ever belongs the same colour, and to the same colour ever belongs the same degree of Refrangibility."

It would be possible, then, to designate any color by a number indicating its degree of refrangibility. Here was the foundation for a science of spectroscopy. Even more important, it was a model of Newton's experimental method. Some belittled Newton by saying that he had really discovered nothing about the "nature" of light. His explanation of colors, they said, was only a "hypothesis." To which Newton firmly replied "that the doctrine which I explained concerning refraction and colours, consists only in certain properties of light, without regarding any hypotheses, by which those properties might be explained. . . . For hypotheses should be subservient only in explaining the properties of things, but not assumed in determining them; unless so far as they may furnish experiments. For if the possibility of hypotheses is to be the test of the truth and reality of things, I see not how certainty can be obtained in any science." It was enough for Newton's purpose to consider light as "something or other propagated every way in streight lines from luminous bodies without determining what that thing is." Of course, he admitted, Huygens was correct in saying that he had not described the mechanism by which colors are made. But that was the virtue and the rigor of Newton's experimental method.

This same rigor would characterize Newton's method when he came to describe the System of the World. As early as 1664, while still an undergraduate, Newton had begun thinking about ways of quantifying the laws of motion of all physical bodies. He had been stimulated, too, by various casual suggestions—Hooke's notion, based not on demonstrative data but on a hunch, that gravitational attraction might decrease as the square of the distance, and Edmund Halley's speculation derived from Kepler's third law, that the centripetal force toward the sun would decrease in proportion to the square of the distance of each planet from the sun. But these were mere suggestions. It was left for Newton to see the universality of the principles, to make the calculations to prove them, and to show that the elliptical orbits of the planets would follow.

In response to a request from Halley, Newton prepared a nine-page "curious treatise, *De motu;* which, upon Mr. Halley's desire, was . . . promised to be sent to the Society to be entered upon their register." This, as we have seen, was Oldenburg's device for ensuring credit to all "first inventors" while inducing communications to the Royal Society. On this occasion, Oldenburg's incentives had paid off. For Halley "was desired to put Mr. Newton in mind of his promise for the securing his invention to himself till such time as he could be at leisure to publish it." Newton's few pages "On the Motion of Bodies in an Orbit," showed that he had already arrived at the crux of his grand theory by demonstrating, among other things, that an elliptical orbit could be explained by suggesting an inverse square force to one focus. In revising the *De motu,* Newton elaborated his first and second law: (1) the law of inertia, and (2) the law that rate of change of motion is proportional to the impressed force.

The power and the grandeur of Newton's system consisted, of course, in its universality. He finally offered one common scheme for terrestrial and celestial dynamics. He had brought the heavenly bodies down to earth, and at the same time provided a new framework, and new limits, for man's grasp on the heavenly bodies. The legend of Newton and the apple is not entirely without foundation. The grand "notion of gravitation" came to him, Newton himself said, "as he sat in a contemplative mood" and "was occasioned by the fall of an apple." He had the bold imagination to think of the apple not simply as falling on his head but as being attracted to the center of the earth. Newton noted that the moon was sixty times as far from the center of the earth as the apple was, and therefore, by the inverse-square law, should have an acceleration of free fall of $1/(60)^2 = 1/3600$ of the acceleration of the apple. By applying Kepler's third law, then, he could test his theory. There were a number of practical difficulties in the way—including Newton's incorrect value for the radius of the earth. But his simple insight had put him on the way to his System of the World. He unified all the physical phenomena on earth with those in the heavens by the generality of his laws, expressed mathematically. For all the motions of earthly and heavenly bodies could be seen, observed, and measured. The grand unifying force in Newton's system, even before gravitation, was mathematics.

Newton's "mathematical way" was a way of discovery. But it was also a way of humility, for the mathematical way was a method of self-discipline as well as an instrument for exploring. The title of Newton's great work, *Mathematical Principles of Natural Philosophy* (*Philosophiae Naturalis Principia Mathematica,* 1687; English translation, 1729) made as plain as he could that he was displacing all the widespread pretensions to reveal the mechanics of nature. Continental reviewers again objected to the narrowness of Newton's stated purpose. He had not explained *why* the physical world behaved as it did but had only provided mathematical formulae. Therefore, they said, what he offered was not really "natural philosophy" at all. Of course, they were quite right again, but at the same time they unwittingly described the new strength of Newton's method. Just as in his *Opticks,* so at the very end of the *Principia,* Book III, "The System of the World," Newton took pains to define the limits of his method and of his achievement. After his concluding paean to the God who "exists always and everywhere," he explained that "We have ideas of his attributes, but what the real substance of anything is we know not," and hence God could be known only "from the appearances of things."

> Hitherto we have explained the phenomena of the heavens and of our sea by the power of gravity, but have not yet assigned the cause of this power. . . . But . . . I have not been able to discover the cause of those properties of gravity from phenomena, and I frame no hypotheses; for whatever is not deduced from the phenomena is to be called an hypothesis; and hypotheses, whether metaphysical or physical, whether of occult qualities or mechanical, have no place in experimental philosophy. In this philosophy particular propositions are inferred from the phenomena, and afterwards rendered general by induction. Thus it was that the impenetrability, the mobility, and the impulsive force of bodies, and the laws of motion and of gravitation, were discovered. And to us it is enough that gravity does really exist, and act according to the laws which we have explained, and abundantly serves to account for all the motions of the celestial bodies, and of our sea.

The most influential of Newton's eighteenth-century disciples, the scientific editor of Diderot's *Encyclopédie,* Jean le Rond d'Alembert (1717–1783), acclaimed Newton for his refusal to play God, seeing nature "only through a veil which hides the workings of its more delicate parts from our view. . . . Doomed . . . to be ignorant of the essence and inner contexture of bodies, the only resource remaining for our sagacity is to try at least to grasp the analogy of phenomena, and to reduce them to a small number of primitive and fundamental facts. Thus Newton, without assigning the cause of universal gravitation, nevertheless demonstrated that the system of the world is uniquely grounded on the laws of this gravitation." Against the pitfalls of common sense, d'Alembert warned that "the most abstract notions, those that ordinary men regard as most inaccessible, are often those that shed the brightest light."

Newton proved so effective an apostle of the bright light of mathematics precisely because he was so acutely aware of the enshrouding darkness. Who but God could penetrate the inmost workings of the universe? Newton's Hermeticism—his feeling for the mystery beneath the unity of the world—grew with the passing years. But throughout his life he saw the limits of the capacity of human reason to encompass experience, which explained, too, his unflagging interest in the Bible and in Prophecy. Newton's experimental and mathematical genius was overcast by a religious and mystical temperament. His copious manuscripts on alchemy (650,000 words) and on Biblical and theological topics (1,300,000 words) baffle Newtonian scholars, who try to fit them into the rational frame of Newton's universe. Without doubt, Newton took the Prophets seriously, exercising all his linguistic learning to seek a common meaning for the mystical terms used by John, Daniel, and Isaiah. But he was wary of priestly pretensions. "The folly of interpreters," he warned, was "to foretell times and things by this Prophecy, as if God designed to make them Prophets." In the Prophetic books God's intention was not to make men Prophets of future events but rather that "the event of things predicted many ages before, will then be a convincing argument that the world is governed by providence." Therefore he applied his sophisticated techniques of astronomical dating to confirm the literal truth of events related in the Bible. Newton never became a thoroughgoing mystic, for he seemed aware of the truth, noted by Roger Fry, that "mysticism is the attempt to get rid of mystery." That Newton never wanted and never dared.

While Newton was widely apotheosized for his mathematical mastery of the world, only a few would sense his awe of the world's mystery—expressed in the line that his mathematics itself drew between man and God. In the next century, both the romantic idealization of Newton and the failure of common sense to encompass his vision were dramatized at a festive literary dinner on December 28, 1817, which Benjamin Haydon (1786–1846), an English historical painter in the grand manner, gave in his studio. Among others there, he reported, were Charles Lamb, John Keats, and William Wordsworth, who "abused me for putting Newton's head into my picture; 'a fellow,' said he, 'who believed nothing unless it was as clear as the three sides of a triangle.' And then he and Keats agreed that he had destroyed all the poetry of the rainbow by reducing it to its prismatic colours. It was impossible to resist him, and we all drank 'Newton's health, and confusion to mathematics.' "

53

Priority Becomes the Prize

THE salute to Newton was a thoroughly modern gesture, for Europe had only lately learned to value the new. "Is it not evident," John Dryden asked in 1668, "in these last hundred years (when the study of philosophy has been the business of all the Virtuosi in Christendom), that almost a new Nature has been revealed to us? . . . more noble secrets in optics, medicine, anatomy, astronomy, discovered, than in all those credulous and doting ages from Aristotle to us?" In this new age of "revelations," honors would be heaped on him who was reputed to be the *first* to unveil a truth of nature. For now the printing press, by speedily spreading word of a new discovery, finally made it possible to define priority. And priority brought a kudos it never could have had before.

Europe's ancient institutions of learning, colleges and universities, had been founded not to discover the new but to transmit a heritage. By contrast, the Royal Society and other parliaments of scientists, with their academies in London, Paris, Florence, Rome, Berlin, and elsewhere, aimed to increase knowledge. They were a witness not so much to the wealth of the past as to what Bishop Sprat called "the present Inquiring Temper of this Age." Robert Boyle put it in a nutshell in the title of his *Essay of men's Great Ignorance of the Uses of Natural Things, or there is scarce any one thing in Nature whereof the Uses to human Life are yet thoroughly understood.*

In earlier times, to possess an idea or a fact meant keeping it secret, having the power to prevent others from knowing it. Maps of treasure routes were guarded, and the first postal services were designed for the security of the state. Physicians and lawyers locked their knowledge in a learned language. The government helped craft guilds exclude trespassers from their secrets. But the printing press made it harder than ever to keep a secret. More than that, the press changed radically, and even reversed, what it meant to "own" an idea. Now the act of publishing could put a personal brand on a newly discovered fact or a novel idea.

We cannot be surprised at Bishop Sprat's defense of the Royal Society:

> If to be the Author of new things, be a crime; how will the first Civilizers of Men, and makers of Laws, and Founders of Governments escape? Whatever now delights us in the Works of Nature, that excells the rudeness of the first Creation, is New. Whatever we see in Cities, or Houses, above the first wildness of Fields, and meaness of Cottages, and nakedness of Men, had its time, when this imputation of Novelty, might as well have bin laid to its charge. It is not therefore an offense, to profess the introduction of New things, unless that which is introduc'd prove pernicious in itself; or cannot be brought in, without the extirpation of others, that are better.

Established tradesmen and artisans of course were suspicious of the new, for "they are generally infected with the narrowness that is natural to Corporations, which are wont to resist all new comers, as profess'd Enemies to their Privileges."

In organizing the Royal Society, the shrewd Henry Oldenburg had seen the new significance of priority. He sensed that members might be reluctant to send their discoveries to the society for fear that others might steal their claims to be the first. So he proposed "that a proper person might be found out to discover plagiarys, and to assert inventions to their proper authors." To protect the rights of priority to investigations still in progress, Oldenburg moved that "when any Fellow have any philosophical notion or invention not yet made out, and desired the same, sealed in a box, to be deposited with one of the secretaries till perfected, this might be allowed, for better securing inventions to their authors." The progress of science would be haunted by the specter of priority. Even the most eminent scientists would seem more concerned to claim the credit than to prove the truth of their discoveries.

The heroic Isaac Newton would embody the spirit of modern science in this as in so many other ways. Soon after his death, Newton's character was idealized as much as his work, and no less misunderstood. The poet William Cowper (1731–1800) described the Godlike Newton:

> Patient of contradiction as a child,
> Affable, humble, diffident, and mild,
> Such was Sir Isaac.

The real Newton was anything but affable. The student who served as Newton's assistant for five years, from 1685 to 1690, declared that in all that time he had heard Newton laugh only once—when someone rashly asked him what benefit there might be in studying Euclid.

Before he was thirty and without the incentive or the rewards of public recognition, Newton was well on the way to his great discoveries. By 1672 he had shaped his theory of fluxions which would be the basis of the calculus, but London booksellers, who usually lost money on mathematical treatises, were not eager to publish it. A passion for priority clouded his later years when he was in a unique position to assert his claims. A concern for priority in his invention of the reflecting telescope, described in Newton's earliest surviving letter, dated February 1669, first brought him into the public community of scientists. The instruments used by Galileo and others before Newton were all "refracting" telescopes, employing lenses to magnify the image and bring the light rays to a focus. But these had to be inconveniently long and suffered from chromatic aberration. Newton's invention, which used concave mirrors instead of lenses, could be much shorter and could produce a greater magnification without chromatic aberration. In the long run they would have still other advantages which did not occur to Newton.

There would be a natural limit to the size of a refracting telescope because a lens can be supported only at the edge and the weight of the lens itself tends to distort its shape. But a mirror can be supported from behind and so can be made much larger without risking distortion. Newton had made and coated the mirrors and the tools for making the mirrors for his telescope with his own hands. "If I had staid for other people to make my tools & things for me," he exclaimed, "I had never made anything of it." His first reflecting telescope, though only six inches long, had a magnification of forty times, which, he boasted, was more than that of a refractor six feet long. When word of Newton's invention reached members of the Royal Society, there was general astonishment, which produced a letter to him in January 1672 from Henry Oldenburg, enclosing a drawing of Newton's telescope:

> Your Ingenuity is the occasion of this addresse by a hand unknowne to you. You have been so generous, as to impart to the Philosophers here, your Invention of contracting Telescopes. It having been considered, and examined here by some of the most eminent in Opticall Science and practise, and applauded by them, they think it necessary to use some meanes to secure this Invention from the Usurpation of forreiners; And therefore have taken care to represent by a scheme that first Specimen, sent hither by you, and to describe all the parts of the Instrument, together with its effect, compared with an ordinary, but much larger, Glasse . . . in a solemne letter to Paris to M.Hugens, thereby to prevent the arrogation of such strangers, as may perhaps have seen it here, or even with you at Cambridge; it being too frequent, that new Invention and contrivances are snatched away from their true Authors by pretended bystanders. . . .

Newton replied promptly, with a show of modesty that would become rarer with the passing years, that he "was surprised to see so much care taken about securing an invention to mee, of which I have hitherto had so little value. . . . who, had not the communication of it been desired, might have let it still remained in private as it hath already done some yeares." Elected a Fellow of the Royal Society the following week, early in February he sent them his first contribution, his paper on the theory of colors fulfilling his hope that "my poore & solitary endeavours can effect towards the promoting your Philosophicall designes."

By a gradual progression, Newton became councillor, and then in 1703 president—in effect, dictator—of the Royal Society for the quarter-century until his death. As his prestige grew, so did his dyspepsia, his unwillingness to give credit to others or share credit for his great discoveries. To assert his primacy in every branch of science that he touched, he marshaled his full powers over what has been called the first scientific "establishment" in the modern world. A martinet in conducting meetings of the society, he never tolerated any sign of "levity or indecorum," and actually ejected Fellows from the meetings for misbehavior. Election to Fellowship, worth honor and money, required his support. When William Whiston, his former assistant in mathematics at Cambridge and his successor in the Lucasian chair, but a man of unorthodox theology, was proposed in 1720, Newton threatened to resign as president if Whiston was elected. In 1714, when Parliament was deliberating on the prize for a way to find longitude at sea, Newton pontificated that no clock would serve the purpose. This probably delayed acceptance of Harrison's clock, which, as we have seen, really did solve the problem. As scientific adviser and pundit, he disposed of coveted government posts—chiefs of observatory and members of scientific commissions—which multiplied with the years. He himself left his Lucasian Professorship of Mathematics for the powerful and remunerative government post as Warden, and later Master of the Mint, during which time his income sometimes came to a then spectacular £4,000 a year. He supervised the great recoinage, hounded out counterfeiters, and seemed to delight in their draconian punishment.

In 1686, when Newton sent to the Royal Society the completed manuscript of Book I of the *Principia,* Robert Hooke immediately claimed that the basic ideas had been plagiarized from his communications to Newton a dozen years before. "Philosophy is such an impertinently litigious Lady," Newton responded to Oldenburg in exasperation, "that a man had as good be engaged in Law suits as have to do with her. I found it so formerly & now I no sooner come near her again but she gives me warning." And his contempt for the presumptuous Hooke grew without bounds. "Now is this not very fine? Mathematicians that find out, settle & do all the business must content themselves with being nothing but dry calculators and drudges & another that does nothing but pretend & grasp at all things must carry away all the invention as well of those that were to follow him as of those that went before." Far from acknowledging Hooke's priority, Newton went back to his manuscript and deleted references to Hooke's work. Halley and others who mildly took Hooke's part so aroused Newton's ire that he threatened to suppress the whole Book III of his great work. They dissuaded him from this act of self-immolation, but Newton still nursed his fury. He kept Hooke as his favorite enemy for the next seventeen years, and to express his pique he refused to publish his *Opticks* or to accept the presidency of the society until after Hooke's death in 1703. The sober verdict of an eighteenth-century French admirer of Newton recognized that Hooke's claims were not entirely without merit but showed "what a distance there is between a truth that is glimpsed and a truth that is demonstrated."

Newton's later years, when he had become the idol of "Philosophic" London, could be chronicled in his acrimonious quarrels with subordinates and his vindictive plots against any who threatened to become his equal. In a first sordid episode he maliciously deprived the unlucky Astronomer Royal, John Flamsteed (1646–1719), of the satisfaction of publishing the scientific products of his lifetime. Though plagued by ill health, Flamsteed had invented new techniques of observation, had improved micrometer screws and calibrations, had spent £2,000 of his own, and finally constructed the best instruments of the age for his work at Greenwich. In a dozen years he made twenty thousand observations, which far excelled Tycho Brahe's in accuracy. But the scrupulous Flamsteed would not yet publish his figures. "I want not your calculations but your Observations only," the imperious Newton badgered Flamsteed. In pique, Newton threatened to drop his own "moon's theory" and blame it on Flamsteed if Flamsteed did not speedily deliver. When the miserable Flamsteed complained that Newton's "hasty artificial unkind arrogant" letters had aggravated his splitting headaches, Newton advised that the best way to cure headaches was "to bind his head strait with a garter till the crown of his head was nummed." The impatient Newton had all Flamsteed's uncorrected observations at the Greenwich Observatory gathered together, then compiled and published them. Appalled to see his lifework garbled, Flamsteed petitioned the Lords of Treasury, managed to buy up three hundred of the four hundred original copies, carefully removed the ninety-seven pages that he had prepared for publication and burned the rest. Flamsteed died before he could complete his work. But his vindication was accomplished by two friends who published his three-volume star catalogue in 1725 which became a landmark in modern astronomy as the first to take advantage of the telescope.

The spectacle of the century in the newly public scientific arena was Newton's battle with the great Baron Gottfried Wilhelm von Leibniz. Now the stakes were one of the greatest scientific priority prizes of any age—the glory of inventing the calculus. The calculus was something that few even of the scientists of that age understood. But the priority issue was easy enough to grasp. Educated laymen recognized that the calculus was an important new way to calculate velocities and rates of change, and that it promised to multiply the uses of scientific instruments and measuring devices. We, too, can understand the priority issue without an expert's knowledge of the calculus. The priority debate, however unedifying, widened the audience for science. What was this "differential calculus" over which great men enjoyed insulting one another in public? The very question was news when the King, his mistress Henrietta Howard, Princess Caroline, and the whole diplomatic corps became interested and discussed ways of settling the dispute.

Newton's antagonist, Leibniz (1646–1716), was himself one of the profoundest philosopher-scientists of modern times. At the age of six he began reading in the copious library of his father, who was a professor of moral philosophy at the University of Leipzig, and by fourteen he was well acquainted with the classics. Leibniz, according to De Quincey, was unlike the great thinkers who were planets revolving in their own orbits, for Leibniz was a comet "to connect different systems together." Before he was twenty-six, Leibniz had devised a program of legal reform for the Holy Roman Empire, had designed a calculating machine, and had developed a plan to divert Louis XIV from his attacks on the Rhineland by inducing him to build a Suez Canal. In 1673, when he visited London on a diplomatic mission, he met Oldenburg and was elected to the Royal Society. Leibniz's European travels put him in touch with Huygens, Spinoza, Malpighi, and Galileo's pupil Viviani. He met the Jesuit missionary Grimaldi, who was about to leave for Peking to become the Chinese court mathematician.

Frederick the Great called Leibniz "a whole academy in himself." Even so, in 1700 the King of Prussia had founded the Berlin Academy. Unlike its counterparts in Paris and London, this was not a spontaneous community of scientific enthusiasts, but was largely the creature of Leibniz himself. The government monopoly of printing and of the newly reformed calendar would be used to fund the academy and its observatory and to make science the property of the whole community. Naturally enough, Leibniz opposed the use of Latin and he championed the vernacular.

> Our learned men have shown little desire to protect the German tongue, some because they really thought that wisdom could only be clothed in Latin and Greek; others because they feared the world would discover their ignorance, at present hidden under a mask of big words. Really learned people need not fear this, for the more their wisdom and science come among people, the more witnesses of their excellence they will have. . . . On account of the disregard of the mother tongue, learned people have concerned themselves with things of no use, and have written merely for the bookshelf; the nation has been kept from knowledge. A well-developed vernacular, like highly-polished glass, enhances the acuteness of the mind and gives the intellect transparent clearness.

When Georg Ludwig, the elector of Hanover, ascended the English throne in 1714 as George I, Leibniz hoped the King would take him along to London as court historian. But the King refused until Leibniz had completed his genealogical history of George's family, and the great Leibniz spent the last two years of his gout-ridden life trying to finish this trivial assignment. When he died in 1716, he had been abandoned by the princes whom he had spent his life trying to please.

For our story the crucial connection in Leibniz's life was his lifelong relation to the Royal Society, first fruitful, but finally fatal. The dramatic climax came with the publication in 1712 of the official report of the august committee of the society that had been appointed to adjudicate the priority dispute between Leibniz and Newton. Technically the occasion was Leibniz's complaint that he had been insulted by John Keill, a Fellow who had accused Leibniz of plagiarizing Newton and claiming to have been the first inventor of the differential calculus.

Though actually charged only with deciding the decorum of Keill's behavior, the committee had seized the opportunity to defend Newton's priority. They summarized the "facts," including numerous conversations and prolific correspondence among the society's Fellows, to show that Keill's statements were not gratuitous insults but simply a recognition of Newton's right to his invention. Through Oldenburg, the committee explained, Leibniz had first been in touch with another Fellow of the Society, John Collins (1625–1683), who had devoted himself to promoting the exchange of mathematical discoveries. Back in 1672 Collins had sent a letter to Leibniz in Paris telling him about Newton's invention of a method of "fluxions" which was essentially what Leibniz now claimed as his own. According to the committee, Leibniz had never done anything more than restate Newton's method, which he had already learned about from Collins' letter, "in which letter the method of fluxions was sufficiently described to any intelligent person." *Commercium epistolicum* (a commerce of letters) was the name they gave to their report, which plainly declared that the opportunities for plagiarism stemmed from the new community of scientific correspondents. So the committee triumphantly convicted Leibniz and awarded the laurel of "first inventor" to Newton. A century and a half after the trial of Leibniz, in 1852 the accomplished mathematician Augustus De Morgan (1806–1871) established that Leibniz had never received the incriminating document but only a copy from which the suggestive passages had been removed.

If the facts had been more widely known, the proceedings would have discredited Newton himself, who at the time was the undisputed dictator of the Royal Society. There was never a direct confrontation between Newton and Leibniz, for all Newton's moves were backstage. Behind the scenes the unsavory role of chief instigator was played by Fatio de Duillier, a half-mad Swiss amateur mathematician and enthusiastic meddler, with whom Newton had a long and curious relationship. Newton had been Fatio's devoted patron, and Fatio, twenty-two years Newton's junior, had occasionally lived with Newton. When the Newton-Leibniz duel held public attention, Fatio had become a religious crank, secretary to a riotous sect of "Prophets" who foretold a second burning of London, for which Fatio had been punished in the pillory at Charing Cross and at the Royal Exchange.

Back in 1699 Newton himself had sent the Royal Society a communication accusing Leibniz of plagiarism. As president of the society, Newton, to humble Leibniz and vindicate his own priority, set up "a numerous Committee of Gentlemen of several Nations" to make an impartial decision on the evidence. The members, all of course appointed by Newton, were five certified Newtonians, plus the Prussian ambassador and a Huguenot refugee. We now know, what was not known at the time, that Newton himself wrote the committee's "impartial" report. Then Newton took the further trouble of writing an anonymous review and summary of the report, which he included in later reprints of the *Commercium epistolicum*. In addition, he was the author of hundreds of other documents "exposing" Leibniz and extolling the originality of his own discovery of the calculus. His most astonishing display of academic overkill was his devotion of the whole of the *Philosophical Transactions* for January and February 1715 (except for three pages) to yet another polemic against Leibniz—and a further backdating of Newton's discovery into the 1660's. Still Newton was not satisfied. Further to humiliate his enemy and to publicize the committee's verdict, he convened a special meeting at the Royal Society to which he invited the whole diplomatic community. Newton "once pleasantly" reported to a disciple that "He had broke Leibniz's Heart with his Reply to him."

The unfortunate Leibniz's Berlin Academy provided no troops nor any arsenal comparable to Newton's Royal Society. Leibniz had hoped to find a champion in the willful Princess Caroline of Anspach, who had accompanied her father-in-law, George I, from Hanover to London, for she was the center of a brilliant salon. After witnessing the sordid quarrel, the philosophical princess, who kept her power in British politics by connivance at her husband George II's amours, concluded that "great men are like women, who never give up their lovers except with the utmost chagrin and mortal anger. And that, gentlemen, is where your opinions have got you." Leibniz died in 1716, before Newton had exhausted his rage. But Leibniz did win a posthumous victory. The mathematical world adopted Leibniz's symbols—the letter *d,* as in *dx* or *dy,* and the long *s* written as $\int$ (the initial letter of *summa*)—and the name *calculus integralis* (which had been suggested to Leibniz by Jakob Bernoulli I in 1690) and these dominated mathematics textbooks into the late twentieth century.

Of course there were priority battles before Newton, and these would become normal afterwards. At the opening of the modern era, Galileo had attacked a host of rivals—one for pretending to have invented the telescopic uses of astronomy "which belongs to me," another for claiming to have preceded him in observing sunspots, others for attempting "to rob me of that glory which was mine, pretending not to have seen my writings and trying to represent themselves as the original discoverers of these marvels," and still another who "had the gall to claim that he had observed the

Medicean planets which revolve about Jupiter before I had" and who then devised "a sly way of attempting to establish his priority." Other notorious quarrels erupted later between Torricelli and Pascal, between Mouton and Leibniz, and between Hooke and Huygens. As the pace of invention and discovery accelerated, so did the bitterness and frequency of priority disputes.

Eighteenth-century Europe saw a vaudeville series of such bouts. Who had first demonstrated that water was not an element but a compound? Was it Cavendish, Watt, or Lavoisier? Each had fervent champions. There was John Couch Adams versus Urbain Jean Leverrier in the case of who first predicted the position of Neptune. Who was first to discover the vaccination against smallpox? Was it really Jenner—or Pearson or Rabaut? As the means of publicity multiplied and with increasing literacy and the rise of the daily newspaper, stakes seemed higher and arguments became more heated. Should the credit for introducing antisepsis go to Lister—or perhaps to Lemaire? The great Michael Faraday (1791–1867), who had worked with Sir Humphry Davy (1778–1829) and had become his intimate, as we shall see, found his election to the Royal Society opposed by Davy (who had earlier fought for his own priority). Davy alleged that William Hyde Wollaston (1766–1828) had preceded Faraday in discovering electromagnetic rotation.

The printing press and the academies made every priority a national victory. Modern rulers in Europe, who had long been patrons of astrologers and alchemists, now became patrons of scientists and technicians. Medieval condottieri had done penance by founding a Balliol College at Oxford or a Trinity College in Cambridge to ensure their entrance into heaven. Modern condottieri founded institutes and prizes. Alfred Nobel (1833–1896) tried to atone for his fortune earned from manufacturing dynamite for warmakers by establishing the prizes first awarded in 1901 for peacemakers and for the great innovators of techno-science. The Nobel prizes, the most coveted of international awards, offer celebrity and money to the winners of the priority race in the sciences. One of the lucky winners, James Watson, in his confession, *The Double Helix* (1968), gave us at long last a frank and unashamed account of how modern scientists scheme for the kudos of priority.

PART TWELVE

CATALOGUING THE WHOLE CREATION

Darwin has interested us in the history of nature's technology.

—KARL MARX, *Capital* (1867)

54

Learning to Look

FOR fifteen hundred years the learned of Europe who wanted to know about nature relied on their "herbals" and their "bestiaries," textual authorities whose tyranny was quite like that of Galen over medicine, and whose poetic delights lured readers away from the outdoor world of plants and animals. Today when we read those guides we understand why medieval Europeans were so slow in learning to look. The pages of the illuminated herbals and bestiaries have never been excelled in charming whimsy or as miscellanies of home remedies.

These sources of medieval botany, the herbals, were the legacy of Dioscorides, the ancient Greek surgeon who had traveled about the Mediterranean with the armies of the emperor Nero. His *De materia medica* (c. 77) surveyed botany mainly as a kind of pharmacology. Physicians went about solemnly trying to match Dioscorides' description of plants he saw on the fringes of the warm Mediterranean with what they found in Germany, Switzerland, or Scotland. Like Galen, Dioscorides had studied Nature, but Dioscorides' disciples studied Dioscorides. He had vainly hoped that his readers would "not pay attention so much to the force of our words, as to the industry and experience that I have brought to bear on the matter." By an alphabetical arrangement earlier writers had separated "both the kinds and the operations of things that are closely related, so that thereby they come to be harder to remember." By contrast, he himself paid attention to where plants grew, to when and how they ought to be gathered, and even to the sorts of containers in which they should be stored. Like other classic authors, he produced few disciples and many exegetes. These treasured his words but forgot his example. He ceased to be a teacher as he became a text.

Yet to the practical-minded of the medieval centuries Dioscorides was delightfully appealing, for he did not distract his readers by theory or taxonomy. Written in Greek, Dioscorides' herbal arranged more than six hundred plants under everyday headings. Which should be sought out for oils, ointments, fats, or aromatics? Which would cure headaches or remove spots on the skin? What fruits or vegetables or roots were edible? What were local sources of spices? What plants were poisonous and what were their antidotes? What medicines could be made from plants?

Countless surviving manuscripts of "Dioscorides" attest his popularity throughout the Middle Ages. The more we read the texts, the less we are

puzzled by Dioscorides' popularity or by the surviving power of his nomenclature. For example, the first item among his "aromatics," in the translation (1655) by John Goodyer:

> Iris is soe named from the resemblance of the rainbow in heaven. . . . The rootes
> under are knotty, strong, of a sweet savour, which after cutting ought to be dryed
> in the shade, & soe (with a linnen thread put through them) to be layd vp. But
> ye best is that of Illyria & Macedonia. . . . The second is that of Lybia. . . . But
> all of them haue a warming, extenuating facultie, fitting against coughs, & extenu-
> ating grosse humors hard to get up. They purge thick humors & choler, being
> dranck in Hydromel to the quantity of seven dragms they are also causers of sleep
> & prouokers of tears & heale the torments of ye belly. But dranck with vinegar
> they help such as are bitten by venemous beasts, and the splenitick and such as
> are troubled with convulsion fitts, & such as are stiff with cold, & such as let fall
> their food.

The berry of the juniper, we learn, is "good for ye stomach, being good
taken in drinck for the infirmities of the Thorax, Coughs, & inflations,
tormina, & ye poysons of venemous beasts. It is also vreticall, whence it is
good both for convulsions, & ruptures, & such as haue strangled wombes."
The common radish "also breeds winde and heates, wellcome to the mouth,
but not good for ye stomach, besydes it causeth belching and is vreticall.
It is good for ye belly if one take it after meate, helping concoction ye more,
but being eaten before, it doth suspend the meate; wherefore, it is good for
such as desire to vomit to eate it before meate." The mandrake root can be
prepared for anaesthesia "to such as shall be cut, or cauterized. . . . For they
do not apprehend the pain because they are overborn with dead sleep.
. . . But used too much they make men speechless."

A thousand years of "Dioscorides" manuscripts shows us what it meant
to be at the mercy of copyists. With the advancing centuries, the illustrations move farther and farther away from nature. The copies of copies grew
imaginary leaves for symmetry, enlarged roots and stems to fill out the
rectangular page. Copyists' fancies became conventions.

Whimsical scribes took clues from the names as much as the properties
of the plants, making botany a branch of philology. From the flowers of the
Narcissus plant emerged tiny human figures, reminiscent of the unlucky
youth who saw and loved his own image everywhere. The "Tree-of-life" was
entwined by a serpent with a woman's head. The "Barnacle-tree" or "Goose
Tree" bore shells that opened and hatched out the barnacle geese found in
northern Scotland.

When the printing press first appeared in Europe, the most useful botanical information was still found in the ancient herbals as expanded and
"improved" by generations of scribes. Printers with a heavy investment in

wood blocks or copperplates were then understandably reluctant to junk them simply because the pictures did not match the words of the text. Even scholars who might have been tempted to look at the plants themselves found it more convenient to compare manuscripts and gloss texts.

Printed herbals quickly became stock items. The *Liber de proprietatiibus rerum* (c. 1470), by an English monk who lived in the thirteenth century, went through twenty-five editions before the end of the fifteenth century. The vernacular opened avenues for facts from all Europe. But the herbal had obvious limits. Of every plant it always asked the same question: How can you amuse me, feed me, salve me, cure me?

In the late sixteenth century the holder of the chair in botany at the University of Bologna was still described as "Reader of Dioscorides." As each generation added its tidbits, seldom distinguished from the original, botanists and pharmacologists were mere commentators. The herbal was a catalogue of "simples," medicines each of which had only one constituent, usually from one plant.

The Italian physician Pierandrea Mattioli (1501–1577) offered the first translation of Dioscorides into a European vernacular. His commentaries in Italian (Venice, 1544) became a publishing phenomenon when it sold thirty thousand copies. Then by translating Dioscorides into Latin and adding synonyms for the plant names in several languages, he helped popularize the work across Europe. More than fifty editions in German, French, Czech, and other European languages made Mattioli's refurbished Dioscorides the ruler of botany for a continent.

What the herbals did for botany the bestiaries did for zoology. They, too, derived from a single ancient original, embroidered over centuries. And during the Middle Ages, they were exceeded in popularity only by the Bible. In our time the printed best seller speedily reaches across space but only seldom reaches out into the generations. In the age of the manuscript the power of a single classic author was deathless. The Empire of the Learned was ruled by an oligarchy of a few chameleon "authors." Classic names were made to serve later generations by countless latent revisions, and the original author became a phantom. The hand of the scribe overruled the author.

The original of the bestiaries took its name from a Greek, Physiologus ("Naturalist"), about whom we know very little. His work, probably written before the mid-second century, appears to have been divided into forty-eight sections, each linked to a text from the Bible. A few facts, embellished by abundant theology, morality, folklore, myth, rumor, and fable, provided zoology for generations. By the fifth century there were translations, besides the Latin, into Armenian, Arabic, and Ethiopian. Later it was among the earliest works translated into the European vernaculars, including Old High

German, Anglo-Saxon, Old English, Middle English, Old French, Proven-
çal, and Icelandic.

The Greek version included some forty animals in a delectable potpourri.
Naturally, the lion, king of the beasts, comes first, and with three salient
facts: he uses his tail to rub out his footprints so hunters cannot follow him;
he sleeps with his eyes open; and the newborn cub remains dead for three
days until the father lion breathes life into it. So, too, the body of Christ
was dead, yet like the newborn lion, He remained awake and ready for
Resurrection on the third day.

The remaining animals—lizard, night raven, phoenix, hoopoe, and
thirty-odd others—carry a heavy baggage of morals. None is more vivid
than the "ant-lion," offspring of the unnatural union of a lion and an ant,
who is doomed to starve because the nature of the ant will not permit it to
eat meat, and the nature of the lion keeps it from eating plants. So, too, none
can survive who try to serve both God and the Devil.

Many "translations" were in verse, because bad verse was more memoriz-
able than good prose. Compounding from Physiologus' work, Pliny and
others pioneered with bestiaries in the new European vernaculars. For
example, the *Bestiare d'amour* of Richard de Fournival delighted readers
at court with the verses of a nobleman urging his lady love to imitate the
turtle dove. But, instead, she imitates the aspis snake and covers her ears
so as not to be seduced by his honeyed words.

"Ask now the beasts," urged Job in a favorite passage of the bestiaries,
"and they shall teach thee; and the fowls of the air, and they shall tell thee:
Or speak to the earth, and it shall teach thee: and the fishes of the sea shall
declare unto thee." Since God himself had named his creatures, the name
of anything was a clue to its meaning. Birds, we are told, are called A-*ves*
"because they do not follow straight roads *(visas)*, but stray through any
byway." "*Ursus* the Bear, connected with the word '*Orsus*' (a beginning),
is said to get her name because she sculptures her brood with her mouth
(ore)."

If we see an uplifting symbolism of divine symmetry, Saint Augustine
himself had declared, we should not worry whether a creature really exists.
There must of course be a sea horse because there is a horse on land, just
as the serpent on land suggests an eel in the sea. And because there is a
Leviathan (a female monster in the sea), there must be a Behemoth (a male
monster on the land).

Myths, unlike facts, were uncorrectable. Who could persuade us to abandon Narcissus, the Phoenix, or the Sirens? Modern authors—Lewis Carroll,
E. B. White, Thurber, Chesterton, Belloc, and Borges—have kept legends
of the animate world alive with their own flights of wit and fancy.

In the herbals and the bestiaries the author and the illustrator were not only different people, they were sometimes separated by centuries. The earliest surviving copy of *De materia medica,* made about A.D. 512, four centuries after Dioscorides' death, offered illustrations copied from those by Krateuas, who had died a century before Dioscorides was born. Commonly scribes wrote the text, leaving space for the illustrator to fill in later, but sometimes the tasks were done in reverse order. Often illuminators could not read the language of the text, and sometimes they could not read at all. Occasionally the master named in the margins the miniature to be copied. Over the centuries different illustrations were used for the same text, and vice versa.

Pliny himself (A.D. 23–79) had noted the difficulties:

> Some Greek writers . . . adopted a very attractive method of description, . . . It was their plan to delineate the various plants in colours, and then to add in writing a description of the properties which they possessed. Pictures, however, are very apt to mislead, where such a number of tints is required for the imitation of nature with any success; in addition to which, the diversity of copyists from the original paintings, and their comparative degrees of skill, add very considerably to the chances of losing the necessary degree of resemblance to the originals. . . .
>
> Hence it is that other writers have confined themselves to a verbal description of the plants; indeed some of them have not so much as described them even, but have contented themselves for the most part with a bare recital of their names, considering it sufficient if they pointed out their virtues and properties to such as might feel inclined to make further inquiries into the subject.

Only a rare few who combined in themselves the talents of both naturalist and artist could transform miscellaneous objects into specimens (from Latin *specere,* "to look at" or "to see"), items not merely written about but shown. The contrast between the schematic designs of the herbals and the true-to-life botanical drawings done about 1500 by Leonardo da Vinci or Dürer is startling. Leonardo himself recalled having made "many flowers drawn from life," and from his renderings of a bramble, a wood anemone, and a marsh marigold modern botanists can unmistakably identify each of the species. Dürer's vivid meadow turf—the random cluster of a dozen different grasses—seen from sod level is said to be the first precise ecological study in botany.

In that Age of Discovery when novelties of all sorts were flooding Europe from distant New Worlds, botanists became discoverers in their own backyards. In one region of Europe, clusters of artists and scientists began collaborating in a variety of new ways, and illustrators lured naturalists out of libraries into the field. As early as 1485 Peter Schöffer, who began as assistant to Gutenberg's associate and successor Johann Fust, had printed an herbal in Mainz, and other popular variations on Dioscorides followed. The modern era in botany was opened by *Living Portraits of Plants* (*Herbarum Vivae Eicones,* 1530)—the joint product of a physician, Otto Brunfels (1489–1534), and an artist, Hans Weiditz—at long last an herbal with illustrations drawn from nature. Brunfels, in the familiar pattern, was destined for the priesthood but turned to medicine, prepared a scholarly medical bibliography, then a new edition of Dioscorides adapted to his own neighborhood. He could not resist including the beautiful pasqueflower, but since it had not been authenticated by Dioscorides and so had no Latin name, he condescendingly labeled it, and others not found in the sacred text, naked orphans *(herbae nudae).* The text was still substantially traditional. But the artist proved bolder than the scholar, and as the title of the book announced, Hans Weiditz had drawn directly from nature. What Leonardo and Michelangelo were doing for the human figure, Weiditz did now for the botanical figure. Of course, faithfulness to the observed specimen would not always please. If it had withered leaves, broken stems, truncated roots, or had been eaten away by insects, just so he drew them.

The courage to look and to draw what was really there was slow in coming. For in this last epoch of herbals the printing press still perpetuated the power of ancient texts. Just as Luther had attempted to reform Christianity by returning to the Bible, so Leonhart Fuchs (1501–1566) urged physicians to return from later commentaries to the original text of Galen, and he produced his own edition (Basel, 1538). Raised in the Swabian Alps, as a boy he would walk through the countryside with his grandfather, who told him the names of flowers. At the university he was taught by the humanist Johann Reuchlin (1455–1522), he read Luther, and became professor of medicine. Then in his herbal, *De Historia Stirpium* (1542; German translation, 1543), he paid heavy tribute in its text to Dioscorides and other ancients. But he boldly departed from ancient visual models. To provide the brilliant illustrations, he had organized a team of artists—one who drew the plants from nature, another who copied the drawings onto the wood blocks, and a third who carved the blocks. The front of the book showed a portrait of each of these "mere" craftsmen.

Far beyond the canon of Dioscorides, the illustrations included woodcuts of four hundred native German plants and one hundred foreign plants. "Each of which," Fuchs' Preface explained, "is positively delineated according to the features and likeness of the living plants . . . and, moreover, we have devoted the greatest diligence to secure that every plant should be depicted with its own roots, stalks, leaves, flowers, seeds and fruits. . . . we have purposely and deliberately avoided the obliteration of the natural form of the plants by shadows, and other less necessary things, by which the delineators sometimes try to win artistic glory." Fuchs' enthusiasm shone through, for "there is nothing in this life pleasanter and more delightful than to wander over woods, mountains, plains, garlanded and adorned with flowerlets and plants of various sorts, and most elegant to boot, and to gaze intently upon them." He still arranged items in alphabetical order.

Fuchs' herbal, which now actually deserved to be called a work of botany, set the standard of plant illustration for modern times, later exciting the admiration of William Morris and John Ruskin. From the New World voyages Fuchs harvested some American plants, notably Indian corn, and posthumously he became the eponym of one of the most beautiful American tropical plants, the fuchsia.

In some ways Hieronymus Bock (1498–1554), the third German father of botany, was even more remarkable. Having first tried to identify the Greek and Latin names with the plants in his part of Germany, he went on, and in his *Neu Kreütterbuch* (1539) he freely described all the plants seen in his neighborhood, and set himself the still novel task of describing local plants in the local language.

All these German fathers of botany were active Lutherans at a time when defying the Church of Rome certainly cost you your professorship, and possibly your life. Their botanical dogma, like the Lutheran dogma, was ambivalent. While they went back to a purified text of their sacred Dioscorides, they also put botanical learning, as Lutherans had put the Bible, into the language of the marketplace.

Reaching far beyond the familiar charms of the German countryside, sixteenth-century Europe was delighted by reports of exotic plants and animals from "the Indies," East and West. New World "facts" did not automatically increase the stock of new knowledge. For sailors, as Shakespeare recounted, enjoyed sensationalizing their experiences—with tales of men whose heads grew beneath their shoulders, or who had no heads, or those like the Patagonians who had a single large foot, or the Labradoreans who bore tails. What followed, the historian Richard Lewinsohn reminds us, was a "Rebirth of Superstition." Out of the Americas, whole new orders of monstrous races and fantastic animals were created. Since it is almost as hard to think up a new animal as to discover one, flimsy facts were grafted onto the familiar creatures of myth and folklore.

The Age of Discovery brought a renaissance of fable. Sea serpents five hundred feet long flourished as never before. Mermen and mermaids were now described in unprecedented detail—tall males with deep-set eyes and long-haired females—hungry for their meal of Negroes or Indians, but eating only the bodily protuberances, the eyes, noses, fingers, toes, and sexual organs. Columbus himself reported his encounter with three Sirens. And, of course, the unicorn's horn was so magically therapeutic that, at the marriage of Catherine de' Medici to the French dauphin, Pope Clement VII himself made the princely gift of one to King Francis I. Doubtful legends were now authenticated by the testimony of Jesuit missionaries, substantial sugar-planters, and sober sea captains. To the figments of medieval fantasy were added the real creatures from every new voyage to the Americas. Those who could not read a Latin text could enjoy the copious printed illustrations.

These opportunities inspired a new generation of encyclopedists of nature. The most remarkable of them, Konrad Gesner (1516–1565), had a genius for grafting the new onto the old. Prodigiously learned in several languages, Gesner was torn between what he had read and what he saw. He was born into a poor Zurich family in 1516, educated himself as a vagrant scholar, and, when he was only twenty, wrote a Greek-Latin dictionary. In the next thirty years he turned out seventy volumes on every conceivable subject. His monumental *Bibliotheca Universalis* (4 vols., 1545–1555) aimed to provide a catalogue of *all* writings that had ever existed in Latin, Greek, and Hebrew. Gesner listed eighteen hundred authors and titles of their works in manuscript and in print, with summaries of their content. Thus he earned his title as the Father of Bibliography. What cartography was to explorers on land and sea, bibliography would be to libraries.

In the library of the Fuggers he came upon an encyclopedic Greek manuscript of the second century which inspired him to become a modern Pliny. Finally his *Historia Animalium*, following Aristotle's arrangement, supplied everything known, speculated, imagined, or reported about all known animals. Like Pliny, he provided an omnium-gatherum, but now added the miscellany that had accumulated in the intervening millennium and a half. A shade more critical than Pliny, he still did not deflate tall tales, as when he showed a sea serpent three hundred feet long. But he circumstantially described whale hunting and offered the first known picture of a whale being skinned for blubber. The enduring influence of Gesner's work came from his feeling for folklore and his power to depict fact and fantasy with equally persuasive vividness.

Within a century the English reader had ready access to Gesner's popular encyclopedia in Edward Topsell's translation, which he called the *History of Four-Footed Beasts, Serpents, and Insects* (1658). There we learn of the Gorgon:

> there ariseth a question, whether the poyson which he sendeth forth, proceed from his breath, or from his eyes. Whereupon it is more probable, that like the Cockatrice he killeth by seeing, then by the breath of his mouth, which is not competible to any other Beasts in the world. . . . By the consideration of this Beast there appeareth one manifest argument of the Creators divine wisdom and Provi-

dence, who hath turned the eyes of this beast downward to the earth, as it were thereby burying his poyson from the hurt of man: and shadowing them with rough, long, and strong hair, that their poysoned beams should not reflect upwards, untill the Beast were provoked by fear or danger. . . .

After the unassailable testimony of the Ninety-second Psalm, he describes how unicorns are sacred because they "reverence Virgins and young Maids, and that many times at the sight of them they grow tame, and come and sleep beside them. . . . for which occasion the Indian and Aethiopian Hunters use this strategem to take the beast. They take a goodly strong and beautiful young man, whom they dress in the apparel of a woman, besetting him with divers odoriferous flowers and spices."

Despite the fantasies of his text, Gesner's thousand woodcuts helped set a new direction in biology. Like the German botanic fathers, Gesner collaborated with artists and provided the most accurate drawing yet of all sorts of creatures, from "the Vulgar Little Mouse," to the Satyre, the Sphinx, the Cat, the Mole, and the Elephant. His illustration for the Rhinoceros, "the second wonder in nature . . . as the Elephant was the first wonder," was made by Dürer. These incunabula of biological illustration began to liberate readers from herbals and bestiaries.

Gesner's work, reprinted, translated, and abridged, dominated zoology after Aristotle until the pathbreaking modern surveys of Ray and Linnaeus, which were not illustrated. His unpublished notes became the basis in the next century of the first comprehensive treatise on insects. For his *Opera Botanica* he collected nearly a thousand drawings, many by himself, but his great work on plants, his first love, he never completed.

He never quite freed himself of his philological obsession. His 158-page book, *Mithridates, or observations on the differences of languages, which have been or are in use among various nations of the whole world* (1555), tried to do for languages what he was already doing for animals and plants. "All" the world's one hundred and thirty languages were described and compared in Gesner's translations of the Lord's Prayer. Incidentally, for the first time, he provided a vocabulary of the Gypsy language.

Gesner found a more characteristically Swiss way to discover nature when he advertised the adventure of exploring the high mountains, which, as we have seen, had so long been a scene of awe and terror. Renaissance Europe saw a brief, if premature, flash of the mountain-adventuring spirit. Petrarch (1304–1374) had led the way near Avignon in 1336 with his ascent of Mt. Ventoux. At the summit he read from the copy of Saint Augustine's *Confessions* that he took from his pocket the caution that people may "go to admire the high mountains and the immensity of the ocean and the course of the heaven . . . and neglect themselves." Leonardo da Vinci, with the eyes of an artist-naturalist, explored Monte Bo in 1511. The Swiss Reformer and humanist Joachim Vadianus (1484–1551), friend of Luther and champion of Zwingli, reached the summit of the Gnepfstein near Lucerne in 1518.

But Gesner was the first European to publish a paean to mountaineering. After his ascent of Mt. Pilatus near Lucerne in 1555, he produced his little classic.

> If you wish to extend your field of vision, cast your glance round about, and gaze off far and wide at everything. There is no lack of lookouts and crags on which you may seem to yourself to be already living with your head in the clouds. If on the other hand you should prefer to contract your vision, you will gaze on meadows and verdant forests, or even enter them; or to narrow it still more, you will examine dim valleys, shadowy rocks and darksome caverns. . . . In truth nowhere else is such great variety found within such small compass as in the mountains; in which . . . one may in a single day behold and enter upon the four seasons of the year, summer, autumn, spring and winter. In addition, from the highest ridges of mountains the whole dome of our sky will lie boldly open to your gaze, and the rising and setting of the constellations you will easily behold without any hindrance; while you will observe the sun setting far later and likewise rising earlier.

Primitive fears were so hard to overcome that two centuries separated Gesner's sallies from the true beginnings of modern mountaineering. Mont Blanc (15,771 ft.), the highest mountain in Europe outside the Caucasus, was not scaled until 1786—by someone who wanted to claim the money-reward offered by a patrician Swiss geologist, Horace-Bénédict de Saussure (1740–1799), twenty-five years before.

55

The Invention of Species

So long as naturalists arranged plants and animals in alphabetic order, the study of nature was doomed to remain bookish and provincial. That order of items would depend, of course, on the language you were reading. The Latin version of Gesner's authoritative encyclopedia opened with *Alces,* the moose, but when translated into German the book began with *Affe,* the ape, while in Topsell's *History of Four-Footed Beasts,* Chapter One described "The Antalope."

Naturalists needed a precise way of naming plants and animals across the language barriers. Even before that, they had to have a common understanding of what they meant by a "kind" of plant or animal. What were nature's units? When pioneer naturalists formulated the concept of "species" they would provide a useful vocabulary for cataloguing the whole creation. In the long run, the new mode of description would open many unanswerable questions. Meanwhile it enlarged the vista of nature's variety. And the quest for a "natural" way of classifying the creation would produce some of the great intellectual adventures of modern time.

In the older popular encyclopedias, such as Topsell's *History of Four-Footed Beasts,* an impenetrable fog enveloped the boundaries between the kinds of animals. Aristotle had described only some five hundred.

A difficulty that we have forgotten lay in the widespread belief in spontaneous generation. Aristotle had written that flies, worms, and other small animals originated spontaneously from putrefying matter. In the seventeenth century the eminent Flemish physician and physiologist Jan Baptista van Helmont (1577–1644?) said that he had seen rats originate in bran and old rags. If animals could arise spontaneously, then it was not feasible to define a species as a creature that reproduced or was reproduced by its own kind.

Only gradually and reluctantly did European naturalists give up this idea. Aristotle's contempt for "lowly" vermin and insects, as we have seen, had been based on his notion that they did not have the differentiated organs found in "higher" animals. Francesco Redi (1626–1697?), a Florentine member of the Accademia del Cimento, who had discovered how snakes produced their venom, was interested in other "lowly" creatures, including insects. After Leeuwenhoek's microscope showed how complex were tiny animals, it was easier for naturalists like his fellow Dutch biologist Swammerdam to argue that these animalcules did not arise by spontaneous generation, but had reproductive organs. And Redi described the parts of insects that produced their eggs. "Flesh and plants and other things . . . putrefiable play no other part, nor have any other function in the generation of insects," he suggested in 1688, "than to prepare a suitable place or nest into which, at the time of procreation, the worm or eggs or other seed of worms are brought and hatched by the animals; and in this nest the worms, as soon as they are born, find sufficient food on which to nourish themselves excellently." Redi had covered putrefying meat with cloth or put it in closed flasks, and so demonstrated that if flies could not reach the meat to lay their eggs no maggots would appear. But he still found some other cases where he suspected spontaneous generation, and the question was to remain alive for two more centuries.

The idea of species would be usefully defined, developed, and applied by biologists long before the notion of spontaneous generation was laid to rest. And the issue was unresolved because it had theological overtones. Radical scientists found the idea of spontaneous generation useful for their natural-scientific explanation of the origin of life, which would have made God's role in the Creation superfluous. Louis Pasteur (1822–1895), the ambitious and hardheaded son of a French tanner, a faithful conservative Catholic and a brilliant experimentalist, saw the matter differently. To him an orderly concept of species was necessary for God's creative work in the Beginning. After acrimonious debate, his simple experiments with fermentation proved the prevalence of microorganisms in airborne dust, and showed that heating and the exclusion of airborne particles would prevent the appearance of vegetation. The successful application of his ideas to "pasteurizing" milk and improving production of beer and wine helped clinch the arguments against spontaneous generation.

When we think of the difficulty of devising a comprehensive system for classifying the whole creation, we are not surprised that the writers of herbals and bestiaries arranged items either alphabetically or according to their human uses. Since the differences between animals are usually more conspicuous than those between plants, the first efforts at general classification were made for animals. Medieval writers derived their first scheme from Aristotle, who had divided the animals with red blood from all others, which he called bloodless. The "blooded" animals were then subdivided according to modes of reproduction (live-bearing or egg-laying) and according to habitat, and the others were subdivided by general structure (weak-shelled, hard-shelled, insects, etc.). Aristotle himself actually used a concept of genus from Greek *genos,* or family; and species from *eidos,* or form, which he seems to have derived from Plato. But for him neither "genus" nor "species" had the sharp definition that they would acquire in modern times. His "genus," or family, designated all groupings larger than the species. Aristotle's rough scheme served European naturalists tolerably well during the Middle Ages, when relatively few novel plants and animals were coming to their notice. They devoted themselves to matching the plants and animals of their region with those described in the ancient texts.

Then in the Age of Discovery countless novelties poured into the European consciousness. How should these be arranged? How could you know whether a particular plant or animal really was new?

Specimens, books, travelers' tales, and newly vivid drawings from nature appeared in profusion and confusion. Encyclopedias like Gesner's piled fancy onto fact. Curiosities from everywhere were jumbled together. For example, a handsomely illustrated volume on the plants and animals of Brazil by the pioneer German illustrator Georg Markgraf (1610–1644) was garbled with William Pies' work on the natural history of the East Indies. Readers were delighted by such potpourris. The word "herbarium" came into use to describe the collection of neatly pressed dried plants piling up in the libraries of noblemen and naturalists. Where should each specimen be placed? How should each one be labeled, organized, or retrieved?

To find a "system" in nature, naturalists first would have to find or make units for their system. This purpose was served by the concept of "species." In the hundred years between the mid-seventeenth century and mid-eighteenth century, more progress was made in cataloguing the varieties of nature than had been accomplished in the whole preceding millennium.

Two great systematizers—Ray and Linnaeus—would accomplish for all plants and animals what Mercator and his fellows did for the planet's whole surface. Just as the map-makers of the earth started from the self-evident boundaries of land and sea, mountains and deserts, the naturalists, too, found self-evident units among plants and animals. Still, as we have seen, even for the earth's surface it was necessary to invent the artificial boundaries of latitude and longitude so others could find their way and all could share the increasing knowledge. Similarly, these naturalists had to supply units that could help others everywhere find their bearings in nature's prolific jungle. Like the "atoms" of the physical system, these "species" would eventually be opened and dissolved, but meanwhile they provided an essential and convenient vocabulary. By the late twentieth century, "species" had become so familiar and so useful that it seemed essential to our thinking about plants and animals, somehow self-evident in the fabric of nature.

In its very beginning the notion of "species" was a labored and controversial product. It was fortunate for the future of biology that John Ray (1627?–1705) invented his definition of species just when he did. Unlike earlier schemes, his applied both to plants and animals and made it possible for his great successor to devise a system for cataloguing the whole creation. At Trinity College, Cambridge, Ray studied classics, theology, and the natural sciences (B.A., 1648), then as a fellow of the college he lectured to undergraduates on Greek and mathematics. Had it not been for the Act of Uniformity, passed by Charles II's Parliament in 1662, he might have remained only another fellow on the college rolls. That Act required clergy, college fellows, and schoolmasters to take an oath accepting everything in the Book of Common Prayer, but Ray would have none of it. Rather than compromise his conscience, he gave up his fellowship.

Another lucky coincidence was Ray's meeting with a wealthy younger member of his college, Francis Willughby (1635–1672), who would make it possible for Ray to spend his life as a private, independent scholar. After a boyhood illness, Ray had formed the habit of country walks, and Ray and Willughby became boon companions, walking the Cambridge countryside together. Ray pursued his scientific interests by describing all the plants he saw, and then went on to survey the plants elsewhere in England. He produced a catalogue of English plants in 1670, incidentally noting variations in proverbs and word usage in different parts of the country, combining the taxonomy of words with that of all other living things. Ray and Willughby together toured the Low Countries, Germany, Italy, Sicily, Spain, and Switzerland, along the way noting the plants. En route they formed a grandiose plan, the sort of youthful pact often made and seldom fulfilled. They would collaborate on a comprehensive *systema naturae*—a description of the whole scheme of nature based on their own observations. Ray would cover the plants, Willughby the animals. This ambitious project was well along when Willughby died in 1672 at the age of thirty-seven.

Meanwhile Ray's letters to Oldenburg had so impressed the Royal Society that not only did they elect him a Fellow but when Oldenburg died in 1677 they offered him the powerful position of Secretary of the Society. But Ray refused, for in his will Willughby had left Ray an annual stipend, and instead of becoming a middleman for other scientists, he preferred to remain an independent naturalist. He moved into Willughby's Middleton Manor, where he revised Willughby's manuscripts and published two substantial treatises, one on birds, and another on fishes, both under Willughby's name.

Then under his own name Ray produced his epoch-making works on plants. His brief *Methodus Plantarum* (1682) offered the first feasible definition of "species," and his *Historia Plantarum* (3 vols., 1686–1704) provided a systematic description of all plants known to Europe at the time. Although Ray started from Aristotle, he went on to develop a more satisfactory arrangement, grouping plants not merely by some single feature like their seed, but according to their whole structure. Following the old axiom that "Nature does not proceed by leaps" *(Natura non facit saltus),* Ray sought out "middle terms," forms that stood between others to fill out the spectrum of the creation. He also improved on Aristotle's general classification of animals, appealing again to affinities of forms. This arrangement has proved useful ever since. Ray went on to survey quadrupeds and serpents, and made the pioneer comprehensive description of insects.

Before Ray's death the grandiose youthful Ray-Willughby scheme for a survey of nature's system based on firsthand observation was near completion. Unlike the alphabetical compendia of Gesner and his predecessors, Ray's work omitted the cherished mythical creatures. Having rid himself of this baggage, and having denied spontaneous generation, he was in a position to define the units of natural life for succeeding generations of naturalists.

Ray's great achievement was his formulation or, more precisely, his invention, of the modern concept of "species." What Newton did for students of physics with his concepts of gravitation and momentum, Ray did for students of nature. He gave them a handle on a system. Like many other world-shaping ideas, his notion was wonderfully simple. Precisely how he came upon it we do not know. But his bold insight and his emphasis must have been stirred by his wide-reaching personal observations in the field. For Ray, finally, the sight of so many different *specimens* suggested the convenience of a concept of *species* (which also derives from the Latin *specere,* "to look at" or "to see.") Unlike his predecessors, he found a system of classification that would serve for both animals and plants.

Others, including Aristotle, had approached the problem by first dividing organisms into large, presumably self-evident, groups, and then subdividing these into smaller and smaller groups. Ray, on the contrary, began with an awe for the uniqueness of individuals and the wonderful variety of "species." As he explained in the Preface to his *Methodus Plantarum*:

> The number and variety of plants inevitably produce a sense of confusion in the mind of the student: but nothing is more helpful to clear understanding, prompt recognition and sound memory than a well-ordered arrangement into classes, primary and subordinate. A Method seemed to me useful to botanists, especially beginners; I promised long ago to produce and publish one, and have now done so at the request of some friends. But I would not have my readers expect something perfect or complete; something which would divide all plants so exactly as to include every species without leaving any in positions anomalous or peculiar; something which would so define each genus by its own characteristics that no species be left, so to speak, homeless or be found common to many genera. Nature does not permit anything of the sort. Nature, as the saying goes, makes no jumps and passes from extreme to extreme only through a mean. She always produces species intermediate between higher and lower types, species of doubtful classification linking one type with another and having something in common with both—as for example the so-called zoophytes between plants and animals.
>
> In any case I dare not promise even so perfect a Method as nature permits—that is not the task of one man or of one age—but only such as I can accomplish in my present circumstances; and these are not too favourable. I have not myself seen or described all the species of plants now known.

For Ray, a species of plants, for example, was a name for *a set of individuals who give rise through reproduction to new individuals similar to themselves.* Among animals the same definition would apply. Bulls and cows were members of the same species because when they mated they produced a creature like themselves.

Ray believed that, as a general rule, each species was fixed and did not vary throughout the generations. "Forms which are different in species always retain their specific natures, and one species does not grow from the seed of another species." As time passed and he studied more and more specimens he saw that minor mutations might be possible. "Although this mark of unity of species is fairly constant," he concluded, "yet it is not invariable and infallible."

Biologists after Darwin uncharitably criticized Ray for his belief in the fixity of species, a proposition that his successor Linnaeus embraced with even more enthusiasm. But in his own day, Ray's insistence on that fixity and continuity of species was a giant step forward. It would make possible an internationally usable catalogue of the whole natural world. His insistence on the power of each species to continue to generate like organisms helped Ray dispose of much baggage that had burdened biologists from antiquity through the age of Gesner. He helped rid scientific literature of the mythical creatures attested in belles lettres and folklore who always propagated more mythical creatures. And he put an indelible question mark beside all "spontaneously generating" creatures. Just as the post-Newtonian world was governed by the laws of physical gravitation, at last biologists were being led into a world governed by the laws of biological generation.

Lyell and other pioneers of geology would introduce uniformitarianism into the history of the earth. Ray brought uniformitarianism into the history of plants and animals. Neither Lyell nor Ray told the whole story, but they both helped open the vistas of time, a new world for evolution and its unsolved problems. Ray was among the first to suggest that the fossil shapes found in mountains and within the earth were not mere accidents but the remains of once living creatures. And he followed through with the possibility that many prehistoric species might have become extinct. Which justified his epitaph (translated by someone from the Latin):

> Nor did his artful labours only shew
> Those plants which on the earth's wide surface grew,
> But piercing ev'n her darkest entrails through
> All that was wise, all that was great he knew
> And nature's inmost gloom made clear to common view.

56

Specimen Hunting

LINNAEUS inherited Ray's mission. His System of Nature, while more comprehensive and more influential than any before, would be built from elements bequeathed by Ray. Sharing a faith in the coherence of nature, Linnaeus would promote Natural Theology as much as Natural Science. He too made "species" his clues to the wisdom of the Creator.

But in their personalities and ways of working Ray and Linnaeus had little in common. Ray, the lonely and humble acolyte of his boon companion and fellow scholar Willughby, wrote mainly from his own observation. Linnaeus, sociable and conceited, was a brilliant teacher, inspiring and organizing legions of specimen hunters to scan the world and send him their findings—for the greater glory of God and of Linnaeus.

Like Ray, Carolus Linnaeus (1707–1778) was intended for the ministry. Born in southeastern Sweden to an impoverished pastor who awakened his love of plants in the parsonage garden, Linnaeus was raised in Stenbrohult, which he called "one of the most beautiful places in all Sweden, for it lies on the shores of the big lake of Möckeln. . . . The church . . . is lapped by the clear waters of the lake. Away to the south are lovely beech woods, to the north the high mountain ridge of Taxas. . . . To the northeast are pine woods, to the southeast charming meadows and leafy trees." He never forgot these infectious charms. "When one sits there in the summer and listens to the cuckoo and the song of all the other birds, the chirping and humming of the insects; when one looks at the shining, gaily coloured flowers; one is completely stunned by the incredible resourcefulness of the Creator."

Yet at school Carolus showed so little interest in theology that his disgusted father was about to apprentice him to a shoemaker. A perceptive teacher persuaded the father to let Carolus try to make his way as a medical student. At Uppsala he substituted for the professor doing demonstrations in the university's botanical gardens. Then in 1732 he was sent by the Uppsala Society of Science on an expedition to mysterious Lapland, to gather specimens and information on local customs. This first strenuous encounter with strange flora and exotic institutions dazzled him with a delight he had never felt so poignantly in neat botanical gardens, nor even in the pages of herbals or bestiaries.

On his return he went to the Netherlands, then a center of medical learning, to qualify himself to make a living as a doctor and also to pursue his botanical ambitions. Within the next three years, even before he was thirty, Linnaeus sketched his grand scheme. His succinct *Systema Naturae* (Leyden, 1735) of only seven folio pages, the first work he published in the Netherlands, was a prospectus for his lifework and for all modern systematic biology. Even before, at Uppsala when he was only twenty-two, he had described the essence of his system to the professor with whom he was living. His New Year's Day greetings then apologized for his inability to offer the customary verse. " 'Poets are born, not made,' I was not born a poet, but a botanist instead, so I offer the fruit of the little harvest which God has vouchsafed me. In these few pages is handled the great analogy which is found between plants and animals, in their increase in like measure according to their kind, and what I have here simply written, I pray may be favorably received." His botanical system was possible because, like Ray, he was not looking at plants alone. But going beyond Ray, he boldly adapted a concept from the animal world for the whole living creation.

Linnaeus was the Freud of the botanical world. With our late twentieth-century freedom to discuss sexuality, we forget the embarrassment in "mixed company" in the pre-Freudian age at public mention of any sexual organs, even though they were only those of plants. In Linnaeus' botany, as in Freud's psychology, the primary fact was sexuality.

Ever since Ovid, poets had played with the metaphor of sexuality in plants. But most people still regarded such suggestions in prose to be perverse, if not obscene. A few naturalists had hinted at and some had dared to demonstrate the phenomenon. The French botanist Sebastien Vaillant (1669–1722), in charge of the Jardin du Roi (now called the Jardin des Plantes), using the peculiarities of the pistachio tree that still stands in its Alpine garden in Paris, had boldly opened his public lectures in 1717 with a demonstration of the sexuality of plants, which awakened the adolescent Linnaeus' interest and set him scrutinizing every plant to count its genital organs.

Some decades before, the essential fact had been revealed by a German botanist, Rudolph Jacob Camerarius (1665–1721), who showed that a seed would not germinate without the cooperation of pollen. But when Linnaeus was a student at Uppsala, the sexuality of plants was still an open and very sensitive question. In the title of his paper, *Sponsalia Plantarum* (1729), he used the discreet language of metaphor—"an Essay on the betrothal of plants, in which their physiology is explained . . . and the perfect analogy with animals is concluded." Just as in the spring the sun animates and enlivens the dormant bodies of animals, so plants, too, he said, wake up from a winter sleep. Plants, like animals, are barren when young, are most fertile in middle years, and waste away in old age. With the microscope, he noted, Malpighi and Nehemiah Grew (1641–1712) had recently shown that plants, like animals, really had differentiated parts. Was it not only logical that they, too, should have organs of generation?

Vaillant had located these organs in the flower, for he said that no fruit was ever produced without a flower. But, the young Linnaeus objected, the botanists who had focused on the corolla or petals were not quite right, because some plants did bear fruit even though they had neither calyx nor petals. The generative organs, Linnaeus ventured, which ought to be the basis of classification were, rather, the stamens and the pistils, whether found on the same or on different plants of the same species. In a cloying passage designed to satisfy the most reverent or the most squeamish, he gives us a clue to the inhibitions of his age. The petals of a flower, he explained, did not directly aid the process of generation. But their attractive shapes and colors, perfumed with appealing odors, had been devised by an ingenious Creator so that the "bridegrooms" and the "brides" of the plant kingdom could celebrate nuptials in their own delightful "bridal beds."

When he arrived in the Netherlands, Linnaeus was already equipped with the data from his field trips and his metaphor of a "sexual system" to make his grand outline. In those seven folio leaves of his *Systema Naturae*, he drew on Ray's notion of species and made each self-generating group of plants a building block. If the self-generating species was basic, it was natural that in Linnaeus' system the generative or "sexual" apparatus of each plant should be the hallmark of classification.

In the details of Linnaeus' argument we begin to see both the boldness of his emphasis on sexuality and why some contemporaries called him salacious. The twenty-three *classes* of flowering plants were distinguished on the basis of the "male" organs (i.e., the relative length and number of the stamens). His twenty-fourth *class (Cryptogamia)*, of the plants like mosses which appeared flowerless, were distinguished into *orders* on the basis of their "female" organs (the styles or stigmas). He made up their names from Greek words with plain sexual and generative overtones, drawing on such Greek words as *andros* (male), *gamos* (marriage), *gyne* (female). He described the class *Monandria* as like "One husband in a marriage," the *Diandria* as "Two husbands in the same marriage." The poppy *(Papaver)* and the linden *(Tilia)*, being *Polyandria*, he observed, showed "Twenty males or more in the same bed with the female." His *Philosophia Botanica* (1751) continued to insist on the calyx as a nuptial bed *(thalamus)* with the corolla acting as a decent curtain *(aulaeum)*. "The calyx," he said, "might be regarded as the *labia majora* or the foreskin; one could regard the corolla as the *labia minora.*" "The earth is the belly of the plants; the *vasa chylifera* are the roots, the bones the stem, the lungs the leaves, the heart the heat; this is why the ancients called the plant an inverted animal." He advised "those who want to penetrate further into the mystery of the sex of plants" to consult his *Sponsalia Plantarum.*

We cannot be surprised that proper professors were troubled by such explicitness. But not Erasmus Darwin (1731–1802), the grandfather of Charles, who soon cast the Linnaean system into a grand epic of heroic couplets, *The Botanic Garden* (1789, 1791). There he described "the Ovidian metamorphosis of the flowers, with their floral harems," the impatient male stamens (belonging to floral beaux, lovers, swains, husbands, and knights) pursuing the recumbent pistils (belonging to virgins, wives, and nymphs). In the lily flower of the *Colchicum* genus:

> *Three* blushing maids [pistils] the intrepid nymph attend
> And *six* youths [stamens], enamour'd train! defend

The flower of turmeric *(Curcuma),* a tropical plant of the ginger family, which Linnaeus had distinguished by its one fertile stamen and its four sterile stamens, was where,

> Woo'd with long care, Curcuma cold and shy
> Meets her fond husband with averted eye:
> Four beardless youths the obdurate beauty move
> With soft attentions of Platonic love.

Other readers did not find it so easy to etherealize Linnaeus. Even an accomplished botanist like the Reverend Samuel Goodenough (1743–1827), a vice-president of the Royal Linnaean Societies who had a plant, goodwinia, named after him, could not conceal his embarrassment at "the gross prurience of Linnaeus' mind. . . . A literal translation of the first principles of Linnaean botany is enough to shock female modesty. It is possible that many virtuous students might not be able to make out the similitude of *Clitoria.*" As late as 1820, even the iconoclastic Goethe was still hoping that young people and women could be shielded from Linnaeus' gross "dogma of sexuality."

The motives behind Linnaeus' sexual system were not mere convenience or prurience. Self-generating species were essential to an all-wise Creator's self-generating nature in which all the units would continue to fit together.

Linnaeus shared both Aristotle's belief in some intelligible underlying order and Aristotle's love of facts. The varied devices by which the Creator had provided for the perpetuation of the system were an awesome spectacle.

Besides his debt to Camerarius, Linnaeus owed most to Andrea Cesalpino, who had directed the botanical garden in Pisa before he became physician to Pope Clement VIII in 1592. A thorough Aristotelian, Cesalpino believed that plants were animated by a vegetable "soul" which both nourished and reproduced them. Their nutrition came entirely from roots in the soil, then up through the stem into the fruit. Cesalpino suggested a classification based on general outward structure—roots, stems, and fruit. He thus avoided altogether the problem of classifying the "lower" plants like lichens and mushrooms, which he believed to lack organs, including those for reproduction, found in higher plants, and which, he explained, sprang by spontaneous generation from putrefying matter. Still, Cesalpino's focus on the general structure of individual plants was a long step forward.

The dominant Aristotelian tradition, as we have seen, had begun from large *a priori* categories based on gross preliminary impressions. Ray's historic departure was to make the *species* his elementary unit. In the modern incremental style, Linnaeus, carrying on with Ray, built up his system from the individual species, which could be scrutinized in specimens. With stamens and pistils as starting points, he used the number and order of stamens to group all plants into twenty-four classes, then according to the number of pistils he subdivided each class into orders. This simple plan was easy to use in the field, and even without a library anyone who could count could classify a plant.

While the "sexual" system provided a simple classifying concept, the nomenclature of biology was still cumbersome, vague, and variant. A growing worldwide community of naturalists would need a common language to be sure they were talking about the same thing. Linnaeus would invent the syntax. Efforts to create other sorts of international language had never had such success. But Linnaeus managed to create an international language, a kind of Esperanto of biology. So he found a universal use for Latin long after it had ceased to be the European language of learning. His "botanical Latin" was based not on classical Latin but on medieval and Renaissance Latin, which he reshaped for his purpose.

In retrospect the binomial nomenclature (e.g., *Homo sapiens,* for genus and species) seems so simple and obvious that it hardly needed to be invented. But before Linnaeus devised his binomial scheme, there was no generally agreed-upon scientific name for any particular plant. The earlier names affixed by different writers aimed to serve both as designation and as description. When more species came to be known and more was known about each plant, names became longer and more confused. Take, for example, the plants of the genus *Convolvulus,* trailing plants of the morning-glory family with funnel-shaped flowers and triangular leaves. In 1576 the French botanist Charles de Lécluse (1526–1609) designated one species as *Convolvulus folio Altheae.* In 1623 the Swiss botanist Gaspard Bauhin (1560–1624) called this same species *Convolvulus argenteus Altheae folio,* which in 1738 Linnaeus amplified to *Convolvulus foliis ovatis divisis basi truncati: laciniis intermediis duplo longioribus,* which by 1753 he had elaborated further into *Convolvulus foliis palmatis cordatis sericeis: lobis repandis, pedunculis bifloris.* And so it went.

Linnaeus came to his solution only gradually, in a search for accurate names that would be usable in the field and handy for the amateur. He did not expect his students on field trips to learn or remember the full Latin description. He did expect them to remember the name of the genus (in the above case, *Convolvulus)* and then in their notes record a number (e.g., "Convolvulus No. 3") referring to the entry for that species in the full list of plants that Linnaeus had published. This gave a hint for a simple binomial system, which could be produced by substituting a word for the number.

The obstacle again was Linnaeus' temptation to make each plant's specific name serve both as a label and as a description. His great simplifying decision was to split these two functions. He would provide only a short, easy-to-remember label. This the student could use when he returned to his library, where it would lead him to a detailed account of the distinguishing features of that species. In the 1740's he tried this for a few plants, but he still stigmatized these as "trivial names" *(nomina trivialia).* To use the species name along with the genus, Linnaeus said, was "like putting the clapper in the bell." Then in his epochal *Species Plantarum* (1753), after twelve months of intensive labor, he supplied such binomial labels for all the fifty-nine hundred species on his list.

Linnaeus wisely realized that it was better to have some handy distinctive name for each species at once than to wait until the perfect word or a thoroughly symmetrical vocabulary could be found. He had to move quickly if he was to accomplish his task at all. Unless he speedily gave some such binomial label to *every* known species, naturalists would be tempted to use the same name for more than one species, which of course would have defeated the whole scheme. His was a monumental task of hasty linguistic invention. He ransacked his Latin for enough terms to make up thousands of labels, sometimes using a single word describing a plant's manner of growth (e.g., *procumbens*), other times using a word for the habitat or for the first discoverer of the plant, or even a Latinized form of a vernacular word. Linnaeus was not too rigorous in the logic of his usage, provided the word was distinctive and rememberable.

When, a few years later, in the definitive tenth edition of his *Systema Naturae* (1758–59) he extended his scheme to animals, Linnaeus showed a similar practical sense. For insects he used specific names designating color or the host plant. To distinguish species of butterflies, he drew on his copious classical learning, and added such epithets as Helena, Menelaus, Ulysses, Agamemnon, Patroclus, Ajax, or Nestor. Then again, in deference to vernacular usage he set up the genus *Felis,* included the lion, tiger, leopard, jaguar, ocelot, cat, and lynx, and designated them by their common Latin names, Leo, Tigris, Pardus, Onca, Pardalis, Catus, and Lynx.

When was there another such colossal feat of name-giving since the Creation? Any parent who has had to name a child can imagine the enormous task of christening that Linnaeus completed in a single year. Within a few decades, even before his death in 1778, his names and his scheme of naming were adopted by his European colleagues. His choices proved themselves over the centuries and would reach across the world. Linnaeus made a world community of naturalists.

The Age of Discovery, meanwhile, had vastly widened the Europeans' vision of nature. From Asia, Africa, Oceania, and the Americas came news of strange plants like the tomato, maize, the potato, cinchona, and tobacco, and new animals like the penguin or "magellanic goose," the manatee, the dodo, the horseshoe crab, the raccoon, the opossum, and countless others.

Linnaeus inspired an unprecedented worldwide program of specimen hunting. His work gave generations of specimen hunters a new incentive to advance science, even at the risk of their lives. No longer would their hard-won finds be relegated to attics or buried in the meaningless jumble of "cabinets of curiosities." Now every plant or animal newly "identified" by Linnaeus' system contributed to a systematic worldwide survey.

Linnaeus himself commanded cohorts of his apostles—his cleverest pupils, "the true discoverers . . . as comets among the stars," who covered the earth. In 1746 his ablest student, Christopher Tärnström, begged to be allowed to go (with free passage on a Swedish East India Company ship) as Linnaeus' emissary to gather specimens in the East Indies. When Tärnström died of a tropical fever on arriving in the Sea of Siam, Linnaeus futilely tried to make amends to the distraught widow and children by naming a tropical genus *Ternstroemia.*

Peter Kalm, another student, had better luck. Linnaeus secured financing for Kalm's costly travels from a group of Swedish manufacturers and from the universities of Uppsala and Åbo. An expedition to lands in the same latitude as Sweden would find new plants to be grown in Sweden for medicine, food, or manufacture. The imported red mulberry, they hoped, would feed silkworms for a whole new industry. These hopes never materialized, but Kalm otherwise proved to be one of the most productive of the specimen hunters. In 1748 after a rough Atlantic crossing, the indefatigable Kalm arrived in Philadelphia, visited his fellow Swedes in Delaware, then with help from Benjamin Franklin and two of Linnaeus' best correspondents, John Bartram and Cadwallader Colden, he explored Pennsylvania and went north to New York and Canada. Linnaeus eagerly awaited the botanical finds, and when Kalm arrived back in Stockholm in 1750, the gout-ridden Linnaeus leaped from bed to greet his adored pupil. Three years later Linnaeus' *Species Plantarum* cited Kalm as his source for ninety species, sixty of them new, and he immortalized Kalm in a whole mountain-laurel genus, *Kalmia*. Kalm's journal, which prophesied American independence, gave one of the most vivid descriptions of colonial life in the New World.

Frederick Hasselquist (1722–1752) was sent with money raised by Linnaeus to Egypt, Palestine, Syria, Cyprus, Rhodes, and Smyrna—all still unexplored by European naturalists. When his expenses exceeded his budget, Linnaeus persuaded the Swedish Senate to make private contributions. And when Hasselquist, still only thirty, died near Smyrna, his creditors refused to release his botanical notes until his debts were paid. Again Linnaeus came to the rescue by inducing the Queen of Sweden to pay off the debts. And when he finally read the journals of his deceased disciple, he was ecstatic. "They penetrate me as God's word penetrates a deacon. . . . May God grant that Her Majesty has them published as soon as possible, so that all the world may taste the pleasure I had yesterday." Linnaeus himself published the *Iter Palaestinum* in 1757, and the world could soon enjoy Hasselquist's discoveries through translations in English, French, German, and Dutch.

To China in 1750 he dispatched another pupil, Pehr Osbeck (1723–1805), as ship's chaplain. "On your return," he wrote, "we will make crowns with the flowers you bring back, to adorn the heads of the priests of the temple of Flora and the altars of the goddess. Your name shall be inscribed on substances as durable and indestructible as diamonds, and we will dedicate to you some very rare *Osbeckia* which will be enrolled in Flora's army. So —hoist your sails and row with all your might; but take heed not to return without the choicest spoils, or we shall invoke Neptune to hurl you and all your company into the depths of the Taenarum." Osbeck heeded the warning, and on his return he delivered to his mentor a rich Chinese herbarium of six hundred specimens.

Nearer home, when the King of Spain requested a Linnaean disciple for a botanical survey of his country, Linnaeus sent "his most beloved pupil,"

Petrus Löfling (1729–1756), who had been living with Linnaeus as companion to his son. Löfling's work in Spain stimulated an expedition to Spanish
South America, with Löfling as chief botanist, aided by two surgeons and
two artists, "to collect specimens for the Spanish Court, the King of France,
the Queen of Sweden, and Linnaeus." But, before Löfling could complete
his mission, he died of a tropical fever in Guiana at the age of twenty-seven.
"Löfling sacrificed himself for Flora and her lovers," Linnaeus lamented,
"they miss him!"

The troubled Linnaeus asked, "The deaths of many whom I have induced
to travel have made my hair grey, and what have I gained? A few dried
plants, with great anxiety, unrest, and care." Still, during the last thirty
years of his life he continued to enlist, organize, and dispatch his apostles
around the world. In 1771, he surveyed his messianic strategy:

> My pupil Sparrman has just sailed for the Cape of Good Hope, and another of
> my pupils, Thunberg, is to accompany a Dutch embassy to Japan; both of them
> are competent naturalists. The younger Gmelin is still in Persia, and my friend
> Falck is in Tartary. Mutis is making splendid botanical discoveries in Mexico.
> Koenig has found a lot of new things in Tranquebar [in south India]. Professor
> Friis Rottböll of Copenhagen is publishing the plants found in Surinam by
> Rolander. The Arabian discoveries of Forsskål will soon be sent to the press in
> Copenhagen.

Linnaeus' worldwide movement gained momentum with the years. Answering a request from England, he sent another favorite pupil, Daniel
Solander (1736–1782), who became his link to the expeditions of the next
centuries. Solander charmed his way up English society, then became librarian to Sir Joseph Banks (1743–1820), who was the European patron of
natural history in the next generation. Banks promoted, organized, and
personally financed natural-history expeditions, and, as we have seen, took
Solander along on Captain Cook's *Endeavour* voyage (1768–71) around the
world. But Linnaeus was disappointed in Solander, who, despite Linnaeus'
schemes, never married his eldest daughter, and then from the round-the-
world voyages "the ungrateful Solander" never sent Linnaeus a single plant
or insect. Banks, who had covered Solander's expenses and had bought
costly equipment, was also disappointed. For he had hoped that Linnaeus
would be willing to come to England to help give names to the finds of the
voyage—twelve hundred new species and one hundred new genera of
plants, with many more animals, fishes, insects, and mollusks.

After Solander's work with Banks it became customary for every exploring ship to carry a naturalist, along with an artist to depict the finds. As
a botanist on his second voyage around the world, Captain Cook chose

another Linnaeus pupil, the young Anders Sparrman (1748–1820), who at the age of seventeen had already gone to China as surgeon on a Swedish East India ship and brought back a treasury of specimens. After returning from Cook's voyage, Sparrman carried his botanical searches into Senegal and the west coast of Africa.

One of the most enterprising apostles was Carl Peter Thunberg (1743–1828), the last of his disciples promoted by Linnaeus himself. At the time the Dutch, on their tiny trading post on the island of Deshima in Nagasaki Bay, were the only Europeans with a foothold in Japan. To catalogue the flora of Japan, Thunberg would have to secure the protective coloration of a Dutchman. Therefore he spent three years in Cape Colony learning Dutch. Incidentally, while he was there he voyaged into the interior, and described three thousand plants, of which about one thousand were new species. In 1775, when he arrived on a Dutch ship at Deshima, the only excursion he was permitted was to accompany the Dutch ambassador on his annual ceremonial visit to the Emperor in Tokyo. Luckily, the young Japanese interpreters on Deshima turned out to be physicians eager to learn European medicine, morsels of which Thunberg exchanged for specimens of Japanese plants. When Japanese servants brought fodder from the mainland for the cattle on Deshima, Thunberg would rummage through it to find specimens for his herbarium. After nine years' absence, Thunberg finally returned to Sweden where he grieved to find that his mentor had died the year before.

In the next generation the unauthorized apostles of Linnaeus were an energetic crew. Pursuant to the custom established by Solander, Sparrman, and Thunberg, the twenty-two-year-old Darwin was enlisted in 1831 as naturalist on H.M.S. *Beagle.* In 1846 the persuasive Thomas Henry Huxley, who had gathered specimens as assistant surgeon on H.M.S. *Rattlesnake* in the South Seas, set a precedent when he secured three years' leave on Navy pay to analyze his finds. The brilliant young Joseph Dalton Hooker (1817–1911), who was carried as assistant surgeon and naturalist on Captain James Clark Ross' expeditions to the Antarctic (1839–43) on H.M.S. *Erebus* (with H.M.S. *Terror*), produced six volumes on polar flora which secured him a Navy commission to study the flora of the Himalayas and Ceylon, and finally made Kew Gardens a world center for botanical research.

The same faith that nourished Linnaeus' quest for a "system" in nature also had convinced him that it was impossible for any man fully to grasp the plan of his Creator. He knew very well that his "sexual" scheme was artificial, only a handy way to file specimens. A strictly *natural* classification would have to group together plants that shared the largest number of attributes.

Linnaeus showed common sense when he seized on Ray's concept of "species" as a useful handle on the whole creation. But he was not above using his theology to validate his vocabulary of convenience. "We can count as many species now," was Linnaeus' most quoted aphorism, "as were created in the beginning." The constancy and permanency of species was, of course, essential to justify the trouble of classification. Why bother to file plants in different species if at any time they could slide into another species or disappear without warning?

As Linnaeus' disciples gathered more thousands of "species," with more examples of hybridization, he began to venture the possibility that in the beginning perhaps not quite *all* species had been created. Perhaps new species could arise later by the combination of the primordial species of one genus with a species of another genus. This opened some chaotic possibilities, and when Linnaeus occasionally speculated on the origins of species, he went off the deep end. Luckily, religious faith and a practical temperament kept him from plaguing himself with origins—probably knowable anyway only by the Creator. *"Deus creavit, Linnaeus disposuit"*—God created, Linnaeus classified—his admirers boasted, with only a hint of blasphemy.

57

Stretching the Past

AMONG the learned in the Europe of their day, it would have been hard to find a sharper contrast to Linnaeus than his aristocratic contemporary Georges-Louis Leclerc, Comte de Buffon (1707–1788). In retrospect they seem allies in the discovery of nature, but in their own time they were notorious antagonists. Perhaps his youth in a poor rural parsonage had led Linnaeus to insist that nature must consist of changeless building blocks "as many as were created in the beginning." Buffon spoke for an urbane world of change. Born into a monied family in Burgundy, where his father was an officer of the bureaucratic nobility, Buffon was educated at a good Jesuit college and the University of Dijon, where he pursued his father's ambition that he become a lawyer. Then, at the University of Angers, he turned to medicine, botany, and mathematics. After a duel he had to leave the university, and he took off on a Grand Tour in the suitable company of the Duke of Kingston and the Duke's tutor, who happened to be a member of the

Royal Society. Returning home, Buffon found that his mother had died, and that his father had remarried and had seized the rich estates which should have come to him from his mother's side. After a bitter quarrel with his father with whom he was never again on speaking terms, he managed to secure for himself the ample estates, including the village of Buffon, which gave him his noble name. The twenty-five-year-old Buffon promptly set himself up as a provincial lord.

Meanwhile he vigorously pursued his scientific interests. In Paris, Buffon first became known for his report to the Navy on the tensile strength of timbers used in ships of war. A paper on probability theory, which brought him *adjoint-mécanicien* membership in the French Academy, was followed by works on mathematics, botany and forestry, chemistry and biology. He used the microscope for his research on the organs of animal reproduction. He translated into French Stephen Hales' *Vegetable Staticks* and Newton's work on the calculus. At twenty-eight, for his impressive attainments he was recognized by the King, who named him superintendent of the royal botanical gardens.

For fifty years Buffon spent spring and summer on his estates in Burgundy, fall and winter in Paris. In the country, rising at dawn, he gave mornings to science, afternoons to business. In his evenings in Paris he charmed the wittiest hostesses in salons, where, as William Beckford acidly recorded, "Zoology, Geology, and Meteorology formed the chief topics discussed, but tautology prevailed over all." After a half-century of this routine he not only was rich from his increased landholdings but had doubled the area and enlarged the buildings of the royal botanical gardens, and had published thirty-six volumes of his *Histoire Naturelle* and scores of important articles on every branch of science. Louis XV made him the Comte de Buffon, Catherine the Great honored him, and he was elected to scientific academies in London, Berlin, and St. Petersburg.

Buffon's fame reached America, which had joined the expanding European community of science. Thomas Jefferson, stationed in Paris in 1785 as American minister to France, had the Marquis de Chastellux deliver to Buffon a copy of his *Notes on Virginia,* just off the press, along with a large American panther skin to contradict Buffon's thesis of the degeneration of animals in the New World. This brought Jefferson an invitation to come discuss natural history and dine in Buffon's gardens. As Jefferson recalled, "It was Buffon's practice to remain in his study till dinner time, and receive no visitors under any pretense; but his house was open and his grounds, and a servant showed them very civilly, and invited all strangers and friends to remain to dine. We saw Buffon in the garden, but carefully avoided him; but we dined with him, and he proved himself then, as he always did, a man of extraordinary powers in conversation."

At the age of forty-five Buffon married a beautiful girl twenty-five years his junior, who died young. Their daughter died in infancy, and their pampered only son (whom Catherine the Great used as an example of geniuses' sons who are imbeciles) was guillotined by Buffon's enemies during the Terror in 1794. After his wife's death Buffon's only liaison was a platonic affair with his "sublime friend" Madame Necker, wife of the French Minister of Finance. When he was bedridden in the last year of his life, she visited him daily. "M. de Buffon has never spoken to me of the marvels of the earth," she wrote, "without inspiring in me the thought that he himself was one of them."

In an age when the sciences had newly gone public, Buffon was a pioneer of popular science, which required a new view of language. Of course, he read Latin, but he wrote in French, which for him was an act of faith—not glossing texts for a learned few but presenting facts to the nation. "Style is the man himself," he declared in his classic *Discours sur le Style* (1753), delivered on his reception into the French Academy. He was suspicious of writers who refined their subtleties, whose thought was "like a leaf of hammered metal, acquiring luster at the expense of substance." Rousseau called him the most beautiful stylist, and his lyrical prose (he wrote no verse) led some to place him among the leading French "poets" of his century.

The thirty-six volumes of Buffon's *Histoire Naturelle* (1749–85), which appeared during his lifetime, supplemented by eight volumes published (1788–1804) after his death, covered every subject in nature from man and birds to cetaceans, fishes, and minerals. For the first time in publishing history, books of popular science were best sellers. His work rivaled Diderot's thirty-five-volume *Encyclopédie* (1751–72), the most successful European publishing venture of the century, which gave its name to the age. Diderot's work was conspicuously collaborative, Buffon's, despite some assistance, was unquestionably his own.

Buffon took aim at the large audience of laymen. In his famous article on the camel a single Proustian paragraph-sentence recaptured the desert (as we can see in this translation by Otis E. Fellows and Stephen F. Milliken):

Try to imagine a country without greenery and without water, a burning sun, a sky always dry, sandy plains, mountains more arid still, over which the eye sweeps in vain and sight is lost without once fixing upon a living object; a dead land, as though stripped bare by the hot wind, offering to the eye only the remnants of bones, scattered stones, outcroppings of rock, upright or fallen, a desert without secrets in which no traveler has ever drawn a breath in the shade, or found a companion, or anything to remind him of living nature: absolute solitude, a thousand times more terrifying than that of the dense forests, for trees are other beings, other life, to the man who sees himself alone; more isolated, more naked, more lost, in these empty and limitless lands, he stares into space, on all sides, space that is like a tomb; the light of day, more melancholy than the shadows of night, is reborn only to shine upon his nakedness and impotence, to let him see more clearly the horror of his situation, driving back the boundaries of the void, extending around him the abyss of the immensity that separates him from the land of men, an immensity that he will attempt in vain to cross, for hunger, thirst, and the scorching heat press upon every moment that remains between despair and death.

Yet his descriptions of some animals were so concise that they were collected to make books for children.

While the stark sexual nomenclature of Linnaeus had simply shocked, Buffon found romance in the sexual activity of his animals. For example, he contrasted the mating of sparrows and of pigeons.

There are few birds as ardent, as powerful in love as the sparrow; they have been seen to couple as many as twenty times in succession, always with the same eagerness, the same trepidation, the same expression of pleasure; and, strange to say, the female seems to grow impatient first with a game that ought to tire her less than the male, but it can please her also much less, for there are no preliminaries, no caresses, no variety to the thing; much petulance without tenderness, movements always hasty, indicative only of a need to be satisfied for its own sake. Compare the loves of the pigeon to those of the sparrow, and you shall see almost all the nuances that extend from the physical to the moral.

Meanwhile, among the pigeons,

tender caresses, soft movements, timid kisses, that become intimate and urgent only at the moment of enjoyment; this moment even, brought back within seconds by new desires, new approaches equally nuanced; an ardor ever durable, a taste ever constant, and a still greater benefit, the power to satisfy them repeatedly, without end; no bad temper, no disgust, no quarrel; an entire lifetime devoted to the service of love and to the care of its fruits.

His work was emphatically not a "system" but a description, "a natural history."

Since the unity that Buffon saw was in the processes of nature, he was wary of nomenclature, whether provided by God or by Linnaeus. It is not surprising that Linnaeus became his *bête noire*. Buffon believed that taxonomy was just a learned technique for making the world seem simpler than it really was. By using stamens to classify plants, Linnaeus had put the veneer of a word over what was really a miscellany. Surely, eyes were given man to distinguish plants from one another, yet Linnaeus' artificial scheme depended on features so minute they could be seen only with a microscope. Buffon concluded that Linnaeus' "system" had actually "made the language of science more difficult than science itself."

Taxonomy and nomenclature, Buffon warned, were only games. His own "true method" was simply "the complete description and exact history of each thing in particular." "One must not forget that these *families* [confidently used by Linnaeus and others] are our creation, we have devised them only to comfort our own minds." To grasp all distinctive features of a particular individual, it is not enough to describe only the individual in hand. We must try to envisage everything about that animal, which means compiling the history "of the entire species of that particular animal . . . their procreation, gestation period, the time of birth, number of offspring, the care given by the mother and father, their education, their instincts, their habitats, their diet, the manner in which they procure food, their habits, their wiles, their hunting methods."

Without any pretense at knowing how many "species" God had created "in the Beginning," Buffon, following Ray's lead, satisfied himself with a purely empirical definition:

> We should regard two animals as belonging to the same species if, by means of copulation, they can perpetuate themselves and preserve the likeness of the species; and we should regard them as belonging to different species if they are incapable of producing progeny by the same means. Thus the fox will be known to be a different species from the dog, if it proves to be the fact that from the mating of a male and a female of these two kinds of animals no offspring is born; and even if there should result a hybrid offspring, a sort of mule, this would suffice to prove that fox and dog are not of the same species—inasmuch as this mule would be sterile.

Mere external resemblance would not prove animals to be of the same species "because the mule resembles the horse more than the water spaniel resembles the greyhound."

Yet he was awed by the very concept of species, and wary of oversimplifying its nuances. His diffidence was much deeper than that of his predecessors. Buffon could not bring himself to believe that "species" provided a key to any divine scheme or a clue to theological truth.

> In general, the kinship of species is one of those profound mysteries of nature which man will be able to fathom only by means of long and repeated and difficult experiments. How, save by a thousand attempts at the cross-breeding of animals of different species, can we ever determine their degree of kinship? Is the ass nearer to the horse than to the zebra? Is the dog nearer to the wolf than to the fox or the jackal? At what distance from man shall we place the great apes, which

resemble him so perfectly in bodily conformation? Were all the species of animals formerly what they are today? Has their number not increased, or rather, diminished? . . . How many more facts we shall need to know before we can pronounce —or even conjecture—upon these points! How many experiments must be undertaken in order to discover these facts, to spy them out, or even to anticipate them by well-grounded conjectures!

The Bible had, of course, disposed of all such troublesome problems in the six days when God created heaven and earth "and every living creature which moveth." Respectable biologists, including Ray and Linnaeus, had made this their point of departure. Since it was an axiom that species could not be either added or subtracted, the precise extent of time since the Creation held little significance for the biologist. Biblical scholarship in the seventeenth century kept biologists focused on those six days of Creation. It seemed both absurd and heretical to suggest that nature had a history. What interested Biblical scholars was the chronology of the Bible in relation to human events.

The Irish prelate James Ussher (1581–1656), an expert in Semitic languages, managed to provide for the first time a generally acceptable Biblical chronology, still found in many editions of the English Bible. A scholar of Trinity College, Dublin, he became a fellow, went to England to collect books for the college library, and then became professor of divinity and archbishop of Armagh. While strident in demanding autonomy for the Church of Ireland, he won the respect of fellow Protestants in England by his scholarly polemics against Rome. In his search for authentic Biblical texts he hired his own agent to gather manuscripts in the Middle East and collected a famous library, which included the Book of Kells. Some of his distinctions between spurious and authentic texts are still accepted by Biblical scholars today. In 1654 he delivered the fruits of his lifetime of scholarship when he declared that the Creation had occurred on October 26, 4004 B.C., at 9:00 A.M.

The precision of this discovery and Archbishop Ussher's prestigious documentation added weight to the widespread belief that the earth and all living creatures had been created within a single week only a few thousand years before the Christian era. This view of the Creation confined biological history to what, by modern geological standards, is a relatively brief time. This brevity itself seemed to confirm the dogma that no species could have been added, nor any have become extinct, and so was a congenial setting for the belief in the fixity of species which made possible Linnaeus' System of Nature.

For geology the brevity of earthly time had an additional consequence, which was in every sense of the word catastrophic. It encouraged a belief in sudden changes, a doctrine known as "catastrophism." Of course everybody could see that weather and climate were still slowly changing the forms of the earth by deepening streams, flooding valleys, and eroding mountains. Herodotus, Strabo, and Leonardo da Vinci had described these processes. But it was generally agreed that in the mere six thousand years since the Creation the flow of water and the crumbling of rocks could not possibly have produced the drastic changes now visible in all the varied landforms. Orthodox naturalists were therefore driven to explain large changes in earth forms by sudden cataclysms, or "catastrophes."

Buffon, not satisfied either by Archbishop Ussher's calculations or by the glib explanations of the catastrophists, plunged into his own study of the earth's dynamism with a naïve experimental enthusiasm. To understand the history of plants and animals, he said, we must first grasp the history of the earth. So Buffon set out to explain how the earth had come into being. Newton, his inspiration in many other ways, had asserted that the six planets, revolving in the same plane in concentric orbits in the same direction, must have been created by God himself. Buffon demanded natural causes, and he came up with his own explanation. "In order to judge what has happened, or even what will happen," he observed, "one need only examine what is happening. . . . Events which occur every day, movements which succeed each other and repeat themselves without interruption, constant and constantly reiterated operations, those are our causes and our reasons."

Buffon's clue for the origin of the earth was Newton's observation that "comets occasionally fall upon the sun." When one such comet collided with the sun, Buffon speculated, fragments of the sun must have been knocked off into space. These liquids and gases (1/650 of the sun's mass) then came together to form spheres revolving in the same direction and in the same plane. Each of them became a planet turning on its own axis, flattened at the poles. And satellites were thrown out.

How did Buffon's new view of the making of the earth affect the extent of historical time? Newton, of course, would not tolerate such an un-Godly account of the Creation. But in the *Principia* Newton had offered some interesting speculations on the rate of cooling of comets. "A globe of red-hot iron equal to our earth, that is, about 4,000,000 feet in diameter," he observed, "would scarcely cool in an equal number of days, or in above 50,000 years." Due to "some latent causes," Newton had ventured, the rate of cooling might be even slower, less even than the ratio of the diameter, "and I should be glad that the true ratio was investigated by experiments." For Buffon, this question held the secret of the age of the earth. If only he could find out precisely how long had been required for the planetary globes to cool down to a habitat suitable for life! He would try.

In his own foundry Buffon cast two dozen globes, one inch in diameter, to be removed white-hot from the furnace. He would then measure the time precisely to "the moment when one could touch them and hold them in one's hand." The answer to his question would come simply by extrapolating that figure to a globe the size of the earth. Even so prosaic an experiment could fire the salacious imagination of these French contemporaries of the Marquis de Sade. As one of Buffon's secretaries recorded, "To determine the epoch of the formation of the planets and to calculate the cooling time of the terrestrial globe, he had resort to four or five pretty women, with very soft skins; he had several balls, of all sorts of matters and all sorts of densities, heated red hot, and they held these in turns in their delicate hands, while describing to him the degrees of heat and cooling." A less sensational report portrayed Buffon himself with one hand holding a watch, the other in a glove cautiously testing the heat of each sphere until he could remove his glove and touch the sphere without being burned.

What Buffon learned in this way about the rates of cooling of spheres he applied to a sphere the size and composition of the earth. And he came up with some bold, theologically dangerous conclusions. "Instead of the 50,000 years which he [Newton] assigns for the time required to cool the earth to its present temperature, it would require 42,964 years and 221 days to cool it just to the point at which it ceased to burn." By further calculations he added to this figure all the years since the earth had cooled to its present temperature, which brought the total age of the earth to 74,832 years.

To his mathematically minded age Buffon was thus able to offer an experimentally verified figure whose precision rivaled the pious calculations of Archbishop Ussher. Modern geologists have, of course, extended this figure into the billions of years. Buffon himself dared observe that "the more we extend time, the closer we shall be to the truth." He had thought of three million years or more, even up to infinity. But he prudently scaled this down, he himself explained, because he did not want to shock readers so much that they might suspect him of pure fantasy. His figure needed to be only enough longer than Archbishop Ussher's to make plausible his modern vista—a world of slow and constant change.

To Buffon the earth no longer seemed the product of one relatively recent Act of Creation. Linnaeus in the ancient taxonomic tradition had focused on the classifiable products of Creation. Buffon would focus on process. The earth would have its own history. Then why not also all of nature, including all the "creatures"?

When Buffon went on from his Theory of the Earth in the very first volume of his *Natural History* (1749) to his *Epochs of Nature* (1779), the fruit of his thirty-year encyclopedic study, he found by a happy coincidence that his vastly extended calendar was divided into precisely *seven* epochs. Which gave a hitherto unsuspected metaphorical meaning to the Book of Genesis. Seven "days" now became seven "epochs."

Buffon's new chronology helped account for many other puzzling facts. In the first epoch the earth and the planets took shape. In the second epoch, as the earth solidified, the great mountain ranges were formed, with their deposits of minerals and "primitive vitreous material." As the earth cooled in the third epoch, gases and water vapors condensed, covering the whole earth with a flood. Fishes and other marine creatures flourished in the deep waters. Chemical processes pulverized the "primitive vitreous material" from the submerged mountains and made sedimentary deposits, which included organic debris like coal. As these waters rushed into the vast subterranean openings left when the earth had cooled, the flood level dropped. In the fourth epoch, when volcanoes erupted, earthquakes shook the earth, and tumbling waters reshaped the lands. In the fifth epoch, still before the separation of the continents, land animals appeared. In the sixth epoch, when the continents separated, the lands received their present shape. Finally, in the seventh, the present epoch, man appeared, heralding a new stage "when the power of man has seconded that of Nature," opening a future of incalculable possibilities.

The residual heat in the globe, a legacy from the sun, explained many things not covered in the Biblical account. For a long period, while the whole earth remained at a tropical temperature, large elephantlike creatures were found in the northern climates of Europe and North America, which incidentally accounted for the huge fossil bones found there. But as the earth cooled, these animals moved south toward the equator. It was this internal heat of the earth that had originally transformed inorganic into organic molecules and so produced the first living creatures. Since these vital powers were proportionate to the heat, the warmer regions of the earth and the warmer periods of history had always produced larger animals.

As animals migrated they adapted to their environments and so produced new varieties. Of the large animals, fewer varieties emerged because they reproduced slowly. But the prolific small mammals, such as rodents and birds, produced countless varieties. The migrations of animals before the separation of the continents explained their distribution across the earth and the fact that only South America has its own fauna.

By opening the gates of time, Buffon opened a new world of change and progress, later to be revealed as a world of evolution. And, incidentally, he opened the way to thoughts of "continental drift." Buffon's heresies, even more obviously than Galileo's, struck at the Creation and the Creator. He invented a whole new category of heresy. If the shape of the earth was so changeful, if old species could become extinct, if new variations could emerge, the world was precariously fluid. Did this not perhaps imply changing ways to salvation, changing sacraments, and even a changing Church?

In 1749, when the first volume of Buffon's *Natural History* appeared, a committee of the theology faculty of the University of Paris demanded that, to avoid their censure, he clarify certain passages in writing. This he did. "I have extricated myself with great satisfaction," Buffon boasted to a friend. They voted 115 to 5 not to censure his work. "I abandon whatever in my book concerns the formation of the earth, and in general all that might be contrary to the narration of Moses," Buffon had written to the committee, "having presented my hypothesis on the formation of the planets only as a pure philosophical speculation." At the same time Montesquieu's *Spirit of Laws* was similarly investigated, but when Montesquieu refused to reply, his work was condemned. Thirty years later, though Buffon included this pious disavowal in his *Epochs of Nature,* a committee of censorship was again appointed, but under pressure from the King they never produced a report.

Whether from piety or from prudence, Buffon steadfastly refused to be embroiled in theological controversy. "I do not understand theology," he explained in 1773, "and I have always abstained from discussing it." Scrupulous in his observance of Catholic ritual, he set up a chapel at the very foundry where he cast the globes with which he revised the Biblical "days" of Creation. He regularly attended confession, and sought the last rites of the Church at his death. But, unlike the pious Newton, Buffon did not allow his religion to stultify his view of the past. And unlike his militant contemporary Baron d'Holbach (1723–1789), he never declared himself the "personal enemy" of God, nor did he believe that one had to be an atheist to "destroy the chimeras which afflict the human race." If Buffon himself would not choose between his parallel faiths in God and in science, the historian today must not choose for him.

By his bold extension of time, Buffon changed the vocabulary of nature from a status world of rigid forms and fixed entities to a changeful world of matter in motion, of fluid, mobile individuals. Nature, no longer the finished product of a beneficent Creator, was now a name for myriad processes. Theology would be displaced by history.

Without Buffon's extension of time there was no room for a history of nature, as the career of Buffon's brilliant, frustrated predecessor had revealed. Nicolaus Steno (1638–1686), like Leonardo da Vinci, was cursed by his own versatility. Born in Copenhagen, the son of a wealthy Protestant goldsmith, he studied medicine. Frustrated in his ambition for a post at the university, he went off to Paris, where he published a treatise on the anatomy of the brain. In Florence the Duke of Tuscany became the patron of his scientific work. A spiritual crisis on All Souls' Day, 1667, led him to convert to Catholicism.

When the Accademia del Cimento assigned him to explore the grottoes at Lake Garda and Lake Como, Steno began his pioneer regional geology, the first of its kind in Europe. He had already explained that "figured stones," which Tuscans called *glossopetri,* or stone tongues, really were not sports of nature but the teeth of sharks that had lived under water there in ancient times. Still only thirty, Steno published in 1669 a revolutionary little book, *Prologue to a Dissertation on how a solid body is enclosed by the processes of nature within another solid body,* which came to be called from its Latin title the *Prodromus.* This book was destined to become a primer of modern geology. Generalizing from his geology of Tuscany, he explained why and how crystals, stones, and fossils were found in strata within the earth.

Steno's radical new insight was that the strata of the earth recorded the history of the earth. With a few simple principles, he transformed the earth's surface jumble into a legible archive. His notion was that the strata found in the earth were originally formed of matter precipitated from water, which then fell to form a sediment at the bottom. In his clear diagram, the first known effort to show a geologic section, he described six successive kinds of stratification. What is below, he said, must normally be older than what is found above. Exceptions occur only when lower layers have been disrupted and then filled in by layers from above. Layers formed by volcanic or chemical means were quite different from those formed by mechanical means. So Steno provided rudimentary definitions of sedimentary, igneous, and metamorphic rocks.

But when he touched the history of the earth, Steno was on dangerous ground. The Bible seemed to say that mountains either had been created in the Beginning by God or had simply grown. Steno began by blandly describing fossils as a class of "solids naturally contained within solids," which included all stony substances of organic origin. Fossilization occurred "where the substances of the shell being wasted, a stony substance is come into the place thereof," which meant that there could be fossils not only of bones and shells but even of plants and soft-bodied organisms. To compress all these processes within six thousand years since the Creation, Steno had to make the six days of the Book of Genesis and Noah's Flood account for more than they could bear. Since there was no history of nature, there could be no prehistory. Therefore the large fossil bones found in the Aretine fields outside Florence could not possibly belong to prehistoric animals, but must be the remains of Hannibal's war elephants.

Steno's *Prodromus* was merely the introduction to a larger work that never came, a foundation on which others could build. In London, Henry

Oldenburg, with his sharp eye for seminal works, promptly translated Steno into English in 1671. Meanwhile the versatile Steno's pioneer work in anatomy had brought him fame. The King of Denmark recalled him to be royal physician and professor of anatomy in Copenhagen. When his Catholic faith made trouble for him, he returned to Florence, and with all the enthusiasm of a convert, he abandoned science. Consecrated a priest in 1675, he diverted his energies to a frenetic ecclesiastical career. Within a year Pope Innocent XI made him a bishop, the vicar apostolic and organizer of Catholic propaganda for all northern Europe. A fanatical propagandist, he even wrote to Spinoza hoping to convert him, but Spinoza never answered. Steno's rabid asceticism hastened his death at forty-eight. He was buried with great ceremony in the Basilica of San Lorenzo in Florence, where we can still see his impressive monument.

It was left to Buffon to open the vistas of modern biology by bringing the whole earth and all its plants and animals onto the stage of history. After Buffon it was harder to believe that anything on earth was changeless. He had glimpsed the "mystery" of species. Now there was time and time to spare for varieties of animals to emerge or become extinct, making the whole world a museum of surprising fossils. By stretching the calendar, Buffon widened the stage for the naturalists' imagination. The creation could be observed not merely as a Linnaean panorama in space, but as a continuous drama in time. "Nature's great workman is Time. He marches ever with an even pace, and does nothing by leaps and bounds, but by degrees, gradations and successions he does all things; and the changes which he works—at first imperceptible—become little by little perceptible, and show themselves eventually in results about which there can be no mistake."

58

In Search of the Missing Link

ONE grand master metaphor dominated, perverted, and obstructed European efforts to discover man's place in nature. This was the simple notion of a Great Chain of Being. The whole universe, European scientists and philosophers explained, consists of an ordered series of beings, from the lowest, simplest, and tiniest at the bottom to the highest and most complex at the top. To the question, "What is man, that thou art mindful of him?" the Psalmist answered (and natural philosophers agreed), "thou hast made him a little lower than the angels, and hast crowned him with glory and honour."

The Chain-of-Being metaphor was pregnant with ambiguities and contradictions. How many links were there in the chain? How different was one link from its neighbor up or down the scale? Answers to such questions presupposed a total knowledge of nature, which was, of course, the exclusive prerogative of the Creator. A figure of speech seemed to tell Alexander Pope in 1734 all that man needed to know of his place in nature.

> Vast chain of being! which from God began,
> Natures aethereal, human, angel, man,
> Beast, bird, fish, insect, what no eye can see,
> No glass can reach; from Infinite to thee,
> From thee to nothing.—On superior pow'rs
> Were we to press, inferior might on ours;
> Or in the full creation leave a void,
> Where, one step broken, the great scale's destroy'd;
> From Nature's chain whatever link you strike,
> Tenth, or ten thousandth, breaks the chain alike.

Since man was infinitely distant from the perfection of his Creator, was there not room above man too for an infinite number of superior beings? Was man only a "middle link" between the lowliest and the highest? If there was indeed a continuous chain, might not man himself differ only infinitesimally from the nearest nonhuman link? And if man partook equally of the material qualities of the beings below him and of the ethereal qualities of those above, was not man condemned to perpetual inner discord? In his unforgettable couplets, Pope observed:

> Plac'd on this isthmus of a middle state,
> A being darkly wise and rudely great,
> With too much knowledge for the sceptic side,
> With too much weakness for the stoic's pride,
> He hangs between; in doubt to act or rest;
> In doubt to deem himself a god or beast;
> In doubt his Mind or Body to prefer;
> Born but to die, and reas'ning but to err; . . .
> Chaos of Thought and Passion all confus'd,
> Still by himself abus'd, or disabus'd;
> Created half to rise, and half to fall,

> Great lord of all things, yet a prey to all;
> Sole judge of Truth, in endless error hurl'd;
> The glory, jest and riddle of the world.

However appealing to poet and metaphysician, the Chain of Being was not much help to the scientist. Though naturalists spoke glibly of the "missing links," they were discouraged from efforts to learn about man from his similarities to the other animals. While the Chain of Being placed man in a continuous chain, it also made him somehow a link uniquely insulated from the forces of nature.

The Chain of Being proved wonderfully flexible and eventually would accommodate an idea of evolution. But at least until the eighteenth century, it described the product and not the process of creation, and was only another way of praising the wisdom and plenitude of the Creator. It described nature in space, and not in time. To discover his place in nature, man would need a sense of history, of how and when all the different species had appeared, and he would need to see how his body was similar to the bodies of the other animals.

Edward Tyson (1651–1708), a prosperous English physician, was well situated and well qualified to open the paths of discovery from natural history to comparative anatomy. He never secured a place alongside Vesalius, Galileo, Newton, or Darwin in the popular pantheon, he shunned controversy and never sought power in the new parliament of science. But what Sir William Harvey was to physiology, Tyson would be to comparative anatomy. Born in Bristol to a wealthy family with a long record of public service, Edward Tyson followed a conventional path—a Bachelor of Medicine degree at Oxford in 1677, then practice in London with his brother-in-law. When he began his anatomical experiments, he became acquainted with Robert Hooke, who illustrated some of his papers and secured his election as a Fellow of the Royal Society in 1679.

As Curator, he was charged with planning demonstrations for the society's regular meetings. He preached the society's modern gospel of incremental science. And he rejoiced at the wealth of facts flooding in from the New World. "New Tracts, new Lands, new Seas are daily found out, and fresh descriptions of unknown Countreys still from both brought in; so that we are forced to alter our Maps, and make anew the Geography of both again. Nor have the discoveries of the Indies more enriched the world of old, than those of Anatomy now have improved both the Natural and Medical Science." But naturalists must not be tempted to slovenly generalizations—"far better a little with accurateness, than a heap of rubbish carelessly thrown together. Malpighi in his Silk-worm hath done more, than Jonston in his whole book of Insects." The patient progress of knowledge of the "lesser" world within must equal that of the "greater" world without, by "taking to pieces this Automaton, and viewing asunder the several Parts, Wheels and Springs that give it life and motion."

"The Anatomy of one Animal," Tyson urged, "will be a Key to open several others; and until such time as we can have the whole completed, 'tis very desirable to have as many as we can of the most different and anomalous." He delighted in Swammerdam's ample account of the Ephemeron or May fly, for life could be understood only by "a *comparative* survey."

> Nature when more shy in one, hath more freely confest and shewn herself in another; and a Fly sometimes hath given greater light towards the true knowledge of the structure and the uses of the Parts in Humane Bodies, than an often repeated dissection of the same might have done. . . . We must not therefore think the meanest of the Creation vile or useless, since that in them in lively Characters (if we can but read) we may find the knowledge of a Deity and our selves. . . . In every Animal there is a world of wonders; each is a Microcosme or a world in it self.

One day when Tyson visited the Tower docks and the Lord Mayor's kitchens in his regular search for unusual fish to dissect, a fishmonger offered him a porpoise. This was the only one of the cetaceans (fishlike mammals lacking hind limbs, including whales and dolphins) found in British waters. It was happy for the future of science that this specimen had lost its way up the Thames.

The Royal Society had expressed a special interest in the anatomy of all rarities, and the porpoise had never been anatomized. Tyson's friend Robert Hooke laid out the society's seven shillings sixpence for the 95-pound "fish," which they took to Gresham College for dissection. There Tyson went speedily about his work, enlisting Hooke to help him make drawings as he went along. Tyson's *Anatomy of a Porpess* (1680) revealed the dangers of classifying animals by their exteriors. John Ray had still classified the porpoise as a fish. "If we view a Porpess on the outside," Tyson observed, "there is nothing more than a Fish." But "if we look within, there is nothing less." Its internal anatomy persuaded Tyson that the porpoise was in fact a mammal, similar to land quadrupeds, "but that it lives in the Sea, and hath but two forefins."

> The structure of the *viscera* and inward parts have so great an Analogy and resemblance to those of Quadrupeds, that we find them here almost the same. The greatest difference from them seems to be in the external shape, and wanting feet. But here too we observed that when the skin and flesh was taken off,

the fore-fins did very well represent an Arm, there being the *Scapula*, an *os Humeri*, the *Ulna*, and *Radius*, and bone of the *Carpus*, the *Metacarp*, and 5 *digiti* curiously joynted. . . .

Tyson's eye for exotic specimens awakened the interest of his colleagues in the Royal Society. They bought an ostrich for him to dissect. He finally offered the society his illustrated dissections (among others) of an American rattlesnake, a Mexican musk-hog, and an opossum, which had been presented to the society by William Byrd of Virginia.

Another accident offered Tyson his opportunity to pioneer on the perilous paths of human origins. An infant chimpanzee which a sailor had loaded on his ship in Angola in southwest Africa had suffered en route an injury that became infected, and it died soon after its arrival in London. Tyson, who had seen the animal while it was still alive, secured the body, and took it to his house for dissection. Lacking refrigeration for his specimen, Tyson had to perform his dissection speedily. By good luck he enlisted as his assistant one of the ablest human anatomists of the day, William Cowper, who helped him make drawings. Their product, published in 1699, was *Orang-Outang, sive Homo Sylvestris: or, the Anatomy of a Pygmie compared with that of a Monkey, an Ape, and a Man*. Just as Vesalius' book had opened human anatomy, this copiously illustrated volume of some 165 pages opened a new era in physical anthropology.

The term "orang-outang," in the Malay language, meant "man of the woods" and in Europe was then being used loosely for all the larger nonhuman primates. The animal that Tyson dissected was not what the modern zoologist would call an orang-outang but an African chimpanzee. This animal, the first anthropoid to appear in European scientific literature, had been noted in 1641 by Dr. Nicolaes Tulp (whom Rembrandt depicted as the teacher in his famous *Anatomy Lesson*). Tyson chose to call his specimen a "pygmie."

What he called it was less important than what he did with it, which was epoch-making. Tyson's anatomy of the orang-outang placed man in a whole new constellation. Just as Copernicus displaced the earth from the center of the universe, so Tyson removed man from his unique role above and apart from all the rest of Creation, for whose nutriment, clothing, and delight plants were created, and for whose service there was a world of animals. Never before had there been so circumstantial or so public a demonstration of man's physical kinship with the animals. Just as Vesalius had detailed and drawn the structure of the human body, so Tyson now detailed the anatomy of what he showed to be man's closest relative among the animals. The implication was plain that here was the "missing link" between man and the whole "lower" animal creation.

Tyson starkly enumerated physical similarities and differences between the chimpanzee and man. Without references to God or speculations about an immortal soul, he listed his conclusions in two columns. One itemized how "The Orang-Outang or Pygmie more resembled a Man, than Apes and Monkeys do," another how it "differ'd from a Man, and resembled more the Ape and Monkey-kind." The forty-eight items of resemblance to man began with "1. In having the Hair of the Shoulder tending downwards; and that of the Arm, upwards," and went through the structural similarities of intestines, colon, liver, spleen, pancreas, and heart. "25. The Brain was abundantly larger than in Apes; and all it's Parts exactly formed like the Humane Brain." Then similarities of teeth, vertebrae, fingers and toes, but finally "whether all the same Muscles in Apes and Monkeys resemble the Humane, could not be determined, for want of a Subject to compare them with, or Observations made by others." The thirty-four anatomical differences from man, and the chimpanzee's resemblances "to the Ape and Monkey-kind" were also listed with technical precision. Having found that the organs of speech and the brain of his pygmie "does so exactly resemble a Man's," he left his readers to puzzle "that there is no reason to think that Agents do perform such and such Actions, because they are found with Organs proper thereunto: for then our Pygmie might be really a Man." Why could man reason, while pygmies could not? Tyson put this question in a new matrix, in the world of physical nature. Just as the heliocentric vista once seen could not be forgotten, so, after reading Tyson, who could believe that man was an isolate from the rest of nature?

Tyson concluded that the chimpanzee more closely resembled man than it resembled the other primates. Man's differences from other animals now became only matters of nuance to be set down on a list. Tyson's expert dissection gave to the theologians' talk of man's "animal" nature a newly precise—and theologically dangerous—meaning. Tyson was on the threshold of physical anthropology.

In the appendix to his *Orang-Outang* he marshaled his copious classical learning to explain how this creature had stimulated reports of satyrs, of men with dog's heads, and of sphinxes—but "they were only a Creature of the Brain, produced by a warm and wanton Imagination, and . . . they never had any Existence or Habitation elsewhere." So he opened the way, too, to cultural anthropology, showing how different peoples gave wild and varied meanings to the same physical phenomenon, to a mere chimpanzee.

Most surprising in the career of so emphatically *physical* an anthropologist was Tyson's pioneer role in treating the vagaries of the human mind. On his way to become the leading English physician of the age, he was elected a Fellow of the Royal College of Physicians, and in 1684 he was named Physician and then Governor to the Bethlehem Hospital. There he earned a place in the pantheon of humanitarians. Bethlehem Hospital, founded in the thirteenth century as a priory for the Order of the Star of Bethlehem, became an asylum for the insane, the first such institution in England. Except for one in Granada, Spain, it was also the first in Europe. When Tyson took charge, "Bedlam" (a common pronunciation of Bethlehem) had long since entered common parlance to mean any place of noise and confusion. There the mentally ill were beaten, shackled, and confined in cells. Bedlam had become so public a spectacle that a staple scene in Restoration comedies showed fashionable people "going to see the Lunatics," as if they were a circus or a zoo. And incidentally, Bedlam was a place of assignation for "lewd or disorderly" persons and for lazy apprentices.

Governors of Bedlam had been reluctant to exclude sightseers, since wealthy "idlers" sometimes took an interest in the institution and made contributions. " 'Tis by the help of such Benefactors," Tyson himself conceded, "that this Hospital is enabled to bear their great Charges." He tried at least to restrict spectators to the more respectable and prohibited all tourists on Sundays.

In a callous age, Tyson was remarkably successful in humanizing treatment of the mentally ill. To change the atmosphere of a jail into that of a hospital, he brought in women nurses, and set up a wardrobe fund to clothe poor patients. "Bedlam" began to become a place not for punishment but for therapy. His great innovation was the postinstitutional treatment of discharged patients, with periodic visits to them at home. During the twenty years that he was the Physician, of 1,294 patients admitted, 890, or some 70 percent, were discharged with their madness cured or relieved. Tyson's reforms survived the centuries and left a permanent mark at Bethlehem and elsewhere. In 1708 the threnodist wrote on his death:

> Great Tyson's Power new Organs cou'd dispense. . . .
> Here ev'n the mental Deprivation cur'd,
> The Man refounded, Light to Souls restor'd,
> The Tyson Art in this Great Cause bestow'd
> Rebuilds ev'n the faln Image of the God.

When Linnaeus later came to place man in his *System of Nature* in 1735, he did not avoid the issue by calling him a fallen angel. Like Tyson, he confessed that he "could not discover the difference between man and the orangutan," and he never did find a single "generic character" to distinguish man from the ape. "It is remarkable," Linnaeus concluded in his twelfth edition, with an irony rare for him, "that the stupidest ape differs so little from the wisest man, that the surveyor of nature has yet to be found who can draw the line between them." "*Homo,*" Shakespeare had said in

Henry IV, Part I, "is a common name to all men." Linnaeus christened man into his binomial system as *Homo sapiens.* He gave *homo* vast new meaning, taking his boldest step when he classified man as a "species," simply another kind of animal. Under Mammalia in his Order of Primates ("Fore-teeth cutting; upper 4, parallel; teats 2 pectoral") Linnaeus placed the human species ("Diurnal; varying by education and situation"), and distinguished these varieties:

> Four-footed, mute, hairy.   *Wild Man.*
> Copper-coloured, Choleric, erect.   *American.*

Hair black, straight, thick; nostrils wide, face harsh; beard scanty; obstinate, content free. Paints himself with fine red lines. Regulated by customs.

> Fair, sanguine, brawny.   *Europeans.*

Hair yellow, brown, flowing; eyes blue; gentle, acute, inventive. Covered with close vestments. Governed by laws.

> Sooty, melancholy, rigid.   *Asiatic.*

Hair black; eyes dark; severe, haughty, covetous. Covered with loose garments. Governed by opinions.

> Black, phlegmatic, relaxed.   *African.*

Hair black, frizzled; skin silky; nose flat; lips tumid; crafty, indolent, negligent. Anoints himself with grease. Governed by caprice.

59

Paths to Evolution

"THE year which has passed," Thomas Bell, eminent president of the Linnean Society of London reported at the end of 1858, "has not, indeed, been marked by any of those striking discoveries which at once revolutionize . . . the department of science on which they bear; it is only at remote intervals that we can reasonably expect any sudden and brilliant innovation." The select Linnean Society (of which Joseph Banks was a founder) had been created in 1788 to preserve the library, herbarium, and manuscripts which Linnaeus had left to his son, and which on his son's death had been bought for them by an English botanist. Despite Bell's observation, the

three papers read to the society on July 1 of that year bore more revolutionary implications than any other offerings to the forum of scientists since Sir
Isaac Newton's day.

Those papers (which came to only seventeen pages in the society's *Journal*), "On the Tendency of Species to form Varieties; and on the Perpetuation of Varieties and Species by Natural Means of Selection," had been
communicated to the society by two of its most accomplished fellows, Sir
Charles Lyell, the geologist, and J. D. Hooker, the botanist. The sponsors
offered "the results of the investigations of two indefatigable naturalists,
Mr. Charles Darwin and Mr. Alfred Wallace. These gentlemen having,
independently and unknown to one another, conceived the same very ingenious theory to account for the appearance and perpetuation of varieties and
of specific forms on our planet, may both fairly claim the merit of being
original thinkers in this important line of inquiry." The three items were:
extracts from a manuscript sketched by Darwin in 1839 and revised in 1844;
the abstract of a letter from Darwin to Professor Asa Gray of Boston,
Massachusetts, in October 1857, repeating his views on species stated in the
earlier manuscript; and an essay by Wallace written at Ternate in the East
Indies in February 1858, which he had sent to Darwin with instructions to
forward it to Lyell if he found it sufficiently novel and interesting.

In later years historians would note July 1, 1858, as the date of the first
public statement of the modern theory of evolution. But at the time the
Darwin-Wallace papers made hardly a ripple. Neither Darwin nor Wallace
was present, and there was no discussion by the thirty fellows who were
there. A scheduled paper with a contradictory thesis was not even given.
The reading of these articles was a rite of priority, required by the new
etiquette of science.

In the progress of the idea of evolution we witness a distinctly modern
phenomenon in the progress of science. Modern times brought new instruments of publicity, the printing press with its new powers of diffusion,
scientific societies with their wider and more public forums. All this meant
a new mobility for scientific ideas and for scientists themselves. Of course,
the new incrementalism of science did not spell an end to revolutions in
thought, but it did change the pace and the character of these revolutions.
Now novel ideas could be introduced piecemeal, unobtrusively, even per-
functorily. And who could tell when one of these ideas might signal a
revolution in thought? On that July day in London the Linnean Society
prepared to publish observations made by Darwin twenty years earlier on
his round-the-world voyage on the *Beagle* alongside complementary observations made by Wallace a few months before in Ternate in the distant
Moluccas.

When Darwin, a young man of twenty-two, had sailed out on December

27, 1831, on the five-year voyage of the *Beagle,* he took with him the just published first volume of Charles Lyell's *Principles of Geology,* a going-away gift from his Cambridge professor of botany. Lyell (1797–1875) would provide the background for all Darwin's thinking about the processes of nature and so make it possible for modern evolutionary thought to bear the name of Darwinism. Lyell's crucial insight, documented with copious evidence in his book, was that the earth had been shaped from the beginning by uniform forces still at work—erosion by running water, accumulation of sediment, earthquakes, and volcanoes. Since such forces through millennia had made the earth what it was in his day, there was no need to imagine catastrophes. This doctrine, christened by the English philosopher William Whewell, came to be known as Uniformitarianism.

Lyell had tried to avoid the shoals of theology and cosmology simply by refusing to discuss the origins of the earth. Speculative theories of a Creation, he said, were unnecessary and unscientific. The implications for plants and animals were obvious. If the present activity of Vesuvius or Etna explained changes in the surface of the earth, could not other forces equally visible today show us how species and varieties of plants and animals had come into being? The Cambridge professor of botany who gave Darwin the copy of Lyell which he read and cherished on the *Beagle* warned him not to believe everything in it. The few other books he took along included the Bible, Milton, and Alexander von Humboldt's travels in Venezuela and the Orinoco basin.

In the mystery story of how Darwin came to his notions of evolution, the voyage of the *Beagle* was, of course, a crucial episode. An essential link in the chain of people and ideas was John Stevens Henslow (1796–1861), the teacher who first inspired the young Darwin with enthusiasm for the study of nature. From the chair of botany the handsome magnetic Henslow single-handedly stirred a botanical renaissance in the university. He initiated field trips to observe plants in their natural habitat and required his students to make independent observations, training a new generation of botanists interested less in Linnaean taxonomy than in plant distribution, ecology, and geography. The Cambridge Botanical Garden became a teaching laboratory.

Henslow's historic accomplishment was to transform the Cambridge playboy Darwin from a listless student of theology into a passionate naturalist. At the age of sixty-seven, Darwin still recalled "a circumstance which influenced my career more than any other":

This was my friendship with Prof. Henslow. Before coming up to Cambridge, I had heard of him from my brother as a man who knew every branch of science, and I was accordingly prepared to reverence him. He kept open house once every week, where all undergraduates and several older members of the University, who were attached to science, used to meet in the evening. I soon got through Fox an invitation and went there regularly. Before long I became well acquainted with Henslow, and during the latter half of my time at Cambridge took long walks with him on most days; so that I was called by some of the dons "the man who walks with Henslow"; and in the evening I was very often asked to join his family dinner. His knowledge was great in botany, entomology, chemistry, mineralogy and geology. His strongest taste was to draw conclusions from long-continued minute observations.

In 1831, when the Admiralty asked Henslow to recommend a naturalist to serve on the *Beagle*'s voyage to map the coasts of Patagonia, Tierra del Fuego, Chile, and Peru and to set up chronometric stations, he recommended his favorite pupil.

Charles was eager to accept. But his father, already irritated by Charles' false start at Edinburgh in the study of medicine, was dead set against any more such casual adventures. "You care for nothing but shooting, dogs, and rat-catching," the elder Darwin had complained, "and you will be a disgrace to yourself and all your family." Now he was determined to keep the vagrant Charles on the path to the clergy, and the dutiful son would not join the *Beagle* without his father's permission. Luckily, Professor Henslow and Charles' uncle, Josiah Wedgwood II, succeeded in persuading Charles' father to let Charles go. "The pursuit of Natural History," Wedgwood argued, "though certainly not professional, is very suitable to a clergyman."

Henslow kept in close touch with his pupil during the five-year voyage of the *Beagle*. They corresponded regularly, and Henslow looked after the specimens that Darwin sent back to London. When the *Beagle* arrived at Montevideo a copy of Lyell's second volume was awaiting, and at Valparaiso on the other side of the South American continent Darwin received the third volume, just off the press. Throughout his trip Darwin was applying Lyell's principles. And at the coral-encrusted rims of submerged volcanic craters in the Indian Ocean, he concluded that the Kelling Atoll had been built up over at least a million years.

The second volume of Lyell went beyond physical geology and applied his Uniformitarianism to biology. Throughout geological time, Lyell explained, new species had been emerging, and others had become extinct. Survival of a species depended on certain conditions of its environment, but geological processes were constantly changing those conditions. Failure in competition with other species in the same habitat might extinguish a species. The success of one prosperous species might crowd out others to extinction. Lyell's survey of the geographic distribution of plants and animals suggested that each species had come into being in one center. Similar habitats on separate continents seemed to produce quite different species equally adapted to their habitats. Environment, species—everything was in flux.

Lyell's interest in these problems had been piqued by the French naturalist Lamarck (1744–1829). But Lamarck, insisting on the inheritance of acquired characteristics, had really abandoned the concept of species. For him a species was only a name for one set of generations while the animal was adapting to its environment. And if every species was infinitely plastic, then no species would ever have to become extinct. While Lyell had kept species as the essential units in his processes of nature, he could not explain how a new species would originate.

The impressionable Darwin was tantalized by Lyell's suggestions. Everywhere in South America he encountered plants and animals he had never seen before. In the Galápagos he was enticed by the variations of bird species on widely separated islands in the same latitude. Meanwhile, Henslow had been so much impressed by Darwin's letters that he had read some of them to the Philosophical Society of Cambridge, and even printed some of them for private distribution. When the *Beagle* returned in 1836, Henslow joined with Lyell in securing for Darwin a grant of £1,000 to help him compile his five-volume report, and then managed his election as Secretary of the Geological Society of London.

During the next few years Darwin, by his own account, saw more of Lyell than of any other man. "His delight in science was ardent," Darwin recalled, "and he felt the keenest interest in the future progress of mankind. He was very kind-hearted, and thoroughly liberal in his beliefs or rather disbeliefs." Still Lyell would be slow in coming around to Darwin's own theories. "What a good thing it would be," the young Darwin had complained to Lyell when older geologists refused to follow Lyell, "if every scientific man was to die when 60 years old, as afterwards he would be sure to oppose all new doctrines." But in his late sixties the courageous Lyell's *Antiquity of Man* (1863) would finally abandon his opposition to evolution and begin to embrace Darwin's views of the origin of species. "Considering his age, his former views, and position in society," observed Darwin, "I think his action has been heroic."

Lyell, twelve years Darwin's senior, and at the height of his fame, remained Darwin's mentor. After the Darwins moved to Down in Kent, the Lyells would come visit for days at a time. As Darwin recalled:

It appeared to me that by following the example of Lyell in Geology, and by collecting all facts which bore in any way on the variation of animals and plants under domestication and nature, some light might perhaps be thrown on the whole subject. My first note-book was opened in July 1837. I worked on true Baconian principles, and without any theory collected facts on a whole-sale scale, more especially with respect to domesticated productions, by printed enquiries, by conversation with skilful breeders and gardeners, and by extensive reading. When I see the list of books of all kinds which I read and abstracted, including whole series of Journals and Transactions, I am surprised at my industry. I soon perceived that Selection was the key-stone of man's success in making useful races of animals and plants. But how selection could be applied to organisms living in a state of nature remained for some time a mystery to me. In October 1838, that is fifteen months after I had begun my systematic enquiry, I happened to read for amusement "Malthus on Population," and being well prepared to appreciate the struggle for existence which everywhere goes on from long-continued observation of the habits of animals and plants, it at once struck me that under these circumstances favourable variations would tend to be preserved and unfavourable ones to be destroyed. The result of this would be the formation of new species.

Here in a nutshell was what Darwin had to add to the thinking about species.

Still, Darwin was "so anxious to avoid prejudice" from the premature exposure of his ideas, that he held back. In June 1842, for his own satisfaction, he penciled a brief abstract of his theory in 35 pages, which he then enlarged in 1844 to another "abstract" of 230 pages. In 1856, when Lyell advised Darwin to expand his treatment, he began at once "to do so on a scale three or four times as extensive as that which was afterwards followed in my Origin of Species."

Then, early in the summer of 1858, as Darwin recorded, all his "plans were overthrown." He received from the Moluccas Wallace's essay "on the tendency of varieties to depart indefinitely from the original type." Wallace asked him, if he thought well of the essay, to send it on to Lyell, and, as we have seen, the scrupulous Darwin did just that. If Wallace's paper was to be published, what would Darwin do with his own labored product of twenty years? Darwin was torn.

Again Lyell, the statesman in the new parliament of science, played a crucial role. Determined to preserve Darwin's claim to priority and at the same time to give Wallace his due, Lyell urged that the three items be promptly offered to the Linnean Society. "I was at first very unwilling to consent," Darwin confessed, "as I thought Mr. Wallace might consider my doing so unjustifiable, for I did not then know how generous and noble was his disposition. The extract from my M.S and the letter to Asa Gray had neither been intended for publication and were badly written. Mr. Wallace's essay, on the other hand was admirably expressed and quite clear. Nevertheless our joint productions excited very little attention, and the only published notice of them which I can remember was by Prof. Haughton of

Dublin, whose verdict was that all that was new in them was false, and what was true was old."

Alfred Russel Wallace (1823–1913), whom history would recognize as co-author of the idea of natural selection, offered a vivid contrast to Darwin. Born into an impoverished family of nine children in Monmouthshire in South Wales, he attended a grammar school for a few years, dropped out at fourteen, and educated himself by reading. As a boy visiting London he frequented the "Hall of Science" in Tottenham Court Road, a workmen's club for advanced teachers where he was converted to Robert Owen's socialism and "secularism," a skepticism of all religions. He supported himself as an apprentice-surveyor with his brother, then read up enough on his own to qualify as a schoolmaster in Leicester. There he had the good luck to meet Henry Walter Bates (1825–1892), who had been working thirteen hours a day drearily apprenticed to a local hosiery manufacturer, but was finding his refuge in Homer, Gibbon, and amateur entomology. Bates and Wallace became fast friends, and joined in beetle-collecting expeditions into the countryside.

A voracious reader, the young Wallace discovered an inspiring assortment of books on science, natural history, and travel, including Malthus' *On Population,* Darwin's journal of the *Beagle,* and Lyell's *Geology.* One of the books that impressed him most was a stimulating book on evolution by another amateur naturalist, Robert Chambers (1802–1871). *Vestiges of the Natural History of Creation* (1844) was so controversial that Chambers had to publish it anonymously to avoid damage to his publishing business, but it went through four editions in seven months and soon sold twenty-four thousand copies. Though condemned as godless by respectable scientists, it irrevocably popularized the ideas of organic and cosmic evolution, and the evolution of species.

Alexander von Humboldt's dramatic personal account of his travels in Mexico and South America emboldened Wallace to enlist Bates on an expedition to gather specimens along the Amazon. Four years (1848–52) of collecting there earned young Wallace a reputation as a field naturalist. On his return voyage to England his ship caught fire and sank, along with his specimens, but he was not discouraged from collecting. He set out promptly for the Malay Archipelago. There and in the Moluccas he spent eight years exploring and gathering specimens, and formulated the theory of natural selection in the paper that Darwin received early in 1858.

If a Greek dramatist had contrived two characters to show how fate could bring men by opposite paths to the same destination, he could hardly have done better than invent Darwin and Wallace. Darwin, the elder by a dozen years, had been dedicated by his wealthy family to a career in the Church. All his life Darwin did his best to follow Lyell's advice "never to get entangled in a controversy, as it rarely did any good and caused a miserable loss of time and temper." Tediously gathering specimens and evidence over two decades, Darwin seemed led to his theory of natural selection almost against his will. The impoverished Wallace, inspired early with a suspicion of religion and all established institutions, was hasty to embrace theories and plunge into controversy. When he was only twenty-two, Chambers' popular *Vestiges* had converted Wallace to an unshakable conviction that species arose through a process of evolution, and his trip to the Amazon was for facts to convince others. By his later trip through the Malay Archipelago covering fifteen thousand miles and gathering some 127,000 specimens, he aimed to gather conclusive evidence. From his arrival there he kept a notebook on evolution, which he called his "Species Notebook." Wallace's essay "On the Law which Has Regulated the Introduction of New Species" (1855) was published three years before the paper he sent to Darwin.

During the 1860's, the very years when the elementary notions of evolution were being publicly tested, Wallace spread himself over the most miscellaneous causes. He became a passionate convert to Spiritualism, pursuing his interest in socialism he was elected the first president of the Land Nationalization Society (1881), and he was an outspoken advocate of women's rights. Curiously, his passion for controversy drew him into the movement against vaccination for smallpox. His pamphlet *Forty-five Years of Registration Statistics, Proving Vaccination to Be Both Useless and Dangerous* (1885) was followed by three days of testimony before the Royal Commission where he argued that more patients died from vaccination than from the disease.

Seeking a wider arena for controversy, Wallace reached into outer space. The eminent astronomer Percival Lowell (1855–1916) argued in *Mars and Its Canals* (1906) that there must have been intelligent inhabitants on Mars, who had made the channels now visible by building a system of irrigation —using water from the annually melting polar ice caps—which created bands of cultivated vegetation. Wallace, though no astronomer, at the age of eighty-four entered the lists. In *Is Mars Habitable?* (1907) he insisted that life could not exist elsewhere in the universe. And twentieth-century evidence has proved that the expert Lowell was probably farther from the truth than the amateur Wallace. Science and reform had produced what Wallace enthusiastically christened *The Wonderful Century* (1898).

The facts of geographical distribution that provided the cautious Darwin with questions supplied the brash Wallace with answers. Seeing natural selection led Darwin away from religious faith. Late in life he recalled that the grandeur of the Brazilian forest had once reinforced his "firm conviction of the existence of God and of the immortality of the soul. . . . But now the grandest scenes would not cause any such convictions and feelings to rise in my mind. It may be truly said that I am like a man who has become colour-blind." "There seems to be no more design to the variability of organic beings and in the action of natural selection, than in the course which the wind blows."

But Wallace's passion for evolution led him more and more toward a belief in a "Higher Intelligence." Increasingly he needed a God to explain what he saw in nature. "I hope," Darwin told Wallace when Wallace's review of Lyell's books in 1869 laid bare his resurgent faith in a God, "you have not murdered too completely your own and my child."

Just as the voyages of Gama and Magellan had been preceded by uncele-brated pioneers on trading voyages across the Mediterranean and by those who inched down around the coast of Africa, so too there were countless pioneers in the voyages toward evolution. But while Columbus knew there was a Japan to be reached, Gama that India was there, the pioneers of evolution were en route to an unknown destination.

To describe amply all who contributed to Darwin's mature theory of evolution would require volumes on the rise of modern biology, geology, and geography. We would have to recount ancient Greek foreshadowings, Saint Augustine's suggestion that while all species had been created by God in the Beginning, some were mere seeds that would appear at a later time, medieval notions of an organic world, Montesquieu's hints of the multi-plication of species from the discovery in Java of flying lemurs, the French mathematician Maupertuis's speculations on the chance combinations of elementary particles, Diderot's suggestions that higher animals may all have descended from "one primeval animal," Buffon on the development and "degeneration" of species, Linnaeus' gnawing doubts that species might not be immutable, the metaphoric fancies of Charles' grandfather Erasmus Darwin on the urges of plants and animals sparked by "lust, hunger, and danger" to develop into new forms—and countless others.

Among earlier contemporaries of Darwin we would have to include Lamarck's bold exploration of the hazy borderland between species and varieties and his evolutionary "tree." Nor could we omit Georges Cuvier's grand systematic arrangement of all classes of the animal kingdom. "These diverse bodies may be looked upon as a kind of experiment performed by nature," Cuvier ventured in 1817, "which adds or subtracts from each of these different parts (just as we try to do the same in our laboratories) and itself shows the results of these additions and subtractions." Many others who, like Cuvier, denied the evolution of species, still detected progress in the sorts of creatures found in the more recent levels of the earth.

Cuvier's *bête noire,* the indomitable Etienne Geoffroy Saint-Hilaire

(1772–1844), took up Napoleon's invitation to join the scientific expedition to Egypt and at the risk of his life collected specimens from the tombs. He translated "evolution" from a word for the embryonic development of the individual into a word for the emergence of species. For Geoffroy, the structural similarity of all vertebrates suggested the evolution of mammals from fishes, and he declared the evolution of the whole animal kingdom. But he said that the innovator, like Christ, must be willing to wear a crown of thorns.

The data for evolution were an unanticipated by-product of a seafaring expedition which had a clearly defined assignment. The *Beagle,* as we have seen, had been sent by the British Admiralty to chart the coast of South America and to fix longitude more accurately by a world-encircling chain of chronological calculations. But the modern parliaments of science—the Royal Society, the Linnean Society, and their counterparts across Europe and the Americas—had made natural history a deliberate forum for the unexpected.

The triumph of evolution was a victory not merely of ideas but of printed matter, which in its European typographic form was a revolutionary new device for spreading grand ideas to the most unlikely places. *An Essay on the Principle of Population* (1798), by Thomas Robert Malthus (1766–1834), which Darwin had read in October 1838, would also catalyze Wallace. In his *Autobiography,* Wallace recalled that when he was a schoolteacher in Leicester in 1844–45 passing many hours in the town library, "perhaps the most important book I read was Malthus's 'Principles of Population,' which I greatly admired for its masterly summary of facts and logical induction to conclusions. It was the first work I had yet read treating any of the problems of philosophical biology, and its main principles remained with me as a permanent possession and twenty years later gave me the long-sought clue to the effective agent in the evolution of organic species." And he recorded vividly the moment when Malthus reappeared on his horizon and changed his life. In January 1858 Wallace had just arrived at Ternate in the Moluccas to collect butterflies and beetles, "bitten by the passion for species and their description, and if neither Darwin nor myself had hit upon 'Natural Selection,' I might have spent the best years of my life in this comparatively profitless work." His thinking had reached a dead end.

I was suffering from a sharp attack of intermittent fever, and every day during the cold and succeeding hot fits had to lie down for several hours, during which time I had nothing to do but to think over any subjects then particularly interesting me. One day something brought to my recollection Malthus's "Principles of Population," which I had read twelve years before. I thought of his clear exposition of "the positive checks to increase"—disease, accidents, war, and famine—

which keep down the population of savage races to so much lower an average than that of more civilized peoples. It then occurred to me that these causes or their equivalents are continually acting in the case of animals also; and as animals usually breed much more rapidly than does mankind, the destruction every year from these causes must be enormous in order to keep down the numbers of each species . . . as otherwise the world would long ago have been densely crowded with those that breed most quickly. . . . Why do some die and some live? And the answer was clearly, that on the whole the best fitted live. From the effects of disease the most healthy escaped; from enemies, the strongest, the swiftest, or the most cunning; from famine, the best hunters or those with the best digestion; and so on. Then it suddenly flashed upon me that this self-acting process would necessarily *improve the race,* because in every generation the inferior would inevitably be killed off and the superior would remain—that is, *the fittest would survive.* . . . I waited anxiously for the termination of my fit so that I might at once make notes for a paper on the subject.

The following two evenings he spent writing the paper that he sent to Darwin by the next post, with the results we have already seen.

Malthus' ideas on population had been a reaction against his father's admiration for the utopian ideas of Rousseau and William Godwin. Though destined for the clergy and actually ordained, the young Malthus at Cambridge had done brilliantly in mathematics. "Population, when unchecked," he gave as the heart of his "principle," "increases in a geometrical ratio. Subsistence increases only in arithmetical ratio." And despite his frequent old-fashioned moralizing, his book had the ring of quantitative social science. Malthus had an eminently practical purpose—to reshape the Poor Laws so that the leaders of England "would not be open to the objection of violating our promises to the poor." And in the long run he would influence economic thinking. Karl Marx learned from him, and John Maynard Keynes would credit Malthus with the idea that effective demand was a way of avoiding depressions. But Malthus' influence on biology was quite unpredicted. The struggle for existence, Darwin explained in the *Origin of Species,* "is the doctrine of Malthus applied with manifold force to the whole animal and vegetable kingdom." The cogency of Malthus' style had much to do with the remarkable impact of his small book, which went through six editions before his death and increased in power with the years.

Publication was often the crux of the matter. Whether readers agreed or disagreed, what mattered was that the published book sparked discussion as it sold copies. When Darwin's *Origin of Species* was offered to the shrewd John Murray (who had published a revised *Voyage of the Beagle* and Herman Melville's tales of the South Seas after several others had refused), he was far from enthusiastic. The cautious Darwin asked Lyell on March 28, 1859, how he should approach Murray:

P.S. Would you advise me to tell Murray that my book is not more *un*-orthodox than the subject makes inevitable. That I do not discuss the origin of man. That I do not bring in any discussion about Genesis, &c., &c., and only give facts, and such conclusions from them as seem to me fair.

Or had I better say *nothing* to Murray, and assume that he cannot object to this much unorthodoxy, which in fact is not more than any Geological Treatise which runs slap counter to Genesis.

Finally, all that Murray objected to were the words "Abstract" and "Natural Selection" in the title. Seeing only the chapter titles, and on Lyell's recommendation, Murray agreed to publish, giving Darwin two-thirds of the net profit.

The Reverend Whitwell Elwin, editor of the prestigious *Quarterly Review,* in a reader's report, which would become a classic in the trade, advised Murray that it was unwise to publish anything that called itself only an "abstract." Since the subject was so controversial, Elwin urged that, instead, Darwin should write a book on pigeons, on which he was known to have some ingenious observations. "Everyone is interested in pigeons," he added. "The book would be reviewed in every journal in the kingdom and would soon be on every library table." Darwin was not persuaded.

A lawyer friend of Murray's encouraged him to print 1,000 copies instead of the planned 500, and the number was raised to 1,250 before publication on November 24, 1859. Until the last moment Darwin feared that Murray was overcommitted, and even offered to pay the cost of his proof corrections. When all copies were taken by booksellers, another 3,000 were printed. The result was beyond expectations. "Sixteen thousand copies have now (1876) been sold in England," Darwin noted in his *Autobiography,* "and considering how stiff a book it is this is a large sale. It has been translated into almost every European tongue, even into such languages as Spanish, Bohemian, Polish, and Russian. It has also, according to Miss Bird, been translated into Japanese and is there much studied. Even an essay in Hebrew has appeared on it, showing that the theory is contained in the Old Testament!" He proudly counted more than 265 reviews, and numerous essays. Darwin attributed the publishing success (not large, for popular novels were equaling Darwin's boasted total in a single year) to his bringing together "innumerable well-observed facts," and to the moderate size of the book, which he said he owed to help from Wallace's essay.

The initial hostile reception of the *Origin of Species,* and especially the ignorant and contemptuous attack by Bishop Samuel Wilberforce, has become proverbial. But contempt rapidly gave way to acclaim. Within a decade of publication, questions for the natural science tripos at Cambridge, instead of asking for "evidence of design" in nature, required an analysis of the concept of the struggle for existence. When even the ill-tempered Bishop Wilberforce reluctantly confessed his error, Darwin's champion, Thomas Henry Huxley, remained unsatisfied. "Confession unaccompanied by penitence . . . affords no ground for mitigation of judgment; and the kindliness with which Mr. Darwin speaks of his assailant, Bishop Wilberforce, is so striking an exemplification of his singular gentleness and modesty, that it rather increases one's indignation against the presumption of his critic." Huxley called Darwin's book "the most potent instrument for the extension of the realm of natural knowledge which has come into men's hands, since the publication of Newton's Principia." "It was badly received by the generation to which it was first addressed, and the outpouring of angry nonsense to which it gave rise is sad to think upon. But the present generation will probably behave just as badly if another Darwin should arise, and inflict upon them what the generality of mankind most hate— the necessity of revising their convictions."

The long-term influence of Darwinism and its fruitful ambivalence for science and religion was embodied in Huxley's invention of the word "agnostic" to describe the limits and the promise of scientific knowledge. Huxley took his clue from Saint Paul's encounter with the Athenians worshipping at an altar inscribed "To the Unknown God." On the urging of twenty members of Parliament, when Darwin died in 1882 he was buried in Westminster Abbey.

BOOK FOUR

SOCIETY

A man alone is in bad company.

—PAUL VALÉRY (1924)

History had to be discovered before it could be explored.
Messages from the past came first through the arts of
memory, then in writing, and, finally, explosively in
books. The earth's unsuspected treasure of relics reached
into prehistory. The past became more than a warehouse
of myth or a catalogue of the familiar. New worlds on land
and sea, resources of remote continents, the ways of faraway peoples, opened visions of progress and novelty.
Society, man's everyday life in community, became a new
and changing scene of discovery.

PART THIRTEEN

WIDENING THE COMMUNITIES OF KNOWLEDGE

. . . building up a library which has no other limits than the world itself.

—ERASMUS, *Adages* (1508)

60

The Lost Arts of Memory

BEFORE the printed book, Memory ruled daily life and the occult learning, and fully deserved the name later applied to printing, the "art preservative of all arts" *(Ars artium omnium conservatrix).* The Memory of individuals and of communities carried knowledge through time and space. For millennia personal Memory reigned over entertainment and information, over the perpetuation and perfection of crafts, the practice of commerce, the conduct of professions. By Memory and in Memory the fruits of education were garnered, preserved, and stored. Memory was an awesome faculty which everyone had to cultivate, in ways and for reasons we have long since forgotten. In these last five hundred years we see only pitiful relics of the empire and the power of Memory.

The ancient Greeks gave mythic form to this fact that ruled their lives. The Goddess of Memory (Mnemosyne) was a Titan, daughter of Uranus (Heaven) and Gaea (Earth), and mother of all the nine Muses. In legend these were Epic Poetry (Calliope), History (Clio), Flute playing (Euterpe), Tragedy (Melpomene), Dancing (Terpischore), the Lyre (Erato), Sacred Song (Polyhymnia), Astronomy (Urania), and Comedy (Thalia). When the nine daughters of King Pierus challenged them in song, the King's daughters were punished by being changed into magpies, who could only sound monotonous repetition.

Everyone needed the arts of Memory, which, like other arts, could be cultivated. The skills of Memory could be perfected, and virtuosi were admired. Only recently has "memory training" become a butt of ridicule and a refuge of charlatans. The traditional arts of Memory, delightfully chronicled by historian Frances A. Yates, flourished in Europe over the centuries.

The inventor of the mnemonic art was said to be the versatile Greek lyric poet Simonides of Ceos (c. 556–468?B.C.). He was reputed also to be the first to accept payment for his poems. The origins were described in the work on oratory by Cicero, who was himself noted for his mnemonic skill. Once at a banquet in the house of Scopas in Thessaly, Simonides was hired to chant a lyric in honor of his host. But only half of Simonides' poem was in praise of Scopas, as he devoted the other half to the divine twins Castor and Pollux. The angry Scopas therefore would pay only half the agreed sum. While the many guests were still at the banquet table a message was brought to Simonides that there were two young men at the door who wanted him to come outside. When he went out he could see no one. The mysterious callers were, of course, Castor and Pollux, who had found their own way to pay Simonides for their share of the panegyric. For at the very moment when Simonides had left the banquet hall the roof fell in, burying all the other guests in the ruins. When relatives came to take away the corpses for the burial honors, the mangled bodies could not be identified. Simonides then exercised his remarkable memory to show the grieving relatives which bodies belonged to whom. He did this by thinking back to *where* each of the guests had been seated. Then he was able to identify by place each of the bodies.

It was this experience that suggested to Simonides the classic form of the art of Memory of which he was reputed to be the inventor. Cicero, who made Memory one of the five principal parts of rhetoric, explained what Simonides had done.

> He inferred that persons desiring to train this faculty must select places and form mental images of the things they wish to remember and store those images in the places, so that the order of the places will preserve the order of the things, and the images of the things will denote the things themselves, and we shall employ the places and images respectively as a wax writing-tablet and the letters written on it.

Simonides' art, which dominated European thinking in the Middle Ages, was based on the two simple concepts of places (*loci*) and images (*imagines*). These provided the lasting elements of Memory techniques for European rhetors, philosophers, and scientists.

A treatise (c. 86–82 B.C.) by a Roman teacher of rhetoric known as *Ad Herrenium,* after the name of the person to whom his work was dedicated, became the standard text, the more highly esteemed because some thought it had been written by Cicero. Quintilian (A.D. c. 35–c. 95), the other great Roman authority on rhetoric, made the classic rules memorable. He described the "architectural" technique for imprinting the memory with a series of places. Think of a large building, Quintilian said, and walk through its numerous rooms remembering all the ornaments and furnishings in your imagination. Then give each idea to be remembered an image, and as you go through the building again deposit each image in this order in your imagination. For example, if you mentally deposit a spear in the living room, an anchor in the dining room, you will later recall that you are to speak first of the war, then of the navy, etc. This system still works.

In the Middle Ages a technical jargon was elaborated on the basic distinction between the "natural" memory, with which we were all born and which we exercise without training, and the "artificial" memory, which we can develop. There were different techniques for memorizing things or words. And differing views about where the student should be when he worked at his memory exercises and what were the best kinds of places to serve as imaginary storage houses for the *loci* and images of memory. Some teachers advised the student to find a quiet place, where his imagined impressions of the *loci* of memory would not be weakened by surrounding noises and passing people. And, of course, an observant and well-traveled person was better equipped to provide himself with many varied Memory-places. In those days one could see some student of rhetoric walking tensely through a deserted building, noting the shape and furnishing of each room to equip his imagination with places to serve as a warehouse for his memory.

The elder Seneca (c. 55 B.C.–A.D. 37), a famous teacher of rhetoric, was said to be able to repeat long passages of speeches he had heard only once many years before. He would impress his students by asking each member of a class of two hundred to recite a line of poetry, and then he would recite all the lines they had quoted—in *reverse* order, from last to first. Saint Augustine, who also had begun life as a teacher of rhetoric, reported his admiration of a friend who could recite the whole text of Virgil—backwards!

The feats and especially the acrobatics of "artificial" memory were in high repute. "Memory," said Aeschylus, "is the mother of all wisdom." "Memory," agreed Cicero, "is the treasury and guardian of all things." In the heyday of Memory, before the spread of printing, a highly developed Memory was needed by the entertainer, the poet, the singer, the physician, the lawyer, and the priest.

The first great epics in Europe were produced by an oral tradition, which is another way of saying they were preserved and performed by the arts of Memory. The *Iliad* and the *Odyssey* were perpetuated by word of mouth, without the use of writing. Homer's word for poet is "singer" (*aoidos*). And the singer before Homer seems to have been one who chanted a single poem, short enough to be sung to a single audience on one occasion. The surviving practice in Muslim Serbia, which is described by the brilliant American scholar-explorer Milman Parry, is probably close to the custom of Homeric antiquity. He shows us how in the beginning the length of a poem was limited by the patience of an audience and a singer's remembered repertoire. Then the achievement of a Homer (whether a he, she, or they) was to combine hour-long songs into a connected epic with a grander purpose, a larger theme, and a complicated structure.

The first manuscript books in the ancient Mediterranean were written on papyrus sheets glued together and then rolled up. It was inconvenient to unroll the book, and frequent unrolling wore away the written words. Since there were no separate numbered "pages," it was such a nuisance to verify a quotation that people were inclined to rely on their memory.

Laws were preserved by Memory before they were preserved in documents. The collective memory of the community was the first legal archive. The English common law was "immemorial" custom which ran to a "time whereof the memory of man runneth not to the contrary." "In the profound ignorance of letters which formerly overspread the whole western world," Sir William Blackstone noted in 1765, "letters were intirely traditional, for this plain reason, that the nations among which they prevailed had but little idea of writing. Thus the British as well as the Gallic druids committed all their laws as well as learning to memory; and it is said of the primitive Saxons here, as well as their brethren on the continent, that *leges sola memoria et usu retinebant.*"

Ritual and liturgy, too, were preserved by memory, of which priests were the special custodians. Religious services, often repeated, were ways of imprinting prayers and rites on the youth of the congregation. The prevalence of verse and music as mnemonic devices attests the special importance of memory in the days before printed textbooks. For centuries the standard work on Latin grammar was the twelfth-century *Doctrinale,* by Alexander of Villedieu, in two thousand lines of doggerel. Versified rules were easier to remember, though their crudity appalled Aldus Manutius when he reprinted this work in 1501.

Medieval scholastic philosophers were not satisfied that Memory should be merely a practical skill. So they transformed Memory from a skill into a virtue, an aspect of the virtue of Prudence. After the twelfth century, when the classic treatise *Ad Herennium* reappeared in manuscripts, the scholastics seemed less concerned with the technology than with the Morality of Memory. How could Memory promote the Christian life?

Saint Thomas Aquinas (1225–1274), his biographers boasted, memorized everything his teachers ever told him in school. In Cologne, Albertus Magnus helped him train his memory. The sayings of the Church Fathers that Aquinas collected for Pope Urban IV after his trips to many monasteries were recorded not from what he had *copied* out but from what he had merely *seen.* Of course he remembered perfectly anything he had ever read. In his *Summa Theologiae* (1267–73) he expounded Cicero's definition of Memory as a part of Prudence, making it one of the four cardinal virtues, and then he provided his own four rules for perfecting the Memory. Until the triumph of the printed book, these Thomist rules of memory prevailed.

Copied again and again, they became the scheme of textbooks. Paintings by Lorenzetti and Giotto, as Frances A. Yates explains, depicted virtues and vices to help viewers apply the Thomist rules of artificial Memory. The fresco of the Chapter House of Santa Maria Novella in Florence provides memorable images for each of Aquinas' four cardinal virtues and their several parts. "We must assiduously remember the invisible joys of paradise and the eternal torments of hell," urged Boncompagno's standard medieval treatise. For him, lists of virtues and vices were simply "memorial notes" to help the pious frequent "the paths of remembrance."

Dante's *Divine Comedy,* with his plan of Inferno, Purgatorio, and Paradiso, made vivid both places and images (following the precepts of Simonides and Aquinas) in an easily remembered order. And there were humbler examples, too. The manuscripts of English friars in the fourteenth century described pictures—for example, of Idolatry in the role of a prostitute—not meant to be seen with the eye, but rather to provide invisible images for the memory.

Petrarch (1304–1374) also had a great reputation as an authority on the artificial memory and how to cultivate it. He offered his own helpful rules for choosing the "places" where remembered images were to be stored for retrieval. The imagined architecture of Memory, he said, must provide storage places of medium size, not too large or too small for the particular image.

By the time the printing press appeared the arts of Memory had been elaborated into countless systems. At the beginning of the sixteenth century the best-known work was a practical text, *Phoenix, sive Artificiosa Memoria* (Venice, 1491), which went through many editions and was widely translated. In that popular handbook, Peter of Ravenna advised that the best memory *loci* were in a deserted church. When you have found your church, you should go around in it three or four times, fixing in your mind all the places where you would later put your memory-images. Each locus should be five or six feet from the one before. Peter boasted that even as a young man he had fixed in his mind 100,000 memory *loci,* and by his later travels he had added thousands more. The effectiveness of his system, he said, was shown by the fact that he could repeat verbatim the whole canon law, two hundred speeches of Cicero, and twenty thousand points of law.

After Gutenberg, realms of everyday life once ruled and served by Memory would be governed by the printed page. In the late Middle Ages, for the small literate class, manuscript books had provided an aid, and sometimes a substitute, for Memory. But the printed book was far more portable, more accurate, more convenient to refer to, and, of course, more public. Whatever was in print, after being written by an author, was also known to printers, proofreaders, and anyone reached by the printed page. A man could now refer to the rules of grammar, the speeches of Cicero, and the texts of theology, canon law, and morality without storing them in himself.

The printed book would be a new warehouse of Memory, superior in countless ways to the internal invisible warehouse in each person. When the codex of bound manuscript pages supplanted the long manuscript roll, it was much easier to refer to a written source. After the twelfth century some manuscript books carried tables, running heads, and even rudimentary indexes, which showed that Memory was already beginning to lose some of its ancient role. But retrieval became still easier when printed books had title pages and their pages were numbered. When they were equipped with indexes, as they sometimes were by the sixteenth century, then the only essential feat of Memory was to remember the order of the alphabet. Before the end of the eighteenth century the alphabetic index at the back of a book had become standard. The technology of Memory retrieval, though of course never entirely dispensable, played a much smaller role in the higher realms of religion, thought, and knowledge. Spectacular feats of Memory became mere stunts.

Some of the consequences had been predicted two millennia earlier when Socrates lamented the effects of writing itself on the Memory and the soul of the learner. In his dialogue with Phaedrus reported by Plato, Socrates recounts how Thoth, the Egyptian god who invented letters, had misjudged the effect of his invention. Thoth was thus reproached by the God Thamus, then King of Egypt:

> This discovery of yours will create forgetfulness in the learners' souls, because they will not use their memories; they will trust to the external written characters and not remember of themselves. The specific which you have discovered is an aid not to memory, but to reminiscence, and you give your disciples not truth, but only the semblance of truth; they will be hearers of many things and will have learned nothing; they will appear to be omniscient and will generally know nothing; they will be tiresome company, having the show of wisdom without the reality.

The perils that Socrates noted in the written word would be multiplied a thousandfold when words went into print.

The effect was beautifully suggested by Victor Hugo in a familiar passage in *Notre-Dame de Paris* (1831) when the scholar holding his first printed book turns away from his manuscripts, looks at the cathedral, and says "This will kill that" (*Ceci tuera cela*). Print also destroyed "the invisible cathedrals of memory." For the printed book made it less necessary to shape ideas and things into vivid images and then store them in Memory-places.

The same era that saw the decline of the everyday empire of Memory would see the rise of Neoplatonism—a mysterious new empire of the hidden, the secret, the occult. This revival of Platonic ideas in the Renaissance gave a new life and a new realm to Memory. Plato had made much of the soul and its "memory" of ideal forms. Now a galaxy of talented mystics developed a new technology of Memory. No longer the servant of oratory, only an aspect of rhetoric, Memory became an arcane art, a realm of ineffable entities. The Hermetic art opened secret recesses of the soul. The bizarre Memory Theater of Giulio Camillo, exhibited in Venice and Paris, provided Memory-places not just for convenience but as a way of representing "the eternal nature of all things" in their "eternal places." The Neoplatonists of Cosimo de' Medici's Platonic Academy in Florence—Marsilio Ficino (1433–1499) and Pico della Mirandola (1463–1494)—built an occult art of Memory into their elusive philosophies.

The most remarkable explorer of the Dark Continent of Memory was the inspired vagrant Giordano Bruno (1548–1600). As a young friar in Naples, he had been inducted into the famous Dominican art of Memory, and when he abandoned the Dominican order, laymen hoped he would reveal the Dominican secrets. He did not disappoint them. For his work *On the Shadows of Ideas, Circe* (1582) explained that Memory-skill was neither natural nor magical but was the product of a special science. Introducing his Memory-science with an incantation by Circe herself, he showed the peculiar potency of images of the decans of the zodiac. The star images, Shadows of Ideas, representing celestial objects, are nearer to the enduring reality than are images of this transient world below. Bruno's system for "remembering" these "shadows of ideas contracted for inner writing" from the celestial images brought his disciples to a higher reality.

> This is to form the inform chaos. . . . It is necessary for the control of memory that the numbers and elements should be disposed in order . . . through certain memorable forms (the images of the zodiac). . . . I tell you that if you contemplate this attentively you will be able to reach such a figurative art that it will help not only the memory but also all the powers of the soul in a wonderful manner.

A guaranteed way to the Unity behind everything, the Divine Unity!

But the everyday needs for Memory were never as important as they had been in the days before paper and printed books. The kudos of Memory declined. In 1580 Montaigne declared that "a good memory is generally joined to a weak judgment." And pundits quipped, "Nothing is more common than a fool with a strong memory."

In the centuries after printing, interest has shifted from the technology of Memory to its pathology. By the late twentieth century, interest in Memory was being displaced by interest in aphasia, amnesia, hysteria, hypnosis, and, of course, psychoanalysis. Pedagogic interest in the arts of Memory came to be displaced by interest in the arts of learning, which were increasingly described as a social process.

And with this came a renewed interest in the arts of forgetting. When Simonides offered to teach the Athenian statesman Themistocles the art of Memory, Cicero reports that he refused. "Teach me not the art of remembering," he said, "but the art of forgetting, for I remember things I do not wish to remember, but I cannot forget things I wish to forget."

The study of forgetting became a frontier of modern psychology, where mental processes were first examined experimentally and subjected to measurement. "Psychology has a long past," Hermann Ebbinghaus (1850–1909) observed, "yet its real history is short." His beautifully simple experiments, which William James called "heroic," were described in *Memory: A Contribution to Experimental Psychology* (1885) and laid the foundation for modern experimental psychology.

Ebbinghaus invented meaningless raw materials for his experiments. Nonsense syllables. By taking any two consonants and putting a vowel in between, he devised some twenty-three hundred rememberable (and forgettable) items, and he put them in series. For his experiments the syllables had the advantage of lacking associations. For two years he used himself as the subject to test the powers of retaining and reproducing these syllables. He kept scrupulous records of all his trials, of the times required for recollection, and of the intervals between his efforts. He also experimented in "relearning." His efforts might have been of little use without his passion for statistics.

Now, Ebbinghaus hoped, not mere sense perceptions (which Gustav Fechner [1801–1887] had already begun to study and to whom he dedicated his work) but mental phenomena themselves could be submitted to "an experimental and quantitative treatment." Ebbinghaus' "forgetting-curve" related forgetting to the passage of time. His results, still significant, showed that most forgetting takes place soon after "learning."

In this unexpected way the inward world of thought began to be charted with the instruments of modern mathematics. But other explorers, in the Neoplatonist tradition, kept alive an interest in the mysteries of memory. Ebbinghaus himself said that he had studied "the non-voluntary re-emergence of mental images out of the darkness of memory into the light of consciousness." A few other psychologists rashly plunged into that "darkness" of the unconscious, but even as they did so they claimed to have invented a whole new "science."

The founders of modern psychology were increasingly interested in forgetting as a process in everyday life. The incomparable William James
(1842–1910) observed:

> In the practical use of our intellect, forgetting is as important a function as
> remembering. . . . If we remembered everything, we should on most occasions
> be as ill off as if we remembered nothing. It would take as long for us to recall
> a space of time as it took the original time to elapse, and we should never get
> ahead with our thinking. All recollected times undergo . . . foreshortening; and
> this foreshortening is due to the omission of an enormous number of facts which
> filled them. "We thus reach the paradoxical result," says M. Ribot, "that one
> condition of remembering is that we should forget. Without totally forgetting a
> prodigious number of states of consciousness, and momentarily forgetting a large
> number, we could not remember at all. . . ."

In a century when the stock of human knowledge and of collective memories would be multiplied, recorded, and diffused as never before, forgetting
would become more than ever a prerequisite for sanity.

But what happened to "forgotten" memories? "Where are the snows of
yester-year?" In the twentieth century the realm of memory was once again
transformed, to be rediscovered as a vast region of the unconscious. In his
Psychopathology of Everyday Life (1904) Sigmund Freud (1856–1939) started
from simple examples, such as the forgetting of proper names, of foreign
words, and of the order of words. The new arts of Memory for which Freud
became famous had both the scientific pretensions of Simonides and his
followers and the occult charm of the Neoplatonists. Of course, people had
always wondered at the mystery of dreams. Now Freud found the dream
world also to be a copious secret treasury of Memories. Freud's *Interpretation of Dreams* (1900) showed how psychoanalysis could serve as an art and
a science of Memories.

Others, stirred by Freud, would find still more new meanings in Memory.
Latent Memory, or the unconscious, became a new resource of therapy,
anthropology, and sociology. Might not the tale of Oedipus record everyone's experience? Freud's own mythic metaphors hinted at our inner inheritance of ancient, communal experience. Carl Jung (1875–1961), more in the
Hermetic tradition, popularized the "collective unconscious." Now Freud,
his disciples and dissidents, as we will see, had again rediscovered, or
perhaps after their fashion reconstructed, the Cathedrals of Memory.

61

Empire of the Learned

THE ancient Roman Empire left a living legacy across Europe. The relics of Roman law defined property, contracts, and crimes for that continent and much of the rest of the world. Memories of political unity encouraged European federalists for centuries. The language of Rome survived, provided the literature of the written book, and created a *European* community of learning. But this legacy that united the culture of Europe also divided the communities of Europe. For all over the continent there were *two-language* communities.

The learned community of the Church and the universities, the community of readers in the Middle Ages, was held together by Latin. So long as Latin was the language of universities there was, at least in the linguistic sense, a single European university system. Teachers and students could move from Bologna to Heidelberg, from Heidelberg to Prague, from Prague to Paris and feel at home in the classroom. Countless ordinary students— along with Vesalius, Galileo, and Harvey—went from one learned community to another. For the first and last time the whole continent had a single language of learning.

But Latin, the bond of the learned, would become a barrier between the learned of each nation and all the rest of their countrymen. Other languages were spoken at home, in the marketplace, and for popular entertainment. Everywhere the populace spoke not Latin but the "vernacular," which meant the local native language (from Latin *vernaculus,* meaning domestic or indigenous, from *verna,* meaning a home-born slave or a native). All across Europe the language of the learned was a foreign language. The curiously cosmopolitan vocabulary of the learned class put still another obstacle in the path of their efforts to understand their neighbors. The consciousness of the common people was provincial and myopic. They could hear the voices only of the living. At the same time the learned were afflicted by a narrow farsightedness. They thought over the heads of their marketplace contemporaries to a special language and literature of faraway and long ago.

Nothing in human nature required that a community be divided in this way. This was an accident of European history which for centuries shaped, directed, and confined the thinking of a continent. As late as the sixteenth century, the German humanist Johannes Sturm (1507–1589), who ran a model grammar school in Strasbourg, wistfully described the unique advantage of the young people of antiquity. "The Romans had two advantages over us," he observed, "the one consisted of learning Latin without going to school, and the other in frequently seeing Latin comedies and tragedies acted, and in hearing Latin orators speak. Could we recall these advantages in our schools, why could we not, by presevering diligence, gain what they possessed by accident and habit—namely, the power of speaking Latin to perfection? I hope to see the men of the present age, in their writing and speaking, not merely followers of the old masters, but equal to those who flourished in the noblest age of Athens and Rome."

Knowledge of Latin was an absolute prerequisite for attending a medieval university. It was not enough to be able to spell out a text laboriously. For all lectures were delivered in Latin and students were required to speak only Latin outside the classroom, a rule enforced by penalties and by informers called "wolves." Perhaps this was a way of discouraging needless talk. At the University of Paris when a student made any request of the rector, the statute required him to present his case without using even a single French word. Before the rise of national vernaculars Latin was the medium of conversation between students from different parts of the country, and was needed for a convivial student life. To help the entering student in Paris, there were Berlitz-style lists of colloquial phrases, where we glimpse the daily round of student life in what to say when you wanted to exchange money, to buy candles and writing materials, wine, or fruit, or pork, chicken or beef, eggs, cheese, or pastry. A handy conversation manual for Heidelberg students in 1480 gave the phrases you would need when you were hazed, when you bought dinner for an upperclassman, when you borrowed money, and, of course, when you wrote home for money. We cannot know whether students would have understood any more of the lectures if these had not been delivered in Latin. But the obstacle of Latin may have been a reason why most university "students" never offered themselves to be examined for any degree.

The Latin of the medieval universities became a richer, more flexible language. Like modern Hebrew, medieval Latin was adapted to daily needs. And this Latin language shaped the thinking of the educated classes across the continent. The "liberal arts"—the prescribed foundation of a "liberal education," i.e., the subjects best suited for *liberi*, freemen—might have been called the "literary arts." For the trivium, the whole curriculum for the Bachelor of Arts degree in the Middle Ages consisted of grammar, rhetoric, and logic, read in the Latin works of ancient Rome. Only for the advanced degree, the Master of Arts, was the student examined in the broader quadrivium, which comprised arithmetic, geometry, astronomy, and music. Fragments of Aristotle and other writers in Greek and Arabic were taught through Latin translations. The Bible, too, was known to the educated classes mainly through the Vulgate (the *editio vulgata,* or "common version"), which was a Latin translation (383–405) based on that by Saint Jerome. In the thirteenth century the faculty of the University of Paris revised and corrected Saint Jerome's work into another Latin translation as the standard version for theological teaching.

The Latin culture of medieval Europe could hardly have prospered without the enthusiasm, the passion, and the good sense of Saint Benedict of Nursia (480?–543?). The father of Christian monasticism in Europe, he was also the godfather of libraries. The preservation of the literary treasures of antiquity and of Christianity through the Middle Ages was a Benedictine achievement. Saint Benedict himself, born of a good family at Nursia, near Perugia, in Umbria, had been sent to school in Rome when the ancient imperial power was in decay and the power of the papacy was on the rise. Troubled by the dissoluteness of the city, he retired for three years to a cave in the Abruzzi hills. When he became known for his holiness, he was invited to become abbot of a monastery, where he disciplined his fellow monks. When a disgruntled friar tried to poison him, he retired again to his cave. But his vision survived. He went on to found in that region alone twelve monasteries with twelve monks in each, all under his direction. He then went south, where, about 529, he founded the abbey of Monte Cassino. Sacked by the Lombards and the Saracens and shaken by earthquake, it still remained the spiritual headquarters of the monastic movement in Europe. It would finally be leveled by aerial bombing in World War II.

The Rule (*Regula*) of Saint Benedict offered a workable compromise between the ascetic otherworldly spirit and the weaknesses of human nature. After a year of probation the young monk vowed obedience to the Rule and lifelong residence in the same monastery. In each monastery the monks elected their abbot for life, and otherwise there was no hierarchy. Saint Benedict's sensible schedule for the monks' daily life spread across Europe, preserving and perpetuating Latin literary culture through the following centuries. According to chapter 48 of his Rule:

> Idleness is the enemy of the soul: hence brethren ought at certain seasons to occupy themselves with manual labour, and again at certain hours with holy reading. Between Easter and the calends of October let them apply themselves to reading from the fourth hour until the sixth hour. . . . From the calends of

October to the beginning of Lent, let them apply themselves to reading until the second hour. During Lent, let them apply themselves to reading from morning until the end of the third hour, and in these days of Lent, let them receive a book apiece from the library and read it straight through. These books are to be given out at the beginning of Lent.

Every monastery needed its library. "A monastery without a library [*sine armario*]," a monk in Normandy wrote in 1170, "is like a castle without an armory [*sine armamentario*]. Our library is our armory. Thence it is that we bring forth the sentences of the Divine Law like sharp arrows to attack the enemy. Thence we take the armour of righteousness, the helmet of salvation, the shield of faith, and the sword of the spirit, which is the Word of God." In each monastery the precentor had the duty of checking out books and seeing that they were returned. It was customary for monasteries to lend their books to other monasteries and even, with proper security, to the secular public. The Benedictine pioneers in "interlibrary loan" provided a kind of public lending library for the learned few.

Special curses were uttered against those who mutilated books or absconded with them. "This book belongs to [the monastery of] St. Mary of Robert's Bridge," warns a twelfth-century manuscript of Saint Augustine and Ambrose, "whosoever shall steal it, or sell it, or in any way alienate it from this house, or mutilate it, let him be forever accursed. Amen." Below this, in the manuscript, now in the Bodleian Library at Oxford, we can read in a fourteenth-century hand: "I, John, Bishop of Exeter, know not where the aforesaid house is, nor did I steal this book, but acquired it in a lawful way."

Wandering clerics and pious travelers entrusted their manuscript treasures to the monastic and cathedral libraries, which vied for the best-collated versions of sacred texts and received substantial fees for the right to copy them. Constantine the African (c. 1020–c. 1087), who spent forty years collecting and translating into Latin scientific treatises from Egypt, Persia, Chaldea, and India, finally settled in Monte Cassino, where he deposited his great collection. When the library of the monastery at Novalesa was destroyed by the Saracens in 905, it was reported to have contained more than sixty-five hundred volumes. Every manuscript copy of a learned work was unique, but one that had been laboriously collated with others had special authority.

Monastic libraries, of course, included the Holy Scriptures, writings of the Church Fathers and commentaries on them. Larger collections, sometimes found in cathedral libraries, would include chronicles like Bede's *Ecclesiastical History,* the writings of Augustine, Albertus Magnus, Aquinas, and Roger Bacon. Their secular books would include Virgil, Horace, and Cicero. Plato, Aristotle, and Galen, among others, were found in Latin translation. Such libraries across Europe were not only the armories of Christian crusaders but the treasuries of European culture. Monkish students who had studied at Paris or Bologna would bring back their lecture notes with the latest interpretations of theology and the classics for their monastery library. It was in these libraries that five books of the *Annals* of Tacitus, the *Republic* of Cicero, and other ancient literary monuments survived.

Not only did the Benedictines collect libraries, but they created them. The "making" (i.e., the copying), like the reading, of books, became a sacred duty, and the scriptorium, the scribes' copying room, was a usual feature of their monasteries. In some ways they were freer to reproduce books than were publishers in the later age of the printing press. Of course, their "publishing" list was limited by orthodoxy and by dogma, but there was no law of copyright and so no royalties to be paid to the author. Their whole stock-in-trade was what the modern publisher would call a backlist. The book was not expected to be, nor dared it be, a vehicle for new ideas carrying messages from contemporaries to contemporaries. Instead it was a device to preserve and amplify the treasured revolving fund of literary works—the sacred Scriptures and their commentators, ancient classics of Greece and Rome, and a few established texts in Hebrew or Arabic.

The age of "authorship" had not yet arrived. When reading a sacred *text,* medieval scholars were quite indifferent to the identity of the author. The writers who were being transcribed had not always troubled to "quote" what they had taken from other writers. Even in an era when students were taught to argue by citing "authority" it was practically impossible, even if it had been thought desirable, to ascribe particular passages to particular authors. Writers of original texts were reluctant to take the credit, or risk the more likely blame, for innovation. In the great age of manuscript books, anonymity was dictated by technology, orthodoxy, and prudence. Even the best modern scholars on the subject cannot devise a satisfactory scheme for arranging these manuscripts in a "bibliography." They must resort to lists arranged not by author but by opening words or other devices. Quotation marks first came into general use with the printed books of Italy and France in the fifteenth and sixteenth centuries. But this kind of punctuation, which guides the reader to the original author, did not receive its modern name or enter standard usage until the seventeenth century.

In the Middle Ages every monastery was its own publishing house, and a monk with writing desk, ink, and parchment was his own publisher. "A book should always be in your hand or under your eyes," advised Saint Jerome. The rich lore surrounding these scriptoria reminds us that long before the making of books was a business it had become a sacred enterprise.

Saint Louis (Louis IX, 1214–1270) insisted that it was better to transcribe a book than to buy the original, because this helped diffuse the Christian gospel. Work in the scriptorium had as much dignity as labor in the fields. "He who does not turn up the earth with the plough," a monk exhorted his fellows in the sixth century, "ought to write the parchment with his fingers." And in the ill-heated hall or cell these fingers were often numb. For fear of fire, artificial light was usually not allowed. Many a monk gave his eyesight to provide the illuminated missals that our own eyes admire.

The sacred work became a penance. Full-time scribes, relieved of routine duties, were allowed to visit the kitchen to melt their wax or dry their parchment. "Jacob wrote a certain portion of this book," we read, "not of his own free will but under compulsion, bound by fetters, just as a runaway and fugitive has to be bound." The Abbot of St. Evroul (c. 1050), a skilled scribe, encouraged his congregation by his tale of a sinful Brother who had been saved by his industry in the scriptorium. At his death the Devil was about to take him to hell. But when the notorious Brother came before the Judgment Seat, God saw the beautiful large folio of holy texts he had transcribed. It was decided that for each letter he had written in the book, he would be pardoned one sin. Since it was a very big book, when the angels reckoned up his sins they found that even after all his sins were forgiven, one letter was left over. The divine Judge then mercifully decreed that the soul of this monk should be allowed to reenter his body on earth so that he could set his life straight. And then not enter life eternal with only one little merit to his name! Another pious chronicler reports an English monk who had been such a devoted scribe that twenty years after his death, when all the rest of his body had gone to dust, the right hand with which he had written his manuscripts remained intact and became a holy relic under the altar of his monastery.

If Saint Benedict was the patron saint of the manuscript book in the Middle Ages, the worldly patron was Charlemagne (742–814). It was a happy coincidence for Western civilization that so effective an administrator was also a devotee of the written word. The shadowy textbook figure who was crowned Holy Roman emperor on Christmas Day, 800, comes to life as the sponsor of bookish culture, reformer of the Latin language and the Roman alphabet. Charlemagne inherited his throne as king of the Franks in 768. A fierce man of ruthless ambition, he overrode the claims of rivals and relatives, subdued the Saxons, conquered Lombardy, and finally organized an empire that included northern Italy, France, and most of modern Germany and eastern Europe. As an ally of the Pope and a passionate Christian, Charlemagne was shocked by the decay of Christian learning. He was dismayed by the crude Latin in the letters he received, even from bishops and abbots. The Carolingian renaissance that he sparked was a Latin renaissance.

When Charlemagne met the personable English monk Alcuin (732–804) in Italy in 781, he persuaded him to come to Aachen (Aix-la-Chapelle) to organize a reform of language and of education. In remote Yorkshire, Alcuin had set high standards that made his cathedral school famous across Europe. Charlemagne agreed, too, that the correct knowledge of Scripture required a correct command of Latin. In his famous edict of 789, written by Alcuin, Charlemagne ordered: "In each bishopric and in each monastery let the psalms, the notes, the chant, calculation and grammar be taught and carefully corrected books be available." Alcuin set the standards at his new school of calligraphy in Tours:

> Here let the scribes sit who copy out the words of the Divine Law, and likewise the hallowed sayings of the holy Fathers. Let them beware of interspersing their own frivolities in the words they copy, nor let a trifler's hand make mistakes through haste. Let them earnestly seek out for themselves correctly written books to transcribe, that the flying pen may speed along the right path. Let them distinguish the proper sense by colons and commas, and let them set the points each one in its due place, and let not him who reads the words to them either read falsely or pause suddenly. It is a noble work to write out holy books, nor shall the scribe fail of his due reward. Writing books is better than planting vines, for he who plants a vine serves his belly, but he who writes a book serves his soul.

Charlemagne's rich library in his palace at Aachen became a cultural center drawing scholarly Christian refugees from the Moors in Spain, and even from the distant islands of Ireland. He ordered every school to have a scriptorium.

Monks adored the holy texts by adorning them, and not only in the metropolitan centers. On Iona (a remote island of the Hebrides off the coast of Scotland) Celtic monks in the monastery founded by Saint Columba in 563 and in the Abbey of Kells in Ireland produced one of the most beautiful books of all time. The Book of Kells, now in the library of Trinity College, Dublin, ornamented the resplendent Latin text of the Gospels in uncial and half-uncial scripts, with vines and leaves in dazzling lapis lazuli. From monasteries in Germany, Italy, and Bulgaria came manuscripts of unexcelled beauty, the work not only of monks but of laymen enlisted in the scriptoria. Some of the best were learned Benedictine nuns, who became famous for the delicate illuminations of the holy texts.

The disciples of Saint Benedict and the scholars of the Carolingian renaissance reformed the very shape of our letters. They improved the function as well as the beauty of our written alphabet by inventing some new forms.

Until then Latin had been written only in capital letters, which was all the
Romans used. No "lower-case," or small, letters are found on the ancient
Roman monuments. The chisel governed the shapes of the letters they
inscribed on stone whose simple dignified forms still mark our cornerstones
and tombstones. When written with a pen on papyrus or vellum, the Roman
letters took on another shape. All the letters were still capitals but the
peculiarities of the pen produced a thin vertical stroke with thicker strokes
on the curves and oblique angles. These so-called "rustic capitals" became
the standard script for books and formal documents. Writing in capital
letters was known as "majuscule." Small, or lower-case, letters were still
unknown. All letters were the same height, confined between a single pair
of horizontal lines.

   Gradually the monks and scribes began experimenting with small letters
of varying shapes. They took clues from the cursive script of business
correspondence. The scarcity of papyrus and the cost of vellum encouraged
them to find a way to write more compactly so they would use fewer sheets.
At the same time the decline of Roman authority dissolved standards in
calligraphy as in everything else. The idiosyncrasies of isolated monasteries
were dividing the culture of Latin Europe.

   When Alcuin joined Charlemagne at Aachen, they naturally made the
reform and standardizing of calligraphy a major concern. To ensure the
accuracy of holy texts it was essential to bring the learned world together.
In this lucky collaboration, Alcuin had the knowledge and the taste to
devise standards, Charlemagne had the administrative power, the organization, and the will to enforce them. At his school of calligraphy in the
monastery of St. Martin's in Tours, Alcuin taught his reformed script. He
had studied ancient monuments and recent manuscripts in search of the
most elegant, most legible, and most writable forms. His capital letters
followed the dignified inscriptions of Augustan Rome. Then, drawing on
the experiments of other monks and on his long experience at York in
supervising the transcription of the famous Golden Gospels, he produced
a standard form for small letters. Alcuin's Carolingian Minuscule proved
successful beyond his dreams. Neat and attractive, easy to write and to read,
it dominated scriptoria and libraries. Seven hundred years later, when
movable type came to Europe, and after only a brief Gothic interlude, the
letters were fashioned on the model of Carolingian Minuscule. Long after
other monuments of Charlemagne's empire have crumbled, the pages of this
book in your hands remain a vivid reminder of the power of the well-
designed written word. What we call the Roman alphabet is really Alcuin's
alphabet.

   During the later Middle Ages, Alcuin's simple legible letters did have some competition. In the eleventh century, the age of the Gothic cathedral, his alphabet was adapted into a script disparagingly called Gothic by the Italian humanists of the Renaissance. This script, now more appropriately called Black Letter, produced a darker page and lent a feeling of solemnity, which explains its survival in the introductory "Whereas" to legal documents and diplomas otherwise printed in Alcuin's Roman. Gutenberg used Black Letter in his 42-line Bible. Happily, the Renaissance would revive the chaste and legible Carolingian Minuscule, which came to dominate the West. Only in Germany and Scandinavia did Black Letter live on, in a new version called Fraktur. Hitler and his Nazis found this pseudo-Teutonic form suited to their taste.

When we look at a manuscript or an inscription before the Age of Charlemagne, we are surprised to see all the letters strung together without space between words, and without periods, commas, or paragraphing. So it was for most of Western history. Before the later seventeenth century "punctuation" meant the pointing of Psalms to indicate pauses for meditation, or the insertion of vowel points in writing Hebrew or other Semitic languages. Our verb "to punctuate" did not appear until the early nineteenth century. Before then "pointing" was the word used for inserting marks on the written or printed page.

With the Carolingian reforms in script came the new practice of separating words by an empty space between them. This helped prevent ambiguities in meaning and so preserve the pure text. The spacing was a symptom, too, that Latin was reaching out to a whole continent of scholars for whom Latin was a foreign language. Scribes in Ireland, England, and Germany felt more secure when they saw the words separated. By the twelfth century, university textbooks used a form of "C" (for *capitulum,* "Chapter") at the beginning of sentences. The title pages of books as late as the sixteenth and seventeenth century show that even skilled printers were dividing words and running on words from one line to another in a way that looks strange today, a relic of days before the separation of words was usual.

After Charlemagne, when punctuation first became common it was an aid for speaking, or for reading a printed text aloud to an illiterate audience. To help the reader follow the principles of elocution, spacing and punctuation marks signaled pauses of different lengths, which incidentally helped a listener follow the meaning. By the later seventeenth century more printed matter was intended for silent reading. Then punctuation came to be governed by syntax and aimed to show the structure of a sentence. Nowadays we punctuate by syntax. Still, in English and other Western European languages a few marks—the exclamation point and the question mark—survive to indicate inflection and intonation.

62

The Duplicating Impulse

"PRINTING," in its origins, meant different things for the West and for the East. In Europe, as we shall see, the rise of printing would mean typography, printing from movable type made of metal. In China and other Asian countries shaped by Chinese culture, block printing was the crucial invention, and the rise of printing meant xylography, printing from blocks of wood. We should be wary, therefore, of generalizing from "printing" in the West to "printing" in the East.

The earliest impulse to printing in China was not to diffuse knowledge, but to ensure the religious or magical benefits from precise replication of a holy image or a holy text. Making repeated images for printing textiles from a carving on wood was an ancient folk art. At least as early as the third century the Chinese had developed an ink that made clear and durable impressions from these wood blocks. They collected the lamp black from burning oils or woods and compounded it into a stick, which was then dissolved to the black liquid that we call India ink, but which the French more accurately call *l'encre de Chine.*

Block printing began to develop during the T'ang dynasty (618–907), when the ruling family tolerated all sorts of religious sects—Taoist and Confucian scholars, Christian missionaries, Zoroastrian priests, and, of course, Buddhist monks. Each of these had its own sacred images and texts. By the early seventh century, the emperor's library held about forty thousand manuscript rolls.

Buddhist monasteries were especially active in experimenting with ways of multiplying images, for the very heart of Buddhism, as historian Thomas Francis Carter observes, was "the duplicating impulse." Just as the faithful themselves were to become replications of the Buddha, so too the devout Buddhist attained "merit" by multiplying images of Buddha and of the sacred texts. Buddhist monks carved images in stone and then took rubbings from them, they made seals, they tried stencils on paper, on silk, and on plastered walls. They made small wooden stamps with handles from which they made primitive woodcuts. Someone had the idea of removing the handle in order to lay the wood block flat on a table with the incised surface up. Then, probably in the seventh or early eighth century, a sheet of paper was laid on the inked block, rubbed with a brush, and so it became possible to make woodcuts larger. But in 845, when foreign religions, including Buddhism, were outlawed in China, forty-six hundred Buddhist temples were destroyed, a quarter-million Buddhist monks and nuns were driven out of the monasteries, and these earliest examples of Chinese printing disappeared.

At the very time when the techniques of block printing were advancing in China, across the neighboring waters the culture of Japan was being transformed under Chinese influence. In the seventh century, strong chiefs like Prince Shotoku (593–622) brought the clans ruled by priest-chiefs into a centralized government on the Chinese model. Shinto, an ancient and multiform nature worship, had been the native religion of the clans. The embassies sent to China by the Japanese rulers used Buddhism, which had originated in India, as the vehicle for importing Chinese ways. Returning students brought some knowledge of the Chinese language along with Chinese literature and arts. Prince Shotoku, mimicking the Chinese emperor, addressed his letter "from the Emperor of the Rising Sun to the Emperor of the Setting Sun." The power of Buddhism was at high tide when the Japanese Empire built (710–84) a splendid capital at Nara, modeled after the Chinese capital of Ch'ang-an (modern Sian), adorned by a 550-ton bronze Buddha (735–49), seventy-two feet high, covered with fifty pounds of gold, still the largest bronze statue in the world.

When the Japanese emperor Shomu abdicated in 749, his throne was taken by his daughter, the nun-empress Koken (718–770). The eloquent chief of the Buddhist hierarchy charmed her by his lectures, then became her personal physician and principal adviser. The Empress put the government in his hands—and presumably her body too. She gave him titles once reserved for the emperor, kept him in the palace, and herself became a fanatical Buddhist.

To avoid recurrence of the smallpox epidemic of 735–37 that had decimated the court, Empress Koken engaged a special corps of one hundred and sixteen priests to drive out disease demons. She recalled from the Buddhist Sutra a text describing how a sick Brahmin had consulted a seer who prophesied that the Brahmin would die within seven days. The Brahmin went to the Buddha himself, offering to become a disciple in return for a cure.

Buddha said to him, "In a certain city a pagoda is fallen. You must go and repair it, then write a *dharani* [charm] and place it there. The reading of this charm will lengthen your life now and later bring you to Paradise." The disciples of Buddha then asked him wherein the power of the *dharani* charm lay. The Buddha said, "Whoever wishes to gain power from the *dharani* must write seventy-seven copies and place them in a pagoda. This pagoda must then be honored with sacrifice. But one can also make seventy-seven pagodas of clay to hold the *dharani* and place one in each. This will save the life of him who thus makes and honors the pagodas, and his sins will be forgiven. Such is the method of the use of the *dharani*. . . ."

Empress Koken, in an unprecedented burst of piety, ordered one million charms, each a single sheet bearing about twenty-five lines of printed text in a tiny wooden pagoda. By 770 the work was completed and the million pagodas were distributed among various temples. Most of these miniature pagodas were three-storied, about four and a half inches high and three and a half inches in diameter at the base, but each ten-thousandth one was seven-storied and each hundred-thousandth one was thirteen-storied. The charms in them are the earliest examples of copper block printing on paper. But apparently they were not effective as medicine, for the Empress died at the age of fifty-two probably of smallpox in the very year when the project was completed.

The next relic of Chinese printing was a more complex project, the *Diamond Sutra* of 868, the oldest surviving printed "book." Selections from the Buddhist scriptures were printed on sheets, each two and a half feet long by a foot wide, pasted together to make a scroll nearly sixteen feet long. These sermons by the Buddha on the nonexistence of all things described the merit gained by all who copied the book, and explained that the Lord Buddha himself would be found wherever there was a copy of this sacred text. Block printing had become a welcome new technique for mass-producing "merit."

Other early printed works, besides sacred Buddhist scriptures, probably included works of Taoist magic, treatises on the divination of dreams, and dictionaries. But handwritten texts embellished by the traditional arts of calligraphy were still generally preferred for sacred matter, and printing supplied those who could not afford better.

Large-scale government printing was to be the predominant and inhibiting form of Chinese printing for centuries to come. Feng Tao, prime minister to the dynasty of Central China which conquered Shu in western China, explained in his official memorial of 932:

During the Han dynasty, Confucian scholars were honored and the Classics were cut in stone. . . . In T'ang times also stone inscriptions containing the text of the Classics were made in the Imperial School. Our dynasty has too many other things to do and cannot undertake such a task as to have stone incriptions cut and erected. We have seen, however, men from Wu and Shu who sold books that were printed from blocks of wood. There were many different texts, but there were among them no orthodox Classics [of Confucianism]. If the Classics could be revised and thus cut in wood and published, it would be a very great boon to the study of literature.

Editing and printing these Confucian classics consumed twenty-one years. In 953, when the director of the National Academy finally presented to the emperor the whole 130 volumes of the Confucian classics, he boasted that now they saw "the universal doctrine made eternal."

Authenticity, not diffusion, remained the purpose. The word *yin* for "print" meant "seal," and so connoted official approval. Until 1064 all private printing of the Confucian classics or anything else was outlawed, and only officially approved texts could be issued.

Block printing made possible the flowering of Chinese culture in the Sung renaissance (960–1127), and the printed Confucian classics revived a Confucian literature. Before the end of the tenth century appeared the first of the great Chinese dynastic histories, a work of several hundred volumes which consumed seventy years. Meanwhile, by 983 the Buddhists had produced something even more spectacular, the Tripitaka, the whole Buddhist canon in 5,048 volumes totaling 130,000 pages, each printed from a separately carved block. The king of Korea received a set from the emperor of China, and when a Buddhist priest brought a set to Japan, there came into the Japanese language the word *suri-hon* for the printed book. Then other sects put their own scriptures into print. The Taoist canon of 4,000 volumes appeared in 1019. Manichaeanism, a religion imported from the West, was authenticated by block printing of its texts. The many Muslims in China during the Sung dynasty seem not to have put their Koran into print, but almanacs and calendars were block-printed for the special use of Moslems.

In China, as in the West, the rise of the arts of printing spelled the decline of the arts of Memory. A Chinese scholar, Yeh Meng-Te (1077–1148), writing about 1130, reports:

Before the T'ang dynasty, all books were manuscripts, the art of printing not being in existence. People regarded the collecting of books as something honorable, and no one had them in large quantity. . . . and students, as a consequence of the great labor of transcription, also acquired great ability and accuracy in reciting them. In the time of the Five Dynasties, Feng Tao first memorialized his sovereign, praying that an official printing establishment might be put in operation. And again in those years of our reigning dynasty called Shun-hua [990–94] officers were commissioned to print the historical records and the annals of the first and second Han dynasties. From that time forth printed books became still more numerous. . . . as students found it easy to obtain books, the practice of reciting was in consequence broken up.

When Marco Polo visited the China of Kublai Khan (1216–1295), he saw nothing worth reporting in their multiplying of sacred texts by block printing. But he did note with astonishment how Kublai Khan by a kind of "alchemy" had made printed paper, in place of precious metals, serve as currency.

> Of this money the Khan has such a quantity made that with it he could buy all the treasure in the world. With this currency he orders all payments to be made throughout every province and kingdom and region of his empire. And no one dares refuse it on pain of losing his life. And I assure you that all the peoples and populations who are subject to his rule are perfectly willing to accept these papers in payment, since wherever they go they pay in the same currency, whether for goods or for pearls or precious stones or gold or silver. With these pieces of paper they can buy anything and pay for anything. And I can tell you that the papers that reckon as ten bezants do not weight as one. . . .
>
> Here is another fact well worth relating. When these papers have been so long in circulation that they are growing torn and frayed, they are brought to the mint and changed for new and fresh ones at a discount of 3 per cent. And here again is an admirable practice that well deserves mention in our book: if a man wants to buy gold or silver to make his service of plate or his belts or other finery, he goes to the Khan's mint with some of these papers and gives them in payment for the gold and silver which he buys from the mint-master. And all the Khan's armies are paid with this sort of money.

What Marco Polo described was an old Chinese institution. By the eleventh century, shortages of metal and the need for more currency had produced a government-supervised system for issuing printed sheets of paper currency, four million in a single year. In the twelfth century the Sung Chinese financed their defense against the Tartars by printing paper currency, and after their defeat they continued to print money for tribute. In 1209 the notes promising to pay off in gold or silver were printed on paper made of silk and pleasantly perfumed, but even their fragrance could not stabilize the currency or stop runaway inflation.

The Sung historian Ma Tuan-lin, who lived through the worst of this inflation, chronicled the familiar consequences;

> After having for years tried to support and maintain these notes, the people had no longer any confidence in them, and were positively afraid of them. For the payment for government purchases was made in paper. The fund of the salt manufactories consisted of paper. The salaries of all the officials were paid in

paper. The soldiers received their pay in paper. Of the provinces and districts, already in arrear, there was not one that did not discharge its debts in paper. Copper money, which was seldom seen, was considered a treasure. The capital collected together in former days was . . . a thing not even spoken of any more. So it was natural that the price of commodities rose, while the value of the paper money fell more and more. This caused the people, already disheartened, to lose all energy. The soldiers were continually anxious lest they should not get enough to eat, and the inferior officials in all parts of the empire raised complaints that they had not even enough to procure the common necessities. All this was the result of the depreciation of the paper money.

Following the example of the more advanced people they had conquered, the Tartars began issuing their own paper currency, and after 1260, when Kublai Khan completed his conquest of China, he made it the regular institution reported by Marco Polo. In Marco Polo's day the notes were still passing at full face value, but in the last years of the Mongols' Yüan dynasty (1260–1368), floods of paper money once again signaled inflation. When the first emperor of the new Ming dynasty (1368–1644) took over, he cut back the paper money in circulation, and finally succeeded in stabilizing the currency.

From its beginnings in China, printing bore this guilty association with unsound currency. For centuries printed paper money appeared to be the only form of printing known to European travelers. A paper-money debacle closer to the West added to the ill-repute of printing. In Tabriz, capital of Mongol-conquered Persia, both Venice and Genoa kept commercial agents during the early years of the fourteenth century. The extravagance of the Mongol ruler Gaikhatu Khan from 1291 to 1295 put pressure on his treasury, which he tried to relieve by issuing paper currency. Block-printed in 1294 in Chinese and Arabic, each of his notes bore the date of the Muslim era, a warning to forgers, and the cheerful prediction that now "poverty will vanish, provisions become cheap, and rich and poor be equal." But the magic did not work. After only a few days of compulsory use of the paper, commerce was disrupted, markets closed, and the Khan's financial officer was reported murdered. All this the Venetians and the Genoese trading with Tabriz could not have failed to observe, and it could hardly have encouraged them to take up printing to solve their fiscal problems.

Others besides Marco Polo, including William of Rubruck, Odoric, and Pegolotti, had noted with admiration how the Great Khan made the bark of trees do the work of precious metals. But this alone seemed not a sufficient inducement to introduce printing in the West. And Westerners had not yet become scholars of the Eastern religions, which could have impressed them with the use of printing for sacred literature. In Europe, though there are records of leather money in the twelfth and thirteenth centuries, no record of paper money appears until an issue in Sweden in 1648.

A more frivolous form of block-printing on paper may have been the vehicle that brought block printing to the West. Playing cards, like dominoes, appear to have originated in China. In the Sung and the Mongol eras complicated card games were being played with what they called "sheet dice" all across China. The fact that the Koran prohibited games of chance may help explain why no mention of playing cards has been found in medieval Arabic literature. But card-playing seems to have been common in the Mongol armies moving westward, and was said to have entered Europe from the land of the Saracens. Printed playing cards somehow leaped across the Arab world to Italy and Western Europe.

Hand-painted cards were still ordered by the rich, but the populace had their cards in print. Printed playing cards, known in Germany and Spain by 1377, had soon become so popular that the alarmed Synod of 1404 forbade card-playing by the clergy. In 1423 Saint Bernard of Siena's invective from the steps of St. Peter's exhorted his listeners to go home, to gather up their cards and burn them in the public square. Even before Gutenberg was printing books, printed playing cards were made in Venice, Augsburg, and Nuremberg, and in 1441 the Council of Venice had to pass a law to protect its domestic printers of cards. The mysterious Master of the Playing Cards (c. 1430–1450) produced an elegant set, of which sixty survive, with a finesse of engraved line that some attribute to Gutenberg himself. Perhaps Gutenberg's later experiments grew out of his efforts to perfect the printing of cards.

Printing was thus first used for a wide variety of everyday purposes long before it was turned to the high cause of knowledge or religion. Textile printing was ancient. Printed fabric found in a tomb of the Bishop of Arles dates from the sixth century. Printed silks survive from an eighth-century palace of Nara in Japan. And other printed textiles of about the same vintage remain from China and Egypt. European textile printers simply pressed a carved block carrying the pigment onto the fabric. In Asia at the same time more sophisticated techniques were used, some to make the dye penetrate the fiber, and others, "resists" or "mordants," to make possible faster dyes in many designs and colors. When Europeans revived the block printing of textiles in the thirteenth century, they still used the rudimentary method of pressing a pigmented block onto a fabric.

When block printing on paper was finally seen in Europe, its uses, its materials, and its technique were all so similar to those long known in China as to suggest their import from there. One of the earliest dated paper block prints made in Europe is an image of Saint Christopher (1423), intended, like the pagodas of Empress Koken, to ward off sickness and death. "In whatsoever day thou seest the likeness of St. Christopher," the inscription read, "In that same day thou wilt from death no evil blow incur." The earliest European block prints used an ink, like that of the Chinese, made of lamp black and dissolved in oil, and the paper itself was of Chinese origin.

The future of printing in East and West and the ease of widening the communities of knowledge depended not only on the technology and the physical materials but on the language itself. The absence of an alphabet in China would never cease to make problems. Long before Europeans, the Chinese experimented with movable type.

After the many-volumed Confucian classics awakened the Chinese to the advantages of printed books, in the tenth century in place of wooden blocks they tried plates of copper. In the early years of the Sung Dynasty a chronicler reported:

> Ever since Feng Tâo started the printing of the Five Classics, all standard works have been printed.
>
> During the period Ch'ing-li [1041–48] Pi Shêng, a man of the common people, invented movable type. His method was as follows: He took sticky clay and cut it in characters as thin as the edge of a copper coin. Each character formed as it were a single type. He baked them in the fire to make them hard. He had previously prepared an iron plate and he had covered this plate with a mixture of pine resin, wax, and paper ashes. When he wished to print, he took an iron frame and set it on the iron plate. In this he placed the type, set close together. When the frame was full, the whole made one solid block of type. He then placed it near the fire to warm it. When the paste (at the back) was slightly melted, he took a smooth board and pressed it over the surface, so that the block of type became as even as a whetstone.
>
> If one were to print only two or three copies, this method would be neither convenient nor quick. But for printing hundreds or thousands of copies, it was divinely quick. As a rule he kept two forms going. While the impression was being made from one form, the type were being put in place on the other. When the printing of the one form was finished, the other was all ready. In this way the two forms alternated and the printing was done with great rapidity.

Three centuries later, during the Mongol dynasty, the Chinese tried casting the separate characters not in ceramics but in tin. Still printers found it "both more exact and more convenient" to carve characters on a large block of wood which was "then cut in squares with a small fine saw till each character forms a separate piece." But the Chinese language had no alphabet, which meant that more than thirty thousand type characters were needed. How could these be stored for easy retrieval? One expedient was to classify the characters into the five tones of the Chinese language, then subdivide them into rhyme sections according to the official Book of Rhymes. With this in mind, printers equipped themselves with revolving tables, each about seven feet in diameter, topped by a round bamboo frame divided into compartments. Even with such aids, the selection of the type for a text would be laborious and the replacement of the pieces for reuse would be tedious.

By contrast to the situation in China, some features of Korean history and geography would give rise to special needs and opportunities. Under the Mongol Empire, Korea's peninsular isolation allowed the government there considerable cultural independence, which was increased by the disintegration of the Mongol Empire. For a brief interlude Koreans were actually the most advanced printers in the world. Printing with wooden blocks in the Chinese manner had been well developed in Korea by the eighth century. By the early twelfth century the kings of the Koryo dynasty had set up a printing office in the national college, and they too were collecting Buddhist documents, not for education but to establish a standard text. A Korean edition (1235–51) of the Tripitaka was sent to the court of the Mongol emperors in the early fourteenth century.

As printing flourished in Korea the scarcity of the woods needed for printing matrices became increasingly troublesome. Although Korea was rich in the pine forests useful for making ink, she was poor in the hard close-grained woods (jujube, pear, or birch) best for printing blocks, and so had to import them from China. Why not try metal? They ingeniously adapted the mold they were already using to cast coins into a novel device for casting type. A character carved in boxwood was pressed onto a trough containing clay to leave the impression for the type. Melted bronze was then poured in through a hole in a plate used to flatten the casting. When the bronze cooled, it left a flat piece of metal type about the size and thickness of a small coin, which in the mid-thirteenth century became standard for printing in Korea. In 1392 an energetic new dynasty established a Department of Books and set up a government type foundry because "the king thought with sadness of the fact that so few books could be printed."

These Korean coinlike metal types presented technical problems of their own. How could they be held firm and smooth while copies were taken from paper rubbed over them? Melted wax or wedges of bamboo strips never worked well. Still, with their primitive technique of movable metal type, the Koreans managed to make several hundred impressions of a number of books.

The crucial new opportunity for Koreans to exploit the advantages of movable type came from innovations in their written language. For centuries the Koreans had written their language only in Chinese ideographic characters. Then King Sejong the Great (1419–1450), of a dynamic new dynasty, to provide a "script for the people" commissioned scholars to devise an alphabet. In 1446 they came up with the new Han'gul of twenty-five letters not based on any existing alphabet.

If Korean scholars and printers had been willing to seize the advantages of their newly devised phonetic alphabet, the future of typographic printing and perhaps also of their science and culture might have been quite different. But they stubbornly hung on to their Chinese characters, or at least the Chinese style, and finally made their own alphabet into a syllabary similar to that of the Japanese. The ironic result was that Korean printing, like the Chinese, still required thousands of different characters.

Unlike their European counterparts, who themselves became a new audience for their product, Korean printers, perhaps because of the number and complexity of the Korean characters, remained illiterate. The bureaucratic concern was for authenticity. The regulations read, "The supervisor and compositor shall be flogged thirty times for an error per chapter; the printer shall be flogged thirty times for bad impression, either too dark or too light, of one character per chapter." This helps explain both the reputation for accuracy earned by the earliest Korean imprints and the difficulty that Koreans found in recruiting printers. In the seventeenth century, when a popular literature did appear in the Korean language, it still circulated in manuscript. Typographic printing with the increasingly scarce copper alloys was confined to official texts that the government wanted to authenticate.

While Korean was the language of the marketplace, the ideographic language of China remained the language of learning. Even more than the learned language in China, the learned language—the Latin of Korea—was isolated from everyday speech. Even today the Chinese written in Korea is said to retain an especially archaic flavor.

For government needs, wood-block printing continued to have advantages over movable type. Wood blocks were cheaper to produce and the calligraphy could be done by the scholars themselves, while movable type required a variety of artisans and complicated processes of casting. Also, the whole blocks were immediately available whenever needed for printing a few additional copies.

Some historians have suggested that these distant Korean experiments a half-century before Gutenberg may have supplied him with crucial clues. But there is no persuasive evidence that Gutenberg ever had word of what the Koreans had done. In Korea itself the pioneer experiments in movable metal type proved a dead end. Korean printers supplied familiar texts to those who already knew them. Most editions counted only two hundred copies, and none exceeded five hundred. Without commercial circulation, there was no incentive to widen the range of titles or increase the numbers printed. There was no effective demand for books printed in the vernacular.

Movable type came to Japan in the sixteenth century from two quite different sources. The first Europeans to arrive in Japan, like the first Europeans in America, came by accident about 1543, when a Portuguese ship was wrecked off the Kyushu coast. The doughty Saint Francis Xavier (1506–1552) went there in 1549 for the task of converting the Japanese. More Jesuit missionaries followed. In 1582 the Jesuit visitor-general Alessandro Valignano persuaded the daimyo of Kyushu to send a delegation to Pope Gregory XIII, and in 1590 they brought back to Japan a printing press along with some European printers. This Jesuit Mission Press was active for twenty years. Its thirty surviving products reveal the familiar Jesuit talent for reaching across cultural barriers. Naturally, most are works of Christian doctrine, but recognizing their limited knowledge of the Japanese language, the Jesuits made no effort at a translation of the Bible. They did print a list designed to appeal to the Japanese—including the classic *Heike monogatari* (1592), sometimes called the Japanese Iliad, collections of Chinese maxims, Aesop's fables (1593), grammars of Latin and Portuguese, a Latin-Portuguese-Japanese dictionary, and a Chinese-Japanese dictionary. The audience for these books could not have been large, for about half the titles were printed in romanized Japanese, which was known to only a few. The Jesuits did use the best European typefaces of the era—by François Guyot, Claude Garamond, and Robert Granjon. With the persecution of Christians in 1611, the Mission Press was moved to Macao, but by then it too had retreated from metal to the old wooden type.

The other source of movable type in Japan was a result of the ambitious enterprises of Toyotomi Hideyoshi (1536–1598), the first Japanese leader to seek an East Asian empire. Among the booty that Hideyoshi's returning Japanese soldiers brought back from their invasion of Korea in 1592 were sets of Korean movable type. These he gave to his Emperor, who ordered them to be used for the new printing of a Chinese classic. In addition, he commanded a new set of movable wooden type for a series of "imperial printings" of Chinese classics (1597–1603), which turned out to be some of the handsomest books ever printed in Japan.

During the next half-century, with an assist from movable type both in bronze and in wood, printing flourished in Japan as never before. There were numerous official editions of Chinese classics, and works of military strategy and of history. The founder of the Tokugawa shogunate, the great Iyeyasu (1542–1616), became an enthusiast for the new technology and ordered thousands of pieces of wooden movable type. In addition, for one work that was never published he commissioned ninety thousand pieces of bronze type. Reams of sacred Buddhist texts and commentaries printed with movable type poured out of Buddhist temples in Kyoto, on Mt. Hiei, and on Mt. Koya.

In Japan commercial printing grew out of the temple publishing. In the new capital city of Edo printing became profitable and certain imprints became famous. Rich physicians supported the publication of medical texts. Popular Japanese classics like the *Ise monogatari* (c. 980) were printed in numerous editions. Using movable type, painters and calligraphers produced works of surpassing beauty on elegant paper of various colors, and some older Chinese and Japanese works appeared in print for the first time. Authors began writing for publication in print.

The earliest movable-type products were in the Chinese script, which required large square pieces to accommodate the complex characters. When more works were printed in the Japanese hiragana or katakana scripts, it was necessary to design a new typeface to accommodate their cursive forms. When we look at the pages of these volumes today, we are amazed that their flowing calligraphy could ever have been reproduced in movable type. Single pieces of type often included two or more of the connected characters.

By the mid-seventeenth century, book publishing in Japan had taken on a new life. Government agencies, Buddhist monasteries, artists working for friends and patrons, and commercial printers had created a wider audience for printed books whose pages were unexcelled in legibility, elegance, and charm.

Then occurred one of the more abrupt breaks in the history of technology. With astonishing suddenness the technique of movable type was abandoned in Japan until the mid-nineteenth century, when it would be reimported from Europe. Economics triumphed over aesthetics. For the Japanese language it proved too costly to continue carving and casting pieces of movable type, but it had required a half-century to discover this expensive fact. Traditional wood-block printing was cheaper and more convenient.

The Japanese never invented a ready technique for multiplying pieces of movable type. Since they had no mold like that which Gutenberg devised for Europe, it was simply easier to provide new carved blocks page by page for each work to be printed. For a society still oriented to its classics, such blocks provided the easiest way to reprint works that were in continuing demand. A negligible number of movable-type editions were printed in Japan in the next centuries. After Iyeyasu banned Christian missionaries by his decree of 1614, Japan remained closed for more than two centuries. During this period Tokugawa culture in the burgeoning cities developed their own ways of educating, informing, and entertaining the people, in the flourishing haiku poetry and the no, bunraku, and kabuki drama. With the abandoned experiments of movable type behind them, the Japanese produced wood-block prints and illustrated books never excelled in Europe.

63

"The Art of Artificial Writing"

WE think of Gutenberg as "the man who invented printing" or at least the inventor of "movable type." But when we commonly identify him with the elegant Bible which was his first major work and which remains a treasure in our great libraries, we obscure his crucial role. For he was not merely a pioneer of the splendid incunabula of his own lifetime. He was a prophet of newer worlds where machines would do the work of scribes, where the printing press would displace the scriptorium, and knowledge would be diffused to countless unseen communities.

Among the heroes of modern history, few are more shadowy than Johann Gutenberg (c. 1394–1468). But while his person is shadowy, his career is not. His work was the climax of enterprise by many others. He put together what others had not, and he risked everything on his try. Much of what we know about Gutenberg comes from the lengthy lawsuits over the financing of his printing establishment and the profits to be made from his invention.

Of course there was printing in Europe long before Gutenberg, if by printing we mean the making of images by pressure. In English "to print" first described the impressing of a seal as in stamping out coins, which makes it understandable that Gutenberg began as a goldsmith. His crucial invention was actually not so much a new way of "printing" as a new way of multiplying the metal type for individual letters. Others before him had thought of carving an image in reverse on wood or metal, then pressing it with color onto textile, vellum, or paper. But usually they printed whole pages, whole designs. Gutenberg broke down the process into its parts. Printing a whole page he saw as the accumulated task of printing individual, frequently repeated letters. Then why not make many copies of each letter, which could be reused as often as needed?

Gutenberg's skill as a goldsmith and metal caster helped him see the problems of the printer which did not afflict the goldsmith crafting a unique piece of jewelry. For example, to make a printed book each cast letter had to be precisely the same *height* as every other. Making the pieces to be "movable" was the very least of it. All examples cast for a given letter had to be *interchangeable*.

Other problems did not appear until Gutenberg seized the opportunity to break down the solid page of type into its several letters. Provided the surface of a wood block was smooth and uniform, after reverse raised letters were carved on it and inked, it would print a uniform legible impression. But if each letter was cast separately, how could they be put together to ensure a uniform flat surface? Gutenberg's crucial invention was his specially designed mold for casting precisely similar pieces of type quickly and in large numbers. This was a machine tool—a tool for making the machines (i.e., the type) that did the printing.

The Roman alphabet, with its small complement of different characters, would make possible the power of interchangeable type and of the printing press in Western civilization. By contrast, as we have seen, the Chinese script with its countless ideograms was not well suited to the opportunities of interchangeable type. For even if multiple copies of each ideogram could be made, how could these thousands be arranged so that the desired one could be speedily retrieved?

In some other ways the Chinese ideograms had advantages for a type cutter. The ideograms were large enough, and interestingly enough varied, to be attractive subjects for carving in wood. Compared with a single letter of the Roman alphabet, the larger size of each ideogram made it easier to align into the form of a page. A letter of the Roman alphabet, on the other hand, was tiny, difficult to grasp in the fingers, apt to slip away. Before Gutenberg could make his invention, the very shape of the Roman letter had to be reimagined and redesigned. He had to see each letter not as flat color on a page but as a little stick to be held in the fingers. He would have to imagine each letter of the alphabet not as a splash of ink but as a stick of type.

The typecasting device that Gutenberg invented now seems simplicity itself. It is a hinged rectangular box open at both ends. One end is closed by inserting a matrix, a flat strip of metal that has been indented by a metal punch leaving the impression of a letter. The box is then stood on that end, and into the open top is poured molten metal. When the metal cools, it has the raised form of a letter at the bottom end, then the hinged mold is opened, and out comes a "stick" of type. By repeating the process, any number of identical, interchangeable pieces of type can be produced. To turn out pieces of type just the right width for the different letters of the alphabet (since "i" is only one-third the width of "w") and yet make them all of uniform height, the box had to be adjustable. Gutenberg managed this by a sliding arrangement that allowed the width of the box to expand or contract to fit the different matrices inserted in the bottom. The mold was enclosed in wood to avoid burning the typecaster's hand. The matrix into which the molten metal was poured would have to be punched with minute precision and to a precisely equal depth at all points. The goldsmith's skill must ensure that the pieces of type inserted in the form made a uniform impression on a page. Gutenberg needed a metal alloy that was easy to melt, that speedily cooled, and that flowed evenly.

Two more problems had to be solved before the pieces of interchangeable type could become effective tools for printing a typographic page. There had to be a way of holding the numerous pieces together and impressing them steadily and heavily on the surface to be printed. Neither the Chinese nor the Europeans had yet used a press for printing. Instead, printing was done by first inking the block, then laying a sheet of paper on the block and brushing on the back of the paper to ensure a uniform impression. In Gutenberg's press, printing was accomplished by an adaptation of the binder's wooden screw press already in use. This was itself perhaps an adaptation of the winepress or the domestic screw press used for pressing linen or extracting oil from olives. Then an ink was needed that would adhere uniformly to pieces of metal type. This would be quite different from what scribes put on their pens to mark parchment or paper. Different, too, from the ink that was used in making impressions from wood blocks. What Gutenberg required was a kind of oil paint. In making his ink, he drew on the experience of Flemish painters who ground their pigment in linseed-oil varnish.

It is no wonder that Gutenberg needed years to solve all these problems and make his solutions work together. Luckily, one problem, how to make a surface on which to imprint his product, had already been substantially solved. Paper was the essential contribution of China to the advance of the book. Other problems would require all the patience, ingenuity, and financial resources that Gutenberg could muster. The leitmotif of Gutenberg's life, amply documented by the records of his numerous lawsuits, was his determination to keep at his work until it was perfected, and meanwhile to keep it secret from competitors. His experiments were costly, and he made many false starts.

Gutenberg's life is written in litigation. Almost all the solid information we have about him comes from the lawsuits against him. We do not even know the exact date of his birth, probably between 1394 and 1399, in the strategic city of Mainz, where the Main River flows into the Rhine. Born Johann Gensfleisch, he took the name Gutenberg from a family estate when the city life was being disrupted by struggles between patrician families like his and the rising guilds. Since Gutenberg's father was connected with the archiepiscopal mint, the young Johann became familiar with the goldsmith's craft. He divided his adult life between Mainz and Strasbourg farther up the Rhine where he had to flee from his enemies the guilds. In our earliest legal notice of Johann he is being sued for breach of promise of marriage. This jilted lady lost her suit, but the case proved costly for Johann, for during the trial he rashly called the Strasbourg shoemaker who testified against him "a poor creature, leading a life of lies and deceit." For this outburst Gutenberg had to pay libel damages of fifteen Rhenish guilders.

In another series of lawsuits we see both Gutenberg's persistence and his desire to keep his inventions secret. One climax came in 1439. Gutenberg the expert goldsmith had taken into partnership three citizens of Strasbourg who invested their capital with him and whom he agreed to instruct in his new process of making the hand mirrors which they intended to sell to pilgrims along the Rhine. But they had miscalculated the year of the pilgrimage, and there would be no market for pilgrim souvenirs. Gutenberg made a new agreement, this time promising to teach his associates an unspecified new secret process in which they would invest heavily. Under the terms of their five-year (1438–1443) contract, if any party died while the agreement was in force, none of his heirs would be allowed to take his place in the partnership. Instead, the heirs would receive a liquidated compensation of 100 guilders. When one of the partners died in 1439, his brothers demanded to take his place, and so share in the partnership secrets. Gutenberg refused. The heirs went to court but lost the suit. All the surviving partners had been sworn to secrecy. During the trial very little was revealed of Gutenberg's invention, but it appeared that the partners had continued to spend large sums on the undisclosed experiments.

The rest of Gutenberg's business career shows him repeatedly soliciting substantial additional capital yet refusing to offer his product for sale until he had perfected the new process to his own satisfaction. Whatever he was doing required costly materials. Meanwhile, when lawsuits arose he regularly instructed his partners to dismantle his experimental machinery so that no one could discover what they were about. All this was still in Strasbourg.

Gutenberg was back in Mainz in 1448 looking for more capital. There he finally found a supporter in the person of Johann Fust, a wealthy lawyer who invested two substantial installments of 800 guilders in Gutenberg's project. After five years Fust still had not received the expected return on his investment. In 1455 Fust sued for the repayment of his money plus compound interest and possession of all Gutenberg's assets. But Gutenberg's aim was not merely to make money. He was determined to find a way to retain the sharp design and brilliant color of illuminated manuscripts while making numerous identical copies. He was in no haste to put an imperfect product on the market.

When Fust won his suit, Gutenberg was ordered to pay him 2,026 guilders and hand over all his materials and equipment, which included the pages and type of the Bible on which Gutenberg had long been working. Fust carried on Gutenberg's enterprises, with the help of his son-in-law Peter Schöffer, who had been Gutenberg's foreman and who therefore knew all the secrets. Schöffer had testified against Gutenberg at the 1455 trial. When sometime before 1456 the "Gutenberg" Bible appeared, it had no colophon. The product of Gutenberg's years of struggle had become the property of the new firm of Fust and Schöffer.

The heavy Gothic type used in Gutenberg's 42-line Bible was not well suited to other sorts of works. But Gutenberg seems to have prepared two more type fonts, which the firm of Fust and Schöffer promptly used for an elegant *Latin Psalter* (1457) and probably also for the *Catholicon* (1460), the reprint of a popular encyclopedia compiled in the thirteenth century. The colophon of the *Catholicon,* written either by Gutenberg or by his successors in the firm of Fust and Schöffer, announced a new miracle:

> With the help of the Most High at whose will the tongues of infants become eloquent and who often reveals to the lowly what he hides from the wise, this noble book *Catholicon* has been printed and accomplished without the help of reed, stylus or pen but by the wondrous agreement, proportion and harmony of punches and types, in the year of the Lord's incarnation 1460 in the noble city of Mainz of the renowned German nation, which God's grace has designed to prefer and distinguish above all other nations of the earth with so lofty a genius and liberal gifts.

Gutenberg must have been a man of monumental persuasive powers, for even after his widely publicized bankruptcy another Mainz official was willing to stake him to a full set of printing equipment. Later, Count Adolph of Nassau, self-styled archbishop of Mainz, who had sacked the city, rewarded Gutenberg, who by then was destitute and nearly blind, with a modest pension, including an annual allowance of corn, wine, and a gentleman's suit of clothing.

We have ample evidence that printing for Gutenberg and his generation was not merely a technology but an art. Bibliophiles agree that the very first printed book in Europe was one of the most beautiful. The technical efficiency of Gutenberg's work, the clarity of impression and the durability of the product, were not substantially improved until the nineteenth century.

It was not dissatisfaction with the work of the best scribes that had stimulated the quest for other ways to reproduce books. The original effort was to find how to multiply manuscripts in larger quantity and at lower costs, but as good as the scribes' and illuminators' best work. The early printers called their craft the art of artificial writing—*ars artificialiter scribendi.*

During the first century of printing the scribes who were practicing the art of "natural" writing and the printers practicing the new arts of "artificial" writing competed for the same customers. The printing press did not at once put the scribes out of business. Nearly as many manuscript books survive from the second half of the fifteenth century, after the invention of printing, as from the first half. Scribes continued to cater to the luxury trade, turning out deluxe volumes for those who could afford them. Certain works continued to be handwritten, especially those in Greek and Latin that had a limited market. In 1481, after eight printed editions of Pliny's *Natural History* had appeared, Pico della Mirandola commissioned his own manuscript copy. Some early printed books were so costly, even in secondhand copies, that it was cheaper to commission your own manuscript. From the fifteenth century and even the sixteenth many examples survive of these manuscript copies of printed books. Some have precisely the same number of lines to a page and even reproduce the printer's colophon.

For a while there was enough business for both scribes and printers. But as the price of printed books declined, scribes began to have trouble finding work. When it became plain that the press was a menace to the calligrapher's craft, the organized scribes and their conservative allies sought laws to protect their monopoly. In 1534 Francis I gave in to their demands and issued an edict suppressing the press in Paris, but this was never enforced. As scribes saw that the printed book was here to stay, they began to cooperate. They themselves started using the press and saved labor by inserting printed sections into their manuscript books. Sometimes, when a printer had failed to run off enough copies of some leaves of a book, scribes were enlisted to fill out the needed text. Printers consulted calligraphers to design their printed pages so they would look like manuscripts.

In the early decades of printing it was risky to commit one's livelihood to so new a technology. While scribes were devotees of an ancient, honorable, and remunerative craft, in those days printers had to be willing to take chances. How long would this new technology last? In fifteenth-century Europe innovation itself was an unfamiliar and suspect idea. Even so, the best-made printed books were already esteemed by aristocratic connoisseurs. Before the fifteenth century was out, the Gonzagas of Mantua, the Medici of Florence, King Ferdinand I of Naples, and the Pope in his Vatican Library had all added printed books to their discriminating collections. Printed books were promptly found, too, in the great libraries of Germany and of Spain. When Columbus' natural son and biographer, Don Fernando Colón, was in London in 1522, he sought out printed books for his own famous library.

In many ways the most interesting period in the history of the printed book is the first century after Gutenberg's Bible, when we can see the ambivalence of cultivated European readers toward the new technology. The old and the new were still in direct competition. Matteo Battiferri of Urbino, a learned doctor and poet, cared enough for the printing press to edit the *Physica* of Albertus Magnus, which was printed in Venice in 1488 and which he dedicated to his father. Yet he took the trouble to illuminate his own vellum copy of *Anthologia Graeca,* printed in Venice in 1494. He had remained such a devotee of the handwritten book that in his own hand he inserted an extra leaf with the explanation that this book was personally "written" and decorated by him. On the printer's colophon he replaced the word "impressum" (printed) by "scriptum" (written). He was not alone among the bibliophiles who wanted their "books" to continue to have the kudos of manuscripts.

Printed books were doctored by eraser and paintbrush to give them the manuscript look, revealing the nostalgia of booklovers who were still irreconcilably attached to the "handmade" product. The "beauticians" of books—binders, illuminators, and rubricators—who flourished in the new age of printing proved that books would never cease to be prized as furniture and works of art. In this competition between the handwritten book and the printed book, who could predict which one would win in the long run? Although in the beginning printing seemed to justify its title as "the art preservative of all arts," a prophet might better have called printing "the art that can revolutionize all the arts." And not only the arts! "He who first shortened the labor of copyists by the device of movable types," Thomas Carlyle could say in 1836, less than three centuries after Gutenberg's Bible, "was disbanding hired armies, and cashiering most kings and senates, and creating a whole new democratic world."

64

Communities of the Vernacular

THE triumph of the printed book soon brought the triumph of the languages of the marketplace, which became the languages of learning across Europe. Vernacular literatures in print shaped thinking in two quite disparate ways. They democratized but they also provincialized. When works of science now appeared not only in Latin but in English, French, Italian, Spanish, German, and Dutch, whole new communities were suddenly admitted to the world of science. Science became public as never before. But when Latin, the international language of the whole European learned community, was displaced by national or regional languages, knowledge tended to become national or regional. Learning garnered from all places and all earlier times was now packaged in parcels that could be opened only by the people of one particular place. As the written, now the printed, word became more popular, literature had a larger ingredient of amusement, fantasy, and adventure. Entertainment had a new respectability.

It is hard to say how many languages or dialects there were in Europe before the age of printing. Scholars nowadays have identified about three thousand languages, excluding minor dialects, which are still spoken. At the end of the Middle Ages there were probably many more. In the twelfth century, as we have seen, when a student from Normandy went to the University of Paris, he would not understand the speech of a student from Marseilles, for there was not yet a standard French. Similar problems plagued students who attended the University of Heidelberg or Bologna or Salamanca or Oxford, for there was not yet a standard German, Italian, Spanish, or English.

The languages spoken in modern Europe, with only a few exceptions, such as Basque and the Uralic languages, belong to the Indo-European family and seem to be derived from a language spoken in northern Europe in prehistoric times, from which seven different branches developed. At the end of the Middle Ages, most languages spoken in western Europe belonged to one of two groups. The "Romance" languages spoken within the former boundaries of the Roman Empire from the English Channel southward to the Mediterranean, and from the Rhine, the Alps and the Adriatic Sea westward to the Atlantic, were derived from Latin, and eventually became French, Italian, Spanish, and Portuguese. The Germanic languages, to the north and eastward from the Atlantic to the Baltic, and from the Rhine and the Alps toward the North Sea and the Arctic Ocean, became Icelandic, English, Dutch-Flemish, German, Danish, Swedish, and Norwegian. In the twelfth century these modern national literary languages were still fragmented into countless local dialects.

We can illustrate the continent-wide emergence of standard national languages by the career of the French language. In 1200 in the present area of France there were still five major dialects, subdivided into numerous minor dialects. Each of these was rooted in the daily life, the folklore and folkways, of its own region. Before there could be a community of French literature, there had to be a standard French language, which was a by-product of both the rise and the fall of the Roman Empire. When the Roman Empire was at its height, only the learned few knew classical Latin. What most people across France heard from Roman soldiers and traders was a rough colloquial version of spoken Latin. With local variants and additions from Celtic and old Frankish, this became a spoken language that surely would not have satisfied Cicero or Alcuin. Then, when the bonds of empire dissolved, the bonds of language weakened. Out of the remnants of Latin, with a local spice of vocabulary, accent, and intonation, different spoken dialects evolved. Without an imperial government, and with meager communications, as time passed these dialects became more and more distinct.

While the Church and the universities still preserved the unity of Latin, everyday speech wandered its own wayward paths. Charlemagne recognized this when he ordered that sermons be preached "in the rustic Romance language." The breakup of Charlemagne's empire brought the vernaculars into official prominence. The first writing in a distinct "French" language was the "Oaths of Strasbourg" of 842, when the oaths of alliance of the armies of Charlemagne's grandson Charles the Bald with the armies of his brother Louis the German were sworn by each in their own vernacular—some in a Teutonic proto-German, others in a Romance proto-French.

In France during the next five centuries the two dialects most widely spoken were the langue d'oïl of the Ile de France and Paris in the north and the langue d'oc of Provence in the south. These were designated by the different words used for "yes" in the two regions. Both regional tongues were producing a rich, mostly oral literature. The northern dialect, and especially the Paris dialect known as Francien, was destined to prevail, which meant that the language of Paris became the language of France. By his Edict of Villers-Cotterêts (1539), King Francis I (1494–1547) made Francien the only official language.

Francis I's strongest ally was the printing press. Within a hundred years after Gutenberg, book publishing was a thriving business not only in Paris but also in Lyons, Rouen, Toulouse, Poitiers, Bordeaux, and Troyes. Forty towns had printing presses. Wherever there was a university, a high court, or a provincial *parlement,* there was an assured market for printed books. As books multiplied, literacy increased, and the literature of the vernacular was enriched. Book buyers came to include not only clerics, lawyers, and government officials, but prosperous merchants and even some city artisans.

On the countryside the spoken word still prevailed. The village social evenings, the *veillées,* brightened the winter months indoors, with some literate person, the traditional storyteller or the schoolmaster, reading familiar tales aloud. The reader, as historian Natalie Z. Davis has shown, was really a translator, because in telling Aesop's Fables or *Le Roman de la Rose* he rendered the printed French into the listeners' spoken dialect. For decades the Church's opposition to reading the Bible in the vernacular was quite superfluous, since few could understand the French anyway.

In cities printing establishments themselves created new readers and writers. Artisans, apothecaries, surgeons, metalworkers, and others came to depend on printed manuals. Artisans would have a book read out to them as they worked. The books read aloud in taverns ranged from the devotional Book of Hours or Lives of the Saints to manuals of arithmetic or metallurgy. These reading groups became prototypes for the secret Protestant gatherings that brought the Reformation.

The national language in France found an eloquent advocate in the remarkable scion of a noble family, Joachim du Bellay (1522–1560). At the age of twenty-seven he wrote the manifesto for the brilliant literary circle known as the Pléiade which he called *The Defense and Illustration of the French Language* (1549). When he met Pierre de Ronsard (1524–1585) they felt kinship in their love of French, and in the extraordinary fact that both happened to be deaf. They naturally found it difficult to make their careers at court, so they dedicated their great talents to cultivating the written word. Inspired by Petrarch's sonnets in Italian, Bellay wrote some of the first love sonnets in French, and his success in the vernacular French would, in turn, inspire English poets like Edmund Spenser.

According to Bellay, all languages were born equal. "They all come from a single source and origin, that is from the caprice of men, and have been formed from a single judgment for a single end, that is to signify amongst us the conceptions and understandings of the mind." The deeds of the Romans seemed so much greater than those of other people not because their language was better but simply because they had so many talented writers. An age that has invented "printing, sister of the muses and tenth of them, and this no less admirable than deadly thunderbolt of artillery, with many other inventions" surely must also be competent to produce a great literature.

Why had science flourished less in France than in ancient Greece and Rome? " 'Tis the study of the Greek and Latin languages. For if the time we consume in learning the said languages were employed in study of the sciences, nature is certainly not become so sterile that she could not bring forth in our time Platos and Aristotles. . . . but as we repent of having left the cradle and become men, we return again to childhood; and in the space of twenty or thirty years we do only the one thing, learn to speak, this man in Greek, that one in Latin, and the other in Hebrew."

Like the Romans, the French should be bold to invent words. "The glory of the Roman people is not less . . . in the amplification of its language, than in that of its frontiers." Before the end of the sixteenth century the upstart French language had effloresced into a brilliant literature—the poetry of Ronsard, the satires of Rabelais (1483–1553), the theology of Calvin, the essays of Montaigne (1533–1592), and the complete Bible in the vernacular.

Vernacular literature promised escape from pedantry. The muckraker of learned monopolies was François Rabelais, himself a man of vast and vagrant learning. After a novitiate in a Franciscan monastery, he became a scholar of Greek and Latin, law and science, tried a Benedictine monastery, studied medicine in Paris, lectured on Galen and Hippocrates at Montpellier, accompanied Bellay's cousin the Cardinal to Rome, edited medical works for the book publishers of Lyons, was patronized by Francis I, was persecuted for heresy, and had his works condemned by the Sorbonne. *Pantagruel* (1532) and *Gargantua* (1534) proclaimed the follies of Greek and Latin pedantry, of astrology, necromancy, traditional medicine, and theology all with a fantastic exuberance and extravagance. The miseducation of Gargantua was entrusted to the great sophister-doctor Tubal Holofernes, who took five years and three months to teach him to say his ABC's by heart backwards. He devoted thirteen years, six months, and two weeks to Latin grammar, then studied the works of Latin eloquence for another thirty-four years and one month so that he could recite them, too, by heart backwards. When his teacher died of the pox, Gargantua's father found "that it were better for his son to learn nothing at all, than to be taught such books, under such schoolmasters, because their knowledge was nothing but all trifle, and their wisdom foppery, serving only to bastardize good and noble spirits, and to corrupt the whole flower of youth."

While French bore relics of a disintegrated empire, the modern German language, creator of a German nation and a rich modern literature, had a far different ancestry. The Romance vernaculars—French, Spanish, Portuguese, and Italian—all had to compete with the vernacular of the Roman Empire and the wealth of Latin literature. The Germanic vernaculars, not the residue of a declining empire but seeds of a rising civilization, commonly had the field to themselves. German arose out of the proto-Germanic group of Indo-European languages, embedded deep in prehistory. In the eighth century, when the earliest written versions of something resembling the modern German language are found, local dialects still prevailed in the marketplace, and there was no standard language across the areas of modern Germany. The dialects fell into two recognizable groups, Low German, or Plattdeutsch, in the lowland north and High German, or Hochdeutsch, in the highlands of the south. It was in the chanceries of the Holy Roman Empire in the fourteenth century that there developed a relatively uniform written language, which gradually took the place of Latin in official records. When Martin Luther set about translating the Bible (1522–34), he chose the High German dialect used by the chancery of the duchies of Saxony, and so provided the norm for modern standard German. He gave dignity to the vernacular while he established a national language. To compete with Luther's Bible, there soon was a Catholic version, also in the vernacular.

Each of the other Germanic languages found itself by a different path. England, too, was a land of many languages. When Gutenberg was printing his Bible, official documents of the English government in London were still written in law-French. Then, barely a century and a half after English became an official language, Shakespeare wrote his plays and the miracle of Elizabethan literature unfolded.

William Caxton (1422–1491) did as much as any man before Shakespeare to standardize the English language. Born in Kent, as a youth of sixteen he was luckily apprenticed to a prosperous textile merchant who became lord mayor of London. When his master died and Caxton was only nineteen, he moved to Bruges, then a center of trade and culture. During the next twenty years Caxton grew rich in the textile business and was chosen governor of the English Nation of Merchant Adventures in all the Low Countries, a powerful trade association. By the time he was fifty he had become financial adviser to Edward IV's sister, Margaret Duchess of Burgundy. Unsatisfied by commercial pursuits, he turned to literature. In 1470 the Duchess encouraged Caxton to complete his translation into English of a popular French collection of stories about Troy. At first he circulated the work in manuscript, but it was in such wide demand that scribes could not provide enough manuscript copies. Caxton went to Cologne to learn the new art of printing, then returned to Bruges, where he set up his own press. The first books off his press, *Recuyell of the Historyes of Troy* (1475) and *The Game and Playe of Chesse* (1476), were the first books printed in English. Eager to print more books in English, he returned to London, where with royal patronage he set up a press. Over the next fifteen years, using his considerable fortune he published some one hundred titles. These books did much to standardize the literary—and eventually, too, the spoken—language of the political and commercial capital into a language for England. The first dated book printed in England was his *Dictes and Sayenges of the Phylosophers* (1477), another translation from the French.

Caxton was faced with a historic decision. Before he translated into "English," he had to decide precisely what he meant by the "English" language. And this was then a more complicated question than we now imagine. When Caxton began publishing, there were almost as many dialects as there were counties in England. The language was as varied and as changing as the twists of the human tongue, and the dialects were not mutually intelligible. Caxton himself illustrated this problem in his tale of merchants who set out from the Thames to sail to the Netherlands. Awaiting a favorable wind, they put in at North Forland on the Kentish coast.

> And one of theym named Sheffelde, a mercer, cam in-to an hows and axed for mete; and specyally he axyed after eggys; and the good wyf answerde, that she coude speke no frenshe. And the merchaunt was angry, for he also coude speke no frenshe, but wolde haue hadde "egges" and she vunderstode hym not. And theene at laste another sayd that he wolde haue "eyren" then the good wyf sayd that she vnderstod hym wel. Loo, what sholde a man in thyse dayes now wryte, "egges" or "eyren"?

To the Kentish housewife in Caxton's day the English spoken by a London merchant sounded for all the world like French. A century later, in Shakespeare's day, that could not have happened. Caxton's work was largely responsible for the change.

For the books that Caxton printed he chose the language of London and of the court. His varied "list" could be the pride of a twentieth-century publisher. He published at least twenty books of his own translations from French, Latin, or Dutch. His list included not only familiar religious works but almost every other kind of book known in his time—chivalric romance, poetry, how-to-do-it manuals, history, drama, theology, philosophy, and morality. His *English-French Vocabulary* (c. 1480) was one of the first bilingual dictionaries. And his encyclopedic *Myrrour of the Worlde* (1481) was the first illustrated book printed in England.

Caxton was the midwife of a flourishing English literature. He published the *Canterbury Tales* and other poems by Chaucer, the poetry of John Gower and John Lydgate, and Sir Thomas Malory's prose version of the Arthurian legend, along with translations of Cicero and Aesop's Fables.

Before Caxton the outcome had been uncertain, and it was conceivable that the literary language of the island might have been some version of French. The fifth-century Germanic invaders of the British Isles had brought with them the West Germanic Frisian language which became Old English. But after the Norman Conquest, French was the official language at court. Only gradually did English displace French. By then, of course, English was full of words of Latin and French origin. In England the establishment of a standard vernacular language had a double significance. It was a victory of the people's language over the Latin language of the learned few, and at the same time it was the victory of a popular vernacular (English) over what in England was an aristocratic vernacular (French). English literature started as a peculiar possession of all the people.

Religion had opened the way and provided strong motives to standardize language in order to spread the good news of Christianity. Calvin's Bible in French and Luther's Bible in German, both early in the career of the printed book, helped establish their languages. In England, too, the Bible called for a common vernacular language. John Wycliffe (1330–1384), hoping to bring its message to the people, had produced an English Bible even before it could be printed. Still, the manuscript copies were themselves numerous enough to make him a dangerous character, who would be condemned by a synod in London and have his works banned in Oxford, but these never reached the wide audience for which he had hoped. By the age of Caxton a common English language and the wonderful vehicle of the printing press had opened the way for a vernacular Bible.

The King James Version of the Bible, besides shaping and invigorating the modern English language, had another rare distinction. It is perhaps the only literary masterpiece ever written by a committee. (Another evidence that it was divinely inspired?) The project was an effort to compromise differences within the Church of England, to bring together Puritans and others. After James I gave it his support, forty-seven approved translators, including the notable Biblical scholars of the day, were organized into six groups. They worked at Westminster, Oxford, and Cambridge on the different parts of the Old and New Testaments assigned to them. When they had completed their assigned parts, each criticized the work of the others. Then a representative group of six, meeting daily at Stationers' Hall in London for nine months, combined their efforts toward publication in 1611. They drew on the latest classical and Oriental scholarship, but willingly followed earlier versions where these were satisfactory. Although there was not one towering literary talent in the lot, their product overshadowed all the other works of literary genius in the language.

$$65$$

Transforming the Book

LANGUAGES would become pathways through space and time. While nations would be held together by their new vernaculars, lone readers could seek remote continents and voyage into the faraway past. From Cicero to Gutenberg, the book, the vehicle of language magic, would be transformed out of all recognition. The modern technical definition of a book, accepted by librarians and by UNESCO for statistical purposes, suggests how much the "book" has changed. A book, they say, is a "non-periodical printed publication of at least 49 pages excluding covers." But for most of history, books did not even have "pages." Our "volume" (from Latin *volvere,* to roll) was first a name for manuscripts in rolls. In early Egypt sheets to write on were made from the papyrus reed that grew in the Nile Delta. The reed was called *byblos,* from the port of Byblos, where it was first found, and whence comes our "Bible" for The Book. Strips from this reed were flattened, then others were laid across them at right angles to make a mat. The surface, when moistened, pounded, smoothed, and dried, was well suited for writing. These sheets were pasted together to make long strips for ceremonial banners in early Egyptian temples. Rolled up, they became a "volume," portable, easy to store, and relatively durable. This was the ancestor of our book.

Other people in other places had, of course, tried many other sorts of writing material. The ancient Babylonians inscribed their wedge-shaped writing on moistened tablets of clay. After those tablets were baked in the Middle Eastern sun, they could carry their messages across the millennia. Before they turned to paper, the Chinese used tablets of bamboo and then sheets of waste silk. In India birch bark and palm leaves were prepared for writing. In Tibet, as we have seen, the smooth shoulder bone of the goat was used for magical inscription. Leather was widely used in the Middle East, and the Burmese wrote on sheets of copper.

The writing material that held the Roman Empire together was Egyptian papyrus. Could they ever have done their business with cumbersome clay tablets? Just as our government is held together by paper, theirs was bound by papyrus. The *Natural History* of Pliny the Elder (A.D. 23–79) described the many grades, from the finest quality in the center of the plant and called "Augustan," and second quality, or "Livian," after the Emperor's wife, to the inferior qualities near the outside rind. For jottings Romans commonly used small wooden tablets sometimes covered with a thin coat of wax. People in the West who had access to the plant generally preferred papyrus for their writing material.

Then, according to tradition, parchment or vellum was invented by King Eumenes II (197–159 B.C.), who wanted to make Pergamum, in Asia Minor, a great center of Greek culture. When his rival Ptolemy VI, king of Egypt, cut off the supply of papyrus, he devised a new technique of cleaning, stretching, and smoothing the skins of sheep and goats which made it possible to write on both sides. This came to be called parchment, after Pergamum, and "vellum" was the name for the especially fine parchment made from the skin of a calf (from Old French *veel*).

Parchment made possible the next and greatest advance in the technology of the book before printing. This idea was so simple that we hardly think of it as an invention. The pages were no longer pasted together and rolled up into a "volumen." Instead they were bound together into a "codex." The name, from Latin *codex*, or *caudex*, for tree-trunk board, or writing tablet, may have been suggested by the form of the Roman notebook made by fastening together sheets of wax-coated wood.

The roll, as we have seen, had many disadvantages. The reader had to unroll the manuscript as he went along and then had to reroll the book before it could be used again, much as a motion-picture film has to be rewound after each showing. In the second century B.C., when this still was the usual form for a book, an average roll might measure 40 feet. Some earlier Egyptian rolls were said to reach a length of 150 feet. The writer of the Apocalypse may have been thinking of such a scroll when he envisioned the sins of Babylon reaching to the heavens. No wonder that the grammarian Callimachus (305–240 B.C.), head librarian at Alexandria, said, "A big book is a big nuisance." After his time smaller rolls became standard. But then a single roll might hold as few as 750 lines, the longest rolls would carry only about two hundred pages of text, and the text of the *Iliad* and the *Odyssey* required thirty-six rolls. Every time a "book" was read—which meant unrolling and rerolling—the text would suffer wear and tear.

We cannot be surprised that quotations in early literature are so variant and so inaccurate. We, too, would naturally rely on our memory rather than unwind a long roll to search for the desired passage. Since every manuscript was unique, there were no numbered "pages," no index, and nothing like the modern title page. The name of the "author" was seldom attached to a roll. The name of the scribe seemed more important, and was more likely to appear somewhere. It was physically as well as intellectually laborious to find the remembered section of any text.

By contrast to the roll, the "codex," bound pages in a form like that we now call a book, was wonderfully convenient. It was handy to use, more durable, more copious in content, and more compact to store. With the codex, eventually there would come a host of reference and retrieval features—a title page, a table of contents, numbered pages, and an index. All these would entice us to "look it up." They would also be finding aids, encouragements to check the precision of the quoted word and the remembered fact.

The parchment codex came into use in the West near the beginning of the Christian era. Modeled on the Roman notebook of wooden leaves, it was naturally first introduced for notebooks or account books. This new format helped the preachers of the new Christian religion emphasize the Good News in their sacred book, in contrast to the roll which was the customary format for the Old Testament or other Jewish books. When used for Christian literature, a manageable codex-book could hold more than one Gospel or Epistle. By the fourth century, pagan manuscripts, too, were appearing in this form. Yet the scroll retained its aura of tradition and long remained in use for solemn and official documents. The Jews still keep the Torah in a roll.

To make a codex, a number of sheets (a "quire") were simply folded and sewn together. Papyrus, which cracked when folded, was not suited to this format. Moreover in a codex both sides could conveniently be read in sequence, and parchment was better suited for writing on both sides. So, before paper was available, parchment was the usual material for codices. Books worth preserving were gradually transferred from the papyrus roll to the vellum codex. The full significance of this Codex Revolution, the first great transformation of the book, would appear only after the invention of paper.

The Chinese had been making a rudimentary paper since about A.D. 105, when Ts'ai Lun, using mulberry, waste fish nets, and rags, made the first known sheet of paper for the emperor. Chinese prisoners of war whom the Arabs had taken at Samarkand introduced them to the arts of papermaking. By 800 the brilliant caliph Harun al-Rashid (764?–809) was having paper made for him in Baghdad. Then through the Arabs paper was brought to Byzantium and across the Mediterranean to Spain, whence it spread over Europe. Even before the invention of printing, manuscripts written on paper were not uncommon, and there were paper mills in Spain, Italy, France, and Germany. Paper still traveled under the respectable old name of papyrus.

During the Middle Ages the "book" which perpetuated Latin culture for the Empire of the Learned had already gone through a long development, and was much improved over that read by scholars in Cicero's time. The first century of printing in Europe saw other elementary changes in design that helped made the book a more compact vehicle of knowledge and discovery.

The pioneer in portable-book design was the great Venetian scholar-printer Aldus Manutius (1450–1515). The Aldine Press, which he founded, was the first modern publishing house. Its list, in Greek, Latin, and Italian, included poetry and reference works. The first printed edition of many Greek and Latin classics appeared under his colophon of the anchor and the dolphin, a symbol of the old Latin proverb "Make haste slowly" (*Festina lente*).

While Gutenberg in the first generation of printers had applied the craft of the goldsmith to make printed books technically feasible, only two generations later Aldus tried to find and reach the market. And he proved that a publisher could prosper by printing elegant, well-designed books. Born into an undistinguished family near Rome, he studied there and became adept at Latin, but very early he fell in love with the Greek language. In 1490 he settled in Venice, whose Biblioteca Marciana held Europe's richest collection of Greek manuscripts, which had been bequeathed to the Republic of Venice by another ardent Greek scholar, Cardinal Bessarion. At the age of forty he made his fateful decision to give up the wandering scholar's life and set up shop in Venice in the risky new business of printing. While the busy sea trade of Venice with the East made it a center of interest in Greek culture, unlike Florence and Milan, the city still had no Greek press. He persuaded the substantial merchant Andrea Torresiani to stake him, then solidified his partnership by marrying Torresiani's daughter.

Aldus Manutius' passion for ancient Greek culture had become a monomania. He made his home into a Greek academy, where the scholars of Venice were expected to speak only Greek. In the mid-1490's, when Aldus began experimenting with Greek fonts, only about a dozen titles had yet been printed in the Greek language. Using his wealthy connections, Aldus flourished. In 1508 Erasmus (1466–1536) reported that he found a printing force of thirty people in Aldus' household, which, as master printer, Aldus was expected to feed.

Unlike Gutenberg, Aldus commissioned others to cast the types of his design, but he still supervised the whole printing operation himself. He gradually printed more and more works in Latin, and then branched out into Italian, with works of Dante and Petrarch. His most ambitious product (1495–97) was a four-volume set of Aristotle in Greek. Aldus' growing list showed that he had chosen the right motto, for he published only items that had already proven themselves as manuscript books.

Before 1500 some one hundred fifty Venetian presses had produced more than four thousand editions, about twice the output of Paris, their closest competitor. Venice accounted for one in seven of all the books known to have been printed in Europe until that time, amounting to about twenty books per capita for the city. Even before the end of the fifteenth century unhappy scribes complained that their city was already "stuffed with books."

But printing was not necessarily the agent of progress. Without the popular Aldine editions and others, Greek philosophy and Greek science could not have had their vogue in the following centuries. The age of incunabula diffused many more works of ancient science than of the new science. In medicine the power of Galen, and in botany the power of Dioscorides, were reinforced by the newly printed bulky texts. Aldus proved to be a resurrection man for Greek thought.

Erasmus, an admirer of the Aldine Press, provided a credo for the publisher in any age:

> However one may sing the praises of those who by their virtue either defend or increase the glory of their country, their actions only affect worldly prosperity, and within narrow limits. But the man who sets fallen learning on its feet (and this is almost more difficult than to originate it in the first place) is building up a sacred and immortal thing, and serving not one province alone but all peoples and all generations. Once this was the task of princes, and it was the greatest glory of Ptolemy. But his library was contained between the narrow walls of its own house, and Aldus is building up a library which has no other limits than the world itself.

The Aldine library would reach even beyond the real world. Thomas More's heroic voyager, Raphael Hythloday, carried in his baggage these handy editions so that he could introduce the Utopians to the wonders of Greek literature.

Not only in Utopia, Aldus' two pioneering innovations—"italic" type and "octavo" size—shaped reading habits. If the black-letter type of Gutenberg's Bible had remained standard, books might never have become compact. For it was not suited to provide the greatest number of legible words on a page. About 1500 Aldus commissioned Francesco Griffo of Bologna to design a more practical font. His radically new typeface was based on the cursive script, then used by the papal chancery, in which the humanists were writing to one another. The narrow slender letters, while lacking the solemnity of the old Gothic script, went well with the roman capitals. The first book printed in the new type was Aldus' octavo-sized edition of Virgil in 1501. Since the Aldine Virgil was dedicated to Italy, the type came to be called "italic." At first it comprised only lower-case letters, and used small roman capitals. Aldus made the new italic type standard for his popular editions of ancient authors. While attractive and readable, it put many more words on the page.

The commercial value of italic type was dramatized in 1502, when the Venetian Senate, while granting Aldus himself a monopoly of publication in Greek, also (over the designer Francesco Griffo's loud protest) gave Aldus the exclusive right to use the italic type for printing in Latin. This is the first known example of a printer seeking a monopoly of all uses of a typeface. But the new type was much too useful for any such monopoly to be enforceable. Griffo and many others went ahead printing in italic, which set the pattern for modern type design.

Aldus' other great innovation also was not quite his own: the octavo-sized, smaller, lighter-weight book that meant the portable book. Even before Aldus, some books in manuscript and in print were smaller than the cumbersome scholar's tomes we see in the familiar portraits of Saint Augustine and Saint Jerome. These smaller books were generally religious texts, meditations, and guides to the order of church services, for prayer still seemed the only occasion for anyone to carry a book outside a church, a monastery, or a scholar's library. The serious scholarly reader would pore over a heavy folio on a solid lectern.

Aldus' vision of the reader was quite different. To make possible his small format, he omitted the lengthy commentaries that in earlier editions had often smothered the text. "Octavo," the word that came to describe this smaller format, originally designated the size of a book made by folding a printer's sheet of paper so that each leaf was one-eighth of a whole sheet. In modern printing jargon it means a page about six by nine inches. Many works that Aldus published in this format had already been printed in the old cumbersome "folios" (made by folding a printer's sheet in half). He liberated the book from the scholar's stuffy study and sent it into the open air.

The learned community warned that popularizing books would vulgarize learning. Even before 1500, a squeamish Venetian man of letters complained that you could not walk along the canal without having cheap books pushed at you "like cats in a bag." "More means worse." "Abundance of books makes men less studious." Corrupt printed versions, they said, were driving out of the market the reliable old manuscript texts.

> Now folks who don't know 'talian
> Will teach you to speak Tullian.

Printing was a whore who ought to be excluded from Venice by law. In 1515 the government of Venice gave the new librarian of the Biblioteca Marciana the impossible task of correcting all literary texts published in the city. Printers' colophons now protested that the volume had been printed *accuratissime*.

The manuscript book had been a kind of sacred object, an aid to religious or legal rituals, to communal memory. Copyright was unknown, and the "author" in the modern sense did not exist. There were special problems of nomenclature when books were commonly composed as well as transcribed by men in holy orders. In each religious house it was customary for generation after generation of monks to use the same names. When a man took his vows, he abandoned the name by which he had been known in the secular world, and he took a name of one of the monastic brothers who had recently died. As a result, every Franciscan house would always have its Bonaventura, but the identity of "Bonaventura" at any time could only be defined by considerable research.

All this, as we have seen, gave a tantalizing ambiguity to the name by which a medieval manuscript book might be known. A manuscript volume of sermons identified as *Sermones Bonaventurae* might be so called for any one of a dozen different reasons, which historian E. P. Goldschmidt enumerates. Was the original author the famous Saint Bonaventura of Fidanza? Or was there another author called Bonaventura? Or was it copied by someone of that name? Or by someone in a monastery of that name? Or preached by some Bonaventura, even though not composed by him? Or had the volume once been owned by a friar Bonaventura, or by a monastery called Bonaventura? Or was this a collection of sermons by different preachers, of which the first was by Bonaventura? Or were these simply in honor of Saint Bonaventura?

When the printed book put an end to these ambiguities, it brought into being the modern author. For as we have seen, manuscript books did not have title pages. Nor did the first printed books. To know what was in the book you had to leaf the pages. The book did not yet advertise its author. With the printed book there soon came the page that announced the name of the author (no longer merely a "writer"), the title and subject, along with the name of the publisher and the printer, the place and date of publication. In future the author would have the credit or blame and some of the profit from his work. The title page also announced a new commercial age in publishing, for here the publisher advertised where copies of the book could be bought. The date showing that the book had recently come from the press incidentally helped make novelty a desired commodity.

Before the title page was invented, at the very end of a volume there was commonly an inconspicuous "colophon" (Greek for "finishing touch") giving the scribe's or the printer's name along with the date and the place where the book was made. Sometimes there was a kind of trademark like Aldus' anchor and dolphin, with an apology or a boast for the quality of the copy. The title page, by contrast, soon became a full-page advertisement for the author and his work. It was not difficult to add an illustration, and from there the way was clear to the increasingly ornamented, baroque title page.

These new features of the printed book served both to standardize and to individualize the items on the market. And the mass production of books paradoxically opened the way for a sharper distinction between the products of individuals. As never before, the individual "author" was encouraged in his efforts at individuality and could be rewarded for his peculiar product. Originality became both respectable and profitable.

For the millions of new readers this meant a new variety of experience and a sharper definition of that variety. For the first time the intellectual menu was properly labeled. Standardizing the written commodity also served the individuality of readers, for books could be produced for their special interests.

The convenience of readers was served in other ways too. For example, manuscript books did not have numbered pages. Scribes using their own contractions competed in the effort to squeeze as many words as possible on each page of costly vellum. Even after the codex displaced the roll, the "pages" were still not standardized and numbered. There was no uniformity between where a given passage appeared in one or another manuscript of the same work. And as we have seen, the first printed books were made to appear as much as possible like manuscripts. Not until 1499, and from the Aldine Press, do we find a book in which each page is numbered in sequence. Well into the sixteenth century, nearly a century after the introduction of typographic printing, many books were still not being paginated, and often the numbering was not correct.

When, following Aldus' example, the numbered page became standard in book design, this apparently trivial innovation made possible some other changes that made the book vastly more serviceable, and appealing to a wider market. The expression "Table of Contents" first came into the English language in a volume Caxton printed in 1481, displaying the framework of the whole book at the beginning or (for Continental readers) at the end. Pagination made it easier, of course, to refer to particular passages, to pursue or verify facts and quotations.

Numbered pages also made it possible for the first time to provide an index to help the book serve varied personal needs. This simple alphabetic finding device is substantially a modern by-product of the printed book, an inconspicuous testimony to individualism and mass production. In the age of the manuscript book there had been rare examples of such efforts, but before pagination, indexes were difficult to make and inconvenient to use. A concordance, or a kind of index, of the Bible, which appeared in 1247, was reputedly compiled by Hugo de St. Caro with the aid of five hundred monks. Manuscripts with some sort of alphabetic index do not appear before the fourteenth century, and the index is by no means customary. Only with the printed book does an index become common. At first it appeared at the beginning of the volume, sometimes with its own title page. By the sixteenth century it was not unusual for printed books to carry indexes, sometimes listing not merely items specifically mentioned by the author but even related subjects and ideas. Before the end of the eighteenth century the value of indexes was widely recognized and readers expected to find them.

In 1878 the Index Society founded in London gave indexers a professional status. *What is an Index?* (1878), by Henry Wheatley, the society's first honorary secretary, reminded readers of the power of the indexer. "Indexes need not necessarily be dry, and in some cases they form the most interesting portion of a book. The Index to Prynne's *Historio-mastix* (1633), unlike the text, is very readable. . . ." And Macaulay, who "knew an author's own words might be turned against himself . . . wrote to his publishers, 'Let no d----- Tory make the Index to my History.' "

The dyspeptic Thomas Carlyle consigned the publishers of any indexless book "to be damned ten miles beyond Hell, where the Devil could not get for stinging nettles." The cause of indexing enlisted the enthusiasm of the great law reformer Lord Campbell (1779–1861), who half seriously proposed that any author who published a book without an index should pay a fine and be deprived of the benefits of the Copyright Act. Speaking for all readers in the early nineteenth century, Isaac Disraeli, Benjamin's father, could "venerate the inventor of Indexes . . . I know not to whom to yield the preference, either to Hippocrates, who was the first great anatomiser of the human body, or to that unknown labourer in literature who first laid open the nerves and arteries of a book."

$$66$$

Books Go Public

IN the multiplying medieval universities there were few institutional libraries, but professors still needed books. These could be had from itinerant book peddlers, an unreliable source over whom the professors had no control. Renting out textbooks, usually at a fixed price per quire, was a valuable privilege that could enrich the university and prevent circulation of heretical texts. The earliest book list of the University of Paris, in 1286, names some one hundred thirty-eight different titles for rent. At Bologna and elsewhere every professor was required to provide the university "stationer" with a copy of his lectures so they could be transcribed and rented or sold. He was called a stationer simply because, unlike the itinerant book peddlers, he stayed in one place. Peddlers still dealt in forbidden books, and it was they who gave wide circulation to John Wycliffe's banned translation of the Bible into English. But the stationer long remained the authorized source of textbooks and writing materials, and also operated a circulating library.

By the mid-fifteenth century, before the full flood of the Renaissance in Italy, the making (i.e., the transcribing) of books was a prosperous secularized industry, centered in university towns. The Florentine bookseller Vespasiano da Bisticci (1421–1498), who collected classical manuscripts for his wealthy patrons, at one time employed forty-five scribes to copy two hundred works for the library that the Medici had founded at Fiesole Abbey. Book publishers were already using block printing to illustrate their books. Some time passed before the universities acquired their own libraries, and then they grew speedily. In the mid-fourteenth century the library of the Sorbonne listed nearly two thousand volumes.

Printing multiplied books in numbers never before imagined. The best estimates suggest that, before Gutenberg, manuscript books in Europe could still be reckoned in the thousands. The population of Europe at the time probably came to less than one hundred million, and most people were illiterate. By 1500 there were probably about ten million printed books circulating (some scholars would double that figure), in addition to the still-growing stock of manuscript books.

The early decades of printing in Europe were marked by a regular increase in the numbers of copies in each printing. Until about 1480, some books had printings of only 100 copies; by 1490 the average had reached 500. By 1501, when the markets were better organized and the price of books had dropped dramatically, historians of printing no longer speak of "incunabula" (first used in 1639, from Latin for "swaddling clothes" or "cradle"), and the number in an average edition had increased to something like a modern figure. Aldus Manutius commonly printed editions of 1,000. During the next century a large edition numbered about 2,000.

Then, as printing itself became an established institution, printers organized their own guilds and tried to limit editions to keep the jobs coming. In England a decree of the Star Chamber in 1587 limited editions to 1,250 with only a few exceptions. About that time the Stationers' Company itself limited editions to 1,500, excepting only such items as grammars, prayer books, statutes and proclamations, calendars and almanacs. In Europe, during the seventeenth and eighteenth centuries a first printing exceeded 2,000 only for Bibles and unusually popular works, such as Voltaire's *Age of Louis XIV* or Diderot's *Encyclopedia*.

A crucial new element in bookmaking was the need and the opportunity for a publisher to estimate the size of the buying public for each book. How many purchasers would there be for still another edition of Cicero, for a legal treatise, for Petrarch's poems, for a work by Erasmus, a herbal, a travel book, or a textbook of astronomy? Who could be sure that more than a few readers would pay for a vernacular translation of the Bible, for a work on natural magic by the suspect Giambattista della Porta, or for Galileo's *Letters on Sunspots.* The mere fact that a book was printed showed that some printer was willing to risk his good money on the chance that hundreds or thousands of readers would pay to share its contents. The act of printing itself became an unprecedented unauthorized declaration of a *public* interest. Of course, governments could license the press or control it in other ways. But the printer's enterprise was a new menace to the oppressive ruler or the inquisitorial priest.

In the heyday of the medieval libraries books had been so valuable that they were chained to their shelf or to a horizontal bar above the desk where they were to be consulted. The symbol of the old library was the chained book. Hundreds of such captive volumes, called *catenati,* can still be seen neatly arrayed on the shelves of the library of Hereford Cathedral. None of the consequences of printing was more far-reaching than the power of the press to free books from these chains. As books became more numerous they no longer rested on their side, as had been the medieval practice, but instead were stood up close together displaying spine, title, and author.

The library of the Escorial near Madrid built in 1584 replaced the old monastic chapel-like bays by shelves that lined the walls, offering a large stock to users who might want to browse. Arranging books in a library became a science. In 1627 Cardinal Mazarin's librarian, Gabriel Naudé (1600–1653), who also served Cardinal Richelieu and Queen Christina of Sweden, wrote the pioneer treatise on librarianship. The Mazarin Library of 40,000 volumes, collected and organized by Naudé, was designed for a great private collector willing to make his treasures available "to all who wish to go there to study." Samuel Pepys followed Naudé's advice in his own elegant library, which still serves scholars in Magdalene College, Cambridge.

The multiplication of books on all subjects challenged philosophers to map the whole terrain of learning. The great German philosopher Leibniz supported himself as a librarian, and helped the dukes of Brunswick-Lüneburg in Hanover arrange their collection of 3,000 volumes. Then he went on to organize the 30,000-volume Ducal Library of Wolfenbüttel, for which he provided one of the first comprehensive alphabetical author catalogues. His new fireproof library design put galleries and shelves around the supporting pillars. But the Duke rejected his plan and constructed the library of wood, with the result that scholars had to shiver in winter because it was too risky to have a furnace. Leibniz saw the library as a congregation of all knowledge, with the librarian as minister keeping the congregation up-to-date and freely communicating. He pioneered in classification schemes, alphabetical finding aids, and abstracts to help the scholar. The library was his encyclopedia.

Leibniz signaled the transition from the royal and ecclesiastical collections for the privileged few to the public library serving everyone. In the next century his visions would be realized in the surprising career of the Italian émigré Sir Anthony Panizzi (1797–1879), a passionate Italian nationalist and energetic man of action. Forced to flee his native Brescello, in the Duchy of Modena, where he had joined a secret society conspiring against the Austrian occupiers, he had been sentenced to death in absentia. He found refuge in England, where he was named the first professor of Italian literature at the University of London. When students did not come, he gave up the honorific post to join the staff of the British Museum in 1831. For the next thirty-five years he dominated and invigorated that place to make it the model of a national library in the modern mode, reaching out to the new reading public.

"What a sad want I am in of libraries, of books to gather facts from!" Thomas Carlyle lamented as he moved from Scotland to London. "Why is there not a Majesty's library in every county town? There is a Majesty's gaol and gallows in every one." The "Majesty's library," the British Museum in London, to which Panizzi came, was ill equipped for Carlyle or for even less irritable scholars. Statuary, fossils, paintings, and maps were cluttered together with books and manuscripts. The extensive private library of George III brought to the museum in 1823 was joined with the old Royal library, and a new building was being built when Panizzi joined the staff. In 1837 Panizzi was appointed Keeper of the Museum's Department of Printed Books, and Principal Librarian in 1856. His fiery temperament was not one to soothe the stuffy trustees, who held a tight rein.

"I want a poor student to have the same means of indulging his learned curiosity," Panizzi told the Parliamentary Select Committee on the Museum in 1836, "of following his rational pursuits, of consulting the same authorities, of fathoming the most intricate enquiry, as the richest man in the kingdom, as far as books go, and . . . Government is bound to give him the most liberal and unlimited assistance." In 1849 Panizzi still boasted that he had "never felt the skin of any reader" and treated them all alike. Of course Carlyle, who was no friend of democracy, considered himself entitled to very special treatment. It also happened that he was hypersensitive to physical discomfort, as he was to almost everything else. Living in Chelsea, he detested the long trip to Bloomsbury to use Panizzi's library, where all the books had to be consulted on the premises, which closed at five o'clock in the afternoon. Naturally, Carlyle became Panizzi's sworn enemy.

Then Carlyle made his clash with Panizzi the occasion for some library innovations of his own. In 1841 he responded to Panizzi's ruthless egalitarianism by organizing the London Library. Carlyle called a public meeting to enlist his rich and noble friends. When the London Library opened in 1841, its five hundred subscribers had access to a collection of 3,000 volumes —and there was no alien radical in charge. The Earl of Clarendon was President, the Prince Consort was Patron, and Carlyle made sure that the first librarian was an amenable Scotsman. Carlyle continued to dominate the London Library, which grew into an unexcelled scholarly subscription library.

Meanwhile, Panizzi was making something quite new of the national library. Under him, librarians ceased to be underpaid clerks. He recruited scholars who were attracted by the secure tenure and the atmosphere of catholic learning. He provided comprehensive catalogues accessible to all, and he enforced the law of legal deposit, which entitled the Museum to a copy of every new British publication. Despite urging by his most respectable patrons, he refused to second-guess the future by collecting only "worthwhile" books on "important" subjects. The grand circular Reading Room of the British Museum was Panizzi's conception, destined to become an exhilarating model for the Library of Congress and other libraries. Devising his own ninety-one *Cataloguing Rules,* he insisted on a complete alphabetical name catalogue, and he refused to print the catalogue until all the library's collection was included. The trustees convened a Royal Commission to bring him to heel, but the Commission's final report in 1850 supported Panizzi.

The public library "in every county town," which Carlyle demanded, was yet to come. Panizzi still required users to present letters of introduction to enter the Reading Room and his books did not circulate. Another Scotsman, Andrew Carnegie (1835–1919), of a temperament very different from Carlyle's, would spread public libraries across a transatlantic continent-nation.

More than three centuries passed after Gutenberg's invention before there was any substantial progress toward admitting the sightless to the world of books. The blind seemed condemned to the age of oral literature. Then, in the era of the French Revolution, a French professor of calligraphy, Valentin Haüy (1745–1822), had the simple idea that the blind might be able to read with their fingers. He designed a simplified italic type in raised letters, which he introduced to pupils in the Royal Institute for the Young Blind, which he founded in Paris in 1785. But he had the perspective of a sighted person who had always seen words in the Roman alphabet. He assumed that the task was simply to emboss the familiar alphabet.

To make the blind at home in the world of written language, they had to be given a system useful for both reading and writing. The eventual solution would be found only by someone imaginative enough to abandon the sighted person's typographic alphabet. An Englishman, T. M. Lucas, following the example of the new systems of shorthand, devised a set of embossed phonetic symbols in which he transcribed the New Testament in 1837. Then James H. Frere (1779–1866), best known in his time for writings on Biblical prophecy, went on to devise a cheap way of embossing the phonetic symbols. He also invented the crucial device of the "return line" —printing the lines alternately, one running from left to right, the next with letters reversed from right to left—so that the reader's fingers could move speedily and accurately from one line to the next.

Finally the problem of finger-reading would be solved by an ingenious sixteen-year-old blind boy, Louis Braille (1809–1852), who had become a student in Haüy's Institute in Paris. Braille had been blinded at the age of three when he accidentally stuck a knife into one of his eyes in his father's leather shop. Then sympathetic ophthalmia made him totally blind. Despite this, he became an accomplished cellist and organist, and when only ten he received a scholarship to Haüy's Institute. Haüy already had achieved a meager success in teaching blind children to read his embossed Roman letters. When Braille entered the Institute, only fourteen books had yet been embossed in Haüy's scheme and these were hardly used. Braille, finding the embossed Roman alphabet unreadable, determined to devise a system that would enable the blind to write as well as read.

The alert young Braille found his clue not in the classroom but in a system proposed for soldiers at night on the battlefield when they needed to communicate but dared not strike a light and so shared the problems of the blind. "Night writing," invented by Captain Charles Barbier, a French artillery officer, used a tiny grill of twelve raised dots. Barbier had grouped and combined these dots in various ways to indicate letters and sounds. The weakness in Barbier's scheme was his twelve-dot "cell," easy enough for a sighted person to envision, but not convenient for the reading fingers and quite. impractical for writing. Braille saw these weaknesses, but he was encouraged by the possibilities. He reduced the "cell" from twelve to six embossed dots, then devised a simple stylo and frame for writing. The system that the sixteen-year-old Braille presented to the astonished principal of the Haüy Institute in 1825 was substantially that still used by the blind today. Braille's 32-page booklet (published by the institute in 1829 in its older system of raised Roman characters) showed how his six-dot scheme could be used for mathematical symbols and musical notes as well as for the alphabet. He also described the stylo and the frame that would enable the blind to write in Braille.

Braille's system was too novel (and too simple.) to be adopted at once. But within twenty-five years it was adopted by the Haüy Institute, then by an international congress in Paris in 1878, and was codified for the English-speaking world in 1932. In 1892 at the Illinois School for the Blind a Braille writing machine was invented. Competing systems have been tried. William Moon, who himself became blind in 1840 at the age of twenty-two, devised a scheme for those who go blind late in life, and the Moon system continues in some use. But Braille was the Gutenberg of the blind. The sightless of the Western world still follow the paths into print invented by this ingenious French boy. In the twentieth century the technology of recorded sound made possible "talking books"—which were among Edison's purposes when he devised the phonograph. Still, no satisfactory substitute has been found for Braille's invention. In the later twentieth century, the Library of Congress, through its National Library Service for the Blind and Physically Handicapped, offered more than 30,000 volumes in the several forms, and every year was transcribing some 2,000 new volumes, and 1,000 current periodicals, into Braille.

67

The Island of Islam

MUSLIM Believers, with some justification, have viewed their worldwide conquest as another of Allah's miracles. Their religion and its Book spread across the world with almost no assist from the printing press. Islam—a religion of the sacred Word—never became a culture of printed books. The refusal of Muslim leaders to adopt the printing press also helps explain many of the features of the modern Arabic-speaking world.

In the later twentieth century Arabic was the everyday spoken language of more than 120 million people—from the Atlantic coast of North Africa eastward to the Persian Gulf. The fifth most widely used vernacular, it remains also the sacred language of 400 million Muslim peoples on all continents. Long before there was a language, much less a literature, of English, French, German, Spanish, or Italian, there was a prolific secular literature in Arabic, with works of enduring value in poetry, history, medicine, astronomy, and mathematics. Paper, the sine qua non of modern printing, as we have seen, came to Europe through the Arabs. It was made in Baghdad in 793 in the reign of Caliph Harun al-Rashid of *Arabian Nights* fame, and then came through Arab Spain in the fourteenth century into Italy, France, and Germany.

Since the Arabic language is alphabetic, we might have expected that it would have been suited for the economies of movable type. Although some letters take different forms depending on their position in a word, writing Arabic requires only twenty-eight easily transcribed letters. Unlike Chinese, it has not been burdened with ideographs. Despite all these advantages, and an uncanny reverence for the written word, the Arabic world resisted the opportunities of the printing press.

What came to be called classical Arabic was spoken by tribes in the northern Arabian peninsula as early as the sixth century, when it had already produced some of its most eloquent long poems. By then the peculiar virtues of the Arabic language—its capacity for rhyme and assonance, the eloquence of Bedouin idiom, a unique stock of prosody and meters and poetic conventions—had already appeared. Rude tribal chiefs were expected to be patrons of poetry, and famous poets were followed about the desert by apprentice "reciters," who in turn became poets in their own right. The Koran overwhelmed the Arabic language in a victory that had no precedent. Revealed to Muhammad (570–632) piecemeal during his life in Mecca and Medina, its canonical text had been fixed by 652 under the caliph Uthman on the basis of the collection of the Prophet's secretary. To establish this text, Uthman had all other versions destroyed.

Henceforth "classical" Arabic was the language of God. No other widely spoken language has been so dominated by a single book. The Koran, according to strict Muslim doctrine, although revealed to the Prophet Muhammad as the word of God, was not "created" by God. The earthly text is believed to reproduce an "uncreated" eternal original in the heavens and is therefore unique in both its divinity and in its perpetuity. Tradition reports that when the Muslim conqueror of Alexandria, Amr ibn al-As (d. 663) entered Alexandria in 642, he asked the caliph Omar (c. 581–644) what he should do with all the books in the Library of Alexandria. The Caliph answered, "If what is written in them agrees with the Book of God, they are not required; if it disagrees, they are not desired. Destroy them therefore." Despite this holy advice, it appears that the conqueror did not burn the library.

The Arabic language was fixed in the Koran. "People loved the Arabs for three reasons," the Prophet said. "I am an Arab; the Koran is Arabic; and the language of the people of Paradise is Arabic." Arabic was not merely the vehicle of religion but the original language of all mankind, given to Adam who first wrote it on clay. Arabic script was not a gradual development. Therefore whatever might be the language of the local marketplace, the prayer addressed to God should be spoken everywhere in His own language, which is Arabic. And so Muslims across the world use Arabic for the prayers they recite five times each day. When a child is born, the Islamic creed (in Arabic of course: *La ilah illa allah; Muhammad rasul allah*) is whispered in his ear. These should be the first words a child learns to speak, and the last words on the lips of a dying man.

No wonder, then, that to imitate the style of the Koran is sacrilege. It is an axiom of Islam that the Koran is untranslatable, and it is forbidden to attempt a translation. "Translations" of the Koran, if by a believer, can be offered only as a kind of exegesis or paraphrase. Therefore Mohammed Marmaduke Pickthall entitles his version in English *The Meaning of the Glorious Koran*.

"The best act of worship by my community," said the Prophet, "is the recitation of the Koran. The best among you is he who learns and teaches the Koran. The people of God and His favorites are those of the Koran." Arabic grammar and lexicography grew up as an aspect of religious worship, as techniques for better understanding the Koran and imitating its rules of Arabic speech. The language of the Koran determined for all time the grammar, the syntax, and even the vocabulary of proper Arabic. Muslims in their everyday speech were expected to follow the rules of language exemplified in the Koran. In Islam, theologians became philologians.

Islam still remains an anachronistic empire of the arts of Memory—a relic and reminder of the power of memory everywhere before the advent of print. Since reciting passages of the Koran was the first sacred duty, a good Muslim child is ideally expected to memorize the whole Koran. When Muhammad himself first uttered the Koran he dramatized this duty. "Each verse of the Koran represents a step to heaven and a lantern in your house." Since vowels were generally not written, it was difficult to distinguish with certainty between the possible meanings in any set of written consonants. But the spoken version could be unambiguous. Therefore memory and recitation preserved the pure Koranic text. Some scholars say that there are today actually fewer significant variations in the commonly used version of the Koran than in the versions of the New Testament.

Not merely the Koran but the Arabic language itself became a sacred vehicle. The Persian historian Al-Biruni (d. 1050?), a classic Muslim authority on mathematics, astrology, and astronomy, rejoiced that all efforts to give a non-Arabic character to the state had failed. They would never succeed "so long as the call to worship continues to sound in their ears five times each day and the clear Arabic Koran is recited among rows of worshippers ranged behind the Imam."

> Our religion and our empire are Arabs and twins. . . . Sciences from all countries of the world have been translated into the language of the Arabs, have been embellished and become attractive, and the beauties of the language have permeated their veins and arteries, even though each people considers beautiful its own language to which it is accustomed and which it uses in its daily business. . . . I would rather be reviled in Arabic than praised in Persian.

Muslim peoples paid a high price for the divinity of their language. Even in the Arabic-speaking world, Muslims lived in a community of two languages. "Classical Arabic" became the sole literary language of the Arab world, the language of the formal written word regimented by the Koran. Modern colloquial Arabic has gone its own several ways into Eastern, Western, and Southern groups of dialects.

Still, classical Arabic, a gift from on high, keeps its dogmatic purity. The vocabulary of the Koran is predominantly of Arabic origin, although modern Orientalists do find there some loan words from Hebrew, Greek, Syriac, and Aramaic (e.g., the words for gospel, law, devil, belief, and prayer). Yet

Muslim dogma holds that the Koran contains no "foreign" words. "Whoever pretends that there is in the Koran anything other than the Arabic language," an eminent ninth-century Muslim philologist proclaimed, "has made a serious charge against God." Any similarity to foreign words, they say, is purely coincidental. Children memorizing the Koran are taught to revere its sounds and not worry over the everyday meaning of each word. The Arabic Koran, like the pilgrimage to Mecca, has remained a translingual bond among illiterate people speaking hundreds of languages.

Just as the Koran revealed in Arabic could not properly be "translated" into any other language, so Believers should transmit the text only in the original handwritten format used by the Prophet's disciples. As we have seen, the Chinese, followed by the Koreans and the Japanese, eagerly enlisted printing to reproduce their sacred classics. In the West, too, the printing press speedily became the vehicle of literature and learning across Europe. Within Christianity the Protestant Reformation made ample use of printed books. But there was nothing like this in the vast and growing community of Islam. The strongest reform movement within Islam, Shi'ism, which expanded in the sixteenth century to become the official faith in Iran and Iraq, with millions of adherents elsewhere, never adopted printing. Within orthodox Sunnite Islam, too, the imamate successfully forbade the use of printing both for the Koran and for other Islamic books. Since all science was nothing but a commentary on the Koran, fear of blasphemy and heterodoxy kept the printing press out of the Muslim world for centuries.

It is not surprising that the Koran in Arabic was printed in Europe long before a Koran was printed in the Muslim community. Less than a century after the Gutenberg Bible, the Arabic text of the Koran was published in Venice in 1530. This was a victory for those who thought the Devil could be combated only by being known. When Peter the Venerable (1092?–1156), abbot of Cluny, visited Toledo in the early twelfth century, he prepared his arsenal for an intellectual assault on Islam. By 1143, his first weapon, a translation of the Koran itself, had been produced by an Englishman, Robert of Ketton.

In 1541 an enterprising printer of Basel, Johannes Oporinus (1507–1568), began setting in type Robert of Ketton's Latin translation. The Basel City Council, following the lead of the Pope, who had ordered the burning of the Venice edition, objected. Against the Pope, Luther argued that knowing the Koran would work to "the glory of Christ, the best of Christianity, the disadvantages of the Moslems, and the vexation of the Devil." The Basel edition did appear in 1542 with prefaces by Luther and Melanchthon. In the Christian West, interest in the Koran grew with the centuries. The first English translation, not from the Arabic but from the French, was made by a Scottish divine, Alexander Ross (1591–1654), who was a student of comparative religion. An Italian priest, Ludovici Marracci, after forty years of study, produced a new Latin translation in 1698. The classic English translation from the Arabic by a lawyer, George Sale (1697?–1736), in 1734, with its helpful introduction, is still widely read. The nineteenth century saw other translations and prolific studies of the Koran in the European languages. Printed Korans spread across the English-reading world. At the suggestion of George Bernard Shaw it was included among the popular Everyman reprints, where it was a "best seller."

Meanwhile Islam remained estranged from printing for its own purposes even while observing the conspicuous advantages of printing for others. Rashīd ad-Dīn (1247–1318), grand vizier of Persia under the Mongols, in his encyclopedic history of the world, told how skillful Chinese calligraphers, overseen by learned scholars, had carved the corrected text of important books in blocks kept in government offices. "Then when anyone wants a copy of this book he goes before this committee and pays the dues and charges fixed by the Government. Then they bring out these tablets, impose them on leaves of paper like the dies used in minting gold, and deliver the sheets to him. Thus it is impossible that there should be any addition or omission in any of their books, on which therefore, they place complete reliance; and thus is the transmission of their histories effected." This prophetic description of "on demand" publishing appears also to be the earliest account of Chinese printed books outside of East Asia. Astonishingly, it never occurred to Rashīd ad-Dīn to have his own works printed. Instead, in his will he left funds to pay scribes to make a copy in Arabic and another in Persian of his complete works every year until there was a set in the mosque of every large Muslim city.

The Muslim community elsewhere was no more receptive than Rashīd ad-Dīn. In the Turkish Empire, Muslim tolerance very early allowed non-Muslim religious communities to operate their printing presses provided they printed nothing in either Turkish or Arabic. Immigrant Sephardic Jews printed a Pentateuch with Commentaries in 1494. Armenians were printing their religious texts by 1568, and the Greeks printed a pamphlet attacking the Jews in 1627. By the end of the sixteenth century, Sultan Murad III (1546–1595) allowed foreigners to trade in imported books. In these ways city-dwelling Turks became acquainted with the products of the printing press within a century after Gutenberg. By the early eighteenth century several libraries had been founded in the Turkish Empire and the exporting of rare books was prohibited.

To establish the first Turkish printing press, the first press within the Muslim world, required a messenger from outside Islam, the Hungarian-

born Ibrahim Müteferrika (c. 1670–1745). As a twenty-year-old student in
Transylvania, he was captured and enslaved by the Turks during their
invasion of eastern Europe. In Turkey he fell into the hands of a cruel
master, and to escape slavery he converted to Islam. He quickly became at
home in the literature of his adopted land, entered the Turkish diplomatic
service, and was named ambassador to the princes of eastern Europe and
the Ukraine. Interested in science, he saw how the printing press would
open paths to progress and as a specimen piece he carved a block of
boxwood for a printed map of the Sea of Marmara in 1719.

For eight years he tried to secure the Sultan's permission to set up a press.
His treatise on *The Means of Printing* (1726) lamented the many copies of
the Koran and other Muslim books that had been destroyed in the Mongol
invasions and during the expulsion of the Moors from Spain. Now inexpensive printed books, while preserving the authentic text, could spread the
true faith. Islam would be rescued from the Europeans who had monopo-
lized the printing of Islamic books. Turks would become champions of
learning for all Islam.

Finally, in 1727, Müteferrika was granted an imperial edict allowing him
to print books. The calligraphers of course protested. They put their ink-
pots, reed pens, and pen sharpeners in a coffin which they paraded to the
site of the proposed press. But when the mufti issued the required religious
permit for "composing letters and words in matrixes so that they can be
imprinted on sheets for reproducing in multiple copies," he expressly forbade the printing of the Koran itself, all books of Koranic exegesis, books
on theology or the Prophetic tradition, and books of law. Ibrahim Müteferrika's press, set up in Istanbul in 1727, was the first in any Muslim country.
For his master printer, he engaged a Jew, along with a staff of fifteen others,
to operate four presses for printing books and two for maps. Molds for Latin
type were imported from Europe. In their short life of eighteen years, his
presses were a promising beginning. They produced altogether some 12,500
copies of seventeen titles, including works of history and geography, astronomy, physics and mathematics, translations from Latin, French, Arabic,
and Persian, Arabic-Turkish and Persian-Turkish dictionaries, maritime
histories, and a book on magnetism. The first illustrated book printed in
Turkey was his edition of a manuscript work of 1583 on the "newly discovered" America. Some of these showed the typographic beauty of fluent
naskhi type equaling the best of later centuries. Ibrahim's presses belatedly
brought news of the invention of the telescope and the microscope to the
Muslim community. With a large backlog of unpublished translations of
Western works, his presses stopped at his death in 1745.

Decades passed before books would be printed again in Turkey. Not until
the Westernizing reform movements (1839–76) of the mid-nineteenth cen-

tury that aimed to secularize education did printed books again become a force in the life of Turkey. Finally in 1874 the Turkish government gave permission to print the Koran, but only in Arabic.

In the rest of the Muslim world, resistance to the printing press and suspicion of its products remained. Muslims have offered various explanations, such as the difficulty of using the Arabic alphabet to print Turkish or other languages of the Muslim community. And they were troubled by fears that would seem trivial to non-Believers, such as horror lest the hog's bristle in the brush used to clean the printing block be allowed to touch the name of Allah.

The story of printing in Muslim Egypt is much the same. When Napoleon arrived in Egypt in 1798, there was still no printing press, nor any newspaper. The town crier or the muezzin, who called for prayers from the minaret, would call out the news. A special class of criers posted along the Nile would alert the people when the river was beginning to overflow and incidentally shout the news bulletins. On his conquering way through Italy, Napoleon had seized the printing machines of the Vatican to accompany him to Egypt. From Italy he brought three compositors and three printers, along with eighteen printers from France. Christened by Napoleon Imprimerie Navale, this printing establishment was boarded on his headquarters ship. While still at sea, the press printed Napoleon's orders for his Army, along with Arabic translations of his proclamations to be distributed by Maltese prisoners whom he had brought for the purpose.

He set up the press in the house of the vice-consul of Venice in Alexandria. Rechristened Imprimerie Orientale et Française, within a day it had printed four hundred more copies of Napoleon's Arabic proclamation. For distribution in the Cairo region, one thousand copies were printed of a pamphlet with statements from Muslim leaders attesting Napoleon's goodwill, his respect for Islam, and his intention to protect all Muslims returning from the *hajj*. And for the French Army came books of Arab grammar with literary exercises. Napoleon himself paid close attention to the press. An old friend had brought his own private press to Alexandria along with the Army, but when the quality of work was unsatisfactory, Napoleon took it over, fired the owner, and transferred the plant to Cairo. When Napoleon made Cairo the center of his operations, he brought his Army press up the Nile to his capital. There it became the target of fanatics, but still Napoleon stayed with the project and issued detailed orders to improve its product. To save his press from the irate mob, he had to move it from place to place.

During the scant three years of Napoleon's occupation, with the help of his presses he opened a new era in Egyptian learning. From his presses came a flood of administrative reports and an informative miscellany. The first daily newspaper in that part of the world, his *Décade Egyptienne,* published in French the news from Europe, reviews of books and concerts, advertisements and poetry, along with articles on Egyptian customs and holidays, and on the annual overflow of the Nile. Napoleon had plans for a similar journal in Arabic.

Even before the first issue of his daily, Napoleon had organized his Institut d' Egypte, a local version of the European academies, which would prove wonderfully productive. To help create a community of Egyptian scientists, Napoleon had brought with him an old friend, the mathematician Gaspard Monge (1746–1818). Although the son of an itinerant knife grinder, he had managed to be admitted to the aristocratic military school of the ancien régime at Mézières. But his lowly birth prevented him from receiving a commission. Condemned to remain a draftsman, in his leisure he invented descriptive geometry, which was to be the basis of modern mechanical drawing. When the Revolution came and his lowly birth was an asset, he was assigned to the committee that devised the metric system and then became Minister of the Navy and the Colonies in 1792. In 1796 Napoleon sent Monge to Italy to select works of art to be confiscated and sold to finance his campaigns. Two years later, setting out to conquer Egypt, Napoleon again enlisted Monge—this time, as an admirer explained, "to offer a succouring hand to unhappy peoples, to free them from the brutalizing yoke under which they have groaned for centuries, and finally to endow them without delay with all the benefits of European civilization."

When the Institut d'Egypte was organized in Cairo on August 21, 1798, Monge was named president, Napoleon vice-president. The Institut made comparisons of the French and the Egyptian systems of weights and measures, studies of vineyards and the date palm, researches on the subterranean movement of water and on irrigation, and surveys of Egyptian aqueducts, of ancient monuments and inscriptions, and of the ruins of ancient cities. It examined an ancient canal supposedly connecting the Mediterranean with the Red Sea, which awakened Napoleon to a possible Suez canal.

The Institut's membership of 165 included medical men working to control the bubonic plague, botanists avid for botanical gardens and museums of natural history, along with entomologists and ornithologists. All these would help compile the elegantly illustrated volumes of the epochal *Description d'Egypte.* Their meetings and their library were open to the public. Napoleon himself asked the Institut: How can the Egyptian ovens be improved for baking bread? Can Egyptian beer be made from some other source than hops? Can the water of the Nile be purified? Does the Egyptian legal system need reform? And finally, what do the people most need?

When Napoleon was forced to leave Egypt, he took the printing presses back with him. The lack of a press in Egypt posed formidable obstacles to public education. Well into the nineteenth century, textbooks were available in manuscript only. The next ruler of Egypt after Napoleon, Muhammad 'Ali (1769–1849), who consolidated his power in 1811, had not learned to read and write till he was forty, but he sent emissaries abroad to learn Western techniques of education, to translate Western books, and to learn the skills of printing. By 1820 he had imported presses, had secured paper and type from Italy, had found skilled workmen, and was operating a government printing press at Bulaq, outside Cairo. His first book, an Italian-Arabic dictionary, came out in 1822, then he provided books for his military academies, for a medical school and a school of music. Of Muhammad 'Ali's many reforms, his printing press was the most potent and the most enduring.

Decades passed before Muhammad 'Ali and his successors could conquer the Islamic fears of the printing press. He did manage to print an edition of the Koran in 1833, but on his death in 1849 the mullahs persuaded his successor, Abbas Pasha, to lock up all printed copies and ban their use. Only later under Said Pasha (1822–1863) were these released for circulation. The first official printed version of the Koran was finally published by the Egyptian government in 1925. But this and even the late twentieth-century versions of the Koran printed in other Muslim countries were generally not made with movable type. Instead they were reproduced by block printing or lithography, which offered visually precise copies of handwritten manuscripts. The latest Pakistani edition of the Koran prints the English text in movable type, but the editor cautiously explains that the Arabic text is "printed from photographic blocks" displaying calligraphy "from the pen of Pir 'Abdul Hamid, with whom I have been in touch and who has complied with my desire for a bold round hand."

68

Toward a World Literature

FROM time to time ingenious and philanthropic spirits have tried to invent a single world language, but no man or government has yet succeeded in inventing the language for a nation, much less the world. The most successful, Esperanto, was designed by Dr. Ludwik Zamenhof, a Polish oculist, in

1887. Aiming to provide a simple and rational second language for all people everywhere, he tried to make Esperanto easy to learn, regular in grammar and pronunciation. Almost a century after its invention this most attractive of prefabricated languages has acquired only about a hundred thousand speakers dispersed over eighty-three countries. Yet even Esperanto was not entirely invented, because its vocabulary is derived from European words, and most are from the Romance languages. The meager success of artificial international languages simply attests to the mystery and vagrancy of language.

The languages of the world, living and dead, number some four thousand. A world community of the spoken, written, and printed word would have to be accomplished by the arts of translation, which could make it possible for anyone through his own vernacular to discover the world's literature.

In preliterate communities and still today, people of different language communities make themselves understood by gesture, facial expression, and tone of voice. And there is no satisfactory substitute for the living translator except to learn the language oneself. As we have seen, Columbus took along on his first voyage a man who could speak Arabic, and so, he hoped, could communicate with the Chinese emperor.

In the world of manuscripts for centuries the arts of translation had helped readers cross the language barrier. The translation by Saint Jerome (340?–420) of the Bible from Hebrew and Greek into the Latin Vulgate was a boon for learned Christendom. Translations of Plato, Aristotle, Galen, Dioscorides, and Ptolemy, and of Arabic manuscripts of mathematics, astronomy, and medicine entered the texture of Western thought.

The printed book further widened the access of learned readers to the literature of distant times and places. Before the end of the fifteenth century, at least twenty works from the Arabic were printed in Europe in Latin translations. While the rise of vernacular languages was narrowing the vision of the literary classes to works in their own national language, printed books were offering them new opportunities to become cosmopolitan. When Francis I made the French of Paris the nation's official language, at the same time he personally paid for translating the classics into French, and classical culture became accessible to Frenchmen who could not read Greek or Latin. In England there were 43 printed editions of classical works in English translation by the mid-sixteenth century, and 119 such editions before 1600. Standard classical authors were the safest investment for both book publishers and book buyers. In Europe, by the end of the sixteenth century, when there were 263 Latin editions of Virgil, there were also 72 translations into Italian, 27 into French, 11 into English, 5 into German, 5 into Spanish, and 2 into Flemish. Some classical authors became better known in translation than in the original. Plato, for example, was widely read in the Latin translation by Marsilio Ficino (five times reprinted in France before 1550), well before the complete Greek text was published in France in 1578.

The literary consciousness of readers was also opened to recent and contemporary authors translated from other vernaculars. Early European favorites included Petrarch, Boccaccio, More's *Utopia,* and Brandt's *Ship of Fools,* along with Machiavelli, Ariosto, Tasso, and the romantic *Amadis de Gaule,* followed by translations of Montaigne's *Essays,* and Cervantes' *Don Quixote.* Works of Spanish literature that we no longer remember proved surprisingly popular in French, English, Italian, German, and Dutch. People across Europe could join the international literary community without reading Latin. There began to be a European literature, available to all readers in translation.

Imagine how provincial we would be if our reading had to be confined to works originally written in our own vernacular! We cannot begin to measure the meaning of translation for civilization. The Renaissance came to England in Elizabethan translations. The flowering of English literature brought John Florio's translation of the *Essays* of Montaigne, Thomas Shelton's *Don Quixote,* Sir Thomas Urquhart's Rabelais, and, as we have already noted, the King James Version of the Bible. Eighteenth-century Englishmen could read Sir William Jones' scholarly translations from Arabic, Hindi, and Persian, which distant Americans promptly included in the Congressional library of their new republic. Shakespeare's works became the subject of an extensive critical literature in German by Lessing, Goethe, and Schlegel, and provided themes for countless authors from Chekhov and Gide to Brecht and Max Frisch, for Verdi's operas, for many ballets and even American musicals. European actors and actresses were expected to prove themselves in Shakespearean roles. Goethe had a similar influence across the continent. Richard Burton's *Arabian Nights* and Edward Fitz-Gerald's *Rubáiyát* opened the world for Victorian readers. Before the end of the nineteenth century literate Europeans would be at home in the great works of their own and other continents, and authors would be writing for a world audience.

Translators are patriots enriching their own national language. Even so, they have seldom received their due. They have too often been dismissed by their beneficiaries with the Italian proverb *Traduttore, traditore* (A translator is a traitor). Some men of letters, with masochistic conceit, have even gloried in the untranslatability of the best works. "Poetry cannot be translated," observed Dr. Johnson, "and therefore it is the poets that preserve the language." "If the translator is a good poet," complained the Irish poet George Moore, "he substitutes his verse for that of the original;—I don't want his verse, I want the original; if he is a bad poet, he gives us bad verse, which is intolerable." Or, as Chaim Bialik puts it more charitably, "Reading poetry in translation is like kissing a woman through a veil." When translators were not contemned, they were often ignored, and so have become the forgotten men and women of letters. The indispensable messengers of a common culture, they take on themselves a linguistic task more complex than that of the original writer, the problem "of adjusting the literal and the literary." We could do worse than define a classic of world literature as a work that has had its greatest influence in translation.

Dictionaries, a modern aid to discovery, began as guides across the barriers between languages before they guided readers and speakers into their own language. The word "dictionary" comes from the medieval Latin, *dictionarium* or *dictionarius,* originally meaning "a repertory of *dictiones,* phrases or words." In Europe dictionaries appeared at first to serve the learned class. "Dictionaries" of antiquity were usually collections, not necessarily in alphabetic order, of words and phrases from the best-known authors. A few appeared in Europe in the thirteenth and fourteenth centuries to help students learn Latin and so they might read the Vulgate version of the Bible. These bilingual dictionaries gave the meanings in one language for the words of another language. Their market was still dominated by Latin scholars reading classical and religious texts.

The first, most successful, and most influential of the early printed dictionaries was a voluminous Latin-Italian dictionary by an Augustinian monk, Ambrogio Calepino (c. 1440–1510), published in Reggio di Calabria in 1502. In successive editions it became increasingly polylingual. By 1590, when Calepino's successors published their edition in Basel, it helped the reader into eleven languages, including Polish and Hungarian. *Calepino* became the Italian word for dictionary. Like "Webster" later in the English-speaking world, "calepin" entered the English language, too, in the sixteenth century and remained current for a century. Calepino's spirit lived on into the eighteenth century, reincarnated in the Italian philologist Jacopo Facciolati's *Dictionary of Eleven Languages* (1718). Surprisingly, the first dictionaries were the most polyglot.

Calepino's success encouraged the enterprising French publisher Robert Estienne (1503–1559) to issue an improved edition with the generous patronage of the bibliophile King. Francis I ordered his friend Robert, of a family of productive scholar-printers, to give the royal library a copy of every book he printed in Greek, and so created what was probably the first national deposit library. In the early sixteenth century the Estiennes made Paris the leader in the Continental book market as Venice had been before. They popularized the "Aldine" kind of book, using roman and italic type in portable octavos. Robert Estienne planned at first to reissue Calepino's dictionary, but then decided instead to revise the whole work. When other scholars would not take on the garguantuan task, he did it all himself, adding some useful features. He drew his vocabulary exclusively from classical authors, he cited other authorities for his meanings, and he supplied ample quotations to show usage. Estienne's *Thesaurus* of Latin appeared in 1531, followed by his Latin-French dictionary in 1538. The only scholarly rival to Estienne's work even now is the *Thesaurus* undertaken in 1894 jointly by five German academies, but after eighty years they had still only reached the letter "O."

Besides pioneering in scientific lexicography, Robert Estienne helped the literate classes in Europe discover the linguistic wealth hidden in their own vernaculars. He came up with handy Latin and French school dictionaries, and he pioneered again with a full-scale dictionary from French into Latin which included technical words. So he helped create a "standard" language for the nation. His practice of basing his approved vocabulary on *"bon autheurs françois,"* first adopted by the French Academy for their dictionary in 1694, still dominates and stultifies French lexicography to this day.

In Venice, commercial meeting place of northern and southern Europe, appeared the first printed bilingual dictionary for the merchant and the common citizen when Adam von Rottweil, an itinerant from Germany, printed his *Vocabolario Italiano-Teutonico* (1477). Then Caxton in London in 1480 printed a concise French-English vocabulary of twenty-six leaves. These are the first known examples of the phrase books that would do so much to help bewildered travelers in later centuries.

Estienne's use of the "best authors" provided the means by which lexicographers set up their standards of correctness for the new national languages. The first comprehensive standard dictionary in a single language, a product of twenty years of work by the Accademia della Crusca, appeared in Venice in 1612 and offered the pattern for other authoritative monolingual European dictionaries. The leader in this project, Leonardo Salviati (1540–1589), used the power of the press to legislate the Tuscan dialect into the standard Italian language. He drew from the great writers of Florence, canonized Dante, Petrarch, and Boccaccio, and so made theirs the language for an Italian nation still three centuries in the future. Some say it was the Italian language that created the nation. The power of Salviati's Tuscan dialect was proved when Manzoni (who had originally written his classic *I Promessi Sposi,* published in 1827, in the dialect of his native Milan) took the trouble to rewrite his book in the dialect established by the Accademia della Crusca two hundred years before. Elsewhere, too, in the *Dictionary of the Spanish Academy* (1726–39), and in Dr. Johnson's *Dictionary* (1755), lexicographers helped establish standard national languages by their choice of examples. These dictionaries, of course, helped natives as well as foreigners discover the growing resource of vernacular vocabularies.

"Standard" English was established empirically and by individuals, by contrast with the product elsewhere of state-supported academies. Since the early English Protestants wanted to help ordinary readers understand the Bible in English, the mainstream of English lexicography, as Allen Walker Read explains, flowed from the lists of words collected to help the devout. One of the first such lists was appended to William Tyndale's English translation of the Pentateuch in 1530. At the same time schoolmasters and spelling reformers wanted to bring some uniformity into the "disorders and confusions" of English spelling, which, they complained, had made any usable dictionary or grammar of the language impossible. "A dictionary and grammar," one of them optimistically boasted in the late sixteenth century, "may stay our speech in a perfect use for ever."

The example of England showed how general enlightenment could come with widespread literacy in a language shared by millions. Roger Ascham (1515–1568), private tutor to Queen Elizabeth, in his *Scholemaster* (1570), one of the first effective critiques of English education, listed the evils of aimless travel abroad (especially to Italy), and urged young men to become masters of their own English language. He even advocated the use of English for teaching the classics.

Another Elizabethan reformer, Richard Mulcaster (1530?–1611), helped provide the printed apparatus. Thirty years of teaching at the prestigious Merchant Taylors' School and St. Paul's persuaded him that teachers, like lawyers and doctors, should be specially trained at the university for their profession. He urged that these schools be open to women, who should also be allowed to go on to the university. And he argued that teachers should respect the differences between individual children, that not age but readiness should determine the curriculum for each pupil, and that the ablest teachers should be assigned to the earliest grades. For all these reforms he found his handle in the teaching of the English language. "I love Rome, but London better," he wrote. "I favour Italy, but England more. I know the Latin, but worship the English." So in his *First Part of the Elementarie* (1582) he compiled a list of some eight thousand words (but without definitions), presumably all the English words in use.

The English dictionary soon became an essential tool for widening education, and word lists became a tool for teaching reading. The first book to go beyond a schoolboy's list and offer explanations in English of English words did not appear until the seventeenth century. Even then the books collected only "hard" words and through etymology still showed a heavy bias in favor of other languages. Such was the first purely English dictionary by Robert and Thomas Cawdrey, schoolmaster father and son, entitled *A Table Alphabeticall, conteyning and teaching the true writing and understanding of hard usuall English wordes, borrowed from the Hebrew, Greeke, Latine, or French &c.* (1604).

The primitive efforts at comprehensive dictionaries for the adult reader were made not by earnest schoolmasters but by hobbyists or hacks. The most notorious of these was by John Milton's nephew and protégé Edward Phillips (1630–1696?), whose work "of doubtful originality, little merit, and great popularity" appeared under the prophetic title *The New World of English Words* (1658). Only after the appearance of professional lexicographers, for whom dictionary-making became a career, were there respectable dictionaries surveying the whole stock of words in the language. The printing press had made this career possible. *A New English Dictionary* (1702), the first such English dictionary aimed at everybody, was by John Kersey the Younger, the first full-time professional English lexicographer.

After Caxton printed the first English book, none of the new national languages flourished more than English. This happened, of course, without benefit of a comprehensive or "authoritative" dictionary. Not until the mid-eighteenth century was there an adequate dictionary of the whole English language. Then Dr. Johnson's *Dictionary* spectacularly demonstrated the power of dictionaries. His work was remarkable not only for its quality and its sudden authority, but as a monument of literary heroism. Five London booksellers signed a contract with the little-known Dr. Johnson in 1746 to compile for them a dictionary of the English language, which he hoped to complete in three years. He enlisted the help of six part-time amanuenses who laboriously copied out illustrative quotations indicated by Dr. Johnson in the best English authors. Johnson himself wrote definitions for 43,500 words, below which these slips of quotations were pasted. "The *English Dictionary,*" he explained in the Preface, "was written with little assistance of the learned, and without any patronage of the great; not in the soft obscurities of retirement, or under the shelter of academick bowers, but amidst inconvenience and distraction, in sickness and in sorrow." Though plagued by ill health and stricken by the death of his wife, he produced the work in two volumes on June 14, 1755, just eight and a half years after he had begun. He legislated standard English into existence—by the power of a printed dictionary—and incidentally provided an unprecedented aid for all explorers of English literature.

For at least a century men of learning had organized to purify, simplify,

and standardize the language. As early as 1664 the Royal Society had
envisioned such a project. In 1711, Alexander Pope, still only twenty-three,
in his *Essay on Criticism* voiced the fear that:

> Our sons their fathers' failing language see,
> And such as Chaucer is, shall Dryden be.

Dr. Johnson, drawing on his 114,000 quotations, applied the new spirit of
incremental science to the ancient world of words. For the common verb
"take," for example, he offered 113 different transitive senses and 21 intransi-
tive senses. He offered 5 senses for "genius," 11 for "nature," 8 for "wit."
"The effect which Newton's discoveries had in mathematics," Dr. Johnson's American disciple Noah Webster (1758–1843) declared, "Johnson's
work had in this world of words."

Yet Dr. Johnson neither lamented nor ignored the organic growth of the
language. His eloquent Preface explained that language was inevitably
changed by conquests, migration, and commerce, and by the progress of
thought and knowledge. "When the mind is unchained from necessity, it
will range after convenience; when it is left at large in the fields of speculation, it will shift opinions; as any custom is disused, the words that expressed it must perish with it; as any opinion grows popular, it will innovate
speech in the same proportion as it alters practice. No dictionary of a living
tongue ever can be perfect, since while it is hastening to publication, some
words are budding and some falling away."

Before Dr. Johnson the best authors had believed that provided the
meaning to the reader was clear, it made no difference how the writer
spelled his words. The basic problem of establishing a uniform English
spelling came from the fact that the alphabet of our language was borrowed
from another language. The Roman alphabet was never designed for English sounds. That alphabet, adapted from the Greek through the Etruscan,
originally had only 20 letters. These became our modern English alphabet,
but lacked *J, K, V, W, Y,* and *Z.* Then the Romans themselves added *K*
(for abbreviations) and *Y* and *Z* (for words they took from the Greek).
These were the 23 letters when the alphabet was first used for English.
Later, to serve the phonetic needs of English, *W* was made by tying together
two *U*'s, and *J* and *V* were introduced for the consonant sounds of *I* and
U, finally producing the 26-letter modern English alphabet.

The community at large went its own way, and the best English authors
spelled according to whim until, by the eighteenth century, printed word
lists and the increasing popularity of the rudimentary dictionaries had
promoted the notion that there could or should be only one way to spell

a word. In 1750 Lord Chesterfield (1694–1773), the paragon of superficial propriety (to whom Dr. Johnson three years earlier had dedicated the Plan for his Dictionary), warned his son: "Orthography is so absolutely necessary for a man of letters, or a gentleman, that one false spelling may fix a ridicule upon him for the rest of his life; and I know a man of quality who never recovered the ridicule of having spelled *wholesome* without the *w.*" Transatlantic Englishmen in America hoped that knowing the proper spelling of standard English would give them the authentic mark of culture. Noah Webster, who began as a schoolmaster, first won fame and fortune by his *American Spelling Book*, which appeared in 1783 and in the next century sold more than sixty million copies. But President Andrew Jackson was reputed to have said that he had no respect for a man who knew only one way to spell a word. American cultural insecurity, which had created the market for Noah Webster's speller, produced a continuing demand for Webster's *American Dictionary of the English Language* (2 vols., 1828), which made his name a synonym for dictionary.

Ironically the works of the most sophisticated English and American dictionary makers in the twentieth century helped liberate the English language from the despotism of dictionaries and sparked the effort to discover the lost treasures of the language in the changing uses of words in the past. Another heroic British lexicographer, James A. H. Murray (1837–1915), conceived the *Oxford English Dictionary* as "the greatest treasure-house of any language in the world."

The Philological Society of London in 1857 began its plans for a historical dictionary. After a number of false starts, in 1879, Murray, then an obscure assistant master of a school, took on the work, gave it shape, and brought more than half the vast project to completion. The purpose was to exemplify every word ever used in English and reveal its changing meaning. Illustrations were collected and copied by thousands of volunteer enthusiasts. Before 1900 these slips exceeded five million. Among these "volunteers" were Murray's eleven children, who joined in the tiresome sorting of the word slips into their alphabetical order. Little Rosfrith, his ninth-born, recalled her father catching her by her pinafore as he passed her in the hall of their Mill Hill House one day (when she had only barely mastered the alphabet) and saying, "It is time that this young woman started to earn her keep." Before his death in 1915 Murray had published nearly half the work —7,207 pages of a total of 15,487. "All the family feel themselves taller by it" was the household motto. Murray's successors brought the work to completion in 1925.

The result, far from establishing a fixed standard, which had been the hope of Dr. Johnson and his predecessors, was to lay out for all to see the responsive, changeful, elusive character of a living world language through several centuries. As Murray explained in his Introduction:

> That vast aggregate of words and phrases which constitutes the Vocabulary of English-speaking men presents, to the mind that endeavours to grasp it as a definite whole, the aspect of one of those nebulous masses familiar to the astronomer, in which a clear and unmistakable nucleus shades off on all sides, through zones of decreasing brightness, to a dim marginal film that seems to end nowhere, but to lose itself imperceptibly in the surrounding darkness.

PART FOURTEEN

OPENING
THE PAST

*Human behavior can be genuinely purposive because only
human beings guide their behavior by a knowledge of what
happened before they were born and a preconception of what
may happen after they are dead; thus only human beings find
their way by a light that illumines more than the patch of
ground they stand on.*

—P. B. MEDAWAR and J. S. MEDAWAR,
The Life Science (1977)

69

The Birth of History

Scholars of India are puzzled by why their culture, so ancient, so rich in sculpture and architecture, in works of mythical and romantic literature, should have been so lacking in critical historical writings. Some suggest that the ancient Indian works of history written in Sanskrit may, for still unexplained reasons, have suffered wholesale destruction. A more plausible explanation is that they never existed. And for this there are ample explanations in Hinduism and the Brahman view.

Hinduism was a religion of cycles. Later religions would be preoccupied with the Creation. They asked when, how, and why the world first came into being, which led to speculation about the purpose of Creation and the end of man. But Hindus were more interested in Re-Creation. A modern view of History would require belief in unique, novel, world-reaching acts. While Hinduism has had many sacred documents, it has no one Sacred Text, no Bible that tells the one true story.

The result was a wonderfully varied and constantly enriching Hindu jungle-garden of truths, but no one path to The Truth. The Hindu lore of cycles carried the Hindu believer far beyond the round of the seasons, far beyond the rhythm of his own birth, life, and death or that of his generation into an unending universe of unending cycles, of cycles within cycles within cycles. The basic cycle, the *kalpa,* is "a day in the life of Brahmâ," who was one of the three supreme Gods. Each *kalpa* lasts 4,320 million earthly years. A "night of Brahmâ" is the same length. A "year of Brahmâ" comprises 360 such days and nights, and Brahmâ lives for one hundred such years.

Each *kalpa* marks another Re-Creation of the world. During each *kalpa*-night the universe is once again gathered up into Brahmâ's body, where it becomes the possibility of still another Creation on the next day. Each *kalpa* contains fourteen smaller cycles, *manvantaras,* each of which lasts 306,720,000 years, when a new Manu, or presiding god, is created and in turn re-creates the human race. Within each *manvantara* there are seventy-one eons or *mahayugas,* a thousand of which comprise a kalpa. Within each *mahayuga* there is a cycle of four *yugas,* each of which is a different age of the world, including, in turn, 4,800, 3,600, 2,400, and 1,200 "years." Each of the four yugas shows a decline in civilization and morality from the yuga just before, until finally the world is destroyed by flood and fire to be prepared for still another cycle of Creation. Change on earth was slower than man could grasp.

The one notable historical work that survives from Sanskrit literature, the long *River of Kings,* by the twelfth-century Kashmiri poet Kalhana, tells us nothing about other parts of India, and carries the moral that man must resign himself to the superhuman forces. *The Ceylon Chronicle* is the story of Buddhism in Ceylon. The main interest of Hindu Indians in their past was not in the rise and fall of historical empires but in the rulers of a mythical golden age. This tantalizes the modern historian trying to chronicle India before the coming of the Muslim kings, for he must piece together his chronology from folklore, from a few scattered monuments, from the writings of foreign travelers. Biographical anecdotes are scarce. The ancient Hindu kings themselves were so persuaded of the evanescence of their works that they did not generally record their achievements on their monuments. The lack of a historical record reveals not merely the Hindu preoccupation with the transcendent and the eternal, but also the widespread sense that social life was changeless and repetitive. Where there was so little difference between past and present the quest for history seemed futile. In a society that did not know change, what was there for historians to write about? When real events were recorded, they were usually transmuted into myth to give them a universal and enduring significance.

After the eleventh century, when the Muslims came to India, records of the Indian past were given new shape. "We tell you the stories of the Apostles," declared the Koran, "which will strengthen your heart, and thus bring you the Truth, an exhortation and a memorial for the believers." For the Muslim, meaning came to events not from what man achieved, but from what God intended. History was not a process but a fulfillment. In India, too, Muslim history became official history, written to celebrate the good ruler. As one of the great Muslim historians of India wrote in the mid-fourteenth century:

> History is the knowledge of the annals and traditions of prophets, caliphs, sultans, and of the great men of religion and of government. Pursuit of the study of history is particular to the great ones of religion and of government who are famous for the excellence of their qualities or who have become famous among mankind for their great deeds. Low fellows, rascals, unfit people of unknown stock and mean natures, of no lineage and low lineage, loiterers and bazaar loafers—all these have no connection with history.

Naturally enough, Muslim history became the history only of Muslims, of the greatest of their prophets, holy men, and rulers. Everywhere it went, Islam brought this Muslim way of filtering the past.

A peculiar genre of Arabic literature, the so-called "battle-day" literature, reaches back before Muhammad. Islam gave a special significance to biography, making all later lives only footnotes to the life of the Prophet. Because in Muslim institutions there could be no novelty, but only the fulfillment of the Koran, Muslim biographies could not have the dignity of new knowledge. History, in the Muslim phrase, became merely a "conversational science" helpful for political wisdom and social skill, a source for illustrations but not for demonstrations. The historian appropriately called himself a compiler of the stories of Muslim crusades and successes.

Since Muhammad himself was the climax of history, there was, of course, no place for the idea of progress. "History," a branch of eschatology, told how all men journeyed toward their Judgment Day. The emphasis on biography increased the temptations of official chroniclers to sycophancy. Their records of events became as suspect as they were fulsome. The biographical dictionary, a characteristic and original creation of the Islamic community, focused on the individual yet did not produce individualism. The historical literature of Islam became an instrument of the faith, not an opener of vistas.

The Muslim emphasis was at its best in Ibn Khaldun (1332–1406), who, in the last stages of the Muslim Empire of North Africa, surveyed the varied fates and opportunities of Muslim community. As Tamerlane's consultant on the sociology of the Muslim-Arab world, he provided for Islam a classic statement like that which Saint Augustine, a thousand years earlier, had provided for Christianity. Unlike Saint Augustine, he saw a destiny unfolded not in time but in space. The earth was not a scene of man's journey toward the City of God, but an arena for conquest by the Prophet's faith. He asked how the varied surface of the planet explained the uneven opportunities for Islam. "The past resembles the future," he concluded, "more than one drop of water another." Ibn Khaldun proved to be the Herodotus and the Thucydides of Muslim historiography. His successors, slow in coming, would not be found in Islam.

Of all modern cultures, the Chinese offer the longest continuous past and the most copious written record of their past. It is all the more remarkable, then, that a modern historical consciousness did not develop in China. The Chinese way of filtering the past, although different from the Hindu, was hardly better designed to awaken people to social change or the power of mankind to transform institutions. Confucianism, rooted in ancestor worship, encouraged record keeping for genealogies. The Confucian consulted the past not to learn how institutions could be changed but, rather, to find the ideal to which they should be restored and for models of virtue to be imitated. Anecdotes collected in the early feudal period were sanctified by attribution to Confucius himself.

Then, in the beginning of the imperial period, in the second century B.C.,

Ssu-ma Ch'ien (145–87? B.C.) set the pattern for historical writing in China for the next two thousand years. His father was Astrologer Royal, or Grand Scribe, of the Han court, with the duty of keeping the calendar and recording official events. When Ssu-ma Ch'ien inherited the office in 108 B.C., the Han dynasty had begun the political unification of all China. He carried on his father's effort to bring together the records of all the Chinese people into a single work. Such a compilation would celebrate the great achievement of the ambitious new dynasty, bent on a New Beginning, marked by a "reform" of the calendar which Ssu-ma Ch'ien helped inaugurate.

A single imprudent word was enough to ruin his life. Once after the general Li Ling had lost a great and bloody battle "the Emperor could find no flavor in his food and no delight in the deliberations of his court." The other generals met in an imperial council to commiserate on the defeat and put the blame on Li Ling. But Ssu-ma Ch'ien considered him a paragon of loyalty and of virtue, and believed the battle had been lost despite Li Ling's bravery. When called into the imperial council, Ssu-ma Ch'ien (by his own account) "took the chance to speak of Li Ling's merits . . . hoping to broaden His Majesty's view and put a stop to the angry words of the other officials." For these ill-considered remarks he was thrown into prison, accused of "defaming the Emperor," for which the penalty was death. "My family was poor," Ssu-ma Ch'ien explained, "and lacked sufficient funds to buy commutation of sentence."

Ch'ien begged a reprieve so that he could finish compiling his history. The Emperor, reluctant to lose so expert and energetic an Astrologer Royal, graciously ordered that instead of being executed, Ch'ien should be castrated. In shamefaced retirement, Ch'ien completed his history, which became the model for all major chronicles of the Chinese past until the end of the imperial period in 1911. Before him, each Chinese state had used its own chronology, with the result that there was no way of knowing which events in the different states were contemporary. Ch'ien brought all these together in a single time series around the chronology of the ruling house of Chou. He also offered a new topical framework divided into five sections: Basic Annals or the lives of men who ruled large areas, Chronological Tables, Treatises on political, economic, social, and cultural themes, Hereditary Houses, and Biographies of important men who were not rulers but who illustrated eminence and virtue. Ch'ien's style made him a classic, but instead of emulating his spirit, disciples imitated his form. His immediate successor, Pan Ku (A.D. 32–92), froze Ch'ien's pattern into a rigid Confucian mold, leaving no leeway for interpretation.

With the reunification of China under the T'ang dynasty in the seventh century, the making of Standard Histories in the genre of Ch'ien, and under the complacent euphemism of "Veritable Records," became a permanent task of the growing national bureaucracy. All versions of the official national history were, of course, government property. For a while they were public, but before long they were secreted in government archives accessible to only a few. Each successive dynasty came to consider it a duty to compile a history of the preceding dynasty, leaving the history of their own time to be written by their successors.

The writing of Standard Histories continued to be governed by contradictory ideals: "truthful record" versus "appropriate concealment," "objectivity" versus "ethical instruction." The whole Chinese past, incorporated into the Confucian tradition, became part of the apparatus of government. A History Office was established in the T'ang dynasty and thereafter controlled all the accessible past. For millennia Chinese history was written by bureaucrats and for bureaucrats.

There was a brief revival of Ch'ien's lively spirit in the remarkable Sung dynasty. But the revival of neo-Confucian orthodoxy under the Ming (1368–1644) kept the past still firmly frozen in Ch'ien's mold. The collapse of the Ming and the conquest by the Manchu did open some cracks, but the few notable efforts to write critical history in the eighteenth century were conspicuous exceptions. Critical techniques in the writing of history and the rise of historical consciousness would have to await the influence of the West.

In this, as in so many other ways, Chinese civilization suffered from its antiquity, its precocity, and its continuity. The greatness of ancient models, the unbroken series of records, and the early effectiveness of central government all reinforced reverence for ancestors and stifled efforts to look at unauthorized vistas of the past or to speculate on what might have been.

The story of history in the West was dramatically different. The kind of history "invented" here would eventually reshape the lives and institutions of the peoples of India, the Middle East, Islam, and China. Western exploration of the past was as momentous as the charting of New World continents and the exploration of the oceans. Here again the story begins in the mystery of the Greek inquiring spirit. For the Greeks made of the past something quite different from that seen from India or China. Greek mythology, the most familiar survival of Greek culture, is not the most distinctive expression of the Greek view of the past.

One of the greatest Greek inventions was the idea of history. The word "history," along with its cognates in European languages, derives through the Latin *historia* from the word *historiē,* which the Greeks used to mean "inquiry," or "knowing by inquiry." Its original meaning survives in the expression "natural history" for inquiry into nature. And this characteristic Greek notion of "inquiry" bore fruit in the sixth century B.C., in the Ionian

Enlightenment. We think of modern scientists as the heirs of that spirit, but so, too, are modern historians. Perhaps it was a by-product of their study of medicine, which observed how the functioning of the body varied with environment, climate, and diet, and so led the Greeks to wonder about the varying ways of communities. Hecataeus of Miletus (c. 550–489 B.C.), one of the first and best-known pioneers of Greek historical writing, compiled genealogies and scrutinized the legends of the great mythical families. "What I write here," he insisted, "is the account of what I considered to be true. For the stories of the Greeks are numerous and, in my opinion, ridiculous." Traveling widely, Hecataeus noted the varieties of customs, and saw a connection between where people lived and how they lived. In this way he helped develop the Greek cosmopolitan awareness of the relation between geography and history.

Before the spread of typographic printing, as we have seen, verse—a convenient and pleasant vehicle of memory—was the customary format for the transmission of countless prosaic facts, skills, and traditions, from the rules of grammar or of morals to the articles of religious faith and the adventures of folk heroes. Poetry, not prose, was the primitive form for storing the community's memory. When Greek writers of the Ionian Enlightenment in the sixth century B.C. began writing about the past in prose, they were noted for their innovation. They came to be known as "logographers," or writers in prose. Figures of transition between the epic poets and the critical historians, they still recounted the careers of gods and heroes and legendary city founders. They were only protohistorians, but they took the first bold steps to free the community's recorded past from the tradition-bound framework of rhythm, verse, and song. And the historical spirit was born in the prosaicizing of experience.

All the Muses were originally goddesses of song, and only later did they become identified with different kinds of poetry, arts, and sciences. Homer, unclear about the number of Muses, sometimes spoke of one, at other times spoke of many. He described them, accompanied by Apollo's lyre, singing to the gods banqueting on Olympus. Hesiod, who wrote after Homer, did distinguish nine Muses, but their specific roles came later. "We know how to tell falsehoods like the truth," the Muses warned Hesiod, "and we know, when we choose, how to speak the truth." Centuries later, Aristotle still gave poetry a higher dignity than history. "It really lies in this: the one describes what has happened, the other what might. Hence poetry is something more philosophic and more serious than history; for poetry speaks of what is universal, history of what is particular." To prefer the particular to the universal, the sobering fact to the exhilarating myth, required both courage and self-denial.

A burgeoning sense of history tempted bold writers very early to bring the gods down to earth. A Sicilian Greek, Euhemerus (c. 300 B.C.), wrote a *Sacred History* from old inscriptions that he claimed had been written by Zeus himself on a golden temple pillar on the island of Panchaea in the Indian Ocean. Euhemerus suggested that the gods were originally real people, heroes or conquerors who were later deified. He claimed that Zeus and his family, for example, were an ancient family of kings of Crete, and he purported to document the whole primitive history of the world since Uranus from those inscriptions. This doctrine, known as Euhemerism, would be used by early Christians to prove that all pagan mythology was a purely human invention, like the Roman apotheosis of emperors.

Elements of the modern historical spirit shone brilliantly in two distinctly Greek literary works. Herodotus and Thucydides, both living in the fifth century B.C., would become the fathers or, more precisely, the godfathers of modern historians.

Herodotus (c. 480–c. 425 B.C.) writing, of course, in prose, was in the new tradition of the logographers. Born in Halicarnassus, an Ionian town on the southwestern coast of Asia Minor which had been ruled by the Lydians and then by the Persians, he had the advantage of coming from the periphery of Greek culture. Far from the settled centers of Athens or Sparta, he was in daily touch with non-Greek peoples. On the Peloponnesus, Greeks might look upon the ways of "barbarians" (i.e., foreigners) with amusement or contempt, but Herodotus, born under barbarian rule, hoped to learn from them.

While the Greeks had an ample store of myths to explain the origins of their own ways, they had no such myths for the Lydians and the Persians. Herodotus planned a survey of the geography and ways of life of non-Greek peoples. Traveling through Asia Minor, the Aegean Islands, Egypt, Syria and Phoenicia, Thrace, Scythia, and eastward all the way to Babylon, he focused on the urban centers. By 445 B.C., when he was in Athens and had become a friend of Pericles and Sophocles, he decided to reshape his ethnographic survey into a history of the Persian Wars (500–449 B.C.), and so revisited the battle sites and routes of the armies. With no written contemporary accounts, no general's memoirs, no war office documents, he had to piece the story together from oral tradition, travel, and observation.

He observed dispassionately the variety of local customs, noting that men naturally preferred the customs into which they had been born. When Darius asked his Greek subjects what he would have to pay them to eat the bodies of their fathers instead of burning them on funeral pyres, no sum could tempt them. He then sent for some Indians, who customarily ate the bodies of their deceased fathers, and asked what would induce them to burn those bodies. But not for any price would they tolerate such sacrilege. Everywhere, Herodotus said, custom is king.

He was freer to speculate about beginnings and sharper in his critical sense than Christian historians would be for the next two millennia. Not boxed in by any hard dogma of the Creation, he could extend historical time back indefinitely. "Nothing is impossible," he ventured, "in the long lapse of ages." If the Nile should reverse its course and flow into the Red Sea, "what is there to hinder it from being filled up by the stream within, at the utmost, 20,000 years?"

In the next generation, Thucydides (c. 460–c. 400 B.C.), in his *History of the Peloponnesian War,* narrowed the focus to political history. We know almost nothing about his life, except that his father had a Thracian name, and that he owned a gold-mining concession in Thrace and was exiled from Athens for twenty years. He, too, had some of the advantages of the out-sider. A young man in 431 B.C., when Athens' next decisive war began, he determined to record its history and stayed with this task for twenty-seven years. At the outset he offered a credo for all later historians:

> He must not be misled by the exaggerated fancies of the poets, or by the tales of chroniclers who seek to please the ear rather than to speak the truth. . . . most of the facts in the lapse of ages have passed into the region of romance. At such a distance of time he must make up his mind to be satisfied with conclusions resting upon the clearest evidence which can be had. . . . Of the events of the war I have not ventured to speak from any chance information, nor according to any notion of my own; I have described nothing but what I either saw myself, or learned from others of whom I made the most careful and particular enquiry. The task was a laborious one, because eye-witnesses of the same occurrences gave different accounts of them, as they remembered or were interested in the actions of one side or the other. And very likely the strictly historical character of my narrative may be disappointing to the ear. But if he who desires to have before his eyes a true picture of the events which have happened, and of the like events which may be expected to happen hereafter in the order of human things, shall pronounce what I have written to be useful, then I shall be satisfied. My history is an everlasting possession, not a prize composition which is heard and forgotten.

Although he outlived the war, his work was never finished. When his book was published posthumously, numerous others tried to complete it.

Herodotus and Thucydides were not followed by other Greek historians of their stature. Historical inquiry in the modern sense, the search for the way it really was, simply to amplify knowledge of the past, did not have wide appeal to the Greeks in their great age. The Ionian Enlightenment, unlike the European Enlightenment of the eighteenth century, was not fertile in works of history, though it did produce a rich imaginative litera-ture and epoch-making writings in biology, mathematics, astronomy, and medicine. The explanation lies partly in the phenomenal Greek genius for poetry, epic, and tragedy, which seemed to sate their emotional needs, and partly in the hypnotic charm for them of philosophic universals like Plato's dazzling ideas. As we have seen, even Aristotle, with his love of specifics, would not dignify history, precisely because it told no more than "what Alcibiades did or had done to him."

70

Christianity Gives Direction

THE great religions of the East, Hinduism and Buddhism, which stretched human vistas into vast and endless cycles far beyond the seasons and the years of an individual life or a generation, brought a refuge from these cycles by helping the individual merge into the All. The Hindu promise was *samsara* (Sanskrit for "migration"), escape from the endless round, not by "life everlasting" but by dissolving the individual into an unchanging anonymous Absolute. Buddhism, too, offered its escape from the "weary reiteration" of life toward *nirvana* (Sanskrit for "blowing out"), the merging of the self into the Universe.

The great religions of the West, also seeking to escape from the animal world of Again-and-Again, found an opposite path. While Hindus and Buddhists sought ways *out of* history, Christianity and Islam sought ways *into* history. Instead of promising escape from experience, these sought meaning in experience. Christianity and Islam were both rooted in Judaism, and all three revealed a dramatic shift from a world of cycles to a world of history.

The Greek gods, timeless on Olympus, had not exhorted people to remember their past. But Judaism was oriented to the past, a historical religion in a sense quite alien to the Hindu, the Buddhist, or the Confucian. "Blessed is the nation," sang the Psalmist, "whose God is the Lord, and the people whom he hath chosen for his inheritance." God's purpose for the Jews was disclosed in the past recorded in Sacred Scripture. By recalling the favors and the tribulations that God had visited on them, Jews discovered and remembered their mission as a chosen people. For Jews, remembering the past was the way to remember their God. Scripture told the story of the world from Creation, and Jewish holidays were celebrations or re-enactments of the past. The Sabbath every week was a reminder of the six days of Creation and God's gift of the seventh day of rest. The Jewish Passover celebrated the coming out from Egypt, marked annually by the Haggadah, the telling of the story. While the Foolish Son of the Passover liturgy saw the Haggadah as a story of what happened to "them," the Wise Son realized that he himself was among those whom the Lord brought out. In this sense, Judaism was emphatically past-oriented but also antihistorical. The Scriptures were read to reinforce what the Jews already knew.

The Jews began and still begin their calendar from the traditional date of the Creation. The historic mission of Israel as the chosen people was established by a particular event, God's Covenant with Abraham. On His side, God agreed to be the God for Abraham and all his descendants, and promised them the land of Canaan, while the people of Israel agreed to worship Him alone and obey His commandments. The Pentateuch, the first five books of the Old Testament, chronicles the making of this historic Covenant and its fulfillment in the delivery of the Laws of Moses on Mt. Sinai. Christian theologians called that the Old Covenant, because they believed that Jesus came to set up a new and better covenant between God and all humankind. This explains "Old Testament" and "New Testament" to describe the two parts of the Bible, for "testament" derives from a Latin mistranslation of a word for "covenant" in the Greek translation of the Hebrew Scriptures.

Both the Creation and the Covenant were more traditional than historical. Although the God of Israel was a universal God, still the religion of Israel, the chosen people, remained tribal. Its laws and customs were substantially confined to the people who were supposed to have a common descent.

Christianity was a historical religion in a new sense. Its essence and its meaning came from a unique event, the birth and life of Jesus. Firmly rooted in the Jewish tradition, Jesus (a Greek version of the Hebrew name Joshua, meaning Savior) was circumcised and confirmed according to Jewish custom, and preached and taught as a wandering rabbi. The basic Scripture of Christianity—the Gospels of Matthew, Mark, Luke, and John—offers chronological biographies of Jesus, with accounts of the life, death, and resurrection.

The very name "Gospel" (from the Old English "godspell," meaning good news, from late Latin *evangelium*) proclaims that this religion is firmly rooted in history, on an unprecedented happening of world significance. The coming of Jesus was the first and greatest news. The Christian calendar therefore commemorates events in the birth and life of Jesus—from the Annunciation (March 25), through Christmas (December 25), Circumcision (January 1), Epiphany (January 6, Twelfth Night; commemorating the baptism of Jesus, the visit of the Wise Men to Bethlehem, and the miracle at Cana), Candlemas (February 2, commemorating the Purification of the Blessed Virgin and the Presentation of Jesus in the Temple), and the Transfiguration (August 6). The Easter holidays similarly commemorate events surrounding the Resurrection. The Christian believer in the uniqueness of these events, naturally enough, counts the years from *Anno Domini*, the coming of his Savior.

The promise of Jesus Christ, the Christian way out of the cycles, was not an escape into some Universal. Rather it was the extension of the uniqueness of the person forever and ever. The Gospels repeatedly promised "that whosoever believeth in him should not perish, but have everlasting life." The Christian ideal was not to escape rebirth but to be reborn and so live on forever in a heavenly afterlife. "Except a man be born again, he cannot see the kingdom of God."

The Christian discovery of history, rooted in the Gospels, was a product of revelation and reason, of crisis and catastrophe. On the night of August 24, 410, as Edward Gibbon recounts, the Goths "directed by the bold and artful genius of Alaric" entered Rome. "At the hour of midnight, the Salarian gate was silently opened, and the inhabitants were awakened by the tremendous sound of the Gothic trumpet. Eleven hundred and sixty-three years after the foundation of Rome, the Imperial city, which had subdued and civilized so considerable a part of mankind, was delivered to the licentious fury of the tribes of Germany and Scythia." Gibbon portrays the sack of Rome for us in some of his most vivid, most persuasive—and more prurient—passages.

The Christian bishop of Hippo, a North African Roman outpost, at that moment happened to be the prodigious Aurelius Augustinus (354–430), an energetic and prolific writer, known to history as Saint Augustine. He would wield a greater influence on Christian thought than any other man between Saint Paul and Luther.

Augustine left a vivid account of his early life in his *Confessions,* which for the twentieth-century psychologist William James was still the classic biography of the converting experience. Augustine's mother had brought him up as a Christian, but when he was sent to be educated at nearby Carthage at the age of sixteen, he abandoned the faith and turned to the study of rhetoric. He dabbled in astrology, which tempted him by its easy prophecies. Before Augustine was twenty he had taken a concubine who had borne him a son. Attracted to the imperial capital, he took them both with him to Rome, hoping to find employment teaching rhetoric. Failing there, he accepted an invitation to lecture at Milan, where he came under the influence of the eloquent Bishop Ambrose. Now the process of his conversion began. Just as the Buddha seeking enlightenment had abandoned his wife and son, so now Augustine dismissed his mistress, mother of his son, who left him reluctantly and in deep grief. After her departure Augustine again was persuaded that continence was impossible for him. He took another concubine, at the same time praying to God, "Give me chastity and continency—only not yet."

Then, one day in a garden in Milan when he was expounding to his pupil Alypius the struggle within him, he was overwhelmed:

> So was I speaking, and weeping in the most bitter contrition of my heart, when, lo! I heard from a neighboring house a voice, as of boy or girl, I know not, chanting, and oft repeating, "Take up and read; Take up and read." Instantly, my countenance altered, I began to think most intently, whether children were wont in any kind of play to sing such words: nor could I remember ever to have heard the like. So checking the torrent of my tears, I arose; interpreting it to be no other than a command from God, to open the book, and read the first chapter I should find. . . . I seized, opened, and in silence read that section [of the Gospel of St. Paul], on which my eyes first fell: Not in rioting and drunkenness, not in chambering and wantonness, not in strife and envying: but put ye on the Lord Jesus Christ, and make not provision for the flesh, in concupiscence. No further would I read: nor needed I: for instantly at the end of this sentence, by a light as it were of serenity infused into my heart, all the darkness of doubt vanished away.

He retired to a monastery, and after his baptism by Bishop Ambrose himself in 387, he returned to Africa. There he became the champion of Christian orthodoxy. In a hundred books, and with letters and sermons, he attacked the leading heretics of his day—Manichaeans, Donatists, Pelagians, Arians. In 395, when only forty, he was consecrated Bishop of Hippo, and stayed there for the rest of his life, for a Church rule forbade the transfer of bishops.

When word of the sack of Rome reached Augustine, he was fully prepared by genius and experience to explain the meaning of Christianity for history and the meaning of history for Christianity. Although he knew Greek only imperfectly, he was a master of Latin. According to Christian theologians, "even if he be not the greatest of Latin writers, he is assuredly the greatest man that ever wrote Latin." Augustine took his cue from the cataclysmic events in Rome on the night of August 24, 410. The Church needed his defense. Many were blaming the Fall of Rome on the rise of Christianity. The religion of Jesus Christ, so recently embraced by Constantine and his followers, was said to be the cancer of the Roman Empire. Would the Eternal City ever have fallen if the Empire had not been "weakened" by Christianity? What did all this foretell for mankind?

In *The City of God* Augustine undertook to answer these questions. He began writing his book soon after he learned of the Fall of Rome and worked at it over the next fifteen years. His foil was the cyclical theory of Plato's *Republic,* according to which the world would last for only 72,000 solar years. The first 36,000 years of the world cycle were a Golden Age, but the second 36,000 years, when the Creator had loosened his control over the world, were an era of disorder, ending in chaos. Then the Deity would intervene and renew the cycle. Augustine's republic, by contrast, existed not in speculation but in history, and his point of departure was the historic events of his time.

Augustine began his massive *City of God* with those very circumstances of the fall of Rome which later engaged Gibbon "to justify the ways of Providence in the destruction of the Roman greatness." And Augustine was impressed by the moderation of the invading barbarians. Never before had the sanctuaries of a conquered people been left unmolested by the conqueror. Since the Romans themselves never spared the temples of those they conquered, history proved that pagan gods could not protect their worshippers. When the Greeks conquered Troy (he quotes Virgil's *Aeneid,* II, 761–67), they used the Temple of Juno for storing the city's treasure until it could be distributed as booty, and for imprisoning the conquered Trojans until they could be sold into slavery. How powerless was that pagan goddess! Even though Juno was "not any vulgar god, of the common sort, but Jupiter's sister, and queen of all the other gods," she still could provide no refuge. How different was the experience of Rome, where the churches were the consecrated memorials of the Apostles of Jesus Christ! "There [in Troy], was freedom lost, here [in Rome], saved; there was bondage shut in, here it was shut out: thither were men brought by their proud foes, for to undergo slavery; hither were men brought by their pitiful foes to be secured from slavery."

This one episode, Alaric's sparing of the Roman churches, set the stage for Augustine's grand Christian interpretation of the past, which gave all history a direction. He collected other examples to destroy any suspicion that the Romans' recent abandonment of their ancient pagan gods was responsible for the barbarian invasions and the sack of Rome. If pagan gods could not provide security in this world, how could they provide bliss in the next? Eternal life surely could not be their gift. Having disposed of the relics of pagan faith, Augustine makes his grand distinction between the two "cities." The City of God, the universal community of the righteous, including God and His angels and all the saints in heaven along with all the righteous on earth, "summoneth its citizens from all tribes, and collecteth its pilgrim fellowship among all languages, taking no heed of what is diverse in manners or laws or institutions." The other city, the familiar city of this world, included all who inhabit the earth and everything that occurs here.

The remainder of his work depicted Augustine's vision of "the beginnings and ends" of the two cities, "the two contrary courses taken by the human race from the beginning, of the sons of flesh and the sons of promise," and finally described "their appointed ends." The end of history was the perfection, glorification, and fulfillment of the City of God, which was not of this world. This notion would dominate Christian thinking about history for the whole Middle Ages.

Augustine did not yet expound a doctrine of progress in our modern sense. He left no place for novelty, for the unexpected good. Still he foreshadowed an idea of progress, of hope for a better life on earth. The Roman Empire, he said, had brought the world together so that Jesus Christ might be born into it, and to provide the Church its opportunity for universal dominion.

To Augustine, a cyclical theory of history was inconceivable, even abhorrent, for it would deny the uniqueness of Jesus Christ and the promise of his Gospel. In his *Confessions* he records his personal struggle against "the lying divinations and impious dotages of the astrologers," who taught that the pattern of events was repetitive, determined by the returning cycles of celestial arrangement. Some of the most eloquent passages of *The City of God* attacked the pagan theory of cycles *(circuitus temporum)*—"those argumentations whereby the infidel seeks to undermine our simple faith, dragging us from the straight road and compelling us to walk with him on the wheel."

He warns us not to misinterpret the wisdom of King Solomon in Ecclesiastes: "The thing that hath been, it is that which shall be; and that which is done is that which shall be done: and there is no new thing under the sun."

> Far be it from the true faith that by these words of Solomon we should believe are meant those cycles by which they [the pagan philosophers] suppose that the same revolutions of times and of temporal things are repeated so that, as one might say, just as in this age the philosopher Plato sat in the city of Athens and in the school called Academy teaching his pupils, so also through countless ages of the past at intervals both the same Plato and the same city and the same school and the same pupils have been repeated, as they are destined to be repeated through countless ages of the future. God forbid, I say, that we should swallow such nonsense! Christ died, once and for all, for our sins.

The other Church Fathers, too, read the prophecies of the Old Testament not as a vision of cycles but as a forecast of the uniqueness of Jesus Christ. The prophecy in Genesis of the coming of a leader who "shall be the expectation of the nations" could refer only to Jesus. "He was obviously the only one," Origen wrote in Alexandria some two centuries before Augustine, "among all his predecessors and . . . among posterity as well who was the expectation of the nations." Jesus Christ had taken mankind off the "wheel." The "finality of Jesus," elaborated by Augustine into a theory of history, would govern Christian thought in Europe for the next thousand years.

While Christianity would be justified in history, its truths could not grow but were simply fulfilled. To the Jewish view of the past the Christians added their own sacred texts. The New Testament, they said, fulfilled the prophecies of the Old. Both Scriptures together were the one God's revelations not merely for a chosen people but for all mankind. While the Gospels were good news for everyone, they were not history in the Greek sense of inquiry, but were verifications of faith. They were both the end and the beginning. The Christian test was a willingness to believe in the one Jesus Christ and His Message of salvation. What was demanded was not criticism but credulity. The Church Fathers observed that in the realm of thought only heresy had a history.

When the literate leaders of Christianity made their record, they were not interested in inquiry. They had no need to seek answers, they had only to document them. During the Christian centuries in Europe these best minds in the Church developed their own techniques for using the past. Origen (185?–254), the precocious Alexandrian Greek, at the age of eighteen became head of the leading Christian theological academy there, and reputedly wrote some eight hundred works. Because he had castrated himself to ensure his purity, he could not be ordained as a priest, but his teachings made him the most influential theologian before Saint Augustine. A genius at allegory, he even managed to discern the outlines of Christianity in the writings of the Greeks, and so gave Christianity the aura of antiquity without endangering faith by historical skepticism. "If the law of Moses had contained nothing which was to be understood as having a sacred meaning," Origen observed, "the prophet would not have said in his prayer to God: 'Open thou mine eyes and I will behold wondrous things out of thy law.' "

When the Mediterranean world began to take up this new religion and the events of Jesus' life receded into the past, it was necessary not only to foresee Jesus in the Scriptures of the Jews but to place all the events of the Bible and the acts of early Christians in the context of the world. This was accomplished by Origen's brilliant successor, Eusebius of Caesarea (A.D. c. 260–340), who sat at the right hand of Emperor Constantine and delivered the opening eulogy of the Emperor at the Council of Nicaea (A.D. 325). For the first time his chronology arranged and encompassed the events of the Chaldean, Greek, and Roman past into the framework of the Bible.

Eusebius provincialized the history of the whole world into a history of Christianity. His calendar of world events managed to incorporate and yet excommunicate the whole non-Christian past:

> Other writers of history record the victories of war and trophies won from enemies, the skill of generals, and the manly bravery of soldiers, defiled with blood and with innumerable slaughters for the sake of children and country and other possessions. But our narrative of the government of God will record in ineffaceable letters the most peaceful wars waged in behalf of the peace of the soul, and will tell of men doing brave deeds for truth rather than country, and for piety rather than dearest friends. It will hand down to imperishable remembrance the discipline and the much-tried fortitude of the athletes of religion, the trophies won from demons, the victories over invisible enemies, and the crowns placed upon all their heads.

The next centuries produced some great works of theology, a copious hagiography of "the athletes of religion," but for a thousand years the spirit of historical inquiry foreshadowed in Herodotus and Thucydides would lie dormant. Christian scholars would share Eusebius' faith in "the incontrovertible words of the Master to his disciples: 'It is not for you to know the times or the seasons, which the Father hath put in his own power.' "

The Christian vision of the past enveloped ancient documents in a haze of allegory and recent actors in an aureole of holiness. History became a footnote to orthodoxy. During the next ten centuries in Europe there were scattered experiments in seeking Christian uses of the past, but they created no tradition of historical inquiry. Saint Augustine did use data from the past to document his *City of God*. The *Seven Books of Histories against the Pagans,* by his disciple Orosius (fifth century A.D.), demonstrated that the evils of post-Christian times could not be blamed on the religion of Christ, for earlier times had suffered even worse calamities. Among the occasional escapees from the prison of Christian orthodoxy were some, such as the Englishman the Venerable Bede (673–735), who oblige us by incorporating at length the documents of their time. Meanwhile the compilers of annals like the Anglo-Saxon Chronicle noted the acts of kings and the careers of churches and monasteries. But the raw materials of history were not history.

The early efforts to create a "national" past for rudimentary nations sometimes followed the pattern of the *Aeneid*. Like Virgil's founders of Rome, other founders elsewhere had been divinely guided. Geoffrey of Monmouth's *History of the Kings of Britain* (c. 1150) traced the descent of British kings back to the Trojans.

Glimmers of modern historical inquiry were found in a few works written by men of affairs. Einhard, companion and adviser of Charlemagne, drew a lively portrait of his hero. Otto, Bishop of Freising, grandson of the Holy Roman emperor Henry IV, was close enough to the throne to give us an intimate glimpse of his nephew Frederick Barbarossa. Still, the full and free exploration of the past was impossible as long as the written record remained above criticism. Sacred "authors" had become "authorities," and medieval chroniclers preferred "authorities" to experience. When the Venerable Bede wrote about Hadrian's Wall, which still stood within sight of where he lived and which he passed every day, instead of describing what he saw he preferred to quote a Roman writer.

71

Revising the Record

CHRISTIAN "history," like Christian "geography," authenticated Church and State. Rulers were quite content to be legitimized by descent from the Trojans or the gods. Revising the ancient views of the earth promised rich rewards. European sovereigns were willing, or even eager, to finance Columbus, Gama, Magellan, or Cabot, to stake out claims on land and sea. But by recharting the past they could only lose. There they preferred to leave well enough alone. Why substitute uncertain facts for authorized legend?

The past, which lay everywhere and nowhere, still was a treacherous no-man's-land with no papal lines to separate competing empires. The prudent sovereign was glad to see his lineage safely shrouded in myth. What might not be uncovered by fearless scholars? The printing press increased the perils. It is not surprising that Cosimo de' Medici exercised a special censorship (1537–74) over historical writing, or that Queen Elizabeth made trouble (1599) for the author who too freely described the dethroning of her predecessor Richard II. A kaleidoscopic past might conjure up visions of a changeful future.

Renaissance Italy would be Europe's first headquarters for exploring the past. What Portugal was for adventurers in geography, Italy was for history. And Florence was the Sagres. While Saint Augustine had drawn the lines of the Christian future, the Italian poet-humanist Francesco Petrarch (1304–1374), in the advance guard of the Renaissance, was a pioneer explorer of the past. He was inspired to a modern sense of history among the same spectacular ruins of the Roman Empire that would inspire Gibbon four centuries later. During the whole Middle Ages those monumental relics had excited very little curiosity in residents, scholars, or travelers. Few questions were asked about who built them or how they were built, and how those ancient people lived. They were simply *Marvels of the City of Rome,* as an anonymous writer of the mid-twelfth century described them, notable only as sites of pagan myth and sacred legend. "There is an arch at St. Mark's that is called the Hand of Flesh," the *Marvels* reported, "for at the time when in this city of Rome Lucy, a holy matron, was tormented for the faith of Christ by the emperor Diocletian, he commanded that she should be laid down and beaten to death; and behold, he that smote her was made stone, but his hand remained flesh, unto the seventh day; wherefore the name of the place is called Hand of Flesh to this day." Ancient buildings themselves were a kind of sacred Scripture.

When Petrarch first visited Rome in 1337, he spent many pleasant hours wandering through the ruins with a Franciscan Brother Colonna for his guide. For him these ruins became clues to strange ways of life, which he reconstructed in a long letter to Colonna, and in a poem about Scipio Africanus describing Rome in its days of glory. The inscriptions on stones became explicit messages from the past, and he scrutinized old manuscripts for other clues. In 1345 in Verona he discovered numerous letters from Cicero to his fellow politicians which transformed the wooden classroom figure into an energetic public man whose observations were the by-product of Roman life. He used coins as a historical source that helped him understand a puzzling passage in Suetonius. Whenever a hoard was newly unearthed, it would be brought to Petrarch for interpretation. He considered his extensive collection of Roman coins a portrait gallery of Roman emperors, and he generously offered a selection to Emperor Charles IV to show him the faces of the Roman Caesars whom he must imitate.

When the Holy Roman emperor Charles IV (1316–1378) was troubled by an old document that purported to exempt "Austria" from his domain, he turned to Petrarch, who proved the purported grant to be a forgery. "I do not know who wrote it," Petrarch concluded in 1355, "but have no doubt that he was not a learned man but a schoolboy, an ignorant writer, a man with the desire to lie but without the skill to do it properly—otherwise he would not have made such stupid mistakes." Petrarch noted that in the counterfeit document Caesar spoke of himself as "we" (although in fact he always referred to himself in the singular), he called himself "Augustus" (although this term was first used only by his successors), and he dated the document "Friday in the first year of our reign" (without any reference to month or day).

Forgery was a prosperous medieval art. Battling feudal lords and upstart kings, eager to substantiate the antiquity of custom, were anxious for the tangible authority of a document. The increasing use of written records increased the need for "authentic" grants, and the Roman-law crime of forgery was limited to cases of property or inheritance. Forgery of documents to support an acknowledged authority was generally considered an act of piety or patriotism. Before falsifying historical documents could have the opprobrium of forgery, it was necessary to believe that the historical past was not a flimsy fabric of myth and legend but had a solid definable reality. The courage to discredit the fictitious past would be a symptom of rising historical consciousness.

The pioneer of modern historical criticism was a peevish character, more fertile in posing embarrassing questions than in supplying comforting answers. Lorenzo Valla (1407–1457), *enfant terrible* of the learned world, was an apostle of Truth in History. He might be better described as a Paracelsus of literary learning, a professional layman, the enemy of pedantry, apriorism, and quiddities. Born in Rome, the son of a lawyer at the papal court, before he was thirty he had outraged the scholarly world. He attacked Stoicism, defended Epicurus, and ridiculed the barbarous Latin used by Bartolus (1314–1357), the revered authority on Roman law. Driven from the University of Pavia, he found temporary refuge in Milan, then in Genoa, before settling down in the south as royal secretary and historian to King Alfonso of Aragon, who was then engaged in staking his claim to the kingdom of Naples.

The political needs of King Alfonso provided the occasion for Valla's most celebrated feat of historical criticism. Against King Alfonso, Pope Eugenius IV was claiming secular authority over all of Italy. The Pope's claim was based on the so-called Donation of Constantine, contained in an old document in which Emperor Constantine the Great (280?–337) purported to grant to Pope Sylvester I (314–335) and his successors temporal dominion over Rome and the whole Western Empire. This was said to be Sylvester's reward for his miraculous cure of Constantine's leprosy, and for converting Constantine to Christianity. During the Middle Ages this unchallenged document had become the most potent weapon in the arsenal of successive popes against kings and emperors. Valla now had an opportunity similar to that seized by Petrarch a century before, to serve both his patron and the cause of history. The occasion was made to order for one of his iconoclastic temperament. His *Treatise on the Donation of Constantine* in 1440 so conclusively proved the document to be a forgery that few champions of papal supremacy ever dared appeal to it thereafter. Drawing on his profound knowledge of changing Latin usage, Valla demonstrated that the document could not possibly be genuine. The forger was so ignorant that he did not know that in Constantine's time a "diadem" was not a gold crown but a coarse cloth, that "tiara" was a word not yet in use. In every line he found flagrant anachronisms—"purple," "scepter," "standard," "banners," and even the word employed for "or"—along with words borrowed from the Hebrew which Constantine's secretaries never could have known.

This was only one of Valla's attacks on citadels of orthodoxy. He disputed the Stoic philosopher Boethius and he revised the interpretations of Scholastic philosophers by simplifying Aristotle's nine categories into three. He criticized the style of Cicero and demonstrated that *Ad Herrenium,* the famous manual of rhetoric and memory commonly attributed to Cicero, was not by him at all. Characteristically, Valla went on to insist that the "Apostles' Creed" could not have been composed by the Twelve Apostles. After the Inquisition convicted him of heresy on eight counts, among them his attempt to revise Aristotle, he would have been burned at the stake if his patron King Alfonso had not come to his rescue.

Valla took the supreme risk by applying his new techniques of critical history to the Bible itself. He criticized Saint Jerome's fourth-century Vulgate, the Latin translation of the Bible that had served as the authorized version during the Middle Ages. And at the insistence of Cardinal Bessarion, who had given his great library to Venice, he even provided *Annotations on the New Testament* dangerous enough to be put on the Church's Index of prohibited books.

Valla's attack on the documentary foundations of papal supremacy still did not prevent the new pope, Nicholas V (1397?–1455), who was a generous patron of arts and letters, from appointing him a papal secretary. The Pope supported his historical writings and commissioned him to translate Herodotus and Thucydides. Valla delivered his final blast in 1457, when he was asked to offer the anniversary eulogy of Saint Thomas Aquinas to the Dominicans at Santa Maria sopra Minerva in Rome. Before an astonished audience he attacked Saint Thomas for logic-chopping and for his "corrupt" style, and pleaded for the simple theology of the Church Fathers.

Valla's personal life was as unorthodox as his scholarship. Although he held numerous ecclesiastical posts, it is doubtful that he was ever ordained a priest. There is no record of his marriage, but he did have a Roman mistress by whom he fathered three children. It was not surprising that so unruly a character should open the Pandora's box of history. The printing press made his writings a time bomb. When finally printed, his *Annotations on the New Testament* (1505) and his attack on the Donation of Constantine (1517, the year when Luther posted his 95 Theses on the church door at

Wittenberg) carried his explosive messages to a wide audience. Erasmus and others following his example brought history into the arsenal of the Reformation.

The critical spirit also served all Christians in their battle against Islam. John of Segovia's translation of the Koran separated the earliest text from later Western additions. *The Sieving of the Koran* (1460), by the versatile Nicholas of Cusa (1401–1464), analyzed the different historical elements that had gone into the sacred book, to show that the surviving text was a product not of divine inspiration but of human events.

The Thirty Years War (1618–48) spawned countless controversies all over Europe between Catholic and Protestant princes claiming jurisdictions based on ancient documents. In France the nobles, building legal cases for their local powers against the threats of an absolute king, engaged in what came to be called the Diplomatic Wars. The science of "diplomatics," essential for modern historical sleuthing, developed in response to these needs. "Diplomatics" has very little to do with diplomacy, but derives from the Greek word *diploma* ("doubled" or "folded") used to describe documents that were usually folded. In ancient Rome such important documents, when engraved on a bronze diptych, were folded shut and sealed, not only for convenience in storage, but to keep the contents secret. The term "diploma" was not much used in the Middle Ages, but Renaissance writers used it generally for an old document, especially one establishing property rights or political authority. Not until the eighteenth century was it used in English to mean an academic certificate.

In 1607 an enterprising Dutch Jesuit, Heribert Rosweyde of Utrecht (1569–1629), drew up an ambitious plan to collect and publish the authentic narratives of the lives of the Christian saints. The latest techniques of philology and textual criticism would be used to separate truth from legend, to purify religious tradition and make hagiography a science. The Jesuit Fathers, who took their assignment seriously, horrified the devout in England when their researches dissolved the legend of Saint George. One of these pioneers of critical history, the energetic Daniel Papebroech (1628–1714), who produced eighteen of these volumes, devised the rules for detecting forged documents, which he then applied to the Benedictine charters to prove that they were spurious. Papebroech asserted that there were no genuine charters surviving from any time before 700. If the Benedictine claims to the French monasteries of St. Denis and Corbie were proven invalid, these properties would probably belong to the Jesuits.

The brilliant Jean Mabillon (1632–1707), who had lately joined the Benedictines, was providentially qualified to defend his Order and at the same time elaborate the techniques of modern textual criticism. Born into a peasant family in Champagne, he received the tonsure at the age of nineteen and spent his early years traveling from monastery to monastery. At St. Rémy he visited the church where kings of France were consecrated for centuries, and as he walked through the cemeteries where the earliest Christians of Gaul were buried he became interested in the testimony of tombstones. During one of his investigations he was so diligent that he was reported to be guilty of "unpaving almost the entire church."

In reply to Papebroech, Mabillon wrote his *De Re Diplomatica* ("On the Study of Medieval Charters," 1681, 1704), which developed diplomatics into a subtle and comprehensive technique for authenticating old documents. The Jesuit Papebroech had cast suspicion on the Merovingian documents because of their peculiar lettering. In reply, Mabillon for the Benedictines explained that over the centuries scripts had changed as much as the events they recorded. He illustrated Latin script from the capital letters of ancient Rome down to the handwriting of the seventeenth century. Surveying the whole range of clues, he opened the "auxiliary" sciences of handwriting (paleography), writing materials, seals (sphragistics), dates (chronology), and vocabulary (philology). Along with his principles for sifting historical evidence, he insisted quite sensibly that the genuineness of a document depended on the consistency of all the clues. Papebroech himself finally had to admit that Mabillon's principles were correct, and Mabillon's book became a classic for future historians. When Mabillon focused his criticism on the legends of popular saints, he risked prosecution. Near the end of his life, after he had undermined the genuineness of the supposed remains of saints in Rome, he had to be defended by Pope Clement XI himself. Even when threatened with the stigma of the Index, he refused to give in with a substantial retraction, which would make him for Lord Acton "as an historian, eminently solid and trustworthy, as a critic the first in the world."

Modern history grew not only in the negative mood but also in positive enthusiasm. The rising glory of the Italian cities and a flourishing vernacular literature in Italian would provide secular subjects for epic narrative. The first modern national anthems were being written as history.

Works about Florence and Italy opened a new chapter in the history of history. The burgeoning city-states engaged official historians to dramatize their struggle for greatness, to celebrate their men and women of *virtù,* and to point the direction of the future. Leonardo Bruni (1368–1444) provided a *History of the Florentine People* (Venice, 1476), which disentangled the story of the city from its legendary past. The greatness of Florence, he said, came from the Florentine Republic and its spirit of liberty. Rome, too, had flourished as a republic and "the Roman imperium began

to go to ruin when first the name of Caesar fell like a disaster upon the
city." The Roman Empire in the West had really ended with the barbarian invasions, for Charlemagne's empire was spurious. And the fortunes
of the Italian cities rose again with their emergence as free republics.
Flavio Biondo of Forlì (1392–1463), while celebrating Florence and Italy,
provided a scheme that would dominate and tyrannize European historical thought for centuries to come. Separating the grandeur of antiquity
from the promise of modern Italy, he made the thousand years after the
seizure of Rome by Alaric into a single "middle" epoch. Sometimes called
the first medieval historian, Flavio Biondo might better be called the first
self-consciously modern historian. For he seems to have been the inventor
of the tripartite framework: ancient, medieval, and modern. Although he
never himself actually used the phrase "Middle Ages" *(medium aevum),*
it was he who gave a new historical coherence to the millenium after the
fall of Rome. Western thought would never recover from his way of slic-
ing the whole European past into a period of ancient glory and a period
of modern rebirth, with a middle era of disintegration and decline in between. European historians preserved these imprisoning categories, which
would even be exported to Asian historians, who recklessly referred to a
"medieval period" in India or in China.

One of these Florentine works of proto-modern history that continues to
entertain and move us is *The History of Florence and the Affairs of Italy,*
by Niccolò Machiavelli (1469–1527). Commissioned by Cardinal Giulio de'
Medici in 1520 to write a history of Florence, Machiavelli drew raw materials from Bruni, Flavio Biondo, and others for the work he delivered in 1525
to his sponsor, who had become Pope Clement VII. *The Prince,* which
Machiavelli had begun writing some years earlier, was a cameo of contemporary history cast in an unfamiliar mold. Machiavelli's *History,* following
Bruni, gave half his work to the period from the barbarian invasions to the
accession of the Medici in 1434. The rest recounted the intrigues and battles
of Florentine Italy until the death of Lorenzo the Magnificent in 1492.
Following Thucydides and Livy, Machiavelli wrote the speeches he thought
appropriate to the characters and the occasions he recounted, and the rise
and decline of the Florentine Republic, the corruption and ruthlessness of
the Borgias, became a political tragedy that climaxed in his own day. His
combination of the critical spirit and the epic form foreshadowed the great
works of modern history.

72

Explorers among the Ruins

A profitable but uncelebrated by-product of the grandeur of ancient Rome was the medieval trade in building materials. For at least ten centuries Roman marble cutters made a business of excavating ruins, dismantling ancient buildings, and digging up old pavements to find models for their own work and materials for new construction. About 1150 a group of them, the School of Cosmati, even created a new mosaic style from the fragments. Marble cutters in their own way continued the more violent and more notorious sack of Rome committed by the Goths in 410, the Vandals in 455, the Saracens in 846, and the Normans in 1084. The despoiling by marble cutters was continuous, quiet, and thoroughly authorized.

The thin slabs of ancient epitaphs were easily adapted into borders and panels or fitted into pavements, which explains why the floors of Roman churches are so richly and so irrelevantly inscribed. It was easier to pry a block from a crumbling ruin or dig it out of the Roman earth than to quarry it fresh from the hills of Carrara. Across Italy the competitive ambitions of rising medieval towns created a demand for new churches that seemed endless. Duomos and campaniles needed heavy stone foundations, thick walls, and monumental arches.

As the industry grew, and as the Roman marble cutters' booty exceeded the needs of the local market, they shipped more and more of their wares abroad on light coasting ships for the new cathedrals of Pisa, Lucca, Salerno, Orvieto, and Amalfi, among others. Pieces of Roman marble can be identified in Charlemagne's cathedral of Aix-la-Chapelle, in Westminster Abbey, and in the churches of Constantinople.

The medieval Roman limeburners prospered by making cement from the fragments of dismantled temples, baths, theaters, and palaces, and from smashed marble ornaments and statues. The Sant' Adriano kilns were devoted to burning the marbles of the nearby imperial forums, the Agosta consumed pieces of the mausoleum of Augustus, while La Pigna was fed by fragments torn from the Baths of Agrippa and the Temple of Isis. Temporary kilns were erected at the Baths of Diocletian, near the Villa of

Livia, at the Basilica Julia and the Temple of Venus and Rome, and remained there until adjacent materials were exhausted. At the Circus Flaminius the whole district was called the Lime Pit. A Vatican document of July 1, 1426, authorized a company of limeburners to demolish the Basilica Julia on the Sacra Via so they could feed their kilns with chunks of travertine, on condition only that the papal authorities receive half the product.

The Renaissance popes who professed enthusiasm for classical culture did little to defend the ancient relics. In fact, demolition of pagan temples and idolatrous statues seemed a pious duty. It was during the reign of Pope Nicholas V (1397?–1455), patron of Valla and other humanists, that many of the most important architectural remains—around the Capitol, on the Aventine and in the Forum, and the Coliseum itself—were denuded. Under Pope Pius II, who issued a bull (April 28, 1462) protecting the ruins of Rome and who even wrote an elegy for them, some of the handsomest remaining monuments were turned into building materials for new construction in the Vatican. Finally, when Pope Paul III (1468–1549) saw that the ancient statues dug up while opening new streets were being thrown wholesale into the kilns, he reinstated the ancient Roman death penalty for anyone destroying such monuments. This was said to increase private collections, but did not substantially impede the large-scale demolition.

Why preserve the clutter from a dead past? There was little interest in the daily lives of pagans and hardly a suspicion of how different everything might have been. Medieval paintings depicted ancient Roman soldiers wearing armor of the Middle Ages. Only gradually did painters begin to note that clothing had changed over the centuries. As we have seen, Petrarch himself had become interested in these changes, and he actually used the peculiarities of Greek clothing to explain a puzzling passage in the *Iliad.* Mantegna (1431–1506) was quite modern when he tried to paint the cult of Cybele in its authentic ancient setting. "The prudent painter," Gilio da Fabriano urged in his book, *The Errors of Painters* (1564), "should know how to paint what is appropriate to the individual, the time and the place. . . . Is it not an error to paint St. Jerome with a red hat, like the one cardinals wear today? He was indeed a cardinal, but he did not wear such a costume, since it was Pope Innocent IV, more than 700 years later, who gave cardinals their red hats and red gowns. . . . All this proceeds from the ignorance of painters."

Marble relics, statues and the remains of public buildings, were still there to be seen and touched. They could make history visible to the "unlettered" populace, Alberti's *non litteratissimi cittadini,* for whom he had to write in vernacular Italian. In the centuries ahead the Acropolis and the Parthenon, the Forum and the Coliseum, the Pyramids and Temples of Karnak, would cease to be mere features of a landscape and become stage sets for the living drama of the past. Then the millions who could not or who would not *read* history could actually *see* history.

The pioneers of Roman archaeology came by many paths. In the fourteenth century Petrarch stigmatized as heirs of the Goths and Vandals all those who were dismantling the ancient grandeur. One of the first to be fascinated by the magic of archaeology, a traveling merchant, Cyriacus of Ancona (1391–1452), sketched monuments and copied hundreds of inscriptions in southern Italy, Greece, and the eastern Mediterranean. "We are able by our art not only to raise from the depths monuments which have been destroyed, but also to bring the names of cities back into the light. Oh what a great, what a divine power has this art of ours!" For Poggio Bracciolini (1380–1459), a survey of Roman ruins provided the obvious introduction to his work *On Changing Fortunes* (1431–48). The successive gates of Rome plainly recorded the growth of the city and provided illustrations for Flavio Biondo's *Rome Restored* (1440–46), which showed Rome to have been an ever-changing city.

Some of the best Renaissance talents were spent on forays into antiquity. Leon Battista Alberti (1404–1472), prototype of Jacob Burckhardt's "universal man," applied his new science of perspective to the surveying and mapping of cities. Using the geometrical principles of perspective, he collaborated with another Florentine, Toscanelli (1397–1482), whose world map inspired Columbus on his first voyage, to make the first modern map of Rome. His advances made possible the conspicuous improvement of European city maps in the following century. Raphael (1483–1520) wanted to use his talents to recapture the vision of ancient glory. When he came to live in Rome in 1509, he was charmed by the ruins and outraged by the daily ravages of the limeburners. Commissioned by Pope Leo X, he had already begun to draw his ideal version of classical Rome when his premature death at the age of thirty-seven ended the project.

In a spectacular display of "urban renewal," the Renaissance popes, aided by artists, architects, condottieri, marble cutters and limeburners, erected elegant new churches and luxurious palazzi. With unconscious irony, when Pope Nicholas V, a celebrated devotee of classical learning, began modernizing the city by broadening old roads and opening new highways, he demolished whatever stood in the way.

The rising sense of history would gradually transform the Roman marble quarry into a vast open-air museum where the unlearned touring public could discover the past. In eighteenth-century England the word "classical," which originally meant simply "first class" or of the highest quality, came to mean specifically a product of ancient Greece or Rome. The Roman column became a symbol of architectural elegance, and "classical" antiquity would be a continent-wide standard of beauty.

The prophet and founding hero of modern archaeology, herald of this ever-widening public significance, was Johann Joachim Winckelmann (1717–1768). The son of a poor shoemaker in Stendal in Prussia, he refused to follow his father's craft. Instead he went to a nearby school where the master was slowly going blind, and the young Winckelmann became his eyes. He never forgot his debt to this master, who awakened his interest in books. Very early, almost as a quirk, he developed a passion for things Greek. At that time German scholars who knew Greek used it mainly for access to the New Testament. At seventeen, Winckelmann went to Berlin to study with one of the few German scholars known for their enthusiasm for Greek literature. At twenty-one he begged his way to Hamburg to secure classical texts from a famous library where holdings were about to be dispersed. Student wanderings took him from his school in Berlin to study theology at Halle and medicine at Jena. While pretending to listen to the long lectures he would secretly read his adored Greek texts.

Raised in threadbare poverty, he would spend much of his life seeking patronage from the rich and powerful. While he served as tutor in the wealthy Lamprecht family, the handsome Lamprecht boy excited in him "a passion which has troubled the peace of my soul." But this was only one of a long series of such unrequited passions. Winckelmann's feeling for the unadorned male figure reinforced his admiration for Greek sculpture.

The ideal secular post for an impecunious scholar in those days was as librarian to some dilettante nobleman, where he could enjoy the pleasant routine of collecting and arranging books, manuscripts, and works of art in a rural mansion. In 1748, when he was thirty, Winckelmann found his opportunity in Count von Bünau's castle near Dresden in Saxony. There he spent seven years helping the Count put together a history of the German Empire. Nearby Dresden, then known as Florence on the Elbe, in its museums and palaces offered as good a sampling of ancient and modern sculpture and painting as could be found outside of Rome or Paris. The city itself was a notorious spectacle of the baroque and the rococo. The filigreed Zwinger Pavillon, made for pageants and public entertainments, the Grosser Garten with its scores of close-packed marble works by the imitators of Bernini, and numerous private collections showed what extravagant modern artists had done to antique motifs. Winckelmann yearned for the pure simplicity of the classic originals.

The Saxon court in Dresden was then a center of Roman Catholic revival.

Winckelmann gave in to these influences, and he never quite forgave himself for it. But for him Rome would be only a way station back to Greece. His *Thoughts on the Imitation of Greek Works of Art in Painting and Sculpture* (1755) appealed from the imitative classicism of Berniniesque Dresden back to the works of the Greeks themselves. With a meager pension from the Elector of Saxony, he went to study in Rome. There an affair with a wealthy painter provided his room and board, and he managed to secure the patronage of cardinals. Beginning as librarian-companion to Cardinal Albani, he became first Librarian and finally Controller of Antiquities in the Vatican. "Cardinal Passionei, a jovial old man of seventy-eight years," Winckelmann boasted, took him "for drives . . . and he always escorts me home in person. When I accompany him to Frascati, we sit down to table in slippers and night-caps; and if I choose to humour him, in our night-shirts too. This may seem incredible, but I am telling the truth."

Despite such distractions, Winckelmann's passion for Greek art never flagged, and he stayed with his ambition to "produce a work in the German tongue, the like of which has never seen the light of day. The History of Ancient Art which I have undertaken to write," he explained, "is not a mere chronicle of epochs, and of the changes which occurred within them. I use the term History in the more extended signification which it has in the Greek language; and it is my intention to attempt to present a system. . . . to show the origin, progress, change, and downfall of art, together with the different styles of nations, periods, and artists, and to prove the whole, as far as it is possible, from the ancient monuments now in existence." Aiming, in Herder's phrase, to trace "the genesis of the beautiful in the art of antiquity," he would show that even among the less known ancients— the Egyptians, Phoenicians, Persians, and Etruscans—art had a history. But he would extol Greek art of the great period in writing so vivid, with examples so persuasive, that Greek art became "classic." Ironically, the "originals" he appealed to were themselves only copies, for original sculptures from the age of Phidias had not yet been discovered, but the copies he had before him were not yet so stigmatized by experts. This "father of scientific archaeology," not for the first time, was basing a "science" on intuition.

"By no people," Winckelmann asserted, "has beauty been so highly esteemed as by the Greeks." The priests who carried the lamb in the procession of Mercury were those to whom the prize of beauty had been awarded. "Every beautiful person sought to become known to the whole people by this distinction, and above all to approve himself to the artists, because they awarded the prize. . . . Beauty even gave a right to fame; and we find in Greek histories the most beautiful people distinguished. Some were famous for the beauty of one single part of their form; as Demetrius Phalereus for his beautiful eyebrows." "Art went still farther; it united the beauties and attributes of both sexes in the figures of hermaphrodites. The great number of hermaphrodites, differing in size and position, shows that artists sought to express in the mixed nature of the two sexes an image of higher beauty; this image was the ideal." "Beauty is one of the great mysteries of nature, whose influence we all see and feel; but a general, distinct idea of its essential must be classed among the truths yet undiscovered." In imagination he assembled all the separate beauties he had observed, "uniting them in one figure . . . a poetic Beauty."

Winckelmann shared the Greek adoration of the human form which made sculpture their great art. While he sneered at the ornate statues that had surrounded him in Dresden, his eulogy of the statue of Laocoön and his two sons crushed by sea serpents became the credo of neoclassicism. "The universal, dominant characteristic of Greek masterpieces, finally, is noble simplicity and serene greatness in the pose as well as in the expression. The depths of the sea are always calm, however wild and stormy the surface; and in the same way the expression in Greek figures reveals greatness and composure of soul in the throes of whatever passions. . . . Laocoön suffers; but he suffers like Sophocles' Philoctetes; his misery pierces us to the soul; but we should like to be able to bear anguish in the manner of this great man."

Only recently there had been excavated outside Naples the vivid remains of Pompeii and Herculaneum, cities suddenly inundated by volcanic ash and lava from Vesuvius in mid-August, A.D. 79. They offered a providential glimpse of the life of ancient Rome, but the excavation, financed by the Bourbon King of the Two Sicilies, had become a secret operation, and to make sketches of the finds was severely prohibited. As superintendent of Roman antiquities, Winckelmann did manage to be admitted to the museum that housed the discoveries. He then wrote his Open Letters describing the objects unearthed, and asserting the right of the whole learned world to receive all messages carried in objects from the past.

With the publication of his *History of Ancient Art* in 1764, incorporating these finds, Winckelmann became an eminent man of letters whose fame reached across the continent. This was one of the first works in the German language to become a classic of European literature. The next year Frederick the Great offered him a post as Royal Librarian. Meanwhile he was being tempted by invitations to Greece. In April 1768 he finally decided to return to Germany, and en route through Vienna he was received by the Empress of Austria. Still, he could not bear to abandon Rome, and he impulsively hastened back by way of Trieste. There at the inn where he spent the night of June 1, 1768, the room next to his was occupied by an "enervated lascivious lackey," Francesco Arcangeli, who had been condemned to death for theft in Austria but was pardoned on condition that he leave the country. As they supped together the boasting Winckelmann showed Arcangeli the gold medals given him by the Empress. Late that evening Arcangeli returned to the room where Winckelmann was correcting proofs of the second edition of his *History of Ancient Art.* Arcangeli strangled him with a rope, then stabbed him to death. Sentenced for his crime, even as he was being broken on the wheel, Arcangeli blamed Winckelmann for seducing him with gold.

"Winckelmann is like Columbus," the adoring Goethe exclaimed, "not yet having discovered the new world, but inspired by a premonition of what is to come. One learns nothing new when reading him but one becomes a new man!" Winckelmann's legacy was a popular movement—incorporating the history of art into the life of art. He, more than any other person, was responsible for elevating Greek and Roman antiquity into a synonym for the "classical."

The great English architect Robert Adam (1728–1792) had come to know Winckelmann on his visit to Rome, but had failed to persuade Winckelmann to accompany him to Greece. The "neoclassic" would become fashionable even before the classic had been properly documented. Adam became famous by embodying the neoclassic ideal in his design for English country houses that included neoclassic details in mantelpieces, window frames, and doorknobs. The enterprising Josiah Wedgwood (1730–1795) built a factory in 1782 which he called Etruria and which incidentally was the first English factory to use steam power. There he turned out plates and cups and vases that would put Winckelmann's ideal on countless middle-class dinner tables. Winckelmann's posthumous spell over Lessing, Herder, Goethe, Schiller, Hölderlin, Heine, Nietzsche, George, and Spengler has been called "the Tyranny of Greece over Germany."

In opening the past, Winckelmann himself was less explorer than discoverer. He awakened Europe to the charms of ancient civilizations, which he only faintly glimpsed. He would entice others to do the exploring. "It is an entirely new and unsuspected world that I am discovering for archaeology!"

73

"To Wake the Dead"

A full century passed before Winckelmann found his Vespucci to show and
tell the world what had really been discovered. Although Heinrich
Schliemann (1822–1890), too, rose from poverty to celebrity, in almost every
other way he was Winckelmann's opposite. Schliemann personally financed
all his exploits. He was his own patron. He imported into archaeology the
enterprise and love of action that had made him his fortune in commerce.
For him, exploring the past became an athletic feat and an adventure in
diplomacy, to feed a newly news-hungry age. And his love of a beautiful
woman helped keep the public spotlight on his digging.

Son of a poor Protestant minister in a village in north Germany, Heinrich
Schliemann's "natural disposition for the mysterious and the marvellous"
was fired by his father's passion for ancient history.

He often told me with warm enthusiasm of the tragic fate of Herculaneum and
Pompeii, and seemed to consider him the luckiest of men who had the means and
the time to visit the excavations which were going on there. He also related to
me with admiration the great deeds of the Homeric heroes and the events of the
Trojan war, always finding in me a warm defender of the Trojan cause. With great
grief I heard from him that Troy had been so completely destroyed, that it had
disappeared without leaving any traces of its existence. My joy may be imagined,
therefore, when, being nearly eight years old, I received from him, in 1829, as a
Christmas gift, Dr. Georg Ludwig Jerrer's *Universal History,* with an engraving
representing Troy in flames, with its huge walls and the Scaean gate, from which
Aeneas is escaping, carrying his father Anchises on his back and holding his son
Ascanius by the hand; and I cried out, "Father, you were mistaken: Jerrer must
have seen Troy, otherwise he could not have represented it here." "My son," he
replied, "that is merely a fanciful picture." But to my question, whether ancient
Troy had such huge walls as those depicted in the book, he answered in the
affirmative. "Father," retorted I, "if such walls once existed, they cannot possibly
have been completely destroyed: vast ruins of them must still remain, but they
are hidden away beneath the dust of ages." He maintained the contrary, while
I remained firm in my opinion, and at last we both agreed that I should one day
excavate Troy.

His mother died when he was nine. Since his father's poverty left him little hope of going to the university, he dropped out of the Gymnasium, where he might have pursued his classical interests, and instead went to the vocational Realschule. At fourteen Schliemann was apprenticed to a grocer and spent five years working from five in the morning till eleven at night, grinding potatoes for the whiskey still, packaging herring, sugar, oil, and candles. He escaped by becoming cabin boy on a ship bound for Venezuela. When the ship was wrecked in the North Sea, he found a post as messenger and later bookkeeper for a trading firm in Amsterdam.

Through these drab years Heinrich never lost his romantic ambition. Determined someday to unearth the real Troy, every spare minute, even as he ran errands or waited in line at the post office, he improved himself by reading. By his own system he acquired a score of languages, never missing an opportunity to learn or to practice what he learned. "This method consists in reading a great deal aloud, without making a translation; devoting one hour every day to writing essays upon subjects that interest one, correcting these under a teacher's supervision, learning them by heart, and repeating in the next lesson what was corrected on the previous day." Within six months, he reports, he had acquired "a thorough knowledge of the English language," as part of the process having "committed to memory the whole of Goldsmith's *Vicar of Wakefield* and Sir Walter Scott's *Ivanhoe.*" In only six weeks devoted to each, he learned to write "and to speak fluently" French, Dutch, Spanish, Italian, Portuguese, and some others. When he traveled through the Middle East, he acquired a practical knowledge of Arabic.

The spoken word interested him most. He never forgot the cadence of spoken Greek, which he first heard when a drunken miller, a dropout from the Gymnasium, came into the grocery store where Schliemann was working and melodiously recited lines from Homer. But he waited till middle life to turn to his beloved Greek. "Great as was my wish to learn Greek, I did not venture upon its study till I had acquired a moderate fortune; for I was afraid that this language would exercise too great a fascination upon me and estrange me from my commercial business."

Schliemann followed an arduous and devious road to fortune. When he was a young man in Amsterdam, Russian merchants were coming there for the indigo auctions. Except for the Russian vice-consul, Heinrich found no one in Amsterdam who knew Russian, and when the vice-consul refused to be his teacher, he employed his usual system in a crash program to teach himself. He hired an old Dutchman to be his audience as he declaimed Russian two hours each evening. When the tenants of his boarding house complained, Heinrich did not change his system but he had to change lodgings twice before he was satisfied with his fluency in the language.

The trading firm where he worked had dealings in St. Petersburg and sent him there as their agent. In St. Petersburg, trading in indigo, dyewoods, and war materials like saltpeter, brimstone, and lead, he soon, to his own surprise, made a fortune. Then he no longer feared the glorious distraction of the classical language. He spent six weeks learning modern Greek and in another three months plunged into the ancient authors. After the Crimean War, he traveled the world pursuing his historical interests. His childhood love was a playmate, Minna Meincke, who had shared his fantasy of pursuing the quest for Troy. And when he was established in business he went in search of her. He did find Minna Meincke, but alas she was already married. In 1852 he made the mistake of marrying a Russian beauty who wanted only his money. She refused even to share his home, much less his archaeological interests. Meanwhile, having become an American citizen by the accident that he was traveling in California when it became a state of the Union, he went to Indiana to benefit from its loose divorce laws and there divested himself of his Russian wife.

Determined not to make the same mistake again, he asked an old friend, his Greek teacher who had since become the archbishop of Athens, to find him a suitable young Greek wife. The Archbishop obliged by suggesting his relative, Sophia Engastromenos, a bright and beautiful seventeen-year-old schoolgirl. In Athens, before he decided to marry Sophia, Schliemann visited her classroom incognito to hear how she recited Homer. Her mellifluous Greek brought tears to his eyes, and clinched his determination to marry her. The polylingual Heinrich, at forty-seven, made Sophia his lifelong student. At their marriage she knew only ancient and modern Greek, but he promised that she would learn four more languages in the next two years. He trudged her about the capitals of Europe and the Near East, expounding history and archaeology, testing her knowledge, pressing her not to lag behind. After an interval of headaches, nausea, and fevers, she survived to become his colleague when he finally began his digging at Hissarlik in 1871. She went into the archaeological trenches, something astonishing at the time for a woman, and managed to direct a crew of Turkish workmen in the excavations.

Schliemann, unlike Winckelmann, believed his vocation to be digging. His proper province was not words but things. But the work he loved required that he supervise laborers who spoke exotic languages. And his gift of tongues, which helped him manage his digs, enabled him to convert doubters and advertise his finds.

A quixotic archaeologist with a beautiful wife directing a hundred and fifty rebellious workmen on the exotic Turkish landscape could hardly prevent his work becoming public, even in those early salad days of the sensational press. The open-air archaeologist would become the public property of the newspaper audience. Now the explorer of the past had to forsake the library and the museum, go to distant places, and move heavy objects up and out into public view. His success would be judged not only by academicians but by impatient millions.

Schliemann could not be budged from his faith that Homer's Troy was at the obscure modern village of Hissarlik in northwestern Turkey on the Asian side just four miles from the mouth of the Dardanelles. When he compared the site with Bunarbashi, several miles south of there, where other scholars had placed it, his conviction became stronger than ever. But Schliemann's chosen site was privately owned. The bureaucratic, autocratic, and corrupt Turkish officials first tried to block him, then blackmailed him before granting the required digging permit. Excavation at Troy would be paid for entirely by Schliemann, who considered it a privilege to spend his fortune in this way. He never complained of the expense, but was prudent and businesslike.

In September 1871, hiring a crew of eighty laborers, he began digging into the mound at Hissarlik. Precisely according to plan, he found layer after layer of cities and fortifications, one below another. He knew that as he dug he was destroying monuments of a more recent era, but his destination was Troy! Twenty-three feet below the surface and stretching down to thirty-three feet, he found ruins of a city he believed to be Troy. He impulsively identified everything he had hoped to find—remnants of the Temple of Athena, the main altar for sacrifices, the Great Tower, houses, and streets —all just as described in the *Iliad*.

In early May 1873, as his workmen were digging out the top of the ancient wall, Schliemann himself spied a shiny gold object. As he recalled seven years later in his own melodramatic account:

> In order to secure the treasure from my workmen and save it for archaeology, it was necessary to lose no time; so, although it was not yet the hour for breakfast, I immediately had *païdos* (rest-time) called. . . . While the men were eating and resting, I cut out the Treasure with a large knife. This required great exertion and involved great risk, since the wall of fortification, beneath which I had to dig, threatened every moment to fall down upon me. But the sight of so many objects, every one of which is of inestimable value to archaeology, made me reckless, and I never thought of any danger. It would, however, have been impossible for me to have removed the treasure without the help of my dear wife, who stood at my side, ready to pack the things I cut out in her shawl, and to carry them away.

Keeping his secret for the moment, he successfully smuggled the gold treasure (eventually nine thousand objects) out of Turkey. His cautions proved justified, for the workman who later found a gold object in the excavations quickly had it melted down by a local goldsmith. The gold, and not Homer's Troy, was what interested Turkish officials. They blocked his further excavations and sued for return of the treasure.

While Schliemann's story of his digging was substantially correct, recent historians raise an eyebrow at this example of how his feeling for drama sometimes overshadowed the fact. His "dear wife," who according to him was catching the treasure in her shawl, appears at that moment to have been not in Hissarlik but in Athens. Still, such trivial embellishments increased public interest in the new romance of archaeology.

Schliemann returned to Greece, where, by intervention of the British prime minister Gladstone and the British ambassador, he had secured permission to dig, and he plunged into another sensational adventure. This time Schliemann was pursuing his hunches about the treasure still to be found at the fabled site of ancient Mycenae. There, he insisted, was the buried treasure of Agamemnon. And there, too, he had been guided by his reading to dissent from the accepted scholarly view. Scholars generally had agreed that the tombs of Agamemnon and Clytemnestra must be outside the walls of the citadel. But Schliemann, a confidant of the ancients, put his faith in Pausanias, the famous second-century traveler who described "the heroes' graves . . . in the midst of the meeting-place." To Schliemann this meant inside the city walls. At Mycenae when he found stelae arranged in a circle suggesting the ancient agora, he began digging. In December 1876 he found the first of five shaft graves. For forty-five days Schliemann and Sophia, their hands numbed by cold and using only fingers, a penknife, and a small shovel, dug within the grave circle.

Their reward—the richest treasure ever yet excavated from the past— was a find of bodies "literally covered with gold and jewels." The faces, distinguishable when unearthed, quickly disintegrated in the air, but each gold mask still had its own character. By intuition, learning, expertise, and good luck, the Schliemanns had found this fabulous prize: "the mask of Agamemnon," gold diadems, gold and silver statuettes, gold sword handles, precious necklaces and bracelets, stone and gold and alabaster vases, goblets of gold and silver, and scores more of dazzling jewels. Not one to hesitate at a moment of drama, Schliemann telegraphed to King George of Greece, "It is with extraordinary pleasure that I announce to your Majesty my discovery of the graves which, according to tradition, are those of Agamemnon, Cassandra, Eureymedon, and their comrades, all killed during the banquet by Clytemnestra and her lover Aegisthus." He declared that no comparable treasure had ever been unearthed. "All the museums of the world taken together," he boasted, "do not have one-fifth as much."

Despite his enthusiasm, faith, and scholarship, Heinrich Schliemann's discoveries were not quite what he thought. He was not off the mark as far as the earlier explorer who aimed for Japan, who thought he had reached

Cathay but had merely discovered America. We now know that the city that Schliemann chose as Homer's Troy from superimposed strata of "the five prehistoric cities" was the wrong one. The spectacular find that he called the treasures of Priam, dug from the second and third layers above bedrock, actually came from a thousand years before Priam. With funds supplied in Schliemann's will, his heir, Wilhelm Dörpfeld (1853–1940), proved Homer's Troy to be the sixth level up from the bottom which Schliemann had cut through in his haste. His conclusions at Mycenae, too, were off the mark. He had not, as he proclaimed, found the tomb of Agamemnon. The tomb he found was many centuries older.

When classical scholars ridiculed his identification of King Priam with his Troy, Schliemann still insisted on "Priam"—"because he is so called by the tradition of which Homer is the echo; but as soon as it is proved that Homer and the tradition were wrong, and that Troy's last king was called Smith, I shall at once call him so." His instinct for the flamboyant, his melodramatic appeal to ancient heroes, awakened the historical curiosity of millions. Even in error, Heinrich and Sophia grandly advanced public knowledge. People everywhere were fascinated by the Schliemanns' courage and determination. The vast watching public came to believe that the earth held relics and messages from real people in the distant past.

Heinrich's contribution to the techniques of field archaeology was not inconsiderable. When twentieth-century archaeologists attack him for destroying en route relics he had not planned to discover, they forget the primitive state of archaeology in his day. In stratigraphy he pioneered by applying to human relics the principles that others had already applied to geology. Homer's *Iliad* was not mere "humanized sun myths," as the oversubtle German scholars were then insisting. Even by his mistakes Schliemann proved the reality of a Homeric civilization by unearthing the pre-Homeric civilization out of which it grew. To the canonical four civilizations: Babylonia, Egypt, Greece, and Rome, he added two more from "prehistory." If there were these two, then why not many more?

Schliemann's successor, Sir Arthur Evans, who took up his clues to unearth still another scintillating civilization at Knossos in Crete in 1900, acknowledged his debt:

> Less than a generation back the origin of Greek civilization, and with it the sources of all great culture that has ever been, were wrapped in an impenetrable mist. That ancient world was still girt round within its narrow confines by the circling "Stream of Ocean." Was there anything beyond? The fabled kings and heroes of the Homeric Age, with their palaces and strongholds, were they aught, after all, but more or less humanized sun myths?
>
> One man had faith, accompanied by works, and in Dr. Schliemann the science

of classical antiquity found its Columbus. Armed with the spade, he brought to light from beneath the mounds of ages a real Troy; at Tiryns and Mycenae he laid bare the palace and the tombs and treasures of Homeric Kings. A new world opened to investigation, and the discoveries of its first explorer were followed up successfully by Dr. Tsountas and others on Greek soil. The eyes of observers were opened, and the traces of this prehistoric civilization began to make their appearance far beyond the limits of Greece itself.

But jealousy of scholarly competitors and the needs of sensational journalism made the glitter of Trojan and Mycenaean gold seem an indictment. Was Schliemann only a mercenary treasure hunter, like others less celebrated? Did he care less to enrich knowledge than to fill his own coffers? Even these accusations had the advantage of focusing public interest on the new worlds of archaeology. But they were unfounded. If Schliemann had not quickly removed the Trojan treasure from Turkey, there would have been little left for historians to study. To the Greek nation he gave all the treasure he unearthed at Mycenae and elsewhere, now splendidly displayed in the museum at Athens. For all his labors and risks, financed by him personally, he had no compensation except his celebrity and the satisfaction of kindling enthusiasm for his beloved Greece.

In the new world of publicity, others did Schliemann's work for him. Back in Winckelmann's time, to be stirred by his enthusiasm for classical Greece you had to read his books. But now, with Schliemann's own shrewd assist, every turn of the archaeologist's spade became news. The reading public did not have to wait for heavy tomes to enjoy the adventures of excavation. Newspaper readers held their breath, watching daily for Schliemann's dispatches to *The Times* of London, the *Daily Telegraph,* and the *New York Times.* The Turkish government's refusal of a permit or the arrogance of a petty official became an international cause célèbre, advertised in letters from Schliemann himself, or in lengthy reports with others' by-lines but later revealed to be his work. Of course, he was made a Fellow of honorary and learned societies, and even the London Grocers' Association invited him to lecture and elected him to membership. His portrait by an artist of the *Illustrated London News* was reprinted across the world, making a trademark of Schliemann's broad forehead and heavy moustache, and reporters inventoried his dandyish wardrobe of fifty suits, twenty hats, forty-two pairs of shoes, thirty walking sticks and fifteen riding crops.

When Dom Pedro II, emperor of Brazil, a devotee of the classics, came with his Empress to visit Turkey, Schliemann, using his fluent Portuguese, toured them through the digs at Hissarlik, after which the Emperor declared himself fully persuaded that this was the true site of Homer's Troy. At Mycenae the Emperor's party was served a sensational luncheon in the depths of the famous Treasury of Atreus, to the delight of the avid press corps. Sophia, of course, added a touch of living romance not commonly found at prehistoric diggings. Heinrich and Sophia became the royal family of archaeology. The young Greek beauty was a welcome variant on the stereotype of fragile Victorian femininity. "The part I have taken in the discoveries is but small," she modestly confessed, "in Troy as well as in Mycenae. I have only superintended thirty workmen." After the upper layer of pebbles was removed from the tombs in Mycenae, "from thence it was exceedingly difficult, because, on our knees in the mud, my husband and I had to cut out the pebbles, to cut away the layer of clay, and to take out one by one the precious jewels."

In London the Royal Archaeological Institute held a special meeting on June 8, 1877, to honor Heinrich and Sophia. The spotlight was on the glowing Sophia, doubly escorted into the hall on the arm of Lord Talbot, the president, and of William E. Gladstone, who had requested the privilege. The address was given by Sophia. Lord Talbot had praised her "as the first lady who has ever been identified in a work so arduous and stupendous, you have achieved a reputation which many will envy—some may emulate —but none can ever surpass." The address of the twenty-five-year-old Sophia dazzled by its erudition and eloquence. In her admiration for Britain she archly confessed that the sin of the ancient Greeks was "envy." Then she read a paean to the Greek sky and the Greek mind, and recalled that the Greek language was so beautiful that "the mere sound filled my husband with wild enthusiasm at a time when he did not know yet a word of Greek." She concluded with "an appeal to the English ladies to teach their children the sonorous language of my ancestors, so that they may be enabled to read 'Homer' and our other immortal classics in the original." At the end she was saluted by a standing ovation. "As I heard and saw the ovation given to my Sophithion by such a notable assemblage," Heinrich wrote, "I could only wonder why the great gods of Olympus had given me this woman as a wife, friend, colleague and lover. My eyes ran with tears so I could barely see."

The press corps that followed the Schliemanns made their own demands. When, as at Tiryns, the digging was slow, the *New York Times* correspondent immediately announced that Heinrich's luck had run out—only days before one of his most spectacular finds, the remains of a palace rivaling that at Troy or Mycenae. In those primitive decades of the daily press, cameras were still cumbersome, and only barely portable. When Schliemann unearthed the remarkably preserved bodies in the tombs at Mycenae, he had no photographer along and so sent urgently for an artist to paint likenesses before the bodies disintegrated. His books on his excavations still contained no photographs, though some of the drawings were copied from

photographs. The first report on archaeological excavation to include pho-
tography was not Schliemann's but the German archaeologist Alexander
Conze's report on his diggings in Samothrace (1873). When we compare
these photographs with the rude line illustrations of earlier reports, we see
how much the camera has increased the vividness of history, making millions eager to see more.

74

Latitudes of Time

FOR a modern sense of history, vivid glimpses of the "serene greatness" of
Laocoön or the golden glitter of Agamemnon's mask were not enough.
Another dimension was needed, what I will call the latitudes of time, vistas
of contemporaneity, a sense of what was going on all over the world at the
same time. This was a much more sophisticated discovery, to be reached
only by devious and surprising paths.

For millennia people dated happenings in their place by the years of reign
of their own king, or by some other fact of local significance. The year A.D.
1900 in the Chinese reckoning was the twenty-sixth year of Kuang-Hsü, the
Brilliant Succession, but in Japan was still called the thirty-third year of
Meiji, the Enlightened Rule. In India, Hindus dated by dynastic eras, but
Buddhists reckoned from the death and Nirvana of Buddha in 544 B.C.
Hindus also used the "Kali" era, a subdivision of the canonical mahayuga
of 4,320,000 sidereal years and the yuga of 432,000 years. Other Indian
schemes in occasional use date from a battle or a calendar reform. All were
complicated by local variations between the lunar and the solar year. Each
ancient civilization—Rome, Greece, Egypt, Babylonia, and Syria—had its
own scheme. The Roman way of dating from the founding of the city had
a vogue elsewhere too. The Muslim calendar, which would reckon from the
Hegira on July 16, 622, as we have seen, was inaugurated only seventeen
years after the event, and still used the lunar year.

In Christian Europe, the modern reckoning—B.C. or A.D.—expressed the
original Christian belief in a unique event, the coming of Christ, which gave
to all history a meaning and direction. But this scheme evolved only gradu-

ally. The Jews had found their unique event in the Creation, and the Christian year 1900 was the Jewish *Anno Mundi* 5661.

Centuries passed after the birth of Jesus before the present system came into use. During the first centuries some Christians dated from the "Indiction," multiples of the 15-year period of imperial tax assessment from the accession of Constantine in 312, others from the Era of Spain (the Easter cycle beginning with the Roman conquest of Spain in 38 B.C.), or from the Era of the Passion (33 years after the Nativity). The inventor of *Anno Domini* was Dionysus Exiguus (c. 500–560), a monk, mathematician, and astronomer, who was trying to figure out how to predict the precise date of Easter, generally agreed to occur on the first Sunday after the full moon or after the March 21 vernal equinox. This meant that in Western Christendom, Easter might fall anytime between March 21 and April 25. Easter has always dominated the Christian Year because it is the date from which all movable feasts are calculated and from which the liturgical year begins.

But the method of predicting Easter in future decades was complicated and a subject of endless controversy. Many European Christians were using a 95-year table when Pope Hilarius devised (461–68) another method. He coordinated the 19-year cycle of the recurrence of New Moons on the same date with the 28-year cycle of the recurrence of days of the week and of the month in the same order, and he came up with a period of 532 years. Dionysus Exiguus set about refining the Pope's figures. In doing so, he rejected the customary use of the accession of Emperor Diocletian in 284 as a base date. Rather than "perpetuate the name of the Great Persecutor," he would "number the years from the Incarnation of our Lord Jesus Christ."

Despite all efforts at compromise, the date of Easter would still divide Western from Eastern Christendom. But Dionysus Exiguus' Christian calendar, numbering the years from the supposed date of the birth of Jesus, except in Islam would come to govern most of the non-Christian world. Dionysus Exiguus' error was only one of detail. He reckoned the birth of Jesus to have occurred in the year 753 from the founding of the city of Rome. Recent Biblical scholars, following the Gospels, generally agree that the Nativity must have occurred before Herod's death, that is, not after 4 "B.C."

In A.D. 525 Dionysus Exiguus proposed to the Pope the use of "A.D." (*Anno Domini,* or the Year of Our Lord) as the standard scheme of dating. Dionysus himself was so little impressed by his own invention that he continued to date his own letters by reference to the "Indiction." Gradually, through the use of Dionysus Exiguus' Easter Tables in Christian Europe, *Anno Domini,* denoting the continuous series of years from the birth of

Jesus, displaced all others. While the system began to be established in the learned world when the Venerable Bede used these "A.D." dates in his *Ecclesiastical History* (731), several centuries passed before "A.D." was generally adopted in Europe. Not until the seventeenth century did scholars begin to use "B.C.," counting years backward from the year of Nativity.

Many ambiguities remained to plague the historian. For example, when did the "year" begin? The numerous competing possibilities included Christmas Day, Lady Day (Annunciation Day, March 25), Easter (a movable feast), and January 1. Our textbooks still bear the mark of these confusions. For example, England's Glorious Revolution, sometimes known as the Revolution of 1688, would by our current reckoning be called the Revolution of 1689, because it occurred on February 13 of that year, but Englishmen at that time did not begin their "new" year until March 25. The date of commencement of the new year changed with the passing centuries. In the eighth century they counted from Christmas Day, but following centuries used Lady Day or Easter, before turning to the modern practice of January 1.

During the Middle Ages in Europe it was common to date legal or official documents not by A.D. but by the regnal year of the ruling king, pope, or bishop, which could add further complications. Since King John happened to come to the throne on Ascension Day (the fortieth day after the Resurrection, i.e., after Easter), which was a movable feast, he began each of his regnal years on that festival day, which varied from year to year. Therefore, some of his regnal years were shorter, and some longer, than our calendar year. King Henry V came to the throne on March 21, 1413, with the result, when the New Year still began on March 25, that each of his regnal years included parts of two A.D. years.

The modern practice of beginning the New Year with January 1 marks a return to pagan practice, for that was when the Roman year began, which explains, of course, why the Church opposed observance of that day. But from the increasing use of almanacs that made their calculations from January 1 and the widespread study of Roman law, January 1 became the general datemark in Europe by the end of the sixteenth century. When Pope Gregory XIII produced his reform of the calendar in 1582, he, too, gave in to the pagan custom. His New Style reckoning created some new complications for the modern historian. Roman Catholic countries soon adopted the sensible Gregorian reforms, but Protestants and the Eastern Orthodox Christians would not follow any pope's rule. During nearly two centuries the British suffered inconvenience rather than live by a papistical calendar, for seasons had long since lost phase with the months.

Finally, in 1751, the freethinking Philip Dormer Stanhope, fourth Earl of Chesterfield (1694–1773), famous for his letters to his son, introduced a bill in Parliament to adopt the New Style (no longer the "Gregorian"!) calendar. By this Act the beginning of the year was moved back from March 25 to January 1, and the day following December 31, 1751 (instead of being January 1, 1751), became January 1, 1752. To correct the error accumulated by the old Julian calendar, the day after September 2, 1752, was to be called September 14. For historians, all this left a legacy of confusion. After 1582, when Old Style was competing with New Style, the British colonies in America generally followed the unreformed British practice of dating, with attendant ambiguities.

It took a communist revolution to persuade the Russians to abandon the Julian calendar, which they did finally in 1919. In Japan, Emperor Meiji, as part of his program of Westernization, on January 1, 1873, finally adopted the Gregorian calendar to be used alongside the old system of reign years. In China a complicated system combined reign-year titles with the lunar year until the republic was founded in 1911. At long last the solar year was adopted, but dates were still reckoned from the founding of the republic. Only in 1949 did the Chinese government go New Style with the Gregorian calendar.

A common time-denominator for the world's events would make it easier to define the latitudes of history, and so discover which events were happening in different places at the same time, and then, too, which of the world's events came before or after others. During most of human history, even in Western Christendom, as we have seen, there was no uniform scheme—in fact, no scheme at all—for dating events in one place in relation to events in another place.

It is hard for us to imagine how insular and fragmentary was the past before scholars around the world established worldwide lines of contemporaneity. Orthodox Christians, spotlighting the events of the Bible, left all the rest of the world in outer darkness. To bring together the events of the Jews, Persians, Babylonians, Egyptians, Greeks, and Romans into a single chronology required superhuman erudition and a willingness to ask embarrassing questions. One of the first to try was the same ambitious cartographer, Gerardus Mercator (1512–1594), who found a way to depict the spherical earth on a flat surface for the convenience of worldfarers on the sea. He also saw the need for a universal chronology to give people their bearings as they explored all past time. In 450 folio pages he produced an ingenious *Chronology . . . from the beginning of the world up to the year 1568, done from eclipses and astronomical observations.* Events among the Assyrians, Persians, Greeks, and Romans were synchronized by the contemporary references to solar and lunar eclipses. This was only the first part of Mercator's grand unfulfilled plan to depict the whole world since the Creation in both spatial and temporal dimensions.

In the age of Copernicus it is not surprising that others too used the new astronomy to illuminate history. The most famous and the most successful of them was the phenomenal French scholar Joseph Justus Scaliger (1540–1609), who was revered as a prodigy and reputed to be, next to Aristotle, the most learned man of all times. According to his admirers, he had "learned Homer complete in twenty-one days." Studying in Paris during the frightful massacre of Protestants on St. Bartholomew's Day (1572), "he sat so intent upon his Hebrew," a fellow scholar reported, "that for some time he heard neither the clash of arms, nor the groans of children, nor the wailing of women, nor the shouting of men. Allured by the marvellous sweetness of these languages, while continually, like a fire, his ardor for learning grew, he acquired in succession Chaldean, Arabic, Phoenician, Ethiopic, Persian, and especially Syriac."

"Phoenix of Europe," "Bottomless Pit of Erudition," "Light of the World," Scaliger drew on philology, mathematics, astronomy, and numismatics to make his Correct System of Chronology, which finally brought together the events of all known antiquity into a single series. While Pope Gregory was proclaiming his reform of the current calendar, Scaliger, too, was using Copernican astronomy to coordinate the numerous ancient calendars. Of course, Scaliger drew fire from all the faithful who believed that "sacred history" should remain esoteric. With this new science of chronology it became possible for the first time to bring together into a coherent narrative the whole European past.

The pious Sir Isaac Newton (1642–1727) devoted the last years of his life to finding ways to use astronomy to confirm Biblical history. As he became more famous he became more religious, and, as we have seen, at his death he left thousands of manuscript pages on theology and chronology. Although some of his speculations would later encourage Buffon to stretch the age of the earth, Newton himself refused to take seriously the possibility that the earth might be much older than the Biblical date (4004 B.C.) fixed by Archbishop Ussher. Newton hoped simply to confirm the Biblical narrative by synchronizing the events in Scripture with those recorded in the chronicles of Egypt, Assyria, Babylonia, Persia, Greece, and Rome. The more eastern and more exotic countries like China, whose chronicles were recently brought to Europe by the Jesuit missionaries, did not yet enter his picture.

While Newton's raw data for the human past were random scraps from dubious sources, his astronomy was brilliantly professional. And his way of using astronomy for historical dating proved to be an advance toward establishing sharper "latitudes" of chronology, so that eventually there could be a single chronological scale for events everywhere. But Newton was by no means the first to understand that this could be done. A century and a half earlier, as we have seen, Mercator and Scaliger too had begun to use astronomical data for a single worldwide chronology. The eminent Polish astronomer Johannes Hevelius had calculated the exact position of the sun in the Garden of Eden at the hour of the Creation, which he fixed at 6:00 P.M., October 24, 3963 B.C. One of Newton's contemporaries, William Whiston, tried to date the comet that had caused the Flood.

For the base event in his chronology, Newton, oddly enough, chose the fabled voyage of the Argonauts. This great scientist erected the whole grand structure of his world chronology on the flimsiest possible foundation—the date of the mythical adventure led by Jason to Colchis in search of the Golden Fleece. The *Argo,* Jason's ship, was said to contain a beam cut from the divine tree of Dodona which could foretell the future. The fleece was guarded by the famous sleepless dragon whose teeth, when sowed, became armed men. Countless wonders attended Jason and his crew of fifty on their celebrated voyage, and not the least wonderful was what Newton managed to make of it.

Newton saw no paradox in choosing a myth as the point of reference for his scientific chronology. He knew that in antiquity the voyage of the *Argo* was assumed to be a historical fact, the trip that first opened the Black Sea to Greek commerce. Like other good Christians who followed the doctrine of Euhemerus, Newton himself believed that the gods of ancient mythology were real heroes who had been deified. If ancient myth was nothing but romanticized fact, then the voyage of the Argonauts must really have taken place, and Newton could fix it in time by its relation to astronomical phenomena.

The peculiar value of the Argonaut date was that it also fixed the date of the fall of Troy, and so, too, of the founding of Rome, since Rome was reputedly founded by Aeneas, who was a refugee from Troy. Newton made much of the fact that the redoubtable Herodotus had written that the time between the voyage of the Argonauts and the fall of Troy was just one generation. If you could date the voyage of Jason, Newton said, then you needed only to define the number of years in a "generation" to provide a precise base line for all Greco-Roman chronology. For many later figures in myth and history traced their descent from one of those who had traveled on the *Argo.*

"The surest arguments for determining things past," Newton wrote, "are those taken from Astronomy." Newton noted that a few historical events —for example, the Peloponnesian War—could easily be dated by the eclipses noted at the time of the events. But eclipses were rare, and none was recorded by the Argonauts. Therefore Newton developed a more sophisticated astronomical technique, which he applied with theological obstinacy. He said that by using data on the annual precession of the equinoxes, which he had computed in the *Principia* at "about 50″ yearly," it would be possible to figure exactly how many years before the present the heavens had a particular aspect.

Newton had read widely in ancient astronomy and he felt deep respect for Hipparchus, the Greek astronomer who had first noted (c. 130 B.C.) the precession of the equinoxes. But, Newton explained, Hipparchus had miscalculated the rate of precession. If Hipparchus' observations of the heavens were reconsidered, then the precise date of the Argonautic expedition would be fixed.

> Hipparchus the great astronomer comparing his own observations with those of former Astronomers concluded first of any man that the equinoxes had a motion backwards in respect of the fixed stars: and his opinion was, that they went backwards one degree in about a hundred years. He made his observations of the equinoxes between the years of Nobonassar 589 and 618: the middle year is 602, which is 286 years after the aforesaid observation of Meton and Euctemon. . . . But it really went back a degree in seventy and two years, and eleven degrees in 792 years. . . . and the reckoning will place the Argonautic expedition about 43 years after the death of Solomon. The Greeks have therefore made the Argonautic expedition about 300 years ancienter than the truth, and thereby given occasion to the opinion of the great Hipparchus, that the equinox went backward after the rate of only a degree in a hundred years.

In this manner Newton constructed his New System of Chronology, by which he dated the principal events of the Greeks, the Persians, and the Egyptians, in relation to the dates of David and Solomon in the Bible. Newton's chronology became a subject of heated international controversy. "The great events of Antiquity," a champion exclaimed, "had long lain like the ruins of some mighty buildings, demolished by the injuries of time, and hid in rubbish, notwithstanding the many attempts made to repair it. But at last we see the noble structure rise, in all its original symmetry, strength, and beauty; every material being restored to its ancient and proper place by the masterly hand of Sir Isaac Newton!" Others called his scheme "no better than a sagacious Romance." But the young Edward Gibbon treated Newton's Chronology with respect. "The name of Newton raises the image of a profound genius, luminous and original," he wrote in his commonplace book in 1758. "His System of Chronology would alone be sufficient to assure him immortality. . . . Experience and Astronomy, this is the thread of Newton's argument."

Newton, a passionate believer in the Biblical prophecies, still pointed toward a practical world chronology based on objective, planet-wide events. In the long run, the kind of base dates that Newton's astronomy recommended would offer usable lines of contemporaneity around the world. People might never agree on the date of the Creation—many would not believe in the Nativity—but all could, and would, share a syntax of history.

Modern chronology came when the old provincial schemes of naming years and epochs by reigning monarchs or dynasties or by magical portents were displaced by a common scheme of numbering. Only very late in the history of our planet did "century" become a widely accepted yardstick of time. In English usage, for example, "century" (from Latin *centuria,* designating a company of one hundred men) originally meant any group of one hundred things, just as Shakespeare has Imogen in *Cymbeline* announce her hope to say "a century of prayers." People still spoke of a "century of years." Not until our "seventeenth century" did the lone word "century" begin to mean one of the successive periods of one hundred years from the beginning of the Christian era. This was a small clue to a great change.

75

The Discovery of Prehistory

IN the eighteenth century, when Buffon stretched the calendar of nature into tantalizing eons, pious Christians still found the Biblical chronology by which Archbishop Ussher had fixed the creation at 4004 B.C. too comforting to abandon. For them the whole course of early history seemed to run from Eden through Jerusalem, and was amply chronicled in the Bible. The ancient events that concerned Christians had occurred exclusively in and around the Mediterranean, and the human heritage was the heritage of Greece and Rome. When Newton made the voyage of the Argonauts the base line for his chronology, he too kept Biblical events in sharpest focus.

But what had happened *before* Bible times? Today we may be amazed that so few Christians asked that question. Yet for believing Christians the question seemed meaningless: What happened *before* history? Before anything really happened? Not until the mid-nineteenth century did the word "prehistory" enter European vocabularies. Meanwhile thoughtful Europeans had somehow excluded from their historical ken most of the earthly past.

Along with the plants and animals and minerals brought back to Europe by missionaries, merchants, explorers, and naturalists came human artifacts for "cabinets of curiosities," familiar features of the households of the rich and the powerful. In the Middle Ages such curious, ancient, and precious objects had been occasionally exhibited too in churches, monasteries, colleges, and universities. In the Renaissance, regal collections including the booty of battle, the gifts of ambassadors, and the works of court artists adorned the palaces of popes and Medicis. So were born the great Vatican collections, the Uffizi and the Pitti in Florence, the Louvre in Paris, the Escorial near Madrid, and others in ducal capitals like Dresden, where Winckelmann was first inspired. These were for the delight of a privileged few.

The eighteenth century in Europe saw a new kind of collection, a novel institution, the *public* museum. The British government pioneered by acquiring the collections of Sir Hans Sloane in 1753, which were opened to the public in 1759. Some private collections, like the Vatican museums, were voluntarily opened to the public. Others, like the Louvre, were seized by revolutionaries for the whole citizenry. Across Europe a new museum public expected to learn, to be delighted, to be entertained. The word "tourist" entered the English language after 1800 for the mobile community of transient spectators. The hopes of breathless museum-explorers were exaggerated by the distances they had traveled.

In the United States and wherever there had not been palaces or royal collections, the public had to start from scratch. New World counterparts appeared: Peale's Museum (1784) in Philadelphia, the Smithsonian Institution (1846) in Washington, and others in South America. Across Asia—in India, Siam, China, Japan—the great collections generally remained the preserve of princely courts or went into the inner sanctums of temples. Only revolutions of one kind or another would open these treasures to public view. From conquered lands—from Egypt, Greece, Rome, and Persia—works of painting and sculpture, and even whole buildings, were transported to the great museums in London, Paris, Amsterdam, or Berlin.

As the European museums grew, at first they showed only the sorts of objects that aristocratic dilettantes had collected for prestige or out of curiosity. Objects of beauty held the spotlight, with occasional items of historical association, such as old crowns, scepters, and orbs, or a rare scientific instrument like an orrery. Objects that were not obviously beautiful or conspicuously strange aroused little interest. Yet, as it turned out, it was precisely these crude anonymous objects that would open prehistoric vistas and give the public a new vocabulary for all history. As we have seen, surviving *objects* had a special power to help people grasp the past. But the buried relics in Rome and Greece simply documented a past familiar from sacred or classical literature. The discovery of prehistory through objects would reach back far beyond the written word and vastly extend the dimensions of human history.

A strange series of coincidences gave the leading role in this discovery to a Danish businessman, Christian Jürgensen Thomsen (1788–1865). Without the erudition of a Scaliger or the mathematical genius of a Newton, he was a man of superlative common sense, richly endowed with the virtues of the dedicated amateur. His passion for curious objects was matched by his talent for awakening the curiosity of the new museum public. Born in Copenhagen, the eldest of six sons of a prosperous shipowner, he was trained for business. He came to know the family of a Danish consul who had served in Paris during the French Revolution, and who had brought back collections purchased from the panicked aristocracy. When young Christian, still only fifteen, helped his friends unpack their treasures, they gave him a few old coins to begin his own collection, and by the time he was nineteen he was a respected numismatist. In 1807, when the British fleet bombarded Copenhagen harbor to keep the Danish fleet from Napoleon, buildings went up in flames, and Christian joined the emergency fire brigade. Working through the night, he rescued the coins of a leading numismatist whose house was hit, and carried them to safety with the Keeper of the Royal Cabinet of Antiquities.

Copenhagen's newly established Royal Commission for the Preservation of Danish Antiquities was being flooded by miscellaneous old objects sent in by public-spirited citizens. The aged secretary of the commission could not face the accumulating pile. It was time for a younger man—and an opportunity made to order for Thomsen, then twenty-seven and known for his own beautifully organized collection of coins. "Mr. Thomsen is admittedly only a dilettante," the bishop on the commission conceded, "but a dilettante with a wide range of knowledge. He has no university degree, but in the present state of scientific knowledge I hardly consider that fact as being a disqualification." Accordingly, young Thomsen was honored with the post of unpaid nonvoting secretary. As it turned out, Thomsen's lack of academic learning equipped him with the naïveté that archaeology needed at that moment.

The dusty shelves of the commission's storerooms overflowed with unlabeled odd bundles. How could Thomsen put them in order? "I had no previous example on which to base the ordering of such a collection," Thomsen confessed, nor had he money to hire a professor to classify objects by academic categories. So he applied the commonsense procedures learned in his father's shipping warehouse. Opening the parcels, first he separated them into objects of stone, of metal, and of pottery. Then he subdivided these according to their apparent use as weapons, tools, food containers, or religious objects. With no texts to guide him, he simply looked at the objects, then asked himself what questions would be asked by museum visitors who saw them for the first time.

When Thomsen opened his museum to the public in 1819, visitors saw the objects sorted into three cabinets. The first contained objects of stone; the second, objects of bronze; the third, objects of iron. This exercise in museum housekeeping led Thomsen to suspect that objects made of similar materials might be relics of the same era. To his amateur eye it seemed that the objects of stone might be older than similar metal objects, and that the bronze objects might be older than those of iron. He shared this elementary suggestion with learned antiquarians, to whom he later modestly gave credit for the idea.

His notion was not entirely novel, but the similar notions found in classical authors were fanciful and misleading. In the Beginning, according to Hesiod, Cronos created men of the Golden Age who never grew old. Labor, war, and injustice were unknown. They eventually became guardian spirits on earth. Then in the Silver Age, when men lost their reverence for the gods, Zeus punished them and buried them among the dead. The Bronze Age, which followed (when even houses were made of bronze), was a time of endless strife. After the brief interlude of a Heroic Age of godlike leaders in their Isles of the Blessed, came Hesiod's own unfortunate Iron Age. Yet worse was still in store for mankind, a future of men born senile, and of universal decay.

Thomsen was not well enough educated to try to fit his museum objects into this appealing literary scheme. He was more interested in objects than in words. There were already "too many books," he complained, and he was not eager to add his own. But finally, in 1836, he produced his practical *Guide to Scandinavian Antiquities,* which outlined his famous Three-Age System. This, his only book, translated into English, French, and German, and spread across Europe, was an invitation to "Pre-History."

It was hard for European scholars at the time to imagine that human experience before writing could have been divided into the epochs that Thomsen suggested. It seemed more logical to assume that stone tools were always used by the poor, while their betters always used bronze or iron. Thomsen's commonsense scheme did not please the pedants. If there was a Stone Age, they scoffed, then why not also an Age of Crockery, a Glass Age, and a Bone Age? Thomsen's scheme, refined but not abandoned by scholars in the next century, proved to be more than an exercise in museum management. It carried the plain message that human history had somehow developed in homogeneous stages that reached across the world. And he arranged the objects in his museum according to his "principle of progressive culture."

Thomsen showed how much was to be learned, not only from those ancient sculptures that embodied Winckelmann's ideal of beauty but even from the simple tools and crude weapons of anonymous prehistoric man. Opening his collections free to everybody, Thomsen offered lively talks about the everyday experience of people in the remote past. A deft lecturer, he would hide some interesting little object behind his coattails, then suddenly produce it at the point in his story when that kind of object—a bronze utensil or an iron weapon—first appeared in history.

Following Thomsen's hints, archaeologists discovered and explored the trash heaps of the past. Their paths into history no longer ran only through the gold-laden tombs of ancient kings, but also through the buried kitchen middens ("middens," from an Old Scandinavian word for muck or dunghill). The first excavation of these unlikely sources was the work mainly of Thomsen's disciple Jens Jacob Worsaae (1821–1885). At the age of fifteen he had become Thomsen's museum assistant and during the next four years spent his holidays digging into the ancient barrows of Jutland with the aid of two laborers paid by his parents. With his athletic temperament and his outdoor enthusiasms he was the ideal complement to the museum-oriented Thomsen. In 1840, when he was only nineteen, using stratigraphy and the field evidence from Danish barrows and peat bogs, he published an article confirming Thomsen's Three-Age theory and assigning prehistoric objects to a Stone Age, a Bronze Age, or an Iron Age. He, too, was suggesting latitudes of time, throughout Denmark and beyond. A dozen years later, in 1853, the Swiss archaeologist Ferdinand Keller (1800–1881), when exploring the lake dwellings of Lake Zurich, concluded that "in Switzerland the three ages of stone, bronze, and iron, are quite as well represented as in Scandinavia."

Some obvious difficulties plagued these prophets of prehistory. How could you stretch human experience to fill the thousands of years of the past opened by Buffon and the geologists? How much neater to fit all pre-Christian history into the comfortable 4004 years B.C. defined by Archbishop Ussher! And then there were new problems created by the geologists, who now revealed that northern Europe had been covered by ice when Stone Age men were living in caves in southern France. To correlate all these facts required a still more sophisticated approach to the early human past. If the Stone Age people of southern Europe advanced northward only after the retreat of the glaciers, then the three universal stages were reached at different times in different places.

To make the Three-Age scheme fit the whole human past in Europe was not easy. The so-called Age of Stone in Thomsen's museum was represented by polished stone artifacts of the kind people would be tempted to send in as curios. Meanwhile, Worsaae, out in the field, was hinting that the Age of Stone was far more extensive and more ancient than was suggested by these skillfully polished stone implements. On the digging sites each object unearthed could be studied not as an isolated curio but among all the remains of a Stone Age community. And these too might provide clues to other Stone Age communities across the world.

Worsaae's opportunity came in 1849, when a wealthy Dane named Olsen was trying to improve his large estate called Meilgaard on the northeast coast of Jutland. Building a road, he sent his workmen in search of gravel for surfacing material. When they dug into a bank a half-mile from the shore, they found no gravel but luckily hit an eight-foot layer of oyster-shells, which was even better for their purpose. Mixed with the shells they found pieces of flint and animal bones. One small bone object two and a half inches long caught their attention. Shaped like a four-fingered hand, it was plainly the work of human craft. Perhaps it had been made for a comb.

Olsen, the proprietor, sharing the popular interest in antiquities which had been stimulated by Thomsen, sent the object to the museum in Copenhagen, where Worsaae's curiosity was aroused. Shell heaps recently turned up elsewhere in Denmark had brought to light flaked flint, odd pottery fragments, and crude stone objects similar to the Meilgaard comb. Perhaps this mound of oyster shells "had been a sort of eating-place for the people of the neighborhood in the earliest prehistoric times. This would account for the ashes, the bones, the flints and the potsherds." Perhaps here, at long last, modern man might visit an authentic Stone Age community. And actually imagine Stone Age men and women at their everyday meals. Worsaae observed that the shells had all been opened, which would not have been the case if they were merely washed up from the shore.

When other scholars disagreed, each with his own theory, the Danish Academy of Sciences appointed a commission. Worsaae, with a zoologist and a geologist, was assigned to interpret these shell heaps found along the ancient Danish shore. These "shell middens," the commission concluded, were really kitchen middens, which meant that now for the first time the historian could enter into the daily life of ancient peoples. Trash heaps might be gateways to prehistory. Such a discovery could not have been made indoors in a museum, but only on the spot in the field. Since the crudely crafted artifacts of the kitchen middens were never polished, unlike the polished stone artifacts of a later Stone Age, they were not likely to be noticed by laymen or sent to a museum. The kitchen middens opened another vast epoch of human prehistory—an early Stone Age, which extended behind the later Stone Age of polished stonework.

Thomsen and his museum collaborators had done their work of publicizing archaeology so well that the question now raised—whether the Stone Age really should be divided into two clearly defined stages—was no longer an arcane conundrum for university professors. The issue was hotly debated in the public proceedings of the Danish Academy. Worsaae's opponents insisted that the shell heaps were only the picnic sites of the Stone Age visitors who had left their best implements elsewhere. The king of Denmark, Frederick VII, who shared the growing interest in antiquities, had excavated middens on his own estate and even wrote a monograph with his interpretation. In 1861, to "settle" the issue, he summoned the leading scholars to a full-dress public meeting at Meilgaard, where he would preside. This royal conclave, no routine academic conference, would be celebrated with the panoply of a coronation. Besides hearing a debate, all those invited would witness the ritual excavation of a new portion of the mound. In the mid-June heat, archaeologists dug into the celebrated mound from eight in the morning till six in the evening, wearing their official "archaeologist's" uniform out of respect for the King. When King Frederick had appointed Worsaae curator of his private collection of antiquities in 1858, he had playfully designated this archaeologist's uniform (high collar and tight-fitting jacket, topped off by a pillbox hat), which was now de rigueur at the diggings.

The lords of surrounding estates entertained the King and his party with banquets and dancing to band music every night. In honor of their royal visitor the neighbors created triumphal arches, and the King was accompanied everywhere by his mounted guard in full livery. A royal welcome to the Old Stone Age!

Early in the meeting it was agreed that Worsaae had won his scholarly point, which now would be proclaimed in royal company and for the whole nation. "I had the especial satisfaction," Worsaae wrote, "of seeing that, among the many hundred stone implements discovered among the oysters, not a single specimen was found with any traces of polishing or of superior culture." And he reported with relish how a human fillip was added to the formal splendor. "Only at the last minute, after we had frequently remarked on this fact, did two polished axes turn up, of a completely different type, which some practical joker had inserted in the heap to cheat us." The practical joker, it was widely assumed, was King Frederick himself.

Seldom has so drab an epoch of history been so splendidly inaugurated. But now, to the royal Danish imprimatur was added the near-unanimous agreement of scholars across Europe. What came to be called the Culture of Kitchen Middens (c. 4000–c. 2000 B.C.) was discovered in due course across the northern European coasts, and in Spain, Portugal, Italy, and North Africa. In southern Africa, northern Japan, in the islands of the Pacific, and in the coastal regions of both Americas, Kitchen Middens cultures seemed to have persisted into a later era. Once identified and placed in the chronicle of human development, the middens provided revealing latitudes of time—and a new vividness for the prehistoric past.

Worsaae, who became professor of archaeology at Copenhagen, and then succeeded Thomsen as director of the museum, is often called "the first professional archaeologist." His mentor Thomsen called him a "heaven stormer." Worsaae accurately praised Thomsen's Three-Age System as "the first clear ray . . . shed across the Universal prehistoric gloom of the North and the World in general." Not in the heavily documented realms of recent history but in the dark recesses of earliest times would mankind. first discover the "universality" of history. The first discovery of the community of all human experience in eras and epochs, the worldwide phenomena of human history, was made when "prehistory" was parsed into the three ages: Stone, Bronze, and Iron. And as Worsaae explored the boundaries between the three ages, he began to raise some profound questions that were explosive for fundamentalist Christians. One of these was the problem, still agitated by anthropologists: independent invention or cultural diffusion?

The disturbing notion, suggested by bold thinkers from Buffon to Darwin —that man had existed long before the Biblical date of Creation in 4004 B.C. —was beginning to be accepted by the scientific community. But the remote antiquity of man was popularized not so much by a theory as by the discovery of a vast and undeniable subject matter, a new dark continent of time, prehistory. More persuasively than a theory, the artifacts themselves seemed to bear witness to a chronology of prehistory that argued the evolution of man's culture.

Gradually, as the word "prehistory" came into use in the European languages, the idea entered popular consciousness. The exhibition in Hyde Park in 1851, which purported to survey all the works of humankind, still gave no glimpse of prehistory. Then, at the Universal Exhibition in Paris in 1867, the Hall of the History of Labor showed an extensive collection of artifacts from all over Europe and from Egypt. The official guide to Prehistoric Walks at the Universal Exhibition offered three lessons from the new science: the law of the progress of humanity; the law of similar development; and the high antiquity of man. In that same year the announcement of the first Congrès International Préhistorique de Paris brought the first official use of the word "prehistoric."

Prehistory entered the curriculum of public education along with the companion ideas of evolution. Charles Darwin's disciple and leading popularizer, John Lubbock (Lord Avebury, 1834–1913), made a European reputation by fitting prehistory into evolution. His *Pre-Historic Times* (1865), which coined the words "Paleolithic" and "Neolithic" for the "Polished Stone Age," was widely read by laymen, who imbibed prehistory and evolution in a single delightful read. *The Origin of Civilization* (1871) drew on widely separated evidences of centers of the Three Ages to argue that the crucial inventions had arisen independently. All of which seemed to support Herbert Spencer's argument that "Progress is not an accident but a necessity. It is a fact of nature."

When Schliemann came to London in 1875, William E. Gladstone saluted him by recalling that when they were growing up, "prehistoric times lay before our eyes like a silver cloud covering the whole of the lands that, at different periods of history, had become illustrious and interesting. . . . Now we are beginning to see through this dense mist and the cloud is becoming transparent, and the figures of real places, real men, real facts are slowly beginning to reveal to us their outlines." The pioneer anthropologist Edward B. Tylor optimistically announced in 1871 that prehistory finally had "taken its place in the general scheme of knowledge"—extending the vistas of human history a thousandfold.

The Three Ages, the worldwide epochs of prehistory, made it easier to imagine other epochs that transcended city, region, or nation. By defining latitudes of history, man had enlarged his view of the world's past and present. The invention of grand historical "Eras," "Epochs," or "Ages" which overreached political bounds would provide time receptacles ample enough to include the whole data of past communities of culture, yet small enough for persuasive definition. Few other concepts have done so much to deprovincialize man's thinking. The Ages of history would dominate (and sometimes tyrannize.) the modern historian, focusing his vision on clusters of past experience—the Great Age of Greece, the Middle Ages, the Age of Feudalism, the Renaissance, the Enlightenment, Modern Industrialism, the Rise of Capitalism, etc.

These notions were to time what the "species" were to nature, a way of classifying experience to make it useful. They were the taxonomy of history. Of course, just as with "species," there was the danger that the label would be taken for the thing, the mere name of an "epoch" might somehow become a force governing interpretation of the events. Still, the advantages of the epochal way of thinking far outweighed the risks. The convenient groupings of men, events, achievements, and institutions helped bring some order into the puzzling miscellany of the past. The six "world periods" *(aetates)* into which the early Church Fathers had divided all time before the coming of Christ were not historical but prophetic and theological. They did not characterize the past, but were categories of prophecy, stages toward the Incarnation.

"The 'spirit of the age,' " John Stuart Mill (1806–1873) explained in 1831, "is in some measure a novel expression. I do not believe that it is to be met with in any work exceeding fifty years in antiquity. The idea of comparing one's own age with former ages, or with our notion of those which are yet to come, had occurred to philosophers; but it never before was itself the dominant idea of any age. Before men begin to think much and long on the peculiarities of their own times, they must have begun to think that those times are, or are destined to be, distinguished in a very remarkable manner from the times that preceded them." The idea of homogeneous epochs in history, he added, was consistent either with the notion of cycles, or with "the idea of a trajectory or progress." Mill plumped for "the progressiveness of the human race . . . the foundation on which a method of philosophizing in the social science has been of late years erected." How could one imagine "progress" without some notion of the coherence of events in each epoch?

Now a host of new influences—museums, archaeological excavations, international expositions, along with the daily press and the periodical press —was spreading a consciousness of history beyond academic circles, preparing people to believe that they lived in an Age of Progress. "There is a progressive change," John Stuart Mill concluded from his study of history, "both in the character of the human race, and in their outward circumstances so far as moulded by themselves . . . in each successive age the principal phenomena of society are different from what they were in the age preceding, and still more different from any previous age."

A vivid reminder of these newly drawn latitudes of time was the invention in the mid-nineteenth century of the "Age of the Renaissance" to describe an era in Europe from about the fourteenth to the seventeenth century. The French nationalist historian Jules Michelet entitled the seventh volume of his History of France *The Renaissance* (1855), and saw the era dominated by "the discovery of the world and the discovery of man." Then the Swiss historian Jacob Burckhardt's *Civilization of the Renaissance in Italy* (1860) offered a classic portrait of the men and institutions that gave the era its character and made it the "mother" of modern European civilization. So a student, confident in the jargon of historical epochs, could characterize Dante as "a man who stood with one foot in the Middle Ages and with the other saluted the rising star of the Renaissance." In our century much of the scholarly debate over the nature of the Renaissance has concerned the latitudes of time: When did the Renaissance begin? Was it the same phenomenon in different parts of Europe?

Two grand assumptions, which lay beneath talk of the Renaissance, shaped future thinking about man's role in all history. First, belief that every age somehow exuded a prevailing spirit—what German scholars called the *Zeitgeist,* what Carl Becker called the "Climate of Opinion"— which favored certain notions and institutions. Second, that within these limits, men had the power to make history. Renaissance men made a Renaissance. If, as Burckhardt explained, they made the state "a work of art," in later ages, too, men could accomplish the unprecedented.

76

Hidden Dimensions: History as Therapy

THE discovery of prehistory had come from the simple effort to arrange artifacts of the distant past into an intelligible order. There was an impenetrable mystery about when and by whom a stone ax had been made, but there seemed no similar uncertainty about man's thought. Ideas were believed to be universal and immutable. Descartes, in his *Discourse on Method* (1637), had insisted on the universality, uniformity, and constancy of man's reason, which he expressed in the familiar proposition, "I think, therefore I am." The world of mind was eternally separate from the physical world of experience and of history. Locke, in his *Essay concerning Human Understanding* (1690), began to relate mind to history by making experience the source of ideas, and knowledge the perception of the agreement or disagreement of ideas. But according to Locke, too, reason and the senses operated constantly and uniformly and the processes of thinking were proof of a universal and eternal mind. Man's ideas, from these various points of view, remained the product of a homogeneous process.

A revolutionary new discovery, or at least a pregnant suggestion, was that man's ideas might be nothing but human artifacts, merely symptoms of changing experience. Then the process by which men acquired what passed for knowledge would not be uniformly rational, nor would that knowledge be universal and changeless. Perhaps some forces other than reason were at work. Did ideas themselves have a history?

A pioneer explorer of this question was the unhappy Italian philosopher Giambattista Vico (1668–1744). The son of a poor bookseller, he had a nearly fatal fall on his head at the age of seven, when doctors predicted that he might become an imbecile. Vico himself said this accident explained his lifelong melancholy. Yet, despite poverty, repeated depressions, and an attack of typhus, Vico from his professorship of rhetoric in Naples managed to finance publication of his writings. Ignored by contemporaries, he came into his own in the late eighteenth century, when Goethe made Vico's "prophetic insights" the basis of his own philosophy of history. In the nineteenth century the eloquent French romantic

Michelet called him "his own Prometheus," and Marx, too, learned much from him.

Born into the generation after Newton that was ebullient with the promise of natural science, Vico declared in his *Principles of New Science . . . concerning the Common Nature of the Nations* (1725) "that the world of civil society has certainly been made by men, and that its principles are therefore to be found within the modifications of our own human mind. Whoever reflects on this cannot but marvel that the philosophers should have bent all their energies to the study of the world of nature, which, since God made it, He alone knows; and that they should have neglected the study of the world of nations . . . which, since men had made, men could come to know." According to Vico, the changing relationship of past peoples to the forces of nature explained their ways of thinking. In the most primitive stage, the Age of the Gods, fearful men were governed by religion and ruled by priest-kings. Then, in an Age of Heroes, to escape the bestial struggle for survival people put themselves under the protection of strong men. "This law of force is the law of Achilles, who referred every right to the tip of his spear." Finally in an Age of Peoples the plebians who had accumulated wealth asserted themselves in "human law dictated by fully developed human reason."

Each stage brought forth its characteristic literature. For example, the Homeric poems were not the product of one talented bard but the unconscious expression of the whole Age of Heroes. "Homer was an idea or a heroic character of Grecian men insofar as they told their histories in song." When poetry gave way to prose in the Age of Peoples, the customs of religion were replaced by written codes that defined rights and privileges. Social classes were not ordained by God but emerged from this progressive development, and brought new ways of thought. The last stage of the cycle, producing ease and luxury, always ended in decadence. The society then relapsed but never quite back to its earlier stage. The progress of humanity was an upward spiral, ascending by the beneficence of a divine Providence.

The crucial novelty in Vico's *New Science* was his treatment of ideas and institutions (excepting only Christianity itself) as mere symptoms of social experience. Man's reason, too, was the product of gradual development. If Vico was correct, then, of course, his own ideas had no absolute validity but were a by-product of the Age of Peoples. He tried to avoid this logical consequence simply by declaring Christianity to be the one true religion for all societies. And Vico's *New Science* would liberate mankind from fears by making people conscious of how their thoughts were shaped. Then they could take charge of their destiny and shape institutions toward desired ends.

Karl Marx (1818–1883) grew up at the end of the century of Adam Smith, James Watt, and Thomas Jefferson, of rising nations, growing colonies, expanding factories, and burgeoning capitalism. He would find hidden dimensions of the past in the forces of production then dramatically exploding in western Europe.

The story of Karl Marx's life, like that of Vico's, is an almost unrelieved tale of personal frustration, flight, and tragedy. Born in Trier in Prussia, he was descended from a long line of rabbis on both sides of his family, an inheritance that biographers recognize in his inclination to dialectics and philosophical debate. His father was a brilliant lawyer, an admirer of Voltaire and an active advocate of a constitution for Prussia. As a practical necessity for a career in the law, he converted to Christianity before Karl was born. Marx's Dutch mother was no intellectual, and spoke German with an accent all her life. She was baptized when Karl was only seven, and he also was baptized about that time, in the year when Heinrich Heine taking the same step called his baptism "an entrance card into the community of European culture." For his university education he followed a familiar German pattern, moving from one place to another according to the appeal of different professors or of the student life. He sowed his wild oats at Bonn, where for drunken rowdiness he actually spent twenty-four hours in the university jail, which would be the only imprisonment of his life. His father insisted that he transfer to Berlin to study law and philosophy. Although two famous German historians, Von Ranke and Von Savigny, were lecturing there at the time, Marx was most influenced by the philosophy of Hegel being preached by a charismatic young *dozent*, Bruno Bauer. He joined the "Doctor's Club" of young Hegelians, who met to debate the social implications of Hegel's idealistic doctrines, which would stay with Marx throughout his life. He seemed unable to forget any theory he had once encountered, and was adept at making it either a foundation or a foil for his own restless philosophizing. A vivid doggerel description of the Marx whom he had not yet met but about whom he heard was composed by Friedrich Engels (1820–1895):

> Who rushes behind with wild bluster?
> A swarthy fellow from Trier, a vigorous monster.
> He walks not, hops not, he leaps on his heels
> And raves, full of rage, as if he wanted to seize
> The broad canopy of heaven, and pull it down to earth,
> His arms extended very wide in the air.
> With angry fist balled, he rants ceaselessly,
> As if ten thousand devils held him by the forelock.
>
> [Translated from the German by Saul Padover]

His doctor's dissertation on the arcane subject of "The Difference between the Democritean and Epicurean Philosophy of Nature" brought him his degree from Jena in 1841. "The glorification of the heavenly body," he explained, "is a cult which all Greek philosophers celebrate. . . . It is the intellectual solar system. Hence the Greek philosophers, in worshipping the heavenly bodies, worshipped their own mind."

In Cologne as editor of a new liberal paper, the *Rheinische Zeitung*, financed by the city's enterprising merchants, Marx advocated various social causes, opposed censorship, and championed freedom of the press, including the freedom to examine such novel notions as communism. Within a year he was forced out, the paper was banned by the Prussian government, and he was on his way to Paris. After a seven-year engagement, in 1843 he married Jenny von Westphalen who was the unflickering gleam of happiness in his life.

In Paris Marx earnestly studied the movements of French and German workers who were organizing a Communist League and a secret society called the League of the Just. He began his collaboration with the twenty-four-year-old Engels, wrote his first works on French politics and economics, and an article calling for the "uprising of the proletariat." He also began to develop his polemics against religion in general, which he unforgettably stigmatized as "the opium of the people." Heinrich Heine, also in Paris at the time, was amused at his "stubborn friend Marx . . . and the rest of the godless, self-appointed gods." When the French government expelled Marx, he fled to Brussels, where he registered as an alien, impetuously took the legal steps to renounce his Prussian citizenship, and at the age of twenty-eight committed himself to a life of exile.

During his stay in Brussels, which lasted three years, he joined with Engels to write the *Communist Manifesto* for the Communist League, which had been meeting in London. In place of the league's old motto, "All men are brethren," Marx substituted his rousing "Proletarians of all countries, unite!" When the liberal revolutions of 1848 broke out in western Europe, Marx fled back to Cologne, where he revived the *Rheinische Zeitung* and attacked both the partisans of representative democracy and their radical opponents. Again banished, he returned briefly to Paris. Expelled once more, in 1849 he arrived in London, his principal habitat for the rest of his life. If he had any home during the remaining thirty-four years of his life, it was the library of the British Museum.

Before he arrived in London, Marx had produced a whole sheaf of polemical pamphlets trying to find his bearings at the same time in philosophy, in history, and in the eruptive politics of his age. His attitude toward violent revolutionary action was vacillating. Although he had exhorted workingmen of all countries to unite, he usually advised against armed rebellion. At least once he and Engels publicly urged that their *Communist Manifesto* should be put aside. He became accustomed to attack both by conservatives who saw him as an agitator for anarchy and by militant socialists who stigmatized him as a capitalist lackey. The constant current in his thought was faith in his own evolving theory of history and his ironic belief that ideas and political movements really could not change the course of history.

Despite poverty and the tragic death of his children, he pursued the dogged research in the British Museum that produced the monumental three-volume *Das Kapital.* He refused to seek regular employment because he would not allow bourgeois society to make him into "a money-making machine." During these years his main financial support was the largess from Engels' Manchester cotton mills and a small family inheritance. His negligible earned income came from occasional articles for the *New York Tribune.*

Marx's economic theory is generally agreed to be an application and critique of the "classical" economic theory of Adam Smith and David Ricardo. But out of his research in the British Museum and his experience of the revolutions of his time came an original theory of history. Instead of explaining social progress as the conscious and unconscious collaboration of social classes, Marx saw the conflict of economic classes as the dynamic force. "The history of all hitherto existing society," the *Communist Manifesto* proclaimed, "is the history of class struggles. . . . In the earlier epochs of history, we find almost everywhere a complicated arrangement of society into various orders. . . . In ancient Rome we have patricians, knights, plebeians, slaves; in the Middle Ages, feudal lords, vassals, guildmasters, journeymen, apprentices, serfs. . . . Our epoch, the epoch of the bourgeoisie . . . has simplified the class antagonisms. Society as a whole is more and more splitting up into two great hostile camps. . . .bourgeoisie and proletariat."

Beneath the theory of class conflict lay his "materialist" faith that ideas were a response to changes in the system of production. Before his time the most influential historians, with a few exceptions like Voltaire and Montesquieu, had focused on the learned, the powerful and the wealthy, on chancellors, princes, and kings, on royal succession and the intrigues of the court, on chanceries and parliaments, and battlefields. They saw Truth battling against Error, Virtue against Vice, Orthodoxy against Heresy. Human Reason was depicted as an autonomous universal faculty dealing in the pure currency of changeless Ideas. Marx shifted the focus to scenes unfamiliar to the literati who had been writing history.

Das Kapital (1867) is a difficult and sometimes pedantic work. Even so, the first of the three volumes, the one published during Marx's lifetime, is widely read. When its first translation from the German into another language appeared in 1872, the Russian censor let it pass because, he observed, "few people in Russia will read it, and still fewer will understand it." There, however, it quickly sold out its first edition of three thousand. The supercilious reviewer of the first English translation (1887), in London's literary *Athenaeum* observed that "Under the guise of a critical analysis of capital, Karl Marx's work is principally a polemic against capitalists and the capitalist mode of production, and it is this polemical tone which is its chief charm."

For the layman who is no economist the book's most intelligible passages are Marx's cameos of social and economic history. For example:

> One of the most shameful, the most dirty, and the worst paid kinds of labour, and one on which women and young girls are by preference employed, is the sorting of rags. It is well known that Great Britain, apart from its own immense store of rags, is the emporium for the rag trade of the whole world. They flow in from Japan, from the most remote States of South America, and from the Canary Islands. But the chief sources of their supply are Germany, France, Russia, Italy, Egypt, Turkey, Belgium, and Holland. They are used for manure, for making bed-flocks, for shoddy, and they serve as the raw material of paper. The rag-sorters are the medium for the spread of small-pox and other infectious diseases, and they themselves are the first victims.

He describes a tile field where a young woman twenty-four years of age made two thousand tiles a day with the assistance of two little girls who carried daily ten tons up the slippery sides of the clay pits, from a depth of 30 feet, and then for a distance of 210 feet. From the Parliamentary Reports themselves he culls telling passages:

> "It is impossible for a child to pass through the purgatory of a tilefield without great moral degradation . . . the low language, which they are accustomed to hear from their tenderest years, the filthy, indecent, and shameless habits, amidst which, unknowing, and half wild, they grow up, make them in after life lawless, abandoned, dissolute. . . . They become rough, foul-mouthed boys, before Nature has taught them that they are women. Clothed in a few dirty rags, the legs naked far above the knees, hair and face besmeared with dirt, they learn to treat all feelings of decency and of shame with contempt. During meal-times they lie at full length in the fields, or watch the boys bathing in a neighboring canal. Their heavy day's work at length completed, they put on better clothes, and accompany the men to the public houses. . . . The worst is that the brickmakers despair of themselves. You might as well, said one of the better kind to a chaplain of Southallfield, try to raise and improve the devil as a brickie, sir!"

Taking off from John Stuart Mill's doubts whether "all mechanical inventions yet made have lightened the day's toil of any human being," Marx shows how modern machinery and steam power actually lengthened the working day, making the conditions of labor ever more intolerable. In a qualifying footnote he observes that in fact machinery had "greatly increased the number of bourgeois well-to-do idlers," then he details the miseries of workers in the cotton mills and mines. He depicts the sufferings of children deprived of all opportunities for education, how women were condemned to work "degrading to their sex," how misleading were coroners' inquests on the causes of death in the mines, how corrupt was the official "inspection"—all of which "by maturing the material conditions, and the combination on a social scale of the processes of production . . . matures the contradictions and antagonisms of the capitalist form of production, and thereby provides, along with the elements for the formation of a new society, the forces for exploding the old one." His facts from government sources were hard to contradict.

Whatever one thought of Marx's revolutionary prophecies, one could not ignore the facts of life on which he had shone the spotlight of his epigrammatic prose. His focus on the conditions of the working classes throughout the whole past was only the most superficial of his influences. More fundamental was his new version of all history and especially of the birth and life of ideas.

Before Marx, the prime movers had been great leaders and great ideas shaping the conditions of life. But for Marx, as he explained in a much quoted passage:

> In the social production of their means of existence men enter into definite, necessary relations which are independent of their will, productive relationships which correspond to a definite stage of development of their material productive forces. The aggregate of these productive relationships constitutes the economic structure of society, the real basis on which a juridical and political superstructure arises, and to which definite forms of social consciousness correspond. The mode of production of the material means of existence conditions the whole process of social, political and intellectual life. It is not the consciousness of men that determines their existence, but, on the contrary, it is their social existence that determines their consciousness.

In 1859, the very year when he wrote these words, Darwin's *Origin of Species* provided him, he thought, with a doubly welcome illustration. The Darwinian struggle for survival seemed only a translation into natural history of the class struggles of all past human history. And the appearance of Darwin's ideas at that time, in the heyday of English capitalism, showed that ideas were symptoms and not causes. While, as we have seen, some were hailing Darwin as a prophet of scientific truth and others were attacking him for blasphemy, Marx saw Darwin's ideas in quite another light.

"The death-blow for the first time to 'teleology' in the natural sciences," Marx exulted, "very important . . . as a natural-scientific basis for the class struggle in history." "It is remarkable how Darwin recognizes among beasts and plants his English society with its division of labour, competition, opening up of new markets, 'inventions,' and the Malthusian 'struggle for existence.' It is Hobbes's *bellum omnium contra omnes,* and one is reminded of Hegel's Phenomenology, where civil society is described as a 'spiritual animal kingdom,' while in Darwin the animal kingdom figures as civil society."

Luckily, in the historical theories of Hegel, Marx had found a perfect foil for his own way of thinking. Just as we can wonder whether Copernicus would have found his framework if the Ptolemaic scheme had not been there ready for the centrality of the earth to be replaced by the centrality of the sun, so we can wonder what Marx might have come up with if Hegel's antithetic scheme had not been there. By tradition and by training a dialectician, Marx thrived on opposition. There was no better example of the dialectic process at work than Marx's own reaction to Hegel and others. Marx's writings are replete with quotations from his spiritual antagonists, usually his former friends, teachers, or associates, against whom he finds his bearings: *The Holy Family* against Bruno Bauer, *The Poverty of Philosophy* against Proudhon, his *Theses on Feuerbach,* and (with Engels) *Anti-Dühring.* While his thought was decisively shaped by Hegel, yet he made Hegel into his Anti-Marx, as he explained in his Preface to *Das Kapital:*

> My dialectic method is not only different from the Hegelian, but is its direct opposite. To Hegel, the life-process of the human brain, i.e., the process of thinking, which, under the name of "the Idea," he even transforms into an independent subject, is the demiurgos of the real world, and the real world is only the external, phenomenal form of "the Idea." With me, on the contrary, the ideal is nothing else than the material world reflected by the human mind, and translated into forms of thought.

Yet Marx's own writings, and the influence of his ideas in the least industrialized parts of the world, would show the limitations of a "materialist" view of history. Marx's pages are often a patchwork of passages from his latest philosophic enemies. It is hard to know what Marx means without reading the works of those he is contradicting, who are only seldom giants in the history of thought.

Despite Marx's hyperphilosophic hyperpolemical style, there is a grandeur, a wit, and a poignancy to his view of history. "Christian Socialism," he says, "is but the holy water with which the priest consecrates the heart burnings of the aristocrat." By his ways of asking he awakens us to ignorance that we had never recognized. He thought he was definitively mapping the whole human past. He was really a discoverer of *terra incognita,* only a Columbus, whose followers liked to think of him as a Vespucci. Mocking the professionally sanctified clichés of historians before him, he was a latter-day Paracelsus. While he outraged by his questions, he could not satisfy by his answers. "The handmill gives you society with the feudal lord; the steammill, society with the industrial capitalist." And other more outlandish, oversimplified but always illuminating connections. His questions opened hidden dimensions of history. After Marx, even non-Marxist historians could never be satisfied by all the old answers.

Marxists came to call the discovery of these hidden dimensions their Science of Society, which they offered as a kind of therapy. To discover the simple truth that "the ruling ideas of each age have ever been the ideas of the ruling class," Marx argued, would eventually liberate the modern proletariat from the ideas of their ruling class and from the illusion that these were universal Truths. Understanding history was not only one way to knowledge, it was the only way. Just as the convert to Jesus was freed from pagan gods, so the convert to Marx would be freed from enslavement to idols fabricated by those who controlled the machinery of production. Saint Augustine had cast Christianity into a historical credo, taking direction from a unique event and moving toward a divine end. "And ye shall know the truth, and the truth shall make you free." Marxists would not dissent from the axiom of Saint John, for Marx, like Saint Augustine and Vico, believed that the cure for man's feeling of powerlessness was knowledge of the true course of history. Again, history had become therapy.

Quite other hidden dimensions of the past were discovered by Sigmund Freud (1856–1939). No vagrant or political organizer, he led a quiet scholar's life in Vienna, where he lived from the age of three. His father, a wool merchant of freethinking liberal political views, had difficulty supporting the family. Like Marx, Freud was born a Jew, but unlike Marx, he never became an anti-Semite. He was an active member of the B'nai B'rith Society and was fond of Jewish anecdotes. In anti-Semitic Vienna, Freud's Jewishness always limited his opportunities and never ceased to affect his thinking. An industrious and successful student, he first inclined to study law. He recalled his interests as he entered the university, in 1873:

> Neither at that time, nor indeed in my later life, did I feel any particular predilection for the career of a physician. I was moved, rather, by a sort of curiosity, which was, however, directed more toward human concerns than toward natural objects; nor had I grasped the importance of observation as one of the best means of gratifying it. My early familiarity with the Bible story (at a time almost before

> I had learnt the art of reading) had, as I recognized much later, an enduring effect upon the direction of my interest. . . . At the same time, the theories of Darwin, which were then of topical interest, strongly attracted me, for they held out hopes of an extraordinary advance in our understanding of the world; and it was hearing Goethe's beautiful essay on Nature read aloud at a popular lecture . . . just before I left school that decided me to become a medical student.

For Freud as for Marx it was a broad humanistic interest that led him to pioneer on the borderlands of science.

The mysterious character of all human experience, as Bruno Bettelheim has observed, was what intrigued Freud and finally captured his energies. This explained, too, why he moved on from treating man's body to treating his soul. Freud began his professional career in a physiology laboratory trying to confirm Hermann Helmholtz's axiom that "No other forces than the common physical-chemical ones are active within the organism." This experience Freud described in his *Autobiography* as "a detour over the natural sciences, medicine, and psychotherapy" from "those cultural problems which had once captivated the youth who had barely awakened to deeper thought." His lifework would be less in the spirit of Helmholtz than of Goethe.

The young Freud's studies at the Sperl Gymnasium in Vienna from the age of nine until seventeen bore a heavy emphasis on Greek and Latin, and he remained a lifelong passionate devotee of the classics. His most influential writings were replete with Greek words and connotations: Eros and Oedipus and Psyche (Greek for "soul"), among many others. From his early years and despite his financial difficulties, Freud's hobby was collecting antique statues, which were, apart from his twenty cigars a day, his only extravagance. When he bought himself a copy of Schliemann's *Troy*, he was delighted by the author's account of his childhood hope to uncover the buried city. This Freud generalized into a lesson for his own life and for the very foundations of psychoanalysis. "The man was happy when he found the treasures of Priam, since the only happiness is the satisfaction of a childhood wish." His own definition of happiness was "the subsequent fulfillment of a prehistoric wish. That is why wealth brings so little happiness: money was not a wish in my childhood." The charm of classical antiquity was a leitmotif in his life, as he noted his delight at acquiring a Roman statue in Innsbruck in 1898 or the pleasures of reading Burckhardt's *Cultural History of Greece.*

The great travel experiences of his life were visits to Rome and to Athens, another focus for his omnivorous interest in origins. Freud, identifying himself with the Semitic Hannibal, likened his difficulties in reaching Rome to the clerical anti-Semitism that had denied him a university professorship in Vienna. When, on the first of many visits, he finally arrived in Rome in 1901, he was entranced by the antiquities of the Vatican Museum, especially the Laocoön and the Apollo Belvedere. In Athens his visit to the Acropolis left him in dazzled disbelief that anything could be so beautiful, a feeling that he never lost. When Freud set out for America, apparently his primary interest, apart from seeing Niagara Falls, was to view the famous collection of Cyprus relics in New York. At the Metropolitan Museum again his focus was on the Greek antiquities. In Vienna, his consulting room and his famous study were lined with cabinets holding his collection. Even the narrow writing desk was cluttered with little statues, mostly Egyptian, which he would replace from time to time with items from his cabinets.

Freud's interest in archaeology, more than a hobby, expressed his quest for our whole unacknowledged inheritance from the past. When in his forties he turned from the world of Helmholtz and neurology to the world of culture and history, he committed himself to the archaeology of the soul, the "psyche." The unexamined strata of experience, both of society and of the individual, were his digging ground. "Every earlier stage of development," he observed, "persists alongside the later stage which has arisen from it." For him our unexcavated memories were the artifacts of human archaeology. Which, of course, was one reason why he gave such importance to recapturing the experiences of childhood.

The central problems of human life, according to Freud, are in its hidden dimensions. "In mental life nothing which has once been formed can perish . . . everything is somehow preserved." Freud saw human frustrations and conflicts arising not from what man had forgotten, but from the embedded memories of which he was not conscious. Could the old arts of Memory now be turned to the service of man's self-discovery? Discovering the past would be not merely a delightful experience but a way of liberation. Psychoanalysis would be a way of curing the self by reviving memories and remembering that they are only memories. Those suffering from hysterics "cannot get free of the past and for its sake they neglect what is real and immediate." The problem of all neurotics was that they were "anchored somewhere in their past." For Freud, knowing one's inner history became therapy.

No "physical-chemical" equations could ever explain human life. For memory was the peculiarly human ingredient, and unless the strata of experience were uncovered, neither a society nor an individual could know itself.

Freud, in his way, too, would be a Paracelsus. The "incurable" ills of the spirit must have their remedies. The professionals would be shaken by the amateur spirit of Freud's *Psyche*-analysis. Freud remained a humanist and man of letters among the test tubes. Immersed in classical literature, he was vividly aware of the beloved of Eros, the mythical Greek girl Psyche, who was so beautiful that the jealous Venus put her to sleep. He never would distill the fertile literary vagueness and ambivalences from his language or his method. When American doctors wanted to restrict the practice of psychoanalysis to trained physicians, Freud strenuously objected to "the obvious American tendency to turn psychoanalysis into a mere housemaid of Psychiatry." He seems to have chosen his favorite terms because of their literary redolence. On more than one occasion he protested against those who would translate his *Ich* (the "I") and his *Es* (the "It") and Psyche-analysis (the analysis of the "soul") itself into a speciously precise jargon of Egos, Ids, and Superegos. " 'Psyche,' " Freud observed as early as 1905, "is a Greek word and its German translation is soul [*Seele*]. Psychical treatment hence means 'treatment of the soul' [*Psyche ist ein griechisches Wort lautet in deutscher Ubersetzung Seele. Psychische Behandlung heisst demnach Seelenbehandlung*]." Ironically, in the United States, where Freud was first publicly acclaimed, his ideas were promptly made a preserve of a medical science and so emptied of the mystery of the prehistoric past that Freud had discovered in everybody.

PART FIFTEEN

SURVEYING THE PRESENT

The known is finite, the unknown infinite; intellectually we stand on an islet in the midst of an illimitable ocean of inexplicability. Our business in every generation is to reclaim a little more land, to add something to the extent and solidity of our possessions.

—THOMAS HENRY HUXLEY on the Reception of the "Origin of Species" (1887)

The eternal mystery of the world is its comprehensibility.

—ALBERT EINSTEIN (1936)

77

"All Mankind Is One"

IN 1537 the great Portuguese cartographer Pedro Nunes, charting the unexpected world in the West, rejoiced in "new islands, new lands, new seas, new peoples; and, what is more, a new sky and new stars." Discovering America brought Europeans face to face with the variety of humankind. At first they were tempted to make the surprising American continents the habitats of the legendary and "monstrous" races who were circumstantially described in Pliny's *Natural History* and who had fascinated and eluded travelers ever since. When Europeans dubbed the natives of the New World "Indians," they not only were committing a geographic error but were announcing their expectations of finding fantastic creatures.

Columbus reported to his surprise and somewhat to his disappointment, that "in these islands I have so far found no human monstrosities, as many expected, on the contrary, among all these peoples good looks are esteemed. . . . Thus I have neither found monsters nor any report of any, except . . . a people . . . who eat human flesh . . . they are no more malformed than the others." These Indians, he reassured the Spanish sovereigns, were "very well built, of very handsome bodies and very fine faces."

While this prosaic reassurance drained the new lands of a legendary charm, the "monstrous races" lived on. Poetry, folklore, and romance repeated old tales of *Anthropophagi* ("man-eaters"), the warlike *Amazons* ("without breasts," women who lived without men, and received their name because they had removed the right breast in order to draw the bow more powerfully), *Cyclopes* ("round-eyes," the one-eyed giants of Homer and Virgil), *Cynocephali* ("dog-heads," who communicated by barking, had huge teeth and breathed flames), *Pygmies* (who braided their long hair into clothing, and warred with the cranes who stole their crops). Then there were *Amyctryae* ("unsociables" who lived on raw meat and whose protruding lip served them as an umbrella), *Antipodes* ("opposite-footed" who lived at the bottom of the world and had to walk upside down), *Astomi* ("mouthless" apple-smellers, who could neither eat nor drink, and could be killed by a bad odor, but lived by smelling, mostly apples), *Blemmyae* (celebrated by Shakespeare as "men whose heads/ Do grow beneath their shoulders"), *Panotii* ("all-ears," whose long ears served for blankets, and, like Dumbo's, could be unfurled for wings), *Sciopods* ("shadow-foot," who had only one great foot, which served as a parasol to protect them from the sun as they lay on their backs).

These and other monstrous peoples inhabited a limbo between theology and fantasy. If, as the Bible explained, all men were descended from Adam, then perhaps these deformities were the punishment of some of Adam's children for their sins or for eating forbidden herbs. "The descendants displayed on their bodies what the forebears had earned by their misdeeds," a twelfth-century German poet explained. "As the fathers had been inwardly, so the children were outwardly."

Recalling Jesus' instruction to the Apostles—"Go ye therefore, and teach *all* nations, baptizing them in the name of the Father, and of the Son, and of the Holy Ghost"—there were tales of missionaries converting dog-headed cannibals in Parthia. Saint Augustine did not refuse the monstrous races a place in his *City of God*.

> Whoever is born anywhere as a human being, that is, as a rational mortal creature, however strange he may appear to our senses in bodily form or colour or motion or utterance, or in any faculty, part or quality of his nature whatsoever, let no true believer have any doubt that such an individual is descended from the one man who was first created.

If these creatures really were human, then they could and should be baptized.

But God had made nothing in vain. "Monstrous" races were so called from the Latin *monstrum* (from *monere,* to warn), meaning a divine portent. There was less agreement on what this omen signified. Since all humankind were descended from Adam in the Garden of Eden, physical deviance from the happy norm preserved in Europe must be explained by degeneration, decadence, or punishment for sin. There was no place in medieval Christian thought for evolving institutions, since all humankind had started at the same time and the whole spectrum of human institutions had been revealed and fulfilled in the Bible. But some people had fallen away.

After the Flood, when the earth was peopled by the sons of Noah, the guilty descendants of Cain or of Noah's son Ham deserved exile and punishment, which still tainted their bodies and their institutions. There were "better" or "worse" in peoples and institutions, but no earlier or later stages of social development. From the Garden of Eden the history of institutions was a one-way street where all roads led backward. Since the Fall there was ample opportunity for decay. But who could improve the Biblical design?

Three times, mankind's original uniform culture had been corrupted into diversity. Cain, punished for murdering Abel, was exiled to the Land of Nod east of Eden, where he and his posterity found strange ways. Later the sons of Noah were dispersed over the earth to go separately. And then again man's single humanity was confused at Babel. Diversity in religion, in language, or anything else was the sign of Cain. During the Middle Ages, when Europeans knew only a narrow range of cultural diversity, belief in the Biblical norm was reinforced by experience.

A revolution in Western thought came late, with the twin discovery that institutions were capable both of novelty not described or foreshadowed in the Bible and of development by which one kind of institution grew out of another. These notions and the idea of progress that came with them were by-products of exploration. The crucial event was the discovery of unexpected continents, which came to be called a New World. Just as in the Middle Ages Christian Europe argued the unity of mankind from the uniformity in the Garden of Eden, modern scientists would find new clues to the unity of the species in the diversity of human ways.

When Columbus reported that the people he encountered were not monsters but only savages, he unwittingly pointed toward a new science of culture. And toward an idea of progress. Extremes of human diversity were no longer exiled to realms of fantasy, for now they could be observed close up. However medieval may have been Columbus' geography, with his circumstantial description of the rivers of Eden, yet when he described the natives he suddenly spoke in the accents of an anthropologist in the field. For he reported their "very handsome bodies and very good faces; the hair coarse almost as the hair of a horse's tail and short; the hair they wear over their eyebrows, except for a hank behind that they wear long and never cut. Some of them paint themselves black (and they are of the color of the Canary Islanders, neither black nor white), and some paint themselves white, and others red, and others with what they have." In his letter to his sovereigns, soon circulated across Europe, as we have seen, Columbus described his encounter with the natives:

> They are so ingenuous and free with all they have, that no one would believe it who has not seen it; of anything that they possess, if it be asked of them, they never say no; on the contrary, they invite you to share it and show as much love as if their hearts went with it, and they are content with whatever trifle be given them, whether it be a thing of value or of petty worth. I forbade that they be given things so worthless as bits of broken crockery and of green glass and lace-points, although when they could get them, they thought they had the best jewel in the world.

When the natives came to greet his ship, they came "in dugouts which are fashioned like a long boat from the bole of a tree, and all in one piece, and wonderfully made (considering the country), and so big that in some came 40 or 45 men. . . . They row with a thing like a baker's peel [a shovellike tool for moving bread into and out of ovens] and go wonderfully and if they capsize all begin to swim and right it and bail it out with calabashes that they carry." "I showed them swords and they grasped them by the blade and cut themselves through ignorance; they have no iron. Their darts are a kind of rod without iron, and some have at the end a fish's tooth and others, other things."

Europeans had not yet associated "race," or levels of humanity, with skin color. Naturally enough, they considered their own color to be the "normal" original color of the human skin. The dark skins of Africans were explained by the burning of the sun in hot climates, and this, of course, affirmed the humanity of African peoples. European experience was still too limited to raise troublesome questions about the correlation of skin color and climate. The Bible was clear enough on the single origin and homogeneous descent of the whole human race. Since all men were descended from Adam and Eve, there was no place for inferiority of genetic endowment. The interesting differences were those of language and religion.

The discovery of America opened intriguing, and then revolutionary, new possibilities. By the eighteenth century, it was plain that there were many species of plants and animals "peculiar to those parts of the world." Jefferson himself noted in 1789 that there was not a single species of terrestrial bird, and he suspected that there was not a single species of quadruped, common to Europe and America. How account for the presence in America of the raccoon, the opossum, the woodchuck, the alpaca, and the bison? If these had been in the Ark, should they not now be found elsewhere too? Some daring naturalists proposed that instead of only a single Creation in the Beginning in the Garden of Eden, there may have been "separate creations" in different parts of the world. Perhaps God had created plants and animal species peculiarly suited to each continental habitat. Then why not also "separate creations" of mankind?

The new problems that the Protestant Reformation created for the Roman Church gave the question of human equality a new urgency. Only twenty-five years after Columbus' landing in America, Martin Luther posted his 95 Theses on the church door at Wittenberg. In Europe by the mid-sixteenth century the Church of Rome was losing millions of souls to the multiplying Protestant heresies. At the same time, by divine Providence the New World suddenly offered its countless pagans for a vast harvest of new believers. And Spanish missionaries were encouraged by their early successes. "Usually the missionaries taught the Indians to read, write, and observe good customs," Alonso de Zorita reported to the Spanish Council of the Indies in 1584. "Many have been taught how to play musical instruments so that they can play in church, while others have been taught grammar and rhetoric. Some have become excellent Latinists and have composed very elegant orations and poetry." An optimistic estimate in 1540 put the number of baptized American Indians at some six million.

Yet the human status of the American Indian—his potential equality in the sight of God—was increasingly disputed. Spanish conquistadores had reasons of their own to insist on the natural inferiority of the Indians, which meant that God had conveniently destined them to be slaves. There were lively debates over the capacities of New World natives. In 1520, Albrecht Dürer was amazed at their artistry when he saw the Indian jewelry and featherwork that Cortés himself had sent to Emperor Charles V for exhibition in Brussels. Cortés, eager to persuade the Pope to legitimize the children whom Cortés had fathered by Indian women, reinforced his request by sending to Rome a band of Aztec jugglers. From the first establishment of the Spanish Council of the Indies in 1524, the humanity of the Indians was disputed.

The last of the Renaissance popes, the notorious Paul III (1468–1549), declared himself sponsor of missionary efforts in the New World. As a young man he had become a caricature of the sensuality of the age. Having exploited his Farnese family connections to become treasurer of the Roman Church, he delighted in the hunt, built the grand Palazzo Farnese in Rome, and fathered four children by a Roman mistress. His patron, the Borgia Pope Alexander VI, created him a cardinal in 1493, but he was not ordained a priest until 1519, when he was past fifty. Then he put his frolics behind him. Pope at sixty-seven, he was the unexpected prophet and organizer of a Catholic Reformation. In the portrait Titian made of him at the age of seventy-five, we can see the vigor that would govern the Church for yet another six years. When the dispute over the humanity of the American Indians reached Rome, Paul III tried to settle the question by his eloquent bull *Sublimis Deus* (1537).

> The sublime God so loved the human race that He not only created man in such wise that he might participate in the good that other creatures enjoy, but endowed him with capacity to attain to the inaccessible and invisible Supreme Good and behold it face to face. . . . Nor is it credible that any one should possess so little understanding as to desire the faith and yet be destitute of the most necessary faculty to enable him to receive it. Hence Christ . . . said to the preachers of the faith whom He chose for that office "Go ye and teach all nations." He said all, without exception, for all are capable of receiving the doctrines of the faith.

Against this mission, Satan had "invented a means never before heard of, by which he might hinder the preaching of God's word of Salvation to the people; he inspired his satellites . . . to publish abroad that the Indians of the West and the South, and other peoples of whom we have recent knowledge should be treated as dumb brutes created for our service, pretending that they are incapable of receiving the Catholic faith. The Indians are truly men."

Even before the Pope's pronouncement, less than twenty years after Columbus arrived in the New World, Spanish settlers had been disturbed by prophetic voices of protest. On the Sunday before Christmas, 1511, when colonists on Hispaniola met for mass in their thatched church in this first Spanish town in the New World, they were shocked by a jeremiad. "In order to make your sins against the Indians known to you," the Dominican friar Antonio de Montesinos declaimed, "I have come up on this pulpit, I who am a voice of Christ crying in the wilderness of this island. . . . This voice says that you are in mortal sin, that you live and die in it, for the cruelty and tyranny you use in dealing with these innocent people. Tell me, by what right or justice do you keep these Indians in such cruel and horrible servitude? On what authority have you waged a detestable war against these people, who dwelt quietly and peacefully on their own land?"

The heroic champion of the Indians was probably also the first man to receive holy orders in America. Born in Seville, Bartolomé de Las Casas (1474–1566) was there when Columbus returned in 1493 from his first voyage. At the age of nineteen, he had a glimpse of the Indians whom Columbus had proudly paraded through the streets, along with colorful New World parrots. When his father returned from service on Columbus' second voyage, he is said to have given an Indian slave to Las Casas, then a student at the University of Salamanca. Las Casas tasted the life of a conquistador when he went to America in 1502, acquired Indian slaves whom he worked in the mines, and built a large estate. For his part in the bloody conquest of Cuba he was rewarded by a further grant of land with more Indian serfs. In Hispaniola when Montesinos delivered his jeremiad, Las Casas remained unmoved, though he was refused the sacraments because he held slaves.

Even after Las Casas took holy orders, about 1512, he remained blind to the plight of the Indians. Then one day in 1514 on his estate in Cuba when he was preparing his Whitsunday sermon for the new settlement of Sancti Espiritus, he was suddenly illuminated. "He that sacrificeth of a thing wrongfully gotten," he read in Ecclesiasticus, "his offering is ridiculous, and the gifts of unjust men are not accepted." Within a few days, repeating the experience of Saint Paul, he was a changed man. Now utterly persuaded "that everything done to the Indians thus far was unjust and tyrannical," he determined, in this his fortieth year, to devote his life toward "the justice of those Indian peoples, and to condemn the robbery, evil, and injustice committed against them."

In his sermon on August 15, 1514, he publicly returned to the Governor all his Indian serfs. For the next fifty years he remained the Indians' most effective champion. Returning to Spain, he defended the Indians in the Barcelona Parliament. Then he persuaded Charles V to sponsor his utopian scheme to build towns where "free Indians" would collaborate with carefully selected Spanish farmers. They would be settled on the Gulf of Paria, between present Trinidad and Venezuela, to provide the model for a new civilization combining the human resources of the Old World and the New. When this scheme failed, he retreated into a Dominican convent in Santo Domingo where he began to write his account of the Spanish in the Indies to enlighten future generations with the wisdom his own age had refused. This manuscript he intended as a work of prophecy.

In 1537, when Pope Paul's *Sublimis Deus* proclaimed the grand principle, Las Casas had already spent twenty years painfully applying his ideals to daily life in the New World. He tried to prove that it was possible to convert the Indians by peaceful means alone, but his ideas were not popular in the Spanish Indies. His "Only Method of Attracting all People to the True Faith" demanded that everything taken from the Indians, including gold, silver, and lands, be given back. Again he tried to demonstrate his unorthodox approach, this time by a new settlement in Guatemala, now part of Costa Rica. When Las Casas returned to Spain, he persuaded Charles V to sign the New Laws declaring that the grants of Indian serfs were not hereditary and requiring Spanish encomenderos to liberate their serfs after one generation. As part of a papal plan that Las Casas himself wrote, he was created bishop of Chiapa expressly to protect the Indians and promote model settlements of Spanish farmers and free Indians. But within two years Spanish settlers sabotaged the scheme and forced Las Casas to return to Spain.

The public climax of Las Casas' struggle provided a spectacle unique in the history of colonization. On April 16, 1550, Charles V, impelled by Las Casas' doubts and accusations, ordered that conquests in the New World be suspended and not be resumed until his theologians had agreed on a just way of proceeding. "In order that all may be done in a Christian fashion," no new conquests would be licensed until the King was informed of how the conquest should be effected. For a while this order was strictly enforced in New Granada, the Chaco, and Costa Rica, policed by the friars against impatient protesting colonists. The moral grandeur of this effort—the ruler of a vast empire refusing to use his power until he was fully satisfied that he was using it justly—would be overshadowed by the brutality of the conquistadores.

Charles V was, of course, declaring faith in the moral judgment of his theologians. They would not provide him a plain and speedy answer, but they did not entirely fail him. His squeamishness was not without effect on the future of the world.

The Spanish colonists, the party of the conquistadores, and other opponents of the New Laws had engaged a weighty champion. Dr. Juan Ginés
de Sepúlveda (1490–1573), a learned humanist and disciple of Aristotle, had
never visited the New World, but he held firm opinions supported by his
ponderous treatise arguing that it was just to war against and enslave the
Indians. By taking on Sepúlveda, whose patron was the powerful president
of the Council of the Indies, Cardinal Garcia de Loaysa of Seville, Las Casas
also took on Aristotle, whose *Politics* Sepúlveda had just translated into
Spanish. Aristotle's proposition that some men are by nature slaves provided the foundation for Sepúlveda's argument. Just as children are naturally inferior to adults, women to men, and monkeys to human beings, so
Indians, he said, were naturally inferior to Spaniards. "How can we doubt
that these people—so uncivilized, so barbaric, contaminated with so many
impieties and obscenities—have been justly conquered by such an excellent,
pious, and most just king as was Ferdinand the Catholic and as is now
Emperor Charles, and by such a most humane nation and excellent in every
kind of virtue?"

To decide between Sepúlveda and Las Casas and make "regulations
which will be most convenient in order that the conquests, discoveries,
and settlements may be made according to justice and reason," on July 7,
1550, Emperor Charles V announced a special congregation of theologians
and councillors to meet in Valladolid, capital of Castile, in August. Las
Casas had already prepared an 870-page "Apologetic History" of the Indians to prove that the American Indians were paragons of reason and
virtue. He marshaled his long experience, embellished by legend and fantasy, under Aristotle's tests of rationality and the good life. In almost
every way, he argued, the Indians were superior to the ancient Greeks
and Romans, and in some respects even superior to the Spaniards. He did
not flatly deny the Aristotelian doctrine of natural slavery, but insisted
that "natural slaves" were a kind of monstrosity, which surely did not
include the Indians.

The fourteen-member council, composed of some of the most learned
and powerful men of the age, took its assignment seriously. Solemnity and
suspense surrounded the Great Debate between the two champions. On
the first day Sepúlveda opened with a three-hour speech summarizing his
book on the inferiority of the Indians. Las Casas followed by reading
verbatim from the 550-page treatise he had prepared specially for the occasion, which the council patiently tolerated for five long days. The deliberations lasted from mid-August to mid-September, when the confused
council members finally asked an eminent jurist of their number to help
them by summarizing the issues. When they reconvened in January 1551,
supposedly to cast their votes, few were prepared to commit themselves.

The lawyers said they needed more time to study the issues, the church-men said they had to prepare for Lent, and two of the members had conveniently been sent by the Emperor to attend the Council of Trent. The only one of their opinions that has survived cautiously concluded that expeditions of conquest were desirable on condition that they be en-trusted to captains "zealous in the service of God and the king who would act as a good example to the Indians, and who would go for the good of the Indians and not for gold."

The council never agreed and so never gave the king a decision. Both parties claimed victory. By every pragmatic test, in the vast arena of the Americas Sepúlveda would prove to be the spokesman of Spanish policy. The conquistadores adored him, sent him presents, and made his works their orthodox defense. Still, he had not won the Battle of Valladolid. Sepúlveda's works were not allowed to be published in Spain during his lifetime, and were not published anywhere until the late eighteenth century. His classic attack on the humanity of the Indians was finally printed in 1892.

Las Casas, the voice of conscience, never wholly suppressed, remained spokesman for the professed doctrine of the Roman Church. Of course, he failed to convert conquistadores into pacifists. But he had put the stamp of the Church on the humanity of the Indians. In 1566, when the King was once again issuing licenses for discovery and conquest, the King felt con-strained to exhort all to obey the laws of a just war. The relatively peaceful conquest of the Philippines after 1570 is sometimes credited to the surviving spirit of Las Casas. When Philip II, in July 13, 1573, proclaimed the law governing future Spanish discoveries and conquests, which remained in force as long as Spain possessed American colonies, he did not follow Las Casas' strict rules of peaceful conversion. But he did order Spanish conquer-ors always to remind the Indians

> that the king has sent ecclesiastics who have taught the Indians the Christian doctrine and faith by which they could be saved. . . . He has freed them from burdens and servitude; he has made known to them the use of bread, wine, oil and many other foods, woolen cloth, silk, linen, horses, cows, tools, arms and many other things from Spain; he has instructed them in crafts and trades by which they live excellently. All these advantages will those Indians enjoy who embrace our Holy Faith and render obedience to our king.

If the Spanish settlers found it necessary to use force against the natives, they must use no more than necessary. Under no circumstances should they enslave the Indians. In a bow to Las Casas, the King banned the word "conquest," in future to be replaced by "pacification."

When Las Casas died in 1566 at the age of ninety-two, he left instructions that his full history of the Indies should be published only after forty years had passed "so that, if God determines to destroy Spain, it may be seen that it is because of the destruction that we have wrought in the Indies and His just reason for it may be clearly evident." Not merely the Spanish but European peoples on all continents would be haunted for centuries by the question debated at Valladolid.

The opportunity to reflect on the variety and unity of mankind which the discovery of America and the outreaching colonies forced on the West was not seized by the peoples of other parts of the world. Islam expanded as an extending empire rather than by colonies-at-a-distance, by conquest and occupation rather than by missionary outposts. Of course, Islam inherited the Biblical baggage of dispersion and original sin and, like Christianity, saw variety only as an evil. But Muslim theology and the hazards of history luckily inoculated Islam against the virus of racism. The solid dogma of the equality of all believers, the spread of Islam across black Africa, the frequent intermarriage with slaves and concubines—all these discouraged any Muslim belief in racial levels of humanity. For Muslims, who would not separate secular from religious life, the all-important distinction was between Believers and non-Believers. Mere variety of social custom, where it did not violate the Koran, seemed insignificant.

For quite opposite reasons, the problem of human equality was not vivid in China. There, where tradition and custom ruled, the best qualities of human life were viewed as products of Chinese tradition and custom. And the China-centric isolationist tradition kept the Chinese from encounters with remote and different peoples. Nor elsewhere in East Asia, in Japan or Korea, do we find anything like Western racism.

Only in India, among the developed cultures, did racial caste become integral with religion. Although the origin of castes is hidden in prehistoric mists, the Hindu caste system may have originated in differences between conquering Aryans and subject Dravidians—which happened to be differences of color. *Varna,* the Hindu word for caste, means "color," but perhaps the original application referred to something other than the color of skin.

78

The Shock of the Primitive

DURING the centuries after Las Casas, European debate over the levels of humanity moved from theology to biology. By classifying all mankind as a single species, *Homo sapiens,* Linnaeus seemed in the mid-eighteenth century to join the party of Las Casas. He gave his own clear answer to the question debated at Valladolid in 1550. But he fogged the issue for European settlers in remote parts when he listed five kinds of *Homo sapiens*—Wild man, American, European, Asiatic, African—"varying by education and situation." Were these different "varieties" of a single human species? If so, what did "variety" mean.

When the assessment of man's capacities moved from religion to science, questions changed from wholesale to retail. Like the earlier movement from cosmology to geography, this too was a move toward incrementalism. Instead of asking the grand monolithic question debated by Las Casas and Sepúlveda about the "nature" of man and his destiny in this and the after life, now they would ask countless questions about the trivia of daily life. Unlike theology texts that were written in a learned language, the data of anthropology were everybody's experience. Focus shifted from human nature to human cultures, from metaphysics to miscellany. Questions of anthropology would be asked and answered not in the library but out in the world. Every human society became a laboratory.

And the New World would be the first, for this new science of mankind. Here large numbers of permanent European settlers lived alongside Stone Age communities. Just as Las Casas had applied Christian theology to his New World encounters with strange people, so in the early nineteenth century observers equipped with new institutions for exchanging the data of science would study the American natives. A strength of this enterprise was precisely that it was new. All inquirers had the naïveté and some had the boldness of amateurs.

The opportunities and the temptations were dramatized in the career of one passionate amateur, Lewis Henry Morgan (1818–1881). Son of a farmer in a frontier village of central New York on the path of the newly constructed Erie Canal, he was an emphatically sociable young man. At school he organized a society for "mutual improvement in useful knowledge," The Erodephecin Society. Graduating from Union College in Schenectady in 1840, he returned to read law in Aurora.

A young lawyer without clients in the doldrums of the business depression that began in 1837, he had ample time for his clubable talents. He founded a secret lodge for sociability and self-improvement which met in an abandoned Masonic Lodge building. Morgan christened his club the Order of the Gordian Knot, for it was an age of classical revival. Ancient Greece and Rome offered models in architecture and synonyms for the best in civilization commemorated in town names like Ithaca, Troy, Delphi, Hannibal, Marcellus, Brutus, Cato, Syracuse, Utica, and Aurora. As his members dispersed from Aurora they formed branches, and within a few years the order had five hundred members in a dozen towns. In 1843 Morgan decided to abandon the ancient classical format for one more distinctly American. "Gordius conceived the mighty enterprise of leading his Phrygian children to this western hemisphere," Morgan explained, "conducted them to Bhering's Strait, thence across to this western world." The Order of the Gordian Knot became The Grand Order of the Iroquois with Morgan as the chief, called Skenandoah, after an Iroquois who had befriended the Americans during the Revolution.

The source of Morgan's first Indian enthusiasm is not clear. Perhaps it started in pure whimsy. But soon enough he proved his serious desire to capture the spirit of the Iroquois. The five tribes of the Iroquois had traded with the first European settlers who moved west, then fought savagely to resist their invasion. Most of the Iroquois had sided with the British during the Revolution, and when the war ended they were forced to surrender their lands in exchange for token payments and confinement on reservations. Morgan's grandfather's six-hundred-acre family farm had been carved from Iroquois land as a reward for his services in the Revolution. In 1843, when Morgan decided to commemorate the Indians' lost cause in his secret society, he knew little about the inner workings of Iroquois life, but this did not prevent his elaborating complicated "Iroquois" rituals. In his solemn initiation ceremony, which he called "Inindianation," the blindfolded candidate was warned "if, in an idle hour, or with careless levity you should dare to lift the veil of Secrecy from our Order and expose it to the Pale-face, a retribution the very thought of which would make you shudder even in the grave will follow quick upon your erring footsteps." At the meetings Morgan's "warriors" wore Iroquois-style leggings and headdress, carried tomahawks, and scorning all who talked "with forked tongue," they spoke in what they believed were Indian figures of speech.

To instruct their fraternity, Morgan boldly invited Henry Schoolcraft

(1793–1864), the nation's leading authority on Indians, whose work would enter American folklore as the basis of Longfellow's *Hiawatha*. Schoolcraft had married a woman of Ojibwa extraction, had negotiated the treaty in which the Ojibwas ceded much of northern Michigan, and then he became Michigan's superintendent of Indian affairs. Addressing the Grand Order of the Iroquois, Schoolcraft urged the Aurora "warriors" not to be satisfied by studying only their European heritage. They should turn to "the history and antiquities and institutions of the free, bold, wild, independent native hunter race. . . . They are relatively to us what the ancient Pict and Celt were to Britain or the Teuton, Goth and Magyar were to continental Europe."

Morgan himself had already determined to study the Indians firsthand. A personable young Iroquoian of the Seneca tribe, whom he had met browsing in an Albany bookstore, gave him his entrée. Ely Parker, son of a chief, had gone to a Baptist mission school, and then had been sent to law school by his tribe so he could defend them against further removal. This time they were threatened with removal beyond the Mississippi. Morgan's Grand Order joined the Iroquois cause, raised money, organized meetings, and signed petitions. Morgan and Parker went to Washington to persuade the Senate Committee on Indian Affairs to give relief from a "treaty" that had appropriated Indian lands worth $200 an acre for a mere $2.50 an acre. The treaty had been signed by a majority of the tribe's chiefs and sachems. In one of the first applications of anthropology to Indian affairs, Parker and Schoolcraft testified that the Indians lived by the rule of unanimity and knew nothing of majority rule. Despite overwhelming evidence of fraud, the Senate refused to abrogate the treaty. Only after another decade of protest the Senate belatedly authorized the Iroquois to buy back their acreage, and appropriated funds for the purpose.

Morgan's trip to Washington persuaded him that Iroquois customs could not long survive. At the same time he had earned the confidence of the Iroquois. On his return, in October 1846, he visited the corn-harvest festival on the Tonawanda Reservation and was adopted into the Hawk Clan of the Seneca tribe. Christened Ta-ya-da-o-wuh-kuh ("One Lying Across"), he was to link Indians and whites. He seized his opportunities, and with passionate nostalgia, a poignant sense of injustice, and omnivorous curiosity, he began gathering data, as he explained, from "those human tablets on which are inscribed the closing events in the career and destiny of the Old Iroquois." From his small-town fraternal beginning Morgan set out on an enterprise of worldwide discovery.

The later career of Morgan's Iroquois friend, Ely Parker, was itself a saga. In Washington on his lobbying trip, the charming Parker had delighted President Polk as a dinner companion. Although Parker had mastered the law, still because he was not a citizen he was refused admission to the bar. Undismayed, he entered Rensselaer Polytechnic Institute to become an engineer, and was hired as superintendent of construction on the government works at Galena, Illinois. There he had the good luck to meet and impress a ne'er-do-well veteran of ten years' army service, Ulysses S. Grant, who was clerking in a leather store owned by his brothers.

When the Civil War broke out in April 1861, Parker's friend Grant had trouble finding himself a suitable army assignment. Parker sought a commission, but Secretary Seward told him that the whites could easily win the war on their own and needed no help from Indians. The indomitable Parker managed anyway to secure a commission as captain of engineers, and soon became Grant's military secretary. At Appomattox Court House when surrender was being negotiated with General Lee the senior adjutant was so nervous that he could not write out the terms. Grant ordered Parker to revise Grant's penciled original and then write clean copies, which became the official documents of surrender signed by General Lee to end the Civil War. Grant commissioned Parker a brigadier general for his gallant and meritorious services, and later as President appointed him Commissioner of Indian Affairs.

When Morgan became a serious student of Iroquois tribal life, he was increasingly troubled by the "merriment and irrelevancy" of the Grand Order. In 1846 he gave up the management, and the society fell apart. But Morgan had already become the nation's leading authority on the Iroquois. He sent his collection of Iroquois objects—mortars and pestles, chisels, knives, tomahawks, kettles, necklaces, pipes, and drums—to Albany for a new Indian museum. Morgan's *League of the Ho-de-no-sau-ne, or Iroquois,* published in 1851, was acclaimed by knowledgeable contemporaries as "the first scientific account of an Indian tribe ever given to the world." In retrospect, too, it is plain that Morgan pioneered a new science of mankind.

Earlier views of the American Indians had been narrowly Christian and Europe-centric. For Spanish conquistadores and Jesuit and Protestant missionaries the Indians were minions of Satan. New England Puritans, with characteristic subtlety, speculated that God had enlisted the savages to keep a new world free from Popery until purified Christianity could occupy the land. Even Morgan's contemporaries who were most friendly to the Indians had not broken the bonds of Christian theology. Biblical history required them to believe that the "savages" of the world had fallen, for their sins, from an earlier state of civilization. Schoolcraft himself pitied the Indians for their "declension from a high type" to a lower type of society. But

Morgan began to see how Iroquois government, tools, domestic architecture, clothing, and language fitted together into a distinctive way of life. He saw no signs of Satan or of an earlier higher civilization from which they had relapsed.

We can follow the stages of Morgan's liberation as he sharpened his appetite for trivial clues to the varieties of human community. *The League of the Iroquois,* still tainted by the vocabulary of Morgan's own culture, tried to fit Iroquois ways into the categories of Aristotle and Montesquieu. But for Morgan even the rudimentary Iroquois organization into "tribes" seemed to mark an advance—"a means of creating new relationships by which to bind the people more firmly together."

The secret of Morgan's intellectual power was his passion for the specific. Unlike Las Casas, Morgan seldom pontificated on the general excellence of Iroquois institutions, focusing instead on the facts of their social organization. In 1856, when he joined a newly founded American parliament of science, the American Association for the Advancement of Science, in Boston, he was encouraged to collect the minutiae of Iroquois laws of consanguinity and descent so he could present them to the assembled scientists.

The AAAS had been founded in 1848 by geologists and naturalists explicitly to nourish the incremental spirit, the democracy of facts against the genteel tradition of "natural philosophy," general science, and the search for scientific panaceas—against that "modified charlatanism which makes merit in one subject excuse for asking authority in others." "The absence of minute subdivision in the pursuit of science, the prevalence of general lecturing on various branches, the cultivation of a literature of science rather than of science itself, has produced many of the evils under which American science has labored, and which are now passing away." True to their love of specifics, the founders had prevented the bequest of James Smithson from being used for a general library of "higher learning," and demanded that it be devoted to "the increase and diffusion of knowledge," which for them meant collecting all possible increments of knowledge. They secured the appointment, as first director of the Smithsonian, of Joseph Henry, famed for his improved design of electromagnets. He would amply fulfill their hopes—for example, by organizing the numerous volunteer weather observers into the nation's first scientific weather prediction service.

Morgan's technical paper on the "Laws of Descent of the Iroquois," delivered to the AAAS in 1856, detailed the Iroquois system of kinship, consanguinity and tribal organization. What made it peculiarly interesting to Europeans was that the Iroquois husband and wife always belonged to different tribes. This, Morgan explained, resulted from a complex system of exogamy and taboo by which children were always assigned to the mother's tribe. Since Iroquois inheritance came through the tribe, the male line was perpetually disinherited. A son could not even inherit a tomahawk from his father, but did inherit all his mother's property. In Iroquois nomenclature, a son referred to all his mother's sisters as "mother," and all those sisters referred to him as their "son." The AAAS audience, finding this bizarre, assumed that it was unique to the Iroquois. Morgan believed he had found a bundle of clues. But clues to what?

When the panic of 1857 required Morgan to go out to Michigan to save his railroad investments, he met a fur trader who was married to an Ojibwa Indian woman. To his delight he learned that the Ojibwa kinship system was similar to the Iroquois's. As he suspected, the Iroquois system was not unique after all. Sparked by the curiosa of kinship classification, a light began to dawn. Morgan recalled that some missionaries' reports suggested similar customs among the remote Micronesian Islanders.

If the Iroquois kinship customs were common to all the American Indians, might this not imply their common descent? And if the same customs were found in the Orient too, then might this not suggest an Asian origin for the American Indian? Linguists had long tried to prove this connection. The reason they had not succeeded, Morgan ventured, was that they had focused on language, which changed rapidly in response to local needs, while "primary" institutions like those of consanguinity were more stable. Here he might have found a reliable link to the distant past, or perhaps to "the imprint of a common mind."

Morgan already had enough evidence to correct the terminology that the most respected historians had transferred from Europe to the American scene. Why should Prescott, in his esteemed *Conquest of Mexico,* have been so puzzled that Montezuma was succeeded on the throne not by a son but by his brother, and then by his nephew? Morgan saw how the discovery of America had opened unexpected vistas of the whole human race. Incidentally, according to a contemporary admirer, he was "charting a new continent of scholarship."

Morgan's efforts to collect data led him to employ a device that was wonderfully adapted to the new world of incremental science. This was the questionnaire. A circular letter or schedule of questions had been tried before by tax collectors and census gatherers. But Morgan's appears to be the first large-scale worldwide effort to gather factual minutiae for scientific purposes. "Questionnaire" does not appear in print in English until 1901.

A century before, the word "statistics" had entered the English language through Sir John Sinclair's 21-volume rural survey, which he called *The*

Statistical Account of Scotland (1791–99). Sinclair had asked the clergy in
each of the 881 Scottish parishes to answer a schedule of more than a
hundred questions. He had then pursued the nonrespondents by twenty-
three successive follow-up letters, aiming to complete "an inquiry into the
state of a country for the purpose of ascertaining the quantum of happiness
enjoyed by its inhabitants and the means of its future improvement." He
tried to induce European governments to follow his example and initiate
their own decennial censuses. Despite his interest in quantitative data,
Sinclair's concern was mainly political and moral. "Are people disposed to
humane and generous actions?" one question went. The answer should
indicate whether the people "protect and relieve the shipwrecked, etc."

Other early efforts in Britain and elsewhere in Europe to gather social
facts on a large scale were reformist in purpose and local in scope. They
aimed to shock readers into a more humane treatment of prisoners, the
mentally ill, or the poor, or to improve sanitation and public health. When
the National Association for the Promotion of Social Science was founded
in England in 1857, it too aimed at education, health, and social reform. In
France and in Germany, the pioneering statistical studies in the social
sciences in the nineteenth century were local efforts to improve health and
morals, to combat the evils of prostitution, and to better the lot of the poor,
of factory workers and rural laborers.

Morgan was on quite another tack. His inquiry aimed to be scientific and
was worldwide. His questions had no obvious practical use. When he returned from his business trip to Michigan, he composed a seven-page
printed questionnaire, with more than two hundred questions on every
aspect of tribal organization, kinship customs and usage—from the name
used to designate a person's father to the name "of the daughter of the
daughter of a brother to the son of the son of the brother's sister." Using
the franking privilege of the congressman from Aurora, Morgan sent out
the questionnaires to missions and federal agencies in the American West.
His covering letter explained that the answers would help "solve the question whether our Indians are of Asiatic origin." Some recipients were too
busy. Others were not interested because they already believed "Mister
Louis Agassiz right in pronouncing them, like the buffalo and grizzly bear,
indigenous." But scores of respondents sent in detailed reports of the Dako-
tas, Shawnees, Omahas, and Pueblos. Morgan's own field trip to Kansas
and Nebraska yielded eleven schedules in eleven languages—nearly all with
clear resemblances to the Iroquois system.

One day Morgan received from a missionary in south India a chart of the
Tamil kinship system identical with that of the Iroquois. He rushed to show
the good news to a scholarly friend, who reported that Morgan had turned

purple with excitement. Now, Morgan saw, it was "imperative to include the entire human family within the scope of the research."

And so he did, with the cooperation of Joseph Henry, the Smithsonian Institution, and the Foreign Service of the United States. Henry printed Morgan's charts under the Smithsonian letterhead and used the Smithsonian mailing privilege to send them around the world. The Secretary of State instructed American diplomats everywhere to cooperate in the study. In January 1860 Morgan's circular went out to all continents, and by spring he had received two hundred completed questionnaires. In 1870, after countless revisions and abridgments to satisfy the cautious Joseph Henry "that its value should be fully established" even before it was published, the Smithsonian Institution finally published Morgan's 600-page book on *Systems of Consanguinity and Affinity of the Human Family.*

Morgan's conclusion, supported by facts from everywhere, was that two basically different modes of reckoning kinship were found in the world, and most of the peoples on the planet could be classified into one or the other. Linguists had never succeeded in establishing such large distinctions, but Morgan now showed that the Indo-European and Semitic nations had one type of kinship system, while other peoples had the other. He then argued that the similarity of the American Indian and the Asian systems of kinship established an Asiatic origin for the American Indians. "When the discoverers of the New World bestowed upon its inhabitants the name of the Indians under the impression that they had reached the Indies, they little suspected that children of the same original family, although upon a different continent, stood before them. By a singular coincidence error was truth." Many anthropologists now do not subscribe to Morgan's thesis, but they still draw on his precious facts about societies that would soon disappear.

Morgan's unprecedented effort produced only a sample of the raw material to be found everywhere for a new science of mankind. But it was an enticing sample. While Las Casas had insisted that mankind was one and equal in the sight of God, Morgan discovered the common experience of all mankind. Primitive peoples, no longer relics of sin or symbols of decadence, became clues to what all mankind had once been. When Morgan read Darwin, at first he resisted the idea of the emergence of species by evolution, but "after working up the results from consanguinity, I was compelled . . . to adopt the conclusion that man commenced at the bottom of the scale from which he worked himself up to his present status." This was Morgan's own version of evolution.

How man "works himself up" was the theme of his *Ancient Society; or Researches in the Lines of Human Progress, from Savagery through Barbarism into Civilization* (1877). He might have called it "A Treatise on Human

Progress," for he showed that everywhere civilization had advanced by similar steps. "Growth of intelligence through inventions and discoveries," "Growth of the Idea of Government," "Growth of the Idea of the Family," and "Growth of the Idea of Property"—these were the means of human advance. Just as Thomsen had parsed prehistory, Morgan went on to describe the three grand epochs of all human development, which were still visible in the arrested development of some societies. He saw them all in the mirror of his America. "The latest investigations respecting the early condition of the human race, are tending to the conclusion that mankind commenced their career at the bottom of the scale and worked their way up from savagery to civilization through the slow accumulations of experimental knowledge. . . . portions of the human family have existed in a state of savagery, other portions in a state of barbarism, and still other portions in a state of civilization . . . so . . . these three distinct conditions are connected with each other in a natural as well as necessary sequence of progress."

Technology and the "arts of subsistence" distinguished the periods and marked the progress of humankind. In *Savagery,* mankind lived by gathering fruits and nuts, learned to fish and to use fire, and invented the bow and arrow. In *Barbarism,* mankind invented the art of pottery, learned to domesticate animals and to cultivate plants, began to use adobe and stone in house building, and finally learned to smelt iron and use iron tools. *Civilization* began with the invention of the phonetic alphabet, and climaxed in all the nineteenth-century wonders:

> The principal contributions of modern civilization are the electric telegraph; coal gas; the spinning-jenny; the power loom; the steam engine with its numerous dependent machines, including the locomotive, the railway, and the steam-ship; the telescope; the discovery of the ponderability of the atmosphere and of the solar system; the art of printing; the canal lock; the mariner's compass; and gunpowder. The mass of other inventions, such, for example, as the Ericsson propeller, will be found to hinge upon one or another of those named as antecedents: but there are exceptions, as photography. . . . With these also should be removed the modern sciences; religious freedom and the common schools; representative democracy; constitutional monarchy with parliaments; the feudal kingdom; modern privileged classes; international, statute and common law.
>
> Modern civilization recovered and absorbed whatever was valuable in the ancient civilizations.

Morgan's own vivid experience of progress had nourished his optimism and made him a prophet and a founder of a science of progress. "The theory of human degradation to explain the existence of savages and of barbarians," he urged, "is no longer tenable. It came in as a corollary from the

Mosaic cosmogony, and was acquiesced in from a supposed necessity which no longer exists. . . . it is without support in the facts of human experience." But in Morgan's day, it was not only Biblical dogma that supported a belief in the decay of human societies. A bundle of secular anti-Christian dogmas of which Rousseau became the prophet had flourished in the eighteenth century and created a cult of the "noble savage." Defying Biblical dogma, these romantic primitivists declared that man, virtuous in his "natural condition," had been corrupted by institutions. As Morgan and his fellow anthropologists collected their data on real-life savages, it was hard to swallow Rousseau's romantic notions.

Ever since the Renaissance, European enthusiasts for science had promoted an *idea* of progress. From Francis Bacon's scheme for *The Advancement of Learning* (1605) to the Abbé de Saint-Pierre's *Observations on the Continuous Progress of Universal Reason* (1737) and Diderot's monumental *Encyclopedia* (1751–72), scholars had proclaimed the inevitable enlargement of human knowledge and the resulting improvement of the human lot. Debate over the relative virtues of "the Ancients" and "the Moderns" exercised the passions of literati and pedants, but the weight of learning was more and more on the side of the moderns. Condorcet's classic *Sketch of a Historical Picture of the Progress of the Human Mind* (1793) announced the certain advance of liberty, justice, and equality.

Morgan found a way to enlist newly gathered facts from everywhere to catalogue the facts of progress. Examples of all the stages, except the "lower status of Savagery, which was the infancy of the human race," could still be found somewhere. America offered only a first opportunity. Collecting data around the earth, a new science of anthropology could show how "mankind commenced their career at the bottom of the scale and worked their way up." The science of anthropology would begin as the science of progress.

Before his very eyes, Morgan saw the contrast between the technology of "barbarism" and the technology of civilization, between communal property and individual property. Karl Marx died before he could write the book he planned to write about Morgan, but Engels incorporated Morgan in the canon of Marxist literature. According to Engels, Morgan had actually anticipated Marx's materialist interpretation, and Morgan's *Ancient Society* was as "necessary" as Marx's *Capital* for understanding the history of civilization. Engels concluded his own work on *The Origin of the Family, Private Property and the State* by quoting Morgan:

Democracy in government, brotherhood in society, equality in rights and privileges, and universal education, foreshadow the next higher plane of society to which experience, intelligence and knowledge are steadily tending. It will be a revival, in a higher form, of the liberty, equality and fraternity of the ancient gentes.

Praise for Morgan by Marx and Engels discouraged Western scholars from recognizing Morgan as a founder of anthropology. But cultivated Europeans were finally transcending the "classical" and "Judaeo-Christian" traditions and were beginning to admit the whole world into the family of civilization.

79

A Science of Culture

THE next grand opening for the European view of civilization was accomplished by another amateur, who also found his first clues in the New World. Edward Burnett Tylor (1832–1917) had the advantage of being an outsider in other senses too. Son of a Quaker brass founder in London, the young Tylor did not attend a "public school" but instead was sent to a Quaker school. At sixteen he went into the family business. Anyway, as a Dissenter he could not have been admitted to the university. So his Quaker inheritance saved him from defining "culture" as the peculiar product of the Greek and Roman classics and the Established Church. The Quaker suspicion of the fine arts saved him too from confining "culture" in the Victorian mold of Matthew Arnold. When he became the first professor of anthropology in Oxford in 1896, he would boast that he had never sat for an examination.

At the age of twenty-three he seemed to be suffering from tuberculosis, and his family sent him traveling for his health. Instead of taking the Grand Tour of the capitals of Europe, which was customary for well-to-do young men in 1835, he went off to America. Vagabonding in Cuba, while on an omnibus in Havana he fell into conversation with a fellow traveler, Henry Christy, a wealthy English banker in his fifties, who also happened to be a Quaker. Christy had already pursued his antiquarian interests in the East and Scandinavia, and was just beginning an American excursion. For Quakers, "ethnology" and the ways of remote peoples had an ethical significance, documenting human brotherhood, and supporting the antislavery impulse. They hoped to enlist anthropology to support Las Casas.

This was not an easy task. In the European West, the words and ideas describing man's social achievements had acquired a eulogistic and self-regarding meaning. "Culture" (from Latin *cultus* for "worship") originally meant reverential homage. Then it came to describe the practice of cultivating the soil, and later it was extended to the cultivating and refinement of mind and manners. Finally, by the nineteenth century "culture" had become a name for the intellectual and aesthetic side of civilization. So Wordsworth deplored any life "where grace of culture hath been utterly unknown."

In Matthew Arnold's familiar phrase, "Culture" was "the acquainting ourselves with the best that has been known and said in the world." This was a most unpromising name for an uncondescending scientific study of all human societies. But Tylor seized on the word and did wonders to empty it of chauvinistic and provincial overtones. For his success in this, and in making "Culture" a neutral term and the focus of a new social science, he is generally agreed to be the founder of modern cultural anthropology. In his own time it was called "Mr. Tylor's Science."

Tylor named his lifelong effort the Science of Culture "to escape from the regions of transcendental philosophy and theology, to start on a more hopeful journey over more practicable grounds." He needed courage to foray into the sacred groves as he did in his epoch-making *Primitive Culture*. "The world at large is scarcely prepared to accept the general study of human life as a branch of natural science, and to carry out, in a large sense, the poet's injunction to 'Account for moral as for natural things.' To many educated minds there seems something presumptuous and repulsive in the view that the history of mankind is part and parcel of the history of nature, that our thoughts, wills, and actions accord with laws as definite as those which govern the motion of waves, the combination of acids and bases, and the growth of plants and animals." The subject of this natural history of society would be Culture, redefined as "that complex whole which includes knowledge, belief, art, morals, law, custom, and any other capabilities and habits acquired by man as a member of society."

Tylor noted that so many eminent thinkers had "brought history only to the threshold of science." "If the field of enquiry be narrowed from History as a whole to that branch of it which is here called Culture, the history, not of tribes or nations, but of the condition of knowledge, religion, art, custom, and the like among them, the task of investigation proves to lie within far more moderate compass." His first casual hint of these opportunities came to him as a young man in Mexico when Christy took him to the ancient obsidian mines. The handworked obsidian prisms found here had formerly been described as maces or handles for weapons, but Tylor showed them to be the cores from which long knifelike flakes had been struck for weapons and tools. Intrigued by this remarkable unfamiliar technology, he was on his way to making the study of technology necessary for the study of society.

Adopting the Italian proverb "All the world is one country," Tylor delighted in the "correspondence" between the ways of widely separated peoples. He avoided the plural word "cultures" and preferred the grand singular, "Culture." For example, he saw "scarce a hand's breadth difference" between an English ploughman, using his hatchet and hoe, boiling his food over a log fire, hearing tales of a ghost in a nearby haunted house, and the similar ways of the Negro of central Africa. Following Linnaeus' example, he set out a taxonomy of society.

> To the ethnographer the bow and arrow is a species, the habit of flattening children's skulls is a species, the practice of reckoning numbers by tens is a species. The geographical distribution of these things, and their transmission from region to region, have to be studied as the naturalist studies the geography of his botanical and zoological species. . . . Just as the catalogue of all the species of plants and animals of a district represents its Flora and Fauna, so the list of all the items of the general life of a people represents that whole which we call its culture.

Instead of digging, like Winckelmann and Schliemann, in classical ruins, or, like Thomsen and Worsaae, sifting the kitchen middens, devotees of this new Science of Culture would discover the past in the ways of living peoples. Tylor's invention was wonderfully simple. To help us "trace the course which the civilization of the world has actually followed," he created a new archaeology of society with his notion of "survivals." "These are processes, customs, opinions, and so forth, which have been carried on by force of habit into a new state of society different from that in which they had their original home, and they thus remain as proofs and examples of an older condition of culture out of which a newer has been evolved." The old Somersetshire woman who still used a handloom inherited from the days before the flying shuttle and still threw her shuttle from hand to hand was not "a century behind the times" but simply a case of survival. These are "landmarks in the course of culture." "When a custom, an art, or an opinion is fairly started in the world, disturbing influences may long affect it so slightly that it may keep its course from generation to generation, as a stream once settled in its bed will flow on for ages. This is mere permanence of culture; and the special wonder about it is that the change and revolution of human affairs should have left so many of its feeblest rivulets to run so long." Sometimes survival passed into revival, as in the case of modern spiritualism. According to Tylor, civilization advanced as much by abandoning the old as by introducing the new.

The older ways still remain the substratum of all modern life. "The past," Tylor wrote, "is continually needed to explain the present, and the whole to explain the part." "There seems to be no human thought so primitive as to have lost its bearing on our own thought, nor so ancient as to have broken its connection with our own life." Following the clues of Lyell's new geology, Tylor brought the uniformitarian idea into social science, which made the living present an uninterrupted avenue into the living past.

To test his doctrine of survival, Tylor rashly entered the most controversial, most passion-ridden arena he could find—Religion. "Animism" was his own word for the minimal form of religion. This he defined as a belief in Spiritual Beings. There appeared to be no tribes of men, Tylor observed, that "have no religious conceptions whatever." Savages saw these spiritual beings in plants, animals, and features of the landscape. And from such elemental notions all religions evolved—through a belief in a future state, to later identification with moral elements, and on toward monotheism. To this notion Tylor devoted more than half his work on Primitive Culture and so incorporated the most sensitive, most sacred notions of his age into his Science of Culture. Could the seeds of Victorian civilization be germinating at that very moment in all the savage tribes of the world? "Animism" was his strongest antidote to British provincialism and complacency. And a clue to countless other paths from Victorian England back to despised savage tribes. While Darwin had dared a flank attack on Christian orthodoxy, Tylor's attack was frontal. His "developmental" approach to humankind was a menacing, and perhaps fatal, blow to the dogmas of Eden, to the sudden revelations of Christian Gospel and a Savior. Was it possible that the great truths of monotheism and Christianity had developed gradually out of the whole worldwide human experience?

Tylor's shocking Science of Culture breathed renewed vigor into the embattled champions of the Christian dogma of human degeneration. Richard Whateley (1787–1863), Anglican archbishop of Dublin, reformer, and apostle of the Irish poor, established himself as a witty defender of the faith by his first book, *Historic Doubts Relative to Napoleon Buonaparte* (1819). He ridiculed David Hume's application of strict logic to Biblical miracles by showing how the same reasoning would raise doubts about the existence of Napoleon. In his popular pamphlet *On the Origin of Civilization* (1855) Whateley next targeted Adam Smith and other partisans of progress. Having described with disgust the polygamous, cannibalistic savage encountered by missionaries, he asked, "Could this abandoned creature entertain any of the elements of nobility?" If savage peoples showed cleverness in the arts, this must be a relic of an advanced civilization from which they have degenerated. Could anyone produce a single example of a primitive people rising to a civilized state, except when aided from the outside by peoples who had not degenerated?

Whateley's degeneration theory was the most popular enemy of Tylor's comparative method and so too of a Science of Culture. Tylor's "progression theory" boldly traced all human history as "the development of culture" and asked "whether we find one recorded instance of a civilized people falling independently into a savage state?" Only a Science of Culture would save man from the pitfalls of piety and tradition. "In discussing problems as complex as those of the development of civilization, it is not enough to put forward theories accompanied by a few illustrative examples. The statement of the facts must form the staple of the argument, and the limit of needful detail is only reached when each group so displays its general law, that fresh cases come to range themselves in their proper niches as new instances of an already established rule." Tylor welcomed questions which he could not begin to answer, but which others had not even begun to ask. To Tylor himself, dominated by the late Victorian idea of "development" (i.e., that all societies had followed a single course of evolution, some more slowly than others), a unilinear view of human progress was especially tempting. It cast a roseate glow over the future of all peoples, and incidentally made all living "primitive" cultures a rich and accessible source for history. Just as Schliemann had some very personal reasons to believe that he had found the true Troy and the relics of Agamemnon's feast, so Tylor and his co-evolutionists were eager to see living savages reenact the "infancy" of their civilization.

Still, Tylor believed himself not the prophet of a dogma but the discoverer of a science. He was happy in the thought that he was opening questions beyond his power to answer. The last twenty-five years of his long life he devoted to organizing and promoting his Science of Culture under the name of Anthropology. The Royal Anthropological Institute with his guidance became a lively parliament of anthropological science. In successive editions of *Notes and Queries on Anthropology* for "the use of travellers and residents in uncivilized lands" Tylor himself collected countless facts and encouraged others to do the same, toward a growing science. When he sensed that unilinear evolution could not explain the varieties of culture, he asked how cultural features might be "diffused" from people to people.

Hoping to illuminate these questions, Tylor persuaded the British Association for the Advancement of Science in 1881 to undertake a long-range study of the little-known tribes of the northwest coast of Canada. That twelve-year field study by Franz Boas (1858–1942) was supervised by Tylor himself, and prepared Boas to become the great reviser of Tylor's science.

As Tylor wrote to Boas in 1895, the time had come for a "reformation" in anthropology.

A frail and precocious child in Westphalia, Boas very early absorbed the liberalism of his freethinking Jewish parents, who had remained loyal to the spirit of the Revolution of 1848. As a young man he studied the natural sciences in several German universities. Then he spent a year exploring Baffin Island and living with Indians in the Canadian Northwest, which awakened his interest in a Science of Culture. At the age of twenty-eight he emigrated to the United States, where he began a whirlwind career in universities, museums, and learned societies which established him, before he was fifty, as the dominant spirit in this new profession in the United States. He helped found the American Anthropological Association, wrote a classic *Mind of Primitive Man* (1911), lectured brilliantly, enlisted a devoted following of students, and as a citizen-spokesman pioneered in applying his antiracist doctrine to American immigration policy.

Boas fulfilled Tylor's hope even more grandly than Tylor could have imagined, for he did more than anyone else to liberate the Science of Culture from the provincial prejudices of Tylor's Britain. In its first phase, "Mr. Tylor's Science" had already done wonders to widen the vision of human society. The doctrine of evolution itself whetted appetites for facts about all the peoples of the earth. The unilinear track of progress seemed to destine all primitive people to the happy climax of Victorian Britain. But Boas could not believe there was only one cultural destination for the whole human race.

If "all mankind is one," if, as Tylor had insisted, all mankind had an equal capacity to develop cultural forms, then there must have been many different tracks and destinations for human progress—as many as the circumstances of geography, climate, language, and historical accidents. Even more than Tylor himself, Boas saw Culture triumphant. The cultural history of each people was unique. All surviving groups of people, Boas argued, had developed equally, but there were as many different ways as there were groups. Also a believer in the primacy of facts, Boas was even more incrementalist than his mentor. Perhaps human nature was too complex, human cultures too diverse, for any simple general solution, even one as grand as "evolution." Perhaps the science of culture would have to grow, not wholesale and speedily—with large organizing ideas like "animism"—but retail, slowly, and piecemeal, by tracing the relations among the elements in one culture and then seeing similar relations reappear in others. While Tylor had opened the vistas to the world of cultures, Boas now opened vistas to the wonderful subtlety within every culture. And to connections between every culture and all its surroundings—its geography, nutrition, diseases, and accidental encounters.

Tylor, as first professor of anthropology at Oxford, believed that he had helped liberate his disciples from an Unholy Alliance between Theology, Classical Studies, and the old-style Natural Sciences. Theology, teaching only the True God, objected to talk of false gods; Classical Studies knew only the cultures of Greece and Rome; the Natural Sciences feared, with some reason, that the new Social Sciences would empty their lecture rooms. Clued by his happy accidental encounter with the New World, Tylor had charted a science that reached beyond English academic orthodoxies. "Mr. Boas' Science" made the culture of every people another New World.

An Expanding Universe of Wealth

To the ancient Greeks "economics" meant the management of a household or a city-state. The standard medieval textbook, Aristotle's *Politics,* explained that "the amount of household property which suffices for the good life is not unlimited." "There *is* a bound fixed," he insisted, for the needs of a household or a state—and the name for this fixed stock was "wealth." This view of economic well-being that long dominated western Europe brought with it some confining dogmas. A "just price" was fixed not by what the traffic would bear but by what the seller ought to ask. "Usury," a name for any interest taken, was frowned on, since money was by its nature supposed to be sterile. In fact, there was a general moral antipathy to "kremastics," the unbounded accumulating of wealth. But there was no economics in the modern sense—no "science" of prices, of supply and demand, of national income, or international trade. Instead, works of moral philosophy that told people how they ought to behave in the marketplace debated and prescribed such questions as the narrow limits of the "just" price. These ways of thought still ruled western Europe in the Age of Discovery.

At the same time, gold and silver, the treasures that could command everything else, seemed the best universal measure of wealth, and became lures for the bold seafarers. Henry the Navigator's sailors were enticed around Cape Bojador by rumors that somewhere in those parts a river of gold emptied into the sea. At the very least they hoped to find a sea route to the African gold mines. Preparing for his first voyage, Columbus marked his copy of D'Ailly's *Imago mundi* at the passages describing the gold and silver, pearls and precious stones, to be found on the Asian shores he expected to reach. The lucky Spanish conquistadores found precious metals in great quantity. First it was gold, but by the mid-sixteenth century new silver mines in Mexico and Peru poured their treasure back to Seville, which for a while dazzled Europe with its prosperity. The myth of El Dorado ("the Gilded" Land) bewitched the imaginations of Spaniards who refused to believe that it would not be found somewhere in America. When they captured Indians, they chose a few to be torn apart by dogs and burned some alive, to force terrified witnesses to reveal the whereabouts of the Gilded Land. And Indians obliged by making up tales that kept the myth alive.

New World supplies of gold and silver proved limited, but Spanish greed was not. The precious metals that flooded into Europe helped bring on a massive inflation, which historians call the Price Revolution. By the year 1600, prices in Spain were nearly four times what they had been a hundred years before. The inflation that spread across Europe disrupted the economy of Spain and hastened the decline of the Spanish Empire.

In western Europe this was the dawning era of the modern national state. Rising European powers competed around the globe for a bigger share of the world's treasure. Queen Elizabeth consolidated England, defeated the Spanish Armada in 1588, and sent out her pirates to seize Spanish treasure wherever it could be found. The nations who were to dominate modern European history organized their policies around the simple ideas that had confined economic thought from the beginning of history: all wealth was limited; one nation's gain was another's loss; your wealth could be increased only at the expense of another's; a larger slice for one nation left smaller slices for others. These assumptions ruled western Europe from the fifteenth till the eighteenth centuries. With stronger armies and more potent navies, your nation could reach for an ever-larger share of the world's treasure.

The "national economy," a notion that grew up in England and France in the seventeenth century, aimed to unify the nation. By breaking up local enclaves, by abolishing local tolls and tariffs, the power of the national government would be increased against rivals on the world scene. The classic doctrine later known as "mercantilism" was stated by a successful English businessman, Sir Thomas Mun (1571–1641), a director of the East India Company. Britons blamed the depression of 1620 on that company's yearly export of £30,000 of bullion to finance their trade. Mun, defending his company before the Standing Committee of Trade, wrote powerful tracts, *A Discourse of Trade, from England unto the East Indies* (1621) and *England's Treasure by Forraign Trade; Or, the Ballance of our Forraign*

Trade in the Rule of our Treasure (1630; published 1664), to reinforce the idea of a national economy and the hypnotic concept of "the balance of trade." He argued that the crucial question was not whether any one company exported bullion but whether the value of the nation's exports *as a whole* exceeded the value of imports. A "favorable" balance of trade meant that bullion was flowing into the country and so the nation was growing richer.

As the modern nations of Europe expanded across the world seeking outposts and colonies in distant parts, they still somehow kept this narrow vision, in myopic search for treasure. Meanwhile they hardly glimpsed the wonderful wider benefits of expanding new communities in America, Asia, Africa, and Oceania. In 1760, after Wolfe captured Quebec and all Canada came under English control, London debated the terms to be imposed on the French. The vast unsettled stretches of unexplored Canada seemed quite useless compared with the tiny sugar-rich islands of Guadeloupe, whose tropical stores could be exported to the world to improve the British "balance of trade." In 1760 Benjamin Franklin, then in London and blessed with a New World vision, argued that Canada would be incomparably more valuable in the long run. He pointed out that in the future the growing Canadian population would increase the market for British products, would strengthen the British Navy by its demand for more British ships, and so add to British strength and well-being. By shutting their eyes to this larger view, the British would lose thirteen American colonies.

The year of the American Declaration of Independence, 1776, by an appropriate coincidence saw the publication of Adam Smith's *Wealth of Nations,* which also in its way was an emancipation proclamation. Just as Jefferson's document proclaimed a new beginning for the politics of the West, so Adam Smith proclaimed a new beginning, a more ample vision, for national economies. Many of Smith's ideas, like Jefferson's, had appeared in the writings of others over the last century. He drew on the ideas of Sir William Petty and John Locke, he owed a debt to Beccaria and Turgot, to the Physiocrats, and especially to his contemporary countrymen David Hume, Dugald Stewart, and Francis Hutcheson. He had borrowed from Grotius and Pufendorf, and even followed some notions of the medieval scholastic moralists. Unlike the work of Newton or Darwin, Adam Smith's was not spectacularly original. All his intellectual raw materials were at hand—the ideas, the historical examples, even many of his most memorable phrases. Master of the lively detail, he illustrated his ideas from Greek and Roman antiquity, from the European Middle Ages, Poland, contemporary China, as well as America. He recounted the policies of those who had ignored the facts of economic life.

The New World stretched the vistas seen from Europe. Prosperous new settlements on an unexploited, unexplored continent inevitably expanded European notions of wealth and material well-being. Definitions from the age of Croesus could not longer serve nations in the age of Franklin and Jefferson. Adam Smith's work proclaimed the freeing of Europe from Old World bonds of economic thought. Expanding Europe required an expanded notion of the wealth of nations. Adam Smith's obvious and conspicuous target was what he called the Mercantile System. He shifted focus from the nation to the world, from the nation to the Wealth of Nations.

> It is not by the importation of gold and silver, that the discovery of America has enriched Europe. . . . By opening a new and inexhaustible market to all the commodities of Europe, it gave occasion to new divisions of labour and improvements of art, which, in the narrow circle of the ancient commerce, could never have taken place for want of a market to take off the greater part of their produce. The productive powers of labour were improved, and its produce increased in all the different countries of Europe, and together with it the real revenue and wealth of the inhabitants. The commodities of Europe were almost all new to America, and many of those of America were new to Europe. A new set of exchanges, therefore, began to take place which had never been thought of before, and which should naturally have proved as advantageous to the new, as it certainly did to the old continent.

Few other subjects in his world panorama so awakened his interest and focused his imagination as did America. But the discovery and settlement of the New World were only a stage in the still wider opening of the world. To try now to organize a great empire simply for the benefit of British merchants and the British "balance of payments" in the home island would be sheer folly.

With uncanny foresight, Adam Smith proposed a scheme of federal union. The American colonists should have representatives in Parliament, "in proportion to the produce of American taxation." Nor would Americans have to fear that the seat of government would always remain across the Atlantic. "Such has hitherto been the rapid progress of that country in wealth, population and improvement, that in the course of little more than a century, perhaps, the produce of American might exceed that of British taxation. The seat of the empire would then naturally remove itself to that part of the empire which contributed most to the general defence and support of the whole."

Adam Smith is commonly awarded his place in the pantheon of economic thinkers as the champion of what he called "perfect liberty," a free competitive economy. But from our perspective he did more than espouse an economic doctrine. He lifted the vision of European man to a new scene. He saw economic well-being not as the possession of treasure but as a

process. Just as Copernicus and Galileo helped raise men above the commonsense fact that the sun circulated the earth, so Adam Smith helped his
generation rise above the specious proposition that a nation's wealth consisted of its gold and silver. And like Copernicus and Galileo, he saw the
whole world and society in constant motion. Just as Lewis Henry Morgan
and Edward B. Tylor would widen the vistas of "culture" to encompass all
mankind, so Adam Smith widened the vistas of "wealth."

The *Wealth of Nations* opens with the homely and familiar example of
a pin factory where the division of labor made it possible for ten people to
turn out forty-eight thousand pins in a single day. "The greatest improvement in the productive powers of labour," he explained, introducing an
illuminating new phrase, was the effect of "the division of labour." This was
"the necessary, though very slow and gradual, consequence of a certain
propensity in human nature which has in view no such extensive utility; the
propensity to truck, barter, and exchange one thing for another." But the
division of labor, the key to human improvement, was limited by "the extent
of the market." Without education there could be no division of labor, and
without division of labor no social improvement.

Adam Smith, a man of sedentary academic temperament, became the
first modern pathfinder for the economic policies of statesmen and businessmen around the planet. Born in 1723, he was the only child in a family in
comfortable circumstances in Kirkcaldy, a coastal village on the northern
shores of the Firth of Forth. His father, a collector of customs, died a few
months before he was born, and Adam Smith remained close to his mother
all her life. So far as we know, no other woman was important to him, and
(as Joseph Schumpeter reminds us) "in this as in other respects the gla-
mours and passions of life were just literature to him." Before his fourth
birthday, when visiting his grandfather on the banks of the Leven, he was
stolen by a band of Gypsies, and it was some time before he could be found.
Would Adam Smith have made a successful Gypsy?

At the Burgh School in Kirkcaldy, one of the best in Scotland, he had
four years' training in the classics. Nearby was the Glasgow ironworks,
called the nailery, which he enjoyed visiting and which would figure in the
opening pages of the *Wealth of Nations.* In 1737 he entered Glasgow College, where he improved his Latin and Greek and came under the influence
of "the never-to-be-forgotten" Francis Hutcheson (1694–1746), the first professor there to lecture in English instead of Latin, who defied the Scottish
Calvinists by teaching a cheerful and beneficent God who governed the
world for what he called "the greatest good of the greatest number."

The Oxford where Adam Smith entered Balliol College as a scholarship
student in 1740 was "steeped in port and prejudice." Comfortably endowed
dons and professors received their salaries "from a fund altogether indepen-

dent of their success and reputation in their particular professions." The colleges showed how "teachers . . . are likely to make a common cause, to be all very indulgent to one another, and every man to consent that his neighbor may neglect his duty, provided he himself is allowed to neglect his own. In the university of Oxford, the greater part of the public professors have, for these many years, given up altogether even the pretence of teaching." They taught him one lesson he never forgot, the fate of any institution that did not depend on the goodwill of its customers. Still he read widely and had time to reflect, and turned from his growing interest in mathematics back to the Latin and Greek classics, which dominated the Balliol library. The only intrusion of university authorities into his education was when they caught him reading David Hume's recently published *Treatise of Human Nature* (1739), but luckily he escaped with only a reprimand and the confiscation of the book. His family hoped he would stay at Oxford for an academic career, but he refused to take the necessary holy orders.

Back in Scotland, he pursued his academic interests under freer auspices. In Edinburgh he gave a series of public lectures on English literature, a novel subject, to an audience of a hundred citizens who paid a guinea each. In 1750–51 he offered a public course on economics, a subject never yet heard of in Oxford's sanctimonious halls. The success of these lectures brought him a chair at the University of Glasgow, first as professor of logic, then as professor of moral philosophy. He was preaching commercial liberty, an idea already stirring Scotland at the time, and it was said that his lectures converted the city to his gospel of free trade.

In the mid-eighteenth century, Glasgow, an ancient provincial Scots city of some twenty-five thousand, was washed by the currents of the future. Situated on both sides of the river Clyde, it had long been a center of religion and education, and of commerce with northern Europe. After the union with England in 1707 Glasgow profited too from the commerce with America. Andrew Cochrane, the Provost of the city, had just founded a Political Economy Club when Adam Smith arrived as professor, and promptly enlisted him as a member. Glasgow merchants, locally known as "tobacco lords," had prospered from the removal of restrictions on their trade with the colonies, and were now agitating against the import duties on the American iron that supplied the Glasgow ironworks. Cochrane's ironworks was then every year importing four hundred tons of iron. It would be a disaster for Glasgow merchants when the American colonies ended their trade in tobacco. Meanwhile, Adam Smith joined with his club members "to inquire into the nature and principles of trade in all its branches, and to communicate knowledge and ideas on the subject to each other." For many facts in the *Wealth of Nations* Adam Smith acknowledged his debt to the practical-minded Cochrane.

Adam Smith's first work, his *Theory of Moral Sentiments* (1759), already showed his knack for simple explanation of complex problems. He described the moral sentiment by a simple figure of speech, an "inner man," an impartial spectator within each of us who passes judgment on everything we do from the point of view of other people. "I consider what I should suffer if I were really you." This, he explained, was quite different from self-love. He had already noticed that to serve society we are "led by an invisible hand."

Smith's fellow Scotsman David Hume facetiously reported from London:

> Nothing, indeed, can be a stronger presumption of falsehood than the approbation of the multitude; and Phocion, you know, always suspected himself of some blunder when he was attended with the applause of the populace.
>
> Supposing that you have prepared yourself for the worst by all these reflections, I proceed to tell you the melancholy news that your book has been very unfortunate, for the public seem disposed to applaud it extremely. It was looked for by the foolish people with some impatience; and the mob of literati are beginning already to be very loud in its praises.

With the bad news of success, Hume noted that he "heard it extolled above all books in the world," and the pundits had already ranked Adam Smith among "the glories of English literature."

The book unpredictably gave Adam Smith his only opportunity to reach out to the larger world of European thought. Incidentally, it led to an annuity that provided him the leisure to write his great work. Among the admirers of Adam Smith's *Moral Sentiments* was Charles Townshend (1725–1767), who, Hume reported from London, "passes for the cleverest fellow in England." It was a splendid irony that the good opinion of the author of the notoriously restrictive Townshend Acts (1767) supported Adam Smith in writing the bible of free trade. The Townshend Acts, violating the traditions of colonial self-government, pushed the American colonists on the irreversible course to revolution. Townshend, who had recently married the widow of the eldest son of the Duke of Buccleuch, was looking for a tutor to accompany his stepson, the young Duke, on the customary Grand Tour of the Continent. Reading Adam Smith's *Moral Sentiments,* he at once decided that the author was his man, and he went straight to Glasgow to persuade his candidate to give up a professorship for this tutorial assignment. Some might have thought the abstracted professor ill suited to guide a young man across the Continent, for while showing Townshend the Glasgow sights, Smith took him to see the great tannery and absentmindedly walked right into the tan pit. The determined Townshend offered Smith a salary of £300 a year, plus traveling expenses while abroad, and then a pension of £300 a year for life. This was an appealing proposition to a Glasgow professor whose income was only £170 a year. There was no pension for overage professors, who had to rely on the sum a successor might pay as the price of the chair.

Adam Smith resigned his Glasgow chair, and in 1764 set out with his charge on their Grand Tour. They were abroad for two and a half years, spending about a year and a half in Toulouse—where Hume's cousin was vicar-general of the diocese—two months in Geneva, and nearly a year in Paris. Toulouse, then a favorite resort of the English, as Florence would be in the next century, offered the most cultivated French society outside of Paris. Without the distractions of the capital, Adam Smith had the leisure to begin to write his big book. During a two-month interlude in Geneva he enjoyed some conversations with Voltaire. Then on to Paris, where Hume himself was secretary at the British embassy. Smith and his charge attended the theater, glimpsed the fashionable salons, and encountered some seminal ideas. The brilliant François Quesnay (1694–1774), King Louis XV's consulting physician who had been installed at Versailles under the patronage of Madame de Pompadour, enlisted Adam Smith in his French version of a political economy club. At sixty, Quesnay had begun writing about economics, on which he had already become the King's pundit.

Quesnay's *Tableau Economique* (1758), aiming to accomplish for social forces what Newton had done for physical forces, invented a whole vocabulary for the new science. Quesnay introduced the notion of economic classes, each with its own flow of products and receipts, he proposed the idea of economic equilibrium, and planted ideas about capital, savings, and investment, which would flower into a vast literature of economic analysis in following centuries. His *Tableau Economique* was first printed in a small edition on the King's private press, but when Madame de Pompadour warned him of the King's certain displeasure with such flighty notions, he let the work reach the large public under the name of the Marquis de Mirabeau.

Quesnay's disciples, first known simply as *les économistes,* became famous as the Physiocrats, offering the first modern model for economics. Their leading ideas were quite simple. A natural law similar to that governing the physical world governed the growth and flow of wealth. The wealth of a society consisted not in its store of gold and silver but in its total stock of commodities, and the best way to increase that stock was to allow the free flow of products in the market without monopolies or tax restrictions. These pioneer economists were appalled by the poverty of the French peasantry in stark contrast to the luxury of the nobles, the tax-farmers and other monopolists. "Poor peasants, poor kingdom," they proclaimed, "poor kingdom, poor king." Their remedy for the nation's ills focused on the plight of the peasants. Improve the techniques of agriculture, remove all blocks to the flow of goods, abolish all existing taxes and all the tax-farmers, and instead establish a single tax on the product of the soil to be collected by honest government officers. When Quesnay was offered a collector-generalship of taxes for his son, he refused. "No," he said, "let the welfare of my children be bound up with the public prosperity," and he made his son a farmer. If Louis XV had only listened to Quesnay, he might have prevented much misery for France and have saved his own grandson from the guillotine.

Salon wits sneered at the Physiocrats' earnest calculations, but Adam Smith felt at home. An economic freethinker himself, as we have seen, he had preached many of the same notions back in Glasgow and had already begun writing his textbook on economic liberty during his idle hours in Toulouse. Now he seized the tiniest clues to what the old régime had done to France. By contrast even to the poor Scottish peasants, French peasants still wore wooden shoes or went barefoot. "In France," he observed, "the condition of the inferior ranks of people is seldom so happy as it frequently is in England, and you will there seldom find even pyramids and obelisks of yew in the garden of a tallow-chandler. Such ornaments, not having in that country been degraded by their vulgarity, have not yet been excluded from the gardens of princes and great lords." Though he saw the people of France "much more oppressed by taxation than the people of Great Britain," he did not yet foresee the coming violence. But like Quesnay, he insisted that economic freedom was essential to improve the condition of the people. Adam Smith said he would have dedicated the *Wealth of Nations* to him had not Quesnay died two years before the book was published.

Unseduced by the charms of the salons, the playhouses, and Quesnay's lively company at Versailles, Adam Smith longed "passionately" to return to old friends in Scotland. His departure was more sudden than he had planned because of the shocking murder in the streets of Paris of the Duke's younger brother, who had also been put in Smith's charge. Stopping briefly in London, where he was elected a Fellow of the Royal Society, he settled back with his mother in the familiar haunts of his native Kirkcaldy. There he spent the next six years, enlivened only by daily walks in the sea breezes of the Firth of Forth and an occasional trip to Edinburgh, writing the *Wealth of Nations.*

In the spring of 1773 Adam Smith went to London with what he thought was a nearly completed manuscript. His manuscript proved less finished than he had imagined, for the next three years in London brought a flood of new facts and ideas. He dined from time to time with Dr. William Hunter the anatomist, Robert Adam the architect, Sir William Jones the linguist, Oliver Goldsmith, Sir Joshua Reynolds, David Garrick, Edward Gibbon,

Edmund Burke, and Dr. Johnson. But he was not always regarded as their equal. "Smith is now of our club," Boswell reported. "It has lost its select merit."

The great question of the hour was, of course, the American rebellion. This was providential for Adam Smith, who believed America a kind of laboratory for the Nature and Causes of the Wealth of Nations. Benjamin Franklin, then Pennsylvania's agent in London trying vainly to avert the separation, boasted, "The celebrated Adam Smith, when writing his *Wealth of Nations,* was in the habit of bringing chapter after chapter as he composed it to himself, Dr. Price and others of the literati; then patiently hear their observations and profit by their discussions and criticisms, sometimes submitting to write whole chapters anew, and even to reverse some of his propositions." The American rebellion dramatized the cause that Adam Smith had defended for thirteen years in Glasgow among the merchants trading with America and the planters who had returned. For the *Wealth of Nations* the American colonies—their settlement, their plight, and their promise—were a source of endless examples. The New World, land of the future, offered a unique opportunity to test the virtues of economic liberty.

The fruit of twelve years of writing and, before that, of at least twelve years of focused thinking on its large subject, the *Wealth of Nations* was finally published in two volumes on March 9, 1776. The publisher, who had paid Smith some £500 for the manuscript, did not lose money. The book sold well from the beginning, and the first edition was exhausted in six months. It was barely noticed by the reviewers, but Adam Smith's friends, London's literary luminaries, praised it privately without stint. They compared it with the first volume of Gibbon's *Decline and Fall of the Roman Empire,* which had appeared just three weeks earlier, on February 17, 1776. David Hume, a loyal Scot, praised Gibbon's volume from Edinburgh with the outburst, "I should never have expected such an excellent work from the pen of an Englishman." Smith's work, he said, required too much thought to be at once as popular as Gibbon's. Still, he prophesied a great future. "What an excellent work," Gibbon himself exclaimed, "is that with which our common friend Mr. Adam Smith has enriched the public! An extensive science in a single book, and the most profound ideas expressed in the most perspicuous language." When a jealous critic carped that the book could not be good because Adam Smith had never been "in trade," Dr. Johnson retorted sententiously that "there is nothing that requires more to be illustrated by philosophy than does trade. . . . A merchant seldom thinks but of his own particular trade. To write a good book upon it a man must have extensive views." Adam Smith's extensive views gave his book a power surpassed by no other modern book. He was the true discoverer of the modern science of economics.

This modern science flourished, opening new paths from all the other sciences to a science of wealth and economic well-being. But, as might have been predicted, Adam Smith's prospectus for exploring the wealth of nations became a chart of orthodoxy. His description of the body economic had much of the appeal that Galen's held for earlier centuries in describing the human body. Brilliant interpreters and elaborators—notably David Ricardo and John Stuart Mill—embroidered Smith's ideas and offered these as established truths. The ambitious Mill's *Principles of Political Economy* climaxed in a chapter "Of the Grounds and Limits of the Laisser-Faire or Non-Interference Principle." There he listed his few exceptions—most conspicuously the enterprise of colonization—to the rule that "Laisser-faire . . . should be the general practice: every departure from it, unless required by some great good, is a certain evil." This "classical" period, lasting for a full century, was followed by a "neoclassical" period pioneered by the Cambridge economist Alfred Marshall (1842–1924), whose *Principles of Economics* (1890) offered a newly persuasive Revised Version of Adam Smith.

Classical economics provided a framework and a vocabulary even for bitter critics of the society it purported to describe. Karl Marx himself, who was both "much more (and much less) than an economist," is placed by historians of economics firmly in the classical tradition. During the heyday of classical economics, most of the leading writers were not full-time professional economists, but were businessmen (like Ricardo or Engels), civil servants (like J. S. Mill), or journalists (like Marx). The word "economics" (in place of "political economy") to describe the subject matter of a new profession came into the English language only in the nineteenth century, and professional associations did not appear until the American Economic Association (1885) and the British Economic Association (1890).

By the mid-twentieth century, just as "classical" physics would describe an outmoded physics, so classical economics came to describe an economic doctrine of the past. For there was a revolution in economics too. The man mainly responsible was one of the most remarkable intellectual phenomena of modern times and, in proportion to his influence, one of the least celebrated. John Maynard Keynes (1883–1946), son of a lecturer on moral science and economics, who was an academic administrator in Cambridge University, hardly had the social background for a revolutionary. Schooled at Eton, he secured traditional grounding there in mathematics and classics, and he thrived under the arcane discipline of that curious place. He even applauded the Etonian version of football, "the present form of legalised ruffianism. . . . the best condition under which one can play the glorious game." At King's College, Cambridge, he was elected president of the undergraduate debating Union, and became a disciple of Alfred Marshall.

Already in Cambridge he had joined the freethinking demimonde of the so-called Bloomsbury group. Their leading spirit was Lytton Strachey, whose mordant wit and irreverence for Victorian sacred cows set the tone of Keynes' own sallies. This group included E. M. Forster, Virginia Woolf, and some of the leading critics and artists, who, for their time, were shockingly tolerant of homosexuality, pacifism, and a bohemian style of life. In the civil service examination, Keynes ranked second in the nation, and after two years in the India Office he returned to Cambridge, where a brilliant *Treatise on Probability* brought him a fellowship at King's College. His catholic tastes in art and in ideas stirred Cambridge gossips, especially when he married the ballerina Lydia Lopokova (some called her a "chorus girl"), who had actually danced the cancan under Massine's direction. Their marriage was long and happy.

Just as the colonial scene in the Age of the American Revolution awakened Adam Smith to a new era in thinking about the wealth of nations, so the tragic scene of Europe after World War I stirred Keynes' reflections. As economic adviser to Lloyd George at the Versailles Peace Conference in 1919, he had an insider's view of the bickering "Big Three." Keynes saw that the narrow nationalism of Lloyd George, the vindictiveness of Georges Clemenceau, and the moralism of Woodrow Wilson were all equally menacing to the prosperity of Europe. He foresaw doom in the unrealistic demands for reparations from the defeated nations. From Paris he reported to his friend the painter Duncan Grant on May 14, 1919:

> I have been as miserable for the last two or three weeks as a fellow could be. The Peace is outrageous. . . . Meanwhile there is no food or employment anywhere, and the French and Italians are pouring munitions into Central Europe to arm everyone against everyone else. I sit in my room hour after hour receiving deputations from the new nations. All ask, not for food or raw materials, but primarily for instruments of war against their neighbours. . . . They had a chance of taking a large, or at least, a humane, view of the world, but unhesitatingly refused it. Wilson, of whom I have seen a good deal more lately, is the greatest fraud on earth. . . . Do write to me and remind me that there are still some decent people in the world. Here I could cry all day for rage and vexation. The world cannot be quite as bad as it looks.

Resigning his post in protest, Keynes left this "nightmare" where peacemakers "gloat over the devastation of Europe," and returned to England.

In the next two months he wrote his *Economic Consequences of the Peace*, which appeared before Christmas and made him famous across Europe and

America. Readers were delighted by his unforgettable caricatures. Clemenceau, he said, "felt about France what Pericles felt of Athens—unique value in her, nothing else matters; but his theory of politics was Bismarck's. He had one illusion—France; and one disillusion—mankind, including Frenchmen, and his colleagues not least." Wilson's "head and features were finely cut and exactly like his photographs, and the muscles of his neck and the carriage of his head were distinguished. But, like Odysseus, the President looked wiser when he was seated; and his hands, though capable and fairly strong, were wanting in sensitiveness and finesse. . . . he was not only insensitive to his surroundings in the external sense, he was not sensitive to his environment at all. What chance could such a man have against Mr. Lloyd George's unerring, almost medium-like sensibility to every one immediately round him? . . . this blind and deaf Don Quixote was entering a cavern where the swift and glittering blade was in the hands of his adversary."

The thrust of his eloquence was that the economy of all Europe—and the world—was one and inseparable. The legacy of a vindictive Versailles would be a contagion of riots and revolutions and dictatorships. "Never in the lifetime of men now living," he concluded, "has the universal element in the soul of man burnt so dimly."

Keynes' bitter predictions would be soon enough fulfilled. Meanwhile, he went back to Cambridge where he remained for the time the star disciple of Alfred Marshall. But Keynes' great strength was a sense of history, a prophetic capacity "to view the world with new eyes. . . . the hidden currents, flowing continually beneath the surface of political history. . . . In one way only can we influence these hidden currents,—by setting in motion those forces of instruction and imagination which change *opinion.* The assertion of truth, the unveiling of illusion, the dissipation of hate, the enlargement and instruction of men's hearts and minds, must be the means."

A Captain Cook in the world of economics, Keynes would also suffer the ardors of negative discovery. While Adam Smith and his classical disciples had been focusing on "wealth" and its causes in the "marketplace," a whole new social phenomenon, a specter, a more negative phenomenon in the world of wealth-oriented theorists, had come into being. Its name was *unemployment.* And this became the center of Keynes' concern. As early as 1924, when unemployment in British mines, shipyards, and factories reached a million, Keynes began to loosen some of the neoclassic dogmas to encompass this growing evil.

He joined Lloyd George (who was no economist) in calling for a large program of public works. *Does Unemployment Need a Drastic Remedy?*

Keynes asked in May 1924. His emphatic answer was to use the Treasury's Sinking Fund "to spend up to, say, £100,000,000 per year on the construction of capital works at home, enlisting in various ways the aid of private genius, temperament and skill." To the objections of his neoclassical colleagues he replied:

> Our economic structure is far from elastic, and much time may elapse and indirect loss result from the strains set up and the breakages incurred. Meanwhile, resources may lie idle and labour be out of employment. . . . We are brought to my heresy—if it is a heresy. I bring in the State; I abandon *laissez faire*,—not enthusiastically, not from contempt of that good old doctrine, but because, whether we like it or not, the conditions for its success have disappeared. It was a double doctrine,—it entrusted the public weal to private enterprise *unchecked* and *unaided*. Private enterprise is no longer unchecked,—it is checked and threatened in many different ways. There is no going back on this. The forces which press us may be blind, but they exist and are strong. And if private enterprise is not unchecked, we cannot leave it unaided.

The Great Depression, beginning in the 1930's, that was marked in the United States by the resounding defeat of Herbert Hoover and the election of the cheerful empiricist Franklin D. Roosevelt, was a worldwide phenomenon. In 1932 there were ten million unemployed in the United States alone. In Keynes' eyes the Depression was marked less by the ancient evil of poverty (i.e., lack of "wealth") than by this modern evil of unemployment. He was shifting the focus of economic theory from the impersonal mechanisms of the marketplace to the spectacle of wasting and despairing human beings.

By 1936 Keynes had elaborated a theory for his new perspective. Considering Keynes' humanistic temperament, it is remarkable that he wrote a book unintelligible to the general public. Adam Smith's book had been written for the literate reader—for those who enjoyed Gibbon's *Decline and Fall*—for there was still no profession of economics. By contrast, Keynes' *General Theory of Employment, Interest and Money* was written only for experts in the new science, and its argument cannot be fairly summarized in a paragraph. But its impact through economists on the popular thinking about economics was plain enough, and it became the most influential work of economic science written in the twentieth century. What was most radical and most revisionary was his new focus on a modern evil that had not seemed to be a significant social phenomenon until his age.

The word "unemployment" does not enter common use in the English language until about 1895, and within forty years Keynes was the first to make the problem central to the whole theory of economics. His seminal book, as his most productive disciple and his definitive biographer R. F. Harrod explained, was "basically an analysis, in terms of fundamental economic principle, of the causes of unemployment." Keynes' conclusion was that a substantially free-market society could be preserved and continuous full employment assured only by the timely intervention of the state with public works or other expedients. He was led to this conclusion by two simple but subtly argued propositions, both radical revisions of laissez-faire dogma. There was no wage so low, he explained, as to lead to full employment. On the contrary, continued reductions of wages would actually increase unemployment. In place of individual demand in the marketplace he offered the crucial notion of "aggregate demand," which was the product not only of individual consumers but of spending by all private investors and government agencies. He gave human expectations themselves a new and leading role in economic theory. In other words, the processes of the marketplace were neither so automatic nor so benignly self-regulating as the classical economists had imagined. To keep a capitalist community fully employed, the "invisible hand" must become visible, and a benign government must control the flow of investment, increasing investment in public works, to ensure that aggregate demand would provide full employment.

Seldom has one scientific book so speedily shaped the policies of government or so widely converted the counsels of government to abandon an inherited orthodoxy. In the United States, Keynes' ideas guided Franklin D. Roosevelt's New Deal, shaped the Employment Act of 1946 requiring the federal government to take measures to keep up employment, and Keynesian policies were pursued by President John F. Kennedy and his successors. At the Bretton Woods Conference in 1944 and in helping to create the International Monetary Fund and the World Bank, Keynes aimed to embody his theories in worldwide institutions to prevent recurrence of another Great Depression. The awakening power of Keynes, like that of seminal thinkers in the other sciences, was not confined to those who understood or accepted his doctrines. His notion of aggregate demand and his proposals for government intervention in the economy led to the collecting of fuller and more accurate statistics on national income in Britain and elsewhere. But most important, his luminous restless mind and his feeling for the human role in the economic world saved the new science of economics from its first orthodoxy.

81

Learning from Numbers

THE pioneer of modern demography, some would say, too, of statistics, was a prosperous London tradesman, John Graunt (1620–1674), an amateur in the world of mathematics. He had no formal training in the subject, but he was bound as apprentice to a haberdasher and became a prosperous man of affairs. "Ingenious and studious," known for his "dexterous and incomparable faculty" of making shorthand notes of sermons, he was deeply pious, experimental in religion, and a peacemaker in the faction-ridden London of the Civil War. Though raised a Puritan, he first converted to anti-Trinitarianism, then to Catholicism. He suffered disastrous losses from the Great London Fire of 1666 from which he never recouped his fortune. A down-to-earth businessman, Graunt was not concerned with the grand estimates of national wealth that occupied the "political arithmeticians" of his day. Yet he was interested in the welfare of his London community. He held numerous city offices, including councilman.

The death toll of the plague years, visible all around him, became the basis for Graunt's interest in demography and statistics. The most obviously distressing fact about the English population was the high death rate during years of plague, some of the worst of which occurred in Graunt's lifetime. During 1625, for example, about one-quarter of the population died. As early as 1527, Bills of Mortality, or lists of the dead, had been occasionally collected in London, and by 1592 these regularly listed the cause of death. During the disastrous plague of 1603 the weekly Bills of Mortality published information gathered by "searchers," "ancient matrons" appointed to view dead bodies to report the cause of death, and to enforce laws of quarantine. Carrying conspicuous red wands of office, these elderly women were notorious for their ignorance of medicine, their thirst for alcoholic beverages, and their obliging readiness, for a sum, to cover up an unpleasant fact like a death from syphilis. Their reports were sold to all interested persons by the parish clerks at 10:00 A.M. on Thursdays at a penny per sheet or by subscription at four shillings per year.

Graunt "knew not by what accident" his thoughts were engaged by the

Bills of Mortality. A practical man, he was puzzled that so many facts so regularly collected had been put to so little use. His friend the pioneer economist William Petty (1623–1687) probably encouraged his curiosity. On February 5, 1662, Dr. Daniel Whistler, a medical doctor, distributed to a meeting of the Royal Society fifty copies of a ninety-page booklet by John Graunt that had come off the press only two weeks before. He proposed Graunt for election to Fellowship, and the Society promptly voted Graunt the honor, which was unprecedented for a mere businessman. King Charles II, endorsing Graunt, urged the Society that "if they found any more such Tradesmen, they should be sure to admit them without any more ado."

The new international community of science was opening its doors. Graunt hoped modestly that his short pamphlet with a long name—*Natural and Political Observations mentioned in a following Index, and made upon the Bills of Mortality . . . With reference to the Government, Religion, Trade, Growth, Ayre, Diseases, and the several Changes of the said City*—might earn him representation in what he called the "Parliament of Nature." His work made no cosmic claims. All he had done was "to have reduced several great confused volumes [of Bills of Mortality] into a few perspicuous Tables, and abridged such Observations as naturally flowed from them, into a few succinct Paragraphs, without any long series of multiloquious Deductions." From "these poor despised Bills of Mortality . . . that ground, which hath laid waste these eighty years," Graunt gained "much pleasure in deducing so many abstruse, and unexpected inferences . . . there is pleasure in doing something new, though never so little, without pestering the World with voluminous Transcriptions."

Graunt was undaunted by the crudity of the available data, and at the very outset offered 106 numbered observations. Refusing to concede that the ineptness of the "searchers" had made their product useless, he was ingenious at squeezing out hypotheses. Even where searchers were known to be tempted, "after the mist of a Cup of Ale, and the bribe of a two-groat fee, instead of one," to list what was really a death from "French-Pox" as a death from "Consumption," he used this fact to add interest to the lists.

After grouping together similar facts from all the seven decades recorded in the Bills of Mortality, he then compared the findings for different groups. Graunt noted, for example, that only 2 persons in 9 died of acute diseases, 70 in 229 of chronic diseases, and only 4 of 229 of "outward griefs" (cancers, sores, broken bones, leprosy, etc.). Seven percent died of old age, while some diseases and casualties kept a constant proportion. Fewer than one in 2,000 was murdered in London, not more than one in 4,000 died of starvation. "The Rickets is a new disease, both as to name, and thing . . . from fourteen dying thereof, Anno 1634, it hath gradually encreased to above five hundred Anno 1660." He probably was unaware that physicians at the time were looking anew at this disease. There were more males than females in England, and though "Physicians have two women Patients to one Man . . . yet more Men die than Women." The fall was the most unhealthful season, but some diseases— spotted fever, smallpox, and dysentery, or "plague in the guts"—were equally menacing all through the year. London was not so healthful now as it once was. While the population of the English countryside would double by procreation only once every 280 years, the population of London would double every 70 years, "the reason whereof is, that many of the breeders leave the Country, and that the breeders of London come from all parts of the Country, such persons breeding in the Country almost onely, as were born there, but in London multitudes of others." He denied the superstition that plague came with the coronation of a king, for in 1660 when Charles II was crowned, there was no plague.

His most original invention was his new way to present population and mortality by calculating survivorship in a "life table." Starting with two simple facts—the number of births that survived to 6 years of age (64 of 100) and the number surviving to age 76 (1 of 100)—he made a table showing the number of survivors in each of the six intervening decades:

At sixteen years 40	At fifty-six. 6
At twenty-six 25	At sixty-six 3
At thirty-six 16	At seventy-six. 1
At forty-six. 10	At eighty. 0

While modern actuaries do not accept his numbers, his table of survivorship opened the modern epoch in demography.

Graunt concluded his pamphlet with a prophetic plea for statistics. "Moreover, if all these things were clearly, and truly known (which I have but guessed at) it would appear, how small a part of the People work upon necessary Labours, and Callings, *viz.* how many Women, and Children do just nothing, only learning to spend what others get? how many are meer Voluptuaries, and as it were meer Gamesters by Trade? how many live by puzzling poor people with unintelligible Notions in Divinity, and Philosophie? how many by persuading credulous, delicate, and Litigious Persons, that their Bodies, or Estates are out of Tune and in danger? how many by fighting as Souldiers? how many by Trades of meer pleasure, or Ornaments? and how many in a way of lazie attendance, &c. upon others? And on the other side, how few are employed in raising, and working necessary food, and covering? and of the speculative men, how few do truly studie Nature, and Things?"

We know of no public national census before the eighteenth century.

Whatever figures there were that revealed a nation's military and economic power were guarded as state secrets, like the maps of newly discovered passages through dangerous waters to distant ports. It seems that the ancient population counts among the Egyptians, Greeks, Hebrews, Persians, Romans, and Japanese were targeted on taxable people and property ("hearths and households") and men of military age. The earliest recorded comprehensive census of a population and its food supply was taken in Nuremberg in 1449, when the town was threatened by a siege. The town council ordered a full count of all the mouths to be fed and an inventory of the food supply, but the results were kept secret and did not become public till two centuries later.

Public numbers are a modern by-product of new ways of thinking about government, about wealth, and, of course, about science. Representative governments have required periodic public censuses of population. The framers of the Constitution of the United States pioneered with their provision (Article 1, section 2) for a national census every ten years. The United States Census of 1790 began the oldest continuous periodic census of a nation, and became a model for the institution elsewhere. Even earlier, the Pennsylvania Constitution of 1776 provided for regular censuses. During the Revolution, the committee of the Continental Congress in 1776 that was delegated to draw up the Articles of Confederation required a census every three years. Although each colony, regardless of population, was to have one vote in the Congress of the Confederation, each state was to be taxed in proportion to its property. The number of inhabitants seemed, in John Adams' phrase, "a fair index of wealth." The familiar compromise in Philadelphia in 1787 between the large and the small states produced a two-chamber Congress with a Senate, in which each state had two votes, and a House of Representatives, where representation was by the numbers of people. The country was growing rapidly by immigration and people were on the move. Without an up-to-date census, how could they know that they were equally represented?

Reasons of national security had not been the only medieval obstacles to publishing data on births and deaths and longevity. The length of different human lives was long believed to be the proper concern only of God. Not until the mid-seventeenth century did the English word "insurance" begin to have its modern meaning. As late as 1783 a French writer boasted that though life insurance was permitted in Naples, in Florence, and in England, it was not allowed in France, where human life was considered too sacred to be a subject of wager.

But ingenious theologians found a way around. John Ray himself had offered some clues in his popular *Wisdom of God* (1691). Following his lead, another member of the Royal Society, William Derham (1657–1735), an expert on clocks who had already proved that nature must be the work of a divine clockmaker, went on, in his *Physico-Theology* (1713), to explain how the facts of population confirmed the divine design. "How is it possible by the bare rules and blind acts of nature," Derham asked, "that there should be a tolerable proportion, for instance, between male and female?" The "surplusage of males," which he calculated at about 14:13, was so "very useful for the supplies of war," for the Navy and other purposes that it must plainly be "the work of the one that ruleth the world." "It is a very remarkable act of divine providence that useful creatures are produced in great plenty and others in less." Derham noted with satisfaction that poisonous reptiles were most plentiful in heathen lands. "Thus the balance of the animal world is throughout all ages kept even and by a curious harmony and just proportion between the increase of all animals and the length of their lives the world is through all ages well but not over stored; one generation passeth away and another cometh." To avoid overpopulation, God had wisely reduced the Biblical age of man first down to 120 and then to 70. "By this means, the peopled world is kept at a convenient stay, neither too full nor too empty."

An even better known champion "of that Order which the Supreme Ruler has chosen and established for the populating of the earth" was J. P. Süssmilch (1707–1767), a chaplain in the armies of Frederick the Great. His popular *Divine Order in the Changes of the Human Race shown by its Birth, Death, and Propagation* (1741) hailed Graunt as "a Columbus," who had discovered a new world of demography.

> We enter on the land of the living step by step and without crowding, and in accordance with certain set numbers bearing a certain set proportion to the Army of the living and the Army of the departing. . . . Note too in this emergence from the void into being there come always twenty-one sons to twenty daughters; also that the whole mass of those coming to the light of day is always a little greater than those returning to dust, and the army of the human race is always growing in set proportions.

Governments should shape their policy to ensure increase because "God Himself has pronounced in favour of a large population."

A half-century later Malthus attacked Süssmilch for his naïve generalizations about the differences between city and country, and for his failure to include the years of epidemic. But the champions of natural theology erased the stigma of sacrilege from the study of human mortality, and an official British census was finally taken in 1801. The numbers, 9 million for England

and Wales plus 1.5 million for Scotland, seemed to prove that God intended man's numbers to increase. But the shocking rate of increase over Gregory King's estimate of 5.5 million in 1688 would be grist for the mills of Malthus and the evolutionists.

Census-taking and a science of statistics grew up together, providing the modern vocabulary of the social sciences, national economy, and international relations. Adolphe Quetelet (1796–1874), born in Ghent, began teaching mathematics at the age of seventeen. As a young man he wrote poetry, collaborated on an opera, served as apprentice in an artist's studio, and painted interesting canvases of his own. He received the first doctorate from the new University of Ghent for a paper on analytic geometry, which made him famous and secured his election to the Belgian Academy. At twenty-three he was named professor of mathematics, and then drew crowds to his brilliant lectures on esoteric scientific subjects. When he proposed establishing a national observatory, the government sent him to Paris to learn from the French experience. There the dynamic Laplace focused his interest on the study of probability. On his return to Belgium he was appointed Astronomer at the new Brussels Royal Observatory. While the astronomical observatory was under construction, the restless Quetelet turned to observing society, and began collecting social facts for a new science of statistics.

While sharing the speculations of French mathematicians and astronomers in Paris, he had felt "the need to join to the study of celestial phenomena the study of terrestrial phenomena, which had not been possible till now." Nor had he lost his artist's interest in the shape and measurement of the human body. In Brussels he began gathering what he called "moral statistics." From the mass of undigested figures he separated out all the statistics on human beings. These included supposedly trivial figures on the physical dimensions of the human body, along with facts on crimes and criminals. "That which relates to the human species, considered en masse," he assumed, "is of the order of physical facts." He noticed, for example, that the number of crimes committed each year by persons in each age group was remarkably constant. Was there perhaps a kind of "budget" for these acts, established by laws of "social physics"? The three sets of figures he selected —for crimes, for suicides, and for marriages, with each classified by age groups—he called "moral statistics," for all these were cases where the individual had a choice of action. Yet in them, too, he found impressive statistical regularities.

Quetelet enlarged "statistics" to mean data about humankind. The earliest known use of the word (German, *Statistik,* a synonym for *Staatswissenschaft,* 1672) had meant a science of the state, or statecraft, and during the eighteenth century it described the study of constitutions, national resources, and the policy of states. Sir John Sinclair, as we have seen, used "statistics" as a name for assessment of "the quantum of happiness" enjoyed by the people of a country and their means of "future improvement." Quetelet came to the subject not from politics or economics but from an interest in mathematics, in probability, and in human norms. In his *Treatise on Man and the Development of his Faculties, An Essay on Social Physics* (1835; English translation, 1842), which made him famous across the Continent, he proposed his original notion of "the average man" *(l'homme moyen).*

From the quantitative data Quetelet had collected on the human body, he concluded that "regarding the height of men of one nation, the individual values group themselves symmetrically around the mean according to . . . the law of accidental causes." This confirmed his notion of "the average man" which for any nation "is actually the type or the standard and . . . other men differ from him, by more or less, only through the influence of accidental causes, whose effects become calculable when the number of trials is sufficiently large." He called his law of accidental causes "a general law, which applies to individuals as well as to peoples and which governs our moral and intellectual qualities just as it does our physical qualities." The average height of men of similar age in a particular nation was the mean around which variations would "oscillate," symmetrically around the mean in the pattern of a binomial or "normal" distribution. Other physical characteristics, he ventured, might follow the same rule, and his theoretical prediction corresponded impressively with the figures on weight and on chest circumference.

In 1844 Quetelet astonished skeptics by applying his notions to discover the extent of draft evasion in the French Army. By comparing his figures for the probable distribution of men of different heights with the actual distribution of height found among 100,000 young Frenchmen who answered the call for the draft, he ventured that about 2,000 men had avoided conscription by pretending to be less than the minimum height. From statistics (1826–31) of the French courts he concluded:

The constancy with which the same crimes repeat themselves every year with the same frequency and provoke the same punishment in the same ratios, is one of the most curious facts. . . . And every year the numbers have confirmed my prevision in a way that I can even say: there is a tribute man pays more regularly than those owed to nature or to the Treasury; the tribute paid to crime! Sad condition of the human race! We can tell beforehand how many will stain their hands with the blood of their fellow-creatures, how many will be forgers, how many poisoners, almost as one can foretell the number of births and deaths.

> Society contains the germs of all the crimes that will be committed, as well as the conditions under which they can develop. It is society that, in a sense, prepares the ground for them, and the criminal is the instrument. . . .

Quetelet was attacked, of course, for using "social physics" to deny the individual's power to choose between good and evil. But he retorted that now, at last, statistics revealed the forces already at work in society and so created "the possibility of improving people by modifying their institutions, their habits, their education, and all that influences their behaviour."

The pious Florence Nightingale (1820–1910), who had been personally called to her work by the voice of God, was the unlikely champion of this new science. She made Quetelet her hero, considered his *Social Physics* her second Bible, and annotated every page of her presentation copy. Since statistics were the measure of God's purpose, the study of statistics became another of her proclaimed religious duties.

"Let us apply to the political and moral sciences," read Laplace's motto on the title page of Quetelet's treatise, "the method founded on observation and mathematics that has served us so well in the natural sciences." For Quetelet the new science of statistics offered nothing less than an international lexicon for a science of improving society. "The more advanced the sciences have become," Quetelet added, "the more they have tended to enter the domain of mathematics, which is a sort of center toward which they converge. We can judge of the perfection toward which a science has come by the facility, more or less great, with which it may be approached by calculation."

An energetic statesman in the new parliaments of science, Quetelet frequently sent his incremental discoveries to the academies and published fragments in their proceedings. His prodigious correspondence with twenty-five hundred scientists, politicians, and men of letters (including Gauss, Ampère, Faraday, Alexander von Humboldt, Goethe, James A. Garfield, Lemuel Shattuck, Joseph Henry, Prince Albert, and King Leopold I of the Belgians) proselytized for his new science of statistics.

Quetelet organized people across Europe and America to gather census data that could serve as "moral statistics." He urged Charles Babbage (1792–1871) to found the Statistical Society of London (1834). Then he made the Crystal Palace Exposition in London in 1851 a forum for international cooperation, which only three years later produced the First International Statistical Congress (1854) at Brussels. As first president, he preached the need for uniform procedures and terminology. Quetelet's influence was crucial during these formative years of the social sciences. International statistics, some said, were Quetelet's own splendid creation. On them Western peoples would base extravagant expectations for the lessons of quantitative data in public health, politics, and education. Meanwhile totalitarian governments would relapse to the age of secrecy.

In the twentieth century, the public numbers would dominate discussions of national welfare and international relations. Such notions as national and per capita income, gross national product, rates of growth and development, developed and underdeveloped nations, and population growth would be a legacy of Quetelet and his disciples. By 1900 the International Statistical Institute, which had been urging publication of all censuses, reported that currently some sixty-eight censuses covered about 43 percent of the world's population. Their proposed world census of population was still in the future.

82

The Infinite and the Infinitesimal

FROM Hiroshima on August 6, 1945, the world received the shocking discovery that man had opened the dark continent of the atom. Its mysteries would haunt the twentieth century. Yet for two thousand years the "atom" had been the most arcane of philosophers' concerns. The Greek word *atomos* meant the smallest unit of matter, supposed to be indestructible. Now atom was a household word, a threat and a promise without precedent.

The first atomic philosopher was a legendary Greek, Leucippus, suspected to have lived in the fifth century B.C. Democritus, his pupil, who gave to atomism its classic form as a philosophy, was so amused at human follies that he was known as "the laughing philosopher." Yet he was one of the first to argue against mankind's decline from a mythical Golden Age and to preach a gospel of progress. If the whole universe consisted only of atoms and void, it was not infinitely complex but somehow intelligible, and there might be no limit to man's power.

In one of the greatest Latin poems, *De rerum natura,* Lucretius (c. 95 B.C.–c. 55 B.C.) perpetuated ancient atomism. Aiming to free people from fear of the gods, he showed that the whole world was made of void and atoms that moved by their own laws, that the soul died with the body, and that therefore there was no reason to fear death or supernatural powers. Understanding nature, he said, was the only way to peace of mind. The Church Fathers, committed to the Christian afterlife, attacked Lucretius, and he was ignored or forgotten during the Middle Ages but became one of the most influential figures in the Renaissance.

So atomism first entered the modern world as a system of philosophy. Just as the Pythagorean symmetry provided a framework for Copernicus, just as geometry enticed Kepler, and as the Aristotelian perfect circle charmed Harvey, so the philosophers' "indestructible" atoms appealed to chemists and physicists. "The theory of Democritus relating to atoms," Francis Bacon observed, "is, if not true, at least applicable with excellent effect to the exposition of nature." Descartes (1596–1650) invented his own notion of infinitely small particles moving through a medium he called the ether. Another French philosopher, Pierre Gassendi (1592–1655), seemed to confirm Democritus and offered still another new version of atomism, which Robert Boyle (1627–1691) adapted to chemistry, proving that the proverbial "elements"—earth, air, fire, and water—were not elementary at all.

The prophetic insights of a Jesuit mathematician R. G. Boscovich (1711–1787) charted paths for a new science of atomic physics. His bold notion of "point-centers" abandoned the old concept of an assortment of different solid atoms. The fundamental particles of matter, he suggested, were all identical, and matter was the spatial relations around these point-centers. Boscovich, coming to these notions from mathematics and astronomy, foreshadowed the increasingly intimate connection between the structure of the atom and the structure of the universe, between the infinitesimal and the infinite.

The experimental path into the atom was charted by John Dalton (1766–1844), a self-educated Quaker amateur, who picked up a suggestive notion from Lavoisier (1743–1794). A founder of modern chemistry, Lavoisier brought atomic theory back to earth when he finally made the atom a useful laboratory concept by defining an "element" as a substance that could not be broken down into other substances by any known method. Born to a family of weavers in Cumberland in the English Lake District, Dalton carried the mark of his modest origins all his life. At twelve he had taken charge of the Quaker school in his village. When he went on to teach in nearby Kendal, in the school library he found copies of Newton's *Principia,* Boyle's *Works,* and Buffon's *Natural History,* along with a two-foot reflecting telescope and a double microscope. There he came under the spell of a phenomenal blind natural philosopher, John Gough, who, Dalton wrote a friend, "understands well all the different branches of mathematics.

. . . He knows by the touch, taste, and smell, almost every plant within twenty miles of this place." He would be celebrated by Wordsworth in his *Excursion.* From Gough, Dalton received his basic education in Latin, Greek, and French, his introduction to mathematics, astronomy, and all observational science. Following Gough's example, Dalton began keeping a daily meteorological record, which he continued until the day he died.

When Dissenters set up their own New College in Manchester, Dalton became professor of mathematics and natural philosophy. In the Manchester Literary and Philosophical Society he found an eager audience for his experiments. To them he offered his "Extraordinary Facts Relating to the Vision of Colours," probably the first systematic work on color blindness, from which John and his brother Jonathan both suffered. "Having been in my progress so often misled by taking for granted the results of others, I have determined to write as little as possible but what I can attest by my own experience." He observed the aurora borealis, suggested the origins of the trade winds, the causes of clouds and of rainfall, and incidentally made improvements in rain gauges, barometers, thermometers, and hygrometers. Dalton's interest in the atmosphere provided the approach to chemistry that led him to the atom.

Newton had expected the smallest invisible bodies to follow the quantitative laws governing the largest heavenly bodies. Chemistry would recapitulate astronomy. But how was man to grasp and measure the movements and mutual attractions of these invisible particles? In his *Principia,* Newton had speculated that phenomena of nature not described in that book "may all depend upon certain forces by which the particles of bodies, by some causes hitherto unknown, are either mutually impelled towards one another and cohere in regular figures, or are repelled and recede from one another."

Dalton went in search of "these primitive particles," seeking some experimental way to encompass them in a quantitative scheme. Since gases were the loosest, most mobile form of matter, Dalton focused on the atmosphere, the mixture of gases that comprised the air, which provided the point of departure for all his thinking about atoms. "Why does water not admit its bulk of every gas alike?" he asked his colleagues in 1803 in the Manchester Literary and Philosophical Society. "I am nearly persuaded that the circumstance depends upon the weight and number of the ultimate particles of the several gases—those whose particles are lightest and single being least absorbable, and the others more, according as they increase in weight and complexity." Dalton had discovered that contrary to the prevailing view the air was not a single vast chemical solvent but a mixture of gases, each of which remained distinct and acted independently. The product of his experiments was his epoch-making *TABLE: Of the Relative Weights of Ultimate Particles of Gaseous and Other Bodies.* Taking hydrogen as 1, he itemized twenty-one substances. Depicting the invisible "ultimate particles" as tiny solid balls, like pieces of shot but much smaller, he proposed to apply to them the Newtonian laws of the attractive forces of matter. He aimed at "a new view of the first principles of elements of bodies and their combinations," which "I doubt not . . . will in time . . . produce the most important changes in the system of chemistry, and reduce the whole to a science of great simplicity, and intelligible to the meanest understanding." When he showed a "particle of air resting on 4 particles of water," like "a square pile of shot" with each tiny globe touching its neighbors, he provided the ball-and-spoke model for organic chemistry in the century to come.

For his popular lectures Dalton invented his own "arbitrary marks as signs chosen to represent the several chemical elements or ultimate particles," displayed in a table of atomic weights. Of course, Dalton was not the first to use a shorthand for chemical substances—the alchemists had theirs. But he was probably the first to use such symbolism in a quantitative system of "ultimate particles." Making an atom of hydrogen his unit, he calculated the weight of molecules as the sum of the weights of component atoms, and so supplied a modern syntax for chemistry. The actual abbreviations using the first letter of each element's Latin name (H_2O, etc.) were designed by the Swedish chemist Berzelius (1779–1848).

The first reception of Dalton's atomic theory was anything but enthusiastic. The great Sir Humphry Davy quickly dismissed his notions as "rather more ingenious than important." But Dalton's ideas, elaborated in *A New System of Chemical Philosophy* (1808), were so persuasive that he was awarded the Royal Medal in 1826. Never forgetting his plebeian origins, he still remained aloof from the Royal Society in London, but was elected without his consent in 1822. Suspicious of the Society's aristocratic, dilettante tone, he felt more at home in Manchester, where he did most of his work, and he joined with Charles Babbage and helped found the British Association for the Advancement of Science to bring science to all the people. Theologically orthodox Newtonians would not believe that God had necessarily made His invisible "ultimate particles" invariable or indestructible. They shared Newton's suspicion that God had used His power "to vary the laws of Nature, and make worlds of several sorts in several parts of the universe."

Dalton's indestructible atom became the foundation of a rising science of chemistry, providing elementary principles—its laws of constant composition and of multiple proportions, its combination of chemical elements in the simple ratio of their atomic weights. "Chemical analysis and synthesis go no farther than to the separation of particles one from another, and their reunion," Dalton insisted. "No new creation or destruction of matter is within the reach of chemical agency. We might as well attempt to introduce a new planet into the solar system or annihilate one already in existence, as to create or destroy a particle of hydrogen." He continued to use the laws of visible heavenly bodies as his clues to the infinitesimal universe. The prophetic Sir Humphry Davy still remained unconvinced. "There is no reason," he said, "to suppose that any real indestructible principle has yet been discovered."

Dalton was only a Columbus. The Vespuccis were still to come, and when they came they would produce some delightful surprises and some terrifying shocks. Meanwhile, for a half-century, Dalton's indestructible solid atom served chemists well and was usefully elaborated. A French scientist, Gay-Lussac, showed that when atoms combined, it was not necessarily in the one-to-one fashion described by Dalton but might be in some other arrangement of simple integers. An Italian chemist, Avogadro (1776–1856), showed that equal volumes of gases at the same temperature and pressure contained equal numbers of molecules. And a Russian chemist, Mendeleyev, proposed a suggestive "periodic law" of the elements. If elements were arranged in the order of increasing atomic weight, then groups of elements of similar characteristics would recur periodically.

The dissolution of the indestructible solid atom would come from two sources, one familiar, the other quite novel—from the study of light and the discovery of electricity. Einstein himself described this historic movement as the decline of a "mechanical" view and the rise of a "field" view of the physical world, which helped put him on his own path to relativity, to new explanations and new mysteries.

On the wall of his study, Albert Einstein kept a portrait of Michael Faraday (1791–1867), and there could have been none more appropriate. For Faraday was the pioneer and the prophet of the grand revision that made Einstein's work possible. The world would no longer be a Newtonian scene of "forces at a distance," objects mutually attracted by the force of gravity inversely proportional to the square of the distance between them. The material world would become a tantalizing scene of subtle, pervasive "fields of force." This was just as radical as the Newtonian Revolution, and even more difficult for the lay mind to grasp.

Like the Copernican Revolution in astronomy, the "Field" Revolution in physics would defy common sense and carry the pioneer scientists once again into "the mists of paradox." If Michael Faraday had been trained in mathematics, he might not have been so ready for his surprising new vision. The son of a poor blacksmith on the outskirts of London, Faraday had to earn his own way from an early age, and when wartime prices were high in 1801, he was said to have lived on a loaf of bread for a week. His parents were members of the Sandemanian Church, a small fundamentalist and ascetic Scottish Protestant sect, which, like the Quakers, believed in a lay clergy and opposed the accumulation of wealth. He attended Sunday meetings regularly and remained an elder until his last years. In his much thumbed Bible the most marked passages were in the Book of Job. He had almost no formal education—"little more than the rudiments of reading, writing, and arithmetic at a common day-school"—but at the age of thirteen he luckily found employment with a friendly French émigré printer and bookbinder, a M. Riebau. At first he delivered the newspapers that Riebau loaned out and then he picked them up to be delivered again.

Among the books that came to Riebau's shop for binding was *The Improvement of the Mind,* by the hymn writer Isaac Watts, whose system of self-improvement Faraday followed by keeping the commonplace book that would eventually become his famous laboratory notebook. One day he received for rebinding the volume of the *Encyclopaedia Britannica* (3d. ed., 1797) that contained the article of 127 double-column pages on Electricity by an erratic "Mr. James Tytler, chemist." Demolishing the prevailing one-fluid and two-fluid theories of electricity, Tytler proposed that electricity was not a material flow at all but a kind of vibration, akin to light and heat. This tantalizing suggestion was the beginning of Faraday's pursuit of science.

In 1810 he began attending public lectures offered by the City Philosophical Society and then Humphry Davy's lectures at the Royal Institution. In December 1811 Faraday had impressed Davy by sending him the neatly bound and beautifully penned notes he had taken at Davy's lectures, along with his request for a post as his assistant. That October, Davy had been temporarily blinded by an explosion in his laboratory, and he now needed an amanuensis. Davy hired Faraday for one guinea a week and the use of two rooms at the top of the Institution with fuel and candles, laboratory aprons, and freedom to use the apparatus. At twenty, Faraday found himself in the laboratory of one of the greatest chemists of the age, where he could experiment at will. A dream come true!

Sir Humphry and Lady Davy rounded off Faraday's education by taking him along on their tour of the Continent in 1813–14, visiting France and Italy, meeting scientists, and sharing the talkative Davy's hopes and doubts. When Faraday returned to England in April 1815, Davy had inoculated him against easy generalizations and had renewed his passion for experiment. Back in the laboratory, he tested heating and lighting oils, and finally discovered benzene. He made the first known compounds of chlorine and carbon, which became ethylene, resulting from the first known substitution-reaction. He pioneered in the chemistry of steel alloys. What eventually proved crucial in his life was the commission from the Royal Society that led him to produce a new "heavy" optical glass with a high refractive index especially useful for experiments in polarized light.

Faraday's sanguine temperament was reinforced by a happy marriage to the sister of someone he had met at the City Philosophical Society. Sarah Bernard never shared the scientific interests that kept him awake nights, but said she was happy to be "the pillow of his mind."

In the new world of priority prizes his early successes even aroused the jealousy of his famous mentor. In 1824, when Faraday was proposed as a Fellow in the Royal Society for his feat of liquefying chlorine, Davy opposed his election and claimed that he himself was entitled to the credit. But Faraday was elected anyway.

Davy had been intrigued by recent theoretical efforts to adapt Newton's ideas to the needs of the chemist in the laboratory. The most attractive of these was Boscovich's "point-center" theory, which had described the atom not as a tiny billiard ball of impenetrable matter but as a center of forces. If the "ultimate particles" of matter should have this character, it might explain the interaction of chemical elements, their "affinities" and the ways of making stable compounds.

Boscovich had limited his radical suggestion to the chemical elements. When Faraday's passion for experiment was focused by chance on the uncharted realm of electricity, he was newly attracted by Boscovich's theory. In 1821 a friend asked Faraday to write for the *Philosophical Magazine* a comprehensive article explaining electromagnetism to the lay public. Current interest had been awakened when, just the previous summer, the Danish physicist Hans Christian Oersted (1777–1851) during the demonstration for an evening lecture showed that a current-carrying wire would deflect a magnetic needle. Following Oersted's clues, Faraday devised a simple apparatus of two beakers containing mercury, a current-carrying wire, and two cylindrical bar magnets. With this he elegantly demonstrated electromagnetic rotation, proving both that a current-carrying wire would rotate around the pole of a magnet and that the pole of a magnet would rotate around a current-carrying wire. Perhaps Faraday began to suspect that somehow surrounding a current-carrying wire there were circular lines of force. And perhaps the forces of magnetism and of electricity were somehow convertible. At this point it was lucky that Faraday was not a sophisticated mathematician. For then he might have followed the conventional path, like that taken by the French mathematical prodigy André Marie Ampère (1775–1836), and have tried to explain electromagnetism simply by a mathematical formulation of Newtonian centers of force. Faraday's naïve vision saw something else.

Without intending it, Faraday had already made the first recorded conversion of mechanical into electric energy. This was, of course, the crucial step toward the electric motor and the electric generator with all their transformations of daily life. Once again a revolution in science would depend on the defiance of common sense. Surprising though it might seem, the power of a magnet, unlike the Newtonian force of gravitation, was not focused in a massy object emanating straight lines of force-at-a-distance. In numerous experiments after 1821 Faraday was beginning to glimpse a bizarre phenomenon, and the possibility that the magnet and the electric current somehow created a "field of force."

Faraday, blessed with the amateur's naïve vision, was not seduced by the revered Newton's mathematical formulae. Faraday's experiments over the next twenty-five years—from his first wires and magnets rotating in beakers of mercury to his grand prophetic outlines of a modern field theory—eventually would open the way to a new vision of the universe. Through it all Faraday would be drawn by his simple Sandemanian faith in the unity and coherence of God's creation.

In 1831, when Faraday learned that Joseph Henry in Albany, New York, had reversed the polarity of electromagnets by reversing the direction of the electric current, he set up his own experiments. He aimed to show how a moving magnet could generate a current of electricity. By an astonishingly simple experiment, which passed an electrostatic discharge through a wet string, he managed to show that static electricity was essentially not different from other kinds, and hence that all known kinds of electricity were identical. Then with his experiments in electrochemistry he showed that the decomposing power of electricity was directly proportional to the quantity of electricity in solution, and hence that electricity must somehow be the force of chemical affinity. Using a piece of blotting paper soaked in potassium iodide, he effected an electrostatic discharge into the air, so disposing of the Newtonian-based theory that electricity, like gravitation, was a force exerted from one "pole" to another. All these were clues to the existence of electric particles and to electric fields—openings toward fields of force with hints of the convertibility of forces and the unity of all phenomena.

By 1838 Faraday had the basis for a new theory of electricity. He developed a whole new vocabulary of such terms as "electrode," "cathode," and "electrolysis." Perhaps, he ventured, electric forces were intermolecular and electricity somehow transferred energy without transferring matter. Wary of using the term "current" because of its mechanical connotations, he described this transfer as a process in which minute particles were put under a strain, which was then conducted from particle to particle.

After a trying five-year interlude when his mind seemed hopelessly fatigued from these early years of relentless experiment, Faraday bounced back for a crucial next step in his chain of experiments. At this moment the young William Thomson (1824–1907), later famous as Lord Kelvin, had been puzzling over the nature of electricity and the difficulty of fitting it into the Newtonian scheme. In August 1845 Thomson wrote to Faraday describing his initial success in giving mathematical form to Faraday's notion of lines of force, and suggesting some further experiments. None of the eminent physicists of the day had been persuaded by Faraday.

But Thomson, then only twenty-one, was open to even wilder possiblities. If there really were lines and fields of forces, might not experiment conceivably prove a kinship between electricity and light? Faraday determined to pursue the outlandish suggestion. At first the difficulties seemed insuperable. "Only the very strongest conviction that Light, Magnetism and Electricity must be connected . . . led me to resume the subject and persevere." On September 13, 1845, Faraday tried passing a ray through a piece of the "heavy glass" with a high refractive index that he had made fifteen years before, and in the field of a strong electromagnet. "There was an effect produced on the polarized ray," he recorded with satisfaction, "and thus magnetic force and light were proved to have relation to each other. This fact will most likely prove exceedingly fertile." He was reassured by finding that the angle of rotation of the ray of light was directly proportional to the strength of the electromagnetic force.

Faraday now found his earlier metaphor of interparticulate "strain" inadequate, and went on to suggest a "flood of power"—the electromagnet being a "habitation of lines of force." From comparing the action of different substances on the passage of magnetic force he contrasted "paramagnetics," which conducted the force well, with "diamagnetics," which conducted poorly. He then showed that his "lines of force" were not polar (directed to the nearest pole) as the old Newtonian theories would have suggested, but were continuous curves. His crucial conclusion, the axiom of modern "field" theory in physics, was that the energy of the magnet was not in the magnet itself but in the magnetic field.

Faraday had sketched the outline of a surprising new invisible world. Among these infinitesimal fields of forces exerted by mysterious minute entities modern physicists would find their New Worlds and their Dark Continents, with secrets of a still wider unity and mystery of phenomena. "I have long held an opinion, almost amounting to conviction," Faraday wrote to the Royal Society in 1845, "in common I believe with many other lovers of natural knowledge, that the various forms under which the forces of matter are made manifest have one common origin; or, in other words, are so directly related and mutually dependent, that they are convertible as it were, one into another, and possess equivalents of power in their action. In modern times the proofs of their convertibility have been accumulated to a very considerable extent, and a commencement made of the determination of their equivalent forces."

The succession of proofs that Faraday had prophesied moved with accelerating momentum in the next century. Communications among scientists were more continuous, and their achievements more collaborative than ever before. Sometimes it became a matter of chance who took (or was given) the credit for taking, the next step. Faraday's discoveries had been the product of an unmathematical mind. But the persuasiveness of the field theory would still depend on its being given a mathematical form. This was accomplished by Faraday's admirer, James Clerk Maxwell (1831–1879), who translated Faraday's "lines" or "tubes" of force into a mathematical description of a continuous field. Just as Newton had given mathematical form to Galileo's insights, so, Einstein noted, Maxwell's equations performed a similar function for Faraday. "The formulation of these equations," Einstein and his collaborator Leopold Infeld called "the most important event in physics since Newton's time, not only because of their wealth of content, but also because they form a pattern for a new type of law." The features of these equations would appear "in all other equations of modern physics." These equations would become a basis, too, for Einstein's own theory of relativity. The next great step after Faraday in the revision of Newtonian physics and the dissolution of the "indestructible" atom came with the discovery of cathode rays, X-rays, and radioactivity. The clues to the electron were followed up by J. J. Thomson (1856–1940), who discovered minute invisible particles of uniform mass, only one eighteen-hundredth that of the hydrogen atom, which till then was the lightest known object. In 1911, Ernest Rutherford (1871–1937) discovered an atomic nucleus for the next generation of physicists to explore, as their predecessors had explored the electron.

The mysteries of the atom multiplied with every new discovery. The limits of mathematics were increasingly disclosed. In the mind of Einstein the unity of phenomena—the quest of Dalton and Faraday—brought "scientific" problems and paradoxes beyond the earlier ken of any but Hermetic philosophers. Just as physicists illustrated their atom by planetary and celestial systems, so the infinitesimal offered clues to the infinite. Time and space came together in a single tantalizing riddle, which led Einstein to conclude that "the eternal mystery of the world is its comprehensibility."

SOME REFERENCE NOTES

These notes will help the reader walk some of the paths of discovery that I have found most rewarding. At the same time I will indicate my heaviest debts to other scholars. I have selected here, for the most part, works likely to be found in a good public library or the library of a college or university. I have omitted many of the specialized monographs and articles in learned journals. More detailed references and the sources of my principal direct quotations are entered on a copy of the manuscript deposited in the Library of Congress in Washington, D.C. The topics below, after a general section, are arranged in the order of the chapters in the book.

GENERAL

While writing this book I have had at my elbow dictionaries and encyclopedias which have helped me into my chosen subjects and at the same time lured me to subjects and people I had never planned to explore. The new *Encyclopaedia Britannica* (15th ed., 1980), which came late in my work, has been a blessing and a delightful eye-opener, with up-to-date bibliographies that have been constantly useful. As a longtime aficionado of dictionaries, reference books, and general treatises, I have found no substitute for owning the basic works. Then there is never an excuse for not pursuing the fugitive thought or checking the puzzling or uncertain fact. Of the indispensable more specialized reference works, I have enjoyed particularly: the monumental *Dictionary of Scientific Biography* (C. C. Gillispie, ed., 16 vols., 1970–80); *A History of Technology* (Charles Singer and others, eds., 5 vols., 1967; and Trevor L. Williams, ed.,

on the twentieth century, 2 vols., 1978); *The International Encyclopedia of the Social Sciences* (David L. Sills, ed., 17 vols., 1968) and its still-useful predecessor, *The Encyclopaedia of the Social Sciences* (Edwin R. A. Seligman, ed., 15 vols., 1930–34); *Encyclopaedia of Religion and Ethics* (James Hastings, ed., 12 vols., n.d.). For the American and the English reader an inexhaustible treasure-house is the *Oxford English Dictionary* (James A. H. Murray and others, eds., 13 vols., 1930) and its supplements (R. W. Burchfield, ed., 1972–).

I am deeply indebted to the phenomenal Joseph Needham for all his works, but especially for his nonpareil *Science and Civilisation in China* (8 vols., and more in progress, 1954–), in addition to his shorter works listed below. No one with the slightest interest in history or in China should fail to savor the delights of Needham, who has achieved one of the great intellectual ambassadorial enterprises of modern times.

Many of the principal texts by the

great discoverers whom I treat are found in the handsome and convenient volumes of *Great Books of the Western World* (an Encyclopaedia Britannica publication, Robert Maynard Hutchins, ed., 54 vols., 1952).

Handy chronologies are found in: *An Encyclopedia of World History* (William L. Langer, ed., 5th ed., 1968); *Chronology of the Modern World* (Neville Williams, ed., 1967); *The Timetables of History* (Bernard Grun, ed., 1975). For geography: *The Times Atlas of World History* (Geoffrey Barraclough, ed., 1978); *The New Cambridge Modern History Atlas* (H. C. Darby and others, eds., 1970); and the concise and inexpensive *Penguin Atlas of World History* (Hermann Kinder and others, eds., 2 vols., 1974–78).

Learned journals, such as *Isis,* a journal of the history of science, *Speculum,* for the Middle Ages, the *Journal of the History of Ideas,* and the *American Historical Review,* files of which will be found in many public and institutional libraries, will, of course, repay exploration in the pursuit of specific people and topics.

BOOK ONE: TIME

Time, which has tantalized philosophers and produced some of their most arcane and opaque writing, has not as a concept much interested the great historians, who have been satisfied to recount its symptoms. The more interesting and intelligible works include: James T. Fraser (ed.), *The Voices of Time* (1966), which surveys the meanings of time for different disciplines, and *Of Time, Passions and Knowledge* (1975); Harold A. Innis, *Changing Concepts of Time* (1952), a historian's view; Wyndham Lewis' suggestive *Time and Western Man* (1957), a literary perspective; Richard M. Gale, *The Philosophy of Time* (1968); R. G. Collingwood, *The Idea of History* (1946), on the threshold between philosophic and historical time.

Part I: The Heavenly Empire

A delightful starting point is O. Neugebauer, *The Exact Sciences in Antiquity* (2d ed., 1969), with George Sarton's lively essay, *Ancient Science and Modern Civilization* (1954). George Sarton's unexcelled and comprehensive two-volume survey, *A History of Science* (1952; 1959), carries the story from ancient Egypt and Mesopotamia to the beginning of the Christian era. Other, more focused works are F. H. Colson, *The Week* (1974); Martin P. Nilsson, *Primitive Time-Reckoning* (1920); F. A. B. Ward, *Time Measurement* (1958); Lawrence Wright, *Clockwork Man* (1969); Peter Hood, *How Time Is Measured* (1969); Kenneth G. Irwin, *The 365 Days* (1963).

On astrology, it is not easy to separate the writings of advocates from those of historians. Useful general works include: Jack Lindsay, *Origins of Astrology* (1971); Ellen McCaffery, *Astrology, Its History and Influence in the Western World* (1970); Christopher McIntosh, *The Astrologers and Their Creed, an Historical Outline* (1969); Eric Russell, *Astrology and Prediction* (1972); Mark Graubard, *Astrology and Alchemy* (1953). A better approach, perhaps, is through the impressive power of astrology in specific times and places, for example: Franz Cumont's little classic, *Astrology and Religion among the Greeks and Romans* (1912); Theodore O. Weden, *The Medieval Attitude toward Astrology, Particularly in England* (1974); Eustace F. Bosanquet, *English Printed Almanacks and Prognostications* (1917); Don Cameron Allen, *The Star-Crossed Renaissance* (1966). The Nostradamus phenomenon, hard to believe but well-documented, can be followed in Edgar Leoni, *Nostradamus: Life and Literature* (1961); and his power over Hitler and other Nazis, in: Louis de Wohl, *I Follow my Stars* (1937) and Sterne, *Krieg und Frieden* (1951); Wilhelm Wulff, *Zodiac and Swastika* (1973); Ellic Howe, *Astrology and Psychological*

Warfare During World War II (1972). The influence of astrology on the fixing of the day and hour for the independence of India in August 1947 is described in Larry Collins and Dominique La Pierre, *Freedom at Midnight* (1975), pp. 181, 196, 228, 341. For other aspects of the relation of astronomy to events on earth, see: Harold Spencer Jones, *The Earth as a Clock* (1939); Theodor Gaster, *New Year: Its History, Customs, and Superstitions* (1955); Ruth S. Freitag, *The Star of Bethlehem: A List of References* (Library of Congress, 1979).

Part II: From Sun Time to Clock Time

The sundial has inspired more sentimental than scientific writing, but some solid facts can be found in: Alice Morse Earle, *Sun Dials and Roses of Yesterday* (1902); Winthrop W. Dolan, *A Choice of Sundials* (1975); Roy K. Marshall, *Sundials* (1963).

The story of clocks and clockmaking has not only interested buffs and collectors but has challenged and inspired some of the best historians of science. No one can read Carlo M. Cipolla's *Clocks and Culture: 1300–1700* (1967) without being stimulated to go on to other books: H. Alan Lloyd, *The Collector's Dictionary of Clocks* (1964); Roger Burlingame, *Dictator Clock: 5,000 Years of Telling Time* (1966); E. J. Tyler, *The Craft of the Clockmaker* (1972); Enrico Morpurgo, *Gli Orologi* (1966) and *L'Origine dell'Orologio Tascabile* (1954). The endless delights are in the detail, which can be explored in: G. H. Baillie, *Clocks and Watches, An Historical Bibliography* (1951); Eric Bruton, *Clocks and Watches, 1400–1900* (1967); Donald de Carle, *British Time* (1947); Herbert Cescinsky, *The Old English Master Clockmakers and their Clocks, 1670–1820* (1938); Maurice Daumas, *Scientific Instruments of the Seventeenth and Eighteenth Centuries* (1972); Ernest L. Edwardes, *Weight-driven Chamber Clocks of the Middle Ages and Renais-*

sance (1965); Robert Silverberg, *Clocks for the Ages: How Scientists Date the Past* (1971); Robert S. Woodbury, *History of the Gear-Cutting Machine* (1958). Silvio A. Bedini's monographs and scholarly articles in the publications of the American Philosophical Society lead us into neglected episodes in the history of timekeeping, for example: *The Scent of Time: A Study of the Use of Fire and Incense for Time Measurement in Oriental Countries* (1963) and (with Francis Maddison) *Mechanical Universe: The Astrarium of Giovanni de' Dondi* (1966).

For the role of timekeeping in earlier ages, the reader has a feast prepared by some of the liveliest scholars of recent decades, to be savored in: Henri Frankfort, *The Birth of Civilization in the Near East* (1956) and (with others) *The Intellectual Adventure of Ancient Man* (1946); Samuel N. Kramer, *History Begins at Sumer* (1981); Georges Contenau, *Everyday Life in Babylon and Assyria* (1954); Svend Pallis, *The Babylonian Akitu Festival* (1926); James H. Breasted, *Development of Religion and Thought in Ancient Egypt* (1912); John A. Wilson, *The Culture of Ancient Egypt* (1951) and *The Burden of Egypt* (1951); Jon M. White, *Everyday Life in Ancient Egypt* (1973); C. M. Bowra, *The Greek Experience* (1957); Jerome Carcopino, *Daily Life in Ancient Rome* (1940); Seyyed Hossein Nasr, *An Introduction to Islamic Cosmological Doctrines* (1978); Jacques Le Goff, *Time, Work, and Culture in the Middle Ages* (1980); H. S. Bennett, *Life on the English Manor* (1974); Victor W. von Hagen, *The Ancient Sun Kingdoms* (1973); Miguel Leon-Portilla, *Time and Reality in the Thought of the Maya* (1973).

On calendars and calendrics in general, see: James C. MacDonald, *Chronologies and Calendars* (1897); Broughton Richmond, *Time Measurement and Calendar Construction* (1956); W. M. O'Neil, *Time and the Calendars* (1975); P. W. Wilson, *The Romance of the Calendar* (1937). And on specific periods: Richard A. Parker, *The Calen-*

dars of Ancient Egypt (1950); Benjamin D. Meritt, *The Athenian Year* (1961); Agnes K. Michels, *The Calendar of the Roman Republic* (1967).

For the rise of the portable clock and timekeeping at sea, start with Derek Howse, *Greenwich Time and the Discovery of the Longitude* (1980). To see how Captain Cook established the utility of Harrison's clock, consult J. C. Beaglehole's suspenseful *Life of Captain James Cook* (1974) and other items under Part VIII, below.

Part III: The Missionary Clock

The most accessible accounts of the remarkable Father Matteo Ricci for the reader of English are Louis L. Gallagher's translation of Ricci's journals, *China in the Sixteenth Century: The Journals of Matthew Ricci: 1583–1610* (1961), and Vincent Cronin, *The Wise Man from the West* (1961). The authoritative modern resources on Ricci are Pasquale d'Elia's *Documenti Originali Concernanti Matteo Ricci e la Storia delle Prime Relazioni tra l'Europa e la Cina* (1949) and his *Storia dell'Introduzione del Cristianesimo in Cina* (3 vols., 1949). A readable introduction to the Eastern world which Ricci entered is Nigel Cameron, *Barbarians and Mandarins* (1976). For the specifically Chinese background, begin with Joseph Needham, Wang Ling, and Derek J. Price, *Heavenly Clockwork; the Great Astronomical Clocks of Medieval China —a Missing Link in Horological History* (1960), and Joseph Needham, *Clerks and Craftsmen in China and the West* (1970) and *The Grand Titration: Science and Society in East and West* (1969); then Dennis Bloodworth, *The Chinese Looking Glass* (1980); Derek Bodde, *Essays on Chinese Civilization* (1981); Nigel Cameron and Brian Brake, *Peking: A Tale of Three Cities* (1965); C. P. Fitzgerald, *China: A Short Cultural History* (1976); Jacques Gernet, *Daily Life in China on the Eve of the Mongol Invasion,*

1250–1276 (1962); Dun J. Li, *The Ageless Chinese* (1965); Shigeru Nakayama and Nathan Sivin (eds.), *Chinese Science* (1973); Jonathan Spence, *To Change China: Western Advisers in China 1620–1960* (1969). On the Christian missionaries, see: Kenneth Scott Latourette, *A History of Christian Missions in China* (1929), and Columba Cary-Elwes, *China and the Cross* (1956). For the wider Asian context, see: Hajime Nakamura, *Ways of Thinking of Eastern Peoples: India-China-Tibet-Japan* (1964), and Donald F. Lach and Carol Flaumenhaft (eds.), *Asia on the Eve of Europe's Expansion* (1965).

BOOK TWO: THE EARTH AND THE SEAS

Part IV: The Geography of the Imagination

Geographers have provided us with some remarkably readable general histories of their subject, notably: Lloyd A. Brown, *The Story of Maps* (1949); Leo Bagrow, *History of Cartography* (1964); John Kirtland Wright, *Human Nature in Geography* (1966); E. G. R. Taylor, *Ideas on the Shape, Size and Movements of the Earth* (1943). On the subject of these chapters and those immediately following, we are fortunate to have C. Raymond Beazley's *Dawn of Modern Geography* (3 vols., 1949), full of lively anecdotes and copious quotations from the sources. For the geographic beliefs and myths of antiquity, start with John F. Blake, *Astronomical Myths* (1877), based on the *History of the Heavens* by the French astronomer and popularizer Camille Flammarion (1842–1925); then C. J. Bleeker, *Egyptian Festivals: Enactments of Religious Festivals* (1967); Jean-Pierre Babard, *La Symbolique du monde souterrain* (1973); E. H. Bunbury, *A History of Ancient Geography* (2 vols., 1879); Richard J. Clifford, *The Cosmic Mountain in Canaan and the Old Testament* (1972); Franz Cumont,

After Life in Roman Paganism (1922); Sir James G. Frazer, *The Golden Bough* (one volume edition, 1922, and numerous editions later); William A. Heidel, *The Frame of Ancient Greek Maps* (1976); Edna Kenton, *The Book of Earths* (1928), a handy anthology of various views of the earth; John H. Rose, *The Mediterranean in the Ancient World* (1969); J. Oliver Thomson, *History of Ancient Geography* (1965); E. H. Warmington, *Greek Geography* (1973). On medieval and early modern geography: Ernest Brehaupt, *An Encyclopedist of the Dark Ages, Isidore of Seville* (1964); Georges Duby, *The Age of the Cathedrals, Art and Society 980–1420* (1981); George H. T. Kimble, *Geography in the Middle Ages* (1938); David C. Lindberg (ed.), *Science in the Middle Ages* (1978); Henry Osborn Taylor, *The Mediaeval Mind* (2 vols., 1930); Paget Toynbee, *Dante Alighieri* (1924); John Kirtland Wright, *The Geographical Lore of the Time of the Crusades* (1965). On the Chinese background, in addition to the works mentioned above by Joseph Needham, see Karl A. Wittfogel, *Oriental Despotism* (1957), and Ulrich Libbrecht, *Chinese Mathematics in the Thirteenth Century* (1973). For articles on almost any of these topics consult the index to *Imago Mundi: A Review of Early Cartography* (founded by Leo Bagrow in 1935 at 's Gravenhage, Netherlands).

On the Sacred Mountain and steps to heaven: André Parrot, *The Tower of Babel* (1955); Raphael Patai, *Man and Temple in Ancient Jewish Myth and Ritual* (1967); I. E. S. Edwards, *The Pyramids of Egypt* (1972); Debala Mitra, *Buddhist Monuments* (1971); Elizabeth B. Moynihan, *Paradise as a Garden: In Persia and Mughal India* (1979); Evrard de Rouvre, *Grands Sanctuaires* (1960).

Part V: Paths to the East

The best introduction to the universal human phenomenon of pilgrimage is Diana L. Eck's brilliant and elegant *Banaras: City of Light* (1982). On European pilgrims and pilgrimages: William C. Bark, *Origins of the Medieval World* (1960); William Boulting, *Four Pilgrims* (1920); John Gardner, *The Life and Times of Chaucer* (1977); Vera and Hellmut Hall, *The Great Pilgrimage of the Middle Ages: The Road to St. James of Compostela* (1966); J. J. Jusserand, *English Wayfaring Life in the Middle Ages* (1950); Alan Kendall, *Medieval Pilgrims* (1970); Thomas D. Kendrick, *St. James in Spain* (1960); Herbert N. Wethered, *The Four Paths of Pilgrimage* (1947). For the texts of their travels, see Arthur P. Newton (ed.), *Travel and Travellers of the Middle Ages* (1968), and Publications of the Palestine Pilgrims' Text Society. For a sample of pilgrims elsewhere see, besides Eck: Pierre Cabanne, *Les longs cheminements: les pèlerinages de tous les temps et de toutes les croyances* (1958); Samuel Beal (trans.), *Travels of Fah-hian and Sung-yun, Buddhist Pilgrims from China to India (A.D. 400 and A.D. 518)* (1964); Maurice Gaudefroy-Demombynes, *Le Pèlerinage à la Mekke* (1923); Oliver Statler, *Japanese Pilgrimage* (1983).

The Crusades have stirred the narrative and analytic talents of modern historians, for example, in Steven Runciman's engrossing *History of the Crusades* (3 vols., 1971), which dramatizes the crusaders without romanticizing them. A concise survey of different attitudes to the Crusades is James A. Brundage (ed.), *The Crusades: Motives and Achievements* (1964), which can be further explored in: Ernest Barker, *The Crusades* (1971); the stimulating works by Aziz S. Atiya, *Crusades, Commerce and Culture* (1962) and *The Crusade in the Later Middle Ages* (1970); Kenneth M. Setton (ed.), the multivolume *History of the Crusades* (1969–). An accessible crusader memoir is Villehardouin and De Joinville, *Memoirs of the Crusades* (1955), with an introduction by Sir Frank Marziale. A classic case study of the problem of recapturing the spoken word in the distant past is Dana C.

Munro, "The Speech of Pope Urban II at Clermont, 1095," *The American Historical Review*, Vol. II (1906), pp. 231–42.

For the story of the Mongols there is in English a scanty but lively and, on the whole, unsympathetic literature: Walter J. Fischel, *Ibn Khaldun and Tamerlane* (1952); René Grousset, *Conqueror of the World* (1966); Walther Heissig, *A Lost Civilization: The Mongols Discovered* (1966); Harold Lamb, *Tamerlane* (1928); Bertold Spuler, *History of the Mongols: Based on Eastern and Western Accounts of the Thirteenth and Fourteenth Centuries* (1972); Leonardo Olschki, *Guillaume Boucher: A French Artist at the Court of the Khans* (1946); John Ure, *The Trail of Tamerlane* (1980).

To follow land travelers to the East, the reader will find the authentic records more inaccessible than the imaginary ones. The narration of William of Rubruck is available in a Hakluyt Society volume (2d Series, No. IV; 1900) reprinted by Kraus (1967), as also is *Mandeville's Travels* (2d Series, No. I; 1953) reprinted by Kraus (1967). And see: Sir John Mandeville, *The Voiage and Travayle of Syr John Mandeville Knight, with The Journall of Frier Odoricus* (Dutton, 1928). Then follow the career of the mysterious forgers and charlatans in: Josephine W. Bennett, *The Rediscovery of Sir John Mandeville* (1954); Robert Silverberg, *The Realm of Prester John* (1972); Vsevolod Slessarev, *Prester John: The Letter and the Legend* (1959).

The Travels of Marco Polo are happily available in numerous inexpensive editions—which leaves no one with a good reason for not tasting his delights. The standard edition in English is that by Col. Sir Henry Yule, *The Book of Ser Marco Polo* (2 vols., 3d ed., 1903). But see Henri Cordier, *Ser Marco Polo* and *The Travels* (Penguin Books, 1967). Yule has also edited numerous other relevant documents for the Hakluyt Society under the title *Cathay and the Way Thither* and it is worth some effort to find and read his "Preliminary Essay on the Intercourse between China and the Western Nations previous to the Discovery of the Cape Route," in the society's publications, Vol. I (1866). For a lively account of twentieth-century travelers in Polo's track, see Jean Bowie Shor, *After You Marco Polo* (1955).

For the larger picture, the reader should come to know Donald F. Lach's monumental and always suggestive *Asia in the Making of Europe* (5 vols., 1965–77).

Part VI: Doubling the World

Historians of the sea have often been passionate sailors themselves, which has given the literature of seafaring history a special vividness. This has also made their writing a potent weapon in the worldwide battle for priority on the seas. See, for example, Emily M. Beck, *Sailor Historian: The Best of Samuel Eliot Morison* (1977). Morison's eloquent partisanship of Columbus and the Spanish sailors is equaled by Armando Cortesão's eloquence and scholarship for the Portuguese, as in *The Mystery of Vasco da Gama* (1973). For a longer perspective, see Lionel Casson, *Ships and Seamanship in the Ancient World* (1971), and Vincent H. Cassidy, *The Sea Around Them: The Atlantic Ocean, A.D. 1250* (1968). A well-illustrated introduction to nautical architecture is Björn Landström, *The Ship* (1961). We are grateful to Dover Publications for providing an inexpensive reprint of A. E. Nordenskiöld's essential *Facsimile-Atlas: to the Early History of Cartography with Reproductions of the Most Important Maps Printed in the XV and XVI Centuries* (1973).

So little is known about Ptolemy that there are no biographies, but see *The Geography of Claudius Ptolemy* (E. L. Stevenson, ed. and trans., 1932). For possible sources of his ideas and the fate of his maps, see: Peter M. Fraser, *Ptolemaic Alexandria* (1972), and R. Walzer, *Arabic Transmission of Greek Thought in Medieval Europe* (1945).

For the larger Portuguese context, see: C. R. Boxer's concise and readable *The Portuguese Seaborne Empire, 1415–1825* (Penguin Books, 1969); Jaime Cortesão, *A Expansão Dos Portogueses No Periodo Henriquino* (n.d.); H. V. Livermore, *A New History of Portugal* (1969); Edgar Prestage, *The Portuguese Pioneers* (1967). On Prince Henry, see: C. Raymond Beazley, *Prince Henry the Navigator* (1895); E. D. S. Bradford, *A Wind from the North: The Life of Henry the Navigator* (1960); Richard H. Major, *The Life of Prince Henry of Portugal* (1868; 1967). A standard source is Gomes Eannes de Zurara, *The Chronicles of the Discovery and Conquest of Guinea,* available in Hakluyt Society publications (C. Raymond Beazley and Edgar Prestage, eds. and trans., 1896). For Gama: Henry H. Hart, *Sea Route to the Indies* (1950); K. G. Jayne, *Vasco da Gama and his Successors, 1460–1580* (1910); the Hakluyt Society Publications, *A Journal of the First Voyage of Vasco da Gama, 1497–1499* (E. G. Ravenstein, ed. and trans., n.d.); Gaspar Correa, *The Three Voyages of Vasco da Gama and his Viceroyalty* (Henry D. J. Stanley, ed. and trans., 1869).

The pride and glory of the Portuguese in their Great Age of Discovery can be heard in the lines of *The Lusiads* of Luis de Camoëns (1524?–1580), the Portuguese Homer, available in the Penguin reprint of an admirable translation (1973) by William C. Atkinson, illuminated by Henry H. Hart, *Luis de Camoëns and the Epic of the Lusiads* (1962).

A stimulating introduction to the role of the Arabs (positive and negative) in this story is Henri Pirenne's *Mohammed and Charlemagne* (1956), which gives a decisive but widely controverted role in early modern history to the Muslim enclosure of the Mediterranean, and George F. Hourani's brilliant *Arab Seafaring in the Indian Ocean in Ancient and Early Medieval Times* (1951). For the wider context: Jacques Berque, *The Arabs* (1964); Bernard Lewis, *The Arabs in History* (1964); I. A. Mayer, *Islamic Astrolabists and Their Works* (1956).

For the role of China on the seas see: in addition to the above-listed works by Joseph Needham, his *Science in Traditional China* (1981), *The Grand Titration: Science and Society in East and West* (1969), and *Within the Four Seas: The Dialogue of East and West* (1969); Hajime Nakamura, *Ways of Thinking of Eastern Peoples: India-China-Tibet-Japan* (1964); C. P. Fitzgerald, *China: A Short Cultural History* (1976), and *The Chinese View of Their Place in the World* (1971); L. Carrington Goodrich, *A Short History of the Chinese People* (1958); Marcel Granet, *Chinese Civilization* (1951); René Grousset, *The Rise and Splendour of the Chinese Empire* (1958); G. F. Hudson, *Europe and China: A Survey of their Relations from the Earliest Times to 1800* (1931); and the admirable collection of documents, Joseph R. Levenson (ed.), *European Expansion and the Counter-Example of Asia, 1300–1600* (1967).

Of special interest for the topics of these chapters is J. J. L. Duyvendak, *China's Discovery of Africa* (1949). For the place of eunuchs in Chinese history, see Marcel Granet, *Etudes sociologiques sur la Chine* (1953), and Taisuke Mitamura, *Chinese Eunuchs* (1970).

Part VII: The American Surprise

Begin with J. H. Elliott's succinct *The Old World and the New, 1492–1650* (1970) or Samuel Eliot Morison's delightful *The European Discovery of America: The Northern Voyages, A.D. 500–1600* (1971) and P. H. Sawyer, *The Age of the Vikings* (2d ed., 1972). On the Vikings in America, see: James R. Enterline, *Viking America* (1974); Joseph Fischer, *The Discoveries of the Norsemen in America . . . their Early Cartographical Representation* (1970); G. M. Gathorne-Hardy, *The Norse Discoverers of America: The Wineland Sagas* (1921); Gwyn Jones, *The Norse Atlantic Saga . . . the Norse Voyages of Discovery and*

Settlement to Iceland (1964). On the Vikings, see also Johannes Brøndsted, *The Vikings* (1973); Gwyn Jones, *A History of the Vikings* (1973); Ole Klindt-Jensen, *The World of the Vikings* (1970); T. C. Lethbridge, *Herdsmen and Hermits: Celtic Seafarers in the Northern Sea* (1950); David M. Wilson and Peter G. Foote, *The Viking Achievement* (1970); Julius E. Olson (ed.), *The Northmen, Columbus and Cabot, 985–1503* (1906), especially pp. 14–66 on Eric the Red; Thorleif Sjøvold, *The Oseberg Find, and other Viking Ship Finds* (1976). Many other relevant articles can be found in the index to *The Mariner's Mirror: The Quarterly Journal of the Society for Nautical Research.*

The Norse sagas are available in Snorri Sturluson, *Heimskringla: Sagas of the Norse Kings* (Samuel Lang trans., 1961), and in Gudbrund Vigfusson and F. York Powell, *Corpus Poeticum Boreale: the Poetry of the Old Northern Tongue* (2 vols., 1883), and, as a special treat, read Paul Taylor and W. H. Auden (trans.), *The Elder Edda* (1969). The role of the Normans in the Old World can be followed in two brilliant essays by Charles Homer Haskins, *Norman Institutions* (1967) and *The Normans in European History* (1959).

A readable layman's introduction to the history of the arts and sciences of navigation is E. G. R. Taylor, *The Haven-Finding Art; A History of Navigation from Odysseus to Captain Cook* (1956), which can be supplemented by: David W. Waters, *The Art of Navigation in England in Elizabethan and Early Stuart Times* (1958); H. L. Hitchins and W. E. May, *From Lodestone to Gyro-Compass* (1953); Frederic C. Lane's invaluable "The Economic Meaning of the Invention of the Compass," in *American Historical Review,* Vol. 68 (1963), pp. 605–17. Thomas Gladwin's *East Is a Big Bird: Navigation and Logic on Puluwat Atoll* (1970) gives a revealing glimpse of how some seafarers still find their way on the ocean without a compass.

The life of Columbus has been vividly recounted by Samuel Eliot Morison (1887–1976), the Naval Historian of World War II, who took the trouble to outfit, man, and sail a vessel like one of Columbus' to recapture his experience of sailing across the ocean. Morison's lifelong love affair with the sea gave his writing an authentic detail and gave his advocacy of Columbus the ring of science. See his *Admiral of the Ocean Sea* (2 vols., 1942), also in an abridged one volume (1942), further abridged in *Christopher Columbus, Mariner* (1955). His *European Discovery of America* (2 vols., 1971, 1974) is a treasure-house of amusing asides, and see also his *Portuguese Voyages to America in the Fifteenth Century* (1965). Other views of Columbus (who has been claimed by nearly every nation and religion of Europe) can be sampled in: Salvador de Madariaga, *Christopher Columbus* (1967); Fernando Colón, *The Life of the Admiral Christopher Columbus by His Son Ferdinand* (Benjamin Keen, trans., 1959). For the wider context, see Américo Castro, *The Structure of Spanish History* (1954). On Columbus' geographic knowledge, see: George E. Nunn, *The Geographical Conceptions of Columbus* (1977); Pierre d'Ailly, *Ymago Mundi* (3 vols., 1930); Peter Martyr D'Anghera, *De Orbe Novo* (Francis A. MacNutt, trans., 2 vols., 1912).

On Vespucci and the naming of the new continents, see: Germán Arciniegas, *Amerigo and the New World: The Life and Times of Amerigo Vespucci* (1955); Frederick J. Pohl, *Amerigo Vespucci, Pilot Major* (1966); John B. Thacher, *The Continent of America: Its Discovery and Baptism* (1896); Louis-André Vigneras, *The Discovery of South America and the Andalusian Voyages* (1976); Arthur P. Whitaker, *The Western Hemisphere Idea: Its Rise and Decline* (1954); Martin Waldseemüller, *The Cosmographiae Introductio: Followed by the Four Voyages of Amerigo Vespucci, and their translation into English; to which are added Waldseemüller's Two World Maps of 1507* (U.S. Catholic Historical Society, 1907). Some light on what Vespucci and others saw (or were said to have seen) and some consequences: Fredi Chiappelli (ed.), *First Images of America: The Impact of the New World on the Old* (2 vols., 1976); W. Arens, *The Man-Eating Myth* (1979); Iris H. W. Engstrand, *Spanish Scientists in the New World, the Eighteenth-Century Expeditions* (1981).

Part VIII: Sea Paths to Everywhere

There is no better introduction to this subject than J. H. Parry, *The Discovery of the Sea* (1974), followed by reading in Antonio Pigafetta, *Magellan's Voyage, a Narrative Account of the First Circumnavigation* (R. A. Skelton [trans. and ed.], 2 vols., 1969). On Magellan, see: F. H. H. Guillemard, *The Life of Ferdinand Magellan, and the First Circumnavigation of the Globe* (1890; 1971); E. F. Benson, *Ferdinand Magellan* (1929); Charles McK. Parr, *Ferdinand Magellan, Circumnavigator* (1964).

A number of readable works give us our bearings in the Age of Discovery and its aftermath: J. H. Parry (ed.), *The European Reconnaissance: Selected Documents* (1968) and *Trade and Dominion: The European Overseas Empires in the Eighteenth Century* (1971); Carlo M. Cipolla's suggestive *Guns and Sails in the Early Phase of European Expansion, 1400–1700* (1965); John F. Meigs, *The Story of the Seaman* (1924); Arthur P. Newton (ed.), *The Great Age of Discovery* (1932); Boies Penrose, *Travel and Discovery in the Renaissance 1420–1620* (1962); David B. Quinn (ed.), *North American Discovery, circa 1000–1612* (1971); G. V. Scammell, *The World Encompassed: The First European Maritime Empires, c. 800–1650* (1981); Sir Percy Sykes, *A History of Exploration* (1961); Louis B. Wright, *Gold, Glory, and the Gospel: The Adventurous Lives and Times of the Renaissance Explorers* (1970). For an introduction to the rich contemporary English seafaring literature, see George B. Parks,

Richard Hakluyt and the English Voyages (1930).

A vivid, up-to-date biography of Sir Francis Drake is Derek Wilson, *The World Encompassed: Francis Drake and His Great Voyage* (1977). See also: Sir Francis Drake, *The World Encompassed, and analogous contemporary documents* (Richard C. Temple, ed., 1969); Christopher Lloyd, *Sir Francis Drake* (1957); James A. Williamson, *The Age of Drake* (1960). As Drake's career revealed, in his age the lines between commerce, piracy, and exploration were seldom clear. See: Henry A. Ormerod, *Piracy in the Ancient World* (1967); Robert Carse, *The Age of Piracy* (1965); Philip Gosse, *The History of Piracy* (1968); Pitman B. Potter, *The Freedom of the Seas in History, Law, and Politics* (1924).

The history of cartography is rich in volumes that will entice the layman to this crossroad of science and the arts. Begin with John N. Wilford, *The Mapmakers* (1981) or J. C. C. Crone, *Maps and Their Makers* (1968) or Norman J. W. Thrower, *Maps and Man ... Cartography in Relation to Culture and Civilization* (1972), and follow Walter W. Ristow's admirable *Guide to the History of Cartography* (1973). See also: for the techniques of mapmaking, David Greenhood, *Mapping* (1951); Walter W. Ristow, *A la Carte: Selected Papers on Maps and Atlases* (1972); R. A. Skelton, *Maps: A Historical Survey of their Study and Collecting* (1975); David Woodward (ed.), *Five Centuries of Map Printing* (1975); Edward L. Stevenson, *Portolan Charts* (1911); A. E. Nordenskiöld, *Periplus* (1897). For a glimpse of how the airplane has revised the map-makers' point of view, and its implications for World War II, see Richard Edes Harrison, *Look at the World* (1944). Specially relevant, in addition to Nordenskiöld's *Facsimile Atlas* (Dover reprint, 1973), are: Gail Roberts, *Atlas of Discovery* (1973); Erwin Raisz, *Atlas of Global Geography* (1944); and Edward L. Stevenson's indispensable *Terrestrial and*

Celestial Globes (2 vols., 1921). Numerous relevant articles can be found in the indexes to the transactions of the Congresso Internacional de Historia dos Descobrimentos and to *Terrae Incognitae,* the annals of the Society for the History of Discoveries.

The New Zealand historian J. C. Beaglehole has given us a grand *Life of Captain James Cook* (1974), which no lover of biography or of seafaring adventure should miss. He also edited *The Journals of Captain James Cook on His Voyages of Discovery* for the Hakluyt Society (4 vols., 1955–67) and *The Endeavour Journal of Joseph Banks, 1768–1771* (2 vols., 1962). A readable short life is *Captain James Cook* (1967), by Alan J. Villiers, who had "sailed a full-rigged ship not much different from his *Endeavour* around the world in as much of Cook's tracks as I dared." For Cook's place in the epic of Pacific exploration and Western settlement, see: J. C. Beaglehole, *The Exploration of the Pacific* (3d ed., 1966); Alan Moorehead's brief and dramatic *The Fatal Impact: The Invasion of the South Pacific, 1767–1840* (1966); Daniel Conner and Lorainne Miller, *Master Mariner: Capt. James Cook and the Peoples of the Pacific* (1978). Robin Fisher and Hugh Johnston (eds.), *Captain James Cook and His Times* (1979), brings together revisionist essays suggesting that Cook was an advance agent of "imperialism," among other evils. A useful cartographic depiction of the illusion of *Terra Australis,* its extent and its migrations, is in Gail Roberts' *Atlas of Discovery* (1973), Chapter 10.

BOOK THREE: NATURE

Part IX: Seeing the Invisible

The reader who is new to astronomy would do well to begin with a clear and concise narrative history, such as J. L. E. Dreyer, *A History of Astronomy from Thales to Kepler* (2d ed., 1953) or Arthur Berry, *A Short History of Astronomy* (1898; 1961), both available in inexpensive paperback reprints. No one should go far into the history of science without reading Thomas S. Kuhn's bold and brilliant *Structure of Scientific Revolution* (1962), with the added fillip of his *Essential Tension; Selected Studies in Scientific Tradition and Change* (1977). I am deeply indebted, as all readers will be, to his *Copernican Revolution: Planetary Astronomy in the Development of Western Thought* (Vintage paperback, 1959). Other suggestive general works include: E. J. Dijksterhuis, *The Mechanization of the World Picture* (1961), a dense book invaluable for reference; Pierre Duhem, *To Save the Phenomena: An Essay on the Idea of Physical Theory from Plato to Galileo* (1969) and *Le Système du Monde* (Vol. I, 1971); Camille Flammarion, *Histoire du Ciel* (1877); Martin Harwit, *Cosmic Discovery: The Search, Scope and Heritage of Astronomy* (1981); Fred Hoyle's lively and speculative *Astronomy: A History of Man's Investigation of the Universe* (n.d.), *Of Men and Galaxies* (1964), and *From Stonehenge to Modern Cosmology* (1972); Antonie Pannekoek, *A History of Astronomy* (1961); H.T. Pledge, *Science Since 1500: A Short History of Mathematics, Physics, Chemistry, and Biology* (1959), which does not skimp technicalities; Charles Singer and C. Rabin, *A Prelude to Modern Science* (1946); A. Wulf, *History of Science, Technology and Philosophy in the 16th and 17th Centuries* (2 vols., 2d ed., 1959) and *... in the 18th Century* (1939). Early sources can be found in: Harlow Shapley and Helen E. Howarth, *A Source Book in Astronomy* (1929); and Thomas Wright (ed.), *Popular Treatises on Science written during the Middle Ages* (1891), which reprints an especially interesting Anglo-Saxon Manual of Astronomy which is an abridgment of that by the Venerable Bede in the eighth century. Some of the principal texts discussed in these chapters—*The Almagest,* by Ptolemy; *On the Revolutions of the Heavenly Spheres,* by

Copernicus; and the *Epitome of Copernican Astronomy* (iv and v) and *The Harmonies of the World* (v), by Kepler—are conveniently and elegantly reprinted in *Great Books of the Western World*, Vol. 16.

On Copernicus, see: Angus Armitage, *Sun, Stand Thou Still* (1947); Josef Rudnicki, *Nicholas Copernicus* (1938; abridged ed., 1943); Edward Rosen (trans. and ed.), *Three Copernican Treatises* (1975); J. Neyman (ed.), *The Heritage of Copernicus: Theories "Pleasing to the Mind"* (1975). The life of a heroic Copernican can be read in Dorothea W. Singer, *Giordano Bruno, his Life and Thought* (1968), including an annotated translation of Bruno's "On the Infinite Universe and World." For Brahe, read J. L. E. Dreyer's vivid *Tycho Brahe* (1890), and consult J. A. Gade, *The Life and Times of Tycho Brahe* (1947) with its bibliography. And meet the magnificent and inscrutable Kepler in Max Caspar, *Kepler* (1939).

Galileo is a never-ending source of scientific melodrama and scholarly controversy. Of the vast literature, some of the works I have found most useful include: Stillman Drake, *Galileo at Work: His Scientific Biography* (1978) and *Operations of the Geometric and Military Compass, 1606* (1978); Ludovico Geymonat, *Galileo Galilei: A Biography and Inquiry into His Philosophy of Science* (1965); Arthur Koestler's engrossing and melodramatic *The Sleepwalkers* (1959); Jerome J. Langford, *Galileo, Science, and the Church* (rev. ed., 1971), and G. de Santillana's vigorously interpretive *The Crime of Galileo* (1955). Galileo's *Dialogues Concerning Two New Sciences* is conveniently available in *Great Books of the Western World*, Vol. 28.

Few topics in the history of science are more redolent of mysticism and religion than optics and the nature of light. A readable and reliable starting point is Vasco Ronchi, *The Nature of Light: An Historical Survey* (1970). Other works of special relevance: David C. Lindberg

and Nicholas H. Steneck, "The Sense of Vision and the Origins of Modern Science," in *Science, Medicine and Society in the Renaissance* (Allen G. Debus, ed., Vol. I; 1972); Henry C. King, *The History of the Telescope* (1955); Edward Rosen, *The Naming of the Telescope* (1947); and numerous articles by Silvio A. Bedini, especially "The Tube of Long Vision (The Physical Characteristics of the early 17th Century Telescope)," *Physis*, Vol. 13 (1971), pp. 149–204.

The history of the microscope, being less clearly tied to cosmology or the heavens, is not invigorated or plagued by the same passions that have infected the study of Galileo and the telescope. A delightful place to begin is Leeuwenhoek himself, who can be followed in his everyday routine through Clifford Dobell, *Antony van Leeuwenhoek and His "Little Animals"* (1932; 1960), which reprints many of Leeuwenhoek's own letters and notes and is happily available in an inexpensive Dover reprint (1960). See S. Bradbury and G. L'E. Turner (eds.), *Historical Aspects of Microscopy* (1967); S. Bradbury, *The Evolution of the Microscope* (1967); Reginald S. Clay and Thomas H. Court, *The History of the Microscope: Compiled from Original Instruments and Documents up to the Introduction of the Achromatic Microscope* (1932); Alfred N. Disney, *Origin and Development of the Microscope, as Illustrated by Catalogues of the Instruments and Accessories in the Collection of the Royal Microscopical Society* (1928); A. Schierbeek, *Measuring the Invisible World, the Life and Works of Antoni van Leeuwenhoek . . . with a biographical chapter by Maria Roosenboom* (1959), and *Jan Swammerdam, 1637–1680, His Life and Works* (1967).

On the telescope and the microscope in China and Japan, in addition to the works cited for earlier Parts, see: Pasquale M. D'Elia, *Galileo in China: Relations through the Roman College between Galileo and the Jesuit Scientist-Missionaries (1610–1640)* (1960); John R. Levenson, *Modern China* (1971); Nathan

Sivin (ed.), *Science and Technology in East Asia* (1977) and *Chinese Alchemy: Preliminary Studies* (1968); Donald Keene, *The Japanese Discovery of Europe, 1720–1830* (1969); G. B. Sansom, *The Western World and Japan* (1951), for the wider context.

Part X: Inside Ourselves

The history of medicine is inseparable from the history of a medical profession. But all professions have been difficult to chronicle because the insiders are usually constructing a defense, and outsiders seldom have the expertise or the jargon to master the specialized subject matter. An interesting introduction to the problems that afflict all professionalized subject matters is Talcott Parsons' article, "Professions," in the *International Encyclopedia of the Social Sciences* (Vol. 12, pp. 536–47; with bibliography), which unfortunately suffers from the author's own sociological jargon. For the wider context of the subject of this Part X, see: Vern L. Bullough, *The Development of Medicine as a Profession,* (1966), on the contribution of the medieval university to modern medicine; Mircea Eliade, *The Forge and the Crucible: The Origins and Structures of Alchemy* (1971); K. J. Franklin, *A Short History of Physiology* (1933); Kenneth B. Keele, *The Evolution of Clinical Methods in Medicine* (1963); Charles Singer, *The Evolution of Anatomy,* on anatomical and physiological discovery until Harvey (1925). For an insight into professional obstacles and the techniques of modern medical discovery, see Claude Bernard's classic *Introduction to the Study of Experimental Medicine* (1927).

Paracelsus' writings are not easily accessible. But see A. E. Waite (trans.), *The Hermetic and Alchemical Writings of Aureolus Philippus Theophrastus Bombast of Hohenheim, Called Paracelsus the Great* (2 vols., 1894), with a biographical introduction; Henry E. Sigerist (ed.), *Paracelsus, Four Treatises* (1941). The best short introduction is

Walter Pagel's article, "Paracelsus," in the *Dictionary of Scientific Biography,* Vol. 10, pp. 304–13, which can be followed by Pagel's *Paracelsus* (1958). See also: Basilio de Telepnef, *Paracelsus: A Genius Amidst a Troubled World* (1945); Henry M. Pachter, *Paracelsus: Magic into Science* (1951); and Allen G. Debus, *The English Paracelsians* (1965), which surveys his posthumous influence in England.

A sample of Galen's writings is conveniently available in *Great Books of the Western World,* Volume 10. A good introduction again is the article by Leonard G. Wilson in the *Dictionary of Scientific Biography,* at Vol. 5, pp. 227–37. The standard edition in English is Galen, *On the Usefulness of the Parts of the Body* (Margaret T. May, trans. and ed.; 2 vols., 1968), usefully supplemented by Owsei Temkin, *Galenism: Rise and Decline of a Medical Philosophy* (1973). There is no more engaging introduction to the mélange of philosophy, psychology, alchemy, astrology, science, and theology that went by the name of medicine before the modern era than Robert Burton's classic *Anatomy of Melancholy* (1624–51), available in many reprints.

On Leonardo as anatomist see: Jean Paul Richter (ed.), *The Notebooks of Leonardo da Vinci* (Dover reprint, 1970); Morris Philipson (ed.), *Leonardo da Vinci, Aspects of the Renaissance Genius* (1966); Erwin Panofsky's stimulating "Artist, Scientist, Genius: Notes on the 'Renaissance-Dämmerung,' " in *The Renaissance* (1962).

For the wider Renaissance perspective, see: Allen G. Debus (ed.), *Science, Medicine and Society in the Reinaissance* (2 vols., 1972); Paul O. Kristeller, *Eight Philosophers in the Renaissance* (1964); George Sarton, *Six Wings: Men of Science in the Renaissance* (1957).

On Vesalius we are fortunate to have a copious, accurate, and readable biography by C. D. O'Malley, *Andreas Vesalius of Brussels, 1514–1564* (1964). Consult the remarkable book, unique in conception, by the American medical pioneer Harvey W. Cushing, *A Bio-Bibliography of Andreas Vesalius* (1962). See: Andreas Vesalius, *De Humani Corporis Fabrica* (1967); L. R. Lind (trans.), *The Epitome of Andreas Vesalius* (1949); and the eyewitness account by a student, Baldasar Heseler, *Andreas Vesalius' First Public Anatomy in Bologna, 1540* (Ruben Eriksson, ed., 1959).

Geoffrey Keynes has given us a monumental and eminently readable *Life of William Harvey* (1966). See also: Kenneth D. Keele, *William Harvey, the Man, the Physician, and the Scientist* (1965); Walter Pagel, *William Harvey's Biological Ideas* (1967); Gweneth Whitteridge, *William Harvey and the Circulation of the Blood* (1971). A selection of Harvey's own writings is reprinted in *Great Books of the Western World,* Vol. 28, and William Harvey, *The Circulation of the Blood and Other Writings* (Kenneth J. Franklin, ed.; Everyman ed., 1963).

Santorio is not easily accessible in English, but begin with M. C. Grmek's article in the *Dictionary of Scientific Biography,* Vol. 12, pp. 101–4, with bibliography. See Ralph H. Major's excellent summary "Santorio Santorio," in *Annals of Medical History,* Vol. 10 (1938), pp. 369–81, and E. T. Renbourn, "The Natural History of Insensible Perspiration: A Forgotten Doctrine of Health and Disease," *Medical History,* Vol. 4 (1960), pp. 135–52; W. E. Knowles Middleton, *A History of the Thermometer* (1966); S. Weir Mitchell, *The Early History of Instrumental Precision in Medicine* (1891).

A good introduction to Malpighi is Luigi Belloni's article in the *Dictionary of Scientific Biography,* Vol. 9, pp. 62–66, supplemented by Joseph Needham, *A History of Embryology* (1934). But there is no competition anywhere else in the history of medicine for the delights of reading and browsing in Howard B. Adelmann's monumental *Marcello Malpighi and the Evolution of Embryology* (5 vols., 1966), which edits and reprints many of Malpighi's writings.

Part XI: Science Goes Public

J. M. Ziman's stimulating essays, *Public Knowledge* (1968) and *The Force of Knowledge* (1976), to which I am much indebted, open the way from the history of science to the characteristically modern organization of science. Of course the history of science in general offers a vast literature, and it is not easy to decide where to begin. The philosopher-mathematician Alfred North Whitehead provides a lucid and literate introduction to the adventures of scientific thought, especially his *Science and the Modern World* (1931) and *Adventures of Ideas* (1933). Other accessible works include: H. Butterfield, *The Origins of Modern Science* (1957); Owen Gingerich (ed.), *The Nature of Scientific Discovery* (1975), a symposium commemorating the 500th anniversary of the birth of Copernicus; Loren R. Graham, *Between Science and Values* (1981); A. R. Hall, *The Scientific Revolution 1500–1800* (1954); F. A. Hayek, *The Counter-Revolution of Science* (1979); Andrei S. Markovits and Karl W. Deutsch (eds.), *Fear of Science—Trust in Science* (1980), a symposium with lively polemical essays by some leading historians of science; Robert K. Merton, *Science, Technology and Society in Seventeenth-Century England* (1970) and *The Sociology of Science* (1973), richly suggestive essays on the frontier between society and scientific thought; Joseph Needham, *The Grand Titration: Science and Society in East and West* (1969); Karl Pearson, *The Grammar of Science* (1957); H. T. Pledge, *Science Since 1500* (1959); Derek J. de Solla Price, *Science Since Babylon* (1961); Cyril S. Smith, *A Search for Structure* (1981), some pathbreaking essays on science, art, and history; Lynn Thorndike, *A History of Magic and Experimental Science* (8 vols., 1923–58), rich in documentation through the seventeenth century.

For the European ideal and the tradition of a scholarly fellowship, see Sir Francis Bacon's *Advancement of Learn-*

ing (1605), his *Novum Organum* (1620), and his *New Atlantis* (1626), all found in *Great Books of the Western World,* Vol. 30, and also in the *World's Classics* of the Oxford University Press (1929).

There are a number of scholarly and readable works on the rise of what I call the Parliaments of Science: Frances A. Yates, *The French Academies of the Sixteenth Century* (1947); Martha Ornstein, *The Role of Scientific Societies in the Seventeenth Century* (3d ed., 1938); Diana Crane, *Invisible Colleges: Diffusion of Knowledge in Scientific Communities* (1972); Harcourt Brown, *Scientific Organizations in Seventeenth-Century France* (1620–1680). (1934). For recent counterparts, see these publications originating from the Congressional Research Service of the Library of Congress and the Hearings and Reports on Migrations and Movements of Scientists: *The Evolution of International Technology* (1970) and *Toward a New Diplomacy in a Scientific Age* (1970).

For Mersenne, begin with A. C. Crombie's article, "Mersenne," in the *Dictionary of Scientific Biography,* Vol. 9, pp. 316–22. All Mersenne materials are surveyed in R. Lenoble, *Mersenne ou la naissance du Mécanisme* (1943). The heart of Mersenne is, of course, in his *Correspondence,* edited by Cornelis de Waard and others (1932–). And see: R. H. Popkin, *History of Scepticism from Erasmus to Descartes* (1964); Frances A. Yates, *Giordano Bruno and the Hermetic Tradition* (1964); and the work promised by A. C. Crombie and A. Carugo, *Galileo and Mersenne: Science, Nature and the Senses in the Sixteenth and Early Seventeenth Centuries.*

The passions of enthusiasts and the suspicions of opponents can be sensed in Bishop Thomas Sprat's *History of the Royal Society* (1667; facsimile reprint, Washington University Studies, 1958). Sprat makes extravagant claims for the organization then still only seven years old, which had hardly begun to show its powers. Like Galileo and Harvey, he wraps himself in the mantle of the an-

cients and "the most ancient author of all others, even Nature herself," but his claim to be a disciple of Francis Bacon, and his defense of novelty, show him marking the new directions. An admirable introduction to Sprat is the article by Hans Aarsleff in the *Dictionary of Scientific Biography,* Vol. 12, pp. 580–87. There, too, in A. Rupert Hall's article (Vol. 10, pp. 200–3) is the best brief introduction to Oldenburg. The atmosphere, passions, and hopes of the early age of the Parliaments of Science can nowhere be better felt than in A. Rupert Hall and Marie Boas Hall (eds.), *The Correspondence of Henry Oldenburg* (12 vols., 1965–). These editors have produced numerous articles in journals of the history of science on Oldenburg and his place in the scientific controversies of his age.

Some of the best mathematicians have been some of the most effective popularizers. A delightful introduction is Alfred North Whitehead, *An Introduction to Mathematics* (Home University Library, 1911). Other volumes appealing to the layman include: W. W. Rouse Ball, *A Short Account of the History of Mathematics* (1960); E. T. Bell, *Men of Mathematics* (1937), a masterly humanization of the story through the biographies of mathematicians; Tobias Dantzig, *Numbers, the Language of Science* (1939); Morris Kline, *Mathematical Thought from Ancient to Modern Times* (1972); David Eugene Smith, *A Source Book in Mathematics* (2 vols., 1929); Lancelot Hogben, *Mathematics for the Million* (1937), a tour de force of economic interpretation, endlessly suggestive; James R. Newman, *World of Mathematics* (4 vols., 1956), an adventurous and witty anthology, and (with Edward Kasner), *Mathematics and the Imagination* (1940); and David Eugene Smith and Yoshio Mikami, *A History of Japanese Mathematics* (1914).

For Simon Stevin a good beginning is the admirable article by M. G. J. Minnaert in the *Dictionary of Scientific Biography,* Vol. 13, pp. 47–51, with bibliography, and D. J. Dijksterhuis, *Simon Stevin: Science in the Netherlands Around 1600* (1970). Stevin's "Art of Tenths" is reprinted in D. E. Smith's *Source Book* (above). On the rise of the decimal system there are useful articles in *Isis* by George Sarton at Vol. 21 (1934), pp. 241–303, and at Vol. 23 (1935), pp. 153–244; and by Dirk Struik at Vol. 25 (1936), pp. 46–56.

The relation of the rise of mathematics to scientific instrumentation and instrument-making can be followed in an attractive variety of works: William Cunningham, *Alien Immigrants to England* (1969), Maurice Daumas, *Scientific Instruments of the Seventeenth and Eighteenth Centuries* (1972); E. J. Dijksterhuis, *The Mechanization of the World Picture* (1969); Derek Howse, *Greenwich Observatory: the Buildings and Instruments* (1975) and *Francis Place and the Early History of the Greenwich Observatory* (1975); Rupert Hall, "The Scholar and the Craftsman in the Scientific Revolution," in *Critical Problems in the History of Science* (Institute for the History of Science, Proceedings, 1959), pp. 3–23; Henri Michel, *Scientific Instruments in Art and History* (1966); W. E. Knowles Middleton's indispensable guides, *The History of the Barometer* (1964) and *A History of the Thermometer* (1966); S. Weir Mitchell, *The Early History of Instrumental Precision in Medicine* (1891).

The problems of devising and agreeing upon standards for weights and measures have aroused political, economic, and patriotic as well as scientific passions. The partisans of the metric system as a minor social panacea are still active today. For the history in the West, see: George Sarton, *A History of Science* (2 vols., 1970) for antiquity; William Hallock and Herbert T. Wade, *Outlines of the Evolution of Weights and Measures and the Metric System* (1906); Henri Moreau, *Le Système Métrique: des Anciennes Mesures au Système International d'Unité* (1975); Edward Nichol-son, *Men and Measures: A History of Weights and Measures Ancient and Modern* (1912); U.S. National Bureau of Standards Special Publications, *U.S. Metric Study Interim Report*, "A History of the Metric System Controversy in the United States" (1971), and *A Metric America: A Decision Whose Time has Come* (1971); Ronald E. Zupko, *French Weights and Measures Before the Revolution* (1978).

To be a qualified biographer of the mystifying and richly contradictory Newton, the historian must have near-Newtonian talents. Richard S. Westfall's *Never at Rest: A Biography of Isaac Newton* (1980) succeeds admirably in using the facts of Newton's life to illuminate his theories and vice versa. The most readable surveys of Newton's thought are the works of I. Bernard Cohen, who has a remarkable talent for interpreting science to the layman. The American reader might begin with his *Franklin and Newton: An Inquiry into Speculative Newtonian Experimental Science and Franklin's Work in Electricity as an Example Thereof* (1956). Cohen has also written the admirable comprehensive article on Newton in the *Dictionary of Scientific Biography*, Vol. 10, pp. 42–101 (to which is appended a suggestive survey of the Soviet literature on Newton by A. P. Youschkevitch). And see Cohen's more extensive *Introduction to Newton's Principia* (1971). There are numerous editions and facsimiles of Newton's *Principia*. A convenient source for the American reader is *Great Books of the Western World*, Vol. 34, which also reprints Newton's *Opticks*.

The bibliography of Newton is vast, and amply surveyed in Cohen's article, above. Alexandre Koyré has given us some elegant essays that lead us from Newton into the wide world of philosophy: *From the Closed World to the Infinite Universe* (1957); *Newtonian Studies* (1965); *Metaphysics and Measurement* (1968). For a refreshing view of some neglected aspects of Newton, see Frank E. Manuel, *Isaac Newton, Historian*

(1963), a searching exploration of Newton's applications of astronomy and Biblical prophecy to history, with its implications for his science, and *A Portrait of Isaac Newton* (1968) emphasizing Newton's early life and psychological problems to produce an even less sympathetic portrait than the familiar one. Of Henry Guerlac's numerous works in this area I am most indebted to his brilliant article, "Where the Statue Stood: Divergent Loyalties to Newton in the Eighteenth Century," in Earl R. Wasserman (ed.), *Aspects of the Eighteenth Century* (1965). For other aspects of Newton, see: A. Rupert Hall, *Philosophers at War: The Quarrel between Newton and Leibniz* (1980), which puts the controversy in the wider context of Continental scientific thought of the era; Sir John Craig, *Newton at the Mint* (1946); Marjorie Hope Nicolson, *Newton Demands the Muse: Newton's "Opticks" and the Eighteenth-Century Poets* (1946), not to be missed by anyone interested in the roots of English Romanticism; J. D. North, *Isaac Newton* (1967); Richard S. Westfall, *Science and Religion in Seventeenth-Century England* (1958).

On the increasing significance of priorities in scientific discovery, see: the sociological writings of Robert K. Merton, especially "Priorities in Scientific Discovery," in *The Sociology of Science* (1973); Lyman R. Patterson, *Copyright in Historical Perspective* (1968); James D. Watson, *The Double Helix: A Personal Account of the Discovery of the Structure of DNA* (1968).

Part XII: Cataloguing the Whole Creation

The most readable scholarly survey of earlier biological thought remains Erik Norkenskiöld, *The History of Biology* (1928), which, for more recent developments, can be supplemented by Url Lanham, *Origins of Modern Biology* (1968); P. B. and J. S. Medawar, *The Life Science: Current Ideas of Biology* (1977),

cogent and lively; Ernst Mayr, *The Growth of Biological Thought: Diversity, Evolution, and Inheritance* (1982), a comprehensive and stimulating interpretation of the history of all biology through a history of ideas of evolution. For the philosophers' view, see: R. G. Collingwood, *The Idea of Nature* (1960). More conventional histories of botany include: the pioneer Richard Pulteney, *Sketches on the Progress of Botany in England from its Origin to the Introduction of the Linnaean System* (1790); Julius von Sachs, *History of Botany* (1890); Ellison Hawke, *Pioneers of Plant Study* (1969); Howard S. Reed, *A Short History of the Plant Sciences* (1942). The gardener and flower lover will find a handsomely illustrated and copiously informative path into history in Alice M. Coats, *Flowers and their Histories* (1968) and *The Plant Hunters* (1969), a history of horticultural pioneers and their quests since the Renaissance. Lucile H. Brockway, *Science and Colonial Expansion* (1979), ties the history of botany through the British Royal Botanical Gardens to the colonies and the empire; and see E. H. M. Cox, *Plant-Hunting in China* (1945). For the relation to the history of printing, see the Catalogue for the exhibit *Le Livre Illustré en Occident* (Brussels, 1977).

On biology in antiquity and the Middle Ages see: Richard Lewinsohn, *Animals, Men and Myths* (1954); Willy Ley, *Dawn of Zoology* (1968); Herbert H. Wethered, *The Mind of the Ancient World . . . Pliny's Natural History* (1968). Pliny's *Natural History* is available in the Loeb Classical Library (H. Rackham et al., trans.; Latin text and English translation, 10 vols., 1942–63).

On herbals, begin with: Agnes Arber's *Herbals, Their Origin and Evolution . . . the History of Botany, 1470–1670* (2d ed., 1970); Frank J. Anderson, *An Illustrated History of the Herbals* (1977); Wilfrid Blunt, *The Art of Botanical Illustration* (1950). And for Dioscorides: Robert T. Gunther (ed.), *The Greek Herbal of Dioscorides* (1959), and

Ben C. Harris, *The Compleat Herbal* (1972). The bestiaries can be discovered in numerous versions, which attest their chameleonlike character: Bishop Theobald, *Physiologus . . . a metrical bestiary printed in Cologne 1492* (Alan W. Rendall, trans., 1928); Edward Topsell, *History of Four-footed Beasts and Serpents and Insects* (reprinted by Da Capo Press, 3 vols., 1967); Jesse L. Weston, ed., "Physiologus," in *The Chief Middle English Poets* (1914), at pp. 325–34; Edward Topsell, *The Elizabethan Zoo*, a selection from Holland's translation of Pliny (1601), and from Topsell's *History* with illustrations in an elegantly printed volume (Godine, 1979); and the characteristically witty translation of a Latin Bestiary by T. H. White, *English Bestiary* (1954). A sketch of Gesner's life with a bibliography is found in the *Dictionary of Scientific Biography*, Vol. 5, pp. 378ff. And see: J. Monroe Thorington, *On Conrad Gesner and the Mountaineering of Theuerdank* (1937), which reprints Gesner's unforgettable paean to the mountains; Claire-Elaine Engel's readable *Mountaineering in the Alps, an Historical Survey* (new ed., 1971) and Engel (ed.), *Le Mont Blanc, vu par les écrivains et les Alpinistes* (1965).

The basic book on John Ray is C. E. Raven's copious *John Ray, His Life and Works* (2d ed., 1950). Ray's complete works are surveyed in G. L. Keynes, *John Ray, A Bibliography* (1951). Ray's *Wisdom of God, manifested in the Works of the Creation* (1691) is available in a 1974 reprint.

The best introduction to Linnaeus is the comprehensive article by Sten Lindroth in the *Dictionary of Scientific Biography*, Vol. 8, pp. 374–81. The definitive biography in English is Benjamin D. Jackson, *Linnaeus: The Story of His Life* (1923), an abridgment of the Swedish work by T. M. Fries; and see Wilfrid Blunt, *The Compleat Naturalist: A Life of Linnaeus* (1971), brief and well illustrated. Of the copious and controversial Linnaean literature, I have found the following most useful: James L. Larson,

Reason and Experience: The Representation of Natural Order in the work of Carl von Linné (1971); Frans A. Stafleu, *Linnaeus and the Linnaeans, the Spreading of Their Ideas in Systematic Botany, 1735–1789* (1971); A. T. Gage, *A History of the Linnean Society of London* (n.d.); W. T. Stearn, *Three Prefaces on Linnaeus and Robert Brown* (1962) and "The Background of Linnaeus's Contributions to the Nomenclature and Methods of Systematic Biology," in *Systematic Zoology*, Vol. 8 (1959), pp. 4–22, to which I am deeply indebted. For a wider context, see: D. Mornet, *Les Sciences de la Nature en France en XVIIIe Siècle* (1911); Helmut de Terra, *Humboldt: The Life and Times of Alexander von Humboldt, 1769–1859* (1955); Douglas Bottong, *Humboldt and the Cosmos* (1973).

A good approach to Buffon is Jacques Roger's article in the *Dictionary of Scientific Biography*, Vol. 2, pp. 576–82, and Otis E. Fellows and Stephen F. Milliken, *Buffon* (Twayne's World Authors Series, 1972). Of special interest here is Arthur O. Lovejoy, "Buffon and the Problem of Species," *Popular Science Monthly*, Vol. 79 (1911), pp. 464–73, 554–67. The best edition of Buffon's writings is the *Oeuvres complètes* (J. L. Lanessan, ed., 14 vols., Paris, 1884–85). *Les Epoques de la Nature* (Jacques Roger, ed., 1962) has extensive critical apparatus. The place of geology in man's enterprise of discovery can be followed in: Ruth Moore, *The Earth We Live On: The Story of Geological Discovery* (1956) and Cecil J. Schneer (ed.), *Toward a History of Geology* (1969). A stimulating study with a sharper focus is Charles C. Gillispie, *Genesis and Geology . . . Scientific Thought, Natural Theology, and Social Opinion in Great Britain, 1790–1850* (1951). The earnest reader may want to venture into Clarence J. Glacken's remarkably suggestive but cryptic *Traces on the Rhodian Shore* (1976), a study of nature and culture in Western thought to the end of the eighteenth century, which follows at length the shifting relations among the idea of a designed earth, the idea of the influence of the environment, and the idea of man as a geographical agent.

Biographies of Louis Pasteur offer an opportunity to compare explanations of the motives of a great discoverer. Would Pasteur have become the Newton of biology if only he had not left his earliest theoretical pursuits? Were fame and the hope for government support temptations away from his most important scientific work? Begin with Gerald L. Geison's admirable article in the *Dictionary of Scientific Biography*, Vol. 10, pp. 350–416, and explore these questions in René J. Dubos' delightful *Louis Pasteur: Free Lance of Science* (1951), which in Chapter VI puts the question of spontaneous generation in the ample context of the facts and passions of the time, or in Emile Duclaux's more technical *Pasteur: The History of a Mind* (1920). The best short biography is by the great man's grandson, Pasteur Vallery-Radot, *Louis Pasteur: A Great Life in Brief* (1958). All these depend heavily on the standard *Life* (2 vols., 1901) by Pasteur's son-in-law and secretary René Vallery-Radot, whose first anonymous essay was disarmingly entitled *Pasteur, histoire d'un savant par un ignorant* (Paris, 1883).

Arthur O. Lovejoy's *The Great Chain of Being* (1936) opened new paths for intellectual history and the history of science, which can be followed in the pages of the *Journal of the History of Ideas*. The basic work on Tyson is M. F. Ashley-Montagu's admirable *Edward Tyson, 1650–1708, and the Rise of Human and Comparative Anatomy in England* (1943). Tyson's *Orang-Outang, sive Homo Sylvestris: or the Anatomy of a Pygmie* (1699) is available in a facsimile, with an Introduction by Ashley-Montagu (1966). For comparative anatomy, see: F. J. Cole, *A History of Comparative Anatomy* (1944); William R. Coleman, *Georges Cuvier, Zoologist* (1964); Stanley M. Garn, *Human Races*

(3d ed., 1971) and (ed.), *Readings on Race* (2d ed., 1968).

There is no better introduction to Darwin than Gavin de Beer's concise, fact-packed, and subtle article in the *Dictionary of Scientific Biography*, Vol. 3, pp. 565–77, with bibliography. Many of Darwin's writings, and especially the *Origin of Species*, are conveniently available in *Great Books of the Western World*, Vol. 49, and many reprints. Darwin comes alive in Gavin de Beer (ed.), *Charles Darwin . . . Thomas Henry Huxley Autobiographies* (1974), and in Paul H. Barrett (ed.), *The Collected Papers of Charles Darwin* (1980), with a Foreword by Theodosius Dobzhansky, a handy one-volume paperback (University of Chicago Press). Two handsomely illustrated accessible paperbacks—Alan Moorehead, *Darwin and the Beagle* (1971) and *The Illustrated Origin of Species* (abridged, with an Introduction by Richard E. Leakey, 1979)—help us see what Darwin saw.

In the popular mind Alfred Russel Wallace has been almost totally overshadowed by Darwin. This generous, erratic, and courageous man begins to receive his due in H. Lewis McKinney's article in the *Dictionary of Scientific Biography*, Vol. 14, pp. 133–40. At greater length, see Lancelot T. Hogben, *Alfred Russel Wallace, the Story of a Great Discoverer* (1918); and he is discovered in his own words in Alfred Russel Wallace, *My Life, A Record of Events and Opinions* (2 vols., 1905).

Into the vast literature on Darwinism, evolution and evolutionism, a good introduction, logically and chronologically, is Thomas H. Huxley, *Evidence as to Man's Place in Nature* (1863). Among the most suggestive other works I have found: Philip Appleman (ed.), *Darwin* (1970); J. W. Burrow, *Evolution and Society, a Study in Victorian Social Theory* (1966); Loren Eisley, *Darwin's Century* (1961), a brilliant narrative summary of the interrelation of pre- and post-Darwin Darwinians; Neal C. Gillespie,

Charles Darwin and the Problem of Creation (1979); Stephen Jay Gould, *Ever Since Darwin* (1977), lively essays on modern natural history; Gertrude Himmelfarb, *Darwin and the Darwinian Revolution* (1959), a perceptive charting of the currents flowing from Darwin; Michael Ruse, *The Darwinian Revolution: Science Red in Tooth and Claw* (1979), chronicles the transformation of the concept of nature; Sol Tax and Charles Callender, *Evolution After Darwin* (3 vols., 1960), papers in celebration of the *Origin of Species*, Vol. 1 (The Evolution of Life), Vol. 2 (The Evolution of Man), Vol. 3 (Issues in Evolution); Scientific American, *Evolution*, essays that appeared in the magazine's issue of September 1978 exploring the far-reaching consequences of molecular biology and other recent developments for the Darwinian concept of evolution.

BOOK FOUR: SOCIETY

Part XIII: Widening the Communities of Knowledge

The pathbreaking books of Frances A. Yates on memory and its place in Western civilization are a delight that no one interested in history should miss. And, incidentally, they remind us of how much remains to be discovered by the bold and imaginative scholar in even the most conventional realms of the past. Begin with *The Art of Memory* (University of Chicago Press paperback, 1966), then to *Giordano Bruno and the Hermetic Tradition* (1964), and *The Rosicrucian Enlightenment* (1972). Also on memory: Frederick C. Bartlett, *Remembering* (1932), a study in experimental and social psychology; M. T. Clanchy, *From Memory to Written Record: England, 1063–1307* (1979); Hermann Ebbinghaus, *Memory: A Contribution to Experimental Psychology* (Dover paperback, 1964); Mircea Eliade, "Mythologies of Memory and Forgetting,"

History of Religion, Vol. 2 (1963), pp. 329–44; Bennet B. Murdock, Jr., *Human Memory: Theory and Data* (1979); Jean Piaget and Bärbel Inhelder, *Memory and Intelligence* (1974). References for Sigmund Freud will be found under Part XIV, below.

On the culture of the written (not printed) word, which is so difficult for us to envisage in our age drenched in printed matter, there are a number of vivid works full of telling detail. A convenient beginning is Charles Homer Haskins' brief *Rise of Universities* (1923) or his *Renaissance of the 12th Century* (1957). Voluminous works that will repay browsing are: the standard Hastings Rashdall, *The Universities of Europe in the Middle Ages* (3 vols., new ed., 1936); George Haven Putnam, *Authors and their Public in Ancient Times* (1894) and *Books and their Makers during the Middle Ages* (2 vols., 1897; reprinted, 1962). Concise and enticing views of the longer perspective are: I. J. Gelb, *A Study of Writing* (1952); Edward Chiera, *They Wrote on Clay* (1938). And see: Stanley Morison, *Politics and Script* (1972), a handsomely illustrated essay on the emergence of new scripts and their social origins; Edward Alexander Parsons, *The Alexandrian Library: Glory of the Hellenic World* (1952), with surprising detail on the Western world's largest ancient library of written books; L. D. Reynolds and N. G. Wilson, *Scribes and Scholars: A Guide to the Transmission of Greek and Latin Literature* (1968), illuminating sidelights on what a scholar in the Age of Writing was apt to know, and why; G. B. Sansom, *The Western World and Japan . . . the Interaction of European and Asiatic Cultures* (1951); Tsuen-Hsuin Tsien, *Written on Bamboo and Silk: The Beginnings of Chinese Books and Inscriptions* (1962); Joyce Irene Whalley, *Writing Implements and Accessories: From the Roman Stylus to the Typewriter* (1975).

The handiest concise studies of the

history of the printed book are S. H. Steinberg, *Five Hundred Years of Printing* (Penguin paperback, 1974), and Lucien Febvre and H. J. Martin, *The Coming of the Book: The Impact of Printing 1450–1800* (1976). A remarkably copious and comprehensive survey of the literature is Elizabeth L. Eisenstein, *The Printing Press as an Agent of Change . . . in Early-Modern Europe* (2 vols., 1979). For terminology and general reference, see Geoffrey A. Glaister, *An Encyclopedia of the Book* (2d ed., 1980), illustrated. Studies of the differential effects of printing in different fields of knowledge are William M. Ivins, Jr., *Prints and Visual Communication* (1973), for biology and the arts; and Stillman Drake, "Early Science and the Printed Book: The Spread of Science beyond the Universities," *Renaissance and Reformation,* Vol. 6 (1970), pp. 43–52. A lively survey of the extent and intensity of the impact of the book is John Carter and Percy H. Muir (eds.), *Printing and the Mind of Man: The Impact of Printing on Five Centuries of Western Civilization* (1967), with succinct essays on individual cataclysmic books, how they came to be published, and whom they reached.

There is no better way to glimpse the kind of effect the printed book did (and did not) have on rural life in Europe than in Natalie Z. Davis' *Society and Culture in Early Modern France* (1975), a model of scholarly eloquence and imagination. Other works on the earliest age of the printed book that I have found especially useful include: Curt F. Bühler, *The Fifteenth-Century Book: The Scribes, The Printers, The Decorators* (1960); Henry J. Chaytor, *From Script to Print . . . Medieval Vernacular Literature* (1974); E. P. Goldschmidt, *Medieval Texts and Their First Appearance in Print* (1943); Sandra Hindman (ed.), *The Early Illustrated Book: Essays in Honor of Lessing J. Rosenwald* (1982), and, with James D. Farquhar, *Pen to Press: Illustrated Manuscripts and Printed Books in the First Century of Printing* (1977); Rudolf Hirsch, *Printing,*

Selling and Reading, 1450–1550 (2d ed., 1974); Henri-Jean Martin, *Le Livre et la Civilisation Ecrite* (1970); Oliver H. Prior (ed.), *Caxton's Mirrour of the World* (1913). For varied sidelights on the later history of printing: Hellmut Lehmann-Haupt, *The Book in America* (1952); F. H. Muir, *Book-Collecting as a Hobby* (1947); Frank A. Mumby, *Publishing and Bookselling . . . from the Earliest Times to the Present Day* (1954); Noel Perrin, *Dr. Bowdler's Legacy, a History of Expurgated Books* (1969); Alfred W. Pollard, *Shakespeare's Fight with the Pirates and the Problems of the Transmission of His Text* (1974); Anthony Smith, *Goodbye Gutenberg, the Newspaper Revolution of the 1980's* (1980); S. H. Steinberg, "Book Production and Distribution," in *Literature and Western Civilization,* Vol. 5 (1972), pp. 509–28; Herbert S. Bailey, Jr., *The Traditional Book in the Electronic Age* (Bowker Lecture, 1978).

The best introduction to the history of paper and papermaking is Dard Hunter, *Papermaking: The History and Technique of an Ancient Craft* (1947). And see: Library of Congress, *Papermaking, Art and Craft* (1968); John Grand-Carteret, *Papeterie et Papetiers de l'Ancien Temps* (1915); Kiyofusa Narita, *Life of Ts'ai Luing and Japanese Paper-Making* (1966).

For a history of the techniques of printing, a good beginning is Warren Chappell, *A Short History of the Printed Word* (1970). And see: Colin Clair, *A History of European Printing* (1976); James Moran, *Printing Presses . . . from the Fifteenth Century to Modern Times* (1978); R. A. Peddie (ed.), *Printing, a Short History of the Art* (1927); Ralph W. Polk, *The Practice of Printing* (1952). An excellent introduction to the history of typefaces and type design is Daniel B. Updike, *Printing Types: Their History, Forms, and Use* (2 vols., 1922); and for an insight into the career of an eminent modern type designer, see Peter Beilenson, *The Story of Frederic W. Goudy* (1965), and Frederic W. Goudy, *Typologia, Studies in Type Design and Type Making* (1977). On the history of printing in eastern Europe, see: Wynar Lubomyr, *History of Early Ukrainian Printing* (1962); George D. Painter and Dalibor B. Chrastek, *Printing in Czechoslovakia in the Fifteenth Century* (1969); Eugene V. Prostov, *Origins of Russian Printing* (1931). Useful articles on the history of books and printing in all parts of the world can be found in the *Quarterly Journal* of the Library of Congress.

For the origins of printing in China, the standard work in English is Thomas F. Carter's scholarly and cogent *Invention of Printing in China and Its Spread Westward* (2d ed., 1955), to which I am much indebted. Other useful works include: C. R. Boxer, *The Christian Century in Japan, 1549–1650* (1951); David Chibbett, *The History of Japanese Printing and Book Illustration* (1977), rich in factual detail and copiously illustrated; Kim Won-Yong, *Early Movable Type in Korea* (1954); Donald Keene, *The Japanese Discovery of Europe, 1720–1830* (1969); Noel Perrin, *Giving up the Gun: Japan's Reversion to the Sword, 1543–1879*, a suggestive cameo of a rare example of "uninvention" or the abandonment of a technological advance, to which Japan's abandonment of movable type in the seventeenth century is an analogue. For the "duplicating impulse" elsewhere and the effect of forms of writing on it, see J. Eric S. Thompson, *The Rise and Fall of Maya Civilization* (1954).

The bibliography on Gutenberg is enormous, but not as rich as we would wish in facts about the man and his career. Much of the writing is speculative. A good beginning is Victor Scholderer, *Johann Gutenberg: The Inventor of Printing* (1970), along with Douglas McMurtrie (ed.), *The Gutenberg Documents* (1941) and *The Invention of Printing: A Bibliography* (1942). The standard work is Aloys Ruppel, *Johannes Gutenberg: Sein Leben und Sein Werk* (3d ed., 1967). For an in-triguing example of the detective problems involved in the earliest history of printing, see Hellmut Lehmann-Haupt, *Gutenberg and the Master of the Playing Cards* (1966), which suggests a close connection between copperplate engraving and the origins of movable type. And for some brilliant latter-day insights into the impact of printing, see Marshall McLuhan, *The Gutenberg Galaxy* (1962).

For the rise of vernacular languages and literature, a readable beginning is Mario Pei, *The Story of Language* (rev. ed., 1966). And see: H. Munro Chadwick and N. Kershaw Chadwick, *The Growth of Literature* (3 vols., 1932–40); David Daiches and Anthony Thorlby (eds.), *Literature and Western Civilization* (6 vols., 1972–76); F. O. Matthiessen, *Translation: An Elizabethan Art* (1965); Edward Sapir, *Culture, Language and Personality* (1956); the ever-stimulating George Sarton, *Ancient Science and Modern Civilization* (1954) and *The Appreciation of Ancient and Medieval Science During the Renaissance 1450–1600* (1955). For French, see: Claude Fauchet, *Recueil de l'Origine de la Langue et Poésie Française* (1938); Joachim Du Bellay, *The Defense and Illustration of the French Language* (1939). For the rise of a popular German literature and a world literature, the Brothers Grimm are especially interesting: Murray B. Peppard, *Paths through the Forest: A Biography of the Brothers Grimm* (1971), and Ruth Michaelis-Jena, *The Brothers Grimm* (1970).

To see some of the effects of the rise of printing on education and classroom methods, begin with a glimpse of the Old Style sketched by *Quintilian on Education*, William M. Smail (ed. and trans., 1938), then on to the career of the too little known John Amos Comenius (1592–1670), pioneer of the illustrated book in the classroom: Will S. Monroe, *Comenius and the Beginnings of Educational Reform* (1971); John E. Sadler, *J. A. Comenius and the Concept of Universal Education* (1966); Matthew Spinka,

John Amos Comenius, That Incomparable Moravian (1943); G. H. Turnbull, *Hartlib, Dury and Comenius* (1947).

For Aldus Manutius, see Martin Lowry, *The World of Aldus Manutius: Business and Scholarship in Renaissance Venice* (1979). And, on indexing, the pioneer guide by Henry B. Wheatley, *What is an Index?* (1878).

A focused and fascinating introduction to the new world of authors, printers, publishers, booksellers, and book buyers opened by the printed book is Robert Darnton's *The Business of Enlightenment: A Publishing History of the Encyclopédie, 1775–1800* (1979) and *The Literary Underground of the Old Regime* (1982), the background for the French Revolution in the literary demimonde. Other varied works rich in detail of this new world: Richard D. Altick, *The English Common Reader: A Social History of the Mass Reading Public, 1800–1900* (1957); R. R. Bowker Company, *Bowker Lectures on Book Publishing* (1957–); Asa Briggs (ed.), *Essays in the History of Publishing* (1974); J. Lough (ed.), *Diderot and D'Alembert, The Encyclopédie* (1954); Robert Escarpit, *The Book Revolution* (1966) and *The Sociology of Literature* (1971); Jack Goody (ed.), *Literacy in Traditional Societies* (1968); John J. Gross, *The Rise and Fall of the Man of Letters* (1969); Mitford M. Mathews, *Teaching to Read, Historically Considered* (1966); Edward Miller, *Prince of Librarians: The Life and Times of Antonio Panizzi* (1967); Jesse H. Shera, *Libraries and the Organization of Knowledge* (1965) and *Foundations of the Public Library* (1965); Siegfried Unseld, *The Author and His Publisher* (1980); *ALA World Encyclopedia of Library and Information Services* (1980).

On the cultural and historical context of the Muslim attitude toward printing, see: Anwar G. Chejne, *The Arabic Language: Its Role in History* (1969); Gustave E. von Grunebaum, *Medieval Islam* (2d ed., 1953), a subtle and elegant introduction; Ibn Hisham Abd al Malak, *The Life of Muhammad* (1955); H. A. R. Gibb, *Arabic Literature* (1963) and his concise *Mohammedanism* (1953); Philip K. Hitti, *Islam, a Way of Life* (1971); Marshall G. S. Hodgson, *The Venture of Islam* (3 vols., 1974); Reuben Levy, *An Introduction to the Sociology of Islam* (2 vols., 1930); Bernard Lewis, *The Muslim Discovery of Europe* (1982); Katharina Otto-Dorn, *L'Art de l'Islam* (1967); F. E. Peters, *Allah's Commonwealth: A History of Islam in the Near East, 600–1100 A.D.* (1974); Maxime Rodinson, *Muhammad* (1980). The Koran is available in numerous English reprints, for example, Mohammed M. Pickthall (trans.), *The Meaning of the Glorious Koran* (Mentor paperback; 1953), and A. Yusuf Ali (trans.), *The Holy Qur'an* (Islamic Center, Washington, D.C., 1978), and see: W. Montgomery Watt, *Bell's Introduction to the Qur'an* (1970) and *Muhammad: Prophet and Statesman* (1964). For some light on the story in Turkey and Egypt, see: Niyazi Berkes, *The Development of Secularism in Turkey* (1964); J. Christopher Herold, *Bonaparte in Egypt* (1962).

From the vast literature on dictionaries and lexicography I suggest that the reader begin with Mitford M. Mathews' brief *Survey of English Dictionaries* (1966). Of the numerous introductions to Dr. Samuel Johnson as master lexicographer, surprisingly, Boswell is less adequate than others, for example, John Wain, *Samuel Johnson,* or James H. Sledd and Gwin J. Kolb, *Dr. Johnson's Dictionary: Essays in the Biography of a Book* (1955). The saga of the *Oxford English Dictionary* is also a suspense story. See William A. Craigie, "Historical Introduction," found in Vol. 1 of the 1933 reissue, and especially K. M. Elisabeth Murray, *Caught in the Web of Words: James A. H. Murray and the Oxford English Dictionary* (1977), a biography of unexcelled wit and charm by Murray's granddaughter.

Part XIV: Opening the Past

A concise and eloquent introduction to the cyclical interpretation of experience is *The Myth of the Eternal Return* (1954), by Mircea Eliade, who helps us see the pervasive significance of religious ideas also in his *Patterns in Comparative Religion* (1958), *From Primitives to Zen: A Thematic Sourcebook of the History of Religions* (1967), and *A History of Religions* (1978).

For a general view of non-Western attitudes to the past, see: Roland H. Bainton et al., *The Idea of History in the Ancient Near East* (1966); John K. Fairbank and Edwin O. Reischauer, *East Asia, The Great Tradition* (1960) and, with Albert M. Craig, *East Asia, The Modern Transformation* (1965); Nancy Wilson Ross, *Three Ways of Ancient Wisdom* (1966). Detailed treatment of individual historians is found in Publications of the London University School of Oriental and African Studies, *Historical Writings on the Peoples of Asia* (3 vols., 1961–62): Vol. 1, C. H. Philips (ed.), *India, Pakistan and Ceylon;* Vol. 2, D. G. E. Hall (ed.), *South-East Asia;* Vol. 3, W. G. Beasley and E. G. Pulleyblank (eds.), *China and Japan;* Vol. 4, Bernard Lewis and P. M. Hold (eds.), *The Middle East.*

Diana L. Eck's admirable *Banaras: City of Light* (1982) reveals Hindu views of the past in the colorful context of Indian geography and daily life. See also A. I. Basham's lively and well-illustrated *The Wonder That Was India* (Penguin paperback, 1954). Texts for Indian attitudes toward the past can be found in the excellent selection by W. Theodore de Bary et al., *Sources of Indian Tradition* (in the series, de Bary [ed.], Introduction to Oriental Civilizations, 1958). And see: Sunuti K. Chatterji, *Languages and Literature of Modern India* (1963); William H. McNeill and Jean W. Sedlar, *Classical India* (1969); Edward Sachau (trans.), *Alberuni's India* (1964); *The Literature of India: an Introduction* (University of Chicago Press, 1974).

A good introduction to the Buddhist view is Edward Conze, *Buddhism: Its Essence and Development* (2d ed., 1953), and Conze (trans.), *Buddhist Scriptures* (Penguin paperback, 1973), or Christmas Humphrey's concise *Buddhism* (Penguin paperback, 1955). See also: H. Fielding-Hall, *The Soul of a People* (1903), on the distinctiveness of Burmese Buddhism; Maurice Percheron, *Buddha and Buddhism* (1957); Melford E. Spiro, *Buddhism and Society: A Great Tradition and Its Burmese Vicissitudes* (1970); William Geiger (trans.), *The Mahavamsa, or the Great Chronicle of Ceylon* (1912).

In China the culture as a whole is obviously and intimately related to distinctive attitudes toward the past. Some of the best scholars have given us readable and enticing essays on this subject, for example: Arthur Waley, *Three Ways of Thought in Ancient China* (1956) and *The Way and Its Power: . . . the Tao Te Ching and its place in Chinese Thought* (1934); Herrlee G. Creel, *Confucius and the Chinese Way* (1960) and *What Is Taoism?* (1970). In more detail, see: Mark Elvin, *The Pattern of the Chinese Past* (1973); Joseph R. Levenson, *Confucian China and Its Modern Fate* (1968). A useful collection of texts is W. T. de Bary et al., *Sources of Chinese Tradition* (2 vols., 1964). The writings of Confucius are available in numerous English translations and reprints, for example: Arthur Waley (trans.), *The Analects of Confucius* (n.d.); Lin Yutang (trans.), *The Wisdom of Confucius* (Modern Library, 1938). More specifically on Chinese historical writing, see: Charles S. Gardner, *Chinese Traditional Historiography* (1938); Ku Chieh-kang, *The Autobiography of a Chinese Historian . . . preface to a symposium on ancient Chinese History* (1931); Burton Watson, *Ssu-Ma Ch'ien, Grand Historian of China* (1958). On the relation of Chinese historical attitudes to politics, see: Charles O.

Hucker, *The Traditional Chinese State in Ming Times, 1368–1644* (1961); Arthur W. Hummel, "What Chinese Historians are Doing in Their Own History," *American Historical Review,* Vol. 34 (1929), pp. 715–24; Jonathan D. Spence, *The Gate of Heavenly Peace: The Chinese and their Revolution, 1895–1980* (1981).

For the Muslim attitudes toward the past a comprehensive introduction is Franz Rosenthal, *A History of Muslim Historiography* (1968). On Ibn Khaldun, see: Ibn Khaldun, *The Muqaddimah: An Introduction to History* (Franz Rosenthal, trans., Bollingen Series, 3 vols., 1958; and in one volume, abridged and edited, in Princeton University Press paperback, by N. J. Dawood, 1969); Muhsin Mahdi, *Ibn Khaldun's Philosophy of History* (1957); Walter J. Fischel, *Ibn Khaldun in Egypt: His Public Functions and His Historical Research, 1382–1406* (1967).

Standard surveys of the rise of historiography in the West are: James Westfall Thompson, *A History of Historical Writing* (2 vols., 1942); C. V. Langlois and C. Seignobos, *Introduction to the Study of History* (1898); James T. Shotwell, *The History of History* (1939). J. B. Bury, the editor of Gibbon's *Decline and Fall* and a brilliant essayist, piquantly introduces us to some implications of the rise of historical writing in his *Selected Essays* (1930) and *The Idea of Progress* (1932), as does Carl Becker in his *Heavenly City of the Eighteenth-Century Philosophers* (1932) and *Everyman His Own Historian* (1935). A lively dissent from Bury is Robert Nisbet's *History of the Idea of Progress* (1980). For a philosopher's perspective, see R. G. Collingwood, *The Idea of History* (1961), which explores the thesis that the modern view of history requires the historian to be in pursuit of "human self-knowledge" and to be concerned with "questions of whose answer the writer begins by being ignorant." For the enticing paths from history to all social philosophy, see Frank E. and Fritzie P. Manuel, *Utopian*

Thought in the Western World (1979), especially Chapter 18, "Freedom from the Wheel."

The texts of Herodotus and Thucydides are available in many modern translations and reprints, for example, in *Great Books of the Western World,* Vol. 6. An admirable introduction is M. I. Finley's *Greek Historians: The Essence of Herodotus, Thucydides, Xenophon, Polybius* (1959), with C. M. Bowra, *The Greek Experience* (1958), Chapter 9. To glimpse the subtleties of interpretation, see Francis M. Cornford's luminous *Thucydides Mythistoricus* (1907; 1971).

Saint Augustine's *Confessions* and his *City of God* are both available in numerous English translations and reprints, notably in *Great Books of the Western World,* Vol. 18. For the place of history and historical thought in Christianity, see Jaroslav Pelikan's grand and delightfully readable *Growth of the Christian Tradition* (5 vols., 1971–83). For an assist in grasping Saint Augustine, see: John N. Figgis, *The Political Aspects of St. Augustine's City of God* (1963); Eugène Portalié, S.J., *A Guide to the Thought of Saint Augustine* (1960); Edward A. Synan, "Augustine of Hippo," and "Augustinism," in *Dictionary of the Middle Ages,* Vol. 1 (1982). On the rise of a historical sense in the West, I have found especially helpful: George Boas, *Essays on Primitivism and Related Ideas in the Middle Ages* (1966), *The Happy Beast: In French Thought of the Seventeenth Century* (1966), and with A. O. Lovejoy, *Primitivism and Related Ideas in Antiquity* (1965); Marc Bloch, *The Historian's Craft* (1963); Jacob Burckhardt, *The Civilization of the Renaissance in Italy* (1944), *The Age of Constantine the Great* (1949); Norman Cohn, *The Pursuit of the Millennium* (1961); Alexander Heidel, *The Babylonian Genesis: The Story of the Creation* (1951); J. Huizinga, *The Waning of the Middle Ages* (1948) and *Dutch Civilisation in the Seventeenth Century and Other Essays* (1968), especially "Two Wrestlers with the Angel," pp. 158–218, on Oswald Spengler and H. G. Wells, and "My Path to History," pp. 244–76; Kenneth S. Latourette, *A History of the Expansion of Christianity* (7 vols., 1937–70); Beryl Smalley, *English Friars and Antiquity in the Early Fourteenth Century* (1960); Lois Whitney, *Primitivism and the Idea of Progress* (1934).

For a pungent taste of the modern spirit in historical writing, the reader should browse in Voltaire, *The Age of Louis XIV* (1935), often reprinted, and Edward Gibbon's *Decline and Fall of the Roman Empire*, in numerous reprints, notably in *Great Books of the Western World*, Vols. 40 and 41. For recent critiques of Gibbon, see the symposium in *Daedalus*, Vol. 105 (1976).

The place of the Renaissance in the rise of the historical sense and historical criticism is explored in two admirably brief essays by Peter Burke, *The Renaissance Sense of the Past* (1969) and *Tradition and Innovation in Renaissance Italy* (1974), and in an eloquent wide-ranging study by Ricardo J. Quinones, *The Renaissance Discovery of Time* (1972), to all of which I am much indebted. See also: Bernard Berenson, *The Italian Painters of the Renaissance* (1932); Ernst Cassirer et al., *The Renaissance Philosophy of Man* (1948); Eric Cochrane, *Historians and Historiography in the Italian Renaissance* (1981); Wallace K. Ferguson, *The Renaissance in Historical Thought: Five Interpretations* (1948); Felix Gilbert, *Machiavelli and Guicciardini: Politics and History in Sixteenth-Century Florence* (1965); Denys Hay, "Flavio Biondo and the Middle Ages," *Proc. British Academy*, Vol. 45 (1959), pp. 97–128; Erwin Panofsky, *Renaissance and Renaissances in Western Art* (1970); Orest Ranum (ed.), *National Consciousness, History and Political Culture in Early Modern Europe* (1975); J. H. Whitfield, *Petrarch and the Renaissance* (1943). Benvenuto Cellini's *Autobiography* illuminates the whole scene and can be read in one of the many modern reprints.

To all the other mysteries of history the practice of archaeology adds the suspense of the treasure hunt that can be sensed in a wealth of scholarly popularizations. C. W. Ceram's *Gods, Graves and Scholars* (1952) has had a well-deserved international popularity and is a place to begin. See also: Geoffrey Bibby, *The Testimony of the Spade* (1962); Glyn E. Daniel, *A Hundred Years of Archaeology* (1950) and *The Origins and Growth of Archaeology* (1971); Leonard Woolley, *History Unearthed* (1963) and *Digging up the Past* (1954). On the destruction and rescue of Roman monuments, see two illuminating fact-packed essays: Rodolfo Lanciani, *The Destruction of Ancient Rome: A Sketch of the History of the Monuments* (1967), and Roberto Weiss, *The Renaissance Discovery of Classical Antiquity* (1969).

On Winckelmann, begin by tasting his fluent and enthusiastic *History of Ancient Art* (Alexander Gode, trans., 2 vols., 1969). And for the commentary: Irving Babbitt, *The New Laokoon, An Essay on the Confusion of the Arts* (1934); Hedwig Weilguny, *Winckelmann und Goethe* (1968); Johann Wolfgang Goethe, *Winckelmann und sein Jahrhundert* (1969). The best approach to Schliemann is through his own writings, *Troy and Its Remains, Narrative of Researches and Discoveries* (1875) and *Ilios: The City and Country of the Trojans, the Results of Researches and Discoveries* (1880), both rich in warm autobiographical detail. For romance, the facts of Schliemann's life need no embroidering, but Schliemann himself did not hesitate to embellish the facts. Resist Emil Ludwig's sensational *Schliemann of Troy: The Story of a Gold-Seeker* and turn to the more reliable and still moving *One Passion, Two Loves: The Story of Heinrich and Sophia Schliemann* (1966), by Lynn and Gray Poole. See also: E. M. Butler, *The Tyranny of Greece over Germany* (1935), a study of the influence of Greek art and poetry over German literature; John Myres, *The Cretan Labyrinth: A retrospect of Aegean Research* (1933). Joseph

Alsop's *The Rare Art Traditions* (1982) offers a stirring original study of the worldwide relation of art collecting to the discovery of art history, the rise of museums, the growth of art forgery, and endless else.

For the dawning of the idea of prehistory, begin with Glyn Daniel, *The Idea of Prehistory* (Penguin paperback, 1971). And see: Grahame Clark, *Aspects of Prehistory* (1970) and *Archaeology and Society* (1965), which awaken us to some surprising implications of an arcane subject; Colin Renfrew, *Before Civilization: The Radiocarbon Revolution and Prehistoric Europe* (1973).

The idea of latitudes of time can be glimpsed in the works mentioned in the Reference Notes, above, for Parts I–III and Part XII. For Scaliger, see his *Autobiography* (George W. Robinson, trans., 1927) and M. Charles Nisard, *Juste Lipse, Joseph Scaliger et Isaac Casaubon* (1899). And for Newton, see especially Frank Manuel, *Isaac Newton, Historian* (1963). See also: G. S. P. Freeman-Grenville, *The Muslim and Christian Calendars . . . tables for the conversion of . . . dates* (1963); John Stuart Mill, *The Spirit of the Age* (1942); Jerome H. Buckley, *The Triumph of Time . . . the Victorian Concepts of Time, History, Progress, and Decadence* (1966).

Vico, a neglected and underestimated figure, offered prophetic insights into the discoveries of anthropology and the worldwide roles of myth and technology, which can be sampled in Thomas G. Bergin and Max Harold Fisch (trans.), *The New Science of Giambattista Vico* (Anchor paperback, 1961). See the same translators' edition of the unabridged *New Science* (1948) and Vico's *Autobiography* (1944). Hayden V. White gives us a good short sketch of Vico in the *International Encyclopedia of the Social Sciences*, Vol. 16, pp. 313–16.

The literature on both Marx and Freud is enormous. Here I mention only a few of the works I have found most helpful for the limited aspect of their ideas that I discuss in my chapters. An interesting discussion of the interrelation of their ideas is found in Stanley Edgar Hyman, *The Tangled Bank: Darwin, Marx, Frazer and Freud as Imaginative Writers* (1962), and in Henri F. Ellenberger, *The Discovery of the Unconscious: The History and Evolution of Dynamic Psychiatry* (1970), especially at pp. 237 ff., 629 ff., and passim.

For Marx's life I have especially profited from Saul K. Padover's balanced and readable *Karl Marx: An Intimate Biography* (1978). The writings of Marx are available in numerous reprints, including *Great Books of the Western World,* Vol. 50. The other principal biography in English is Franz Mehring's defensive *Karl Marx: The Story of His Life* (1936). A convenient selection of the writings of Marx and Engels, including some journalistic and fugitive pieces is Emile Burns (ed.), *A Handbook of Marxism* (1935). The article by Robert S. Cohen, in the *Dictionary of Scientific Biography,* Vol. 15: Supplement I, at pp. 403–17 is interesting for a number of reasons, not the least of which is that it appears that this lone social "scientist" (together with Friedrich Engels) had to be added in a supplement in order to make a *Dictionary of Scientific Biography* acceptable and available for translation in the Soviet Union. It is notable that Western governments have exerted no pressure for the inclusion of Moses or Jesus! For a perspective on Marx's view of history, it is worth reading Georg W. F. Hegel, *Lectures on History,* and Karl R. Popper's brilliant essays attacking historical determinism, *The Poverty of Historicism* (1957) and *The Open Society and Its Enemies* (2 vols. 1971), which climaxes in the case against Marx and his followers.

On Freud, the best beginning is still Ernest Jones' authorized *Life and Work of Sigmund Freud* (3 vols., 1953–55; abridged in the one-volume Anchor paperback, 1963). Bruno Bettelheim's humane and delightful *Freud and Man's Soul* (1983) should come next. I have found the copious writings of Paul Roazen helpful beyond measure—especially his *Freud and His Followers* (1976) and *Freud, Political and Social Thought* (1968). See also: Jacques Barzun, *Clio and the Doctors: Psycho-History, Quanto-History and History* (1974); Ronald W. Clark, *Freud: The Man and the Cause* (1980), copious on the social context; O. Mannoni, *Freud* (1971), a lively short narrative; Frank J. Sulloway, *Freud, Biologist of the Mind* (1979). Freud's writings have been often reprinted, for example, in A. A. Brill (ed. and trans.), *The Basic Writings of Sigmund Freud* (Modern Library Giant, 1938), and in *Great Books of the Western World*, Vol. 54 (the final volume!).

Part XV: Surveying the Present

For the wider background of Las Casas' thought, the vigorous and cogent works of Lewis Hanke are basic, especially *The Spanish Struggle for Justice in the Conquest of America* (1965) and *All Mankind Is One . . . the disputation between Bartholomé de las Casas and Juan Ginés de Sepúlveda* (1974); and see his *Do the Americas Have a Common History? A Critique of the Bolton Theory* (1964). A Library of Congress publication makes some sources available in a handsome, beautifully edited facsimile: Helen Rand Parish (ed.), *Las Casas as a Bishop* (1980). For pre-Columbian anthropological thought and its later currents, see: John B. Friedman, *The Monstrous Races in Medieval Art and Thought* (1981) and Margaret T. Hodgen, *Early Anthropology in the Sixteenth and Seventeenth Centuries* (1964). And for modern tendencies: Stanley Diamond (ed.), *Culture in History* (1960); Theodosius Dobzhansky, *Mankind Evolving* (1970); Pierre L. van den Berghe, *Race and Racism* (1967) and his excellent article, "Racism," in the *Encyclopaedia Britannica* (15th ed.), Vol. 15, pp. 360–66; Timothy Raison (ed.), *The Founding Fathers of Social Science* (1969).

Morgan's life is summarized in two brief biographies: Carl Resek, *Lewis Henry Morgan, American Scholar* (1960), and Bernhard J. Stern, *Lewis Henry Morgan, Social Evolutionist* (1931). His *Ancient Society* is still readable and thought-provoking (1877; Belknap Press reprint, Leslie A. White, ed., 1964). To follow the influence of Morgan, read Friedrich Engels' brief *Origin of the Family, Private Property, and the State* (1902), often reprinted.

A good introduction to Tylor is R. R. Marett, *Tylor* (Modern Sociologists series, 1936). Tylor's works, like Morgan's, are still eminently readable and suggestive, for example, his *Primitive Culture* (2 vols., 1871; reprinted 1929) and *Anthropology: An Introduction to the Study of Man and Civilization* (1896; abridged with an Introduction by Leslie A. White, 1960). And see: Margaret T. Hodgen, *The Doctrine of Survivals* (1936); and articles by George W. Stocking, Jr., "Matthew Arnold, E. B. Tylor and the Uses of Invention," *American Anthropologist*, Vol. 65 (1963), pp. 783–99 and "Franz Boas and the Culture Concept . . . ," Vol. 68 (1966), pp. 867–82.

Robert L. Heilbroner's *Worldly Philosophers* (5th ed., 1980) is a masterpiece of popularization, challenging all other historians of ideas to match this combination of clarity, vividness, and scholarship. The comprehensive work is Joseph A. Schumpeter's monumental *History of Economic Analysis* (1954), edited from manuscript by Elizabeth S. Schumpeter, rich in detail and subtle and charitable in judgments. For deeper background, see Eli Heckscher's definitive *Mercantilism* (2 vols., 1933). Other readable works include: Robert Lekachman, *A History of Economic Ideas*; Erich Roll, *A History of Economic Thought* (rev. ed., 1954). The standard biography of Adam Smith is John Rae, *Life of Adam Smith* (1895; reprinted with a valuable introduction by Jacob Viner, 1965). See also: Otto Mayr, "Adam Smith and the Concept of the Feedback System . . . ," *Technology and Culture*, Vol. 12 (1971), to which I am much indebted; William R. Scott, *Adam Smith as Student and Professor* (1937). Adam Smith's writings have been often reprinted, for example, in the Modern Library (Edwin Cannan, ed., 1937), and in *Great Books of the Western World,* Vol. 39.

For John Maynard Keynes, begin with Robert Lekachman's eminently readable *Age of Keynes* (1969) or the standard biography by R. F. Harrod (1951). Keynes himself writes with some of the flair of Adam Smith and with even more wit. Those who may find it difficult to follow his *General Theory of Employment, Interest and Money* (1936) or his *Treatise on Money* (2 vols., 1930), will enjoy *The Economic Consequences of the Peace* (1920), *Essays in Persuasion* (1931), and *Essays in Biography* (1933).

For the discovery of demography and the quantitative dimensions of society, see: James Bonar, *Theories of Population from Raleigh to Arthur Young* (1931); F. N. David, *Games, Gods and Gambling* (1962); Ian Hacking, *The Emergence of Probability . . . early ideas about Probability, Induction, and Statistical Inference* (1975); John Koren, *The History of Statistics* (1918); American Economic Association, *The Federal Census* (1899); Carroll D. Wright, *The History and Growth of the United States Census* (1900). John Graunt's *Natural and Political Observations . . . upon the Bills of Mortality* has been reprinted by Arno Press (1975). See: Robert Kargon, "John Graunt, Francis Bacon, and the Royal Society: The Reception of Statistics," *Journal of the History of Medicine and Allied Sciences,* Vol. 18 (1963), pp. 337-48; Charles F. Mullett, *The Bubonic Plague and England* (1956). For Quetelet, see: Frank H. Hankins, *Adolphe Quetelet as Statistician* (1968), and Quetelet's *Sur l'Homme et le Développement de ses Facultés ou Essai de Physique Sociale* (1836), translated as *A Treatise on Man and the Development of his Faculties* (reprint, 1968).

From the vast literature on the history of physics and chemistry I was first stimulated to an interest in this subject by Edwin A. Burtt's *Metaphysical Foundations of Modern Physical Science* (1927), still useful and suggestive, and by the writings of Alfred North Whitehead (see above, Part XI). I have found Lancelot Law Whyte, *Essay on Atomism: From Democritus to 1960* (1961) an indispensable outline and guide into the subject. For a witty introduction to relativity, especially designed for the layman, see George Gamow, *Mr. Tompkins in Paperback* (1967), which we owe to the editorial imagination of C. P. Snow. Other works especially useful include: Andrew G. Van Melsen, *From Atomos to Atom, the History of the Concept Atom* (1952); Selig Hecht, *Explaining the Atom* (1954); Leonard K. Nash, *The Atomic-Molecular Theory* (Case 4, Harvard Case Histories in Experimental Science, 1950); Banesh Hoffman, *The Strange Story of the Quantum* (1959); Arnold Thackray, *Atoms and Powers . . . Newtonian Matter-Theory and the Development of Chemistry* (1970). Lucretius' "On the Nature of Things" *(De rerum natura)* is reprinted in *Great Books of the Western World,* Vol. 12; a selection of Lavoisier's works is in Vol. 45.

For the wider background I have found the writings of Gerald Holton especially stimulating because of the sharp focus of his examples and his constant tie to the grand themes: *Thematic Origins of Scientific Thought: Kepler to Einstein* (1973); *The Scientific Imagination: Case Studies* (1978), and, with others, the *Harvard Project Physics Readers* (1975), and *Albert Einstein, Historical and Cultural Perspectives* (1982). The lay reader should share my pleasure in C. P. Snow, *The Physicists* (1981), which has much of the suspense of a novel, and Heinz R. Pagels' tantalizing *The Cosmic Code, a Quantum Physics as the Language of Nature* (1982).

On Dalton's life and work there is an array of readable works: Frank Greenaway, *John Dalton and the Atom* (1966); Elizabeth C. Patterson, *John Dalton and* the *Atomic Theory* (1970); Henry E. Roscoe, *John Dalton and the Rise of Modern Chemistry* (1895); Henry E. Roscoe and Arthur Harden, *A New View of the Origin of Dalton's Atomic Theory* (with an Introduction by Arnold Thackray, 1970); C. S. L. Cardwell (ed.), *John Dalton and the Progress of Science* (1968).

The standard biography is L. Pearce Williams' copious *Michael Faraday* (1964), a delight to read. And see: John Tyndall, *Faraday as a Discoverer* (1961). A good selection of Faraday's writings along with some of Lavoisier's is in *Great Books of the Western World,* Vol. 45. Other biographies of special interest for the history of chemistry and physics include: Lewis Campbell and William Garnett, *The Life of James Clerk Maxwell* (1882), with a selection from his writings; Dorothy Michelson Livingston, *The Master of Light: a biography of Albert A. Michelson* (1973); Robert J. S. Rayleigh, *The Life of Sir J. J. Thomson* (1942); Robert Reid, *Marie Curie* (1974); George P. Thomson, *S. J. Thomson and the Cavendish Laboratory in his Day* (1964); J. J. Thomson, *Recollections and Reflections* (1936).

For an introduction to Einstein, I have enjoyed: Jeremy Bernstein, *Ein-stein* (1973); Ronald W. Clark, *Einstein: The Life and Times* (1971); Albert Einstein and Leopold Infeld, *The Evolution of Physics from Early Concepts to Relativity and Quanta* (1938). See Harry Woolf (ed.), *Some Strangeness in the Proportion* (1980), a centennial symposium on Einstein, and for a readable recent assessment of Einstein's impact beyond physics, Gerald Holton and Yehuda Elkana (eds.), *Albert Einstein: Historical and Cultural Perspectives* (1982), the centennial symposium in Jerusalem.

Some works on the wider context that I find interesting include: Ginestra Amaldi, *The Nature of Matter: Physical Theory from Thales to Fermi* (1966); Max Born, *The Restless Universe* (1951); Fritjof Capra, *The Tao of Physics* (1977); Freeman Dyson, *Disturbing the Universe* (1979); A. S. Eddington, *The Nature of the Physical World* (1928); Herbert Friedman, *The Amazing Universe* (1975); Stanley L. Jaki, *The Relevance of Physics* (1966) and *The Road of Science and the Ways to God* (1978); Robert Jastrow, *Red Giants and White Dwarfs* (1967) and *Until the Sun Dies* (1977); Daniel J. Kevles, *The Physicists: The History of a Scientific Community in Modern America* (1978).

ACKNOWLEDGMENTS

My debts for this book go back almost as far as I can remember, at least to my first visit to Florence a half-century ago, and to my first reading of Oswald Spengler and Edward Gibbon. The book has been the pleasure of my private hours over the last fifteen years. Unlike most of my earlier books, it has not been tested on or shared with colleagues or students or research assistants or lecture audiences. However, a number of friends have offered me their insights, have given me suggestions, or have read parts of the manuscript. They have saved me from errors of fact, but have often not shared my interpretations or my emphases. It is a pleasure to thank them. They include: Silvio A. Bedini of the National Museum of American History, The Smithsonian Institution, Washington, D.C.; Simon Michael Bessie of Harper and Row, Publishers; Dr. Charles A. Blitzer, President and Director of the National Humanities Center, Research Triangle Park, North Carolina; Subrahmanyan Chandrasekhar, Morton D. Hull Distinguished Service Professor of Astrophysics, The University of Chicago; Dr. Elizabeth Eisenstein, Alice Freeman Palmer Professor of History, University of Michigan; Dr. Ivan P. Hall of the Japan-American Friendship Commission, Tokyo; Dr. O. B. Hardison, Director of the Folger Shakespeare Library, Washington, D.C.; Dr. Chauncy D. Harris, Samuel N. Harper Distinguished Service Professor of Geography, The University of Chicago; Professor Sandra Hindman, Department of the History of Art, The Johns Hopkins University; Dr. Gerald Holton, Mallinkrodt Professor of Physics and Professor of History of Science, Harvard University; Sol Linowitz of Washington, D.C.; Dr. Edmund S. Morgan, Sterling Professor of American History, Yale University; Dr. Jaroslav Pelikan, Sterling Professor of History and Religious Studies, Yale University; Dr. Edmund D. Pellegrino, John Carroll Professor of Medicine and Medical Humanities, Georgetown University; William Safire of the *New York Times;* Dr. Emily Vermeule, The Zemurray-Stone-Radcliffe Professor, Department of Classics, Harvard University; Dr. Paul E. Walker, Executive Director, American Research Center in Egypt; and my sons, Paul Boorstin, Jonathan

Boorstin, and David Boorstin. I owe the title of this book to Paul Boorstin.

At every stage in preparation of the manuscript, the assistance, scrupulous accuracy, and discrimination of Genevieve Gremillion have been essential. Her warm friendship and her devotion to the enterprise have been my rare good fortune and have contributed immeasurably to this book.

Robert D. Loomis, vice-president and executive editor of Random House, has had an intuitive grasp of my hopes for this book from the beginning. His patience, his critical intelligence, his sense of what this book should (and should not) try to be, his enthusiasm and encouragement, have helped over many years. For me he has become the ideal of how a publishing editor can guide an author.

But this book would not have been possible without the cheerful companionship, intimate collaboration, intellectual stimulus, editorial scrutiny, and poetic vision of my wife, Ruth F. Boorstin. She has been, as always, my principal and most penetrating editor. For this book, which in its writing has been more private than any I have written before, her creative, catalytic and inspiring role has been crucial. To dedicate the book to her is a conspicuous understatement. My debt to her is beyond words. She has once again been the indispensable companion of discovery, and remains for me the most delightful discovery of all.

Grateful acknowledgment is made to the following for permission to reprint previously published material:

Edward Arnold Publishers Ltd: Excerpts from the series "Documents of Modern History," *The Renaissance Sense of the Past*, by Peter Burke. Edward Arnold, London, 1969.

Basic Books, Inc.: Excerpts from *Life and Work of Sigmund Freud*, by Ernest Jones. Originally published by Doubleday-Anchor Books, Copyright © 1963 by Ernest Jones.

Blackwell Scientific Publications Ltd: Excerpts from the *The Circulation of the Blood*, by William Harvey, Kenneth J. Franklin, ed. Everyman's Library Edition, 1963.

Cambridge University Press: Excerpts from three sources. *Heavenly Clockwork; The Great Astronomical Clocks of Medieval China*, Joseph Needham, Cambridge, 1960. *John Ray, His Life and Works*, second edition, quoted and translated by C. E. Raven, Cambridge, 1950. *Science and Civilization in China*, volume III, by Joseph Needham, Cambridge, 1959.

Cornell University Press: Excerpts from *Marcello Malpighi and the Evolution of Embryology*, 5 volumes, Cornell University Press, 1966.

Dodd, Mead & Company, Inc.: Excerpt from *My Life*, by Alfred Russel Wallace, 2 volumes, Dodd, Mead & Company.

Hakluyt Society: Excerpt from *Cathay and the Way Thither*, Henry Yule, ed., revised by Henri Cordier. Copyright © The Hakluyt Society, London.

G. K. Hall & Co.: Three excerpts from Buffon, *Histoire Naturelle*, translated by Fellows and Milliken, Copyright © 1972 by Twayne Publishers, Inc., and reprinted with the permission of Twayne Publishers, a division of G. K. Hall & Co., Boston.

Lewis Hanke: Excerpts from *The Spanish Struggle for Justice in the Conquest of America*, by Lewis Hanke. Little Brown, 1965, pp. 17, 21, 80, 121, 123, 129, 131.

Harcourt Brace Jovanovich, Inc., and Macmillan and Company Ltd, London: Excerpts from *Economic Consequences of the Peace*, by John Maynard Keynes. Harcourt, Brace & Howe, 1920.

Harcourt Brace Jovanovich, Inc.: Excerpts from *Antony Van Leeuwenhoek and His 'Little Animals,'* by Clifford Dobell, from Translations of the Philosophical Society, Harcourt, Brace, 1932.

Harper & Row: Excerpt from an article translated by Ralph Major, "Santorio Santorio," in *Annals of Medical History*, volume 10, New York, Paul B. Hoeber, publisher, 1938.

Harvard University Press: Excerpts from *The Medieval Mind*, by Henry Osborn Taylor, Fourth Edition.

Hodder & Stoughton Ltd: Excerpts from *The Great Age of Discovery*, Arthur P. Newton, ed., University of London Press, 1932.

Little, Brown & Company: Excerpts from *Christopher Columbus, Mariner*, by Samuel Eliot Morison. Copyright © 1955 by Samuel Eliot Morison. Reprinted by permission of Little, Brown & Company in association with the Atlantic Monthly Press.

Macmillan & Company Ltd, London: Excerpts from *Life of John Maynard Keynes,* by Roy F. Harrod. Macmillan, London, 1951. Used by the permission of Macmillan, London and Basingstoke.

McGraw-Hill: Excerpts from *Karl Marx, An Intimate Biography,* by Saul Padover, McGraw-Hill, 1978.

The Medici Society Limited: Excerpts from *The Confessions of St. Augustine,* translated by E. B. Pusey, Medici Society, London, 1930.

Methuen & Co., Ltd: Excerpts from *Geography in the Middle Ages,* by H. T. Kimble, Methuen, London, 1938.

Octagon Books: Excerpts from *Amerigo Vespucci, Pilot Major,* by Frederick J. Pohl, Octagon, 1966.

Oxford University Press: Excerpts from two sources. *The Norse Atlantic Saga,* quoting "The Greenlander's Saga," by Gwyn Jones, Oxford, 1964. *Autobiographies of Charles Darwin and Thomas Henry Huxley,* Gavin de Beer, ed., Oxford, 1974.

Penguin Books Ltd: Short excerpts from *Marco Polo: The Travels,* translated by R. E. Latham. Penguin Classics, 1958. Copyright © 1958 by Ronald Latham. Reprinted by permission of Penguin Books Ltd.

Popular Science : Excerpt from Buffon, *Histoire Naturelle,* translated by Arthur O. Lovejoy, "Buffon and the Problem of Species," by Arthur O. Lovejoy, 1911. Reprinted with permission from Popular Science, copyright 1911.

Arthur Probsthain: Excerpts from *China's Discovery of Africa,* by J. J. L. Duyvendak, Lectures at University of London. Copyright 1949 Arthur Probsthain.

Random House, Inc.: Specified excerpts from *China in the Sixteenth Century: The Journals of Matthew Ricci: 1583–1610,* by Matthew Ricci, translated by Louis J. Gallagher, S.J. Copyright 1953 by Louis J. Gallagher, S.J. Reprinted by permission of Random House, Inc.

Charles Scribner's Sons: Excerpts from *Dictionary of Scientific Biography,* volumes VII, IX, and XI are quoted with the permission of Charles Scribner's Sons. Copyright © 1973, 1974, 1975 American Council of Learned Societies.

University of California Press: Excerpts from *Andreas Vesalius of Brussels,* by C. D. O'Malley. University of California Press, 1964.

University of Michigan Press: Excerpt from *Galileo, Science, and the Church,* by Jerome J. Langford. Copyright © 1966, 1971 The University of Michigan. Reprinted by permission of University of Michigan Press.

Walker & Company: Excerpts from *The European Reconnaissance: Selected Documents,* J. H. Parry, ed. Published by Walker & Co., 1968.

John Wiley & Sons, Inc.: Excerpts from *The Invention of Printing in China,* Second Edition, by Thomas Francis Carter, revised by L. C. Goodrich. Ronald Press, 1955. Reprinted by permission of John Wiley & Sons, Inc.

INDEX

ABOUT THE AUTHOR

DANIEL J. BOORSTIN, the Librarian of Congress Emeritus, directed the nation's library from 1979 to 1987. He had previously been director of the National Museum of History and Technology and senior historian of the Smithsonian Institution, Washington, D.C. At the University of Chicago, where he was the Preston and Sterling Morton Distinguished Service Professor of American History, he taught for twenty-five years.

Dr. Boorstin was a Rhodes Scholar at Balliol College, Oxford, winning a coveted "double-first," and was admitted as a barrister-at-law of the Inner Temple, London. He has been visiting professor of American History at the University of Rome, at Kyoto University, at the University of Puerto Rico, and at the University of Geneva. He was the first incumbent of the chair of American History at the Sorbonne, and Pitt Professor of American History and Institutions and a Fellow of Trinity College, Cambridge University, which awarded him its Litt.D. degree.

Born in Georgia and raised in Oklahoma, Dr. Boorstin received his B.A. with highest honors from Harvard and his doctor's degree from Yale. He is a member of the Massachusetts Bar and has practiced law. Before going to Chicago in 1944, he taught at Harvard and Swarthmore. He has lectured widely within this country and all over the world.

The Americans, his most extensive work, is a trilogy with a sweeping new view of American history, revealing through the story of our past some of the secrets of the distinctive character of American culture. The third volume, *The Americans: The Democratic Experience* (1973), was a main selection of the Book-of-the-Month Club and won the Pulitzer Prize. Dr. Boorstin has received numerous other awards, including the Bancroft Prize for *The Americans: The Colonial Experience* (1958) and the Francis Parkman Prize for *The Americans: The National Experience* (1965). His books have been translated into the European languages as well as Chinese and Japanese.

Dr. Boorstin is married to the former Ruth Frankel, who has been editor for all his books. The Boorstins have three sons.